THE GEOLOGY OF WESTERN EUROI

The Geology of Western Europe

M. G. RUTTEN

Professor of Geology

State University of Utrecht, Utrecht, The Netherlands

ELSEVIER PUBLISHING COMPANY

AMSTERDAM LONDON NEW YORK, 1969

ELSEVIER PUBLISHING COMPANY
335 JAN VAN GALENSTRAAT
P.O. BOX 211, AMSTERDAM, THE NETHERLANDS

ELSEVIER PUBLISHING CO. LTD.
BARKING, ESSEX, ENGLAND

AMERICAN ELSEVIER PUBLISHING COMPANY, INC.
52 VANDERBILT AVENUE
NEW YORK, NEW YORK 10017

LIBRARY OF CONGRESS CARD NUMBER: 68–15624
STANDARD BOOK NUMBER 444–40710–3
WITH 303 ILLUSTRATIONS AND 20 TABLES

PRINTED IN THE NETHERLANDS

To the faculty of the
Department of Geology,
University of Michigan,
who suggested the writing
of this book

Preface

INTRODUCTORY

This book is a sequel to a course on the geology of western Europe, given at the University of Michigan during my tenure of the Netherlands Visiting Professorship in 1957–1958. Only then did I realize how scattered the literature on European geology is, and how strong, moreover, is the influence of the language barrier.

Also, although the primary material, the rocks and their history, is pretty much the same in the Old and in the New World, I was struck by the difference in outlook of many earth scientists in both areas. This has led, in many cases, to a quite different approach, if not to differences in philosophy.

An important factor in this difference is that European geology is older, but that there have, on the other hand, been fewer European geologists than their American counterparts. European geology, although the youngest of natural sciences, reaches back for more than two centuries (GEIKIE, 1962) as against hardly a century of American geology. On the other hand, European geological societies number their members by the hundreds, the American by the thousands.

A major result of these differences in the development of our science is that, on the whole, it was slower in Europe and more inclined to the formulation of all-embracing theories. Whereas, on the other side of the ocean, it was quicker, if not explosive, and leaned more to the descriptive.

In Europe, from the onset, geology has been in the hands of relatively few, academically minded, professionals and amateurs. Its mineral wealth had been tapped for over two milennia, and in the earlier geologic studies basic information on the history of the earth was sought. Only relatively recently has geology in Europe been applied to the exploration for new mineral deposits or for deposits of grades so low that they had formerly been passed by.

The early geologists were not only interested in the immediate questions of, say, a small-scale geological map, but also (and often much more) in the general background, such as, for instance, the origin of the earth, or that of granites, or in the reality of the Deucalian Flood. They extrapolated happily; one of the results being the establishment of various schools of geology, each dominated by a gifted "maître", who normally was at loggerheads with his equally gifted opposite numbers in other schools.

Although the controversies are not so sharp any more as during the days of the Neptunists and the Plutonists, it is still worth while, in trying to understand a given author's papers, to know the school to which he belongs.

Another result of these differences in the development of our science in the Old and in the New World, is the different philosophy of its stratigraphy. In Europe the emphasis has always been on "how old?". Patient fossil collecting, often by gifted local amateurs, has in most cases supplied more or less reliable answers. In America, on the other hand, emphasis has lain pretty well on "what is there?". European stratigraphy therefore is chronostratigraphy, American stratigraphy is lithostratigraphy. And even though, in more recent years, and mainly under American influence, lithostratigraphy has gained some ground in Europe, it still is always correlated with the stage names of time-stratigraphy. An irksome influence in European stratigraphy has moreover been the strong influence both of various schools and of nationalistically minded geological surveys, which has resulted in a great number of different local names for the same, or for more or less the same, epochs.

ABOUT THIS BOOK

With these considerations in mind, the main object in writing this book has been to produce an overall picture of the geology of western Europe, in which stress is laid on present controversies related to the habits of contemporary geological thinking.

Of necessity subjective in choice of subject material and in its treatment, I have, on the other hand, always tried to indicate when and how certain "knowledge" and "facts" ought to be qualified. If, on the grounds of personal preference, I have leaned over too much to a certain way of thinking, I have nevertheless tried to include in the bibliography enough literature representative of the opposite camp.

It stands to reason that it is impossible, in a book this size, to give a detailed description, with maps and sections, of all of western Europe. Moreover, the days of E. Suess, in which a single geologist could write about *Das Antlitz der Erde* (SUESS, 1885, 1888, 1901), are over. Western Europe alone would require a large team of regional specialists, with at least several editors to cement their necessarily fragmentary text.

Bearing in mind the success of my father's *Voordrachten over de Geologie van Nederlandsch Oost-Indië* (RUTTEN, 1927), and BAILEY's 1935) *Tectonic Essays: Mainly Alpine*, I have tried to limit myself to the discussion of certain areas and of certain problems only. For a more complete coverage the reader is referred to the regional literature. I hope that in this way the main aspects of European geology can be highlighted, while anyone interested in a certain area will still find his way by means of the more important references to the regional literature.

The area thus treated in this text is limited to certain parts of continental western Europe. Iceland, though culturally belonging to Europe, has been left out, because geologically it belongs to the Atlantic Ocean. Great Britain has been omitted, because there exist so many reviews and guide books in English that it seemed unnecessary to rehash this information.

Review literature on the whole of this area is scarce, and in fact limited to VON BUBNOFF (1926, 1930, 1935, 1936). Apart from that much information is contained in the stratigraphy texts of BRINKMANN (1960) and GIGNOUX (1955). The latter author also treats problems of more general aspect, especially for the Alps. There are many more descriptive texts of smaller regions, which will be cited in the relevant chapters.

As to a general map, we are at present far better off, because we have the exellent new tectonic map of Europe 1:2,500,000 (BOGDANOFF, 1962; BOGDANOFF et al., 1964). The accompanying text is, however, rather disappointing. Not only does it suffer from the conglomeratic character of most multi-author volumes, but, moreover, the authors being amongst the grand old men of their respective countries, their presentation is decidedly old-fashioned.

Besides the tectonic map of Europe, which should be in the hands of anyone interested in the geology of this continent, a necessary tool is a set of good dictionaries. To complete their information of a more general aspect, the *Geological Nomenclature* of SCHIEFERDECKER (1959) is strongly recommended. This lists equivalent geological terms in English, French, German and Dutch, often supplemented by a short commentary on controversial terms. For anyone wishing to study the literature of western Europe through the original publications, it is well-nigh indispensable.

Similar to the problem of what areas and what problems of western Europe to treat in this text, I have had to make a choice of what part of the literature had to be cited in the bibliography, and what had to be left out. I have tried to select those papers which give an up-to-date picture of the state of our knowledge, and only included earlier references, when this was necessary to understand the development of certain ideas.

In this way a kind of instantaneous picture of our knowledge has been aimed at. Not that I think that the older literature is superseded, and hence its knowledge superfluous, but the inclusion of more complete lists of regional literature would have swollen the bibliography beyond all bounds.

Anyone interested in a certain region or in the development of a certain line of thought, can easily compile for himself a more complete bibliography of such more detailed subjects from the literature cited in the papers listed here. He will, moreover, find that the snowball effect resulting from the consultation of earlier and still earlier papers is not as large as might be expected. The reason for this lies in the fact that the earlier really classic papers are few in number, and cited by every later worker in the same field.

Parenthetically, when he really takes the trouble to immerse himself in such a study of the older literature, he will come to the conclusion that the better of the early workers in any particular field, have already "seen anything there is to see".

The reason for this is that most of these grand old masters of geology were extremely sharp observers. They always produced valuable descriptions and the dogma and controversies based on these descriptions, which often only came later, can be separated easily from the descriptive part of their memoirs. As an example one of the most important of such a succession of theories based on the same, or nearly the same rocks, the development of our ideas on the Alps has been treated in some detail in Chapters 9 and 10.

As to the illustrations, apart from some generalized maps which had to be newly drawn to suit this text, I have used figures taken from the existing literature. Although some of these had to be re-drawn, and some of the names adapted to conform to the use in my text, I still hope in this way to convey the "feel" of the original publications. Because the areas described by various authors do not coincide with the divisions used in this book, there is often considerable overlap in illustrations taken from different sources. If the attentive reader will find discrepancies between such illustrations, this may serve as a warning against the differences of opinion existing on many aspects of the geology of western Europe.

A WORD ABOUT GEOLOGICAL TRAVEL

Assuming that many colleagues from the other side of the Atlantic are interested in geological field trips in western Europe, a word for them about geological travel seems appropriate. There is a happy maxim: no matter how a trip is planned in Europe, it is easy to take in some of its geology while touring, because every scenic route is interesting for its geology too. Even if accompanied by a non-geological family, it should therefore be possible to arrive at a travel scheme in which culture and cathedrals can be satisfactorily blended with scenery and geology.

In such a case it is, however, necessary to have good road maps, besides the geological maps. One should not try to travel geologically on any scale less than 1:200,000, if such maps are available. An added advantage is that they indicate the routes off the beaten tracks, which, even in Europe, can be delightfully tranquil in summer.

Road maps are only rarely distributed free at gas stations. And even when this is the case, they only show the main roads. One needs, of course, maps on which hills and mountains, rivers and railroads are indicated, together with the size of towns, historic monuments and what not. The pioneers in the manufacture of such geographical road maps are the Michelin tyre company. Their maps, covering most of Europe, are still the best to be found.

Personally I have a preference for the Michelin Guidebooks, several of which have English editions, too. There are more elaborate guides, but the general information on history, geography, and often on geology, found in the Michelin guides is very succinct, and somehow tailored to our motorized period.

The cost of such travel in western Europe varies enormously according to the style of travelling. One may rough it, riding a bicycle built for two, and sleeping in hay lofts. The next step is camping, which has become a major tourist industry because of the paid vacations of one month most European workers now enjoy. There are many official camping grounds, with sanitary accomodations of a sort. Except along the sea coast where too many people crowd into the littoral, however, it is almost always possible to find a private spot in a meadow or on a river bank. Most of the higher mountains have free meadows for grazing cattle, well suited for camping. In view of the temperate European climate, one should, however, not try to camp without a tent.

The next step is, of course, hotels. Which is so varied a proposition, both as to comfort and as to price, that I had better leave this field to the professional travel guides and agencies.

There is a point to make, however, about the quality of drinking water and milk. It seems many Americans are under the impression that water is always polluted and milk never pasteurized or sterilized. Although I am afraid to pull a prop from underneath the soft drink industry, which profits enormously from this superstition, I should like to state that tap water is always drinking water, when not otherwise indicated, and that milk sold in bottles is always either pasteurized or sterilized, depending on the form of the bottles. Those who subscribe to the widely spread superstition cited above, and even put it in print, ought to remember that Louis Pasteur was a Frenchman.

ACKNOWLEDGMENTS

Many are the geologists of western Europe who have helped me in some way or another, such as by permitting the reproduction of their figures, or by supplying new illustrations, or by helping in the search for literature. Having come to the end of this book — for a preface is always written as an afterface — I want to thank them most heartily for their cooperation.

Thanks are due to the staff of the Geologisch Instituut der Rijksuniversiteit in Utrecht. This applies in particular to Mr. F. J. L. Henzen, who prepared most of the illustrations.

Thanks are also due to Dr. E. Ten Haaf and Professor M. E. Kauffman who read part, and to Professor A. J. Pannekoek, Mr. P. van der Kruk, and Mr. H. Visscher, who read all, and quite thoroughly polished the original typescript.

May then this text serve the purpose of the dissemination of the geology of western Europe and may the reader enjoy as much its reading, as I did its writing.

Bunnik, May 2nd, 1966

REFERENCES

BAILEY, E. B., 1935. *Tectonic Essays: Mainly Alpine*. Clarendon, Oxford, 200 pp.

BOGDANOFF, A. A. (Editor), 1962. *Carte Tectonique Internationale de l'Europe 1:2,500,000*. Nauka, Moscow.

BOGDANOFF, A. A., MOURATOV, M. V. and SCHATSKY, N. S. (Editors), 1964. *Tectonique de l'Europe. Notice Explicative pour la Carte Tectonique Internationale de l'Europe au 1:2,500,000*. Nauka, Moscow, 360 pp.

BRINKMANN, R., 1960. *Geologic Evolution of Europe*. Enke, Stuttgart, 161 pp.

GEIKIE, A., 1962. *Founders of Geology* (reprint of 2nd ed., 1905). Dover, New York, N.Y., 486 pp.

GIGNOUX, M., 1955. *Stratigraphic Geology*. Freeman, San Francisco, Calif., 759 pp.

RUTTEN, L. M. R., 1927. *Voordrachten over de Geologie van Nederlandsch Oost-Indië*. Wolters, Groningen, 839 pp.

SCHIEFERDECKER, A. A. G. (Editor), 1959. *Geological Nomenclature*. Noorduijn, Gorinchem, 523 pp.

SUESS, E., 1885. *Das Antlitz der Erde. I.* Tempsky, Wien, 778 pp.

SUESS, E., 1888. *Das Antlitz der Erde. II.* Tempsky, Wien, 703 pp.

SUESS, E., 1901. *Das Antlitz der Erde. III.* Tempsky, Wien, 508 pp.

VON BUBNOFF, S., 1926. *Geologie von Europa. I. Einführung, Osteuropa, Baltischer Schild*. Borntraeger, Berlin, 322 pp.

VON BUBNOFF, S., 1930. *Geologie von Europa. II. Das Ausseralpine Westeuropa. 1. Kaledoniden und Varisciden*. Borntraeger, Berlin, 691 pp.

VON BUBNOFF, S., 1935. *Geologie von Europa. II. Das Ausseralpine Westeuropa. 2. Die Entwicklung des Oberbaues*. Borntraeger, Berlin, 441 pp.

VON BUBNOFF, S., 1936. *Geologie von Europa. II. Das Ausseralpine Westeuropa. 3. Die Struktur des Oberbaues und das Quartär Nordeuropas*. Borntraeger, Berlin, 468 pp.

Contents

CHAPTER 1

The Main Divisions

THE NECESSITY OF CLASSIFICATION

In studying the geology of western Europe it is first of all necessary, I think, to bring some simple order into the multitude of facts and names. A sort of classification is needed, grouping this bewildering mass of facts into some kind of a system. A system which, of course, will always remain an oversimplification, but which will, nevertheless, serve as a frame of reference for the more detailed descriptions that follow.

Western Europe, though so important an area in the eyes of its inhabitants, really is no more than a peninsula of the much larger Eurasian continent, of which it forms only a small part. This fact is nowadays easy to realise by studying one of the beautiful geologic maps of the U.S.S.R. Is it necessary, one might ask, to dwell upon a system of main divisions for so small an area? Would not a better course be to go directly into a regional description of the various units which make up this peninsula?

Many factors, however, make such an introduction of main divisions an absolute necessity. To cite a few, there is first the fact that this peninsula of western Europe contains a highly varied geology. It is, moreover, the region where numerous classic notions of the science of geology originated. To these two factors of a geological nature must be added the fact that the literature on the geology of western Europe was of necessity grafted upon topographic nomenclature developed since Roman times.

The first factor, the varied geology, does not only apply to time, but also to other aspects such as facies and structure. Rocks occur from the Precambrian to the Holocene in quite different facies, metamorphosed or not, and thrown moreover into quite varied pictures of tectonic styles by the successive orogenies western Europe has known. Also many geological phenomena of comparable nature are found scattered all over this peninsula whilst they were separated by unrelated areas, which makes it all the more difficult to see a general pattern.

The second factor, the fact that many of the classic notions of geology started in western Europe, has not been very propitious for an easy understanding of the main lines of its structure. Most of these notions, dating back to the era before the advent of the railway, the car or the airplane, were of necessity based on local phenomena. Later their proper usage was often blurred by unwarranted extrapolation, from local features to general law.

As to the last-mentioned factor, there are in Europe an immense number of named physiographical features, which are often cut by political divisions. Lowlands and uplands, moors, mountains and rivers all carry names which are fairly constant. The political divisions, on the other hand, the duchies, counties, provinces, states and what not, have been constantly shifting their boundaries during the centuries. Perhaps because of the anxiety instilled by this unstable situation, the temporal powers that be have often induced the usage of political divisions in geological descriptions, thereby greatly complicating an easy comprehension of the general structure of western Europe.

So, although, when seen at large, when looked upon for instance, from the centre of the Eurasian continent, western Europe is but a small peninsula, it is not so for the man within. To the geologist working in one of its many countries, western Europe is not unimportant in areal extension, but is, on the other hand, too wide an area for anyone to get properly acquaintanced with. An area which contains, moreover, a rabble of local names, in which it is difficult to find one's way; more or less hermetically sealed as is everyone to a certain extent by political and by language barriers.

This soliloquy will not, I hope, make our readers reluctant to study the geology of western Europe, as being a hopeless matter. But I needed it to excuse myself for introducing the following main divisions of western Europe. Divisions which have, as stated already, been necessarily oversimplified, to arrive at a convenient arrangement. But if we start without this oversimplification, we will be in danger of becoming stuck immediately in a morass of unrelated or even contradictory local notions and names, and of losing sight of the main lines altogether.

THE THREE MAIN DIVISIONS

We can distinguish in western Europe three main types of terrain. These are, (*1*) the lowlands together with the low plateaus, (*2*) the uplands, and (*3*) the mountain chains (Fig.1).

Nature, as we all know, has no sharp limits. So it is better not to apply exact altitudinal figures to these regional divisions. There is distinct overlap in mere altitude above sea level, between what is still a low plateau or already an upland, and between upland and mountain chain. The Norwegian Alps, for instance, are just as impressive as, and even higher than, Mount Olympus. We may even extend this comparison because the first are called Jotunheimen or House of Gods, a name indicating a similar veneration as that due to Mount Olympus. Nevertheless, the Norwegian Alps will be classed here with the uplands, Mount Olympus with the Alpine mountain chain.

My friend Professor A. J. Pannekoek of Leyden, thinks there is too much schematization in the divisions given above. He proposed the following scheme:

(*1*) Lowlands: (*a*) epicontinental, (*b*) rifts, (*c*) intra-Alpine.

(*2*) Uplands: (*a*) with subhorizontal Mesozoic and Cenozoic cover; (*b*) with basement uncovered.

(*3*) Mountain chains.

In his classification, therefore, my "low plateaus" are classed with his "uplands", because, geomorphologically, they are so similar. The reason I still prefer my system is not that it is better, but that it is simpler, and I intend to schematize as far as possible.

It follows that, to have any meaning at all, the main division into lowlands and low plateaus, into uplands, and into mountain chains, must be based not only on topography,

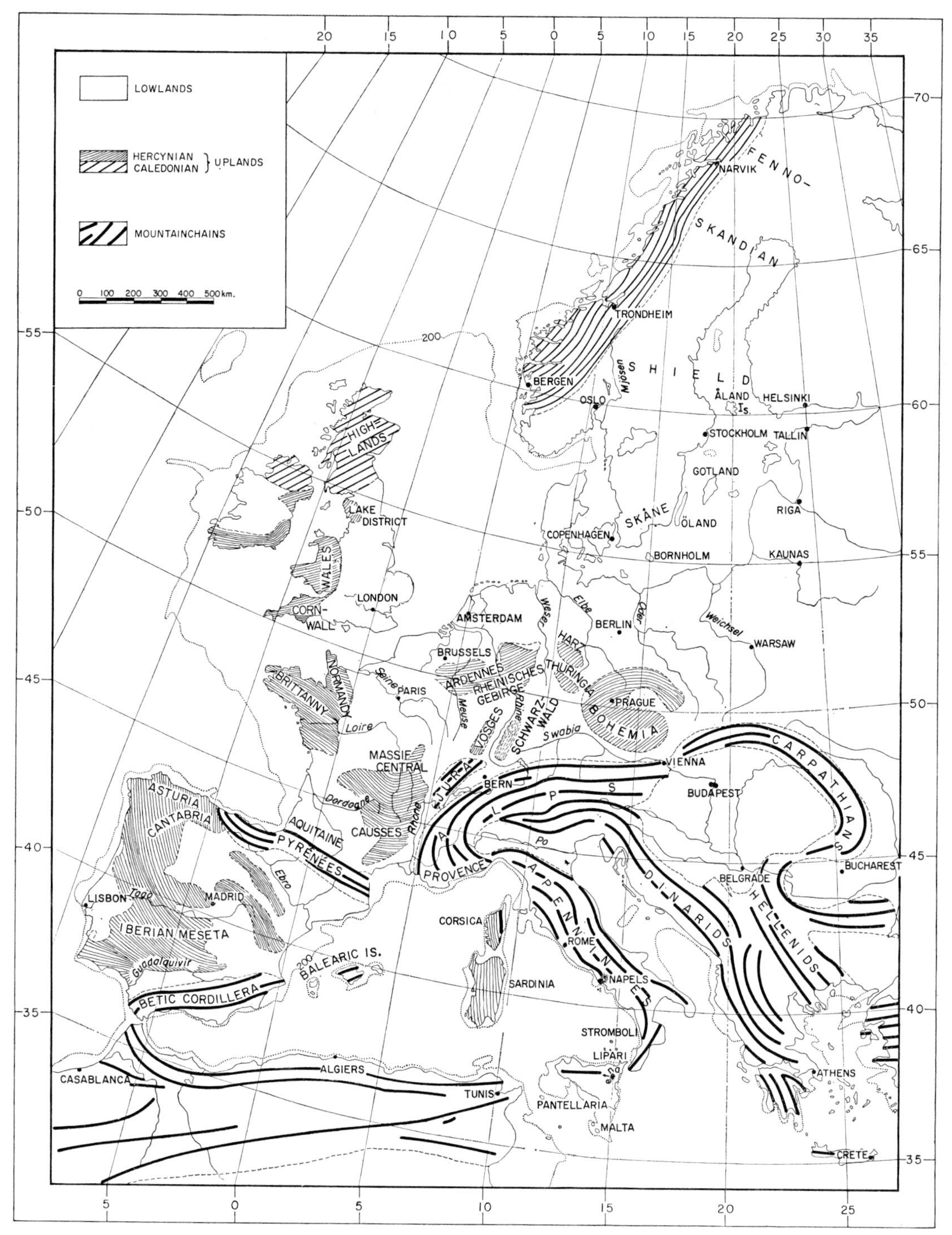

Fig.1. Generalized tectonic map of western Europe, showing the three main divisions.

but also on geological structure. Although each of these main divisions is rather complex it is normally possible to classify a certain region quite rigorously into this simple scheme, if only one takes note too of its geological structure.

Our next step consequently will be to explore the geological context of each of the three main physiographic divisions. The first group, the lowlands and low plateaus are quite complex. The second group, the uplands, show remarkable uniformity in their later history, being crustal blocks which have risen in rather recent times, thereby exposing the basement. It is very complex, however, in the nature of the basement exposed. The third group, that of the mountain chains, is the most coherent geologically, being related strictly to the Alpine orogeny.

LOWLANDS AND LOW PLATEAUS

The first group is the most complex, morphologically, because here I have thrown together lowlands and low plateaus, and geologically, because these regions show quite a variation in structure too. To begin with, lowlands have been formed in areas where positive crustal movements have been so slow that erosion could cope with them, resulting in a more or less base leveled, so-called "stable" region. Another type of lowlands is formed in areas with negative crustal movements, either slow or quick, where sedimentation was able to offset crustal subsidence. The low plateaus are similar in general to the latter group of lowlands, the difference being that they form sedimentation areas where subsidence and sedimentation stopped some time ago, whereupon uplift took place.

THE LOWLANDS

To return to the lowlands proper, we find that the first type, due to slow positive crustal movement balanced by erosion, is represented only by the Fennoscandian Shield and those areas of western Russia in which non-disturbed Paleozoic overlies the Precambrian basement. The latter area, of course, hardly belongs to western Europe, but is cited to round off the picture.

The shield itself rose so far that erosion could expose the Precambrian basement, whilst in the more southerly regions a cover of the Paleozoic sediments is still preserved. This of course makes the Fennoscandian Shield look very different from its Paleozoic aureole on a geologic map, but the difference is not essential. Morphologically, both are lowland regions, formed by peneplains virtually eroded down to base level.

The Fennoscandian Shield and its southerly cover of undisturbed Paleozoic is very similar to the Canadian Shield with its surrounding Paleozoic aureole of the Stable Interior of North America. But in Europe this simple pattern has survived only in a zone reaching from Finland southwards into the Baltic states and western Russia. Going west from Finland, we run into the morphologically different region of the Scandinavian uplands. Geologically most of this still belongs to the old shield, but its westerly part is formed by the mountanous Caledonides of Norway and Sweden, classed here with the uplands. Eastwards from Finland the simple relationship of a Precambrian shield covered by an aureole of undisturbed Paleozoic, is bounded by the Ural Mountains, another upland region exposing a basement dating from the Hercynian orogeny.

In contrast to the first type of lowlands, comprising the Fennoscandian Shield and its Paleozoic aureole, and characterized by slow positive crustal movements, the second type of lowlands, formed by negative crustal movements offset by sedimentation, is found scattered all over western Europe.

As examples let me cite the plains of Holland, northwestern Germany and western Belgium, the Upper Rhine valley, the delta of the Rhône in southern France, the basin of the Guadalquivir in southwestern Spain, or that of the Po in northern Italy. These indicate both the geographical scatter and geological variety of this group. The latter is perhaps less selfevident than the former, and requires some evaluation.

The lowlands of Holland, northwestern Germany and western Belgium are in an epicontinental area that has been stable since the Hercynian orogeny. The Upper Rhine valley and the Rhône delta are formed by active graben. The basins of the Guadalquivir and the Po are regions with strong crustal subsidence, related in a general way to the post-orogenetic stage of the Alpine orogeny.

In the lowlands of Holland, sinking rates, though they may have been heavy in certain regions and for short periods, were generally small all through the Mesozoic and the Cenozoic, and interrupted by many periods of non-subsidence or even of slight uplift and erosion. Against the history of this typical epicontinental lowland region, the two last examples cited are more localized basins of strong subsidence, which, moreover in the last three examples, started only rather recently. In general, the history of these basins began as such only during the Late Tertiary, after the main phase of the Alpine orogeny. Their cumulative subsidence during this relatively short geological time span is, however, impressive. It is of the order of many kilometers; 16 km of Late Tertiary and Quaternary being, for instance, often cited for the Po Basin.

THE LOW PLATEAUS

The second group of our first division comprises the low plateaus. Their structure is in general comparable to that of the lowlands of the epicontinental type. The difference is that in the low plateaus the crustal subsidence with its accompanying sedimentation did not continue up to the present time, but stopped somewhere in the past. After that some uplift has taken place, followed by erosion and often by some continental sedimentation.

Although some erosion has taken place, considerable time often elapsed between the point at which subsidence and sedimentation stopped and the beginning of the uplift. During this time crustal movements were either absent or remarkably slight. The uplift is often quite recent, resulting in a raised peneplain or peneplains, with a rejuvenated river pattern.

The Paris Basin is perhaps the most classic example of these low plateaus, but many others could be cited — for instance the Aquitaine Basin in southwestern France; the plateau of Swabia in southern Germany, classic ground for Jurassic stratigraphy; the area between the lowlands of northern Germany and the uplands of central Germany; most of Denmark and of Scania, the southernmost province of Sweden.

Just as with the lowlands, we find considerable geographic scatter. Also we find quite a varied history, when going from one region to the other, which is caused by the fact that the various low plateau areas represent individual epirogenetic crustal segments.

The variations in the history of the individual low plateau regions are, however, limited by two boundary conditions — one being that all sedimentation is post-Hercynian, the other that these areas are not disturbed, or only very slightly disturbed tectonically. The sedimentary history of a low plateau might have started in the Permian, or only begun at some time during the Mesozoic. It might contain important nonconformities, or be rather complete, superficially at least. But there is a strong disconformity always between the pre-Hercynian basement and the sedimentary cover. Moreover, tectonics are almost always of the regional tilting and block faulting type only. One apparent exception, found all over the low plateaus of northern Germany, is due to halokinesis or salt tectonics, and not so much to strong primary crustal movements.

THE UPLANDS

If we now turn our attention to the second item in our main division of the regional geology of western Europe, that is to the uplands, we do again find strong geographical scatter coupled with quite a varied geologic history in this group which, when seen superficially, seems so similar.

To the uplands I have assigned all areas where the basement crops out, with the exception of the lowlands of the Fennoscandian Shield on the one hand and the mountains of the Alpine system on the other. "Basement" in this context is meant to include all pre-Hercynian rocks — that is, everything up to the Middle or Upper Carboniferous, dependent upon the time the Hercynian orogeny did occur locally.

This group therefore covers, for instance, the Scandinavian and Scottish uplands and parts of Ireland, as well as the Lake District, the Ardennes and the Rheinisches Schiefergebirge, and the Harz Mountains. Further south we find Normandy and Brittany, the Massif Central, the Vosges, Schwarzwald and Bohemia. Still further south most of the Iberic Peninsula, i.e., the sierras which form the backbone of the Meseta of Portugal and Spain, also belongs to this division.

The genetic difference between the low plateaus of the former divisions and the uplands is slight. It is dependent only upon the balance between uplift and erosion. Regions where the carapace of relatively undisturbed post-Hercynian sediments of Permian, Mesozoic or Cenozoic age is still present, are classed with the low plateaus. Where this cover has been removed by erosion, so that the underlying pre-Hercynian rocks crop out, we speak of uplands. This small difference in the balance between uplift and erosion results, however, in a strikingly different geology and as such it is justified to use it as a boundary between two of the main divisions. The difference in geology, that is to say in surface geology, is so striking, that the upland regions as defined here stand out on any geologic map of Europe. Especially on the most strongly schematized black and white maps, the uplands are seen as readily recognizable darker blobs. There may be differentiation within the upland regions, but more often than not Paleozoic, igneous and Precambrian rocks are just indicated together, surrounded as they are on all sides by a cover of post-Hercynian sediments, easily differentiated according to their age.

VARIATION IN EARLY GEOLOGY OF THE UPLANDS

The similarity in representation on generalized and outline maps tends to obscure the considerable variation there is in geologic history between the different upland regions, and often even within a single region. We will of course encounter this variation in due time, when studying several of these upland regions separately. However, I should like to press home already at this early stage that the overall similarity of the various upland regions of western Europe has nothing to do with their earlier geologic history. Instead it is a product of quite recent crustal movements. These movements may locally have started early after the Hercynian orogeny, but they are much younger in general.

The earlier, pre-Hercynian, history of the basement rocks cropping out in the uplands may be mainly concerned with Precambrian times; or with the Caledonian or the Hercynian orogenetic cycles; or with several of these combined. That part of their history during which they acquired their present upland character, is, on the other hand, much younger. It is not even connected with the aftermath of the Hercynian orogeny, but is Late Tertiary to Quaternary.

This later history is often difficult to date exactly, but it is quite commonly referred to as "Plio-Pleistocene". But in some cases, notably for the Ardennes and the adjoining parts of the Rheinisches Schiefergebirge, it has been proved that only the Quaternary is responsable for the upland features, produced by an uplift beginning with the Pleistocene.

So it follows that after the Hercynian orogeny a long period of general quiet characterized western Europe north of the Tethys. Relatively strong post-orogenetic vertical movements still occurred during the Permian and parts of the Triassic. We find their influence translated into the deposition of continental clastics of the various series of "New Reds". But all through the main part of both the Mesozoic and the Cenozoic, crustal movements were relatively unimportant. Then, in Late Tertiary and Quaternary times, upward movements affected large but disconnected regions. Where these remained relatively small, and where erosion was not too forceful, we now find the low plateaus. But where uplift and erosion have been relatively strong, the uplands formed.

UPLANDS AND LOW PLATEAUS

It follows that the difference between the low plateaus and the uplands is largely a geologic one, whereas there is considerable overlap in mere altitude. This has been stated before, but just to be explicit, let me add two examples. Erosion will of course have been more active near the sea and here, even if the recent epirogenetic uplift has not been too fast, the pre-Hercynian basement becomes uncovered at a low level. The peninsula of Brittany, consequently, though not reaching 500 m in altitude, belongs to the uplands. Contrarily, in areas situated far from the sea, and also where the post-Hercynian cover is largely calcareous, leading to karst erosion, remnants of the younger sedimentary cover may still be found at considerable altitude. South of the Massif Central in southern France, we find such a region in the scenic Causses. Mesozoic limestone plateaus are preserved here, though uplifted more than 1 km. In this text they are classed with the low plateaus, because of their geological structure.

SIMILARITY IN LATER EVOLUTION OF THE UPLANDS

The superficial similarity of the uplands, is, as we saw, dependent entirely upon the similarity of their later history, e.g., upon upward crustal movements of quite recent times. This similarity is consequently expressed mostly in their geomorphologic habit. All uplands are plateau mountains with their earlier uplifted peneplains still largely intact. These peneplains are normally complex in nature, being formed during the long times between the last local orogenetic phase, be it the Hercynian or an older one, and the later Tertiary when the younger history started. Tilting during the upward movements is also often well expressed in the position of these uplifted peneplains. All river systems are typically re-juvenated, being inherited from earlier courses. These superimposed or epigenetic rivers are now deeply incised in narrow V-shaped valleys. But it is difficult to reconstruct the earlier pattern of the rivers flowing on top of the peneplains before the uplift, because of the many river captures.

Everywhere in the uplands then, we find the same morphology: an undulating plateau, formed by a former peneplain, or, more often, by a system of former peneplains, cut by narrow, deep river valleys. From the valley floor the steep valley may look mountainous indeed, but once climbed, there are no more mountains to be seen. The most widely known of these rivers is, of course, the middle course of the Rhine with its many castles perched on steep valley walls. However, even the Lorelei, rising 130 m from the water's edge in a sheer cliff, shows a flat top and is connected by a broad shoulder formed by a Late Pliocene river terrace to the undulating plateau of the non-dissected former peneplain.

Only in uplands which have been glaciated during the Ice Ages of the Quaternary has glacial erosion modified this general picture. In central Europe, for instance in the Massif Central, in the Vosges and in the Schwarzwald, glaciation affects only the highest parts of the uplands. The main feature there are cirques and wide U-shaped valleys. In Scandinavia, of course, most of the country was covered by the ice and extensive moraines formed, while a well-expressed monadnock topography is characteristic of many parts of northern Scandinavia.

ABSENCE OF STRUCTURAL CORRELATION BETWEEN OLDER AND YOUNGER STRUCTURES

Before finishing this introduction on the uplands of western Europe I should want to stress the fact that in general there is no parallelism between the earlier structures of the basement and the main directions of the Plio-Pleistocene uplift. The later uplift, with its tilting and faulting, normally cuts across the earlier structural lines, either obliquely or more or less at right angles, as the case may be. One consequently gets the impression that even where there is some parallelism locally, this is only accidental. There seems to be no control by the earlier structures on the younger uplift.

If we compare this with North America, the European uplands in this respect are similar, for instance, to the Ouachita and Ozark mountains, whereas the parallelism between earlier basement structures and younger uplift exhibited by the Appalachians has no counterpart in western Europe.

THE MOUNTAIN CHAINS

As stated already, our third division, that of the mountain chains, is much more uniform than the two earlier divisions, both in geographical location and in geological history.

Geographically, it comprises the Betic Cordillera in southeastern Spain and the Pyrenees between Spain and France and their westerly continuation; further parts of the Provence in southern France and all of the Alps: French, Swiss, Austrian and Italian. Included also are the Jura Mountains in eastern France and northwestern Switzerland, the Carpathians, several Mediterranean islands, the Apennines of Italy, and most of the Balkan. This may look like quite a list, but on any map of Europe the geographical coherence of this chain of mountains is immediately apparent.

As a counterpart to this geographical coherence, there is also a general geological uniformity. All mountain chains of western Europe were folded during the Alpine orogeny. The orogenetic cycle, in the main, started its sedimentary phase after the Hercynian orogeny, in what is now southern Europe. The main geosyncline was part of the famous Tethys. The main orogenetic phases of the Alpine system fall in the Early and Middle Tertiary, whereas post-orogenetic sediments of the molasse type are common from the Oligocene or the Miocene onwards.

There is, of course, great diversity within the Alpine system. But this does only apply to variations in stratigraphy and structure within the boundaries set above. As to stratigraphy, there are, of course, differences between epicontinental, miogeosynclinal and eugeosynclinal development, or, to take another main difference, between those areas where the Triassic developed epicontinentally, the so-called German facies, and those where geosynclinal conditions already prevailed in the Triassic, the Alpine facies. As to structure in relation to stratigraphy, there are variations in time in the main orogenetic phases, the Pyrenees and the Provence having been folded earlier than most of the other parts of the chain. Or one might cite the influence of earlier orogenetic movements, to which for instance the famous Gosau nonconformity in Austria is due.

Moreover, quite apart from time, there are great variations in structure in a purely geometrical way. When speaking of the Alps, the structural picture which comes to mind immediately is that of nappes. But other structural types occur too, down to autochthonous or almost autochthonous simple folds in, for example, the Jura Mountains, the Provence or the Dolomites. And even within those parts of the chain which exhibit nappe structure, an amazingly great difference exists between epidermis structures formed by post-Hercynian sediments which slipped over the basement, and more deeply folded structures, exhibiting a *tectonique de fond*, where the pre-Hercynian basement became involved in the cores of the individual nappes.

Important as these variations may be, they are, however, not more than variations upon a single theme: that of mountain chains belonging geologically to one and the same orogeny — the Alpine one.

THE "GIPFELFLUR"

Just as has been said for the uplands and the low plateaus, a point to be made is that mere altitude above sea level is not sufficient a parameter to distinguish between the main divisions. Although most of the mountain chains of the Alpine system qualify for our normal mental picture of the Alps, viz. a mass of high, rugged peaks, not all of it is that high. The Jura and Provence mountains for instance are lower than many uplands and even lower than some areas classed here with the low plateaus. Their morphology is, however, that of typical mountain chains, and this is corroborated by their geological structure. There is not the slightest danger of confounding the strongly folded post-Hercynian strata of any stretch of the Alpine system with either the pre-Hercynian basement rocks of the uplands or the undisturbed post-Hercynian cover of the low plateaus.

When we now look at these mountain chains of the Alpine system, we find, here too, the importance of uplifted former erosion surfaces (formed, according to various authors, as pediments). Another feature, however, is still more significant for the stronger rise of most mountain chains, viz. the summit level or "Gipfelflur." From afar, one always sees how the individual tumble of jagged peaks, so impressively irregular at close sight, approaches a strikingly even upper limit. This summit level passes over the tops, which are, in most cases, residual mountains of some former erosion surface. Of course, many individual peaks have already been eroded below this level. Also there are wide range undulations of the summit plain itself. And lastly, some individual peaks or a group of peaks, a massif, may individually stand out above it, being either exceptionally high residual mountains, or resulting from quite young vertical uplift.

But as an overall feature this summit plain is a very real thing indeed. It is strikingly apparent on all general pamphlets of the Alps, say on the scenic panoramas of parts of the Swiss and Austrian Alps available in every tourist centre. Everytime one is situated at a distance from the chain large enough to get a general outline, it just jumps to the eye. It is just as apparent in the other mountain chains of the Alpine system. A classic example is, for instance, found in the Pyrenees, where the view from Pau is so scenic that it brought the French poet Lamartine to the statement that what Naples is to the sea, Pau is to the land:

"Pau est la plus belle vue sur terre,
comme Naples est la plus belle vue sur mer".

MOUNTAIN CHAINS AND UPLANDS

Morphologically the difference between the uplands and the mountain chains of the Alpine system lies in the fact that in the latter erosion has attacked the uplifted earlier erosion surfaces to a far greater extent. Whereas in the uplands most of the former peneplain is still present, resulting in plateau mountains cut only by narrow deep river valleys, most of the erosion surfaces (which were far from complete peneplains) have gone in the mountain chains. The valleys are very much deeper, so that often the valley walls intersect in sharp ridges. Consequently very little has been preserved of the successive uplifted erosion levels. This gives us something like a negative print of the upland morphology in the mountain chains.

The reason for this difference in the morphology of uplands and mountain chains is two-fold. For one thing, the mountain chains normally have risen much higher than the uplands. Moreover this uplift started earlier. As to the first factor, the former peneplains of the uplands are now generally situated about 1 km above sea level. Rarely do they rise to an altitude of more than 1.5 km. This differs greatly with the normal figure of between 3 km and 4 km for the altitude above sea level of the Gipfelflur. This of course provides erosion with a much larger amount of potential energy in the mountain chains. The second factor, the time during which erosion was able to do its work, is important too. It is generally thought that the Gipfelflur passes over residual mountains belonging to a Miocene relief, the uplift having started in Late Miocene time. This is considerably earlier than the Late Pliocene onset of the uplift in most of the uplands.

In the mountain chains erosion consequently was able to remove the main part of the rock mass below the former erosion surfaces, leaving only fragments, from which the pinnacles of the high peaks rise up to point out their former position. In the uplands, on the contrary, erosion has only just started and has only just managed to cut narrow and deeply incised river systems more or less down to their present base level.

RECAPITULATION

I hope the foregoing short discussions are sufficient to characterize the three main divisions of the geology of western Europe, so that in the next chapters some regions can be treated in more detail. The main divisions, as defined above, can be followed on any good physiographical map of Europe. LOBECK (1951) has, however, given a somewhat different picture of the main lines of the geology of western Europe, very easy to read through his superior draughtsmanship. A comparison of his text with mine will show considerable differences. Fresh insights into the geologic history, gained in later years, have tended to substitute a more genetic picture for the earlier one. As a superficial index to the main divisions used, Fig.1 has been prepared. Moreover, Fig.2 is included, because it gives a wealth of topographical information important in relation to the structural units of Europe, although BOGDANOFF (1963) uses a system of divisions which slightly differs from mine. Most of my "low plateaus", for instance, are classed by him as "Hercynian, covered by younger sediments". This hardly influences the choice of the structural units recognized.

GENERALITIES

Before proceeding with a more detailed description of various separate regions, that is before closing this chapter, I should like to discuss two general topics. One is the importance of former peneplains in this overall picture of the geology of western Europe. The other is the well diffused but invalid theory of a periodic accretion of the European continent, from the early core of the Fennoscandian Shield outwards.

THE IMPORTANCE OF PENEPLAINS

So let us look first into the importance of peneplains. One of the surprises of a stay in the U.S.A., was to notice the particularly low value many of our American colleagues now give to the notion of the peneplain, because, after all, peneplains are an American invention. Perhaps the reason for this slight esteem is that, as so often happens, the pendulum has swung too far. After planations all and sundry had been called peneplains, with or without reason, the other extreme seems to have been reached now, and there is a tendency to deny the possibility of a relatively stable situation of the earth's crust over times long enough to achieve peneplanation.

To illustrate this, let me cite just two examples. The Schooley Harrisburg "peneplains", that have so long served as showpieces for a cyclic relief development, have now been swept away by the school of the equilibrium theory (compare, e.g., HACK, 1960). The gravel-covered Harrisburg "peneplain" would, moreover, hardly ever have been classed as such in Europe, and its former designation as a peneplain has obscured the meaning of that notion.

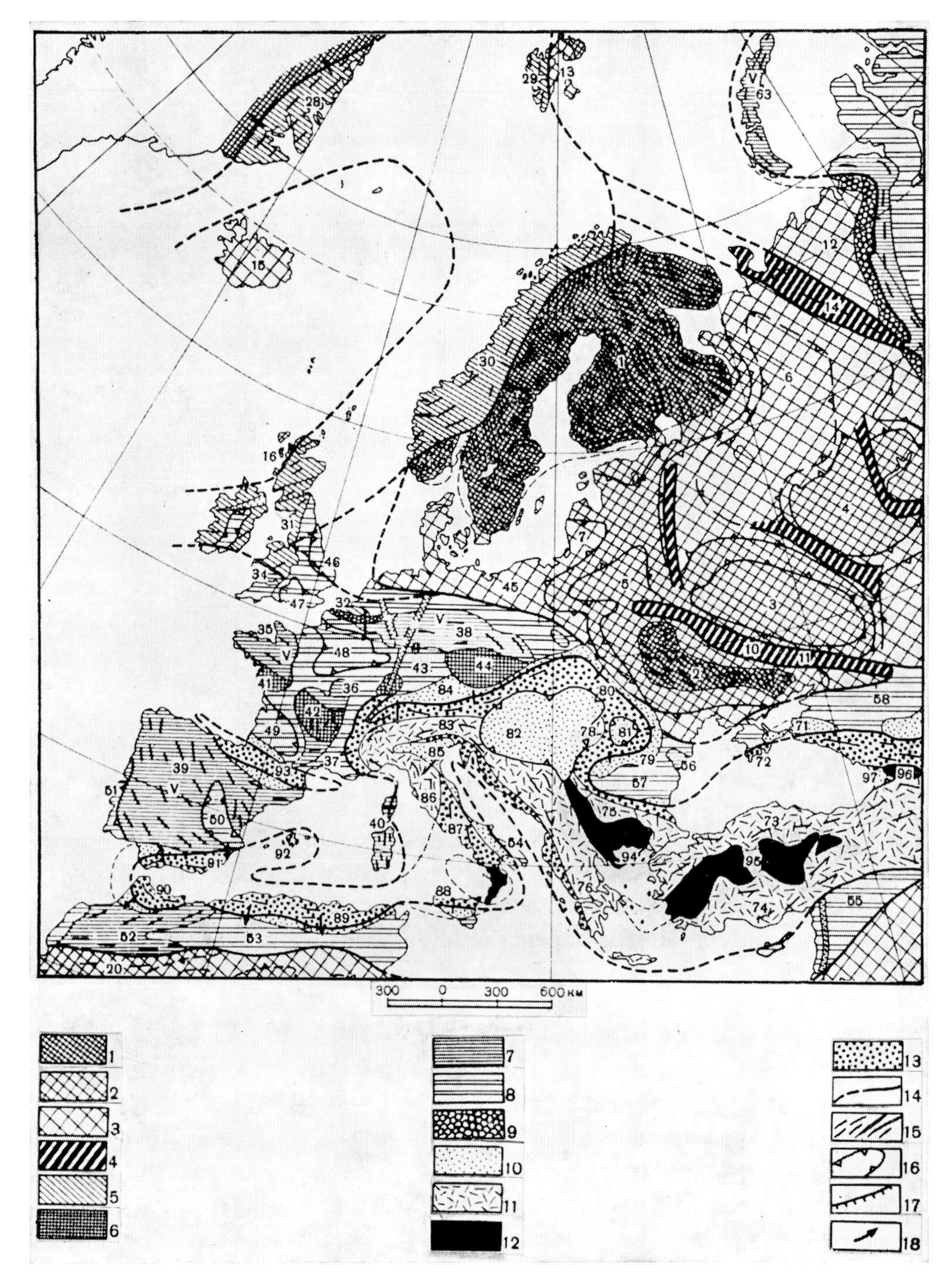

Fig.2. Schematic structural map of Europe.

1–4: The basement and its cover. *1* = Precambrian shields; *2, 3* = Precambrian shields under a thinner and thicker cover of relatively nondisturbed sediments; *4* = younger furrows on the shield.

5 = fold belt of the Caledonides.

6–9: Hercynian Europe. *6* = older cores in Hercynian structures; *7* = Hercynian fold belt exposed; *8* = Hercynian fold belt covered by younger sediments; *9* = foredeep.

10–13: Alpine Europe. *10* = post-orogenetic molasse basins; *11* = eugeosynclinal areas; *12* = stable areas; *13* = miogeosynclinal areas.

14 = limits of fold belts; *15* = general strike in fold belts; *16* = limits of epicontinental subareas; *17* = main rift zones; *18* = direction of "Vergenz" in fold belts.

Legend continued on p.13

Another extreme standpoint is well expressed in a paper by FAIRBRIDGE (1961), an author always thought-provoking through his clear language. Repeating a question first put by Wheeler, he rhetorically asks what is so sacrosanct about sea level. "There is nothing sacrosanct about sea level, in Europe either", we might answer. Sea level, at least the relative position of sea level versus continent, was never stable. Even in the most quiet moments of the history of the epicontinental basins, transgressions and regressions were actively proceeding over wide areas.

There is, however, a definite and very important quality about sea level. That is that we find erosion above it and marine sedimentation below it. It is not the stability of sea level over millions of years, or even over thousands of years only, which is important. Nor is it the "average sea level", a figure difficult to assess anyhow.

What matters is relative crustal stability during certain periods of the earth's history, and relative crustal unrest during others. Relative in this context means measured as against the power of erosion and/or sedimentation in the same period. During times of relative crustal stability, vertical crustal movements can be offset by erosion and sedimentation, whereas this is not normally the case during periods of relative crustal unrest. The point is, whether the effect of the vertical crustal movements in a given period is smaller or larger than the effect of the exogenic reactions of erosion or sedimentation they give rise to.

Observation teaches us that in the geologic history of western Europe there have been several periods of relative crustal stability, alternating with more active periods. For our main division the most conspicuous of the relatively quiet periods is the last one, including the Permian, the Mesozoic and most of the Neozoic. Vertical crustal movements were of geosynclinal nature within the Alpine system, of epicontinental nature elsewhere during this last long quiet period. The most important of the more active periods of vertical crustal movements for our classification again is the youngest one. That is the Plio-Pleistocene uplifts occurring localized outside the Alpine system and the general post-Miocene uplift of the Alpine system itself.

We only have to look around to be convinced of the reality of peneplains. We see peneplains everywhere. We see earlier peneplains which have become uplifted and dissected, even

---

Numbers on the map (several numbers are missing, because the original map covers a larger area than that which is reproduced here):

1–20: The basement and its cover. *1* = Fennoscandia; *2* = Ukraine; *3* = Voronej High; *4* = Volga High; *5* = Belorussian High; *6* = Moscov Basin; *7* = Baltic Basin; *10, 11* = Donbass–Donetz Basin; *12* = Bolchezemelskaia Basin; *13* = Spitzbergen High; *14* = Timan chain; *15* = Iceland volcanics; *16* = Llewisian; *20* = Anti-Atlas High.

28–32: Caledonides. *28* = East Greenland; *29* = West Spitzbergen; *30* = Skandinavia; *31* = Great Britain; *32* = Brabant Massif.

34–44: Hercynian Europe. *34* = Ireland and Wales; *35* = Armorican Massif; *36* = Morvan; *37* = Montagne Noire and environments; *38*=Ardennes, Rheinisches Schiefergebirge, Harz, Thuringia, etc.; *39*=Iberian Meseta; *40* = Corsica and Sardinia; *41–44* = older cores; *41* = Vendée; *42* = Massif Central; *43* = Vosges and Schwarzwald; *44* = Bohemia.

45–63: Younger basins. *45* = Polish–German plains; *46, 47* = London Basin; *48* = Paris Basin; *49* = Aquitaine Basin; *50* = Mancha Basin; *51* = west Portugal Basin; *52, 53* = basins in the Atlas; *56, 57, 58* = Dobrudja, Walachia and south Russia Basin; *63* = Novaya Zemlya.

71–93: Alpine Europe. *71* = Indolo-Kubanskaya depression; *72* = south Crimea; *73* = north Turkey; *74* = Taurus Mountains; *76* = Hellenides–Dinarides; *78, 79, 80, 81* = Carpathians; *82* = Vienna Basin; *83* = Alps; *84* = molasse basin; *85* = Po Basin; *86, 87* = Apennines; *88* = Sicily; *89, 90* = Atlas; *91* = Betic Cordillera; *92* = Balearic Isles; *93* = Pyrenees. (After BOGDANOFF, 1963.)

up to the stage where only their ghost structures are recognizable at some distances above the Gipfelflur of the Alps. We do, however, note that not every peneplain is perfect. Not every erosional cycle reached its final stage everywhere, especially not in the mountain chains. So the notion of peneplain has to be supplemented by that of residual mountains, "monadnocks" and "mosors" or "Fernlinge". We must bear in mind that multicycle peneplains do often exist. Nor must we forget that earlier peneplains may have become tilted or warped during their uplift. But these are more or less details, varying in every low plateau, upland and mountain chain, and important only for their local history.

These qualifications apart (for a more detailed discussion, see PANNEKOEK, 1967), we conclude that we are very happy with this notion of the peneplain, or peneplain systems, a notion which is of considerable aid in tracing the main divisions of the geology of western Europe.

REJECTING CONTINENTAL ACCRETION

The second point to be made before closing this chapter is about the non-validity of the idea of gradual or periodic accretion of the continent of western Europe.

Of course, the Fennoscandian Shield is the oldest part of Europe, as far as it is exposed now, whereas the influence of the youngest orogenetic cycle, the Alpine, is felt only in the south, with Caledonian and Hercynian areas in between. This led to the above-mentioned theory.

Like so many other notions on geotectonics, this one was most forcibly expressed by Professor H. Stille. By sheer impact of his own great influence, and by the spreading of the gospel by his many pupils, the picture reproduced here as Fig.3 became generally accepted as proof for this accretion of Europe with each succeeding orogenetic phase. The Fennoscandian Shield — Fennosarmatia in the nomenclature of Stille — and the hypothetical old shield of Eria — since foundered in the North Atlantic Ocean — together formed the Precambrian core of Primeval Europe, his "Ur-Europa". Across this Ur-Europa the Caledonian geosyncline of "Palaeo-Europa" developed, and further southward the subsequent areas of the Hercynian and Alpine geosynclines, of "Meso-Europa" and of "Neo-Europa" (STILLE, 1924).

One aspect of this picture is true. That is that the Fennoscandian Shield was no more affected since the Cambrian by any of the younger orogenetic movements during all later times and only suffered block-faulting and tilting. In the same way Stille's Palaeo-Europa was not affected by either the Hercynian or the Alpine orogenies; nor was Meso-Europa folded anew by the Alpine orogeny.

But although this aspect of the picture is true in that there have been definite limitations on the area of the younger orogenies, it is not the same as saying that these areas were never continental before that orogeny.

Precambrian crustal rocks are found, for instance, below the Caledonian nonconformity in Scandinavia. So in the Precambrian there was no difference between "Eria", and Palaeo-Europa or Fennoscandia. It was all continent. The difference only started in Late Precambrian time, when the Caledonian geosyncline developed across the earlier continent, while the remainder of the Fennoscandian Shield remained unaffected.

Similarly, although we do not find evidences of the Caledonian orogenetic cycle further

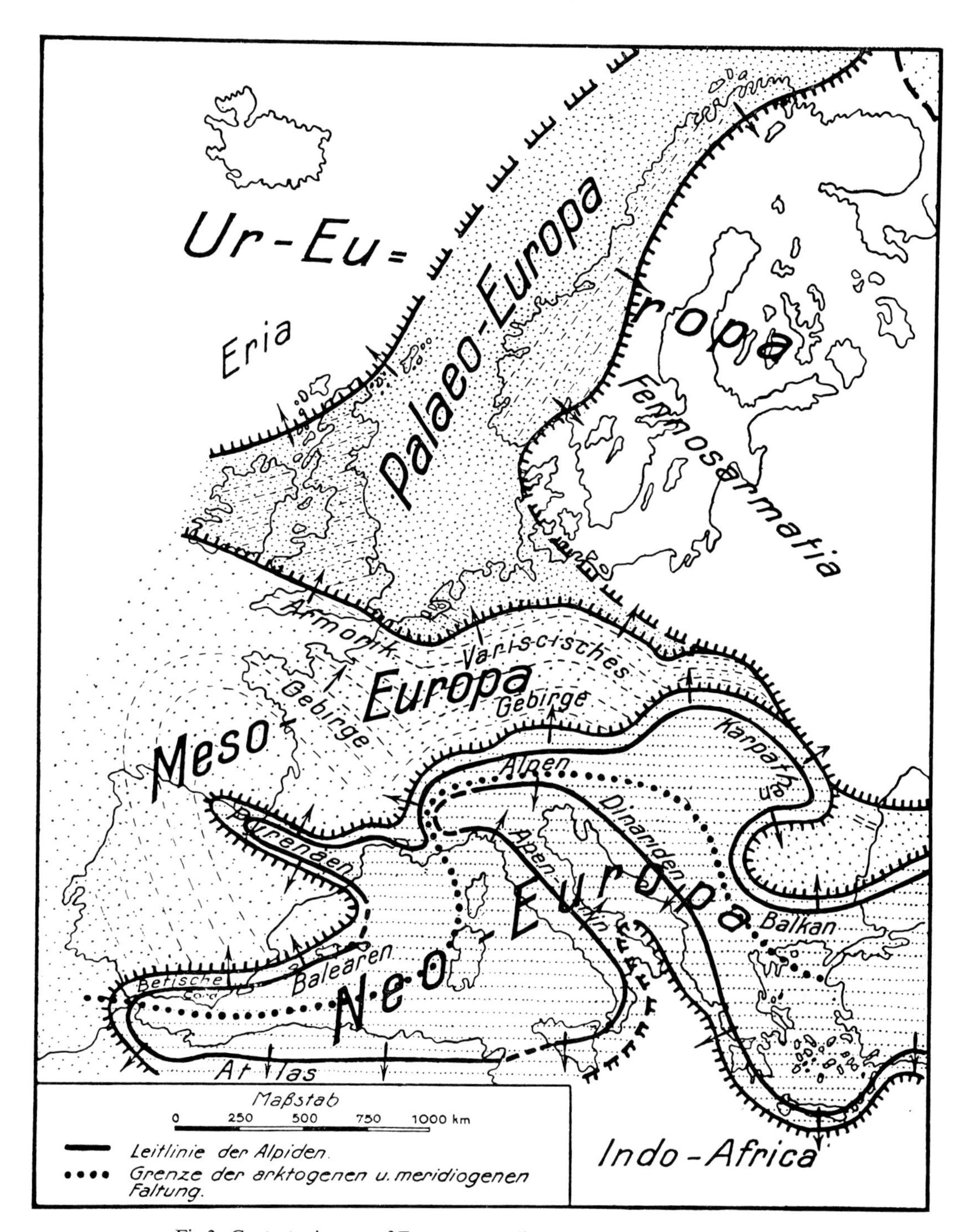

Fig.3. Geotectonic map of Europe, according to Stille. (After STILLE, 1924.)
— Main structural line of Alpine Europe ... borderline between the arctogenetic and meridiogenetic foldings.

south than the Ardennes, for instance, in Brittany and in the Massif Central, Precambrian and Early Paleozoic sialic series abound there in the midst of Stille's Meso-Europa. Even in southern Europe, clastic series of several kilometers thick in the Montagne Noire and in the Pyrenees, and ranging in age from Late Precambrian to Lower Carboniferous, indicate the existence of continental crust that far south at that early date. In the same vein, sialic rocks deformed by the Hercynian orogeny are well known in the basement of many of the Alpine nappes, indicating that in Stille's Neo-Europa too, a continent existed far earlier than the Alpine orogenetic cycle.

It follows that former continents have been demonstrated with certainty all over western Europe in the basement of the younger geosynclines, as far as these basement rocks are still recognizable. There is no evidence at all for a gradual or episodic accretion of the European continent to be drawn from Stille's delightfully simple schema. The picture of Ur-Europa, Palaeo-Europa, Meso-Europa and Neo-Europa only indicates the youngest orogenetic cycle active in a given area. They correlate with one or other of the Precambrian orogenies and with the Caledonian, Hercynian and Alpine orogenies respectively. It is quite apparent that there has been migration of the younger geosynclinal belts, but this is not accretion of continental crust. This distinction, we may conclude, has not always been clearly understood by people making use of Stille's schema.

REFERENCES

Bogdanoff, A., 1963. Sur certaines problèmes de structure et d'histoire de la plate-forme de l'Europe. *Bull. Soc. Géol. France*, 7 (4): 898–911.

Fairbridge, R. W., 1961. Eustatic changes in sea level. *Phys. Chem. Earth*, 4: 99–185.

Hack, J. T., 1960. Interpretation of erosional topography in humid temperate regions. *Am. J. Sci.*, 258-A: 80–97.

Lobeck, A. K., 1951. *Physiographic Diagram of Europe*. Geographical Press, New York, N.Y., 1 sheet.

Pannekoek, A. J., 1967. Generalized contour maps, summit level maps and streamline surface maps as geological tools. *Z. Geomorphol., N.F.*, 11: 169–182.

Stille, H., 1924. *Grundfragen der vergleichenden Tektonik*. Borntraeger, Berlin, 443 pp.

Von Bubnoff, S., 1926. *Geologie von Europa. I. Einführung, Osteuropa, Baltischer Schild.* Borntraeger, Berlin, 322 pp.

Von Bubnoff, S., 1930. *Geologie von Europa. II. Das Ausseralpine Westeuropa. 1. Kaledoniden und Varisciden.* Borntraeger, Berlin, 691 pp.

Von Bubnoff, S., 1935. *Geologie von Europa. II. Das Ausseralpine Westeuropa. 2. Die Entwicklung des Oberbaues.* Borntraeger, Berlin, 441 pp.

Von Bubnoff, S., 1936. *Geologie von Europa. II. Das Ausseralpine Westeuropa. 3. Die Struktur des Oberbaues und das Quartär Nordeuropas.* Borntraeger, Berlin, 468 pp.

CHAPTER 2

Northern Europe: Fennoscandian Shield and Caledonides

INTRODUCTION

To start with our regional studies, it so happens that we can follow normal procedure in geology and begin with the oldest known. That is, within the limits of this book, northern Europe; all of Finland and most of Norway and Sweden. Within this broad area, a number of geologic units are easily distinguished (Fig.4). First, of course, is the Precambrian core, the Fennoscandian Shield. Next, the zone of the Caledonides, which, though much of its material is of Precambrian age, attained its present structure mainly during the orogenetic cycle which ended with the Silurian. The epicontinental counterpart to the geosynclinal Caledonides is the Paleozoic cover of the Fennoscandian Shield, exposed mainly along its southern rim. In between, preserved by later graben tectonics, the Oslo area in Norway presents us with isolated remnants of Early Paleozoic of a facies transitional between the geosynclinal Caledonides and the stable epicontinental area, followed by an interesting sequence of Permian volcanics. The southern part of Scania, the Swedish province of Skåne, with its much younger cover of Cretaceous, falls outside this chapter. Geologically it belongs to the lowlands of Denmark and northern Germany.

The hard rock geology of northern Europe consequently is one of old and very old rocks. But at the same time the area is also a classic region for glacial geology, having been covered almost completely by the ice sheets of the successive Ice Ages of the Quaternary. This peculiarity of the geology of northern Europe, this contrast between very old and very young, is mirrored in the specialization of northern geologists. The hard rock geologist is mainly a petrographer, while the glacial geologist is interested in pollen and varved clays. In either domain, however, northern geologists have been leaders in European geology. In petrography, let us mention the names of Sederholm, originator of most of our present ideas on migmatization and granitization, and Eskola, *pater intellectualis* of the mineral facies concept. As to glacial geology, let us remember that both the use of varved clays for absolute dating and the use of pollen for relative dating and tracing the climatic history originated in northern Europe, with De Geer and Von Post respectively.

The geology of northern Europe has drawn much attention during the 21st Session of the International Geological Congress, organized by the five countries forming "Norden" (the North): Iceland, Norway, Sweden, Finland and Denmark, and held in Copenhagen in 1960. Field trips were arranged in all five countries, and most of the crucial points of their geology have been described in the *Guides to Excursions*. These guide books, although ex-

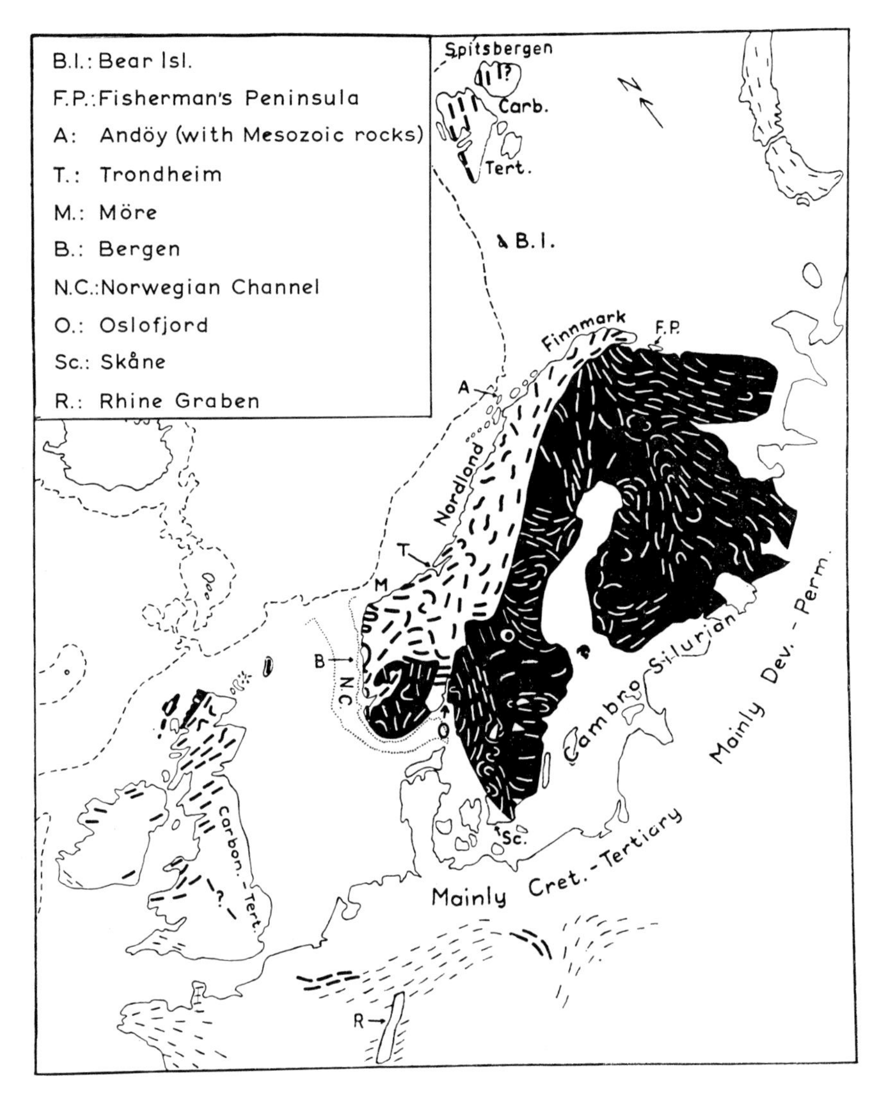

Fig.4. Northwestern Europe with main structural trends of Precambrian (white lines), Caledonian (thick black lines) and Hercynian orogenies. (After HOLTEDAHL, 1960.)

cellent when one is interested in special areas or special topics, do not, however, offer an easy way towards a generalized picture of the regional geology. They are best suited to the geologist who has already a basic understanding of the general structure. Luckily, at the time of, or just before the congress, regional descriptions were also published on the geology of Norway, Sweden and Finland, so that we now have up-to-date general as well as detailed information on the geology of northern Europe. The most complete of these general works is the volume on Norway (HOLTEDAHL, 1960), accompanied by a 1:1,000,000 geological map. For Sweden we have maps on the same scale, one for bedrock (MAGNUSSON, 1958a) and one for the deposits of the Quaternary (LUNDQVIST, 1958), accompanied by short descriptions (LUNDQVIST, 1959; MAGNUSSON, 1960b). The former has, however, no bibliography, for which we must turn to the *Lexique Stratigraphique International* (MAGNUSSON, 1958b). For Finland there is the short description of the hardrock geology by SIMONEN (1960a), accompanied by a geological map 1:2,000,000. The older map by GAVELIN and

MAGNUSSON (1933), which covers all of Scandinavia, Denmark and the Baltic states, still gives a good general picture and is a suitable wall map.

There was no special section on regional geology of Norden during the Copenhagen congress, but several papers pertinent to this chapter were read in the sections *IX*–"Precambrian Stratigraphy and Correlations" and *XIX*–"Caledonian Orogeny".

Local and regional geology has, of course, been treated extensively in the publications of the various geological surveys and geological societies of the countries concerned. References to special topics can mostly be found through the *Guides to Excursions.*

An exhaustive and modern presentation on the Precambrian, that is of the Scandinavian Shield, by A. Noe-Nygaardfor Denmark, T. F. W. Barth and Reitan for Norway, P. Geijer for Sweden and P. Eskola for Finland, is formed by the first volume of K. Rankama's series on the Precambrian (RANKAMA, 1963). In this volume too, our knowledge up to about 1960 has been summarized.

THE FENNOSCANDIAN SHIELD

The Precambrian core of northern Europe is called Fennoscandian or Baltic Shield. In the literature, the term "Baltic Shield" is, however, often used in a wider sense, the epicontinental cover of Paleozoic being included. Although, of course, this southerly rim also forms a stable area, which has not been affected by tangential movements since the beginning of the Paleozoic, one had better, in my opinion, reserve the designation of "shield" to the area where Precambrian rocks outcrop. The southerly rim of covering Paleozoic will then be comparable to the Stable Interior of North America.

In studying the literature of the Fennoscandian Shield, one is in danger of drowning in a flood of detailed investigations of ore bodies and their surroundings. The Precambrian is known for its mineral richness, and a large part of the literature is devoted to such local and detailed studies. To make matters worse, all attempts at a more general view of the history of this vast area are hampered by the lack of dependable stratigraphic criteria. If one follows the difficulties which Precambrian geology presents, even in drawing up the most simple and schematized stratigraphic outlines, one will realize the enormous help fossils give in all later formations. This is, of course, stated explicitly in every textbook on geology. But the average geologist who works in younger formations quite naturally tends to forget that it is so. Following the history of our science in an area like the Fennoscandian Shield then will be an eye opener.

Historically, the quest for stratigraphic criteria in the Precambrian of Fennoscandia is seen to comprise three main steps. At first it was thought that "older" correlates directly with "more strongly metamorphosed". After the fallacy of this postulate had become evident, the main attention turned towards the unravelling of the succession of Precambrian orogenetic cycles. Lastly, but only after the last war, absolute dating became available. Although the principle of a number of successive orogenetic cycles into which the Precambrian history can be divided still holds, absolute dating has nevertheless shown that many regional and local correlations stand in need of revision. Consequently, we have not advanced much since 1952, when Von Bubnoff stated that no synopsis could do more than just give the state of the problems encountered (VON BUBNOFF, 1952).

THE OROGENETIC CYCLE CONCEPT

Realization of the fact that "older" is not necessarily the same as "more strongly metamorphosed" came with the notion that the origin of most of the crystalline rocks of the Fennoscandian Shield is not igneous and abyssal, but supracrustal, both sedimentary and volcanic. In these newer views, which were fostered mainly by Sederholm, the degree of metamorphism is more or less irrelevant. On the other hand, once abstraction was made of their present state of metamorphism, it became apparent that these ancient supracrustal rocks could be classed in definite orderly series. Each of these corresponds with the development of an orogenetic cycle, such as had become known from the younger geologic history. Just as in these later ones, each of the ancient orogenetic cycles too, begins with a geosynclinal phase, characterized by strong sedimentation, often accompanied by volcanics, normally of a basic composition. Next follows the orogenetic phase, with folding and with intrusion of synkinematic — primorogenic (VON BUBNOFF, 1952) — well-differentiated granites. This again is followed by the late orogenetic phase with its suite of less differentiated, and generally more acid granites (serokinematic plutonites of ESKOLA, 1960).

As an example the Precambrian rocks of the Stockholm area may be cited. In this area supracrustal series of the Svecofennian and Gothian cycles are found, which have been variously metamorphosed to gneisses, migmatites and granites. The order of their evolution is strikingly parallel in each cycle. To illustrate this we can best compare two stratigraphical

TABLE I

EVOLUTION OF THE SVECOFENNIAN CYCLE IN SOUTHERN SWEDEN
(After Lundegårdh, in: GAVELIN and LUNDEGÅRDH, 1960)

	Subdivision	*Tectonic activity*	*Rock types*
	Late Sveco-fennian	repeated tectonic activity and migmatization	development of porphyroblasts and/or veins in older rocks, acid granites, aplite, pegmatite
		development of joints	dikes of metabasites and porphyries, plutonic peridotite and anorthosite, ultrabasic gabbro
	Middle Sveco-fennian	decreasing orogeny	acid, intermediate, and porphyritic granites[1]
		strong tectonic activity,	basic granite = granodiorite tonalite diorite, metabasites[2] gabbro lherzolite, ultra-mafic gabbro
Early Svecofennian	Mälar Series[3]	strong volcanism	sedimentary gneisses (most frequently veined), schists, graywackes, intermediate and basic volcanics, porphyritic intrusions, quartzite
Early Svecofennian	Leptite-hälleflint Series[4]	increasing tectonic activity	reddish and red gneisses (frequently veined), leptites, hälleflints, porphyrites, quartzite, limestone, iron ore

[1] In part supracrustal rocks granitized in situ.
[2] Mainly metamorphic supracrustal rocks.
[3] In Finland known as the Bothnian.
[4] In Finland known as the Svionian.

TABLE II

EVOLUTION OF THE GOTHIAN CYCLES IN SOUTHERN SWEDEN

(After Lundegårdh, in: GAVELIN and LUNDEGÅRDH, 1960)

Sub-div.	*Tectonic activity*	*Rock types*
Late Gothian	repeated tectonic activity and migmatization	development of microcline pophyroblasts in older rocks, acid granite (fine-grained), pegmatite
	decreasing orogeny	microcline granites (Kroppefjell Granite, part of the red Småland Granites, etc.) intermediate granites porphyritic granites and granodiorites (Askim Granite, Filipstad Granite, etc.)
	strong orogeny	basic granites = granodiorites and tonalites (Åmål Granite, gray Småland Granite, etc.) diorite gabbro plutonic ultra-basites
	increasing orogeny	sedimentary rocks (mainly quartzites) and volcanics (Åmål Series, Småland Porphyries, Västervik and Vestanå Series, etc.)
Early Gothian	migmatization	development of veined gneiss, pegmatites
	strong orogeny	granites plutonic basites and ultra-basites
	increasing orogeny	sedimentary rocks (mainly graywackes) and volcanics (Stora Le–Marstrand Series, Blekinge Coastal Gneiss, etc.)

tables, in which the tectonic activity is correlated to the metamorphics formed during these cycles (Table I, II).

Of course, this is only possible when field relations are quite clear. Pitfalls abound when one tries to apply this concept. The sedimentary facies of a given series, instead of its present metamorphism, becomes of paramount importance, for by this facies it can be placed in a given phase of an orogenetic cycle. But during the long times of Precambrian history several orogenetic cycles have left their mark on the Fennoscandian Shield, and it is often all too easy to correlate a facies belonging to a certain phase of an orogenetic cycle with a quite similar facies elsewhere, which in reality, however, belongs to an earlier or later orogenetic cycle. Moreover, migmatization and anatexis of earlier supracrustal rocks may have made them very similar to later intrusives, which in their turn often are none other than older supracrustal rocks, in which granitization went so far that they became mobilised.

These difficulties apart, however, it must be stated that the introduction of this concept of successive orogenetic cycles to solve Precambrian stratigraphy has been very successful. It has not only been the leading concept of all later studies of the Fennoscandian Shield, but it has also been successfully applied to other old shields of the earth. Even more important, to my mind, is the fact that it has greatly widened the horizon of our historical outlook in geology. It has helped to abolish the earlier complacent distinction between the Precambrian and all later history from the Cambrian onwards, between "basement" and "historical geology". The fact that actualistic extrapolation of the concept of the orogenetic cycle far back into the rocks of the "basement" proved fruitful, made us really aware that the old shields are not so very different from the younger areas of our earth. It made us aware too, of course, of the very long periods represented by these basement complexes.

In the newest development, absolute dating has often resulted in considerable shifting of units from one orogenetic cycle to another. But it has not invalidated the concept of successive orogenetic cycles as the main element in Precambrian stratigraphy. It offers a refinement, by which earlier mistaken correlations can be amended. Also it is providing dates which enable us to stagger the successive cycles properly in time.

Up to now, however, absolute dating still suffers from two adverse factors. One is the paucity of dates as yet available, the other is the widespread regeneration, which has occurred at several periods. As to the first factor, POLKANOV and GERLING (1960), for instance, report on only 480 dates for all of the Fennoscandian Shield, including dates from Russia. This will, of course, improve when more dates become available over the years.

The other factor, the regeneration, is much more of a nuisance. This results in an apparent rejuvenation, during a younger cycle, of rocks formed during earlier cycles. MAGNUSSON (1960a, 1965) was first aware of the important errors in dating and stratigraphic correlation this may lead to. Much more detailed field work, combined with petrographic and petrochemical studies, related to intensified dating programmes, will be required, before the influence of these periods of regeneration can be evaluated properly.

During this evolution of geological concepts in the Fennoscandian Shield, the earlier nomenclature of local and regional rock series has largely been retained. Most of the older names survived the vicissitudes of the newer concepts in geology. The rocks are still there, of course, though their interpretation may have changed. This makes the newer literature often difficult to follow for an outsider. Often only the new interpretations are stressed, and the reader is left in doubt about what all these rock series, with their local names, stand for in reality, what they look like in the field or under the microscope. The detailed evaluation of the literature would be beyond the scope of this book, but the subject is brought up, because it presents itself immediately to anyone who wants to arrive at a more detailed picture.

Here, it should suffice to cite summarily the stratigraphy arrived at by applying the concept of successive orogenetic cycles, together with the main differences this picture has undergone through absolute dating.

The stratigraphy based on successive orogenetic cycles contained the following divisions (VON BUBNOFF, 1952):

(*5*) Non-metamorphic Precambrian (Jotnian).
(*4*) Gotokarelian cycles.
(*3*) Svecofennian cycles.
(*2*) Marealbidian cycles.
(*1*) Saamid cycles.

The oldest of these orogenetic cycles must, of course, have developed on a still older basement. This has, however, proved to be extremely elusive. No outcrop of this oldest basement has as yet been found.

Quite apart from their interest for the unravelling of the history of the Fennoscandian

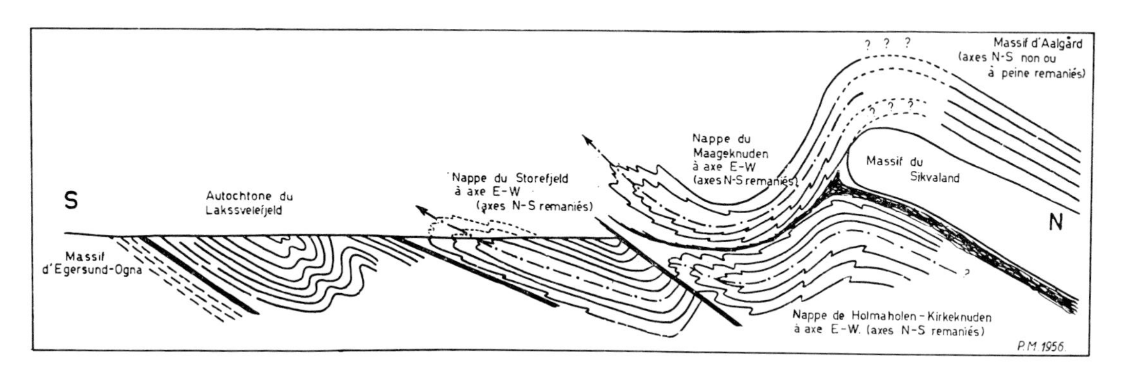

Fig.5. Section through the catazonal gneiss nappes north of Egersund, southern Rogaland, Norway. The feather-like folds with subhorizontal axes within the nappes show, in the field, a similar re-folding, mirrored on the axial plane, and without any drawing out of the reversed limbs, down to the order of decimeters, or even centimeters. (After MICHOT, 1960.)

Shield, these earlier orogenies are important for the study of the deeper levels of the crust. Thanks to prolonged erosion of rising crustal blocks, one now finds rocks that were formed or metamorphosed in the catazone at the surface in part of the Fennoscandian Shield exposed.

In this respect the investigations in southern Rogaland of Professor Michot of Liége, at the southern tip of Norway, must be cited (MICHOT, 1960,). It has been shown that the tectonic style of these deeper zones of a fold belt differ markedly from that found in its higher levels. We do not find the tectonic style of the series of open synclines and anticlines, with steep axial planes, and combined into synclinoria and anticlinoria, such as, for instance, found in the Ardennes (Chapter 5). Nor do we find the subhorizontal thrust plate nappes, such as in the Caledonides (this Chapter) or the Helvetides (Chapter 11). Instead, one finds in the catazone an extremely supple style of recumbent folds with horizontal axial planes. Although nappe thrust plates occur too, the material within a nappe is folded and re-folded conformably into the minutest details, without any apparent drawing-out of the inversed limbs, indicating that it always flowed and did almost never fracture (Fig.5).

Quite apart from this tectonic aspect of the catazonal "Stockwerk" of a fold belt, it is important to remember that we are here at what was formerly the very base of the continental crust. Michot holds that in those levels continental crust is formed from material of the upper mantle. Continental accretion would in this case occur basally, not laterally, such as has been held by so many tectonicians. The details of the petrographic studies of Michot can not be gone into here. They have been cited only to stress the importance of the Fennoscandian Shield for our insight in geologic questions in general.

ABSOLUTE DATING

Absolute dating has now established, as a main difference with the earlier views cited above, that of the "Gotokarelian cycles" the Gothides are much younger than the Karelides. Moreover, the Karelides, which in their geology are more similar to the Gothides than to the Svecofennides, seem to have about the same age as the latter. Consequently, the Karelides and Svecofennides, though strongly different in geology, are now thought to belong to the same orogenetic cycle or group of cycles. They seemingly represent regional variations in facies of the same orogeny.

As such, they form an indication of the extent of the differences in facies which may occur regionally, during the same orogenetic cycle. Also it follows that they indicate the caution which must be exercized in applying the concept of the successive orogenetic cycles.

We may, just to realize the extent of these differences, briefly enumerate them. First, there is a difference in geographical location. The belt of Svecofennides, or Svionides as they are sometimes called, extends from central Sweden, around Stockholm across the Botnian Gulf into southern Finland. The Karelides are found from northern Sweden through northern and eastern Finland to the province of Karelia, which now forms part of Russia, bordering Finland on the east. The Svecofennides and the Karelides consequently occupy different areas altogether. Apart from that, to quote SIMONEN (1960a, b, c, 1963), the Svecofennides

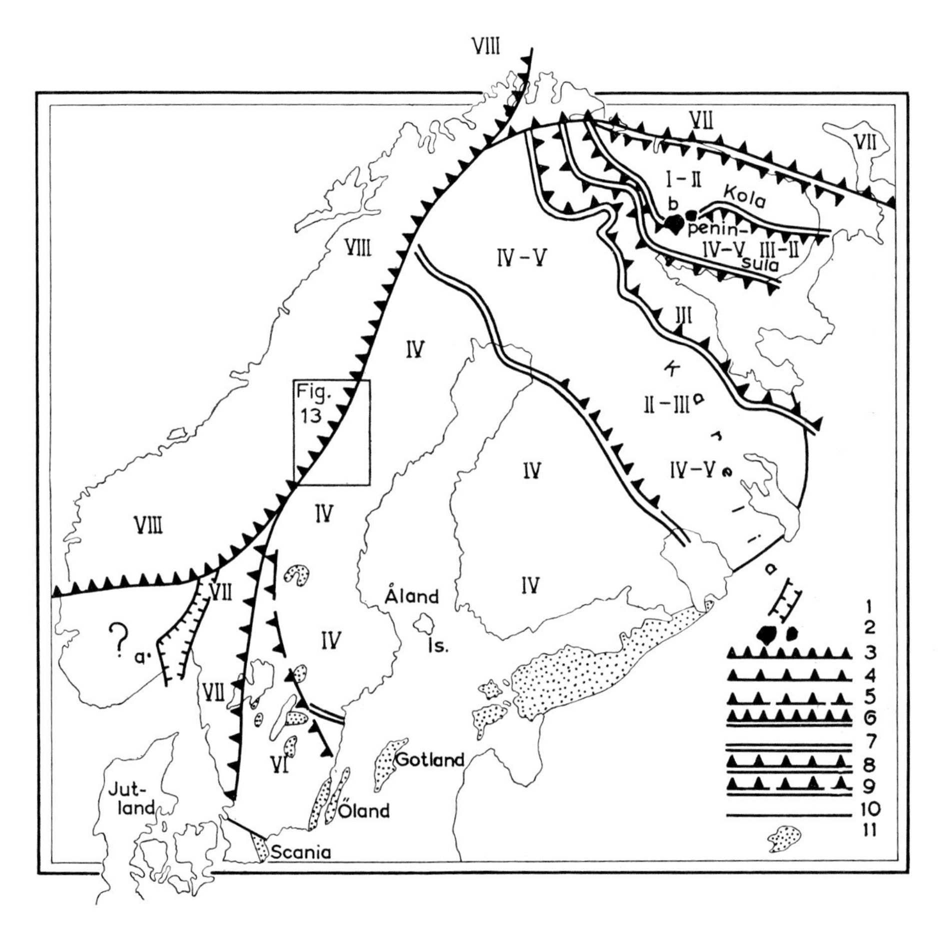

Fig.6. The Fennoscandian Shield.

1 = Oslo graben; *2* = alkaline intrusions (*a* = Paleozoic Fen area; *b* = Karelian Kola Peninsula intrusives); *3* = boundary of the Caledonides (dentition directed towards the folded belt), and *VIII* = Caledonides; *4* = boundary of the Ripheides, and *VII* = Ripheides; *5* = boundary of the Gothides, and *VI* = the Gothides; *6* = boundary of the Svecofennides–Karelides, *IV* = Svecofennides, and *IV–V* = the upper and lower Karelides; *8* = boundary of the intermontane region of the Belomorides, and *III* = Belomorides; *9* = boundary of the Saamides, and *II* = the Saamides; *I* = the Katarchean; *10* = boundary between the outcropping Precambrian and the cover of younger sediments; *11* = epicontinental Paleozoic. Rectangle indicates area of Fig.13.
(Mainly after POLKANOV and GERLING, 1960.)

TABLE III

OROGENETIC CYCLES OF THE FENNOSCANDIAN SHIELD[1]

Sweden, outside the Caledonides[2] (MAGNUSSON, 1960a)	*Finland* (SIMONEN, 1960)	*Baltic Shield* (POLKANOF and GERLING, 1960)
		Caledonides 620(!)–500
Eocambrian		
Jotnian	Jotnian 1620(?)	
Subjotnian		Ripheides (Baltic) 1021–865
1100–900		
↑ Dalslandian[3] (Bohus and Karlshamn Granites)		
Gothian 1300		Gothides 1400–1265
	Rapakivi Granites 1620	Rapakivi Granites, alkaline intrusions 1640–1610
1800		
↑ Karelian	Karelian and Svecofennian 1850–1750	Karelides and Svecofennides 1870–1640
Svecofennian or Svionian		
Pregothian		Belomorides 2100–1900 (Kola Peninsula and Skelefte, Sweden: Marealbides)
	Prekarelian ~ 2600	Saamides: Upper 2490–2100
		Saamides: Lower 2870–2500
		Katarchean: Upper granite 3100–3060
		Katarchean: Lower granite 3500–3200

[1] Ages in million years.
[2] Vertical arrows indicate regeneration of earlier cycles.
[3] Also southern Norway (Barth, in: HOLTEDAHL, 1960).

show much stronger metamorphism than the Karelides. The Svecofennides contained more clays and arenaceous rocks, now metamorphosed to phyllites, micaschists, quartz–felspar schists and gneisses, and basic volcanics now metamorphosed to amphibolites. The Karelian rocks, on the other hand, originally contained much more pure quartz sandstones and carbonates, now represented by quartzites and dolomites, but metamorphosed basic volcanics are also present.

These differences in facies between the Karelides and the Svecofennides, which have about the same age, point to an even more detailed analogy of this Precambrian orogenetic cycle with younger ones. The Svecofennides would in this case represent the eugeosyncline, the Karelides the miogeosyncline of one single cycle. We should bear in mind, however, that

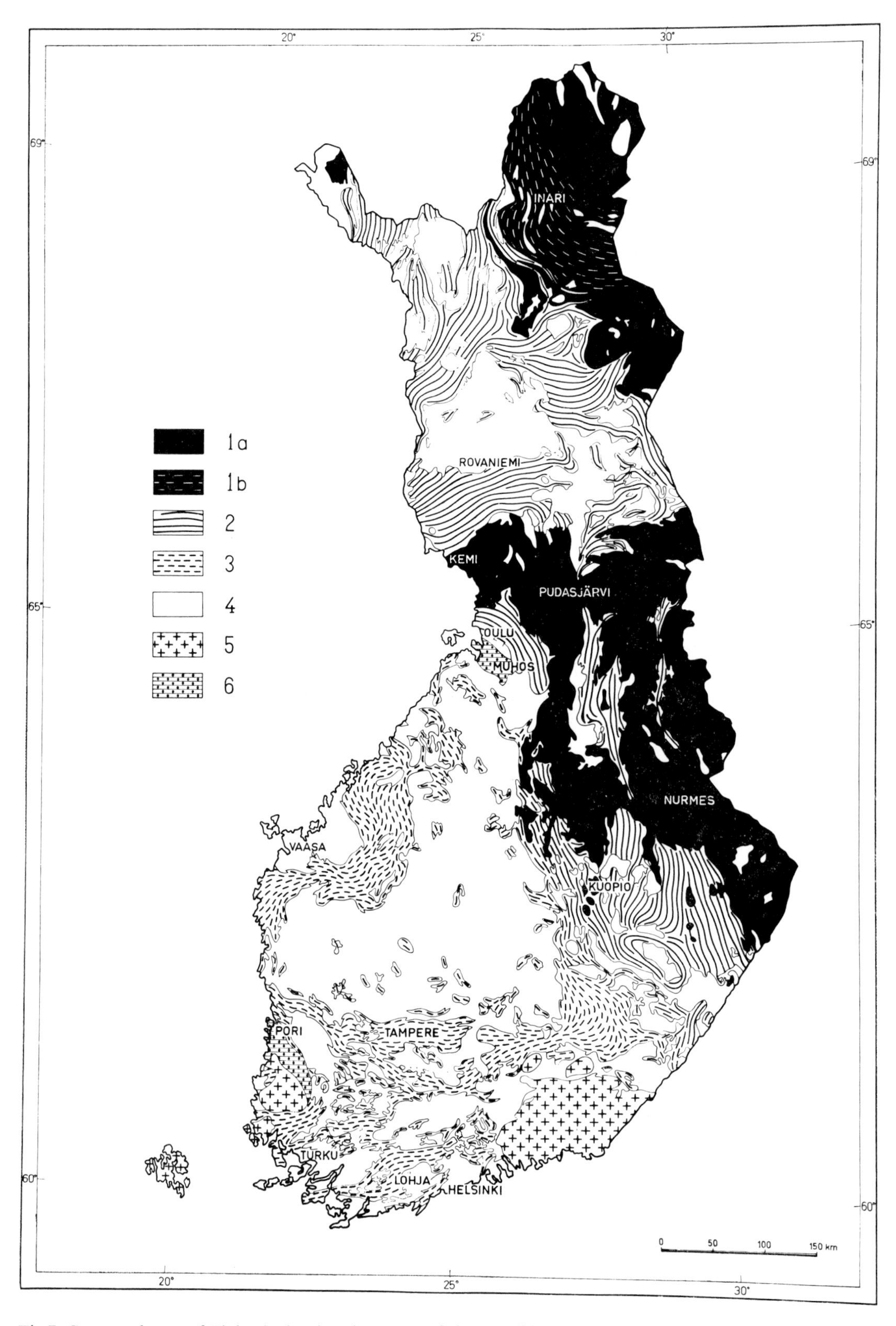

Fig.7. Structural map of Finland, showing the extent of the Karelides in the northeast, the Svecofennides in the southwest.

1–4: "Grundgebirge", rocks of the Svecokarelian or earlier orogeneses. *1a* = granite–gneiss; *1b* = granulite; *2* = Karelian schist zone; *3* = Svecofennian schist zone; *4* = syn-orogenetic plutonites; *5, 6*: Precambrian, younger than "Grundgebirge". *5* = an-orogenetic Rapakivi intrusives; *6* = Jotnian sediments. (After SIMONEN, 1963.)

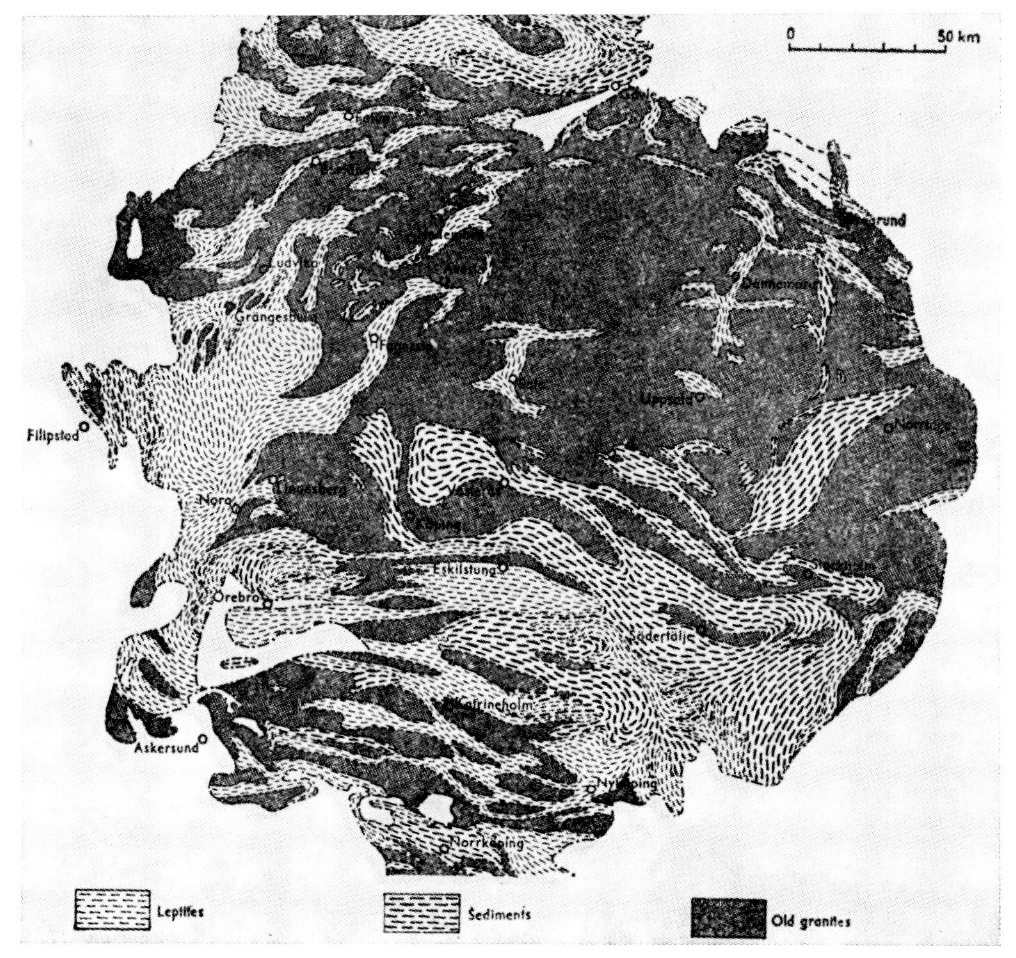

Fig.8. Outline of the Precambrian geology of central Sweden. Compare the absolute dates on Fig.9. (After MAGNUSSON, 1965.)

our knowledge of the Precambrian is still rudimentary, while, on the other hand, not every one of the younger cycles shows such a marked distinction between eugeosynclinal and miogeosynclinal development. But the results so far have been encouraging, and more absolute dating will surely lead to a much more detailed understanding of this area.

We cannot go further into other details of the modern viewpoints, but must refer to Table III, where the main stratigraphic elements of the Fennoscandian Shield, as they now appear, have been summarized. A synthesis of the absolute dates of the Fennoscandian Shield and its geologic structure, mainly based on Russian literature, is given by HOFMANN (1962). Its general structure follows from Fig.6. For a summary on the Precambrian geology of Finland, the reader is referred to SIMONEN (1960a, b, c, 1963), and to Fig.7; for that of Sweden to MAGNUSSON (1960b; 1965), and to Fig.8, 9.

To finish this short description of the Fennoscandian Shield, special mention must be made of the youngest members of the Precambrian, the non-metamorphic Jotnian and Sub-Jotnian systems. These series are found mainly in the Dalecarlia province[1] of central

[1] Dalecarlia is named Dalarna in Swedish, and this is the name most commonly found on the maps. Administratively, however, the former province is now called Kopparberg Lan. The ancient mining area of Kopparberg is, to make matters worse, not included in Kopparberg Lan, but lies to the south of it.

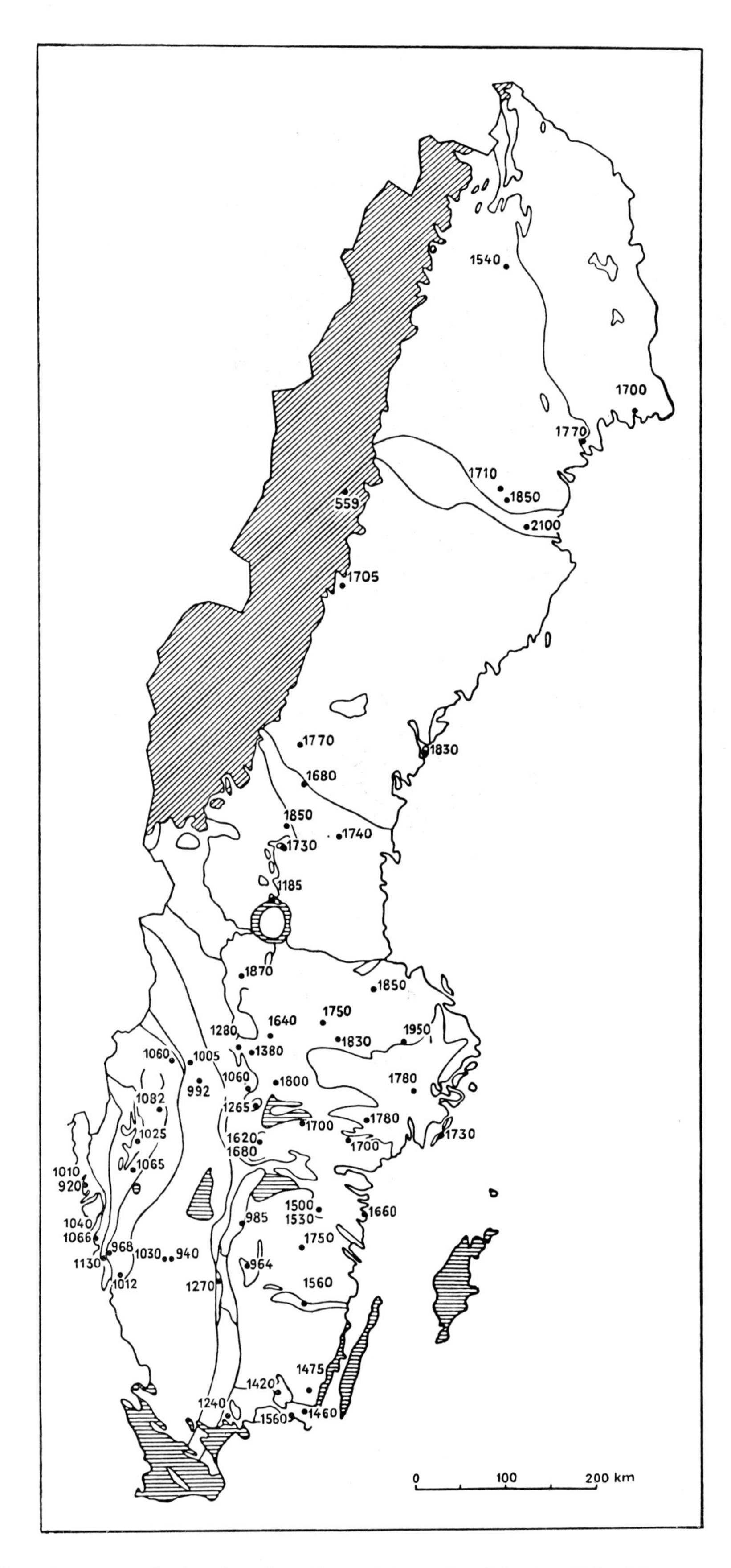

Fig.9. Potassium–argon absolute dates from Precambrian rocks of Sweden. (After MAGNUSSON, 1965.)

Sweden and consist for the larger part of red clastics, overlying porphyries now believed to be ignimbrites (HJELMQVIST, 1956, 1966; RUTTEN, 1966). This, the "Oldest Red" of Sederholm, unconformably covers all earlier metamorphic series of the Precambrian. It is a typical post-orogenetic series of continental clastics and acid volcanics, comparable to the "Old Red" and "New Red" series which follow upon the Caledonian and Hercynian orogenies respectively. To the west are found the Sparagmite Series, as its geosynclinal equivalent. This is a thick Late Precambrian to Eocambrian series of clastics, which forms the base of the geosynclinal sedimentation of the Caledonian orogenetic cycle in northwestern Sweden and in Norway (p. 30).

THE CALEDONIDES

The Caledonides form a continuous belt throughout Scandinavia, stretching from southwestern Norway, continuing along both sides of the Norwegian–Swedish border, all the way up to Lapland. They are quite similar in their history to the Caledonides of Scotland (BAILEY and HOLTEDAHL, 1938)[1]. The literature on the Caledonides of Scandinavia is unfortunately split up into Norwegian and Swedish papers, because the international border divides the chain in a rather erratic way, but see also STRAND (1961). In Sweden the Caledonides are often referred to as "The Mountain Chain".

The early history of the Caledonides of Scandinavia is more or less obscure, as it starts with geosynclinal sedimentation in Late Precambrian times. This is non-fossiliferous, and therefore not easily dated. It forms, however, the lower part of the sedimentary cycle which continues into the Silurian, so its age will not be so much higher than that of the base of the Cambrian. These Late Precambrian series have been called Eocambrian by W. C. Brögger in 1904 (HOLTEDAHL, 1958; NEUMANN and PALMER, 1958). Sedimentation started at different times in various basins along the strike of the future Caledonide fold belt and became general only during the uppermost Eocambrian.

The three younger systems which make up the geosynclinal period of sedimentation of the Caledonides, Cambrian, Ordovician and Silurian, have been taken together as Silurian in most of the older literature, while this complex is designated as Cambro-Silurian in more recent times. Up to the 1960 Copenhagen Congress, the Ordovician normally ranks as the lower part of the Silurian system.

Folding of the Caledonides has, at least in some parts of the chain, taken place in various phases, as yet not well understood. It was mainly completed at the end of the Silurian, before the deposition of the Downtonian. This stage[2], together with overlying Devonian sediments, is developed in a typically postorogenetic red bed facies, which shows posthumous folding only. During the main folding the whole belt was intensively tectonized and numerous and extensive nappes developed. These are, in general, thrust towards the southeast; the outer nappes largely overriding the rim of that part of the Fennoscandian Shield which remained stable.

[1] See also the recent review by STÖRMER, (1967).

[2] The position of the Downtonian as either uppermost Silurian or lowermost Devonian is doubtful (GIGNOUX, 1955, p. 72, 99).

SPARAGMITE SERIES

The early geosynclinal sedimentation during Late Precambrian times is represented by the famous Sparagmite Series. With the exception of the Dala Sandstones of Sweden, Late Precambrian sedimentation was restricted almost exclusively to the area of the later fold belt.

The relationship between more limited subsiding basins in which the earlier beds of the Sparagmite Series were deposited, against the far wider spread of their later sedimentation, is indicated in Fig.10. It should be remembered, however, that sparagmite stratigraphy can only be established along the eastern border of the foldbelt, in the outer nappes of central Sweden and in the Lake Mjösa area north of Oslo, and also farther north, in Finmark. Elsewhere, the series probably developed a much thicker accumulation but it has become more strongly metamorphosed, and the tectonization up to now has obscured the true stratigraphic relations.

Two features of the Sparagmite Series have always drawn particular emphasis: (*1*) the arkosic nature of many of its beds; and (*2*) the intercalation of layers of till, which offer proof of Late Precambrian Ice Ages.

A stratigraphic column of the classical Lake Mjösa area is found in Table IV. For more detailed descriptions and other sections compare HOLTEDAHL (1960, pp.111–126) and Table VI.

The arkosic nature of the Sparagmite Series is more or less restricted to its lower part, which was laid down in localized basins. Here gray and red arkoses, the typical sparagmite rock, dominate, though alternating with quartzites, shales and some limestones. The upper part of the series, which has a far wider distribution, shows predominance of pure quartzites. It is thought that the earlier arkosic members, the sparagmite proper, are derived from nearby granitic sources, whilst the material of the more generally distributed quartzitic upper beds was more widely transported and hence weathered more fully before deposition.

The tillite intercalations, in their most typical form, consist of coarse boulderclays. These are unsorted, boulders of widely varying size being quite irregularly scattered in a fine-grained matrix. Apart from the classical Lake Mjösa area, tillites have now also been

TABLE IV

STRATIGRAPHY OF THE SPARAGMITE SERIES AT LAKE MJÖSA, NORTH OF OSLO
(After HENNINGSMOEN and SPJELDNAES, 1960)

Approximate thickness (m)	*Stratigraphy*	
	arenaceous shale with thin basal conglomerate (Lower Cambrian.)	
Min. 200	Ringsaker Quartzite Vardal Sparagmite	The Quartz-Sandstone Formation
50	Ekre Shale (red and green)	
350	Moelv Conglomerate, (tillite) with transition to: Moelv Sparagmite (red, coarse, in part conglomeratic)	
100	Biri Limestone (and shale)	
150	Biri Conglomerate	
	thin horizon with shale and limestone	
600	Brøttum Sparagmite (dark grey, with shales) (base not known)	

found in northern Norway and in western Sweden. They form regularly interstratified layers, which can be followed over considerable distances. Such former ground moraines indicate the existence of vast, widespread ice caps. It is probable that various tillites occur, at successive levels, indicating repetitive glaciation during a definite period of the Late Precambrian.

We will meet with more or less comparable series of Late Precambrian clastic sediments in the Brioverian of the Armorican Massif (pp.133–136).

CAMBRO-SILURIAN

From the Cambrian onward, after a stratigraphic break in the lower part of the system, sedimentation occurred over a much wider area. For these younger series of the Caledonian sedimentary cycle a marked difference exists between the very thick geosynclinal facies in the Caledonides proper and the thin epicontinental series found outside the fold belt. Transitions between both facies occur, for instance in the autochthonous Cambro-Silurian along the eastern border of the mountain chain, and in the isolated Oslo area.

The Paleozoic era began with a widespread marine transgression, which, although in general of Cambrian age, shows rather wide variations in time of onset in different parts of Scandinavia. The Cambrian seas transgressed over an earlier peneplain, the "sub-Cambrian peneplain" of the Scandinavian literature (Fig.10).

The genesis of this earlier peneplain is sometimes not well understood. ASKLUND (1960), for instance, tells us that "a great change took place in Scandinavia at the beginning of the Cambrian period, when the Lower Cambrian transgression had the power to plane down the earlier coastal platform, resulting in a peneplain". Now the Cambrian sequence does not show any influence of such an abrading power of the Lower Cambrian transgression. As we will see in the next chapter, the Cambrian consists mainly of argillaceous and carbonate rocks, with some sands. Conglomerates are rare, and, when found, often not situated at the base of the series. These conglomeratic intercalations, as well as the sandy members, do not result from the abrading action of the Lower Cambrian transgression. Instead they may be due to temporarily stronger crustal movements between the sedimentary basin and the erosional hinterland during Cambrian sedimentation, through which some coarser material became available by intensified erosion, and was deposited as a coarser intercalation in the sedimentary series.

Against this idea of the abrading power of the Lower Cambrian transgression (which incidentally would imply that the sub-Cambrian peneplain is an abrasion surface and not a peneplain), we must stress the fact that a very long time elapsed after the close of the last orogenetic cycle of the Ripheides and before the onset of the Cambrian transgression. All through the continental periods of the Subjotnian and Jotnian peneplanation took place. The so-called sub-Cambrian peneplain is in all probability an old and complex surface. It was well planed down before the Lower Cambrian sea moved over it with hardly any abrasion due to the transgression.

The "great change in Scandinavia" did not take place at the beginning of the Paleozoic, but much earlier, with onset the Late Precambrian sedimentation of the Sparagmite Series. A similar thing happened at the beginning of the Paleozoic, where its younger part (the Varegian of Asklund) shows a much wider distribution than the Precambrian Sparagmites.

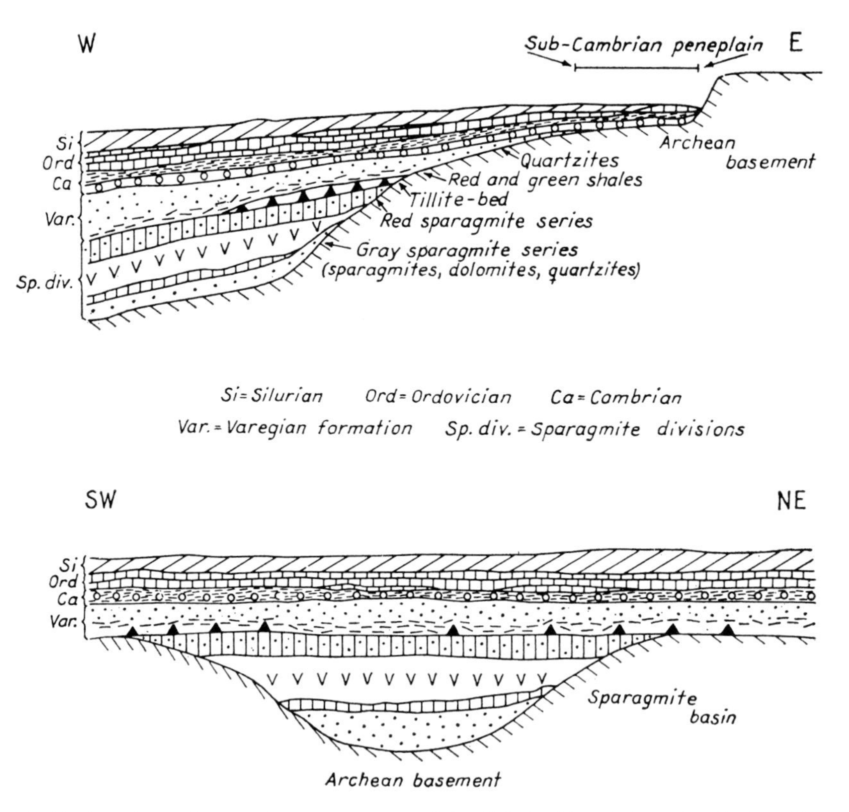

Fig.10. Schematic ideal sections of the sedimentary series of the Caledonides of Scandinavia. The Varegian is the equivalent of the upper sparagmite series of authors. Like the Paleozoic, it shows definite onlap over earlier series. There is often a nonconformity at its base. The tillite layers are normally found at this nonconformity, but may occur slightly higher in the series. (After ASKLUND, 1960.)

But it was not a completely new theme, but only a wider area of distribution, occurring after a sedimentary break, within a cycle of deposition which started with the Sparagmites.

Another well-publicized feature is the "Central Scandinavian Torso" (Fig.11). It is stated that the Cambrian never covered this "torso", a belt some 200 km wide between the epicontinental sedimentary basin around the Baltic and the geosynclinal basin of the Caledonides. Along its western edge the existence of the former Cambrian coast line is inferred from present-day topography, which becomes more rugged to the east of the Cambrian outcrops (see Fig.13).

There is, however, no indication of littoral conditions in the Cambrian sediments themselves. Moreover, in many places in Scandinavia a planed down surface, whether the sub-Cambrian peneplain or not, continues evenly outside the present area of the Cambrian sediments. On top of this, it must be remembered that for the Ordovician and Silurian no "torso" is indicated. (Compare the paleogeographic maps of Thorslund, in: MAGNUSSON, 1960b.)

All through the Cambro-Silurian alternating regional tilting movements took place, resulting in repeated regressions, transgressions and diastems of irregular areal distribution. Moreover important vertical post-Caledonian movements occurred, as witnessed by the erosional remnants of beds in Old Red facies of the Devonian, now found at very different altitudes as erosional remnants scattered over Scandinavia.

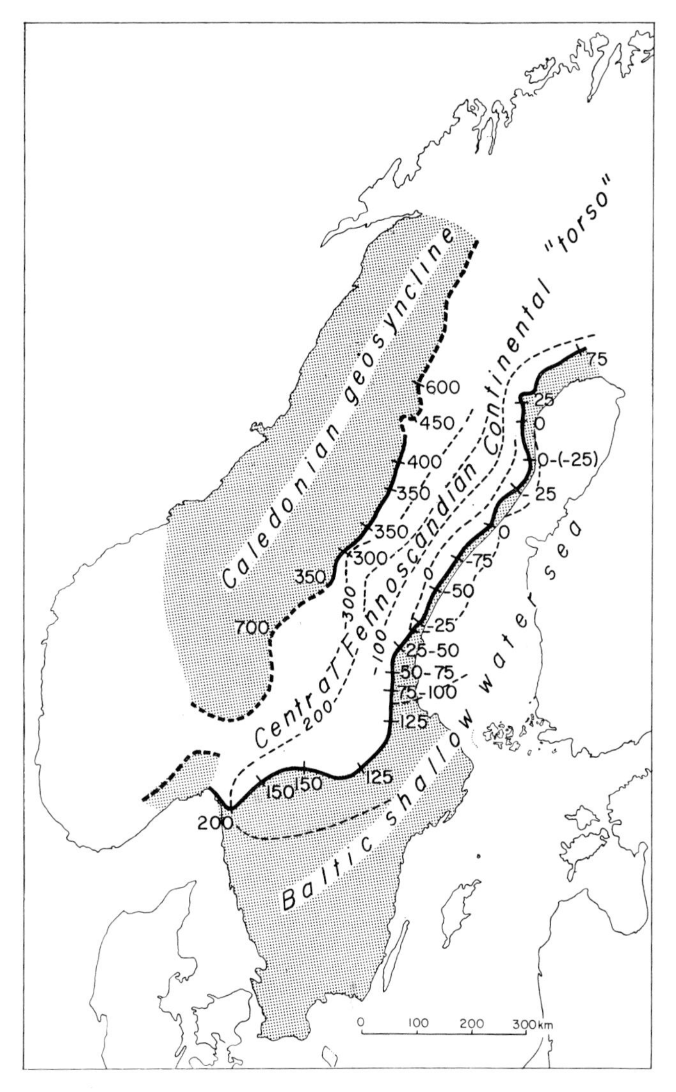

Fig.11. Scandinavian Early Cambrian paleogeography. According to Asklund the thick curves indicate the highest level of the Early Cambrian sea, coinciding with the limits of the Early Cambrian, or sub-Cambrian peneplain, whilst the thin broken lines indicate the present deformation of this old sea level surface. The figures indicate its present altitude above sea level, so the Quaternary deformation by ice loading is included. (After ASKLUND, 1960.)

It seems more probable therefore that the "torso" resulted from removal by erosion of the Paleozoic sequence in those areas where post-Caledonian uplift was relatively strong. The sub-Cambrian peneplain, uncovered by the removal of the Paleozoic sequence, was then also attacked by erosion, resulting in the present-day relief. The importance of post- (or late-) Caledonian deformation of the sub-Cambrian peneplain is indicated for instance, by Strand (in: HOLTEDAHL, 1960, fig.52), who gives a maximum present altitude of 1,800 m for this surface in southern Norway.

Turning now from the sub-Cambrian peneplain to the Cambro-Silurian, there is, as stated, a strong difference between the epicontinental series outside the Caledonides, and the geosynclinal series belonging to the foldbelt. The former, a predominantly carbonate sequence, will be treated in the next chapter. It is known in great detail and is often very fossiliferous. The latter is much less known. It forms great thicknesses of clastic rocks, often with many volcanics, and also with many thin-bedded limestone intercalations. It is moderately to strongly metamorphosed. Scattered fossils, found all over the area, prove that all three lower Paleozoic systems are represented (HOLTEDAHL, 1960, fig.45).

These series have, however, been studied in detail only in localised areas, two of the best known being those of the Bergen Arcs (C. F. KOLDERUP and N. H. KOLDERUP, 1940), and the Trondheim volcanic facies (CARSTEN, 1960). It would lead me to too many scattered descriptions, to cite them more fully here. The reader is therefore referred to the literature for further information, first of all, of course, to the relevant chapters in *Geology of Norway* (HOLTEDAHL, 1960).

STRUCTURE OF THE CALEDONIDES

The Caledonian folding has affected the Cambro-Silurian together with the Eocambrian sediments, and often with considerable portions of the Precambrian basement. A striking characteristic of this belt is the prevalence of nappe structures. As such, the Caledonian fold belt is much more comparable with the Alpine than with the intervening Hercynian fold belt in Europe. In contrast to the Alpine fold belt, the Caledonides have undergone much more intensive erosion. Consequently their orogenetic structure is exposed at a much lower tectonic horizon. As a result, we find in the Caledonides predominantly the subhorizontal thrust planes of the lower nappes, their frontal parts having long since been eroded.

In the Caledonides of Scandinavia certain regional differences may be found, just as they are found in, say, the Alps. Along their eastern border there is a prevalence of easily distinguishable nappe structures, mainly formed by thin, extensive, subhorizontal slices of Eocambrian, or of Cambro-Silurian, or both. In southern Norway, on the other hand, nappes of much more strongly metamorphosed supracrustals, of gneisses and granites of unknown age, form the overthrust masses. Further west, mainly in the coastal districts of southern and central Norway, nappe structures are less prominent, though strong metamorphism has taken place here also. This regional structure is schematically illustrated in Fig.12.

It is again not feasible to enter into too much detail, so only a few examples of these various regions will be given here. An example of the area with many thin, superposed nappes along the eastern border of the fold belt as found in central Sweden is described by ASKLUND (1960) (see Fig.13, 14). Other, more northerly, sections are those of DURIETZ (1960) (see Fig.15), and that from Narvik eastwards by Kulling (in: KULLING and GEYSER, 1960; Fig.16). An excellent microtectonic investigation is that by LINDSTRÖM (1962).

A well-studied example of the big sub-horizontal nappes of strongly metamorphic crystalline rocks is that of the Bergsdalen nappes, including the Jotunheimen and Hardangerjökulen districts, as studied by KVALE (1960) (see Fig.17, 18). These are well exposed along the railroad from Bergen to Oslo, near Finse, where they form the imposing monadnocks rising steeply above the general peneplain. As is the case in most fold belts with nappe structures, the amount of overthrusting is disputed, but even the minimum figures stand at "thrusting in the order of 10–30 km, perhaps as much as 50 km in an east-southeast direction" (OFTEDAHL, 1961).

To the west of the Bergsdalen nappes are the famous Bergen Arcs (C. F. KOLDERUP

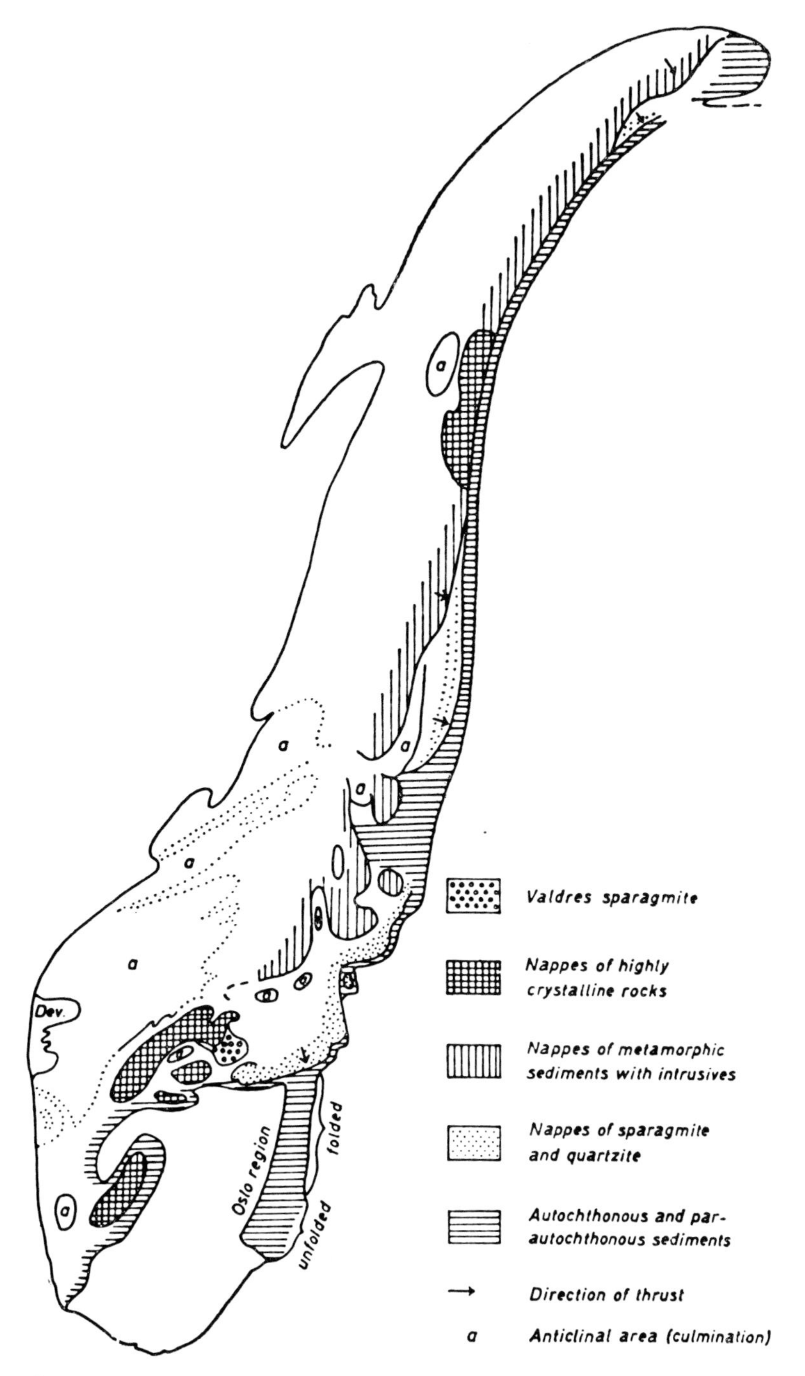

Fig.12. The main tectonic features of the Caledonides of Scandinavia. (After HOLTEDAHL, 1960.)

and N. H. KOLDERUP, 1940; Fig.17–19). As seen from the map, the arcs proper are but the southeasterly closure of a much larger composite synclinal structure, which stretches from Bergen towards the northwest. This structure is, in turn, cut off obliquely to the west by the more strongly metamorphosed Western Gneisses of the coastal islands. The question whether the composite synclinal Bergen structure is autochthonous, or whether it forms part of a nappe, has not yet been decided.

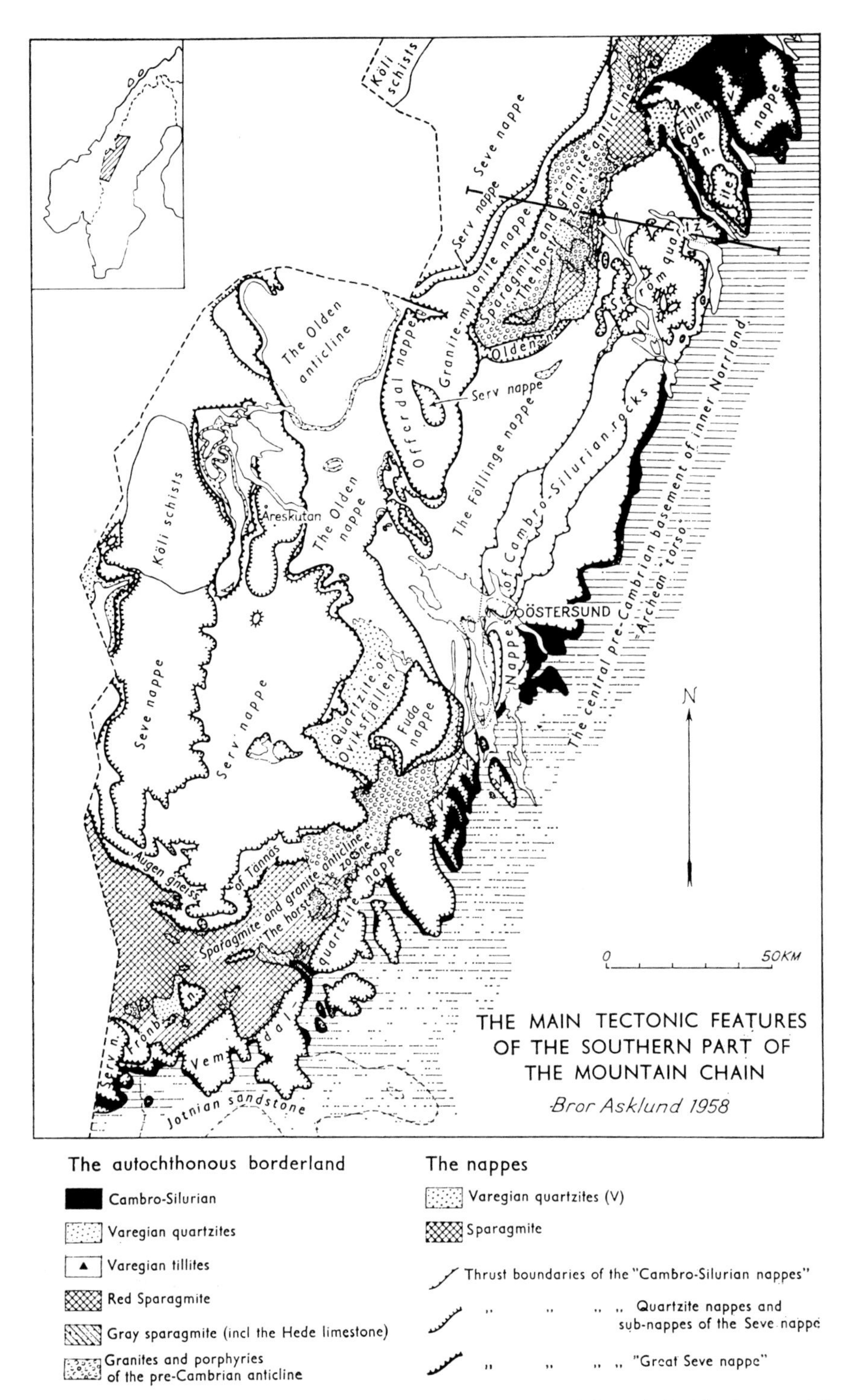

Fig.13. Tectonic map of the nappe structures along the eastern front of the Caledonides in central Sweden. (After ASKLUND, 1960.)

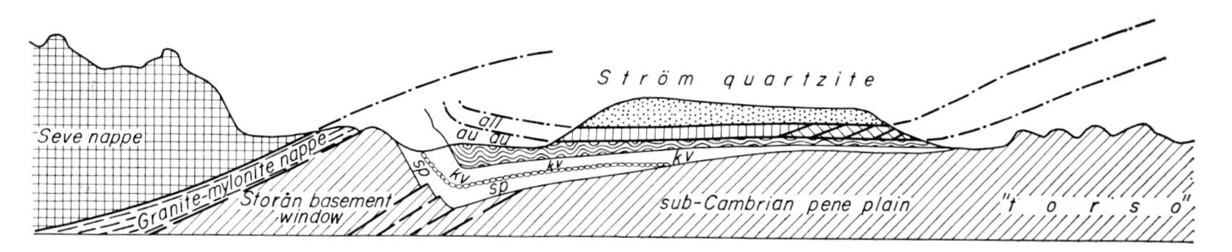

Fig.14. Schematic section through the nappe structures along the eastern front of the Caledonides in central Sweden. Compare Fig.13 for approximate location of section. (After ASKLUND, 1960.)

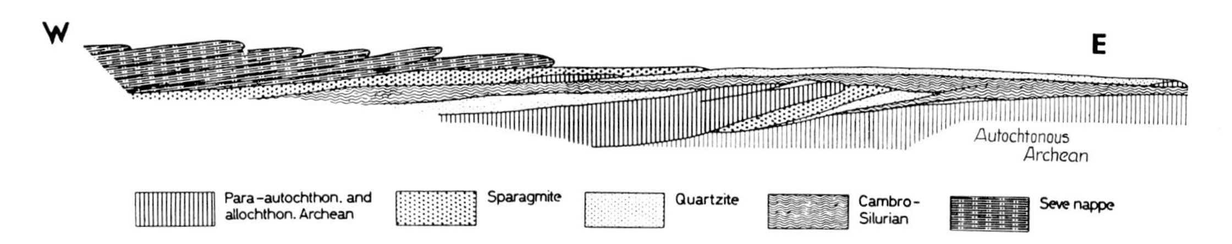

Fig.15. Schematic section through the front range of the Caledonides in northern Sweden, showing the extreme thinness of individual overthrust slices and the low dip of overthrust planes. (After DURIETZ, 1960.)

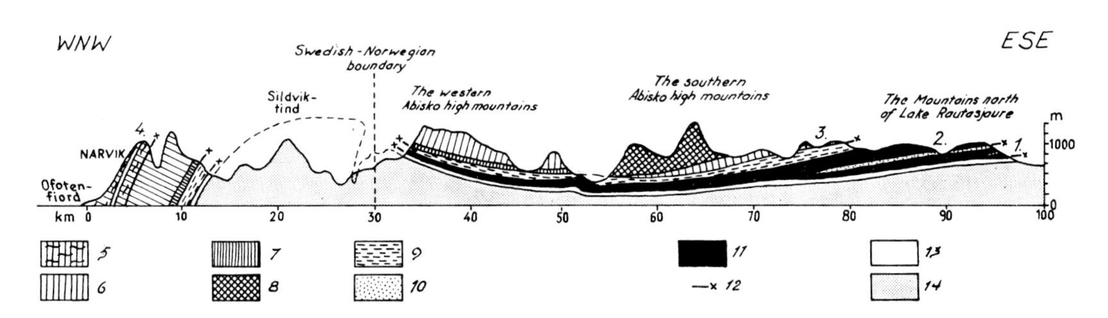

Fig.16. Section through the northern tip of the Caledonides, east of Narvik.
1 = Rautas nappe complex; *2* = Abisko nappe; *3* = Seve-köli nappe; *4* = Rödingsfjäll nappe (?); *5* = schists rich in granite intrusions; *6* = mica-schists and mica-gneiss; *7* = marble, mostly calcite-marble; *8* = amphibolite; *9* = banded sericite-quartzite, schists and dolomite; *10* = quartzite, slate and dolomite; *11* = cataclastic Archean granite and syenite of thrust units; *12* = thrust; *13* = autochthonous Lower Cambrian; *14* = Archean substratum of the Caledonides. (After Kulling in: KULLING and GEYER, 1960.)

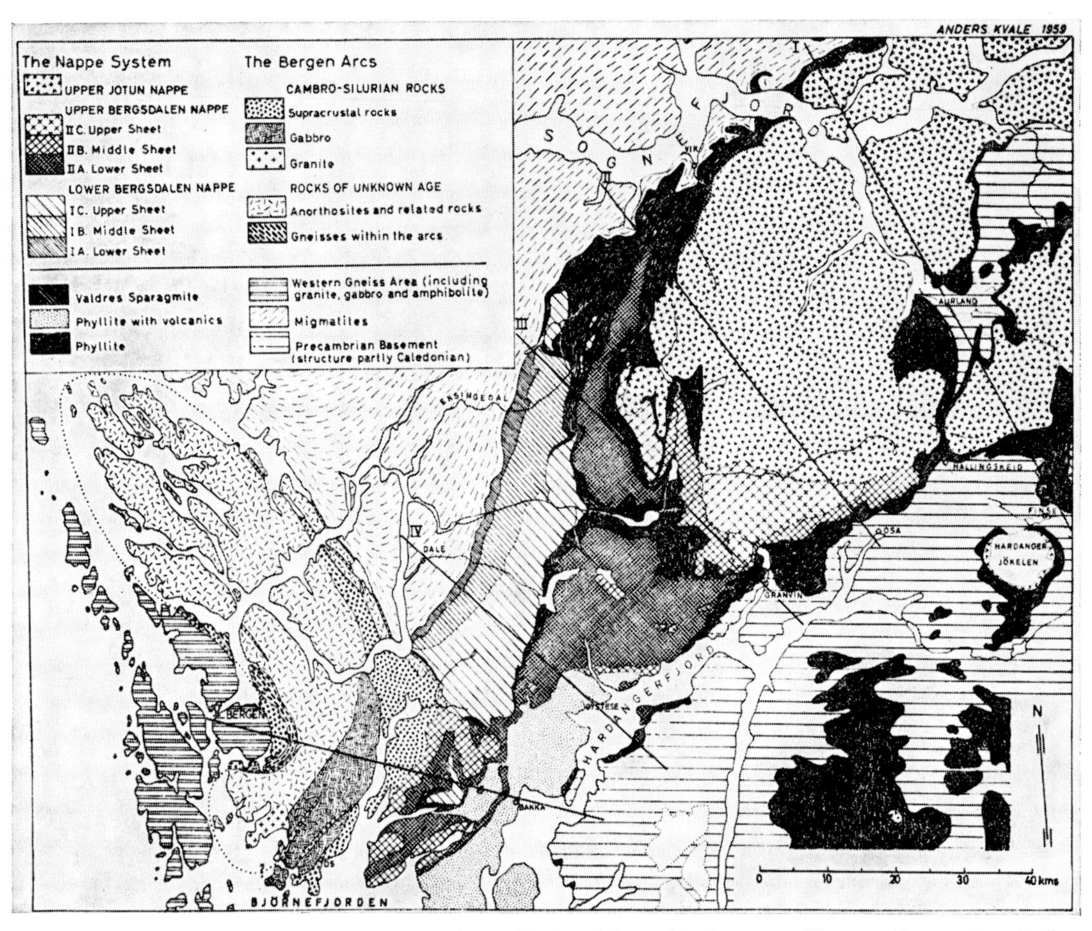

Fig.17. Tectonic map of the area between Hardangerfjord and Sognefjord, western Norway. Bergen Arcs in the southwest and Bergsdalen nappes in the northeast. (After KVALE, 1960.)

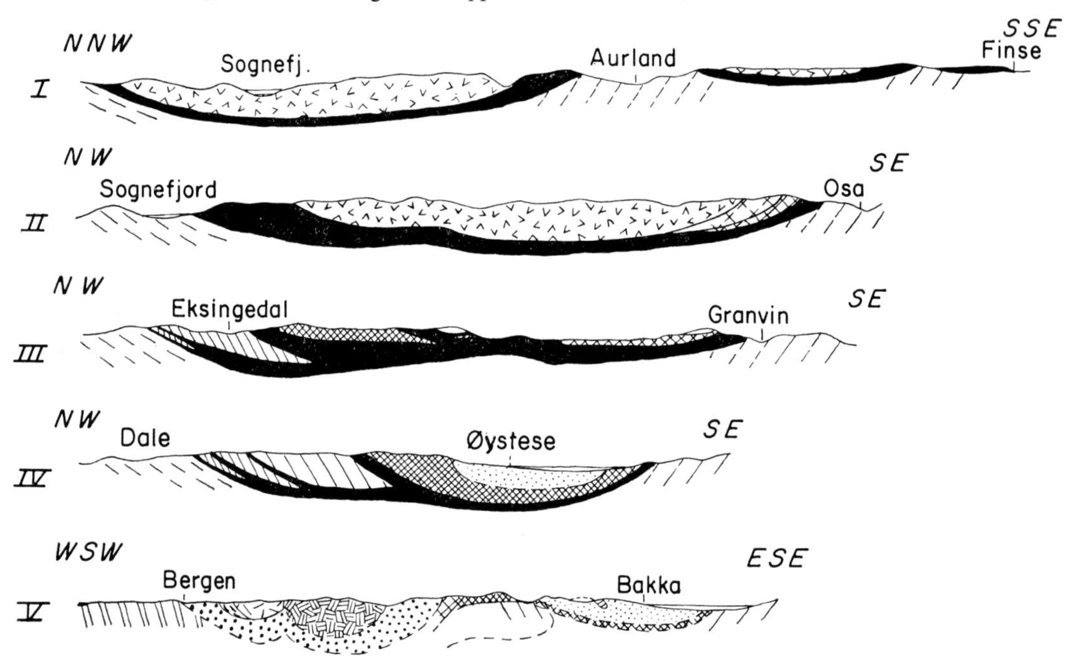

Fig.18. Sections through the Bergsdalen nappes and the Bergen Arcs. Compare Fig.17. (After KVALE, 1960.)

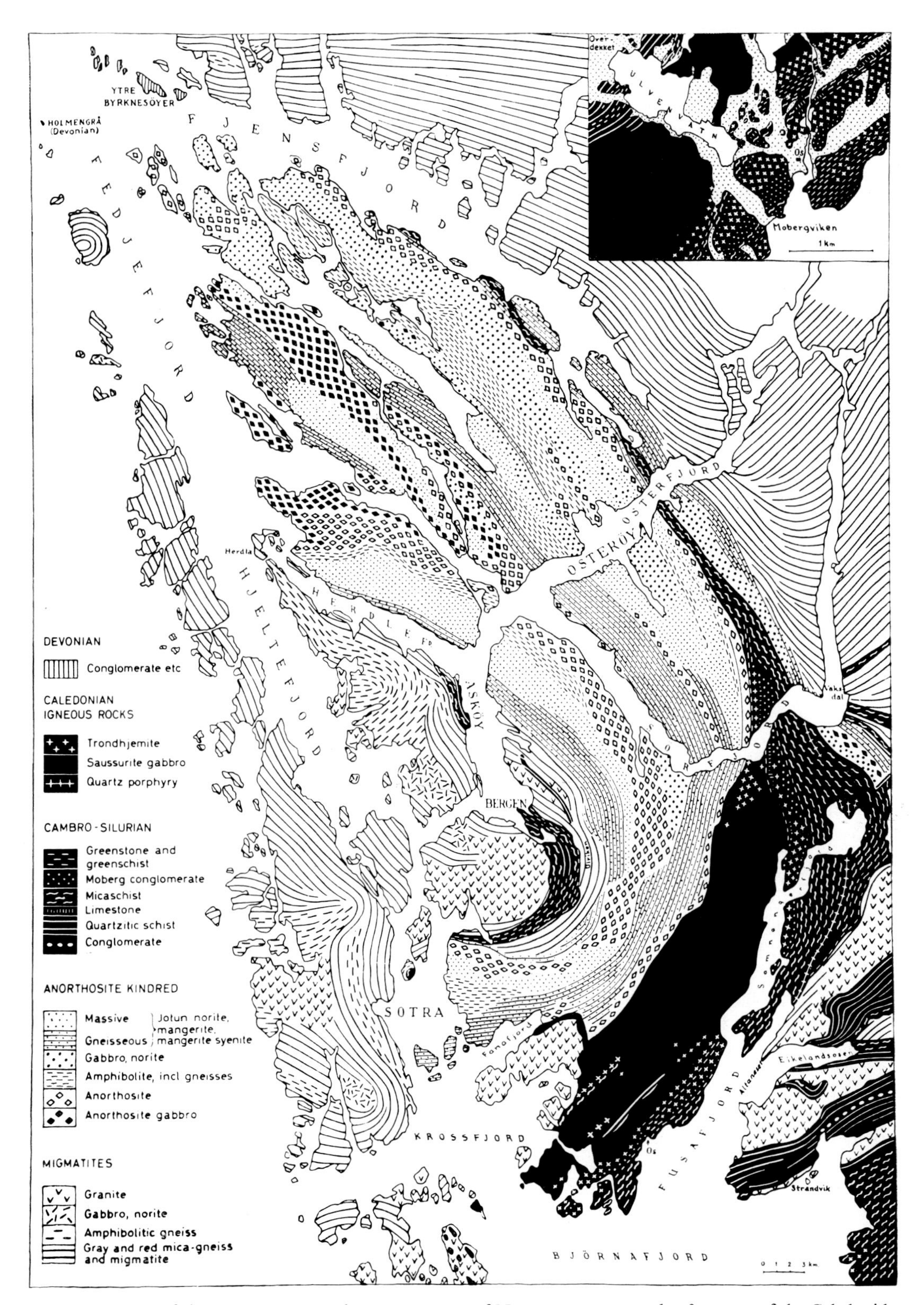

Fig.19. Structure of the Bergen Arcs on the western coast of Norway as an example of an area of the Caledonides which possibly is autochthonous. Compare Fig.17 and section *V* of Fig.18. (After C. F. KOLDERUP and N. H. KOLDERUP, 1940.)

REFERENCES

ASKLUND, B., 1938. Hauptzüge der Tektonik und Stratigraphie der mittleren Kaledoniden in Schweden. *Sveriges Geol. Undersökn., Ser. C*, 417: 99 pp.

ASKLUND, B., 1958. Le problème Cambrien–Éocambrien dans la partie centrale des Calédonides suédoises. *Centr. Natl. Rech. Sci., Colloques Intern.*, 76: 39–52.

ASKLUND, B., 1960. Studies in the thrust region of the southern part of the Swedish mountain chain. *Intern. Geol. Congr., 21st, Copenhagen, 1960, Guides to Excursions*, A 24, C 19: 60 pp.

BAILEY, E. B. and HOLTEDAHL, O., 1938. *Northwestern Europe - Caledonides. Handbook of Regional Geology.* Akademie Verlag, Leipzig, 76 pp.

CARSTEN, H., 1960. Stratigraphy and volcanism of the Trondheimsfjord area, Norway. *Intern. Geol. Congr., 21st, Copenhagen, 1960, Guides to Excursions*, A 4, C 1: 22 pp.

DURIETZ, T., 1960. Tectonic conditions in the front range of the Swedish Caledonian in central Norrland. *Sveriges Geol. Undersökn., Årsbok*, 53: 57 pp.

ESKOLA, P., 1960. Granitentstehung bei Orogenese und Epeirogenese. *Geol. Rundschau*, 50: 105–123.

GAVELIN, S. and LUNDEGÅRDH, P. H., 1960. Development of gneisses and granites in southern Sweden. *Intern. Geol. Congr., 21st, Copenhagen, 1960, Guides to Excursions*, A 28, C 23: 36 pp.

GAVELIN, S. and MAGNUSSON, N. H., 1933. *Geologisk Oversiktskarta över Norden, 1:1,000,000.* Generalstabens Lithogr. Anstalt, Stockholm.

GIGNOUX, M., 1955. *Stratigraphic Geology.* Freeman, San Francisco, Calif., 759 pp.

HENNINGSMOEN, G. and SPJELDNAES, N., 1960. Paleozoic stratigraphy and paleontology of the Oslo region, Eocambrian stratigraphy of the sparagmite region, southern Norway. *Intern. Geol. Congr., 21st, Copenhagen, 1960, Guides to Excursions*, A 14, C 11: 29 pp.

HJELMQVIST, S., 1956. On the occurrence of ignimbrite in the Precambrian. *Sveriges Geol. Undersökn., Ser. C*, 542 *(Årsbok, 49)*: 1–12.

HJELMQVIST, S., 1966. Berggrundskarta över Kopparbergs Län, 1:200,000. *Sveriges Geol. Undersökn., Ser. Ca*, 40.

HOFMANN, J., 1962. Zum präkambrischen Kristallin Kareliens und der Halbinsel Kola. *Geologie (Berlin)*, 11:397–415.

HOLTEDAHL, O., 1944. On the Caledonides of Norway. *Skrifter Norske Videnskaps Akad. Oslo, Mat. Naturw. Kl.*, 4: 31 pp.

HOLTEDAHL, O., 1952. The structural history of Norway and its relation to Great Britain. *Quart. J. Geol. Soc. London*, 108: 65.

HOLTEDAHL, O., 1958. La Sparagmite Formation (Kjerulf) et l'Éocambrien (Brögger) de la Péninsule Scandinave. *Centr. Natl. Rech. Sci., Colloques Intern.*, 76: 33–38.

HOLTEDAHL, O. (Editor), 1960. *Geology of Norway — Norg. Geol. Undersøk.*, 208: 540 pp.

KOLDERUP, C. F. and KOLDERUP, N. H., 1940. Geology of the Bergen Arc system. *Bergens Museums Årsbok, Naturv. R., Skrifter*, 20: 137 pp.

KULLING, O. and GEIJER, P., 1960. The Caledonian mountain chain in the Torneträsk-Ofoten area, northern Scandinavia. The Kiruna iron ore field, Swedish Lapland. *Intern. Geol. Congr., 21st, Copenhagen, 1960, Guides to Excursions*, A 25, C 20: 76 pp.

KULP, J. L. and NEUMANN, H., 1961. Some potassium–argon ages on rocks from the Norwegian basement. *Ann. N. Y. Acad. Sci.*, 91: 469–475.

KVALE, A., 1953. Linear structure and their relation to movement in the Caledonides of Scandinavia and Scotland. *Quart. J. Geol. Soc. London*, 109: 51.

KVALE, A., 1960. The nappe area of the Caledonides in western Norway. *Intern. Geol. Congr., 21st, Copenhagen, 1960, Guides to Excursions*, A 7, C 4: 43 pp.

LINDSTRÖM, M., 1962. Beziehungen zwischen Kleinfaltenvergenzen und anderen Gefügemerkmalen in den Kaledoniden Skandinaviens. *Geol. Rundschau*, 51: 144–180.

LUNDQVIST, G., 1958. Karta över Sveriges jordarter, 1:1,000,000 (Quaternary deposits of Sweden). *Sveriges Geol. Undersökn., Ser. Ba*, 17.

LUNDQVIST, G., 1959. Description to accompany the map of the Quaternary deposits of Sweden. *Sveriges Geol. Undersökn., Ser. Ba*, 17: 116 pp.

MAGNUSSON, N. H., 1958a. Karta över Sveriges berggrund, 1:1,000,000 (pre-Quaternary rocks of Sweden). *Sveriges Geol. Undersökn., Ser. Ba*, 16.

MAGNUSSON, N. H. (Editor), 1958b. *Lexique Stratigraphique International. I. Europe. 2c. Suède.* Centr. Natl. Rech. Sci., Paris, 498 pp.

MAGNUSSON, N. H., 1960a. The stratigraphy of the Precambrian of Sweden outside the Caledonian mountains. *Intern. Geol. Congr. 21st, Copenhagen, 1960, Rept. Session, Norden*, 9: 132–140.

REFERENCES

MAGNUSSON, N. H. (Editor), 1960b. Description to accompany the map of the pre-Quaternary rocks of Sweden. *Sveriges Geol. Undersökn., Ser. Ba*, 16: 177 pp.

MAGNUSSON, N. H., 1965. The Precambrian history of Sweden. *Quart. J. Geol. Soc. London*, 121: 1–30.

MICHOT, P., 1960. La géologie de la catazone. *Intern. Geol. Congr., 21st, Copenhagen, 1960, Guides to Excursions*, A 9: 54 pp.

NEUMAN, R. B. and PALMER, A. R., 1958. Critique of Eocambrian and Infracambrian. *Intern. Geol. Congr., 20th, Mexico, 1956, Rept, El Sistema Cambrico*, 1: 427–435.

OFTEDAHL, C., 1961. On the genesis of the gabbroic rock bodies of the Norwegian Caledonides. *Bull. Geol. Inst. Univ. Uppsala*, 40: 87–94.

POLKANOV, A. A. and GERLING, E. K., 1960. The Precambrian geochronology of the Baltic Shield. *Intern. Geol. Congr., 21st, Copenhagen, 1960, Rept. Session, Norden*, 9: 183–191.

RANKAMA, K. (Editor), 1963. *The Precambrian. I. Fennoscandian Shield.* Interscience, New York, N.Y., 279 pp.

RUTTEN, M. G., 1966. The Siljan ring of Paleozoic, central Sweden: a posthumous ringcomplex of a Late Precambrian Dala Porphyries caldera. *Geol. Mijnbouw*, 45: 125–136.

SIMONEN, A., 1960a. Pre-Quaternary rocks in Finland. *Bull. Comm. Géol. Finlande*, 191: 1–49.

SIMONEN, A., 1960b. Petrographic provinces of the plutonic rocks of the Svecofennides of Finland. *Intern. Geol. Congr., 21st, Copenhagen, 1960, Rept. Session, Norden*, 13: 28–38.

SIMONEN, A., 1960c. Precambrian stratigraphy of Finland. *Intern. Geol. Congr., 21st, Copenhagen, 1960, Rept. Session, Norden*, 9: 141–153.

SIMONEN, A., 1963. Alter und allgemeine Charakteristik des finnischen Grundgebirges. *Geol. Rundschau*, 52: 250–260.

STÖRMER, L., 1967. Some aspects of the Caledonian geosyncline and foreland west of the Baltic Shield. Quart. *J. Geol. Soc. London*, 123: 183–214.

STRAND, T., 1961. The Scandinavian Caledonides, a review. *Am. J. Sci.*, 259: 161–172.

STRAND, T., STÖRMER, L. et SIMONEN, A., 1956. *Lexique Stratigraphique International. I. Europe. 2a, 2b. Norvège—Finlande.* Centr. Natl. Rech. Sci., Paris, 101 pp.

VON BUBNOFF, S., 1952. *Fennosarmatia.* Akademie Verlag, Berlin, 450 pp.

CHAPTER 3

Northern Europe: Epicontinental Deposits, the Oslo Region, Post-Glacial Rise

INTRODUCTION

Continuing our analysis of northern Europe, this chapter will contain three different subjects. First, a short survey of the epicontinental cover of the Cambro-Silurian in Sweden. Second, an account of the Oslo region, a graben in which both an interesting series from the Cambro-Silurian, and much later Permian volcanics and subvolcanics have been preserved. The Oslo region Cambro-Silurian forms a nice intermediary, both as to its sedimentation and as to its tectonic position, between the geosynclinal facies of the Caledonides and the epicontinental cover of Sweden. The Permian igneous rocks are renowned in petrography as representatives of a mildly alkaline suite, and whose volcanic and volcanotectonic character has attracted recent attention. The third subject is the post-Glacial isostatic rebound from the unloading of the melting ice cap. This is a classic example, and is found in most textbooks. Nevertheless it should be included in a regional description of northern Europe.

EPICONTINENTAL CAMBRO-SILURIAN

Deposits of Cambro-Silurian age, non-metamorphosed and nondisturbed except for slight tilting and some faulting, form a definite turning point from the earlier Precambrian rocks of the Fennoscandian Shield. In continental Sweden they are preserved only in small disjunct areas by favour of graben tectonics or the cover of a dolerite cap (Fig.20). Along the eastern coast they are continuous from Kjalmar, over Öland and Gotland, across the Baltic towards Estonia.

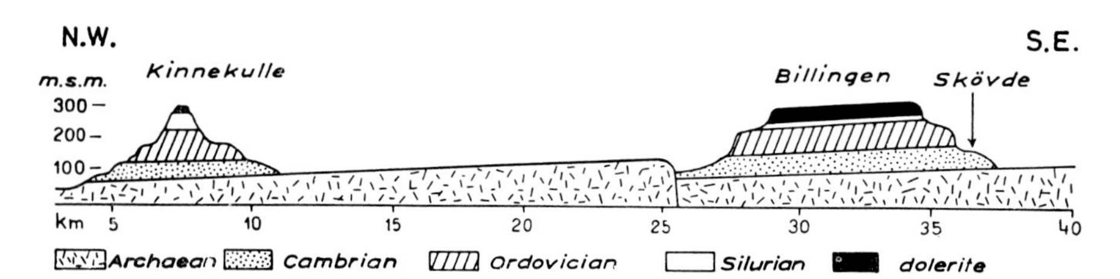

Fig.20. Section across the erosional remnants of Lower Paleozoic of Mount Kinnekule and Mount Billingen central Sweden. (After THORSLUND and JAANUSSON, 1960.)

The series is formed mainly by a succession of shales and limestones. These show the strong lateral variation typical for epicontinental deposits in a shallow sea. These variations are due to local tilting movements, to local variations in the supply of terrigenous sediments, and perhaps to local variations in other causes. These variations seem to be haphazard in space and time.

The series is highly fossiliferous, and as such has attracted many detailed stratigraphic analyses. Although rather complete at first sight, with most biostratigraphic zones more or less well established, the sequence is found to be very incomplete when studied in detail. This has been graphically expressed for the first time, by the great number of diastems indicated by TJERNVIK (1958) in his measured sections. The interested reader will find general accounts in GIGNOUX (1955), BRINKMANN (1960), Thorslund (in: MAGNUSSON, 1960b); and in the *Lexique Stratigraphique International* (MAGNUSSON, 1958b). More detailed descriptions will be found in two of the *Guides to Excursions* of the 21st Session of the International Geological Congress, those by THORSLUND and JAANUSSON (1960) and by REGNELL and HEDE (1960).

BRINKMANN (1960) cites a thickness of 150 m for the epicontinental Cambrian in Scania, as against 55 m in central Sweden. A drastic change is found in the Silurian, with 100 m in an incomplete development in central Sweden, thickening to about 500 m in Gotland. In Scania, however, the Silurian reaches a thickness of over 1,000 m, probably up to 1,500 m, in a pure shale facies. This local, strong, temporary subsidence, compensated by rapid sedimentation, was situated in a truly cratonic environment, far away from the geosyncline of the Caledonides. It serves to show that possibly one of our most cherished tenets, that of the fundamental difference between craton and geosyncline, between stable and mobile, is over-emphasized or over-schematized.

Apart from the detailed stratigraphies and fossil correlations, a couple of points are of more general interest. These are the alum shales and the kolm of central Sweden, the *Orthoceras* limestones of the Ordovician, and the Silurian reefs of Gotland.

ALUM SHALES AND "KOLM"

Alum shales occur in almost every Cambro-Silurian sequence in southern Scandinavia. They are a facies of shales with a varying amount of biogenic material. This may only consist of very fine-grained pyrite, disseminated through the rock, but often there is more of it, the rock becoming a typical black shale. In central Sweden the organic content is locally so high, that oil shales are found.

Alum shales are restricted neither to the Early Paleozoic, nor to Scandinavia. It is a facies widely found, for instance in the Devonian of the Hercynian geosynclines, or in the Lias of the post-Hercynian basins. The name is derived from their former great economic importance as raw material for the manufacture of alum. This went out of use so long ago, that it is almost forgotten now. So, instead of designating "only" a lithofacies, the name originally had the meaning of alum ore, and we should look into this matter a little more. Alum, $KAl(SO_4)_2 \cdot 12H_2O$, served as the most important bleaching agent for the medieval textile industry. It was extracted, often by means of human urine, from the ashes resulting from roasting black shales. In the process the aluminum and potassium of the clay minerals were attacked by sulphurous acid formed from the finely disseminated pyrite or other biogenic material, and recombined to form alum. The only requisite for shales to be considered alum shales, consequently, is that they must be very fine grained and contain a certain amount of disseminated pyrite of organic matter. Moreover, the name alum shales is, it stands to reason, normally applied only to such shales as found around early centres of industry. Many similar blackish shales now go under other aliases, only because they were described after the period when such shales were of economic importance as a source for alum.

In the Middle and Upper Cambrian of central Sweden the alum shales locally grade into bituminous shales through an increase of organic matter. These have long been quarried as oil shales, with local grades of over 5.4% oil in sections over 10 m thick. Also, in a newer development, they are burnt in bulk for the production of "*ytong*", a porous, very light building material.

Within these black shales lenses may occur of a sort of coal, the "kolm". Both shales and kolm have in recent times acquired importance for their uranium content. Apart from the economic importance, the latter has permitted absolute dating of these deposits (COBB and KULP, 1961). The age of the post mid-Upper Cambrian, that is the faunal zone of *Peltura scarabaeoides*, on Mount Billingen now stands at a minimum of 500 million years.

Another general aspect of the alum shales of Scandinavia is that during Ordovician and Silurian they generally carry graptolites, often in great quantities. One is accustomed to interpreting the graptolite facies as a deep water deposit. This is emphatically not the case in this area. Not only have the deposits been formed on a stable platform, but their very regular alternation with limestones of neritic facies, coupled with a great number of stratigraphic discontinuities and smaller diastems, indicate a deposition in very shallow water. It is quite possible that the black graptolite shales represent sedimentation in shallow basins with hypertrophic bottom conditions. In such areas only a planktonic or pseudoplanktonic fauna could survive, drifting at the sea's surface.

ORTHOCERAS LIMESTONE

The *Orthoceras* limestone (Fig.21) of the Ordovician is a striking building stone, used all over Scandinavia and adjacent areas. Either in its grey variety, used as flooring or wall coating, or in the more sophisticated "red marble" facies, the long slender *Orthoceras* cones catch the eye everywhere in buildings in northern Europe.

It is quarried and mined in central Sweden and in Öland. Its stratigraphy is known in detail, but the paleoecology of these deposits does not seem to have attracted much interest. This is the more surprising, because in the province of Dalecarlia in central Sweden a bioherm facies of the Ordovician is also found (Thorslund, in: MAGNUSSON, 1960b, p.91; THORSLUND and JAANUSSON, 1960, pp.23–34). A comparison between the paleoecology of these reefs of the Ordovician; the quite different facies of the *Orthoceras* limestone, and the somewhat younger Silurian reefs of Gotland seems well worth while. Especially because the Ordovician Siljan Reefs in Dalecarlia seem to be related to localized posthumous volcano-tectonic movements of a Late Precambrian caldera (RUTTEN, 1966).

One of the best exposures of the *Orthoceras* limestone is found in northern Öland all along the sea cliff or "klint" of the Littorina Sea.

Fig.21. *Orthoceras* limestone; Ordovicium.

SILURIAN REEFS OF GOTLAND

In contrast to the *Orthoceras* limestone, the Silurian reefs of Gotland have always attracted much attention as one of the oldest zoogene biohermal formations known. They are moreover especially well exposed along the beautiful coast of the island. In the areas of Silurian reef growth this coast is normally developed as a steep cliff, or "klint". These "klints" have been formed by the rising of Scandinavia owing to unloading of the melting ice cap in post-Glacial times. The coastal "klints" of Gotland result from uplift since the

TABLE V

SILURIAN STRATIGRAPHY OF GOTLAND[1]

Gotland		*Scania*	*Great Britain*
stratigraphy	*thickness (m)*		
13 Sundre Limestone	10	Lower Öved–Ramsåsa	
12 Hamra Limestone	40	Formation	
11 Burgsvik Sandstone and Oolite	50		
10 Eke Group	15		Ludlowian
9 Hemse Group	100	*Colonus* Shales	
8 Klinteberg Limestone	100		
7 Mulde Marl	25		
6 Halla Limestone	15		
5 Slite Group	100		
4 Tofta Limestone	10	*Cyrtograptus* Shales	Wenlockian
3 Högklint Limestone	20		
2 Upper Visby Marl	10		Llandoverian
1 Lower Visby Marl	10		

[1] Compare Fig.22.

Littorina Sea, while further inland, remnants of the earlier Ancylus Sea cliffs are found. There is a good correlation between the distribution of the Silurian reefs in the "klints" and present topography. All northern headlands show relative predominance of reef structures, which evidently were much better able to withstand erosion than the normal stratified limestones.

The literature on Gotland, mainly of a paleontological and stratigraphical nature, is extensive. The detailed maps 1:50,000 and their descriptions, summarized in MUNTHE et al. (1924), and also in HADDING (1933, 1941) are basic. Other papers are by HADDING (1950), JUX (1957), RUTTEN (1958) and the *Guides to Excursions* by REGNÈLL and HEDE (1960).

Gotland is formed by a slab of Silurian sediments, about 500 m thick and shows an overall dip of about 0.5° towards the southeast. They belong to a mixed facies of alternating marls and limestones, with very minor amounts of sand (Table V; Fig.22). Reefs occur in most of the limestone members (Fig.23, 24).

Although different types can be recognized, the most striking feature of all reefs on the

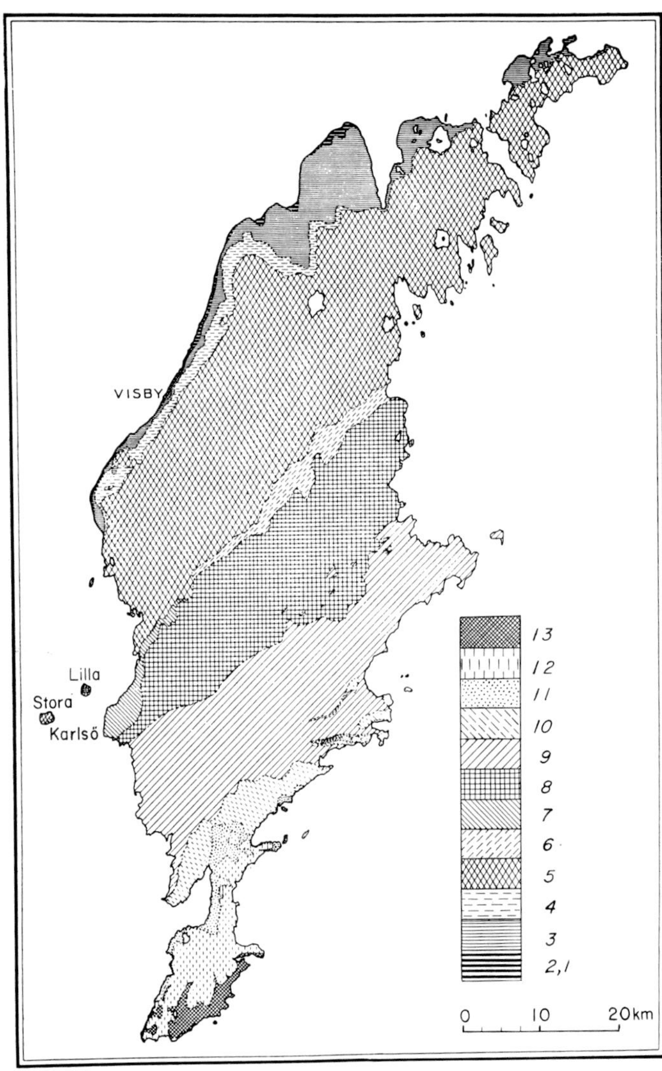

Fig.22. Map of Gotland. Numbers of formations correspond to Table V. Bioherms occur in formations *2*, *3*, *5*, *6*, *8*, *9*, *12*, *13*, and, more rarely, in formations *10* and *11*. (After REGNÈLL and HEDE, 1960.)

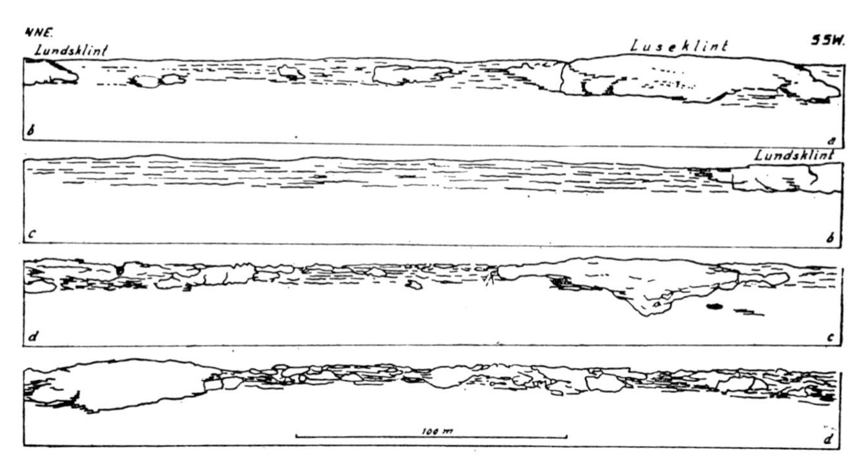

Fig.23. Section through 1.5 km of coastal "klint" north of Visby, Gotland. Showing the many small disjuncted bioherms intercalated in the stratified Högklint Limestone. (After B. Boekschoten, in: RUTTEN, 1958.)

main island is their small size. Thickness is not more than 20 m, normally less than 10 m, while lateral dimensions of over 100 m are rare. Moreover most of the reef bodies contain many smaller local discontinuities, due to hiatuses in growth. The transition towards the surrounding stratified limestones is sharp; individual bioherms having only a narrow envelope of reef detritus. Another feature, peculiar to most of the reefs of Gotland, is their loose structure. This stems from the fact that the bioherms are built mainly of globose or discoidal colonies of stromatoporoids and corals, without much support from branching corals which in other bioherms aid in the construction of a much more massive rock. It is, therefore, rather difficult, in a general study, to distinguish between actual bioherm and reef detritus. Possibly this difficulty has led JUX (1957) to quite erroneous correlations and conclusions. For an analysis of reef growth in space and time in Gotland based on a more thorough field study, the reader is referred to MANTEN (in preparation).

It follows that these Silurian reefs are neither the fossil counterpart of our present-day

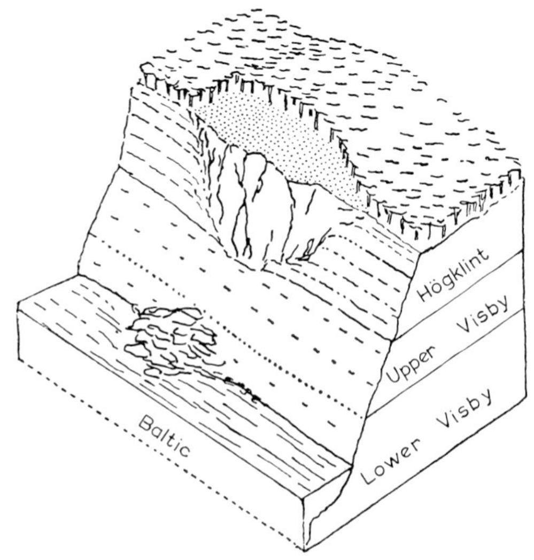

Fig.24. Schematic block diagram of a Högklint Limestone bioherm along the coast north of Visby, Gotland. Present surface of the bioherm, covered by grass in contrast to the wooded stratified limestone, approximately coincides with original surface. (After B. Boekschoten, in: RUTTEN, 1958.)

reefs, of atolls and barrier reefs; nor of other fossil reefs from geosynclinal basins. From the present-day reefs they differ in that they have not had to grow upward rapidly for 100 m or even more, to offset post-Glacial eustatic rise of sea level. From the fossil bioherms in geosynclinal basins, such as the Devonian reefs of the Ardennes (p. 102) they differ in not having had to contend with much stronger crustal subsidence.

The Silurian reefs of Gotland are typical examples of reef growth in a shallow epicontinental sea characterized by slow crustal subsidence. This limited the possibility of reef growth, because during the periods of limestone formation the growing reef could of course not grow any higher than contemporary sea level. The growing reef consequently never projected very much over the floor of the surrounding shallow sea, and did not produce much detritus. On the other hand, during the ensuing period of marly sedimentation, terrigenous supply of fine clayey material prohibited the growth of reef organisms. During the latter periods only biostromes could develop temporarily, often on a base of extensive *Halysites* flats. But these never reach more than a few tens of centimeters in thickness.

A rather different type of reef has developed on the Karlsöarna, the two Karls Islands, off the west coast of Gotland (Fig.25). These reefs are much larger and, moreover, surrounded by a thick and extensive mantle of reef rock. The massive rock of the main reefs is almost exclusively formed by stromatoporoids; the well-bedded mantle, which shows a regular primary radial dip of around 20°, consists predominantly of crinoid detritus. The reef rock mantle thus is not formed from detritus derived from the actual bioherm, but built up by crinoids which found a favourable biotope all around its slope.

The Karls Islands bioherms possibly could reach greater dimensions because during their formation crustal movements were somewhat stronger and more irregular, as is evidenced by rapid facies transitions in the Slite Group.

Both of the Karls Islands show strong glaciotectonics along their southern shores. The weight of the ice cap folded and squeezed the underlying marls into the overlying more massive reef limestone and reef rock. Of course, the fact that the Karls Islands exist at all today as erosional remnants is due to the resistance of their reef limestones to glacial erosion.

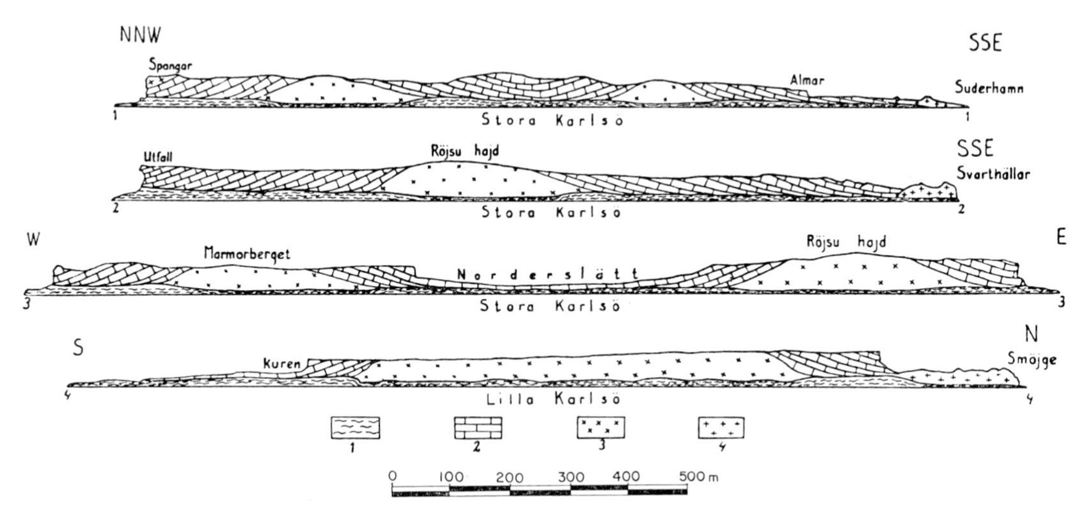

Fig.25. Sections through Stora Karlsö and Lilla Karlsö, off the west coast of Gotland. *1* = marls; *2* = well bedded reef rock, mainly crinoidal, showing radial primary bedding; *3* = central stromatoporoid bioherms; *4* = lateral corraligene bioherms. (After G. Y. Sondaar, in: RUTTEN, 1958.)

THE OSLO REGION

The Oslo region (Fig.26) is formed by a graben, some 200 km long, averaging about 40 km wide, and stretching N 20°E from the mouth of the Oslofjord, past Oslo, towards Lake Mjösa. The downthrow is estimated to be between 2 km and 3 km in its southern part, diminishing northwards. In the graben Cambro-Silurian sediments and Permian volcanics are preserved, which outside the graben have been eroded, exposing the Precambrian basement all along both sides of the graben (Fig.27). The age of the graben, consequently, is post-Permian, but not further datable. It is probable, however, that the massive Permian volcanism, which in part is thought to be of fissure eruption type, is related to the onset of graben tectonics.

The Oslo graben can be associated with similar graben structures also trending approximately N–S further south. These are covered by younger sediments in Denmark and northern Germany, but seem to be present in the basement. Then follow the Hessen and Upper Rhine graben; the Belfort wind gap and the Rhône valley. Together with the Oslo graben they form one of the characteristic lineaments of western Europe, the Mediterranean –Mjösa line of H. Stille. The age of the more southerly graben structures is variable; their structure is, in general, composite. Although faulting normally began in Permian times, it continued intermittently up to the Tertiary or even to Recent times. It is thus possible that the Oslo graben also took a rather long time before it arrived at its present state.

The literature on the Oslo region is varied and extensive. Good summaries are found in *Geology of Norway* (HOLTEDAHL, 1960) and in several of the *Guides to Excursions*, notably those by HENNINGSMOEN and SPJELDNAES (1960) and by OFTEDAHL (1960). Most of the elements of the Oslo region geology can be studied in the immediate vicinity of the town of Oslo, if not in the town itself. An excellent guidebook to this more restricted area is HOLTEDAHL and DONS (1957); with map 1:50,000 and detailed road logs.

The two main elements of the Oslo region geology arouse interest within quite different groups of geologists. The Cambro-Silurian sediments are remarkable for their stratigraphy, paleontology and tectonics; the Permian being known mainly for the petrography of its igneous rocks and for vulcanological questions.

CAMBRO-SILURIAN IN THE OSLO GRABEN

The Cambro-Silurian is notably developed in the same mixed alum shale-and-limestone facies as the comparable rocks of Sweden. Of course, there is no correlation of a single lithofacies all the way through. But the fact that the Cambrian is almost wholly developed in the alum shale facies is in keeping with the Swedish development. The main difference is that there is more of it in Oslo. This does not apply so much to the Cambrian, with about 50 m thickness for Middle and Upper Cambrian, the Lower Cambrian not being present. But then we find 300 m of Ordovician and 510 m of Silurian. On top of that follows a new element, the Downtonian, developed in a slightly folded red bed facies at least 500 m thick (Fig.28).

A good impression of this repetition of detrital and carbonatic series is gained from Fig.29 representing a facies analysis of the Cambro-Silurian of the Oslo region by SEILACHER and MEINSCHNER (1965).

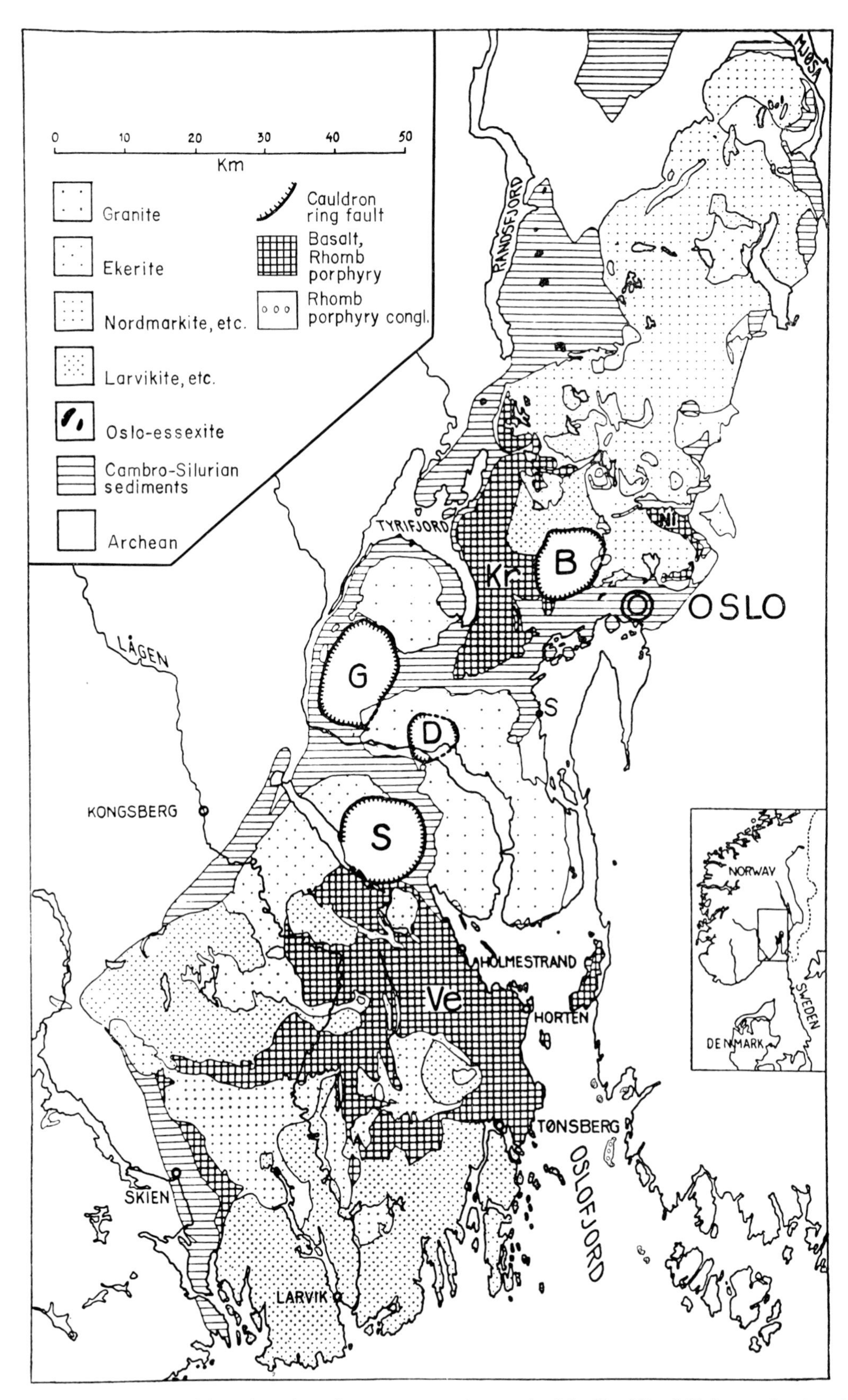

Fig.26. Generalized map of the Oslo region. The Ramnes area, just south of the *Ve* of Vestfold, is now also thought to be formed by supracrustal rocks, ignimbrites and rheo-ignimbrites (RUTTEN and VAN EVERDINGEN, 1961). *Ni* = Nittedal area; *Kr* = Krokskogen area; *Ve* = Vestfold; *B* = Baerum cauldron; *G* = Glittrevann cauldron; *D* = Drammen cauldron; *S* = Sande cauldron; *S* = Slemmestad. (After OFTEDAHL, 1952.)

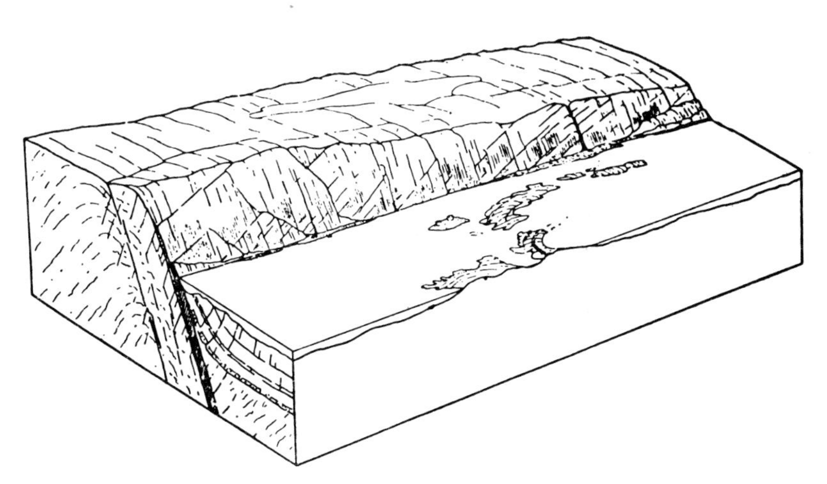

Fig.27. View of the eastern boundary fault of the Oslo graben along Nesodden Peninsula, looking southeast. Islands in the fjord are formed by Ordovician, the much higher mainland by Precambrian basement. (After H. Cloos, in: HOLTEDAHL and DONS, 1957.)

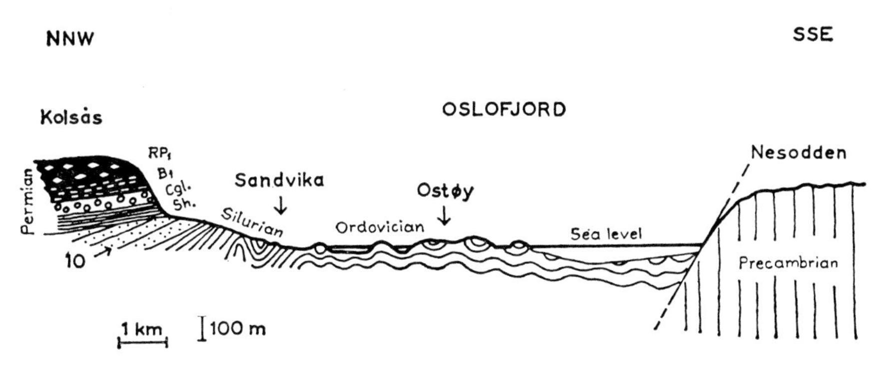

Fig.28. Schematic section from Kolsås to Nesodden, across the Oslofjord. At left Kolsås with dipslope formed by rhombporphyry RP 1, overlying Permian basalt B 1 and Permian conglomerates and shales, exposed at the southern cuesta. Underneath follow the slightly folded red beds of the Downtonian (*10*), nonconformably overlying strongly folded Silurian and Ordovician. Along Nesodden Peninsula the eastern graben fault throws up the Precambrian basement. (After HENNINGSMOEN and SPJELDNAES, 1960.)

These authors assumed that much of the a-faunal alum shales were formed as deep-water deposits, whereas the coquina shales and the limestones indicate shallow water. Together with the alternation of these sediments as depicted in Fig.29, this excludes, in their view, a single phase of regular sinking of the basin. Instead, they conclude that the basin had sank rapidly at first, following the Cambrian transgression. During the Ordovician subsidence must then have stopped almost completely, and the basin was filled. During the Silurian strong subsidence set in again, but this time it was offset by stronger sedimentation, so that the basin remains from shallow to very shallow during this period (Fig.30).

This conclusion rests, however, on the fallacious asumption that the a-faunal alum shales should form only beneath the depth where benthos thrives. From the Dutch Wadden Sea we know, on the other hand, that quite similar deposits may form at high tide level. Moreover, an appreciable amount of organic material, so characteristic for the Cambro-Silurian alum shales of Scandinavia, may originate in this environment from the development of reed marshes.

In the case of the Cambro-Silurian of the Oslo region, pseudomorphoses of halite crystals are even known to occur in the Middle Cambrian alum shales near Slemmestad in the southeastern part of the district. These indicate the existance of temporary eva-

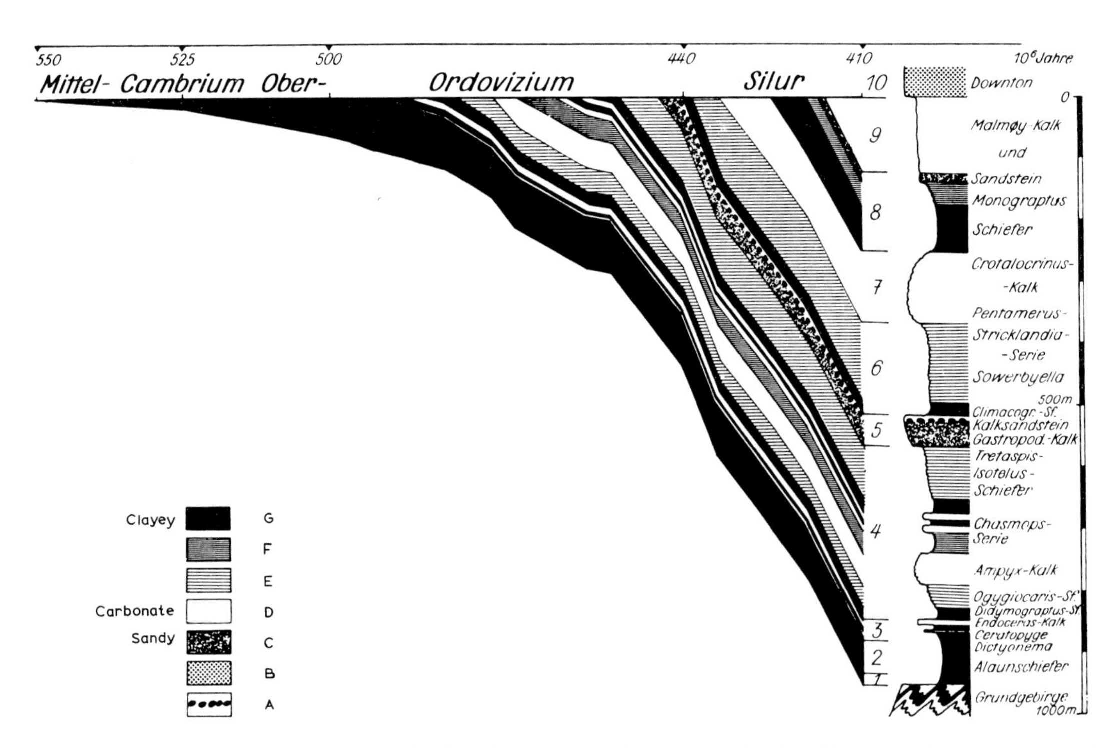

Fig.29. Sedimentation of the Cambro-Silurian of the Oslo region. Accumulated sediment thickness is presented against isotopic age. Under the assumption that all sediments shown were formed at or about sea level, this graph also presents an oscillogram of the Oslo region. G = clayey; D = carbonate; C = sandy.
(After SEILACHER and MEINSCHNER, 1965.)

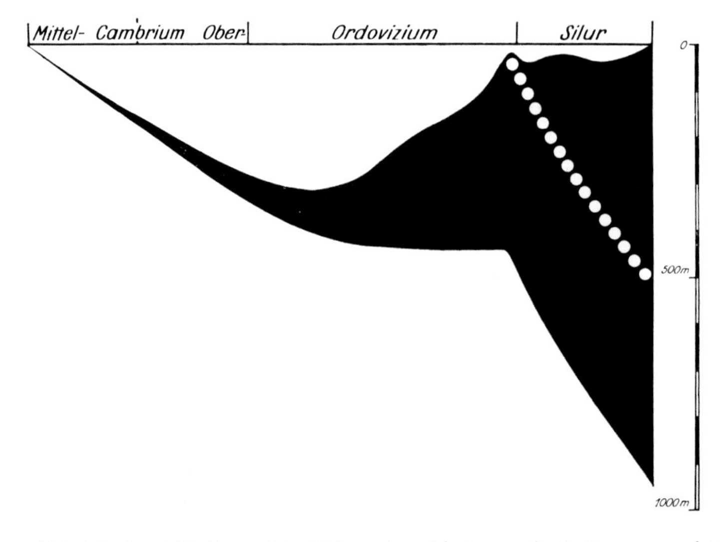

Fig.30. Oscillogram of the Cambro-Silurian of the Oslo region. Black area indicates accumulated sediment thickness. Upper white area indicates contemporary depth of the basin. White dots indicate the position of the conglomerate of zone *5* (Fig.29), at the base of the Silurian.

The diagram has been constructed under the assumption that the alum shales of the Cambrian represent deep water sediments. (After SEILACHER and MEINSCHNER, 1965.)

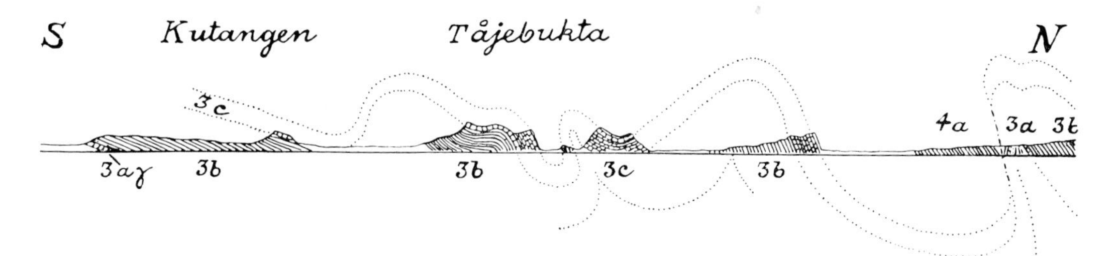

Fig.31. Section through the Cambro-Ordovician along the westcoast of the Oslofjord, north of Slemmestad, showing Jura Mountain type of epidermis folding. (After HOLTEDAHL and DONS, 1957.)

poritic conditions, sustaining a shallow water, instead of a deep water origin of these alum shales.

The facts assembled in SEILACHER and MEINSCHNER's (1965) study are very interesting, because they draw on so many and so varied a combination of both the organic and the inorganic characters of the sediments studied. Their conclusion seems, however, a little apodictic. The same sedimentary series could just as well have been formed in a basin which remained shallow to very shallow during all of its history, and in which rhythmic variations of cyclothem type of subsidence and sedimentation have more or less continuously occurred.

We see here a transition from the epicontinental Cambro-Silurian of Sweden towards the geosynclinal facies of the Caledonides. A similar transition can be found in the tectonics of the Oslo region Cambro-Silurian. For, apart from the lowermost beds exposed along the southern rim, the Oslo sedimentary series is quite distinctly folded.

The direction of this Caledonian folding within the Oslo graben averages some N 60°E near Oslo, turning to east–west further north. As such it shows typically Caledonian directions, and is cut off obliquely by the younger graben faults.

The type of folding is quite similar to that found in the Jura mountains. It consists of tight asymmetric or overturned anticlines, often upthrusted, and wide, shallow synclines. The "Vergenz" or overturn is generally towards the south, which is consistent with the southerly overthrusting of the Caledonian nappes farther north.

Just as the comparable folding of the Jura Mountains, the Caledonian folding of the Cambro-Silurian in the Oslo graben is a typical epidermis folding. The rigid Precambrian basement did not move. A difference with the Jura Mountains lies in the nature of the zone

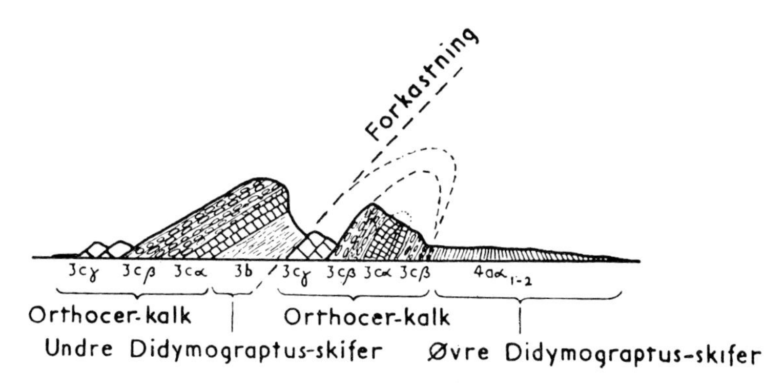

Fig.32. Section through a typical narrow overturned an upthrusted anticline, forming the southern promontory of Bygdöy Peninsula, Oslo City. (After HOLTEDAHL and DONS, 1957.)

of "décollement". In the Jura this is formed by evaporites of the Triassic, but in the Oslo region no evaporites, no rock salt or gypsum, are present. Their role is taken over by the alum shales. The lowermost Cambrian still adheres to the Precambrian basement, dipping about 20°N with the sub-Cambrian peneplain. Then, within the 50 m of Cambrian alum shales, tectonization is strongly developed. Above that, from the lowest sequence of alternating shale and limestone beds of the Ordovician upward, everything is folded.

This style of epidermis folding is illustrated in the sections of Fig.31, 32. Because of the differential erosion of the alternating limestone and shale beds, tectonic structure has a strong influence on the morphology of the Oslo region. This is quite clear, for instance, in the Bay of Oslo proper, where all skerries are aligned according to the direction of the Caledonian folding.

PERMIAN PLUTONIC ROCKS AND VOLCANICS

The second aspect of the geology of the Oslo graben, for many geologists the primary interest of the area, lies in the Permian plutonic rocks and volcanics. Before starting their analysis, we have to say a few words about the many names used. Because the Oslo volcanics seemed to directly overly Downtonian red beds, they were dated as Devonian in all older literature. Only after the finding of continental fossils of Permian age by O. Holtedahl in 1931 was it recognized that the uppermost horizons of these red beds belong to the Permian, nonconformably overlying Downtonian red beds. The overlying volcanics, basalts and rhombporphyries, are consequently now dated as Permian and with them the whole related suite of volcanics and plutonic rocks.

TABLE VI

PRINCIPAL IGNEOUS ROCK TYPES IN THE OSLO REGION AND TOTAL AREA AT PRESENT SURFACE
(After C. Oftedahl, in: HOLTEDAHL, 1960)

Magma group	*Plutonic rocks*		*Extrusive rocks*	
	rock type	*area (km^2)*	*rock type*	*area (km^2)*
Gabbroic	"Oslo-essexite" (gabbros, kauaiite, bojite)	15	basalt and trachy-basalt	220
Monzonitic	kjelsåsite	201	rhomb porphyries	1,160
	larvikite, etc.	1,705	tuff	25
	?		trachyte, rhyolite, welded tuff, explosion vents	55
Nepheline monzonitic to syenitic	lardalite to foyaite	65		
Syenitic	alkali syenite, nordmarkite	1,400		
Granitic	ekerite	821		
	biotite granite	840		
Total area		5,047		1460

TABLE VII

CHEMICAL COMPOSITION OF TYPICAL PERMIAN PLUTONIC OSLO ROCKS
(After C. Oftedahl, in: HOLTEDAHL, 1960)

	Olivine gabbro	*Kauaiite*	*Kjelsåsite*	*Larvikite*	*Lardalite*	*Alkali syenite*	*Nordmarkite*	*Ekerite*	*Granite*
SiO_2	46.22	46.80	57.44	57.80	54.55	59.63	63.36	76.23	77.17
TiO_2	2.12	2.78	1.42	1.15	1.40	1.03	0.96	0.27	tr.
Al_2O_3	14.70	11.60	17.30	18.82	19.07	18.47	17.00	11.35	12.09
Fe_2O_3	1.34	4.27	3.48	1.60	2.41	1.39	2.26	1.88	1.22
FeO	9.82	9.73	4.61	3.50	3.12	2.92	1.55	0.32	0.07
MnO	0.20	0.22	0.26	0.14	0.17	0.13	0.12	0.20	tr.
MgO	10.04	8.44	1.42	1.48	1.98	1.06	0.77	0.00	0.05
CaO	10.68	10.89	6.20	3.72	3.15	2.35	1.01	0.43	0.65
BaO	0.07	0.04	—	0.17	—	0.16	0.05	—	—
SrO	—	—	—	0.13	—	—	—	—	—
Na_2O	2.41	2.61	4.63	6.48	7.67	5.90	7.00	4.71	3.88
K_2O	1.12	1.30	2.96	3.97	4.84	5.61	5.58	4.57	5.05
H_2O-	0.04	0.09	—	0.02	—	0.07	0.13	0.09	—
H_2O+	0.80	1.01	0.08	0.64	0.72	0.70	0.32	0.30	0.24
P_2O_5	0.28	0.31	0.50	0.55	0.74	0.29	0.19	tr.	tr.
CO_2	0.13	0.20	—	0.10	—	0.20	0.12	—	—
Cl	—	—	0.03	0.05	ca	tr.	—	—	—
F	—	—	0.03	0.04	0.12	0.03	tr.	—	—
S	0.05	0.05	—	0.03	—	—	—	—	—
FeS_2	—	—	0.15	—	—	0.08	—	—	—
Total	100.02	100.34	100.51	100.39	99.94	100.02	100.42	100.35	100.42

Also, in earlier literature, the district is called the Christiania district, which name was changed when, in 1924, the capital of Norway was re-christened back to its original name of Oslo. Lastly we find in the nomenclature of the Oslo rocks the influence of W. C. Brögger, formerly Norway's grand old man in geology. Among all his qualities Brögger had the tendency to be what in systematics is now called a splitter. The early nomenclature of the Oslo rocks is therefore somewhat top-heavy. BARTH (1945) established a nomenclature more in keeping with international usage, which, in the main, has been followed since.

Table VI lists the various types of Oslo rocks in the present nomenclature, together with the total area they occupy. In general we are dealing with a mildly alkaline suite, ranging in composition from monzonitic to granitic, and accompanied by minor amounts of gabbroic rocks. Chemical analyses of the principal rock types are given in Table VII, while their distribution in the Oslo region is indicated in a schematic way in Fig.26. Their petrographic relationship is expressed in Fig.33.

There is, at present, in the Oslo graben a marked predominance of plutonic over volcanic rocks. It is generally assumed, however, that the two groups are intimately related. The plutonic rocks are not deep-seated batholiths, but of a subvolcanic nature. Since the Permian, strong erosion has taken place. The present surface of the Oslo region consequently is situated at a considerable depth, estimated at several kilometers, below the former surficial volcanics. It follows that the Oslo region, with its predominance of plutonic rocks

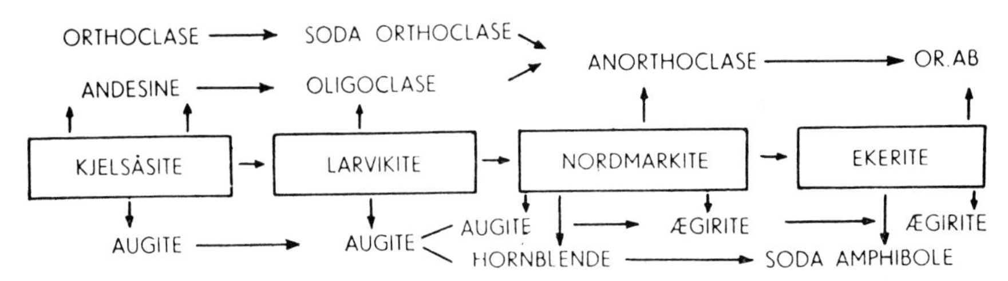

Fig.33. Petrographic relationship of main Permian plutonic Oslo rocks. (After BARTH, 1954.)

over volcanics, offers for inspection a much deeper level than can be studied in any of the younger volcanic areas. In this vein, the comparison between calderas, ring dykes and the Oslo cauldrons made by REYNOLDS (1956) may be mentioned.

The later graben faults indiscriminately cut off all Permian structures. There is no indication whatsoever where the original boundaries of the volcanic district were situated. As there is no sign of any decrease in volume of the Permian rocks towards the present graben limits, it is thought that the original volcanic district was really much larger than the portion now preserved within the graben. Actually several large dikes of rhombporphyry, probably feeder dikes of the Permian eruptives extend for over 10 km beyond the present graben. Instead of the present graben area of 200 km by 40 km, figures for the original volcanic district of 200 km by 100 km, or even of 300 km by 150 km are thought likely. Present volume of the volcanics is estimated at 300 km^3, while a figure of 6,000 km^3 for the original amount of volcanics may well be possible.

RHOMBPORPHYRY FLOWS AND IGNIMBRITES

The main feature of the Oslo volcanics is the series of syenitic flows of rhombporphyries. These rocks are named after the rhomb-shaped sections of their plagioclase phenocrysts (Fig.34), a variety with rectangular shapes being called rectangle porphyries. It has been possible to build up a detailed stratigraphy of the rhombporphyry flows by using only the habit of the plagioclase phenocrysts, which varies from unit to unit. Some of these units are formed by successive flows, others by one single massive flow of rhombporphyry. From the stratigraphic analysis it follows that most units, even single flow units, can be followed over considerable parts of the Oslo region, if not over the whole area. The areal dimensions of single flows or of units of several consecutive flows thus are amazingly large whereas their total thickness is of the order of 2 km.

Most rhombporphyry flows have been faulted and slightly tilted by subsequent cauldron or graben tectonics. They show beautiful dipslope-and-cuesta morphology, as is seen for instance in the RP 1 cover of Kolsås near Oslo (Fig.28, 37). Together with the Caledonian strike of the folded sediments of the Early Paleozoic, the dipslope-and-cuestas of the rhombporphyries are the dominant geologic control of the morphology in the surroundings of Oslo.

The rhombporphyry flows are not, however, the only Oslo volcanics. Volcanism started with basal basalt flows, and the sequence contains intercalations of basalt lavas at various levels. The volcanic stratigraphy is presented in a simplified version in Table VIII.

The basalts occur as thin lava flows, characterised by scoriaceous tops. There are normally several lava flows composing one basalt horizon. Petrographically they belong

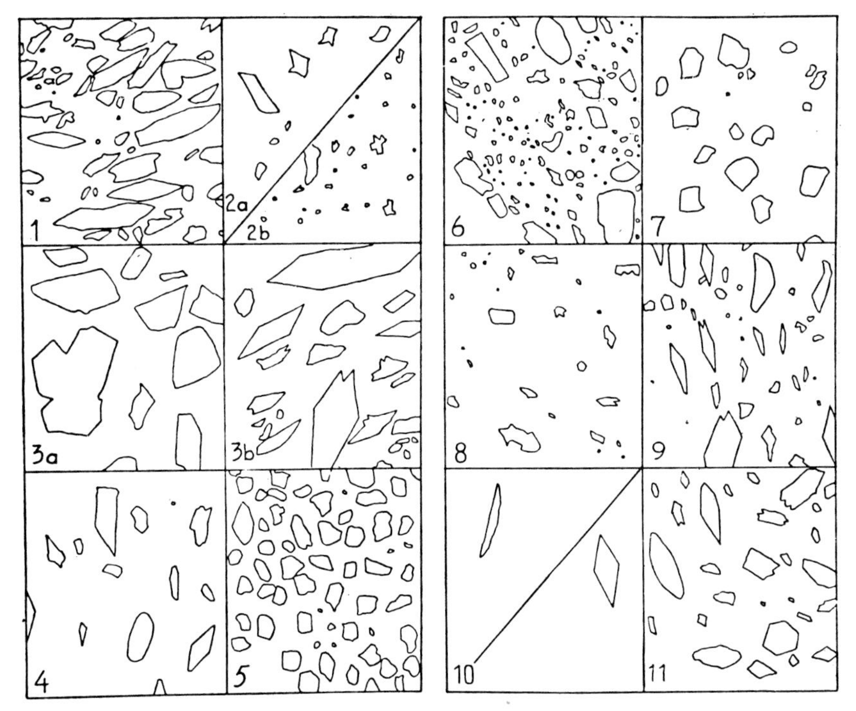

Fig.34. Habit of the plagioclase phenocrysts of the various rhombporphyry types RP 1–RP 11; Krokskogen area near Oslo. (After C. Oftedahl, in: HOLTEDAHL, 1960.)

TABLE VIII

SIMPLIFIED STRATIGRAPHY OF THE PERMIAN OF THE OSLO REGION[1]

Permian of the Oslo region
Basalts B_5
Rhombporphyry flows RP_{13}–RP_{17}
Rhombporphyry flow RP_{12}, tuff and tuff breccia
Basalts B_3
Rhombporphyry flows RP_{10}–RP_{11}
Basalts B_2
Rhombporphyry flows RP_1–RP_9
Basalts B_1
Conglomerates
Shales
~~ (unconformity)
Red beds of the Downtonian

[1] This stratigraphy has since been completed for the younger volcanics in the southern part of the Oslo region (see OFTEDAHL, 1967).

mainly to the trachybasalts of Rittmann's nomenclature. They have been wrongly called essexites by BRÖGGER (1934), a name which still survives in the present "Oslo-essexites" of authors who, mainly for historical reasons, do not like to drop Brögger's term altogether.

The rhombporphyries are in general of the same chemical composition as the plutonic kjelsåsite. They do show, however, a wide range of variation. More acid members occur, such as the syenite prophyries of Vestfold, the equivalent of nordmarkites. Tuffs and tuff breccias may even range to granitic composition.

The feldspar phenocrysts of the rhombporphyries are oligoclase or andesine, An 20–35, but often show alkalic selvages. Their apparent monoclinic symmetry, so readily visible in the typical rhombs from which the rock derives its name, is due to fine albite twinning. Earlier designations as "Na-microcline" or "anorthoclase" seem to be erroneous (OFTE-. DAHL, 1948).

In contrast to the basalts, the rhombporphyries are not typical lava flows. They are thick — RP 1, for instance, is over 100 m thick — massive, uniform in texture, and show faint but very even flow structure, coupled with an extremely wide and regular areal distribution. Their characteristics remain constant over the whole area of a given flow or unit. Altogether this indicates extreme fluidity upon eruption, a character not proper to acid lava flows.

On the other hand, these characteristics all fit in well with those of ignimbrite or ash flow eruptions. It is thought with REYNOLDS (1954) that in this type of eruption a fluidization process took place somewhere in the eruptive vent. A two-phase system of ash fragments, every single one surrounded by its own hot gas film, or even a three-phase system, with additional lava droplets also surrounded by their proper gas films, erupted at low pressure and by its extreme mobility flowed out at high speed. For further discussion of the general properties of ignimbrites or ash flows, the reader is referred to SMITH (1960) or WEYL (1961), in which the relevant literature from all over the world is cited. This mode of eruption for the rhombporphyry flows, first discussed in the field in the Oslo region, has since been more widely accepted (VAN EVERDINGEN, 1960).

It must be stressed that no single characteristic correlates a certain volcanic with the ash flow mode of eruption; not even welding. For instance, I think that the ignimbrites described by OFTEDAHL (1957) from the Baerum cauldron — his "welded tuffs" and "tuff breccias" (compare Fig.34) — are not ignimbrites but airfall tuffs, though they show welding. They are found in an irregular series, with rapid variations, both vertically and laterally, and as such are typical of airfall tuffs. For the normal, massive, uniform rhombporphyry flows, Professor Oftedahl has now, however, also accepted the ignimbritic or ashflow mode of eruption (personal communication, 1962).

A further refinement of the concept of ignimbrite eruptions is that of the rheo-ignimbrite (RITTMANN, 1958). It is thought that when an ash flow comes to rest and degasses, it may sometimes still contain so much heat that its base temporarily melts and so forms a pseudo lava. Such rheo-ignimbritic structures possibly exist in the Permian Ramnes Volcano in Vestfold (RUTTEN and VAN EVERDINGEN, 1961). Here it is believed that a central exposure of granular kjelsåsite, within an aureole of prophyric rocks, is not plutonic. Instead it is seen as but the vent filling of a volcano erupting in ignimbritic mode. When the eruption stopped, the vent filling degassed, subsequently melted, and upon cooling recrystallized as a granular rock resembling plutonic rocks in structure. To what extent other "plutonic"

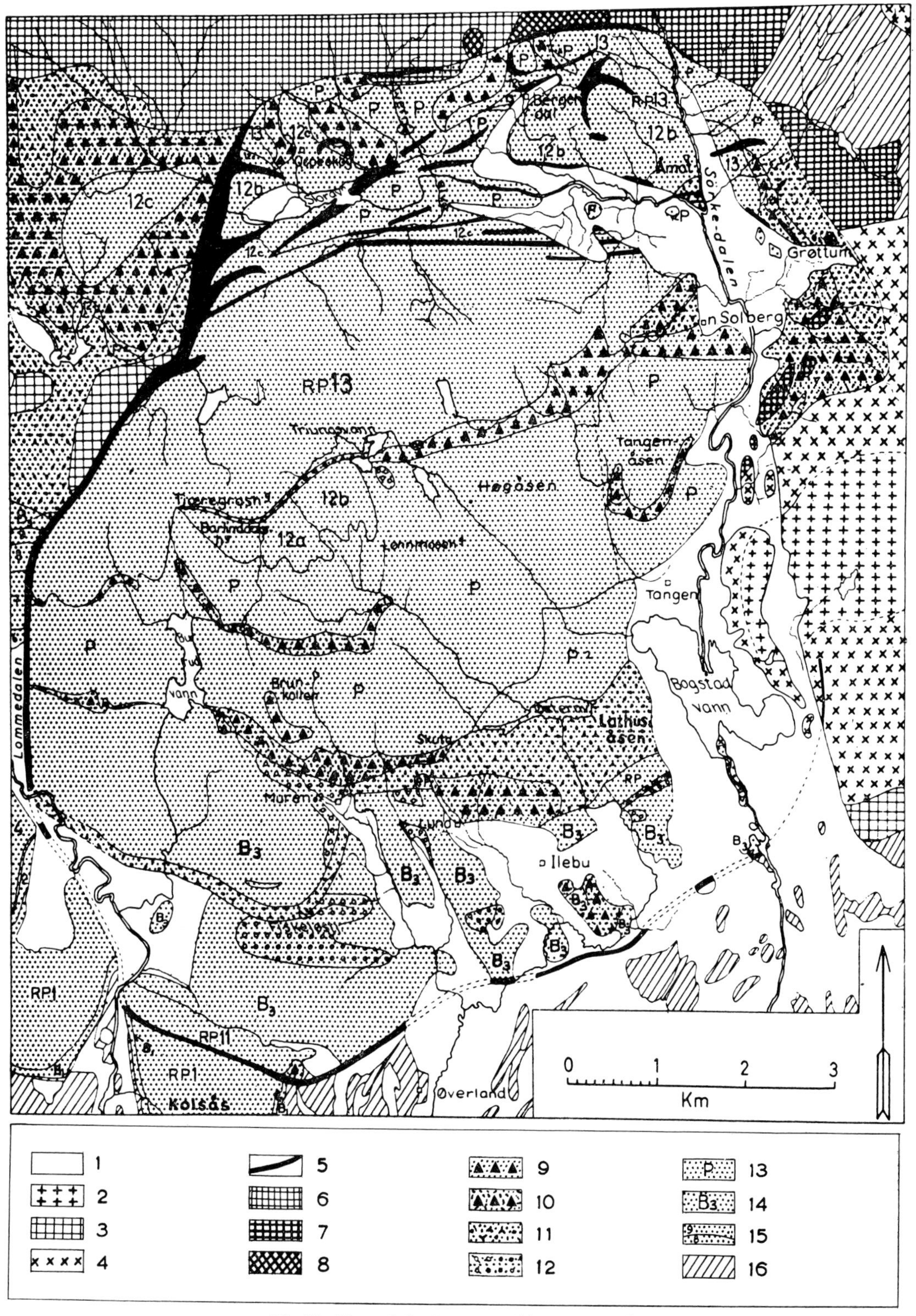

Fig.35. Map of the Baerum caldera near Oslo.
1 = Quaternary deposits; *2* = aplite-granite; *3* = monzonitic hybrid (akerite); *4* = syenites (nordmarkite, etc.); *5* = ring dike, mostly syenite porphyry; *6* = monzonitic hybrid; *7* = Sörkedalite; *9* = agglomerate, welded; *10* = agglomerate, welded tuff; *11* = welded tuff (felsite porphyry); *12* = tuff with conglomerates, etc.; *13* = porphyries, monzonitic to syenitic; *14* = basalts, with stratigraphic number; *15* = rhombporphyries, with stratigraphic number; *16* = Cambro-Silurian sediments. (After OFTEDAHL, 1953.)

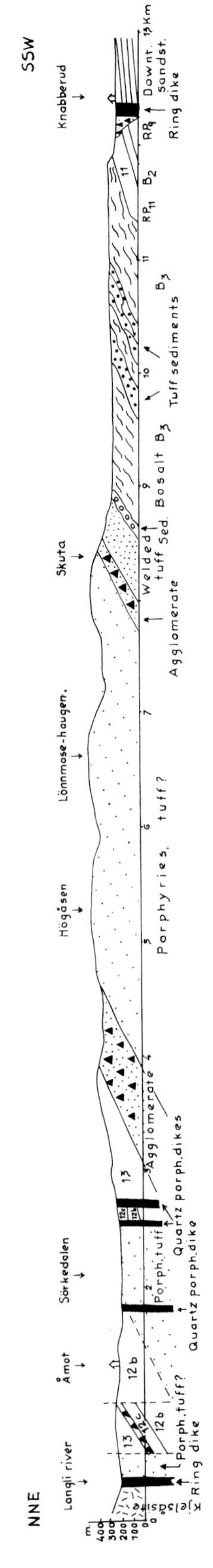

Fig.36. North-northeast–south-southwest section through the Baerum caldera near Oslo. Compare Fig.35. (After OFTEDAHL, 1953.)

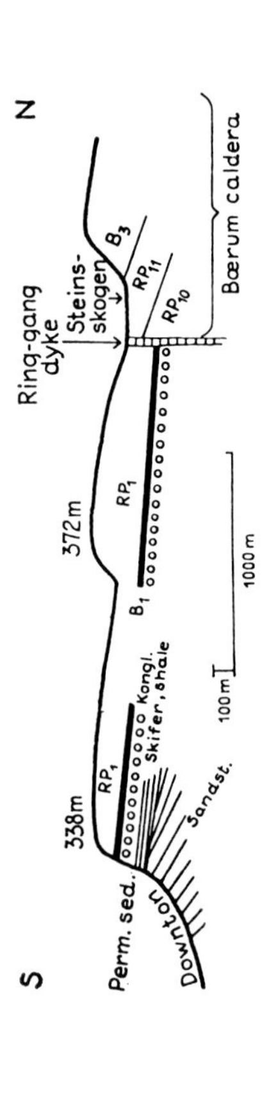

Fig.37. Section through Kolsås and the southern rim of the Baerum caldera. Tilted position of RP 1, outside the cauldron. Ring dike and RP 11 and B 3 within the cauldron. (After HOLTEDAHL and DONS, 1957.)

subvolcanic rocks of the Oslo region might have a similar origin cannot, at present, be estimated. Rheo-ignimbritic structures also occur locally at the bottom of several rhombporphyry flows but these still await detailed description.

CAULDRONS

Another feature of the Oslo region is the occurrence of cauldrons. It is thought that the cauldrons of the Oslo region are calderas, cut at a rather deep level. The main Oslo cauldrons are indicated in Fig.26, while their size and subsidence is listed in Table IX.

TABLE IX

STATISTICS OF THE MOST IMPORTANT CAULDRONS IN THE OSLO AREA
(After C. Oftedahl, in: HOLTEDAHL, 1960)

Cauldrons	*Diameter (km)*	*Mean vertical subsidence (m)*	*Volume of subsidence (km³)*
Bærum	12 × 8.5	1,000–1,500	80–120
Drammen	7	ca. 500	ca. 17
Glitrevann	16 × 10	1,500	180
Sande	12	500–800	55–80
Alnsjø (remnant)	15(?)	1,500–2,000	—
Øyangen (remnant)	ca. —8	3,000–4,000(?)	—

In their most typical development the cauldrons are circular or elliptical in ground plan, and surrounded by a narrow sub-vertical ring dyke. This structure is well shown by the cauldron nearest to Oslo, the Baerum cauldron (Fig.35–37). Along its southern rim the syenitic ring dyke, between 1 m and 2 m thick only, separates rhombporphyry flow RP 1 outside the cauldron from RP 11 and B 3 within the cauldron, which are approximately 1,000 m removed in stratigraphic sequence. In other cauldrons this simple plan may be somewhat obscured, for instance by a drag of the inner part of the cauldron along part of the circumference, in which case no ring dyke will develop, or by the intrusion of younger plutonic masses within the cauldron. But the general structure remains the same.

ORIGIN OF OSLO ROCKS

After this rather detailed analysis of the igneous Permian rocks of the Oslo region, some words as to their origin must be added. The rocks are thought to be anatexitic, deriving from a re-melting of the Precambrian metamorphic series such as form the border on both sides of the Oslo graben. The main factor responsible for this anatexis is thought to be hot gases ascending along the incipient fault zone of the later Oslo graben (BARTH, 1954). This is in accordance with CLOOS' (1939) tenet "Hebung–Spaltung–Vulkanismus" (rising, fissuring and volcanism; see Chapter XVIII). In Oslo, of course, we again would be at a much deeper level than in any of the examples cited by Cloos.

The theory of an anatexitic origin of the Oslo rocks is mainly based on two lines of reasoning. The first notes the predominance of syenitic over basaltic rocks, even under the assumption that, through gravitational differentiation, relatively more basaltic rock will be found at still deeper levels. It is thought impossible that so much acid rock should be derived through fractional differentiation from a basaltic mother liquid. On the other hand, there is a close similarity between the mean composition of the Oslo rocks and the average of the Precambrian found outside the graben.

For the calculation of the mean composition of the Oslo rocks, a prism was assumed 200 km long, 40 km wide and 30 km deep. One further assumption is that this "Oslo prism" contains one-third of basaltic (= Oslo-essexitic) rocks and two-thirds of syenitic rocks, because one expects to find relatively more basalt at deeper levels, owing to gravitational differentiation. Calculated in this way, the resemblance in chemical composition between both areas, at first sight so strikingly different in geology, is convincing indeed (Table X).

In detail there is of course the question whether the anatexitic processes took place in a closed system, or in an open system, with "apport" of cations. And also the more general question of the origin of alkali rocks. But this would lead us too far into details. The main thing to retain is that the Oslo rocks presumably originated through anatexis of the surrounding Precambrian, a view now generally accepted.

To finish the story of the Oslo region, let us cite two figures determined from Oslo rocks and each of them used as a calibration point in general geology. First, a Permian granite and a nordmarkite, dated as post-Middle Autunian, gave absolute ages of 259 million years and

TABLE X

COMPARISON BETWEEN THE IDEAL COMPOSITION OF THE PRE-EXISTING CRUST FORMED BY PRECAMBRIAN METAMORPHICS, AND THE OSLO IGNEOUS ROCKS

(After BARTH, 1954)

	Oslo prism[1]	*Pre-existing crust (=average igneous rocks)*	*Ions added to or subtracted from the pre-existing crust to give the Oslo prism*	
			+	—
Si	54.0	55.8		1.8
Ti	1.1	0.7	0.4	
Al	18.2	17.0	1.2	
Fe^{3+}	2.2	2.2		
Fe^{2+}	3.0	3.0		
Mn	0.1	0.1		
Mg	2.9	4.9		2.0
Ca	4.8	5.2		0.4
Na	8.9	7.1	1.8	
K	4.5	3.8	0.7	
P	0.3	0.2	0.1	
Sum cations	100.0	100.0	+4.2	—4.2
F	0.2	0.1	0.1	—
OH	4.5	7.3	—	2.8
O	156.8	157.2	—	0.4
Sum anions	161.5	164.6	+0.1	—3.2

[1] Composed of ⅓ Oslo-essexite and ⅔ average subvolcanic Oslo rock.

260 million years, respectively (KULP, 1961). Second, the position of the Permian south pole, calculated as a mean from 538 samples of Permian volcanics is situated at the present location of 157°E 47°N (VAN EVERDINGEN, 1960).

POST-GLACIAL UPLIFT OF SCANDINAVIA

As a sort of postcript to the geology of northern Europe, the post-Glacial upbending due to unloading by the melting ice cap must be mentioned. It is not that it is unimportant. On the contrary, this experiment in isostasy undertaken by the earth has attracted such wide attention that it is a classic in geologic literature, and thus it is hardly necessary to mention it here. Just to refresh the memory, let us recall that, owing to a combination of uplift in the central ice cap area and a downward recoil along its margin, coupled with the eustatic rise of the sea level, the Baltic was temporarily dammed twice. The history, in short, is that of the succession of the Baltic Ice Lake, the Yoldia Sea, the Ancylus Lake and the Littorina Sea, which is the precursor of the present Baltic. Total uplift of the ancient shore lines of the Baltic Ice Lake reaches more than 250 m above present sea level (see, e.g., BRINKMANN, 1959, 1960; GIGNOUX, 1955). Uplift continues at present in Finland at a rate varying from 1 mm/year in the southeast to 9 mm/year in the northwest (KÄÄRIÄINEN, 1953).

In detail this picture is more intricate, because the uplift has not been uniform, but has in the past shown definite hinge lines (SAURAMO, 1954, 1955).

The post-Glacial uplift of Scandinavia has been used to calculate the constant of viscosity of the plastic mantle of the earth. This is possible, because flow of mantle material towards the rising area must have sustained this rise after a short initial period of elastic reaction of the crust (VENING MEINESZ, 1937, 1964; HEISKANEN and VENING MEINESZ, 1958). The viscosity of the mantle has been estimated at 10^{22} poises.

From the assumption of a compensation by viscous flow in the mantle, it also follows that the time of relaxation is inversely proportional to the size of the area deviating from isostatic equilibrium. In other words, relaxation time, multiplied by diameter, is constant. This constant too has been determined from the rate of post-Glacial rise of Scandinavia. For the two-dimensional case, in which the area is thought to be infinite in one direction, the formula is:

$$t_r \cdot L = 6.3$$

Here t_r is the relaxation time during which an initial deviation from isostatic disequilibrium is reduced to 1/e its initial value, expressed in units of 1,000 years, while L is the breadth of the area, expressed in units of megameters or 1,000 km.

REFERENCES

BARTH, T. F. W., 1945. Studies on the igneous rock complex of the Oslo region. II. Systematic petrography of the plutonic rocks. *Skrifter Norske Videnskaps-Akad. Oslo, Mat. Naturv. Kl.*, 8: 104 pp.

BARTH, T. F. W., 1954. Studies on the igneous rock complex of the Oslo region. XIV. Provenance of the Oslo magmas. *Skrifter Norske Videnskaps-Akad. Oslo, Mat. Naturv. Kl.*, 4: 20 pp.

BRINKMANN, R., 1959. *Abriss der Geologie. II. Historische Geologie*, 8. Aufl. Enke, Stuttgart, 360 pp.

BRINKMANN, R., 1960. *Geologic Evolution of Europe*. Enke, Stuttgart, 161 pp.

BRÖGGER, W. C., 1934. Die Eruptivgesteine des Oslogebietes. VII. *Skrifter Norske Videnskaps-Akad. Oslo, Mat. Naturv. Kl.*, 1: 1–147.

BRÖGGER, W. C. and SCHETELIG, J., 1923. *Geologisk Oversiktskart over Kristianiafeltet, 1:250,000*. Norges Geograph. Opmåling, Oslo.

CLOOS, H., 1939. Hebung–Spaltung–Vulkanismus. *Geol. Rundschau*, 30: 401–519; 637–540.

COBB, J. C. and KULP, J. L., 1961. Isotopic geochemistry of uranium and lead in the Swedish kolm and its associated shales. *Geochim. Cosmochim. Acta*, 24: 226–249.

GIGNOUX, M., 1955. *Stratigraphic Geology*. Freeman, San Francisco, Calif., 759 pp.

HADDING, A., 1927. The pre-Quaternary sedimentary rocks of Sweden. I, II. *Lunds Univ. Årsskr., Avd. 2*, 23 (5): 171 pp.

HADDING, A., 1933. The pre-Quaternary sedimentary rocks of Sweden. V. On the organic remains of the limestones, a short review of the limestone forming organisms. *Lunds Univ. Årsskr., Avd. 2*, 29 (4): 93 pp.

HADDING, A., 1941. The pre-Quaternary rocks of Sweden. VI. Reef limestones. *Lunds Univ. Årsskr., Avd. 2*, 37 (10): 132 pp.

HADDING, A., 1950. Silurian Reefs of Gotland. *J. Geol.*, 58: 402–409.

HADDING, A., 1959. Silurian algal limestones of Gotland. Indicators of shallow waters and elevation of land, some reflections on their lithological character and origin. *Lunds Univ. Årsskr., Avd. 2*, 56 (7): 25 pp.

HEISKANEN, W. A. and VENING MEINESZ, F. A., 1958. *The Earth and its Gravity Field*. McGraw-Hill, New York, N.Y., 470 pp.

HENNINGSMOEN, G., 1956. The Cambrian of Norway. *Intern. Geol. Congr., 20th, Mexico, 1956, Rept., El Sistema Cambrico*, 1: 45–58.

HENNINGSMOEN, G. and SPJELDNAES, N., 1960. Paleozoic stratigraphy and paleontology of the Oslo region, Eocambrian stratigraphy of the Sparagmite region, southern Norway. *Intern. Geol. Congr., 21st, Copenhagen, 1960, Guides to Excursions*, A 14, C 11: 29 pp.

HOLTEDAHL, O. (Editor), 1960. Geology of Norway. *Norg. Geol. Undersøk.*, 208: 540 pp.

HJELMQVIST, S., 1966. Beskrivning till berggrundskarta över Kopparbergs Län. *Sveriges Geol. Undersøkn.*, 40: 217 pp. (with English summary).

HOLTEDAHL, O. and DONS, J. A., 1957. Geological guide to Oslo and district, with map 1:50,000. *Skrifter Norske Videnskaps-Akad. Oslo, Mat. Naturv. Kl.*, 3: 86 pp.

JUX, U., 1957. Die Riffe Gotlands und ihre angrenzenden Sedimentationsräume. *Stockholm Contrib. Geol.*, 1 (4): 41–89.

KÄÄRIÄINEN, E., 1953. On the recent uplift of the earth's crust in Finland. *Veröffentl. Finn. Geodät. Inst. Helsinki*, 42: 106 pp.

KULP, J. L., 1961. Geologic time scale. *Science*, 133: 1105–1114.

MAGNUSSON, N. H., 1958a. Karta över Sveriges berggrund, 1:1,000,000 (pre-Quaternary rocks of Sweden). *Sveriges Geol. Undersökn., Ser. Ba*, 16.

MAGNUSSON, N. H. (Editor), 1958b. *Lexique Stratigraphique International. I. Europe, 2c. Suède*. Centr. Natl. Rech. Sci., Paris, 498 pp.

MAGNUSSON, N. H., 1960a. The stratigraphy of the Precambrian of Sweden outside the Caledonian mountains. *Intern. Geol. Congr., 21st, Copenhagen, 1960, Rept., Session Norden*, 9: 132–140.

MAGNUSSON, N. H. (Editor), 1960b. Description to accompany the map of the pre-Quaternary rocks of Sweden. *Sveriges Geol. Undersökn., Ser. Ba.*, 16: 177 pp.

MANTEN, A. A., in preparation. *Middle Paleozoic Reefs of Gotland*. Elsevier, Amsterdam.

MUNTHE, H., HEDE, J. E. and VON POST, L., 1924. Gotlands geologi. *Sveriges Geol. Undersökn., Årsbok*, 18 (3): 130 pp.

OFTEDAHL, C., 1946. Studies on the igneous rock complex of the Oslo region. VI. On akerites, felsites and rhomb-porphyries. *Skrifter Norske Videnskaps-Akad. Oslo, Mat. Naturv. Kl.*, 1: 51 pp.

OFTEDAHL, C., 1948. Studies on the igneous rock complex of the Oslo region. IX. The feldspars. *Skrifter Norske Videnskaps-Akad. Oslo, Mat. Naturv. Kl.*, 3.

OFTEDAHL, C., 1952. Studies on the igneous rock complex of the Oslo region. XII. The lavas. *Skrifter Norske Videnskaps-Akad. Oslo, Mat. Naturv. Kl.*, 3: 64 pp.

OFTEDAHL, C., 1953. Studies on the igneous rock complex of the Oslo region. XIII. The cauldrons. *Skrifter Norske Videnskaps-Akad. Oslo, Mat. Naturv. Kl.*, 3: 108 pp.

OFTEDAHL, C., 1957. Studies on the igneous rock complex of the Oslo region. XVI. On ignimbrite and related rocks. *Skrifter Norske Videnskaps-Akad. Oslo, Mat. Naturv. Kl.*, 4: 21 pp.

OFTEDAHL, C., 1959. Volcanic sequence and magma formation in the Oslo region. *Geol. Rundschau*, 48: 18–26.

REFERENCES

OFTEDAHL, C., 1960. Permian igneous rocks of the Oslo graben, Norway. *Intern. Geol. Congr., 21st, Copenhagen, 1960, Guides to Excursions*, A 11, C 7: 23 pp.

OFTEDAHL, C., 1967. Magmen-Entstehung und Lava-Stratigraphie im südlichen Oslo-Gebiete. *Geol. Rundschau*, 57: 203–218.

RAMSAY, W., 1930. Changes of sea-level resulting from the increase and decrease of glaciations. *Publ. Mineral. Geol. Inst. Helsinki*, 53: 52 pp.

REGNÈLL, G. and HEDE, J. E., 1960. The Lower Paleozoic of Scania; the Silurian of Gotland. *Intern. Geol. Congr., 21st, Copenhagen, 1960, Guides to Excursions*, A 22, C 17: 87 pp.

REYNOLDS, D. L., 1954. Fluidization as a geological process and its bearing on the problem of intrusive granites. *Am. J. Sci.*, 252: 577–614.

REYNOLDS, D. L., 1956. Calderas and ring-complexes. *Verhandel. Koninkl. Ned. Geol. Mijnbouwk. Genoot., Geol. Ser.*, 16: 355–379.

RITTMANN, A., 1958. Cenni sulle colate di ignimbriti. *Boll. Acad. Gioenia Sci. Nat. Catania*, 4 (10): 524–533.

RUTTEN, M. G., 1958. Detailuntersuchungen an Gotländischen Riffen. *Geol. Rundschau*, 47: 359–384.

RUTTEN, M. G., 1966. The Siljan ring of Paleozoic, central Sweden: a posthumous ringcomplex of a Late Precambrian Dala Porphyries caldera. *Geol. Mijnbouw*, 45: 125–136.

RUTTEN, M. G. and VAN EVERDINGEN, R. O., 1961. Rheo-ignimbrites of the Ramnes Volcano, Permian Oslo graben. *Geol. Mijnbouw*, 23: 49–57.

SAURAMO, M., 1954. Das Rätsel des Ancylussees. *Geol. Rundschau*, 42: 197–233.

SAURAMO, M., 1955. Land uplift with hinge-lines in Fennoscandia. *Ann. Acad. Sci. Fennicae, Ser. A*, 3 (44): 25 pp.

SEILACHER, A. and MEINSCHNER, D., 1965. Fazies-Analyse im Paläozoikum des Oslogebietes. *Geol. Rundschau*, 54: 596–618.

SIMONEN, A., 1956. Cambrian sediments in Finland. *Intern. Geol. Congr., 20th, Mexico, 1956, Rept., El Sistema Cambrico*, 1: 91–96.

SMITH, R. L., 1960. Ash flows. *Bull. Geol. Soc. Am.*, 71: 795–842.

STRAND, T., STÖRMER, L. and SIMONEN, A., 1956. *Lexique Stratigraphique International. I. Europe. 2a, 2b. Norvège-Finlande.* Centr. Natl. Rech. Sci., Paris, 101 pp.

THORSLUND, P. and JAANUSSON, V., 1960. The Cambrian, Ordovician and Silurian in Väster-Götland, Närke, Dlarna and Jämtland, central Sweden. *Intern. Geol. Congr., 21st, Copenhagen, 1960, Guides to Excursions*, A 23, C 18: 51 pp.

TJERNVIK, T. E., 1958. On the Early Ordovician of Sweden. *Bull. Geol. Inst. Univ. Uppsala*, 36: 107–284.

VAN EVERDINGEN, R. O., 1960. Studies on the igneous rock complex of the Oslo region. XVII. Palaeomagnetic analysis of Permian extrusives in the Oslo region, Norway. *Skrifter Norske Videnskaps-Akad. Oslo, Mat. Naturv. Kl.*, 1: 80 pp.

VENING MEINESZ, F. A., 1937. The determination of the earth's plasticity from the post-Glacial uplift of Scandinavia; isostatic adjustment. *Kon. Ned. Akad. Wetenschap., Proc., Ser. B*, 40: 654–662.

VENING MEINESZ, F. A., 1964. *The Earth's Crust and Mantle.* Elsevier, Amsterdam, 124 pp.

WESTERGARD, A. H., 1922. Sveriges olenidskiffer. *Sveriges Geol. Undersökn., Ser. Ca*, 18: 205 pp.

WETZEL, W., 1947. Sedimentpetrographische Studien an den Kambro-Silurischen Ablagerungen des Billingen. *Z. Deut. Geol. Ges.*, 99: 139–149.

WEYL, R., 1961. Mittelamerikanische Ignimbrite. *Neues Jahrb. Geol. Paläontol., Abhandl.*, 113: 23–46.

CHAPTER 4

Hercynian Europe: Generalities

HERCYNIAN AND VARISCAN

Together with the Scandinavian and Scottish Caledonides, Hercynian Europe forms the second main division of the physiography of western Europe, that of the uplands. Hercynian Europe embraces all areas where the youngest orogeny is the Hercynian and which are at present not covered — or no longer covered — by post-Hercynian strata.

Hercynian is the term used in the Romanic languages. It is equivalent to *variszisch*, Variscan, in German. To complicate matters, *herzynisch* is used, in German, not for an orogenetic period, but for a direction of blockfaulting. Consequently, in German, *variszische Orogenese* and *herzynische Richtung* are two completely different things. The *Richtung*, the direction, applies to faults which are younger than the *Orogenese*, and thus post-Hercynian according to the Romanic nomenclature. Moreover the *Richtung* strikes NW–SE, more or less perpendicularly to the strike of the fold belt in Germany. Even German authors (KNETSCH, 1963) are now willing to follow the Romanic use of the term hercynian, but in all older literature in German and German-influenced areas, from Spain to Finland, *variszisch* is used instead of Hercynian.

In this matter, Cloos' lamentation (CLOOS, 1948, pp. 290, 291; compare p. 119) still seems very appropriate:"There came the 'tectonic directions': Erzgebirgic, Rhenic, Hercynic; a facilitation for the inventor and the specialist, but not for the reader from foreign countries. And how to call the intermediate directions, which unruly earth nowhere leaves out. Perhaps 'Franconic' and 'anti-Franconic'? Or 'flat' and 'steep' Rhenic or Hercynic, etc.? In which 'flat' and 'steep' do not mean 'flat' and 'steep', but 'more easterly' and 'more northerly'. And what about it, when, for instance, 'Erzgebirgic' is meant as a direction, but 'Sudetic' as a time phase?"

COE (1963) recently has complicated matters still more, by using Hercynian "only in a time sense", as against Variscan "in a geographical sense [only?] for the whole fold belt". Moreover, he uses Armorican in a separate geographical sense "for the western part to the Central Massif", forgetting that within the Armorican Massif itself the northern part is said to have variscan, the southern armorican direction (see p. 132).

As to the literature on Hercynian Europe, this is, of course, scattered widely in thousands of publications. A general idea, mostly of the stratigraphy, can be found in VON BUBNOFF (1930), GIGNOUX (1955) and BRINKMANN (1960), cited in the general introduction, and further in volume I, 4 and I, 5 of the *Lexique Stratigraphique International.* Because central Europe, Germany in particular, but also Belgium and France, contain so many elements of Hercynian Europe, regional works on these countries also offer a good introduction to the subject. In particular FOURMARIER (1934), BERTRAND (1946), DORN (1960), and KNETSCH (1963) may be cited. Further pertinent literature will be cited under the more detailed descriptions.

THE HERCYNIAN UPLANDS

As already explained in Chapter 1, the Hercynian uplands do not constitute a uniform area or a single chain of mountains. They are separate crustal blocks, which may be of rather large size and composite structure. Their structure is due mainly to the Hercynian, but in part, and always to a lesser degree, also to older orogenies. In contrast to their internal structure, the blocks owe their present position as uplands to much younger movements, mainly of Plio-Pleistocene age. Moreover these are of a vertical character only.

In enumerating the Hercynian uplands, we may start from the area south of lowlands of Holland and northern Germany, which separate the Caledonian uplands of Scandinavia and the Fennoscandian Shield in the north from Hercynian Europe further south.

South of these lowlands follow the Ardennes in Belgium (Chapter 5), which are contiguous to the much larger Rheinisches Schiefergebirge in Germany (Chapter 6). This in turn is bordered on its southeastern side by two units which are more or less separate topographically, but which none the less form a geological entity, i.e., Hunsrück and Taunus. Further east, across the low plateaus of central Germany, a much smaller separate unit is to be found, the Harz Mountains (DAHLGRÜN et al., 1925; SCHRIEL, 1954; SCHWAN, 1956).

Southwest of the uplands just enumerated, another large unit is formed by the peninsulas of Brittany and Normandy (Chapter 7), together forming the "Massif armoricain" of the French. Its southeastern border is formed by the Poitiers Straits, which connect the low plateaus of the Paris Basin with those of the Aquitanian Basin.

The Hercynian basement, masked by its Mesozoic cover in the Poitiers Straits, towards the east crops out again in the big, composite Massif Central upland (Chapter 8). The latter is cut in a north–south direction by the "Sillon houiller", a narrow graben structure with thick, post-Hercynian coal deposits. In addition to the main area of the Massif Central a southern outlier is present, the Montagne Noire. This area differs in its basement geology, which is in many aspects more like that of the Pyrenees.

Towards the northeast the Massif Central upland is bordered by another low plateau, the Langres Straits, connecting the Mesozoic of the Paris Basin with that of the Jura Mountains. Across these straits follow the Vosges Mountains, which in turn are bordered by the narrow north–south trending lowland of the Upper Rhine graben. To the east of this graben the Hercynian basement then crops out again in the Black Forest or Schwarzwald Mountains (Fig.38).

Still further towards the east again follow the extensive low plateaus of central and southern Germany, which in their turn are bordered by the Hercynian uplands of Bohemia. These, together with their northern prolongations into Thuringia and the Erzgebirge, are beyond the scope of this volume (VON GAERTNER, 1950; WEBER, 1955; ZOUBEK, 1960; PIETSCH, 1962).

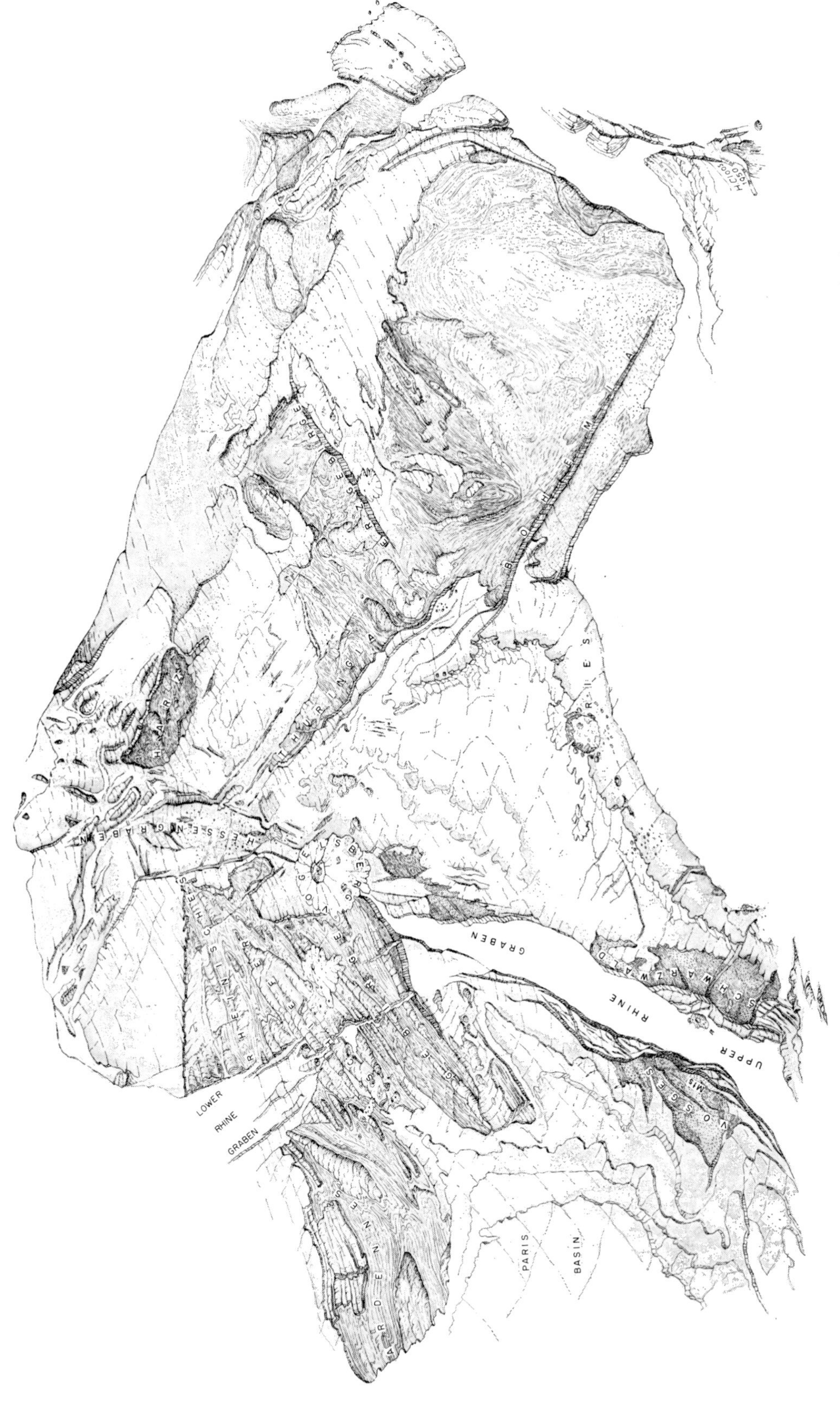

Fig.38. Block diagram of Germany. Viewed from the south, it illustrates the relation between Hercynian uplands, the low plateaus and the younger blockfaulting movements and volcanism, which are discussed in Chapter 18. (After CLOOS, 1955.)

A further group of Hercynian uplands is found south of the Pyrenees and comprises many of the "sierras" and part of the "mesetas" of the western half of the Iberian Peninsula, of western Spain and of Portugal (LOTZE, 1945). This area, complex in structure, and with wide regional variations, will also not be treated in this volume, owing to space restrictions.

Apart from the Hercynian uplands enumerated above, Hercynian and older rocks also occur in the core of the Alpine mountain chains. They may be found either as slightly metamorphosed sediments, such as in the axial zone of the Pyrenees, or as strongly metamorphosed crystalline rocks, as in the central massifs and in the crystalline cores of the Penninic nappes of the Alps. These will, however, not be treated in these chapters, which are concerned only with Hercynian areas that were not affected by the Alpine orogeny.

VARIATIONS IN STRUCTURE IN THE HERCYNIAN UPLANDS

As was stated already in Chapter 1, the Hercynian and earlier history of the uplands of western Europe shows strong variation. It seems well, at this stage, to indicate in a general way the nature of these variations in the uplands enumerated above.

The Ardennes, the Rheinisches Schiefergebirge, together with the Hunsrück and Taunus, are characterized by a very prominent development of the Devonian. The "Schiefer", the slate which gave its name to the largest part of this unit, is mainly of Lower and Middle Devonian age. There is practically no influence of Hercynian plutonic rocks in this unit. The quite prominent, though local, volcanism of Late Cenozoic age on both sides of the Rhine has, of course, nothing to do with the Hercynian history of these uplands, but is related to much younger blockfaulting. In the Ardennes the geologic history is exposed somewhat more completely than in the easterly units. Underlying Caledonian structures crop out in a number of nuclei. And Lower Carboniferous is widely included in synclinal structures in their northern half.

The Harz Mountains further east offer a new element, which we will also meet with in the more southerly units, i.e., a number of plutonic cores of granite surrounded by gneiss and crystalline schists.

The Armorican uplands are characterized by the occurrence not only of strata belonging to the Hercynian but also of older rocks. The base of the Paleozoic sedimentary cycle includes as its lowermost unit the Brioverian. This is a local stage name for a thick series of clastic sediments of Late Precambrian age, which are comparable to the sparagmites of the Scandinavian Caledonian geosyncline. In addition to the sediments, there are large areas of plutonic rocks, mostly granites, and of crystalline schists and gneisses. Some of these are clearly Hercynian in age, but others are definitely older, while there are, of course, many of undefined age.

The general strike in the fold belt of the Armorican uplands is west-southwest–east-northeast in its northern part, in northern Brittany and in Normandy. This lines up with a similar general strike in the Ardennes and in the Rheinisches Schiefergebirge. In southern Brittany the general strike is, however, west-northwest–east-southeast, in line with the western part of the Massif Central across the Poitiers Straits. The limit between these two contrasting parts of the Armorican uplands is formed by an east–west trending graben structure, in which deposits of the uppermost Carboniferous are found. The character of this feature is, as yet, not well understood. It seems, however, in a way to be comparable to the "Sillon houiller" of the Massif Central.

The Massif Central is characterized by a predominance of plutonic and metamorphosed rocks, almost to the exclusion of sediments. The latter are still found in a zone along its western border, the Bas Limousin. They belong to the Early Paleozoic, and may exhibit epizonal metamorphism. Some elements are mined as roofing slates.

Almost everywhere else the Massif Central consists of crystalline schists and gneisses, of granites and related rocks. These rocks, the "séries crystallophylliennes" and the "roches cristallines" of the French literature, are nowadays commonly interpreted as metamorphic and ultrametamorphic sediments. The Sederholm concepts of migmatization and granitization have found their earliest application outside Fennoscandia in the French Vosges Mountains and in the Massif Central under the influence of JUNG (1927).

There are definite indications of several successive cycles of migmatization and granitization in the Massif Central. These may be apparently correlated with successive orogenetic cycles. But in the absence of non-metamorphosed sediments, and hence of fossils, the latter have as yet not been dated. A program of absolute dating is under way at the University of Clermont Ferrand, so we will know much more about this in the near future.

The strike of the metamorphic rocks in the western half of the Massif Central is west-northwest–east-southeast, forming a direct continuation of the southern part of Brittany. In the eastern part it is southeast–northeast, pointing directly towards the Vosges Mountains. It follows that, in a very simplified way, the general strike of the Hercynian structures form a big V across France. This V structure is played up rather strongly in all regional literature, but there has never been a serious attempt to explain it.

The transition from one limb of this V to the other, from the west-northwest trending structures in the western half of the Massif Central to the northeast trending structures in its eastern part, is exceedingly sharp. It is marked, moreover, by the "Sillon houiller", the Carboniferous furrow. This is a narrow north–south trending graben structure, rarely more than 2 km wide, in which locally thick coal seams have accumulated. These date from the uppermost Carboniferous — Late Westphalian and Stephanian — and consequently are post-orogenetic to the Hercynian orogeny. The "Sillon houiller" gives the impression of being related to a transcurrent fault. But if so, it was an older fault whose horizontal movements ended at the time of the Hercynian orogeny, or even before. The present graben structure offers proof only of vertical subsidence, without a possibility to accommodate larger horizontal movements. Of course, the present graben structure might follow a cicatrice of a former transcurrent fault, which still formed a line of crustal weakness.

Apart from the "Sillon houiller", there are many other graben structures cutting through the Massif Central. All of them are wider and not as sharply cut as the "Sillon houiller". They give the impression of graben structures of a more normal nature. Several of these are more or less contemporaneous with the "Sillon houiller", and responsible for local coal basins such as those of St. Étienne, Alais and Autun. Others are of much younger Tertiary age, such as the Limagne graben between Clermont-Ferrand and Vichy. The border faults of the latter structures in many spots are responsible for the mineral springs which form the base for famous watering places.

Related to these younger blockfaulting movements is the Late Cenozoic volcanism.

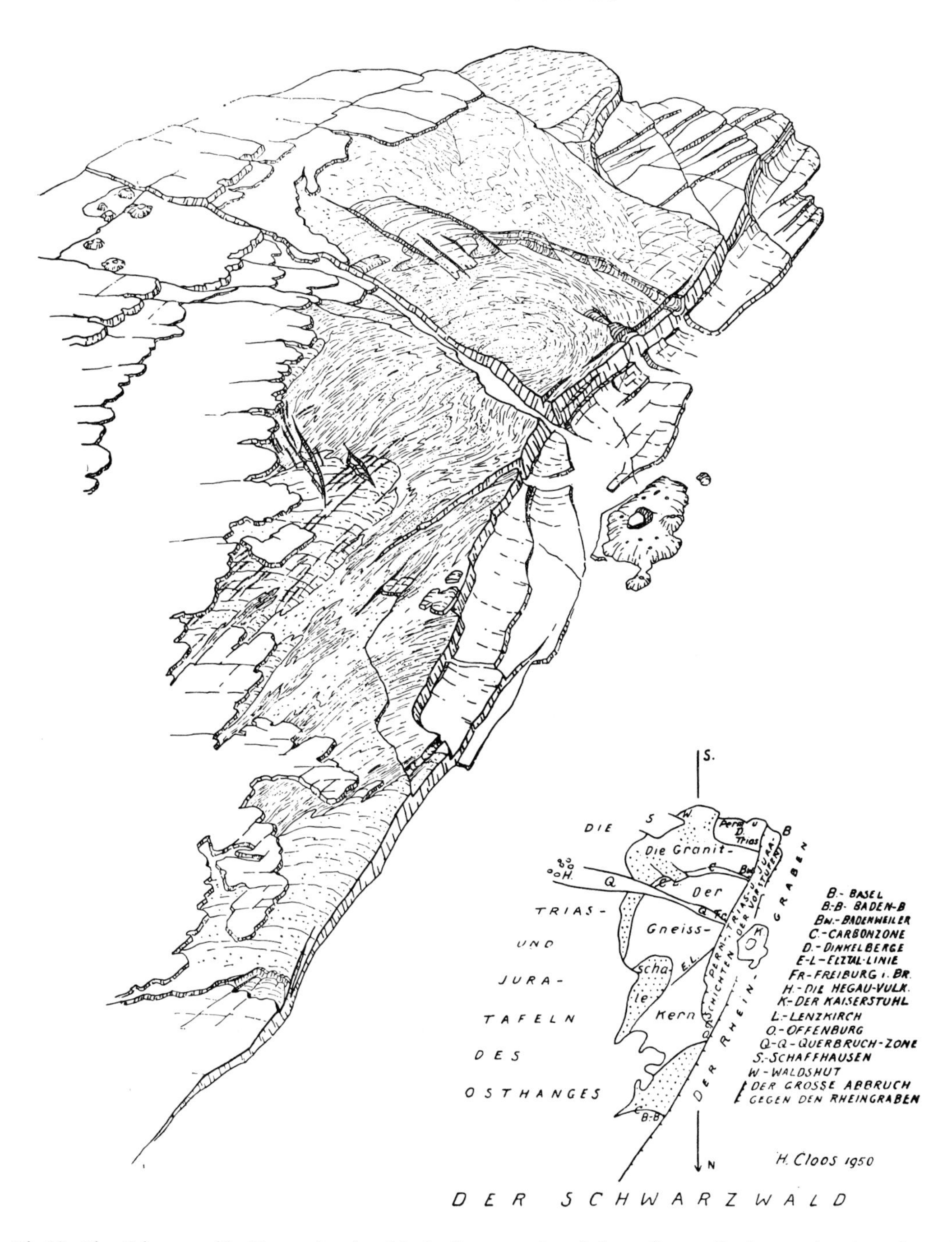

Fig.39. The Schwarzwald. Upper drawing block diagram, viewed from the north. Lower drawing schematic geological map. The Hercynian basement crops out in an oval beneath the Mesozoic of the low plateaus of southern Germany. Its eastern face is formed by the faults bordering the Upper Rhine graben. *K* = the young Kaiserstuhl Volcano. (After CLOOS, 1952.)

This has resulted in a very substantial number of smaller and larger volcanoes and necks, not only in the Auvergne, but also further south and southeast. Along a north–south direction with the Auvergne volcanism are situated the big central volcanoes of Mont Doré and Cantal, a trend which can be followed down to Agde on the Mediterranean. Towards the southeast we find the volcanism around Le Puy in the Velay province, which can be followed down into the present Rhône valley, to the Montagne des Coirons.

The southern outlier of the Massif Central, the Montagne Noire, is quite unrelated in its geologic structure. Although plutonic cores occur in this area too, sediments dominate again. These mainly belong to the Early Paleozoic, with a strong development of Cambrian, Ordovician and Silurian. Their facies is comparable to that found in the Pyrenees. Moreover (a relative rarity in Hercynian structures) it seems that nappes are widely present.

Northeast of the Massif Central then follow the Vosges Mountains (JUNG, 1927; VON ELLER, 1961) and the Schwarzwald or Black Forest (CLOOS, 1952; MEHNERT, 1958). The Hercynian structure of these two units is very similar. They are now separated by the Upper Rhine graben, but this is a young, truly post-Hercynian element only. It might well be a rejuvenation of an older structural zone (ILLIES, 1962; THÉOBALD, 1963), but at present it just cuts across the single Hercynian structure of Vosges Mountains and Schwarzwald. Vosges Mountains and Schwarzwald also show a marked predominance of plutonic rocks over sedimentary rocks. Moreover, in the Schwarzwald migmatization has been largely superseded by granitization (Fig.39). It follows that within the plutonic rocks there is a marked predominance of granitic rocks over gneissic rocks, when compared with the Massif Central. This explains why magmatism has retained so strong a foothold in the Schwarzwald, when migmatism was already "à la mode" in the Massif Central and the Vosges.

GERMAN CLASSIFICATION OF HERCYNIAN EUROPE

The enumeration of the units of Hercynian Europe as given above is that of a western European. We start in the west and follow the various trends eastward. However, in the German literature the description of Hercynian Europe normally starts in the east and gradually works westward. This is, perhaps, in part due to the fact that so many of the old masters of German geology (let us cite only Von Bubnoff, Cloos and Stille) started their investigations in eastern Germany. But it most certainly also stems from the fact that we have in eastern Germany and the neighbouring parts of Czechoslovakia that large and composite Hercynian unit formed by Bohemia, the Sudeten Mountains, the Erzgebirge and Thuringia. The different elements of this area show very strong variations in history and structure. It is tempting to extrapolate what has been found there, and also the underlying principles one thinks to have found there, to the various units of western Europe (STILLE, 1951; KÖLBEL, 1963).

Bohemia, the type area for the *Moldanubische Zone* (from the rivers Moldau and Donau), is characterized by a great amount of Precambrian metamorphosed sediments and a general prevalence of gneisses and granitic batholites. The fossiliferous Middle Cambrian containing the classical "primordial fauna" of Barrande, and forming the type of the local Barrandian stage, unconformably overlies the Precambrian. This major non-

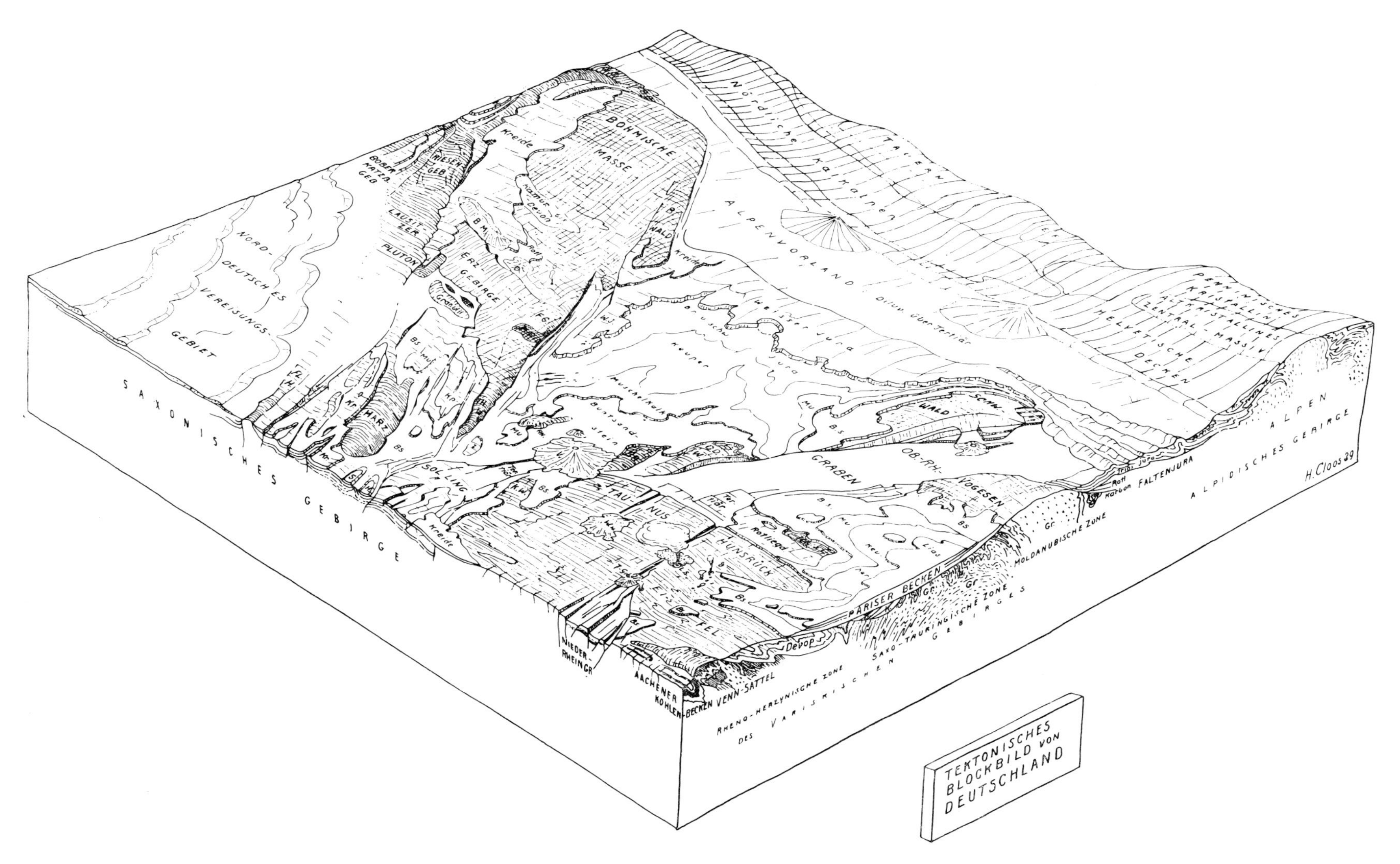

Fig.40. Block diagram of Germany. Viewed from the north it illustrates the three successive zones of Hercynian Europe in Germany (*Rheno-herzynische Zone*, *Saxo-thüringische Zone* and *Moldanubische Zone*), the Jura Mountains (*Faltenjura*), the molasse basin (*Alpenvorland*) and the Alps.

TH.W. = Thüringer Wald, Thuringia; *SP* = Spessart; *OD* = Odenwald; *SCHW.W.* = Schwarzwald. (After CLOOS, 1940.)

conformity, indicative of an orogenetic period, has been classed by Stille as the Assyn tian orogeny. It might well correlate with one or other of the Precambrian orogenies on the Fennoscandian Shield, but this is a fact difficult to ascertain, because nothing is known about its age. Consequently Stille's Assyntian orogeny, although upgraded by that author to represent the first orogeny of Hercynian Europe, has not found as much acceptance as the names of the younger orogenies in his codex. In the German literature it is, however, consistently employed.

Although the youngest main orogeny in Bohemia has been ascribed to this Precambrian Assyntian phase, which has left the fossiliferous Middle Cambrian undisturbed, strong migmatization and granitization has persisted in other areas of Bohemia until Hercynian times. It follows that Bohemia itself is a composite element and thus could hardly serve as the type area for a single structural zone (Fig.40).

North of the Moldanubicum of Bohemia follows the Saxothuringicum or *Saxo-thüringische Zone*, formed mainly by the Sudeten Mountains and the Erzgebirge. The Precambrian basement hardly crops out in this zone and consequently the oldest orogeny found is the Caledonian. Towards the north this zone is followed in turn by the Rhenohercynicum or *Rheno-herzynische Zone*. The most easterly unit of this zone is the Harz Mountains.

The German viewpoint is that this zonation found in eastern Germany can be followed westward. The Moldanubicum correlates with the Schwarzwald, the Vosges Mountains and the Massif Central. The Saxothuringicum can be followed westward into the Odenwald and Spessart, two minor Hercynian units situated on the northeastern extremity of the Upper Rhine graben. Further west the Saxothuringicum submerges under the Paris Basin, leaving ample space for the theories of the geotectonician. The Harz Mountains can be correlated westwards to the Rheinisches Schiefergebirge and the Ardennes.

This grand picture of a tripartite zonation of Hercynian Europe rests, according to my mind, on altogether too scanty data. Moreover one often gets the impression that stratigraphy and paleography of the Hercynian units in Germany are put into a straight jacket so that they will conform with the grand picture. When followed in detail, this picture becomes much less straightforward. To cite only one instance, the facies of the eastern end of the Rheinisches Schiefergebirge is variously indicated as "Saxonic" or "Bohemic", although it does not form the link between these zones, which are supposed to be parallel, instead of continuations of each others (p. 111). Compare also BEDERKE (1962) on the time of folding of the Harz Mountains.

The picture is, however, basic to all German literature. It is therefore hardly possible to understand German literature on Hercynian Europe, without knowing the meaning implied by the terms described above.

TIME OF FOLDING

Just as elsewhere in Europe, the ideas among geologists on the time of folding during Hercynian orogenesis have been strongly influenced by the theory of Professor H. Stille of Göttingen (after 1932: Berlin), who postulated worldwide, strictly contemporaneous orogenetic phases, sharply limited in time. In the Stille codex six tectonic phases together compose the Hercynian orogeny (Table XI).

TABLE XI

THE TECTONIC PHASES OF THE HERCYNIAN OROGENY ACCORDING TO H. STILLE

Phase	*Age*
Pfälzic	between Permian and Triassic
Saalic	within the Rotliegendes (Lower Permian)
Asturic	between Westphalian C and Stephanian (within the uppermost Carboniferous)
Erzgebirgic	between Namurian and Westphalian (within the Upper Carboniferous)
Sudetic	between Visean (uppermost Lower Carboniferous) and Namurian (lowermost Upper Carboniferous)
Bretonic	between uppermost Devonian and lowermost Carboniferous

The wholehearted approval of these ideas, not only from the influential Berlin school, but also from a substantial number of other European geologists, has led to deplorable habits of sloppy dating in many tectonic studies. "Because these phases have been proven to be worldwide and synchronous, dating is relatively easy. If a nonconformity of any sort is, for instance, younger than uppermost Devonian and older than Namurian, it is datable as Sudetic, because there have been no other tectonic movements during this time span". Such has been, in short, the philosophy underlying much of tectonic dating in Europe; philosophy so widely accepted that one gets the impression it is often applied unconciously. There is no excuse for doing so.

Stille's idea of worldwide, strictly synchronous orogenetic phases has met with considerable criticism in later years. But though the fallacy of this line of reasoning has been stressed many a time (GILLULY, 1949; L. M. R. RUTTEN, 1949; SPIEKER, 1956; M. G. RUTTEN, 1962; SIMPSON, 1963) the names of the Stille codex are still widely used.

Moreover, and even much worse, many authors, though aware of the difficulties of a strict application of Stille's theory, nowadays use his names in a much vaguer way. It is, of course, so much easier to designate a given nonconformity as "Asturic", than having to describe it as "occurring sometime between Westphalian and Autunian". Used in this way, the names of the Stille codex only indicate very vague dates and do, moreover, not imply strict contemporaneity, a usage quite opposed to that of Stille and his school. This naturally leads to a considerable muddle. To be quite sure, one has consequently to ascertain beforehand what meaning is guiding a given author, when he uses the names of the Stille codex.

Because of the fact that the latter are well anchored in European geology they are at present widely employed. And since it is relatively easy to learn a short list of names by rote, I think they will be used for many years to come. I have the impression, however, that the names of the Stille codex are serving more and more in their second, improper, meaning as a vague indication of some period not too precisely datable, instead of in their proper, original, precise meaning.

For the purpose of this very general narrative the exact date of a certain event is not so important as is its nature and its role within the frame of the Hercynian orogeny. The names of the Stille codex will therefore be used here in their more recent, vague — and of course improper — meaning. It then becomes possible to schematize the events of the Hercynian orogeny as follows:

(*1*) The *Bretonic phase*, at the transition from Devonian to Carboniferous, has not been active all over Hercynian Europe. Where its influence was felt, it mostly resulted in a minor disconformity within the geosynclinal series. It is essentially composed of pre-orogenetic movements (pp.140–141).

(*2*) The *Sudetic phase*, at the transition from Lower Carboniferous to Upper Carboniferous, and in many places extending rather high up into the Upper Carboniferous, stands for the main Hercynian orogenetic movements of central Europe. As such, its effects are known from western and central France (Armorican Massif, Massif Central, exterior chains of Western Alps (cf. Fig.41), Saar Basin) into eastern Germany and the Sudeten Mountains, on the border of Czechoslovakia and Poland, from which this phase was named.

(*3*) The *Erzgebirgic phase* (from the Erzgebirge on the border of eastern Germany and Czechoslovakia) is not so well understood. It is perhaps a somewhat younger substitute in eastern central Europe of the Sudetic phase farther west.

(*4*) The *Asturic phase*, uppermost Carboniferous (that is Upper Westphalian and Stephanian) was the main orogenetic phase in a zone surrounding the region affected by the Sudetic phase. It is found from northwestern Spain, where it took its name from the Asturian Mountains, to Britain, northwestern France, Belgium, The Netherlands and the Ruhr Coal Basin, in a wide arc around central Europe. But we meetit again, surprisingly, in the inner zone of the Western Alps, in the Briançonnais zone.

(*5, 6*) The *Saalic* and *Pfälzic phases* have found their most important expression in those parts of central Europe where the main orogenetic movements developed during the Sudetic and/or the Erzgebirgic phases. They are truly post-orogenetic, with a prevalence of blockfaulting movements. Such folding as is found, is due to draping of the sedimentary cover over fault blocks in the basement or to settling in graben structures.

Absolute ages of Hercynian granites from the Massif Central, the Vosges and the Schwarzwald range from K/Ar ages of 296–330 million years (FAUL, 1963). The older ones presumably date from the Sudetic phase. But there is only one locality which has been exactly dated as such, i.e., the Gien-sur-Cure Granite of the Massif Central of which only the Rb/Sr age of 334 million years is available. Erzgebirge granites give unaccountably higher ages (329–390 million years).

HERCYNIAN PALEOGEOGRAPHY

It follows that, in a schematic picture, it is possible to divide Hercynian Europe into a central and a peripheral part. In the central part the main Hercynian orogeny was "Sudetic" — sometime between Lower Carboniferous and middle Upper Carboniferous — whereas in the peripheral part it was "Asturic" — meaning sometime during the uppermost Carboniferous.

The paleogeography of these two distinct phases of the Hercynian orogeny is listed here according to their present position. We have ample reason to state this fact purposely. As is well known, our ideas about mantle, crust and continents are in a fluid state at present. They are transitory from fixistic to mobilistic views, due mainly to the effects of paleomagnetic measurements.

In our case there is to be considered not only the wholesale continental drift of continents as they are today. We must also have in mind the possibility of relative movements between parts of the present day continents during a major orogeny. We will see how parts of the present Alpine foldbelts probably drifted over very large distances relative to "Meso-Europa", during the geosynclinal and early stages of the Alpine orogenetic cycle (pp.211–212).

It is therefore most probable that the elements of the mosaic now forming Hercynian Europe were in quite different positions relative to each other at the time of folding (RUTTEN, 1964, 1965). But paleomagnetic analysis of the Carboniferous and earlier systems is at present not yet sufficiently advanced to supply us with a reliable paleogeography. Thus, for lack of data, we still have to list the Hercynian paleogeography according to the present position of its elements.

HERCYNIAN COAL BASINS

The distinction arrived at above, between a central part of Hercynian Europe which was folded earlier than the peripheral part, is important in view of the many coal basins of western Europe (Fig.41).

In the Sudetic part the geosyncline in general maintained marine facies up to the time of folding. Coal formation only started as a post-orogenetic process, from the Upper Westphalian onwards. It developed in limited graben structures, and there are marked differences in the lithologic columns of these various basins. Deposition remained continental throughout. This facies of the coal basins of central Europe consequently is called the limnic coal facies. Although the individual basins are of small extent, strong subsidence occasionally led to the deposition of very thick seams of coal, of 10 m and over. It follows that the economic importance of the basins of the limnic facies has been great. Their reserves are, however, limited and several of them are already exhausted.

Incidentally, this explains why the term "Anthracolithicum" was proposed for the Permo-Carboniferous by the French stratigrapher Émile Haug. With the exception of the Pas-de-Calais Basin in the northwest and the Briançonnais in the southeast, all French coal basins belong to this type. Coal deposition only starts in the uppermost Carboniferous and generally extends into the Lower Permian, the Autunian of Autun in central France. In this environment there is consequently no main break between the Carboniferous and Permian systems, which seemed a valid reason to unite them into the "Anthracolithicum".

In the Asturian part of central Europe geosynclinal conditions persisted right up to the Upper Westphalian. Here it was the geosyncline itself which gradually switched over to continental facies sometime during the Upper Namurian or the Lower Westphalian. Coal formation took place in swamps of very wide extent, evidently adjoining the open sea. Now and then transgressions interrupted the continental regimen and hence this geosynclinal belt is indicated as the paralic coal facies. As is to be expected, the areal extent of

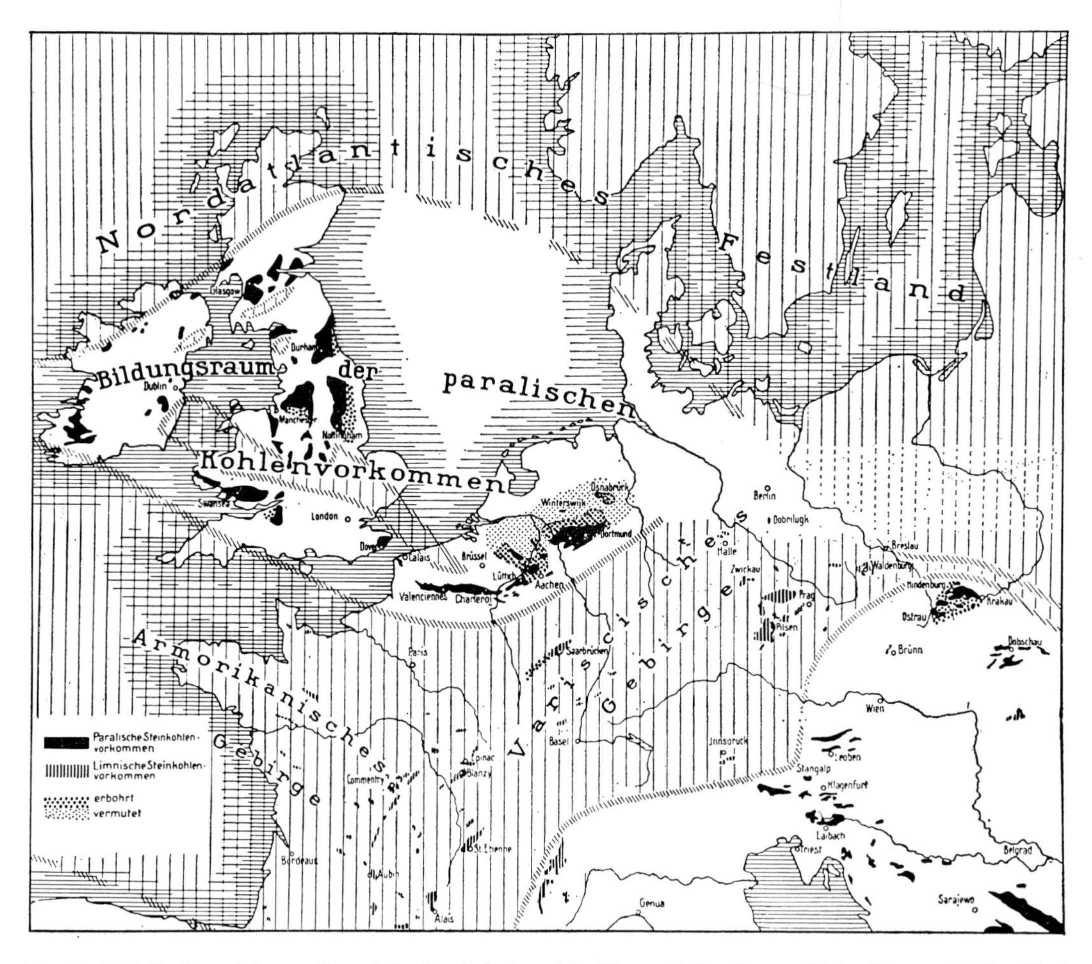

Fig.41. Distribution of the paralic and the limnic facies of the Upper Carboniferous. (After KUKUK, 1938; with the Briançonnais in the French Alps interpreted as belonging to the paralic facies.)

individual coal seams is very large, but they rarely reach 2 m in thickness. In contrast to the limnic coal facies in the individual graben structures, the reserves of the coal belt in paralic facies are enormous. Moreover, in the lowlands along the North Sea, coalification gases were trapped under Triassic salt layers far to the north, where they gave rise to gas fields of the Slochteren type in the northern Netherlands and underneath the North Sea, being amongst the largest accumulations of natural gas known (PATIJN, 1963, 1964).

REFERENCES

BEDERKE, E., 1962. Das Alter der Harzfaltung. *Neues Jahrb. Geol. Paläontol., Monatsh.*, 1: 24–27.

BERTRAND, L., 1946. *Histoire Géologique du Sol Francais. II.* Flammarion, Paris, 365 pp.

BRINKMANN, R., 1960. *Geologic Evolution of Europe*. Enke, Stuttgart, 161 pp.

CLOOS, H., 1940. Ein Blockbild von Deutschland. *Geol. Rundschau*, 31: 148–153.

CLOOS, H., 1948. Gang und Gehwerk einer Falte. *Z. Deut. Geol. Ges.*, 100: 290–303.

CLOOS, H., 1952. Der Schwarzwald. *Mitt. Naturforsch. Ges. Schaffhausen*, 24: 1–6.

CLOOS, H., 1955. Ein Blockbild von Deutschland. *Geol. Rundschau*, 44: 480.

COE, K. (Editor), 1963. *Some Aspects of the Variscan Fold Belt.* Manchester Univ. Press, Manchester, 163 pp.

DORN, P., 1960. *Geologie von Mitteleuropa*, 2. Aufl. Schweizerbart, Stuttgart, 488 pp.

REFERENCES

DAHLGRÜN, F., ERDMANNS DÖRFFER, O. and SCHRIEL, W., 1925. Harz. *Sammlung Geol. Führer*, 30 (1): 228 pp; 30 (2): 306 pp.

FAUL, H., 1963. Age and extent of the Hercynian complex. *Geol. Rundschau*, 52: 767–781.

FOURMARIER, P., 1934. Vue d'ensemble sur la géologie de la Belgique. *Ann. Soc. Géol. Belg., Mém.*, 1934: 200 pp.

GIGNOUX, M., 1955. *Stratigraphic Geology*. Freeman, San Francisco, Calif., 759 pp.

GILLULY, J., 1949. Distribution of mountain building in geologic time. *Bull. Geol. Soc. Am.*, 60: 561–590.

ILLIES, H., 1962. Oberrheinisches Gebirge und Rheingraben. *Geol. Rundschau*, 52: 317–332.

JUNG, J., 1927. Contribution à l'étude des Vosges hercyniennes d'Alsace. *Mém. Carte. Géol. Alsace Lorraine*, 2: 481 pp.

KNETSCH, G., 1963. *Geologie von Deutschland und einigen Randgebieten*. Enke, Stuttgart, 386 pp.

KÖLBEL, H., 1963. Der Grundgebirgsbau Nordostdeutschlands im Gesamtrahmen der benachbarten Gebiete. *Geologie (Berlin)*, 12: 674–682.

KUKUK, P., 1938. *Geologie des Niederrheinisch-Westfälischen Steinkohlengebietes*. Springer, Berlin, 706 pp.

LOTZE, F. (Editor), 1945. Zur Geologie der Iberischen Meseta. I. *Geotekton. Forsch.*, 6: 92 pp.

MEHNERT, K. R., 1958. Die geologische Entwicklung des Schwarzwald-Grundgebirges unter Berücksichtigung neuer absoluter Altersbestimmungen. *Z. Deut. Geol. Ges.*, 110: 2–3.

PATIJN, R. J. H., 1963. De vorming van aardgas tengevolge van nainkoling in het noordoosten van Nederland. *Geol. Mijnbouw*, 42: 349–358.

PATIJN, R. J. H., 1964. La formation de gaz due à des réhouillifications dans le nord-est des Pays Bas. *Congr. Intern. Stratigraph. Géol. Carbonifère, Compte Rendu, 5, Paris, 1963*,2: 631–646.

PIETZSCH, K., 1962. *Geologie von Sachsen*. Deuticke, Berlin, 870 pp.

RUTTEN, L. M. R., 1949. Frequency and periodicity of orogenetic movements. *Bull. Geol. Soc. Am.*, 60: 1755–1770.

RUTTEN, M. G., 1962. Strata, movement and time. *Congr. Avan. Études Stratigraph. Géol. Carbonifère, Compte Rendu, 4, Heerlen, 1958*, 3: 603–608.

RUTTEN, M. G., 1964. Paleomagnetism and paleogeography of the Carboniferous. *Congr. Intern. Stratigraph. Géol. Carbonifère, Compte Rendu, 5, Paris, 1963*, 1: 255–260.

RUTTEN, M. G., 1965. Some recent paleomagnetic work carried out in The Netherlands. *Phil. Trans. Roy. Soc. London, Ser. A*, 258: 53–58.

SCHRIEL, W., 1954. Die Geologie des Harzes. *Veröffentl. Niedersächs. Amt. Landesplanung Statistik*, 49: 308 pp.

SCHWAN, W., 1956. Gliederung und Faltung des Harzes im Raum und Zeit. In: F. LOTZE (Editor), *Geotektonisches Symposium zu Ehren von Hans Stille*. Enke, Stuttgart, pp. 272–288.

SIMPSON, S., 1963. Variscan orogenic phases. In: K. COE (Editor), *Some Aspects of the Variscan Fold Belt*. Manchester Univ. Press, Manchester, pp. 65–73.

SPIEKER, E. M., 1956. Mountainbuilding chronology and nature of geologic timescale. *Bull. Am. Assoc. Petrol. Geologists*, 40: 1769–1815.

STILLE, H., 1951. Das mitteleuropäische variszische Grundgebirge im Bilde des gesamteuropäischen. *Geol. Jahrb., Beih.*, 2: 138 pp.

THÉOBALD, N., 1963. Évolution tectonique post-hercynienne de la région vosgéso-schwarzwaldienne. In: M. DURAND DELGA (Editor), *Livre à la Mémoire du Professeur Paul Fallot. II.* Soc. Géol. France, Paris, pp. 159–177.

VON BUBNOFF, S., 1930. *Geologie von Europa. II. Das ausseralpine Westeuropa. 1. Kaledoniden und Varisciden*. Borntraeger, Berlin, 690 pp.

VON ELLER, J. P., 1961. Les gneiss de Sainte-Marie-aux Mines et les séries voisines des Vosges moyennes. *Mém. Carte. Géol. Alsace Lorraine*, 19: 160 pp.

VON GAERTNER, H. R., 1950. Probleme des Saxothuringikums. *Geol. Jahrb.*, 65: 409–450.

WEBER, H., 1955. *Einführung in die Geologie Thüringens*. Deut. Verlag Wiss., Berlin, 42 pp.

ZOUBEK, V. (Editor), 1960. *Tectonic Development of Czechoslovakia*. Nakl. Českosl. Akad. Prague, 226 pp.

CHAPTER 5

Hercynian Europe: The Ardennes

INTRODUCTION

The Ardennes and the Rheinisches Schiefergebirge together form the most northwestern of the main units of Hercynian Europe, with the exception of Great Britain. Geographically the Ardennes form only a small western appendix of the main unit, and would therefore not seem to be so important. They do show, however, the Hercynian structures more completely than does the larger Rheinisches Schiefergebirge and consequently it seems best to begin our description in the west, with the Ardennes.

Moreover this area forms one of the classic grounds of European geology. In the main part its description goes back to the old master, J. Gosselet, professor of geology of the University of Lille in northern France. Gosselet and his contemporaries had, in the last decades of the 19th century, the advantage of many fresh outcrops where the railroads were being built. As is often the case on classical ground, we now find many of these outcrops completely overgrown or walled in, or at best blackened by the soot from countless railroad engines. These are the type localities for a substantial number of stages of the Devonian, named by Gosselet and contemporaries from the Ardennes.

Excellent descriptions of the geology of the Ardennes can be found in the *Livrets-Guide* of the 13th Session of the International Geological Congress in 1922 (CONGRÈS GÉOLOGIQUE INTERNATIONAL, 1922). The geology has since been synthesized, in 1934, by FOURMARIER, the stratigraphy and paleontology by MAILLIEUX in 1933. Detailed information on all stratigraphic names can be found in the two relevant volumes of the *Lexique Stratigraphique International* (PRUVOST, 1957; WATERLOT, 1957). Newer developments are reviewed in the *Prodrome d'une description géologique de la Belgique* (FOURMARIER, 1954). Although containing an excellent coloured map 1:500,000, this, however, is not an easily digested volume. A good introduction can be found in the short survey article by BEUGNIES (1964). Modern views on the stratigraphy and paleontology — here taken together under the catchword sedimentology — of parts of the Devonian and the Carboniferous can be found in the excellent *Livrets-Guide des Excursions* of the 6th International Congres of Sedimentology. Unfortunately, these mimeographed guide books have not have a wide distribution (CONGRÈS INTERNATIONAL DE SEDIMENTOLOGIE, 1963). Two more points of general interest in these newer developments are, first, the detailed analysis of Frasnian reefs by LECOMPTE (1954, 1957, 1961). And, secondly, the study on the famous Theux window by GEUKENS (1959), which seems to indicate that the formerly accepted nappe structures do not exist.

Geographically, the Ardennes form the upland region between the Sambre–Meuse valley in the north and the Mesozoic cover of the Paris Basin in the south (Fig.42, 43). Towards the west the Hercynian structures plunge beneath the cover of Upper Cretaceous and Tertiary sediments, only to reappear again in the small-sized amygdale of the Boulonnais, situated on the Channel coast north of Boulogne-sur-Mer (PRUVOST, 1924; BOUROZ, 1960). Towards the east the Ardennes pass into the Eifel without any clear demarcation. Geologically the boundary between the Ardennes and the Eifel had best be drawn at the transverse depression extending northwards from the "Gulf of Luxemburg", or Trèves Embayment, the *Trierer Bucht* of the German literature. Because of a slightly smaller amount of younger uplift, considerable areas of the post-Hercynian Mesozoic cover are still preserved in this transverse depression. Geographically the boundary is, however, normally drawn somewhat more to the west, following the political boundary between Belgium and Germany.

Structurally, the Sambre–Meuse valley, which follows a long and narrow, complicated Hercynian syncline, the Namur syncline, in which the coal measures are preserved, also belongs to the Hercynian Ardennes. North of this valley follows a low plateau, the Brabant

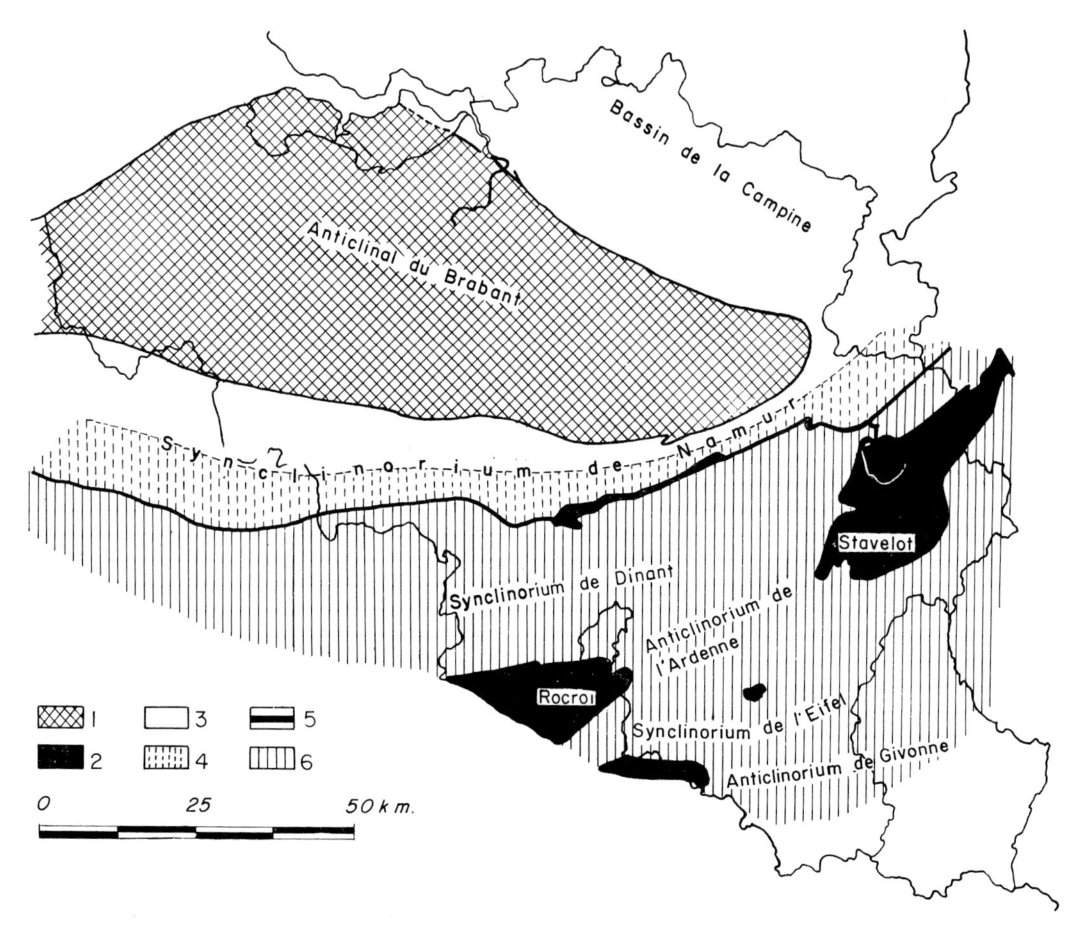

Fig.42. General view of the Paleozoic structure of Belgium.
1, *2* = Caledonian Massifs (*1* = non-affected, and *2* = retectonised by the Hercynian orogeny); *3* = non-folded post-Caledonian or post-Hercynian cover; *4* = Namur syncline; *5* = faille du Midi (= faille Eifélienne); *6* = Ardennes proper. (After FOURMARIER, 1954.)

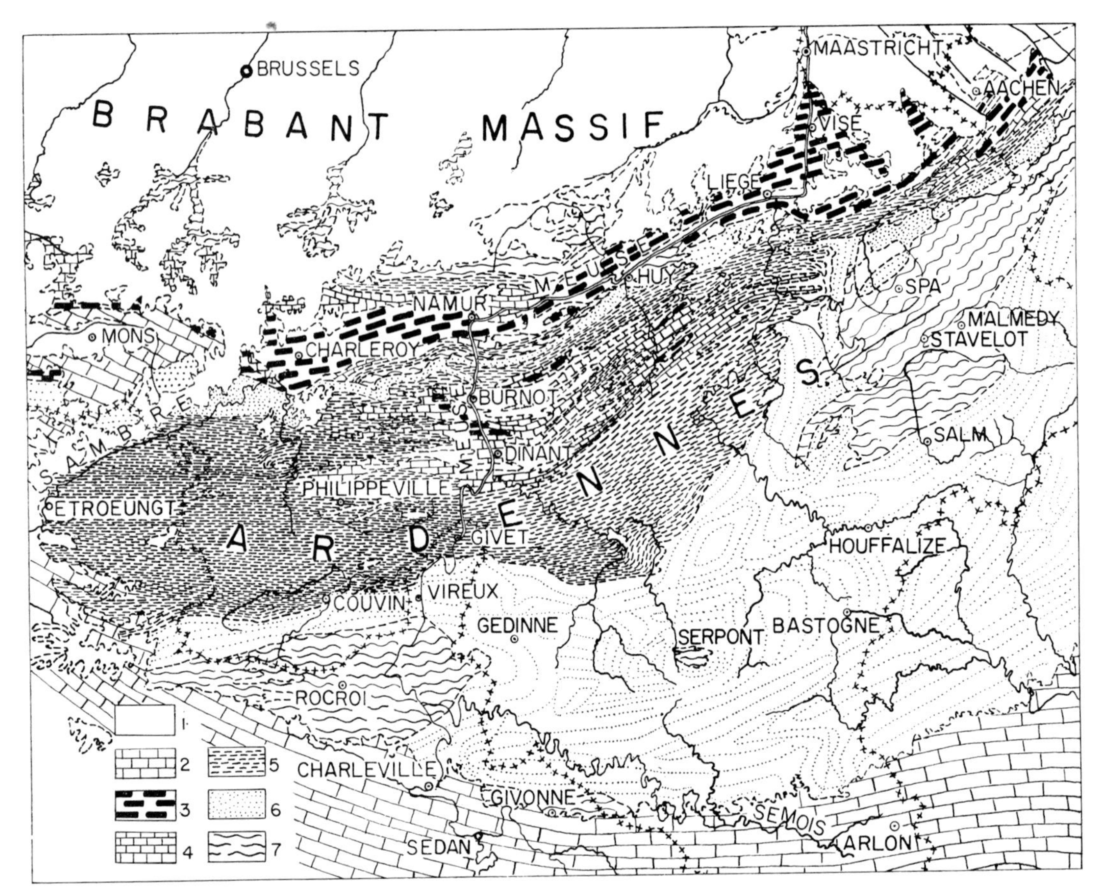

Fig.43. Generalized geologic map of the Ardennes.
1 = Cenozoic; *2* = Mesozoic; *3* = "Houiller" (Upper Carboniferous); *4* = Dinantian (Lower Carboniferous); *5* = Upper and Middle Devonian; *6* = Lower Devonian; *7* = Cambro-Silurian. (After FOURMARIER, 1934.)

Massif. This is a Caledonian structure unconformably covered by younger formations, since the Upper Devonian which have not been folded (Fig.42).

South of the Sambre–Meuse valley the structures of the Ardennes override the coal measures along a major overthrust fault. This overthrust, called the "Faille du Midi" in the west and the "Faille Eifélienne" in the east, became known to coal miners at an early date. Shafts sunk in the Lower Carboniferous, or even in the Devonian, encountered productive coal measures beneath this overthrust. It is through Gosselet's description of the "Faille du Midi" that L. Bertrand was led to postulate the nappe structures in the Swiss Alps. This overthrust consequently is a very important one, not only for the structure of the Ardennes, but also for the development of geological theory (Fig.44.)

In what I like to think of as the type section through the Ardennes, that is the section along the Meuse valley south of Namur, the situation is a little more complex. The trace of the "Faille du Midi", accompanied by a zone of strongly tectonized Ordovician and Silurian about 0.5 km wide, runs several kilometres south of the Sambre–Meuse syncline. Consequently, an additional element is visible. The intervening structure, called the Condroz anticline, shows a Devonian stratigraphy quite distinct from that of the main part of the Ardennes. It is therefore treated separately from the Ardennes in Belgian literature (KAISIN et al., 1922).

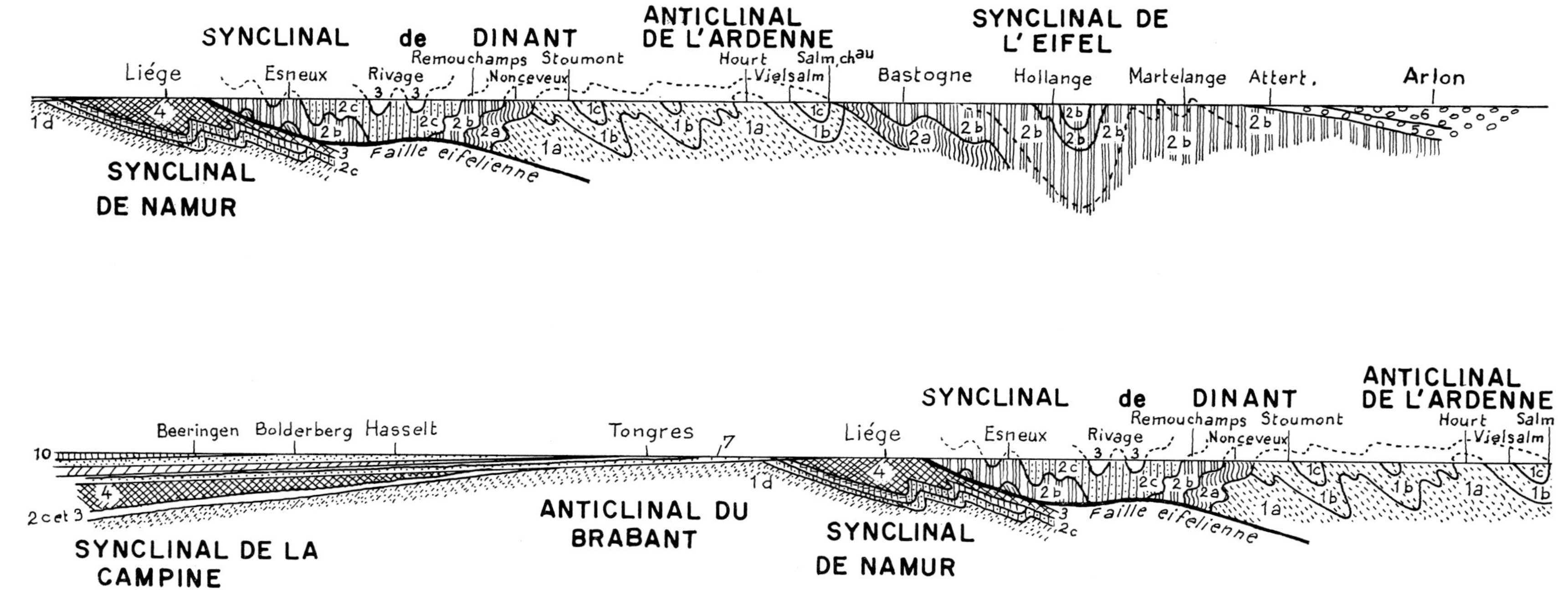

Fig.44. Generalized section through the Ardennes and the Brabant Massif, across Arlon and Liège. *1*: Cambrian. *1a* = Devillian; *1b* = Revinnian; *1c* = Salapian; *1d* = Ordovician and Silurian. *2*: Devonian. *2a* = Gedinnian; *2b* = Coblencian; *2c* = Middle and Upper Devonian. *3*, *4*: Carboniferous. *3* = Dinantian; *4* = Namurian and Westphalian. *5*: Triassic. *6*: Jurassic. *7*: Cretaceous. *8*: Eocene. *9*: Oligocene. *10*: Miocene and Pliocene. (After LOHEST and FOURMARIER, 1922.)

Within the Ardennes proper the Caledonian basement crops out in the domes of Rocroi, Stavelot, Givonne and the minuscule Serpont Massif (GEUKENS and RICHTER, 1962). The Givonne Massif (GEUKENS, 1962) is also quite small and but imperfectly known. It is situated along the southern border of the Ardennes and partly overlain by the transgressive series of the Jurassic of the Paris Basin. The Rocroi and Stavelot Massifs are more or less aligned along the Hercynian strike. They are generally held to form an anticlinal area, separating the two synclinoria of Dinant in the north and of the Eifel in the south.

As can readily be seen from Fig.43, 44 both Devonian and Lower Carboniferous crop out in the Dinant synclinorium, as against a prevalence of Lower Devonian in the Eifel synclinorium. The latter deepens eastwards, however, and in the Eifel synclines are developed with classical exposures of Middle and Upper Devonian.

THE OLD-RED CONTINENT AND THE TRANSGRESSION OF THE DEVONIAN SEA

BORDER POSITION OF THE ARDENNES

In relation to their northwesterly position in Hercynian Europe, the Ardennes are situated just south of the border between the Old Red continent, the stable part of Europe, and the adjoining Devonian geosynclinal sea. This marginal position has led to considerable variations in thickness, and also in facies, between the southerly and northerly Devonian series of the Ardennes. And, even more strongly, between the development of the Devonian in the Ardennes and that on the southern border of the Brabant Massif (Fig.45).

LOWER DEVONIAN

In the Ardennes proper the transgression over the Caledonian basement started everywhere during the lowermost stage of the Devonian, the Gedinnian. There is some indication that the transgression was a little later in the north than in the south, but fossils are scarce in the slaty facies of this series, and it is not quite clear in how much this interpretation rests upon fossil evidence or on wishful thinking.

The transgression has been remarkably tranquil. The well-advertised basal conglomerate hardly anywhere amounts to more than a number of arkose beds a couple of meters thick. The Caledonian structures must thus have been well base-levelled before the advent of the Devonian transgression.

All of the Lower Devonian is developed in detrital facies, being formed of terrigenic material, i.e., clay and sand. There is a definite tendency towards coarser deposits further north, indicating the nearness of the Old Red continent.

One of the most interesting facts, paleogeographically, is that the coast line between continent and geosynclinal sea remained by no means constant during the Early Devonian. Continental intercalations in the normally marine sedimentation of the Devonian in the Ardennes show that it sometimes shifted southwards over considerable distances, presumably when sedimentation overwhelmed geosynclinal subsidence.

The most important of these oscillations occurred at the end of the Early Devonian.

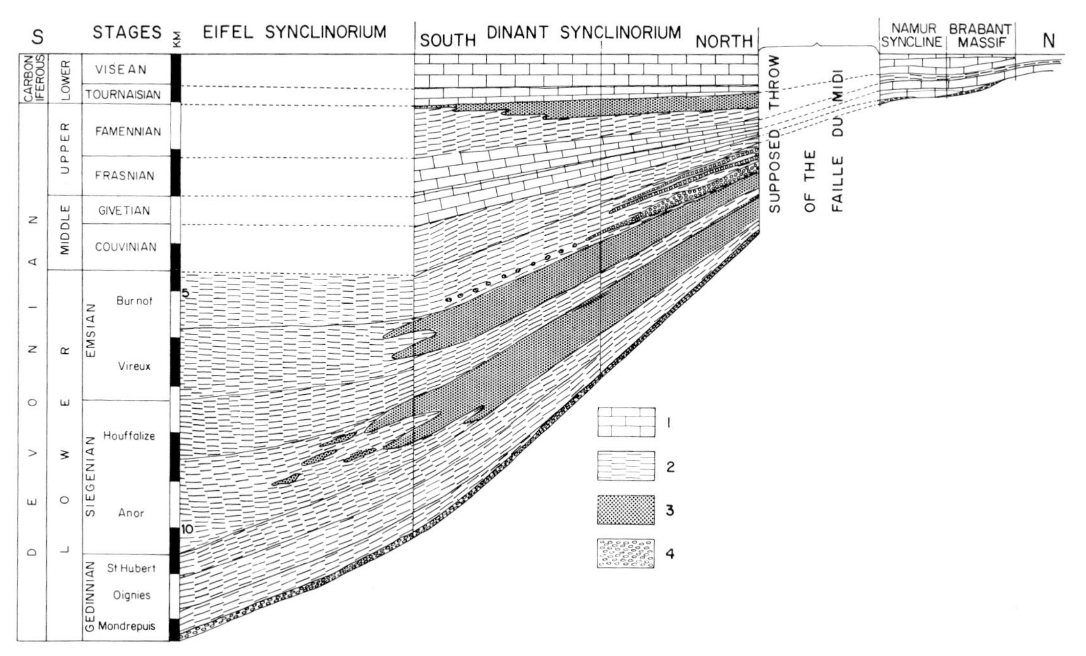

Fig.45. Generalized stratigraphic section through the central Ardennes.
1 = limestone; *2* = shale; *3* = sandstone; *4* = conglomerate. (After FOURMARIER, 1934.)

Redbeds — shales alternating with sand and coarse conglomerates — form the typical Burnot Series. This continental intercalation reaches about 1 km thickness in the northern Ardennes and peters out southwards. Intercalations of red, continental shales can, however, be followed down to Vireux, at the extreme southern limit of the Dinant synclinorium.

The total thickness of the Lower Devonian diminishes strongly, and supposedly more or less gradually, from 7.5 km in the south to 2 km in the north. This variation has often been interpreted as if the Devonian sea transgressed very gradually northwards during the Early Devonian. It has also been stated that the southern part of the geosyncline represents a deep water facies as compared with the northern part.

Neither of these statements is valid. As to the first idea, it must not be forgotten that the Gedinnian, the lowermost stage of the Devonian, is everywhere transgressive in the Ardennes proper. The Devonian sea consequently spread without much delay over all of this area. In regard to the water depth within the geosynclinal sea, there are very few detailed sedimentological studies available in this respect, so nothing can be said about it as yet.

What we seem to have instead, is a clear example of regional variation in the rate of geosynclinal subsidence, compensated by variation in the rate of sedimentation.

MIDDLE AND UPPER DEVONIAN

With the advent of the Middle Devonian the supply of terrigenous clastic material became sharply reduced and calcareous sedimentation became dominant. In the Ardennes we of course know this only for the northern part, because the Middle Devonian has not been preserved in the southern Eifel synclinorium. But where the latter deepens towards the east, we find a similar development in the Eifel and also further east in the Rheinisches

Schiefergebirge, so this seems to be a rather general feature. It indicates a slower rate of vertical crustal movements. This holds good both for the hinterland, where fewer material became available for erosion, and for the geosynclinal sea, where the limestone sedimentation was able to maintain a shallow water environment.

The Frasnian Stage of the Upper Devonian saw a repeated, rather sharp, alternation of shale and limestone deposition. It is during the latter that the beautiful reefs developed. The uppermost Devonian Stage, the Famennian, then saw a return of the detrital sedimentation, which, in its turn was followed by the predominantly calcareous facies of the Dinantian or Lower Carboniferous.

For the Middle and Upper Devonian we find a northward thinning of the strata similar to that of the Lower Devonian; from a thickness of 3.5 km in the south to 1 km in the north. In this case also we do not find any indication of a marked change of water-depth in this direction during sedimentation. Reefs of the Frasnian, for one thing, are found with minor differences both in the south and in the north. Only the Couvinian shows a decidedly more continental development, with intercalations of redbeds, shales and conglomerates, in the north, as against a purely marine facies in the south. This situation is comparable to that during the Burnotian.

CARBONIFEROUS

In contrast to the Devonian, the Lower Carboniferous, about 1 km thick, shows no marked variation in thickness from south to north. The predominance of its calcareous facies indicates overall quietness.

The quietness of the sedimentary environment of the Lower Carboniferous has recently been stressed more fully by research of Professor P. Michot of Liège and his school. They found that the major part of the Visean is characterized by extremely regular marine rhythmic series. Starting with an analysis of the Visean 2b (Michot et al., 1963; cf. Conil and Pirlet, 1963) and the Visean 3b (Pirlet, 1963) between Liège and Namur, similar rhythms, or even the exactly equivalent sequences, have now been found, not only much further west in the Namur syncline, but also in the Dinant synclinorium. Moreover, it seems probable that other horizons of the Visean will, on further analysis, show rhythmic sedimentation too.

Each sequence is mainly characterized by a lower and coarser bed that is organodetrital and zoogenic, and an upper cryptogranular bed that is phytogenic. Several other factors are found added to this basic rhythm in individual sequences. These are often persistent enough to be used to distinguish individual sequences over all of the area studied so far.

Individual sequences normally reach a thickness of several meters, rarely of 10 m to 20 m. They have been follwed, up to now, over distances of the order of 50 km. Apart from the fact that they are fully marine they are quite similar to Coal Measure cyclothem rhythms.

There was variation in geosynclinal subsidence during the Lower Carboniferous, but of a pattern different from that found for the Devonian. Stronger subsidence, in this case coupled with sedimentation in deeper water, characterized by turbites, is in general found towards the east. This is the Culm facies of the Lower Carboniferous, characteristic for the Rheinisches Schiefergebirge (p.114).

To complete the picture, let us state that during the Upper Carboniferous the trend will become reversed. At that period the thickest sediments are found in the paralic facies of the Coal Measures, to the north of the Ardennes and the Rheinisches Schiefergebirge; in the Namur syncline and in the Ruhr Basin.

SOUTHERN BORDER OF THE BRABANT MASSIF

Returning to the Devonian, we run into a quite different picture, when crossing the Namur syncline northwards, from the Ardennes towards the southern border of the Brabant Massif. Here the transgression only started in the Middle Devonian, with the Couvinian, or even with the Givetian. Further, only thin, truly epicontinental, deposits have been laid down with a maximum thickness of 500 m for the Middle plus the Upper Devonian, and a similar thickness for the Lower Carboniferous. The stratigraphy of the Devonian in the Condroz anticline is similar in facies to that of the southern border of the Brabant Massif, which seems a valid reason indeed to separate this structure from the Ardennes proper.

NAPPE STRUCTURE IN THE NORTHERN ARDENNES?

Fig.45 has been so constructed that extrapolation of the northward thinning of the Devonian in the Ardennes proper exactly correlates with the thin epicontinental facies of the Devonian on the southern border of the Brabant Massif. This rests upon the assumption of P. Fourmarier that the variation in the geosynclinal facies found in the Ardennes proper

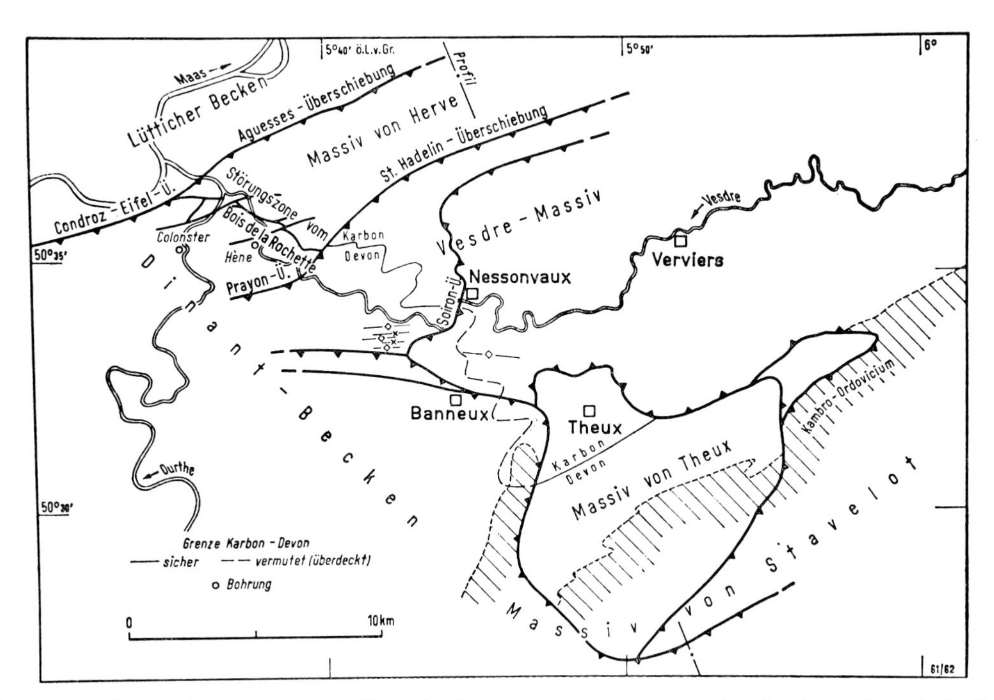

Fig.46. Outline of the structure near Theux in the northeastern Ardennes. See also the section of Fig.47. (After GEUKENS, 1962.)

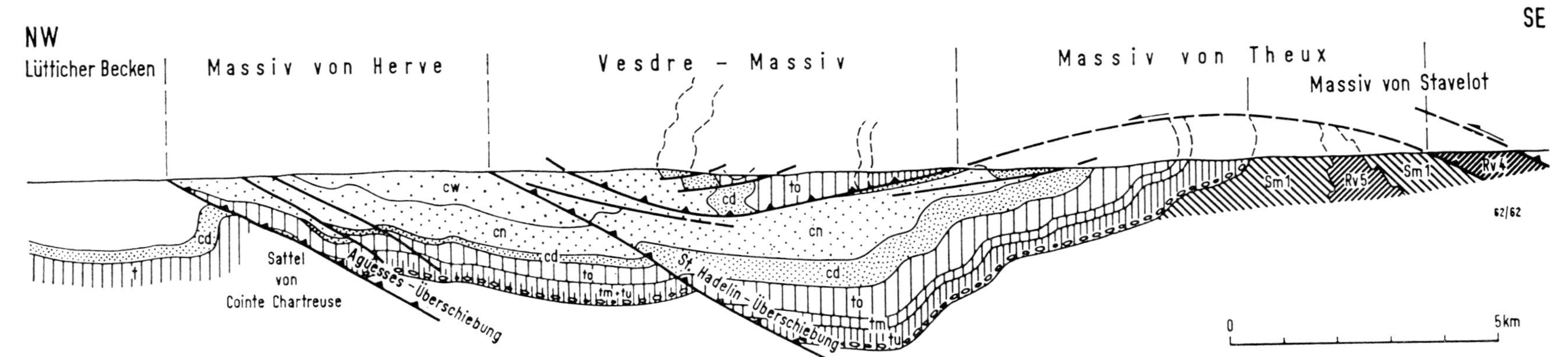

Fig.47. Section through the northeastern part of the Ardennes, across the so-called Theux window, from the Caledonian Stavelot Massif to the coal measures of the Liège Basin (compare Fig. 46). Note local overthrust block of the Vesdre.

cw, *cn* = Westphalian and Namurian, Upper Carboniferous; *cd* = Dinantian, Lower Carboniferous; *to*, *tm*, *tu* = Upper, Middle and Lower Devonian; *Sm* = Salmian, Ordovician; *Rv* = Revinian, (?) Cambrian. (After GEUKENS, 1962.)

can indeed be extrapolated linearly thus far northwards; even to the epicontinental area of the Brabant Massif. This extrapolation would then give us the horizontal amount of the northward overthrust of the Ardennes on the main northern fault zone, called rather indefinitely the "Faille du Midi", the "Charriage de Condroz" or the "Faille Eifélienne", at an estimated 15 km. Consequently, in all paleogeographic reconstructions by Fourmarier, we find southern Belgium cut off from northern Belgium along this major fault line and shifted 15 km to the south.

The fallacy of assuming linear variations in thickness from a geosynclinal area towards an adjoining epicontinental block seems evident. We thus really have no data whatsoever upon which to base an estimate of how broad this zone of transition originally was. It might have been quite narrow, being no more than the expression of a dislocation in the basement, which separated the then stable part of Europe from the remobilized part in the south.

The idea of a broad, nappe-like overthrust structure along the northern front of the Ardennes did correlate well with Fourmarier's interpretation of thrust structures farther to the northeast, near Theux, as a tectonic window in a thrust plate. The Theux window, the only example of nappe structure of an Alpine type in the northern units of Hercynian Europe, has been quite famous in the literature. More recent studies by GEUKENS (1959) seem, however, to disprove completely the window character of the Theux structures.

The impression of a tectonic window is created by the intersection of two flat-lying overthrusts with different directions. Consequently the structure is not a tectonic window with the autochthonous peeping through a major nappe, but formed by a coincidence of local tectonic features. In connection with the main overthrust along the northern Ardennes other local overthrust blocks are known (Fig.46, 47) but these are of local importance only. There is no reason to any longer assume the existence of a major nappe along the northern thrust front of the Ardennes.

STRUCTURE

The general type of the structure of the Ardennes has already been indicated in Fig.44. Apart from the Caledonian massifs on the one hand and the coal measures in the Namur syncline on the other, it is characterized by rather simple, open folds and relatively few faults. As is evident from Fig.43, individual anticlines and synclines in the Dinant synclinorium are normally rather narrow. Moreover there often is pronounced axial plunge and "en échelon" arrangement along the strike.

In more detail we may distinguish between four main structural styles in the Ardennes, i.e., those of: (*1*) the Caledonian massifs; (*2*) the Devonian; (*3*) the Lower Carboniferous; and (*4*) the Coal Measures of the Upper Carboniferous.

The difference in tectonic style between the Caledonian massifs and the series belonging to the Hercynian orogenetic cycle is mainly due to the fact that the strata of the Caledonian massifs suffered twice the vicissitudes of a major orogeny. The differences between the various Hercynian units are probably due mainly to the rather strong variations in lithofacies, resulting in quite different reactions to the same orogenetic forces. We may see in these differences a kind of "Stockwerk tectonics", although, due to limited height of individual exposures, we rarely find these different tectonic styles actually preserved on top of each other.

CALEDONIAN MASSIFS

The beds in the Caledonian massifs are practically nonfossiliferous. Only locally, in slates in the Stavelot Massif, graptolites (*Dictyonema flabelliformis*) are found, defining the Tremadocian or lowermost Ordovician. The stratigraphy commonly used in the thick detrital series of slates and quartzites is based on lithology only. A more greyish-black group is conventionally distinguished as the "younger" Revinian from a more greenish-black and "older" Devillian. Both series are named from towns on the Meuse, in the westerly Rocroi Massif. Here also is the town of Fumay, where violet-coloured series crop out, which at times have been taken as the type of a third lithostratigraphic division.

Apart from the graptolite shales in the Stavelot Massif, the type of the local stage of the Salmian, all series are assigned to the Cambrian. There is, however, a quite obvious possibility that they are Precambrian in part. In that case they should be the counterpart of detrital series of similar facies and age both in the north and in the south. That is, of the Sparagmites of Norway, the Brioverian of Brittany, and the oldest beds of unknown, but also possibly Precambrian age in the Pyrenees and the Montagne Noire.

The Revinian in the Rocroi Massif contains dikes of plutonic rocks, both quartz porphyries and diabases, which have not been found in the Devillian. This might be construed as an indication that the conventionally accepted stratigraphy is the wrong way up and that the Revinian in reality is older than the Devillian. The only other criterion is the pseudo fossil *Oldhamia*. Although rather heavily played up in the literature, this too is probably indicative only of a certain facies. Recent mapping has, however, affirmed the commonly accepted lithostratigraphy, according to BEUGNIES (1962, 1963).

The sediments in the Caledonian massifs show light epizonal metamorphism, which in places has led to the formation of roofing slate. The Revinian quartzites, very hard and resistant to abrasion, and moreover easily recognizable from their small included pyrite cubes, are about the best known and the most widely distributed erratics from the Ardennes found in Pleistocene deposits in The Netherlands.

The tectonic structure of the Caledonian massifs can best be studied in the classic section along the Meuse valley through the Rocroi Massif (GOSSELET, 1888; KAISIN et al., 1922; WATERLOT, 1937, 1945, 1958; ANTHOINE, 1938).

Waterlot's section, reproduced here as Fig.48, serves well to typify the tectonic style. This is characterized by a general imbricate structure, with a predominantly southerly dip of both strata and overthrust planes. Overturned isoclinal folding may be present, but proof for such structures, either from the stratigraphy or from top and bottom features, is scarce. There is a strong antithesis between ANTHOINE (1938), who favours nappe structures and larger overthrusts, and the other authors who think more in terms of overturned folds cut by faults. The absence of key beds makes any structural interpretation hazardous. Moreover, apart from the exposures along the Meuse valley and some of its tributaries, the Rocroi Massif forms a plateau well covered by a thick mantle of weathering products, where anybody's fantasy may well run loose.

As for the tectonics of the Stavelot Massif, which shows somewhat stronger metamorphism in its southeastern part, the reader is referred to GEUKENS (1962).

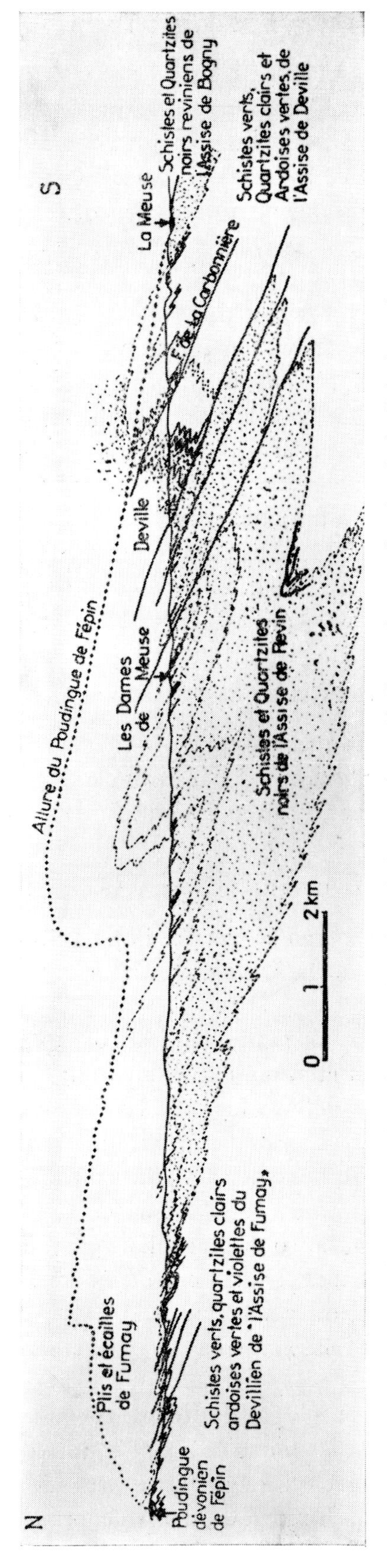

Fig.48. Section along the Meuse river through the Rocroi Massif. The Fépin unconformity of Fig.49 is situated at the extreme left. (After G. Waterlot, in: FOURMARIER, 1954.)

FÉPIN NONCONFORMITY

Returning to more simple facts, the classical unconformity of Fépin deserves to be mentioned. It is situated along the eastern valley wall of the Meuse at the northern boundary of the Rocroi Massif, and is formed by the transgressive Gedinnian over the Caledonian series. The Gedinnian, represented by its basal coarse to conglomeratic sandstone, forms a simple, open, synclinal fold, asymmetric towards the north, with a subhorizontal northern and a subvertical southern limb. Underlying Revinian slates generally dip rather regularly southwards (Fig.49).

This exposure not only forms one of the classic examples of a major nonconformity in Europe, but it also beautifully shows the differences in reaction to the same orogenetic strain by different rocks. This applies to the way in which the Revinian was reactivated during the Hercynian orogeny. At that time the Gedinnian was not yet folded nor was it metamorphosed, whereas the Revinian slates had already been both folded and metamorphosed. If we try to reconstruct the pre-Hercynian structure at Fépin by simply folding back the Gedinnian together with the underlying Revinian towards the horizontal, we run into space problems in the latter. There has evidently been a quite different reaction to the Hercynian stress. Whereas the Gedinnian was folded in concentric type folding, the Revinian slates must have moved along pre-existing cleavage planes in a type of shear folding.

STRUCTURE OF THE DEVONIAN

As indicated above, the Devonian is normally characterized by a regular, open style of folding, without many complications. As an example, Fig.50, of part of the Devonian along the Meuse north of Dinant, is given without further comment.

STRUCTURE OF THE LOWER CARBONIFEROUS

The Lower Carboniferous, as can be seen for instance from Fig.51, normally shows a much closer style of folding than the Devonian, with serrated series of tight, subvertical anticlines and synclines.

Presumably this is due to the alternation of series of thin-bedded limestones with thin-bedded marly limestones, which not only results in strong variations in competence, but also leads to a high degree of mobility along the bedding planes.

Characteristic for the Lower Carboniferous are the acute minor folds superimposed in many places upon the anticlinal and synclinal structures. These minor folds are of the type called "plis en chaise". They are a couple of meters to some ten meters across. Their axial planes are invariably inclined, so that a subhorizontal "lower" limb and subvertical "upper" limb alternate. Beneath the subhorizontal limb of one such minor fold then follows the subvertical limb of the next lower minor fold, and vice versa. Cross sections through one and one half of such minor folds consequently quite strikingly resemble a chair, hence their local name (Fig.52).

It is difficult to think of such folds as due to horizontal tangential compression. They are reminiscent of the cascade folds found in alpine tectonics and it seems more sensible to suppose that they too originated through gravity tectonics.

The regional overthrust which separates the Ardennes proper from the Namur syn-

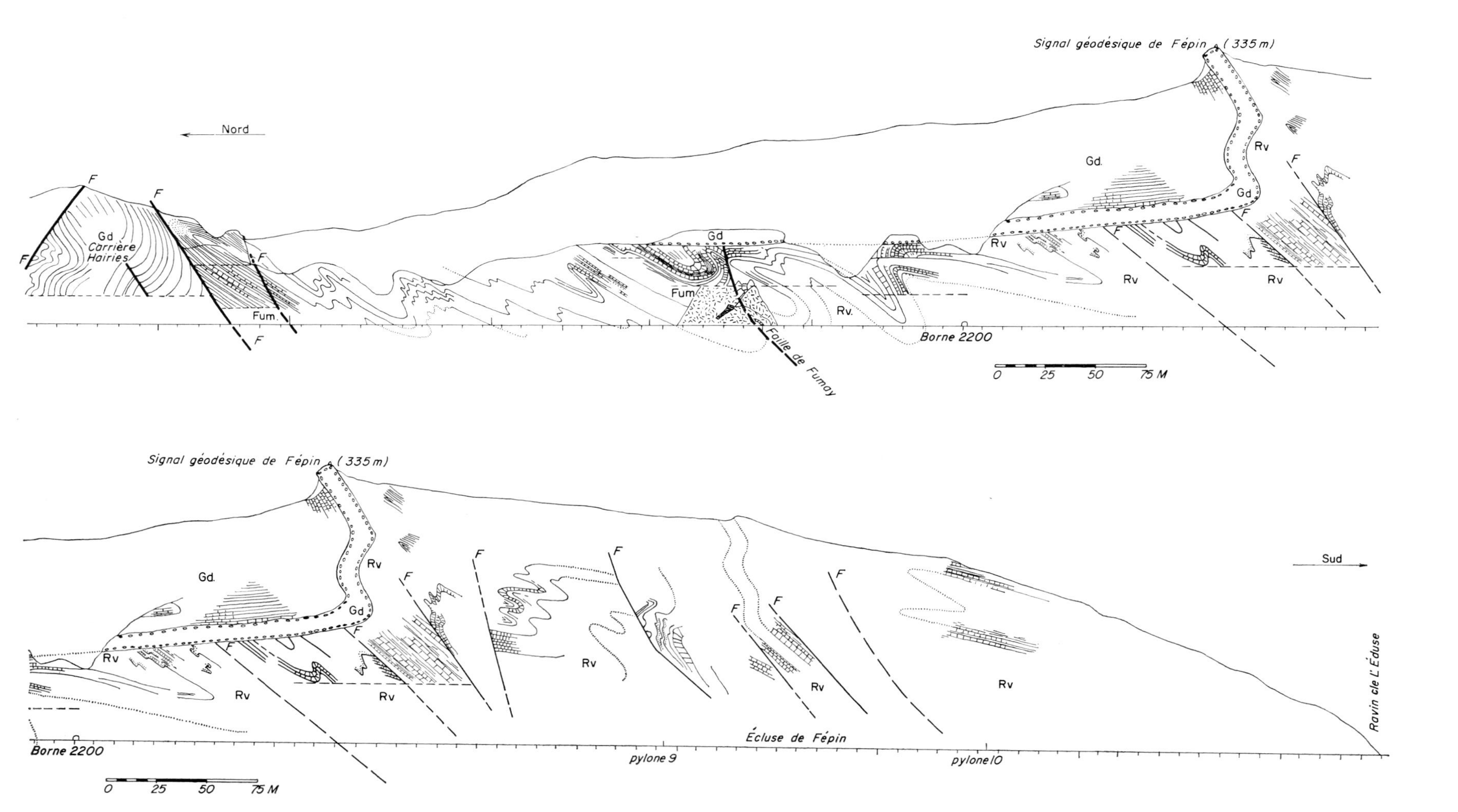

Fig.49. The Fépin nonconformity along the eastern Meuse valley at the northern limit of the Rocroi Massif. Basal Gedinnian Grès d'Anor (*Gd*) nonconformably overlies (?) Cambrian slates of the Revinian (*Rv*) and Fumay Series (*Fum*). (After ANTHOINE, 1938.)

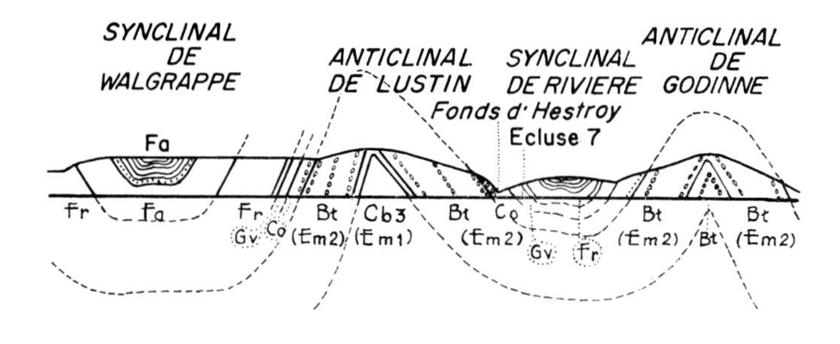

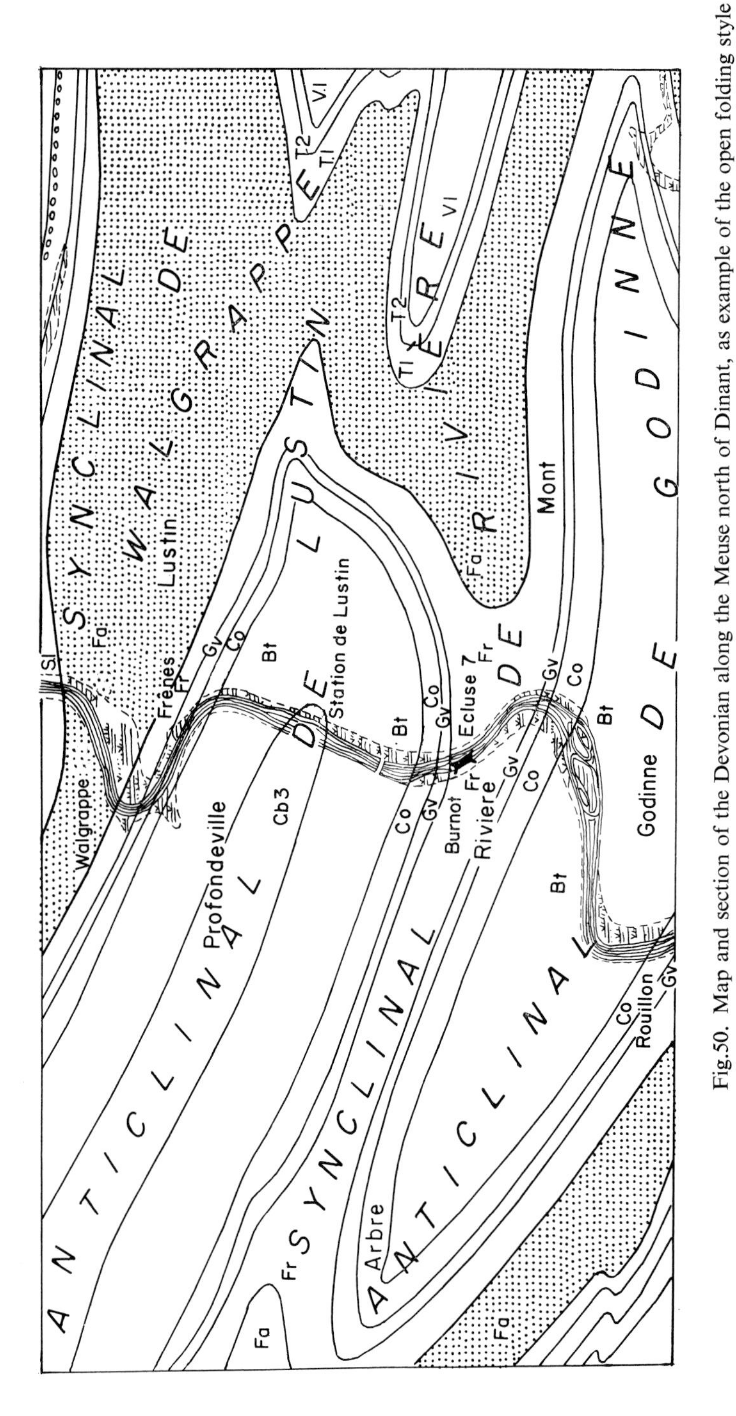

Fig.50. Map and section of the Devonian along the Meuse north of Dinant, as example of the open folding style of the Devonian.

V_2, V_1 = Upper and Lower Visean; T_2T_1 = Upper and Lower Tournaisian; *Fa* = Famennian; *Fr* = Frasnian; *Gv* = Givetian; *Co* = Couvinian; Em_2 = Upper Emsian, (with *Bt* = Burnotian); Em_1 = Lower Emsian (= Cb_3 = Upper Coblencian of authors). (After KAISIN et al., 1922.)

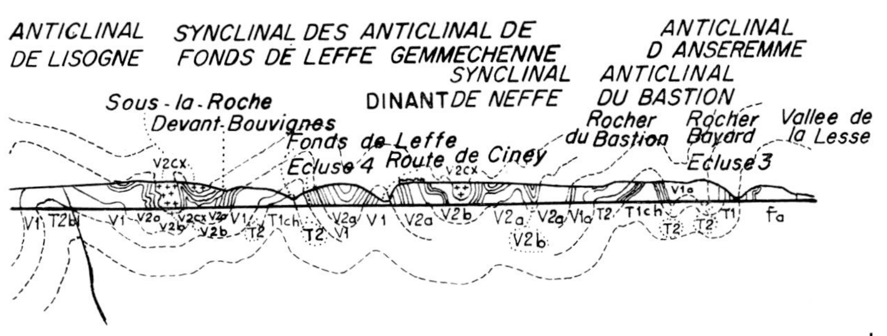

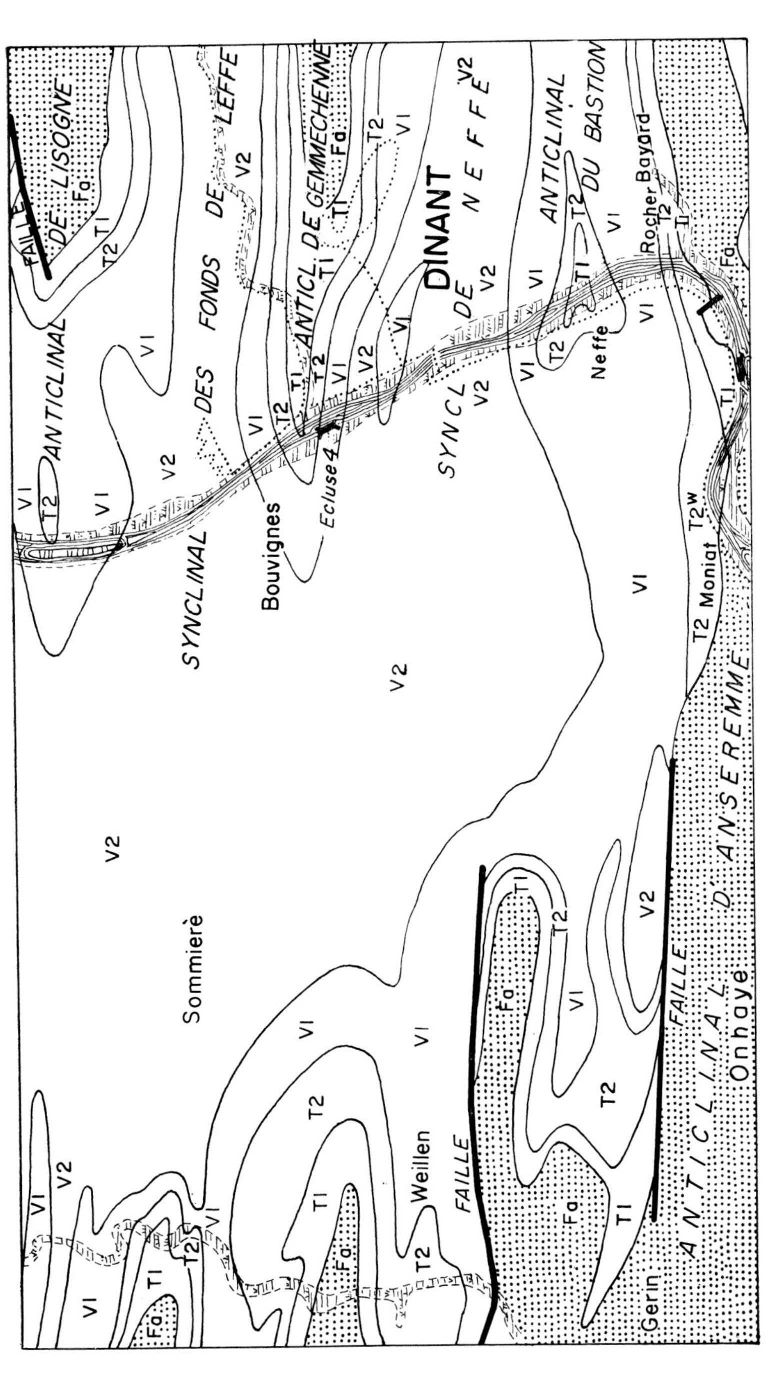

Fig.51. Map and section of the Lower Carboniferous along the Meuse near Dinant as example of its narrow, complicated style of folding, which is actually far more complex in detail. This in part is due to the many "plis en chaise", for instance well visible at the Dinant railroad station. For abbreviations of stratigraphic units, compare Fig.50. (After KAISIN et al., 1922.)

Fig.52. "Pli en chaise" in Dinantian limestone. Moresnet, northeastern Belgium.

cline, the "Faille du Midi" or "Faille Eifélienne", has been mentioned already. It is not everywhere as simple in structure as is often assumed. Further east, where the Namur syncline opens into the Liège and Aachen Basins, this overthrust splits into several parallel structures, which may show considerable displacements of local thrustblocks (Fig.47). The intercalation of the Condroz anticline between the Ardennes and the Namur syncline, has also already been mentioned.

STRUCTURE OF THE UPPER CARBONIFEROUS

The Upper Carboniferous of the Namur syncline shows a tectonic style quite different from that of the other units. It consists of the detrital series of the Namurian and the Westphalian, built up by several kilometers of shales and sandstones with intercalated coal seams, regularly exhibiting cyclothemic sedimentary rhythms. Its structure is characterised by a series of subhorizontal overthrusts cutting through a sequence which at first sight looks deceptively simple and tranquil (Fig.53).

Individual overthrusts often lose themselves in the bedding planes, where the amount of thrusting can no longer be estimated. As is the case in many coal measure series, there has been, however, considerable bedding plane slippage. Most coal seams for instance show signs of movement, either along their roofs, or along roofs and floors and also along clayey intercalations.

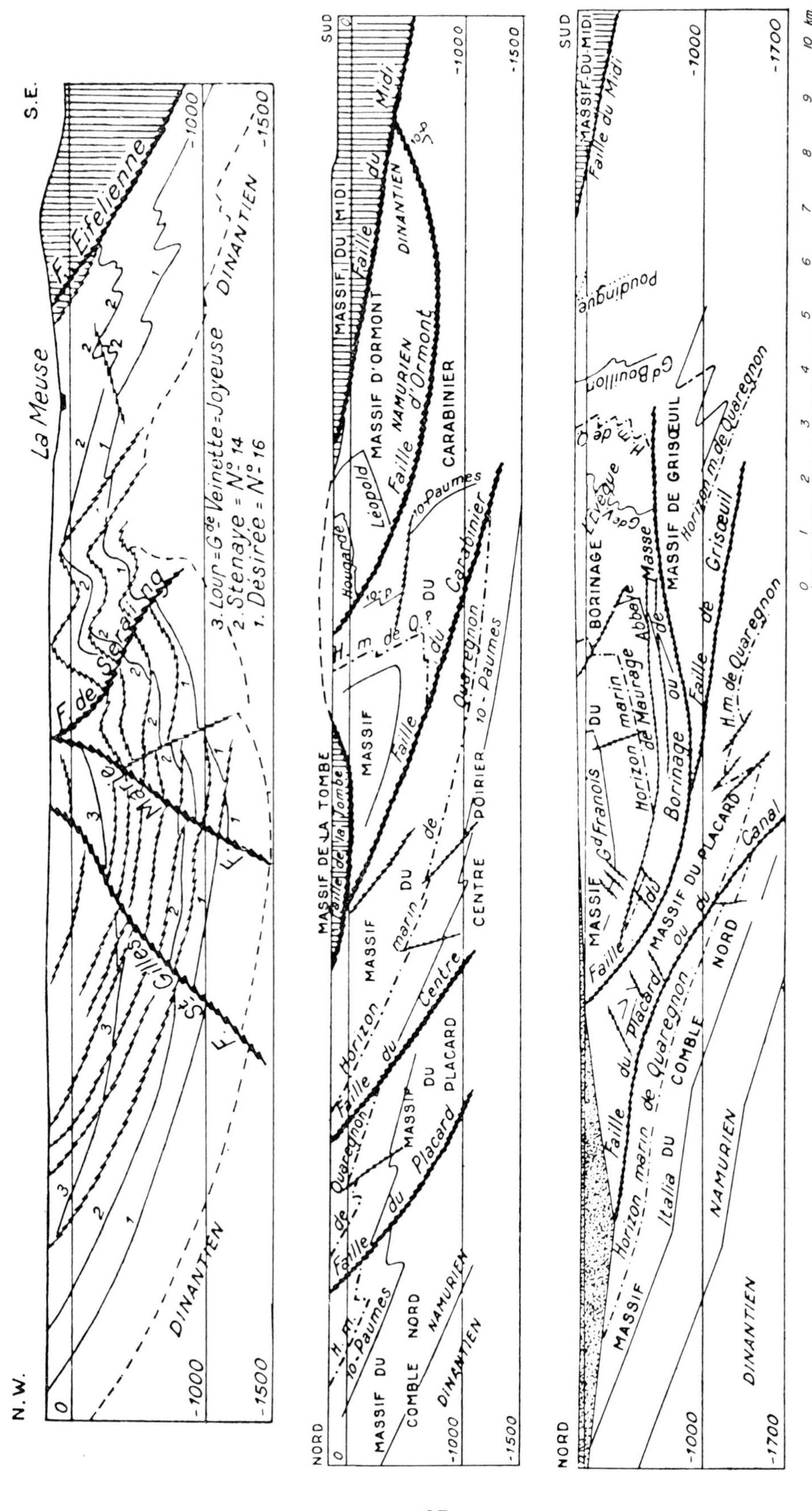

Fig.53. Sections through the Coal Measures in the Namur syncline. (After FOURMARIER, 1954.)

In a general way the overthrusting is seen to die out northwards. There the tectonized series of the Namur syncline gradually changes into the smooth sedimentary mantle of the southern flank of the Brabant Massif, with a regular southerly dip of only 5–10°.

TYPE LOCALITIES OF ETROEUNGT AND VISÉ

A note might be added here on the type localities of Etroeungt and Visé, the one situated at the extreme western plunge of the Ardennes, the other at its extreme north-eastern limit, on the Meuse, north of Liège. The Etroeungt Stage forms the transition from Devonian to Carboniferous. A unanimous conclusion as to which of the two systems it must be ultimately assigned, has, I think, not yet been arrived at amongst top-level paleontologists (SIMPSON, 1962) .Visé is the type locality for the uppermost part of the Lower Carboniferous.

The reason for mentioning these type localities here under one heading is not their dissimilar stratigraphic value, but the characters they have in common, i.e., their lithological aspect and their general unsuitability as a type locality.

Etroeungt forms a synclinal structure of limestones of the (?) lowermost Carboniferous within shales of the Famennian, whereas Visé is an anticlinal structure of limestones within the shales of the Upper Carboniferous. Both form the first limestone outcrops, when one comes from the surrounding lowlands, and have been quarried since early times. In both localities this has led to the establishment of large collections of fossils, which at that early date was the only thing required for a type locality.

For more modern stratigraphical studies, both localities are, however, both equally unsuitable. Not only are the surrounding shale members badly exposed, but in both cases the difference in competence between the limestone and the surrounding series has resulted in considerable overthrusting. There is, consequently, no good stratigraphical sequence to be found at either of these two ancient type localities, a fact which does not facilitate modern stratigraphical studies.

TIME OF FOLDING AND THE "GRANDE BRÈCHE" OF THE VISEAN

The northwesterly position of the Ardennes in Hercynian Europe implies a question as to the time of folding within the Ardennes, i.e., Sudetic or Asturic (p.74). For the northern part of this Hercynian unit, that is for the Namur syncline, there is no such question. The Coal Measures of the Upper Carboniferous belong to the geosynclinal, that is the paralic facies, and were folded some time after the Middle Westphalian. They typically belong to the Asturic phase of the Hercynian orogeny.

At most there has been some vertical uplift, accompanied by tilting between Lower and Upper Carboniferous. In the Namur syncline, this has resulted in a slight disconformity between Lower and Upper Carboniferous. Between Namur and Liège, from west to east, a progressive onlap of the Namurian from Visean 3bα to Visean 3aα is found. That is a vertical difference of 70 m, over a distance of 16 km. Actually, the nonconformity is

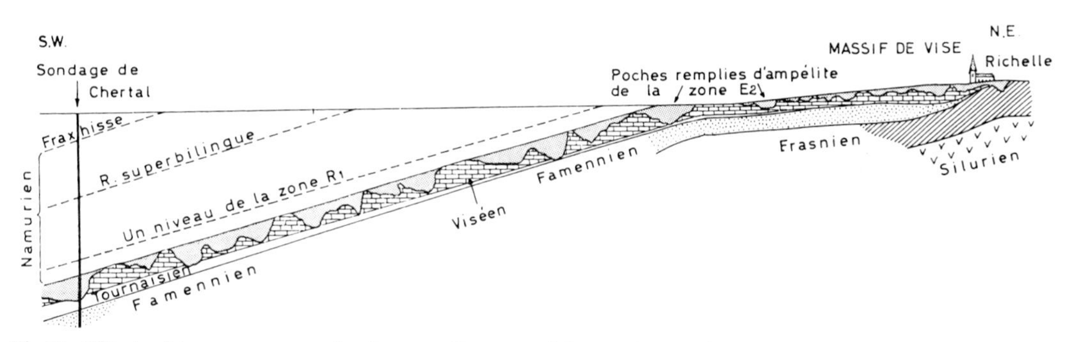

Fig.54. Effects of temporary emersion between Lower and Upper Carboniferous in the Namur syncline near Visé. Development of karst erosion in the Visean limestones. (After GRAULICH, 1963.)

situated slightly lower. For the uppermost Visean, the Vec, only a couple of meters thick, already belongs to the sequence of the Namurian (PIRLET, 1964).

Further east, near Visé, karst erosion could even develop on the Visean limestones during the regression between Lower and Upper Carboniferous (GRAULICH, 1963; Fig.54).

Further south, in the Ardennes proper, the situation is, however, less clear (WATERLOT, 1945). With the exception of the small Anhée syncline, halfway between Dinant and Namur, where quite strongly tectonized lowermost Namurian is preserved, the youngest strata found belong to the Visean, or uppermost Lower Carboniferous. It is thus well possible that the Ardennes proper, or at least their southern part, were already folded during the Sudetic phase of the Hercynian orogeny. The total absence of postorogenetic sediments of Upper Carboniferous age in the Ardennes makes this assumption impossible to prove. It cannot be more than a well-considered guess.

One indication of stronger crustal movement during the later part of the Lower Carboniferous can be found in a breccious facies, for instance in the "Grande Brèche" of the Upper Visean. This is a coarse sedimentary limestone breccia. Presumably it is the result of calcareous turbidites, consisting of lime mud and of angular fragments derived from earlier, already consolidated, limestone beds. It locally shows well-developed graded bedding.

The "Grande Brèche" is found in the Upper Visean all over the Dinant synclinorium, for instance in the Fonds de la Leffe, north of Dinant (cf. PIRLET, 1964; Fig.51). But it is also found, intriguing though this may be, in the Condroz anticline and on the southern flank of the Brabant Massif, which are within the realm of the Asturic phase of the Hercynian orogeny. Consequently, although the "Grande Brèche" can certainly be used to advocate the fact that stronger crustal movements have taken place at the end of the Lower Carboniferous, it cannot, just as certainly, be used as proof of a Sudetic folding phase. All over the Dinant synclinorium it might very well be no more than a repercussion of a Sudetic folding taking place much further south.

TANGENTIAL COMPRESSION VERSUS GRAVITY TECTONICS

The generally accepted picture of the forces which have formed the Hercynian fold belt of the Ardennes is that of tangential forces originating somewhere to the south. This view is substantiated by the overall tendency towards northwards overturn in the Ardennes. Even if the nappe structures along the northern front of the Ardennes are not as large a feature as has been supposed, the fact remains that the main overthrust shows a southerly dip. Similar southerly dips are generaly found in the low-angle overthrusts of the Namur geosyncline. And also, much more distinctly, in the rather scarce overthrusts within the Ardennes proper. The imbricate structure in the Caledonian massifs, lastly, points the same way.

It can thus be concluded that there is an overall northward "Vergenz" in the Ardennes. This is explained by the resistance, thought to have been offered by the Brabant Massif to a tangential thrust from the south. The latter, the rim of the stable, consolidated part of "Ur-Europa", is thought to have formed the buttress, on which wave after wave of the Hercynian foldings came to break.

Contrary to this generally accepted picture, VAN LECKWIJCK (1956) has tried to show that there is not the slightest evidence that the Brabant Massif already formed an elevated or even a stable area during the Upper Carboniferous. While following the stratigraphy of Namurian and Westphalian coal measures, from the Namur syncline along the eastward plunging nose of the Brabant Massif towards the Campine Basin situated on its northern flank, he found no indication whatsoever that the present Brabant Massif influenced Upper Carboniferous sedimentation or that it was expressed in the paleogeography of that time. Instead of a stable area of positive elevation as it is now, it was during the Upper Carboniferous an unstable area of geosynclinal subsidence just as the Namur syncline and the Ruhr Basin.

According to VAN LECKWIJCK (1956) the rise of the Brabant Massif is a much younger feature. This view has since been fully supported by PATIJN (1963), cf. Fig.55. It consequently cannot have formed a buttress during Hercynian orogeny. This demolishes one of the props on which the conventional theory of tangential compression rests. It seems therefore much more sensible to admit gravity as the moving force behind the Hercynian tectonics in the Ardennes.

Several other features can also be better explained by gravity tectonics than by tangential compression. Such as, to mention only what has been cited in this text, the gradual northward dying out of the overthrusts in the Namur syncline, and the "plis en chaise" structures in the Lower Carboniferous. In the Namur syncline one would, in the tangential compression model, expect the strongest tectonisation to have occurred on the southern flank of the Brabant Massif. It is there that all of the Ardennes are thought to have been pushed upon the buttress of stable Europe. But instead of structures reminiscent of a breakwater during a severe storm, one finds a gradual dying out of all tectonic effects. In the "plis en chaise", attention has been already drawn to their overall similarity to series of cascade folds, which form one of the typical indications for gravity tectonics.

Up until now Van Leckwijck has had very few adherents and the classic picture of tangential compression as the main force during Hercynian orogeny is certainly still the most widely accepted. But we will see in our discussions of Alpine Europe, how at this date time seems to work for the ideas of gravity tectonics. So we might at present well leave these views to the able assistance of time, and see how things will develop in the future.

Fig.55. Genetic sections across the Brabant Massif. Vertical scale exaggerated. Showing how the Brabant Massif only developed since post-Permian time. It can, consequently, not have acted as a buffer during the Hercynian orogeny. *1*. Late Devonian time. *2*. Late Westphalian time. *3*. Triassic time. *4*. Jurassic time. *5*. Early Cretaceous time. *6*. Late Cretaceous time. *7*. Present time. (After PATIJN, 1963.)

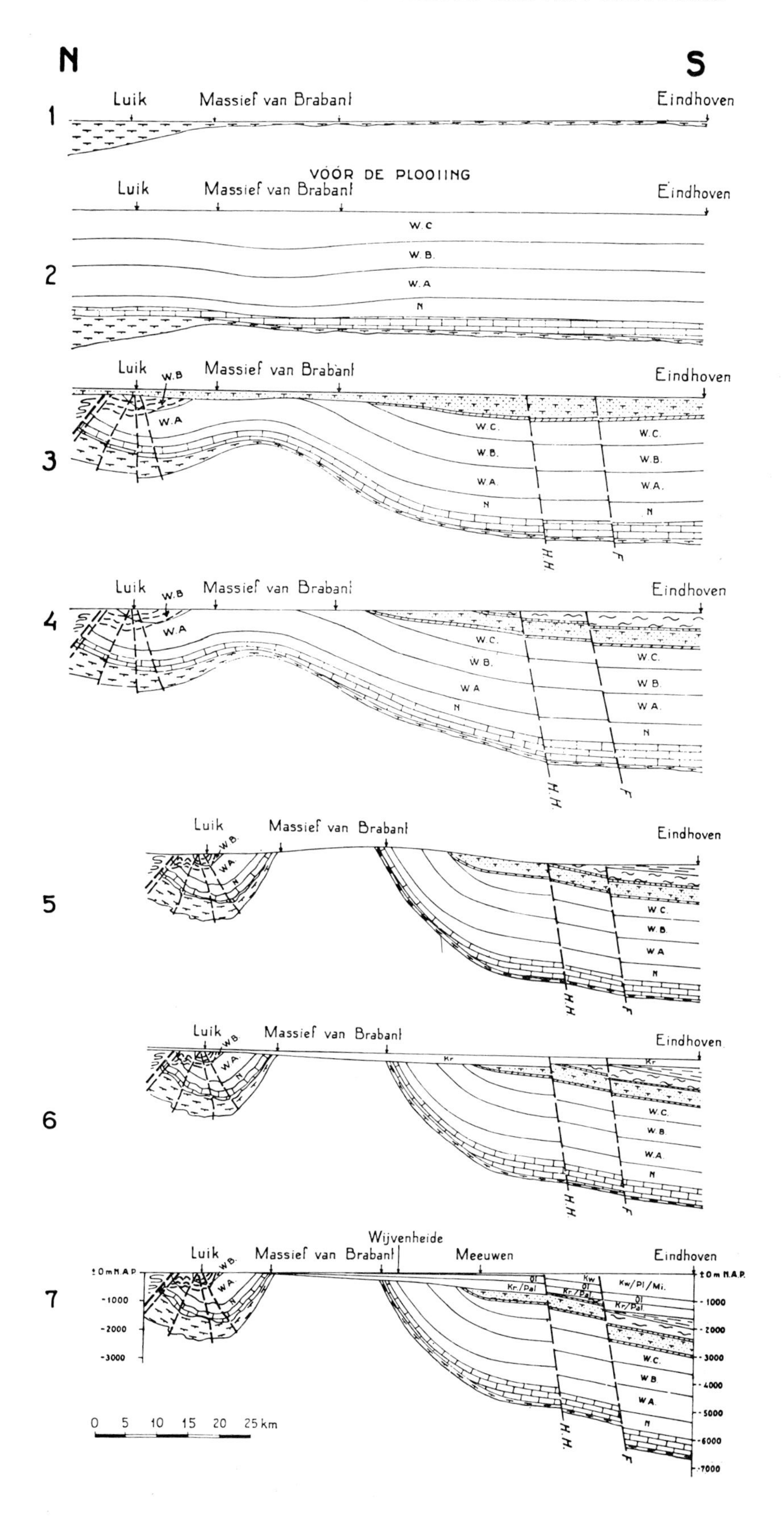
N
S
1
Luik
Massief van Brabant
Eindhoven
VÓÓR DE PLOOIING
2
W.C
W.B.
W.A
N
3
4
5
6
7
Wijvenheide
Meeuwen
±0m N.A.P.
-1000
-2000
-3000
-4000
-5000
-6000
-7000
Kr/Pal
Kw/Pl/Mi.
0 5 10 15 20 25 km
H.H.
F

DEVONIAN REEFS

Before leaving the Hercynian Ardennes we must still mention one very interesting facet of its geology, i.e., the Devonian reefs. They are best developed in the area between Philippeville and Couvin, where they have attracted the attention of geologists from the earliest times. In later years they have become much more intensively known through the painstaking work of Professor M. Lecompte of Brussels (LECOMPTE, 1954, 1957, 1961).

Fig.56, in a very old drawing indeed, dating back to GOSSELET (1888), schematically shows the position of these reefs north of Couvin, whereas Fig.57 shows their position in several stratigraphical columns. The reefs occur in three horizons within the *F2*, which paleontologically is the Middle Frasnian, but which as to actual thickness forms its main part. Two limestone members in the lower part of the *F2* (*F2c* and *F2g*), intercalated in shales, locally give rise to reefs (*F2d* and *F2h*). Moreover in the upper shale member *F2i* a younger period of reef formation (*F2j*) occurs.

All reefs show a similar helmet form, with flat base and globular top. Evidently they started growing on a more or less even sea floor, and thereupon quickly outgrew the surrounding sedimentation. The *F2d* and *F2h* reefs grew into surficial depths and show the influence of wave action. Reef detritus has been piled up around the living reef. The *F2j* reefs remained in (slightly?) deeper water during all of their growth period and are not found influenced by wave action.

There are several ecological correlates to this difference in water depth of the *F2d*

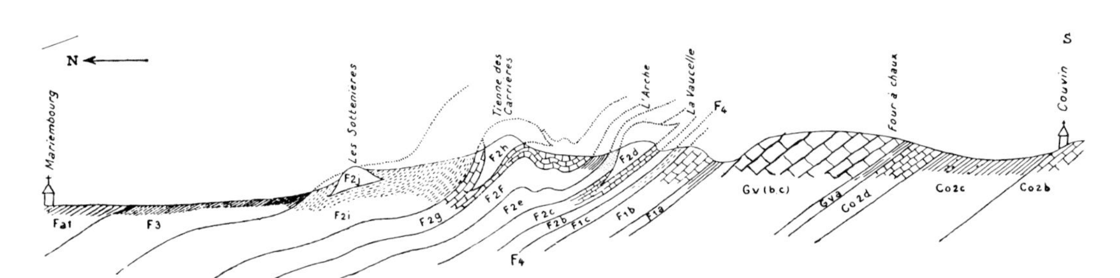

Fig.56. Section through the Devonian north of Couvin, 1:40,000. Note three successive periods of reef formation in the Frasnian (*F2d*, *F2h* and *F2j*).
Co = Coblencian; *Gv* = Givetian; *F* = Frasnian. (After KAISIN et al., 1922.)

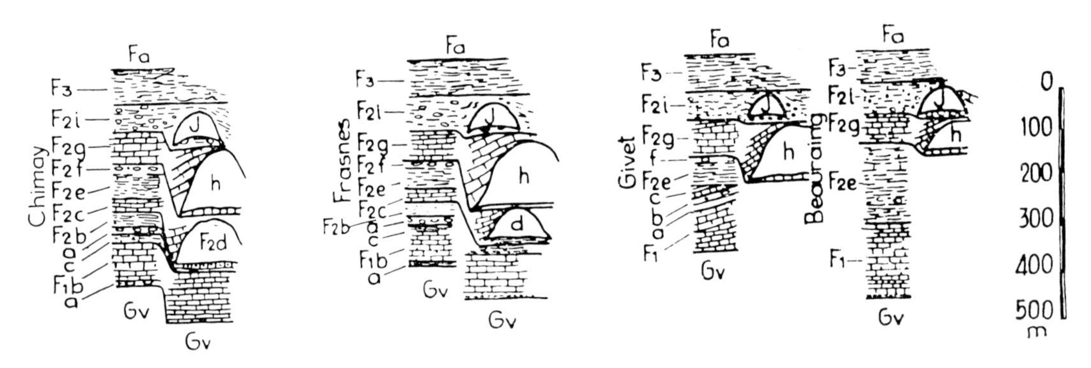

Fig.57. Stratigraphical sections through several Devonian reefs. Note general helmet form.
Gv = Givetian; *F* = Frasnian. (After FOURMARIER, 1954.)

and *F2h* reefs, as compared with the *F2j* reefs. The most important is perhaps that corals were able to flourish, not just to survive, in deeper water than stromatoporoids. This conclusion of Lecompte has since been found to be valid in the Silurian reefs of Gotland also (p.48). For a more detailed analysis of the various ecological relations, the reader is referred to the publications mentioned above.

But in contrast to the Silurian reefs of Gotland, the Devonian reefs of Belgium may reach quite large dimensions. Diameters of several hundreds of meters and heights of up to 100 m are common.

The reddish limestone of the *F2j* reefs is exploited for marble and as such is found all over western Europe. Notably, most tops of the little round tables, so characteristic for old-fashioned cafés, come from the Belgian *F2j* reefs. These tops are slabs from the cylindrical cores extracted out of the first big corner holes that must be drilled into the marble before blocks can be cut. Sheaves carrying the cutting wire are lowered into these holes. The astounding similarity of all these old-fashioned cafés, somehow related to our image of France and Romanic culture, depends largely on the exactly similar size, everywhere, of all these little round tables. This in turn is dependent upon the standardized size of the bore holes in which the wheels carrying the wires needed to cut the marble are lowered, an indication how far reaching the influence of trifling things and Gallic thrift can be.

DEVONIAN RHYTHMS AND COAL MEASURE CYCLOTHEMS

It is evident that the alternation of series of shales and limestones in the Frasnian indicates a sedimentary rhythm in relation to geosynclinal subsidence. The sea was shallow during the deposition of the limestone series and decidedly deeper during shale deposition. This can be deduced from the scanty dwarf faunulae which characterize the shale members. Moreover, during the formation of the limestone series, supply of terrigeneous material was non-existent, or almost non-existent, whereas it was active during the deposition of the shales. There is consequently a positive correlation between the deepening of the geosynclinal sea and a stronger supply of detrital, terrigeneous material. This indicates stronger crustal movements; both a renewed rising of the hinterland which made available the detrital material through renewed erosion, and a contemporary stronger subsidence of the geosyncline.

If we now compare these rhythms with those other well-known rhythms, the cyclothems, we may note two distinctions:

(*1*) Individual cycles of these Devonian rhythms are much thicker than individual cyclothems normally are. This thickness is of the order of 100 m for the Devonian rhythms, as against 10 m for an average cyclothem.

(*2*) The Devonian sequences were deposited on the seaward side of the hinge line between geosynclinal subsidence and hinterland rising, whereas the cyclothems of the coal measures were deposited to the landward of that line (RUTTEN, 1956). Or, in other words, in the Devonian rhythm the sea deepened in the sedimentary area concurrently with the influx of more terrigeneous material, whereas in the coal measure cyclothem a regression coincides with the deposition of the sandy member, there also indicative of stronger hinterland rising.

POST-HERCYNIAN HISTORY

PENEPLANATION

Following in the wake of the Hercynian orogeny, erosion immediately started. It is probable that peneplanation had already been mainly accomplished during the Permian. Remnants of post-orogenetic Permian beds deposited nonconformably over folded Paleozoic rocks can be taken as indication for this assumption, but most deposits of that time have since been carried away. Of such local remnants the Malmédy Conglomerates, whose age has recently been confirmed by paleomagnetics (DE MAGNÉE and NAIRN, 1963), are the most important.

All during the Mesozoic and the Tertiary the area retained a lowland character and the rate of its vertical movements, either up or down, must have been small. Nothing is known of a possible cover on the Ardennes of Triassic, Jurassic or Lower Cretaceous, because no remnants of these formations have ever been found. Further east, on the transverse depression north of Luxemburg, extensive remnants of Buntsandstein are known, however. Scattered localities with remnants of marine Upper Cretaceous make it probable that the Senonian transgression, whose deposits now envelop the Ardennes from the north, the west and the south, has actually spread all over it. The Tertiary, on the other hand, left quite frequent evidence in the form of continental sand deposits nonconformably covering the folded Paleozoic.

FORMATION OF UPLANDS

The transition from lowland to the present upland position of the Ardennes occurred only very late in geologic history. This is a feature common to all uplands of Hercynian Europe, as has been stated in Chapter 1. However, in most uplands actual dating of this transition is difficult, and it is generally assigned rather vaguely to the Plio-Pleistocene. In the Ardennes and the Rheinisches Schiefergebirge, this uplift can, on the other hand, be dated quite nicely. It took place at the transition from Pliocene to Pleistocene. Because of the importance for all of the history of western Europe, we will go into some detail on this question.

It is quite possible that the Ardennes and the Rheinisches Schiefergebirge showed rising tendencies long before the beginning of the Pleistocene. MACAR (1954, 1965), among others, has defended this thesis on the basis of some fifteen high levels of planation found all over the Ardennes.

But these earlier rising movements must have been on a smaller scale and their influence has been offset by contemporaneous erosion. Apart from the possibility of subdued residual hills on the divides, far from the rivers, no topographically higher areas have developed from these earlier movements. We can be sure of this for two quite different reasons. For one thing, no fresh detrital material from the Paleozoic of the Ardennes ever reached the surrounding areas of Tertiary deposition. These deposits are predominantly fine-grained sands and clays. When coarser material is found, this was not derived from the Paleozoic core of the Ardennes — such as, for instance, the pebbles derived from Senonian flint, found in the Miocene, and the famous pebbles with silicified oolites found in the Pliocene to the north of the Ardennes and the Rheinisches Schiefergebirge.

The latter form the second argument against the existence during the Pliocene of an elevated area where Ardennes and Rheinisches Schiefergebirge are now, because they were transported across these areas from a southern origin to their present northern position. The gravels of the Pliocene are well sorted and contain pebbles of only the least weatherable materials, quartz and quartzites and similar siliceous rocks. They could have been derived, perhaps in a number of steps of erosion and deposition, from almost anywhere. The only index pebbles found are the silicified oolites, occurring at a rate of about $3^0/_{00}$, and, much rarer still, silicified Jurassic brachiopods. Both can be traced back to the oolitic limestone of the Middle Jurassic or Dogger Series ("Hauptrogenstein"), presumably exposed, then as well as now, along the limit between Paris Basin and Vosges Mountains (Fig.58).

Geographers have taken hold of this fact. Moreover, they were impressed by the capture of the upper Meuse valley by the Moselle south of the Ardennes at Toul in France, one of the classic facts of European physical geography. So they speak of an arch-Meuse, still possessing the upper Moselle as a tributary, responsible for the transport of these silicified oolites across the Ardennes and the Rheinisches Schiefergebirge during the Pliocene. Consequently the pebbles are called the "traîneau mosane" (MACAR and MEUNIER, 1955).

I think this is extrapolating earlier drainage patterns beyond our real state of knowledge. The capture of the upper Meuse by the Moselle is a much younger feature. It occurred at or slightly before the Riss Glacial (TRICART, 1952), that is during the Middle Pleistocene. The valley of the Meuse at that time was already deeply incised in the uplifted peneplain. This feature is still very evident in the dry valley near Toul, which connects the valleys of the present Moselle and Meuse, and forms the severed limb of the former Meuse. It has been nicely preserved, and the valley of the upper Meuse with it, because with most of its water gone by the capture of the Moselle, there was no longer an erosive force to speak of, and the valley floor was fossilized.

The deposition of the silicified oolites, on the other hand, took place during the Pliocene; extending perhaps into the Early Pleistocene, depending on the paleobotanic interpretation of the Reuver clays near Venlo on the Meuse.

Moreover the early Meuse, which still possessed the present upper Moselle as its main

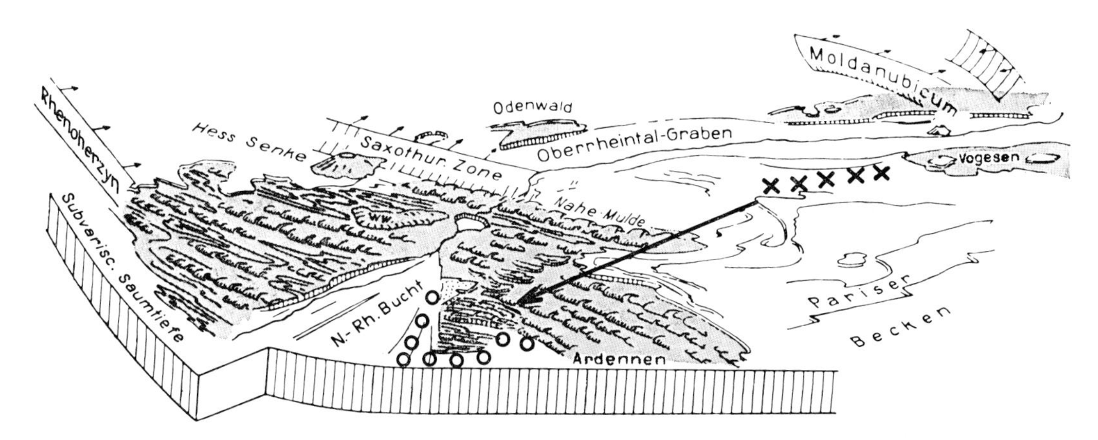

Fig.58. Diagram showing the relation between the outcrops of the Dogger oolites on the eastern side of the Paris Basin with the deposits of silicified oolites in the Pliocene north of the present Ardennes and Rheinisches Schiefergebirge.
Crosses: outcrop of Dogger. Circles: Pliocene with silicified oolites.

tributary, flowed westward to the south of the Ardennes, and belonged to the drainage system of the early Seine. This is the "Meuse de Mézières", or the "Meuse lorraine" of some authors (PISSART, 1961) And, lastly, the silicified oolites are not only found in the area of the present Meuse, but in that of the present Rhine as well. It seems safer to state that at that time no topographic obstacle formed by the Ardennes and the Rheinisches Schiefergebirge existed as yet for the Rhine area (cf. QUITZOW, 1959). The Rhine, at that time, had an extremely wide flood-plain, only slightly lower than the peneplain. It transparted only the least weatherable heavy minerals (mainly tourmaline, zircon and rutile), taken from the deeply weathered spoils of the peneplain. A river, or several rivers, were thus able to transport materials right across these areas.

As a corollary, coarse pebbles of the Paleozoic core of both the Ardennes and the Rheinisches Schiefergebirge are found in the Earliest Pleistocene gravel terraces of Meuse and Rhine. Both their coarseness and their sudden appearance in the surrounding basins of sedimentation indicate an exceptionally strong and quick uplift of these uplands, beginning with the Quarternary. This has at the same time cut off a further supply of silicified oolites from the south.

The typical uplands geomorphology of these areas, with their broad plateaus formed by the uplifted peneplain, and the narrow, deep river valleys of an extremely young character, are of course quite in line with the ideas set forth above, of a major uplift that was both strong and young.

REFERENCES

ANTHOINE, R., 1938. Contribution à l'étude du massif cambrien de Rocroi. *Mém. Acad. Roy. Belg., 2e Sér.*, 61 (4): 201 pp.

ASSELBERGHS, E., 1946. L'Éodévonien de l'Ardenne et des régions voisines. *Mém. Inst. Géol. Univ. Louvain*, 14: 598 pp.

BEUGNIES, A., 1962. Compte rendu de la session extraordinaire 1961. *Ann. Soc. Géol. Belg., Bull.*, 85: 1–84.

BEUGNIES, A., 1963. Le massif cambrien de Rocroi. *Bull. Carte Géol. France*, 270: 166 pp.

BEUGNIES, A., 1964. Essai de synthèse du géodynamisme paléozoïque de l'Ardenne. *Rev. Géograph. Phys. Géol. Dyn., Sér. 2*, 6: 269–277.

BOUROZ, A., 1960. La structure du Paléozoïque du nord de la France au sud de la grande Faille du Midi. *Ann. Soc. Géol. Nord*, 80: 101–112.

CONIL, R. and PIRLET, H., 1963. Sur quelques foraminifères du Viséen supérieur de la Belgique (Bassins de Namur et de Dinant). *Bull. Soc. Belge Géol., Paléontol., Hydrol.*, 72: 1–15.

CONGRÈS GÉOLOGIQUE INTERNATIONAL, 1922. *Livrets-Guide — Congr. Géol. Intern., 8e, Bruxelles, 1922.* Vaillant-Carmanier, Liège.

CONGRÈS INTERNATIONAL DE SEDIMENTOLOGIE, 1963. *Livrets-Guide des Excursions C–D, E–F, G, I–J — 6e Congr. Intern. Sedimentol. (Hollande–Belgique)*, Brussels.

DE BETHUNE, P., 1952. Geologie der steenkoolafzettingen van België. *Tech. Tijdschr. Unie Ingenieur., Leuven*, 80 (16): 37 pp.

DE MAGNÉE, I. and NAIRN, A. E. M., 1963. La méthode paléomagnétique. application au poudingue de Malmédy. *Bull. Soc. Belge Géol. Paléontol. Hydrol.*, 71: 551–565.

FOURMARIER, P., 1934. Vue d'ensemble sur la géologie de la Belgique. *Ann. Soc. Géol. Belg., Mém.*, 4: 200 pp.

FOURMARIER, P. (Editor), 1954. *Prodrome d'une Description Géologique de la Belgique.* Soc. Géol. Belg., Liège, 826 pp.

GEUKENS, F., 1959. Het pseudotektonisch venster van Theux. *Mededel. Akad. Wetenschap.*, 21: 1–8.

GEUKENS, F., 1962. Überblick über die tektonischen Beziehungen zwischen dem Massiv von Stavelot, dem Vesdre-Massiv und dem Massiv von Herve. *Fortschr. Geol. Rheinland Westfalen*, 3: 1145–1154.

GEUKENS, F. and RICHTER, D., 1962. Problèmes géologiques dans le massif de Serpont (Ardennes). *Bull. Soc. Belge Géol. Paléontol. Hydrol.*, 70: 196–212.

GOSSELET, J., 1888. L'Ardenne. *Mém. Carte Géol. France*, 1888: 881 pp.

GRAULICH, J. M., 1963. La phase sudète de l'orogène varisque dans le synclinorium de Namur à l'est de Samson. *Bull. Soc. Belge Géol. Paléontol., Hydrol.*, 71: 181–199.

KAISIN, F., MAILLEUX, E. and ASSELBERGHS, E., 1922. Traversée centrale de la Belgique. *Congr. Géol. Intern., 13e, Bruxelles, 1922, Livret-Guide*, A: 90 pp.

LECOMPTE, M., 1954. Quelques données relatives à la genèse et aux caractères écologiques des "récifs" du Frasnien de l'Ardenne. *Inst. Roy. Sci. Nat. Belg. — Vol. Victor van Straelen*, 1: 153–181.

LECOMPTE, M., 1957. Les récifs dévoniens de la Belgique. *Bull. Soc. Géol. France*, 6 (7): 1045–1068.

LECOMPTE, M., 1961. Faciès marins et stratigraphie dans le Dévonien de l'Ardenne. *Ann. Soc. Géol. Belg., Bull.*, 85: 17–57.

LOHEST, M. and FOURMARIER, P., 1922. Traversée orientale de la Belgique. *Congr. Géol. Intern., 13e, Bruxelles, 1922' Livret-Guide*, A 1: 46 pp.

MACAR, P., 1954. L'évolution géomorphologique de l'Ardenne. *Bull. Soc. Belge Géograph.*, 78.

MACAR, P., 1965. Aperçu synthétique sur l'évolution géomorphologique de l'Ardenne. *Géographie (Brussels)*, 17: 1–11.

MACAR, P. and MEUNIER, J., 1955. La composition lithologique des dépôts de la "Traînée mosane" et ses variations. *Ann. Soc. Géol. Belg., Bull.*, 78: 61–88.

MAILLIEUX, E., 1933. *Terrains, Roches et Fossiles de la Belgique*. Musée Roy. Nat. Hist. Belge, Brussels, 217 pp.

MICHOT, P., CONIL, R. and GRAULICH, J. M., 1963. Sédimentologie des formations viséennes du synclinorium de Namur, dans la vallée de la Meuse. *Congr. Intern. Sédimentol., 6e, Bruxelles, 1963, Livret Guide Excursions, G* (1): 22 pp.

PATIJN, R. J. H., 1963. Het Carboon in de ondergrond van Nederland en de oorsprong van het Massief van Brabant. *Geol. Mijnbouw*, 42: 341–348.

PIRLET, H., 1963. Sédimentologie des formations du Viséen supérieur, V3b dans la vallée du Samson (Bassin de Namur, Belgique). *Ann. Soc. Géol. Belg., Mém.*, 86: 1–41.

PIRLET, H., 1964. La sédimentation rythmique de la partie inférieure du V3a dans le Bassin de Namur; les relations entre le Dinantien et le Namurien de Namèche à Moha. *Ann. Soc. Géol. Belg., Bull.*, 86: 461–468.

PISSART, A., 1961. Les terrasses de la Meuse et de la Semois. La capture de la Meuse lorraine par la Meuse de Dinant. *Ann. Soc. Géol. Belge, Bull.*, 84: 1–108.

PRUVOST, P., 1924. A synopsis of the geology of the Boulonnais. *Proc. Geologists' Assoc. Engl.*, 35: 29–59.

PRUVOST, P., *Lexique Stratigraphique International. I. Europe. 4aII. France, Belgique, Pays-Bas, Luxembourg: Paléozoïque supérieur*. Centr. Natl. Rech. Sci., Paris, 224 pp.

QUITZOW, H. W., 1959. Hebung und Senkung am Mittel- und Niederrhein während des Jungtertiärs und des Kwartärs. *Fortschr. Geol. Rheinland Westfalen*, 4: 389–400.

RUPKE, J., 1965. The Esla nappe, Cantabrian Mountains, Spain. *Leidse Geol. Mededel.*, 32: 1–74.

RUTTEN, M. G., 1956. Devonian reefs from Belgium: relation between geosynclinal subsidence and hinterland erosion. *Am. J. Sci.*, 254: 685–692.

SIMPSON, S., 1962. The Devonian–Carboniferous boundary and the problem of the British and Ardennes–Rhineland successions from the upper Famennian to the Namurian. *Congr. Avan. Stratigraph. Géol. Carbonifère, Compte Rendu, 4, Heerlen, 1958*, 3: 629–633.

TRICART, J., 1952. *La Partie Orientale du Bassin de Paris. Étude morphologique*. Thesis, Sorbonne, Paris, 471 pp.

VAN LECKWIJCK, W. P., 1956. Tableaux d'une aire instable au Paléozoïque supérieur: la terminaison orientale du Massif du Brabant aux confins belgo-néerlandais. *Verhandel. Koninkl. Ned. Geol. Mijnbouwk. Genoot., Geol. Ser.*, 16: 252–273.

VAN VEEN, J., 1965. The tectonic and stratigrafic history of the Cardaño area, Cantabrian Mountains, northwest Spain. *Leidse Geol. Mededel.*, 35 (1966): 45–103.

WATERLOT, G., 1937. Stratigrafie et tectonique du massif cambrien de Rocroi. *Bull. Carte Géol. France*, 195: 53 pp.

WATERLOT, G., 1945. L'évolution de l'Ardenne au cours des divers phases des plissements calédoniens et hercyniens. *Bull. Soc. Géol. France*, 15: 3–44.

WATERLOT, G., 1957. *Lexique Stratigraphique International. I. Europe. 4aI. France, Belgique, Pays-Bas, Luxembourg: Antécambrien–Paléozoïque inférieur*. Centr. Natl. Rech. Sci., Paris, 432 pp.

WATERLOT, G., 1958. Le Cambrien de l'Ardenne. *Intern. Geol. Congr., 20th, Mexico, 1956, Rept., El Sistema Cambrico*, 1: 161–184.

CHAPTER 6

The Rheinisches Schiefergebirge

INTRODUCTION

Towards the east the Ardennes pass into the Rheinisches Schiefergebirge.[1] The name is used here in a geological, that is in its widest sense, to comprise all of the unit formed by the Hercynian upland. It consequently includes areas such as Eifel, Sauerland, Kellerwald, Hunsrück and Taunus, which topographically form more or less separate units (Fig. 59).

The northern and the southern border of the R.S. are formed by the carapace of transgressive post-Hercynian formations, mostly Mesozoic in age. Its eastern border is a major fault zone, which separates the R.S. from the Hessen graben (KOCKEL, 1958). The latter is the direct northward continuation of the Upper Rhine graben, and as such forms a link in the Mjosa–Mediterranean zone (p. 49). Towards the west there is, as noted in Chapter 5, a gradual transition towards the Ardennes, which geologically form a part of the same Hercynian structure. As an upland unit it is, however, almost cut off by the depression in the western Eifel, where the cover of younger formations in the Lower Rhine graben and in the Luxemburg Gulf almost meet (cf. W. SCHMIDT and SCHRÖDER, 1962). On top of the R.S. numerous, though always very local, representatives of the younger, Cenozoic volcanism are found. Of these, perhaps, the Siebengebirge near Bonn and the many prominent volcanoes in the Eifel are the best known, but the Westerwald is by far the largest example. Except for these younger volcanics, the R.S. forms one solid block of Hercynian structures.

If we now compare the R.S. with the Ardennes, there is overall similarity. Both mainly belong to the Hercynian orogenetic cycle. Their geosynclinal period started with the Devonian and continued during the Carboniferous, while the main orogenetic phases are found either at the end of the Lower or at the end of the Upper Carboniferous.

Within this framework we find, however, quite a series of differences. To begin with, the R.S. is much broader, about two times as broad as the Ardennes. This is because it is built up by a larger number of individual structural units than the Ardennes. Both the more southerly and the more northerly of these additional units of the R.S. dip westwards under the Mesozoic cover, whereas only the central part of the R.S. continues into the Ardennes.

[1] In this chapter abbreviated to R.S.

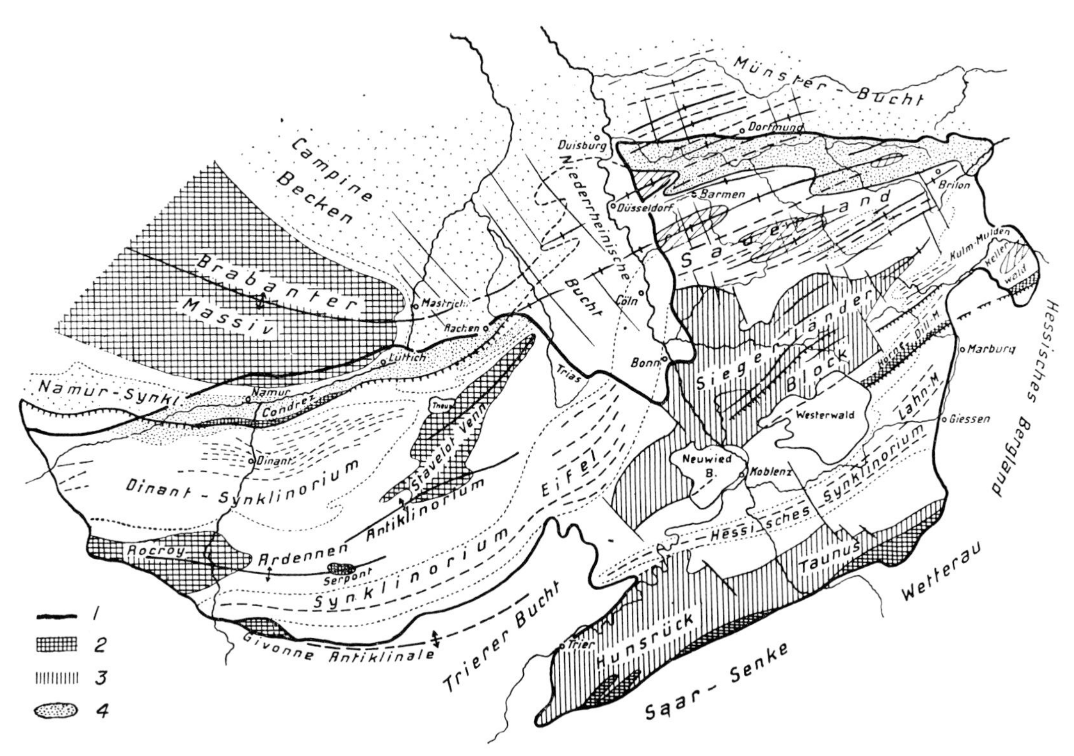

Fig.59. Structural outline of the Ardennes and the Rheinisches Schiefergebirge. *1* = limit of Hercynian upland; *2* = the Brabant Massif, Caledonian massifs in the Ardennes, and local pre-Devonian in the Rheinisches Schiefergebirge; *3* = "stable blocks"; *4* = Upper Carboniferous of Namur syncline and Ruhr Basin. (After VON BUBNOFF, 1930.)

In spite of this broadness, the R.S. shows, however, a more monotonous structure than the Ardennes. For one thing, it lacks the Caledonian massifs[1]. Pre-Devonian rocks are only found locally; at the base of a number of main overthrusts in the northeast, in the Sauerland (BEYER, 1941; JAEGER, 1962) and in a zone, not yet well understood, which accompanies the southern border of Hunsrück and Taunus (BIERTHER, 1953; STENGER, 1961). Moreover the Carboniferous, which forms such an important part of the Dinant synclinorium in the Ardennes, occurs only along the northern limit of the R.S.

Schematically, we may, as a consequence, say that the R.S. mainly consists of a monotonous series of tightly folded clastic sediments, predominantly of Lower and Middle Devonian age. In this series slates, and to some extent sandy slates, dominate, which amply justifies the name of *Schiefergebirge* (= "Slate-mountains").

A feature not found in the Ardennes is the initial, geosynclinal volcanics of the R.S., both of the Devonian and of the Lower Carboniferous. This distinction makes it difficult for those who like rigid classifications. For are not the Ardennes, without initial geosynclinal volcanics, typically miogeosynclinal, whereas their direct continuation, not a parallel zone, the R.S., exhibits eugeosynclinal characteristics?

[1] Of course the Stavelot Massif crosses the political boundary between Belgium and Germany. Its easterly plunging nose is called Venn Sattel (W. SCHMIDT, 1952; THOMÉ, 1955) in the latter country. But geologically it forms part of the Ardennes.

THE DEVONIAN OF THE RHEINISCHES SCHIEFERGEBIRGE

COMPARISON WITH THE ARDENNES

Before these points can be discussed in more detail, it must be noted that in crossing from the Ardennes to the R.S. several of the internationally accepted stage names are replaced by German synonyms. The Couvinian is replaced by the local *Eifel*, although the first name has a long and honorable priority, whereas the Upper Devonian is divided into local stages. Table XII summarizes these differences and adds some more names of supposed orogenetic phases, local refinements of the Stille codex. More detailed information may be found in KUTSCHER and H. SCHMIDT (1958).

During the Devonian the geosynclinal history of the R.S. is much less simple than that of the Ardennes. Instead of a continent to the north and a geosynclinal sea to the south, with subsidence increasing generally southwards, the R.S. geosyncline consisted of a number of relatively stable blocks, alternating with more strongly subsiding zones. The southern border of the Old Red continent, on the other hand, is nowhere exposed, though its influence is shown by reddish marine sediments in the northern part of the R.S.

The main stable blocks during the Devonian are those of the Siegerland and of the Hunsrück and Taunus. The main subsiding areas during that time were the Sauerland (H. SCHMIDT and PLESSMANN, 1961) and the Moselle or Hessen synclinoria (Fig. 59).

VARIATIONS IN SUBSIDENCE

The story is, however, much more complicated, because rates of subsidence have varied widely from one stage of the Devonian to the other. This can, for instance, be seen

TABLE XII

STAGES OF THE DEVONIAN IN THE ARDENNES AND THE RHEINISCHES SCHIEFERGEBIRGE, AND SOME SUPPOSED OROGENETIC PHASES

	Ardennes		*Rheinisches Schiefergebirge*	*"Orogenetic phases"*
Upper Devonian	Famennian		Dasberg	~~~~ Marsic
			Hemberg	
			Nehden	
	Frasnian		Adorf	
Middle Devonian	Givetian		Givet	~~~~ Meggen
				~~ Brandenberg
			Eifel	
	Couvinian			
Lower Devonian	Coblencian	Emsian	Koblenz = Ems	~~~~ } pre-Sideritic
		Siegenian	Siegen	~~~~ }
	Gedinnian		Gedinne	

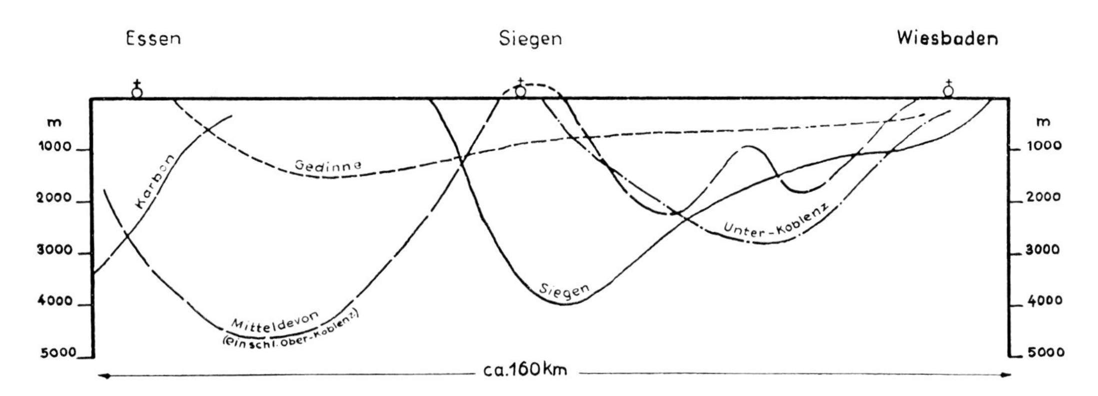

Fig.60. Composite stratigraphic section across the Rheinisches Schiefergebirge, from Essen to Wiesbaden, showing variations in thickness of parts of the Devonian and of the Carboniferous. (After KEGEL, 1948.)

in the schematic north–south section through the R.S., from Essen to Wiesbaden of Fig.60. Further information on these variations of geosynclinal subsidence, with isopach maps for the various units of the Devonian, is presented in KELLER (1948), W. SCHMIDT (1952) and PILGER (1952) for the R.S. and by H. SCHMIDT (1962) for the R.S. and the Harz.

In contrast to the Ardennes, we find that during the Gedinnian subsidence was still rather limited all over the R.S. The first really geosynclinal subsidence sets in during the Siegenian, in the area around Siegen on the Sieg river. That is in the zone which will form a relatively stable block, and even undergo some erosion, during the Middle Devonian. This type area of the Siegenian has lately come in for much detailed investigation (HESSISCHES LANDESAMT FÜR BODENFORSCHUNG, 1960).

Thereupon, from the end of the Lower Devonian and during the Middle Devonian, the future Sauerland and Moselle synclinoria became the strongly subsiding zones. Development of the Upper Devonian, mainly preserved in the northern part of the R.S., was a much more localised affair.

This picture is, of course, grossly schematized. Moreover, the facies of the different units of the Devonian show strong differences along the strike. For instance, in relation to the Lower and Middle Devonian of the southeastern part of the R.S., and also for the Upper Devonian in its northeastern part, the literature indicates that it "shows variations toward", or "is developed in" Bohemic or Saxonic facies. One gains the impression that the celebrated "Rheinische Fazies" is only found in its true development in the cross section along the river Rhine.

It must be remembered that there is, in the literature, a single Rheno-Hercynic zone in tectonics (p.74), comprising both the R.S. and the Harz. On the other hand, there are separate facies provinces for the Devonian, the Rhenish facies in the R.S. and the Hercynic facies in the Harz Mountains (ERBEN, 1962; H. SCHMIDT, 1962).

These variations in the rate of geosynclinal subsidence have been construed as proof for intra-Devonian orogenetic phases. Such evidence is, however, of a character no more convincing than for instance: "In the Lahn syncline uppermost Middle Devonian with pebbles overlies Lower Emsian. About 700–800 m of sediment has been eroded. The angular nonconformity might well attain 10°–15°" (PILGER, 1952, p.199). This, to my mind, is carrying the Stille concepts altogether too far. It should be kept in mind that

Stille himself stressed the dual mode of tectonic movements, in which epirogenetic vertical movements form a counterpart to orogenetic tangential movements. In a geosyncline in which syn-sedimentary tilting has led to the strong variations in thickness described above, local uplift and erosion is to be expected. And in an area as strongly tectonized as the R.S., the assumption of an intra-Devonian angular unconformity should be based on more than intraformational disconformities.

DEVONIAN SEDIMENTOLOGY

The clastic series of the Devonian have formerly been interpreted as deposits of a shallow, or even a tidal or wadden sea. A detailed analysis of a small area of the Famennian in the Ardennes has led VAN STRAATEN (1954), easily the best informed geologist of the present day Dutch Wadden Sea, to maintain that this is not the case. According to Van Straaten the Condroz Sandstone investigated do not show the characteristics of a tidal flat area and must have formed in a deeper environment. Although stated only for a small part of the Upper Devonian, the similarity in facies of the Famennian, at least all over the Ardennes, is so conspicuous, that a tentative extrapolation of this result seems justified.

On the other hand it has been established that many, if not most, of the greywackes occurring in the Devonian of the R.S. are turbidites (HELMBOLD, 1952, 1958; PLESSMANN, 1964). This is an indication that their sedimentation probably has taken place at an altogether deeper level. This gives no indication whatsoever as to the exact depth of the geosynclinal sea.

The well-known and beautifully preserved pyritic fauna of the Bundenbach Slate and other fossiliferous localities in the Hunsrück Schiefer, at the transition from Siegenian to Emsian in the Lower Devonian, also indicates deep water environment. It consists of Coelenterata, some brachiopods and molluscs, worms, early arthropods, cystoids, asteroids, crinoids and fish (KUHN, 1961). Here again it is difficult to assess at what depth sedimentation actually took place.

Consequently we can conclude that most of the clastic facies of the Devonian in the R.S. was laid down in a deep-water environment. And also that here is a field wide open for further sedimentological and paleoecological studies, which should in future indicate with somewhat greater precision what the actual environment has been.

In contrast, the limestone facies of the R.S., developed as in the Ardennes in parts of the Middle and the Upper Devonian (KRÖMMELBEIN et al., 1955; LECOMPTE, 1961), is a typical shallow water facies. Although considerably less studied than its Belgian counterpart, it is known to contain both bioherm and biostrome structures (JUX, 1960).

As is always the case, not everybody seems convinced of such a new interpretation. Note that, for instance, ERBEN (1962) in a recent publication still speaks of "near-shore regions with high percentages of terrigenous, coarse to middle grained sandy material" and of "off-shore regions lacking, or almost lacking terrigenous material". Or, to cite a more controversial case, JUX (1964) even thinks to have found wadden sedimentation with indications for tidal runs in the Lower Devonian of the Bergischer Land, east of Cologne. His descriptions of the sedimentary series contain, however, all the characteristics of a turbidite sequence, which may suffice to indicate the difficulties still inherent in sedimentological interpretation.

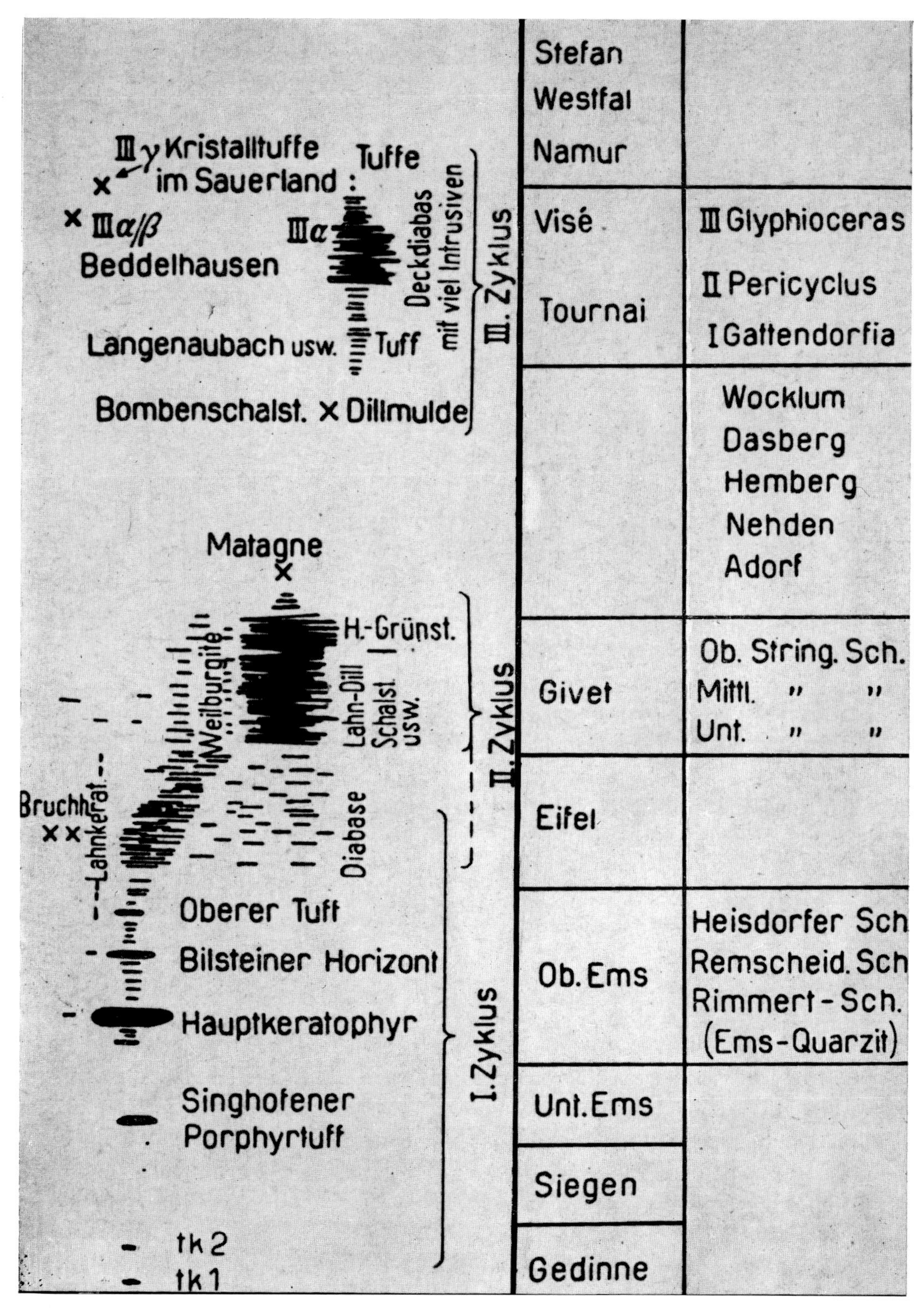

Fig.61. Stratigraphy and sequence of the three cycles of initial magmatism in the Rheinisches Schiefergebirge. (After PILGER, 1952.)

INITIAL MAGMATISM IN THE RHEINISCHES SCHIEFERGEBIRGE

The next point of main interest in the geosynclinal history of the R.S. is its initial magmatism. It is important here to distinguish between acid igneous rocks found around the eastward plunging nose of the Stavelot Massif in the northwestern part of the R.S., and basic rocks found in its southeastern part, mainly in the Lahn and Dill river areas (KEGEL, 1922, 1934).

The acid rocks, formerly considered to belong to the Cambrian, are now thought to be of Gedinnian Age. These rocks, commonly called Venn–Porphyre (W. SCHMIDT, 1955), are excluded from the rocks belonging to the initial magmatism of the Hercynian orogenetic cycle. From their petrographic character (roughly tonalitic), their stratigraphic position (following hard upon the Caledonian orogeny), and their far northern position (near the border of the Old-Red continent), they must rather be considered as a post-orogenetic feature of the Caledonian orogenetic cycle. This tallies also with the relatively small subsidence during the Gedinnian, indicating that Hercynian geosynclinal conditions started rather late in this area.

Quite another picture is presented by the magmatism in the Lahn and Dill province. Here lavas, pillow lavas and sills abound. Petrographically they range from the more acid, sodium syenitic, keratophyres to diabases and weilburgites (LEHMANN, 1952). Diabase is used here in the early German petrographic sense, as denoting a greenstone, a strongly altered volcanic or subvolcanic rock of a composition akin to that of a basalt. The alteration normally is a spilitization and many of these so-called diabases in reality are true spilites.

The spilitization, the frequent occurrence of pillow lavas and of predominantly glassy rocks (German: Schalstein), all indicate submarine geosynclinal eruption (HENTSCHEL, 1960). These have occurred in three more or less separate consecutive cycles (Fig. 61). The first is mainly formed by keratophyres, the second and third by diabases and weilburgites.

CARBONIFEROUS OF THE RHEINISCHES SCHIEFERGEBIRGE

The main point of interest in the Carboniferous lies in the change in facies of the Lower Carboniferous between the Ardennes and the R.S. In the Ardennes a limestone facies is found for all of the Dinantian or Lower Carboniferous. In the R.S., on the other hand, the Tournaisian or Lower Dinantian is mainly developed as siliceous shales, whereas the upper part of the Dinantian, the Visean, is replaced in most of the R.S. by a clastic facies, the Culm (PAPROTH and GRAULICH, 1958; PAPROTH and TEICHMÜLLER, 1961); cf. Fig. 62, 63, 64.

In the older literature we find the limestone facies interpreted as bathyal, the clastic Culm facies as neritic or even litoral. The calcarenites and calcilutites of the Ardennes were compared with modern *Globigerina* oozes; the sandstones and shales of the Culm with shallow water or with tidal flat deposits.

As is the case with the Devonian, recent sedimentological thinking, together with some sedimentological investigation, has completely reversed this interpretation. The limestones

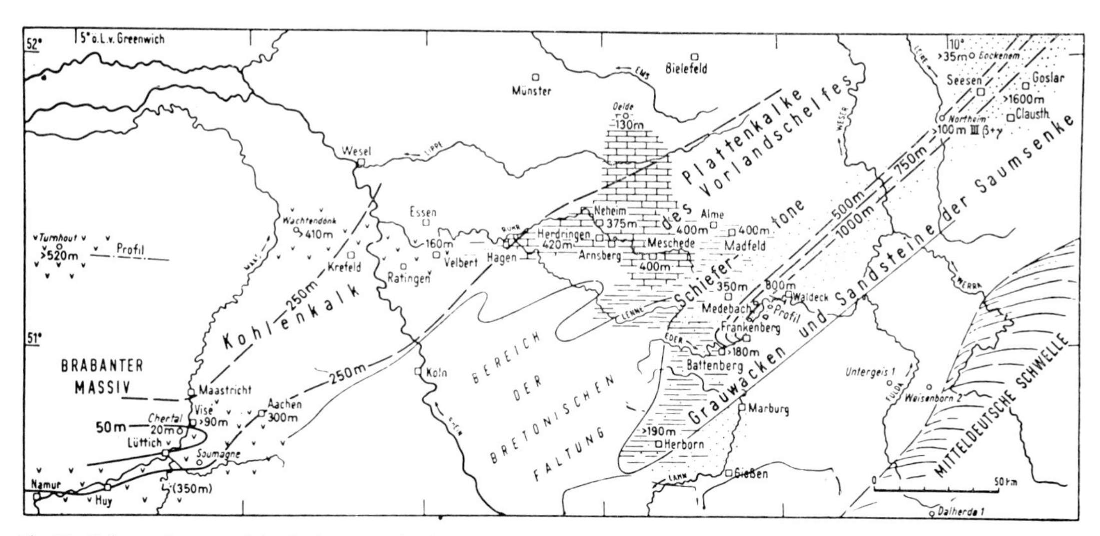

Fig.62. Schematic map of the facies zones in the Lower Carboniferous of the Ardennes and the Rheinisches Schiefergebirge.
Kohlenkalk stands for the carbonate facies of the Upper Carboniferous; *Plattenkalk* is a well bedded series of limestone alternating with shale. (After PAPROTH and TEICHMÜLLER, 1961.)

are now thought to have originated in quite shallow water. And the clastic Culm deposits, on the other hand, are thought to represent a relatively deep-water turbidite series (KUENEN and SANDERS, 1956).

Although the turbidite character for the Culm seems to be far less evident than for the greywackes of the Devonian, this nevertheless is today the generally accepted version.

The development of the Upper Carboniferous is similar to that found in the Namur

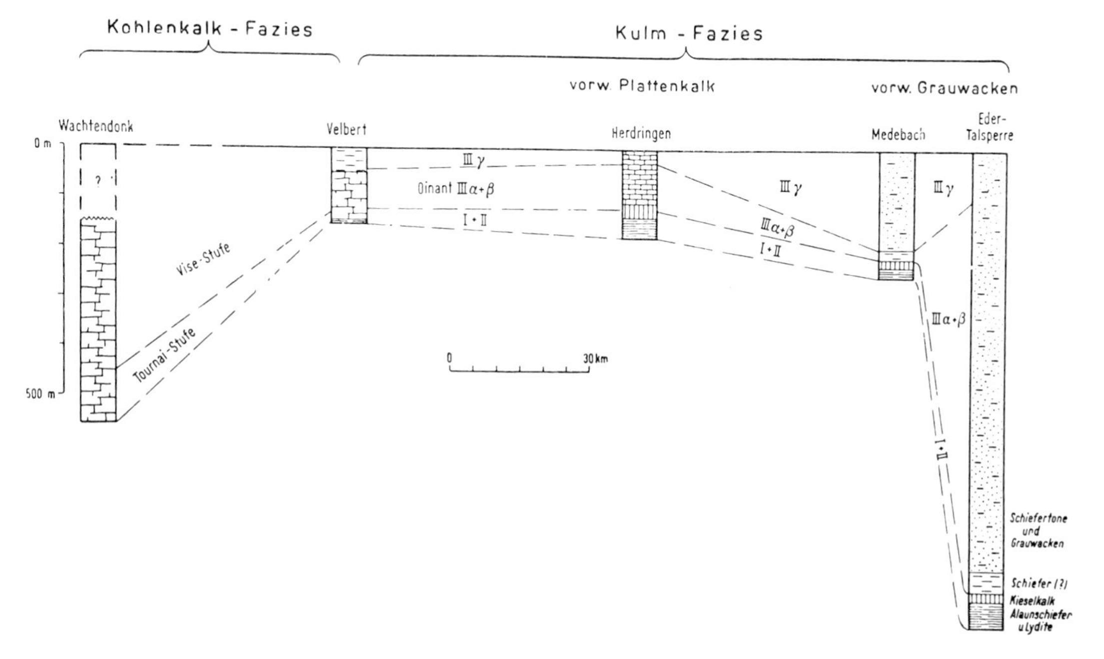

Fig.63. Schematic stratigraphic sections of the Lower Carboniferous in the Ardennes and the Rheinisches Schiefergebirge. (After PAPROTH and TEICHMÜLLER, 1961.)

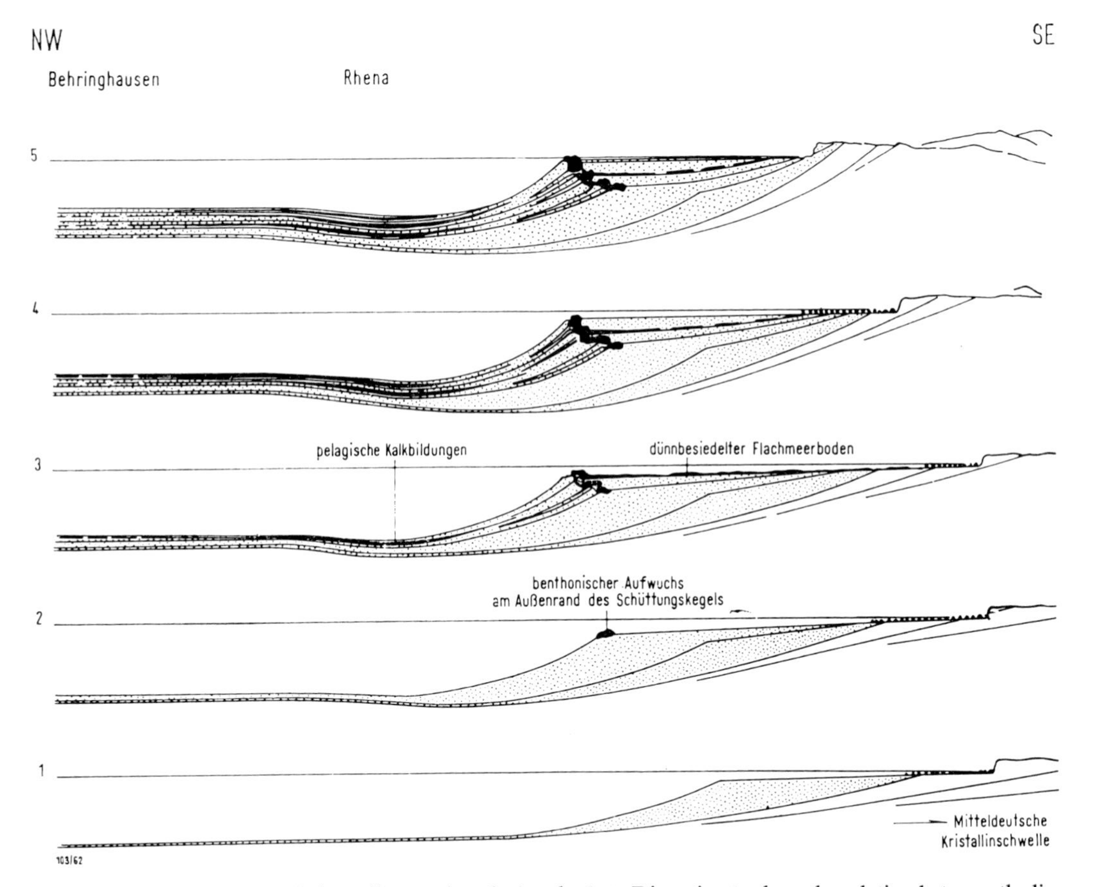

Fig.64. Schematic diagram of the sedimentation during the Late Dinantian to show the relation between the limestone facies in the Ardennes (*Kohlenkalk*) and the clastic facies of the Culm in the Rheinisches Schiefergebirge. (After TEICHMÜLLER, 1962.)

syncline further west. It is exposed along the northern border of the R.S., and can be followed underground in the collieries of the Ruhr Basin. We will come back to its mode of origin in the last section of this chapter.

TECTONICS

SINGLE STOCKWERK; MONOTONY

The tectonics of the R.S., in their overall picture, are extremely monotonous. As follows from Fig.65, 66, we have to deal with a large number of tightly folded parallel anticlinal and synclinal structures of varying size. Upthrusts are common, but there are no major overthrusts of a nappe type. Several "highs" and "lows" (*Blocke* and *Mulden*) are distinguished, but the differences between these areas are secondary.

A more varied picture is found along the northeastern border of the R.S., where slates and limestone series alternate (RICHTER, 1963). But the section through the R.S. exposed along the Rhine belongs to a single tectonic Stockwerk. Nothing is left of a higher Stockwerk of the Upper Carboniferous, which may, or may not, have covered part of all of

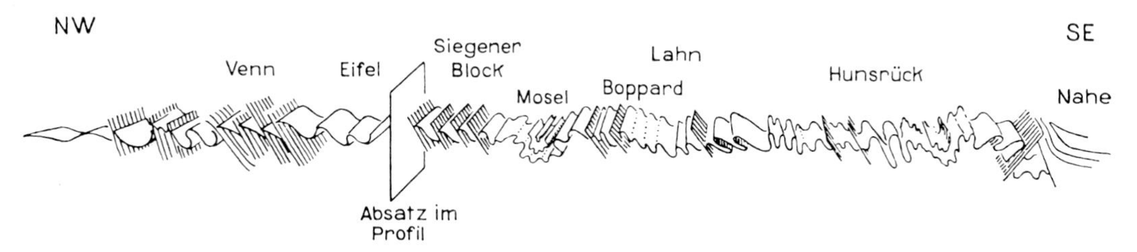

Fig.65. Schematic cross-section through the Rheinisches Schiefergebirge. (After KNETSCH, 1963.)

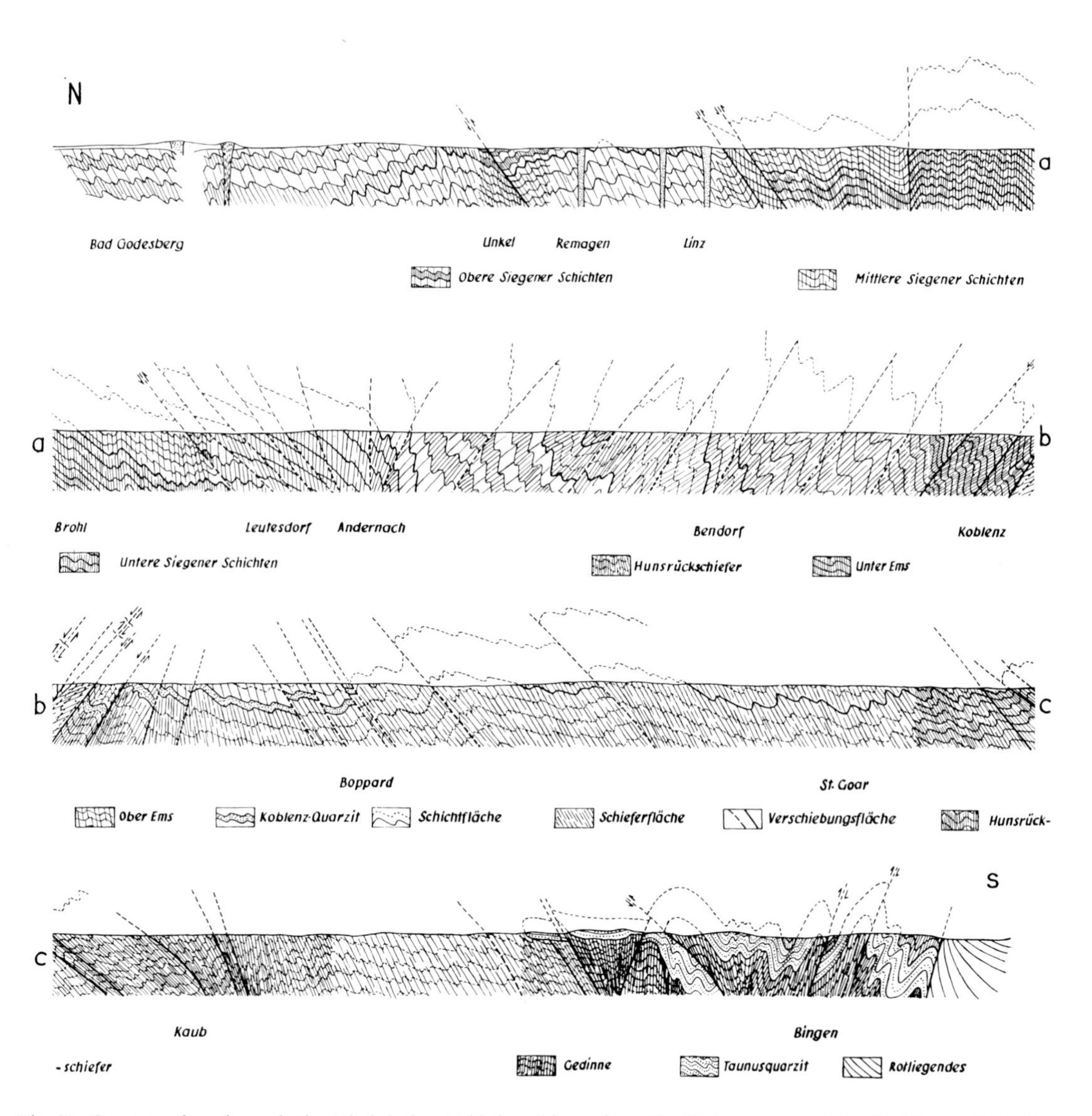

Fig.66. Cross-section through the Rheinisches Schiefergebirge along the Rhine, summarizing bedding, schistosity and thrust planes. The *Mosel Mulde*, or Moselle synclinorium, is centered near Boppard. The section begins in the north in the upper left at Bad Godesberg, and continues to Bingen in the south. (After HOEPPENER, 1955.)

the R.S. Nor is anything exposed of a lower Stockwerk, of the Caledonian foldbelt or of possible migmatitic or plutonic structures.

In view of this monotonous succession of anticlinal and synclinal structures in the R.S. it is not surprising that workers in this area have turned their attention to other facets of the tectonic picture. On the one hand one finds a more generalizing attitude, a tendency to treat a group of individual structures as a whole. On the other hand, detailed investigation of minor structures within a limited area, or even in single folds, has awakened considerable interest. From the first springs the use of notions such as "Vergenz "and "Faltenspiegel", from the second the exhaustive studies of microtectonics and schistosity.

"VERGENZ"

Vergenz or overturn indicates the main direction of tectonic activity. It consequently is non-existent, or zero, in symmetric structures. Vergenz may be expressed only in the asymmetry of a fold structure. But it is normally found in a larger number of connected features, such as overthrusts accompanying the core of an asymmetric anticline — the French "pli-faille" — isoclinal and imbricate structures, minor folds, and of course nappe structures. All indicate Vergenz or overturn towards a certain direction. In the R.S., as in many other foldbelts, it is found that the Vergenz remains the same for quite a number of successive parallel structures. Moreover, it often shows a gradual transition from one direction to another. The foldbelt of the R.S. can thus be divided into broad zones with either a northward or a southward Vergenz, separated by zones where a subvertical, fan-like arrangement prevails.

On the cross-section through the R.S. along the Rhine (Fig.66), the following main units may be distinguished according to their Vergenz. First a northern zone with northward Vergenz. This shows gradual transition to the Siegen high north of Brohl. Then follows rather sharply a zone with southward Vergenz, which forms the northern flank of the Moselle synclinorium. The centre and southern flank of this synclinorium show a number of zones of alternating Vergenz, separated by narrower fan like zones. The southernmost zone near Bingen is formed by a more irregular, more strongly tectonized area, which is the equivalent of Hunsrück and Taunus.

"FALTENSPIEGEL"

The parallel series of tightly folded structures in the R.S. has induced the notion of *Faltenspiegel*, a sort of enveloping contour perpendicular to the axial planes of a series of individual folds. The Faltenspiegel can be very useful in generalized sections in which individual folds cannot be indicated. Instead of schematizing by drawing in only a certain part of the individual structures, the Faltenspiegel indicates the position of their axial planes to which it is perpendicular. Its use comes in very handy in those areas where the original bedding has been more or less completely obscured by schistosity. According to theory, the schistosity is parallel to the axial planes, and hence the Faltenspiegel is perpendicular to the schistosity. This is an easy way out, when it becomes too difficult, or too time-consuming, to unravel the actual folding from ghost structures. It carries, of course, one danger, because what one actually does is measure the "Schistositätsspiegel", which is also the Faltenspiegel by implication only, a fact often slurred over.

Although the Faltenspiegel consequently is a very useful concept in a generalized picture, it does not do justice to those regions in which wider and/or shallower individual structures alternate with the tight folding normally found in the R.S. If only the axial planes of these wider structures are approximately parallel to the axial planes of the tightly folded series, the Faltenspiegel will remain the same. In this case differences in tectonic style become obscured by the use of the Faltenspiegel concept.

MICROTECTONICS

"Came the blissful era of the stress ellipsoid and its less glorious, because too schematic, application to every available map and section, which led to a side-tracking, rather than to an explanation (of tectonic questions). Came the shift of tectonics into the well lit, round field of the microscope, in which the so vertiginous expanses of nature are helpfully limited, in which questions are reduced to sharply formulated rules, promising firm results even to the beginner and the desperate, who seem, however, mistakenly felt exempted from the obligation (still) to question Nature (herself) actively and originally, and to try to understand Nature's answers when delivered in her own language." (CLOOS, 1948, p.290).

Speaking of microtectonics in this chapter and not in the preceeding one is incorrect in regard to the extensive work on schistosity of FOURMARIER (1923, 1952, 1953a, b, c) in the Ardennes. On the other hand, microtectonic investigations have since been pursued on a much more comprehensive basis and in much more detail in the R.S. These were started under the gifted inspiration of H. Cloos and have since his death been pursued under the direction of R. Hoeppener.

For unity of presentation, we had therefore better discuss microtectonic investigations both in the Ardennes and in the R.S. in this section.

Fourmarier's main interest has been schistosity or flow cleavage, as against fracture cleavage, and its relation to folding. His main theses are, first, that schistosity develops parallel to axial planes, and second, that it can only develop under a certain thickness of overburden. The schistosity is thought to arise through dimensional orientation ("Einregelung") of mica flakes on the schistosity planes, as a result of differential movement along these planes ("Durchbewegung"). The thickness of the overburden required for schistosity to develop is estimated at a minimum of 4,000 – 6,000 m.

In the R.S., BORN (1929) had expressed a different opinion about the close parallelism of schistosity and axial planes, which was corroborated by VAN WIJNEN (1953) for the Luxemburg Ardennes. Instead of a dynamic flow process orienting the mica flakes, static recrystallization perpendicular to stress according to the "principle of Riecke" seemed more probable (RUTTEN, 1955). This idea incidentally accords well with the old German term for schistosity, i.e. "Druckschieferung" or pressure cleavage.

I have been privately told since that BORN's (1929) field evidence had proved insufficient, a thing never published, "because he was such a nice man". DE SITTER (1956), on the other hand, has stated that the general laws and all accessory features cited by Fourmarier have been confirmed by so many authors that they now belong to the ordinary arsenal of the tectonician. According to him, parallelism of cleavage and axial planes is so evident that we need not bother about exceptions, and that axial plunge can even be measured from the line of intersection of bedding and cleavage.

CLOOS AND HIS SCHOOL

The matter thus threatened to develop a dogmatic character, just about the time when the modern investigations of Cloos' students began to throw more light on the subject. Accordingly it is best to retrace the steps of the Bonn group in the field of microtectonics.

From the onset Cloos stressed the need of an analysis, in maps and sections, as detailed as possible — statistical if feasible — of all tectonic features of a given structure. Not only is it necessary to present schistosity and folding both in detailed sections and in plan, but thrusting, striae, fractures, diaclases (or joints), lineation and all other characters found in a given structure or structures must be taken into account simultaneously.

The title of CLOOS' (1948) classic study of the structures at Schuld on the Ahr river in the R.S., "Gang und Gehwerk einer Falte" (Mode of development and processes of motion of a fold), clearly indicates this approach to tectonic analysis. Two figures taken from this publication (Fig. 67, 68) may further illustrate this emphasis on detailed and comprehensive analysis of all tectonic features. This same approach is also found in the work of his pupils (HOEPPENER, 1953, 1955, 1956, 1957, 1960; KUBELLA, 1951; ENGELS, 1959; JANKOWSKY, 1955).

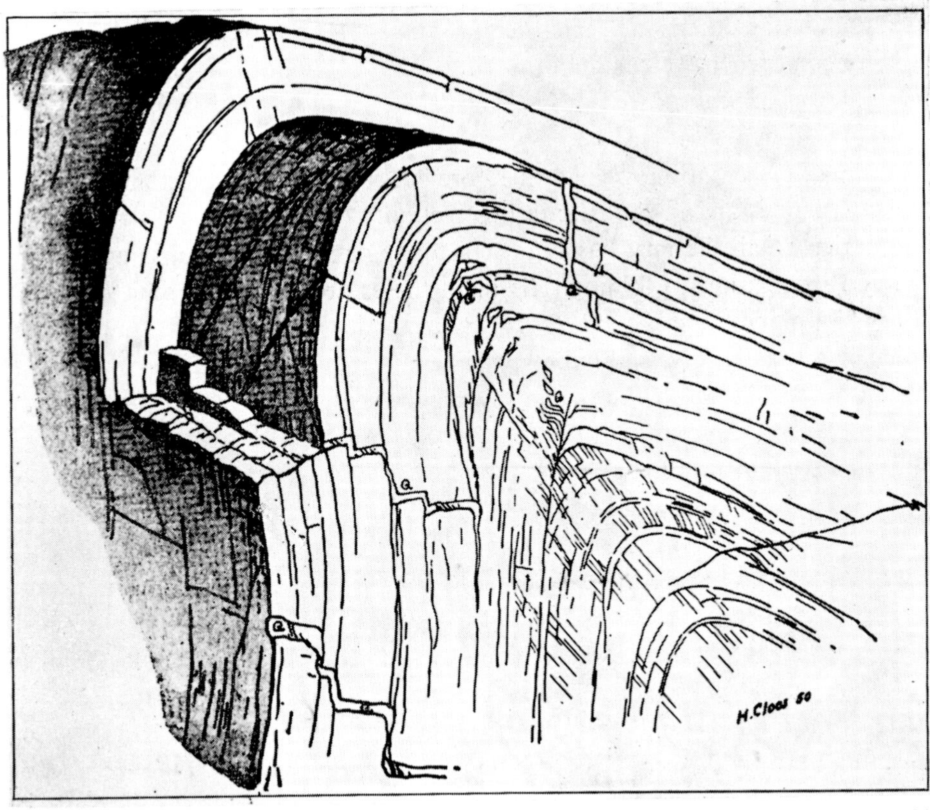

Fig.67. Small, asymmetric anticline near Altenahr on the Ahr river. Grey: bedding planes with striae. White: cross section. Graywackes with quartz veins showing later displacement by bedding plane slip, covering slates with schistosity. (After CLOOS, 1948.)

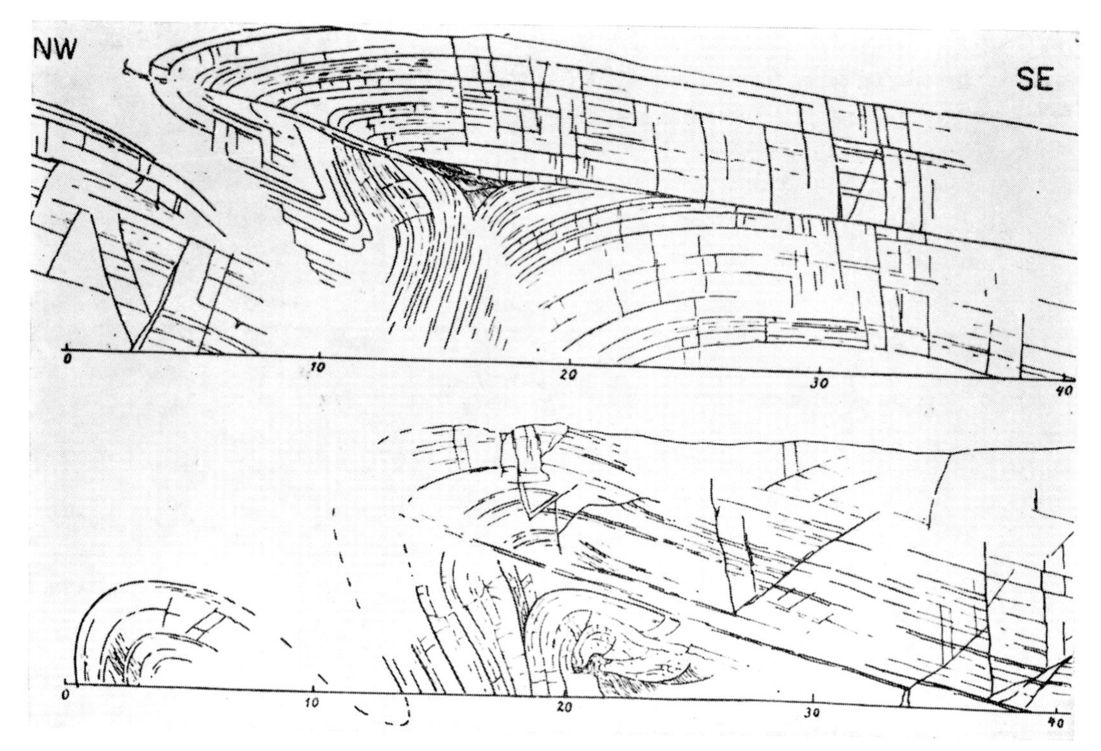

Fig.68. Two parallel detailed sections through the small anticline of Schuld on the Ahr river. Showing relation of asymmetric fold, thrust in the core fracture cleavage and minor fractures. As is generally the case in the northern part of the Rheinisches Schiefergebirge, the "Vergenz" is northwards. (After CLOOS, 1948.)

The main result of the microtectonic studies of the Bonn group apparently lies in the stress laid on the dynamics during the development of the structures. It became clear that, in the case of the asymmetric structures characterizing the R.S., rotation always accompanies deformation. Interrelation of the various tectonic features consequently is determined by the time at which they developed during the tectonic evolution. Inversely, this often offers a means of relative dating of these features.

Rotation, a major theme during the development of the tectonics of the R.S., may be found on many different scales. A small-scale example is the rotation of the fracture cleavage toward fold hinges. The fracture cleavage, developed perpendicular to the bedding planes in a very early stage in competent beds, will rotate through a sort of flowage ("Biegegleitung") within these beds, without further fracturing.

Further constraint will lead to bedding-plane slip, mainly resulting in a relative forward movement of the upper limbs of asymmetric structures ("Faltungsvorschub"). This results in a larger-scale rotation, affecting the outline of the whole fold. Inhomogeneities within each bed now stay relatively undisturbed, but earlier inhomogeneities across the bedding are offset at the bedding planes through this bedding plane slip (for instance, the displaced quartz veins in Fig.67).

Although this is difficult to pinpoint, schistosity will begin to develop about that time in the shales. It forms a single set, parallel to the axial planes in true shales, but two sets, symmetrical to the axial planes, in somewhat sandier material (JANKOWSKY, 1955). Parallelism between axial planes and the single set schistosity or the plane of symmetry of the two sets of schistosity exists, however, in contemporaneous folding only. When earlier folds have rotated considerably before the onset of schistosity, the latter will develop according to the new stress field, and will have no relation to the axial planes of the earlier folding (HOEPPENER, 1956).

Schistosity introduces such an important element of anisotropy that later movements may make intensive use of it. One of the common features are thrust planes developing knick zones following the schistosity (Fig.69). But also the later, and larger, vertical

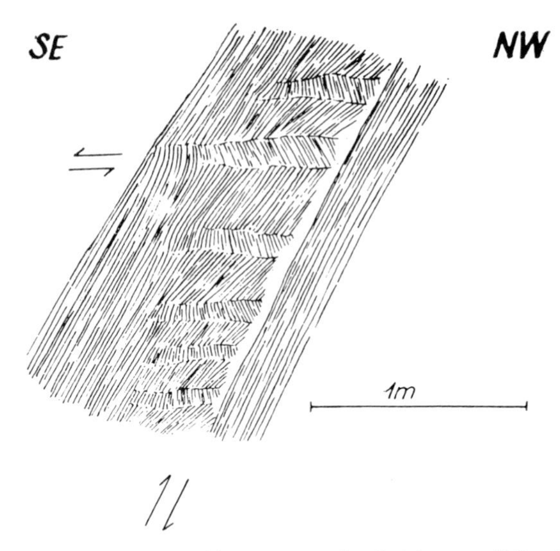

Fig.69. Upthrust in schistose slates, with accompanying knick zone. (After HOEPPENER, 1955.)

movements may result in quite different patterns depending on whether they happen to occur cross-wise or parallel to the earlier schistosity.

The schistosity itself may also be rotated during further tectonic evolution. Fan-wise arrangement of the schistosity may develop in this way. If, however, the schistosity has rotated so far in relation to the stress direction that it cannot function any more as the easiest plane of translation, movement may start along a new set of planes of different orientation. A sort of crenelation will then form on the primary schistosity which will ultimately develop into a second schistosity (Fig.70, 71). This is the "microplissement" of FOURMARIER (1952, fig.1).

In this way a detailed history of the dynamics of folding has been arrived at for small areas in the R.S. It is thought that the results of these test cases will have a wider application over larger parts of the R.S.

One more general result is that there is apparently not one single factor which causes schistosity to form. Mere thickness of overburden cannot have been the only influence, because schistosity occurs in many areas where in all probability the overburden has always been thin. The material in which it develops, more sandy or more purely clayey, is important, as is the amount of stress. The most important factor, however, might well be the length of time during which the material has been under stress, as schistosity takes time to develop by slow solid state processes.

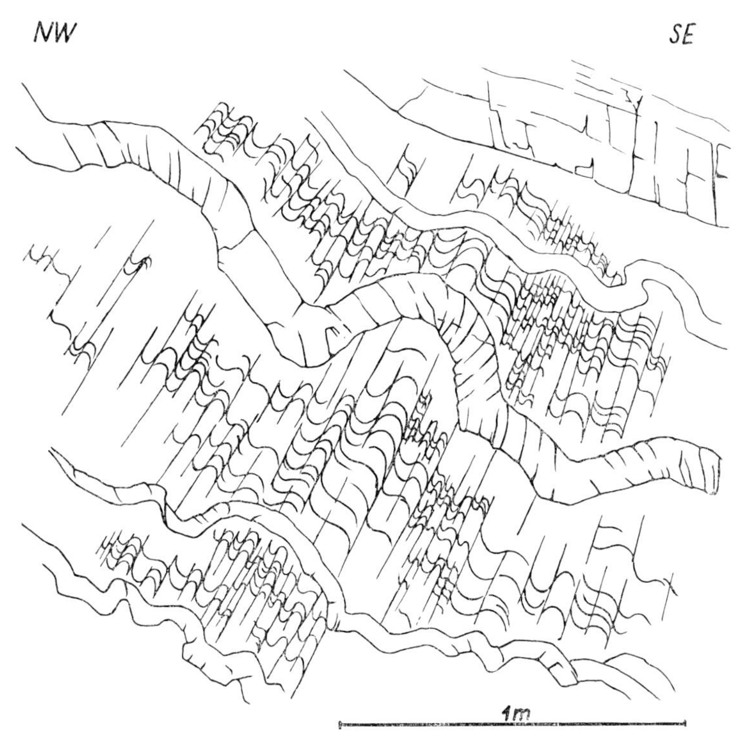

Fig.70. Bedding and fracture cleavage in competent beds; first and second schistosity in slates. (After HOEPPENER, 1956.)

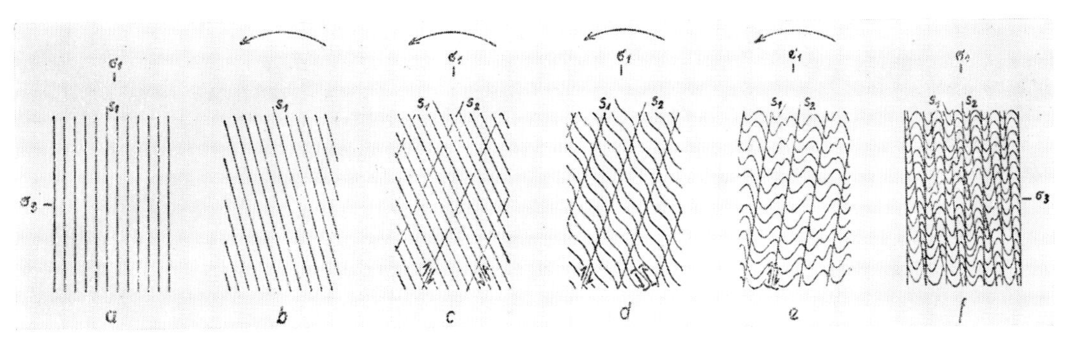

Fig.71. Development of second schistosity, after rotation of the first schistosity. (After HOEPPENER, 1960.)

"ACHSENRAMPE"

Another general feature that has transpired, is the occurrence of narrow transverse zones of axial updip ("Achsenrampe"; Fig. 72). These are not related to single foldstructures, but strike N 20°E, obliquely across the foldbelt. They are thought to represent late orogenetic or post orogenetic movements of a more general character. This same N 20°E direction is also found, both earlier than the folding in geosynclinal sedimentary facies boundaries, and later than the folding in block faulting and ore veins. It is therefore supposed that it indicates recurrent movements of the basement. The "Eifeler Nord-Süd Zone", between the Lower Rhine graben in the north and the Luxemburg Gulf in the south (p. 81) also belongs to this same direction.

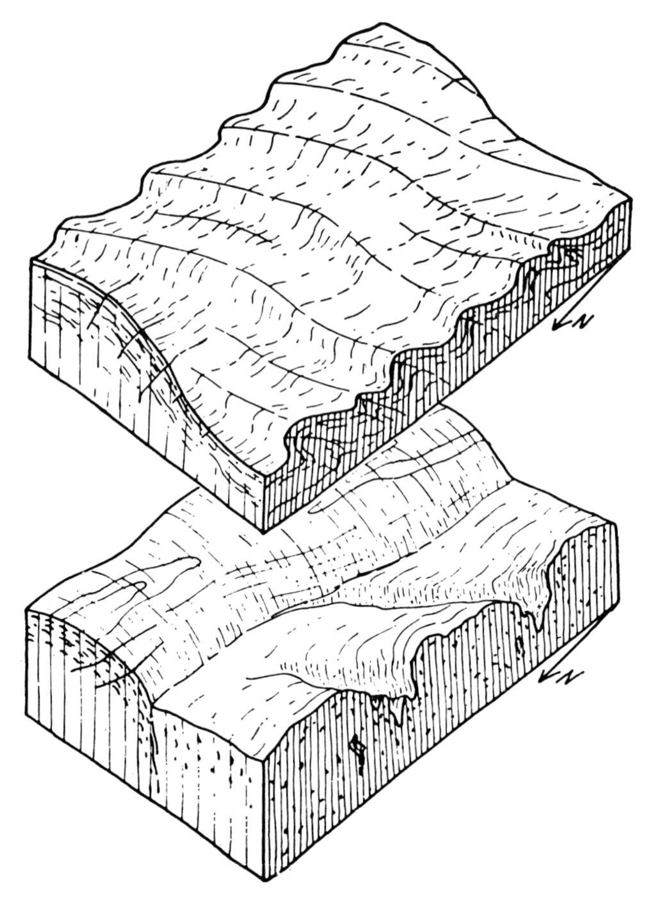

Fig.72. Schematic block diagram of a zone of axial updip (*Achsenrampe*). Upper part is taken from the Rheinisches Schiefergebirge, lower part from the Odenwald, in which the crystalline basement is exposed, with major faults that are supposed to cause the formation of *Achsenrampe* in the sedimentary cover of the Rheinisches Schiefergebirge. (After CLOOS, 1940.)

THE MOSELLE SYNCLINORIUM

I should like to close this exposé of the tectonics of the R.S. with a short survey of the Moselle synclinorium. It is here that microtectonics have been of great value in unravelling the geologic history (HOEPPENER, 1957).

Returning to Fig.66, the cross-section of the R.S. along the Rhine, it must be noted that the zone with southward "Vergenz", which forms the northern flank of the Moselle synclinorium, is not simply a symmetrical counterpart to the northern zone of the R.S. with northward "Vergenz". Surprisingly, individual structures in either zone are not mirror images, but are of similar build. They show long southern flanks and short northern flanks, indicating primary northward "Vergenz" in both zones. The difference between the two zones is thought to have been caused by a later subsidence of the Moselle area, in which individual structures of the second zone were secondarily rotated far enough to acquire southward "Vergenz".

Farther southwest (Fig.73) a quadripartite zonal structure of the Moselle synclinorium is found, in which four parallel zones, each with conformable "Vergenz", are separated by narrower zones with a fan-like structure. According to HOEPPENER (1960), the main theme is rotation in different sense of older features, during the further tectonic evolution of the Moselle synclinorium. This is illustrated in Fig.74. After embryonic folding, primary schistosity is seen to begin to develop in the south, in the future zones *III* and *IV*. These were still similar at that time, both being characterized by initial northward "Vergenz". During time *4* primary schistosity also developed in the north, in zones *I* and *II*. At the same time secondary schistosity, obliquely cutting the strongly rotated primary schistosity, developed in zones *III* and *IV*. Further constraint led to the formation of thrust faults

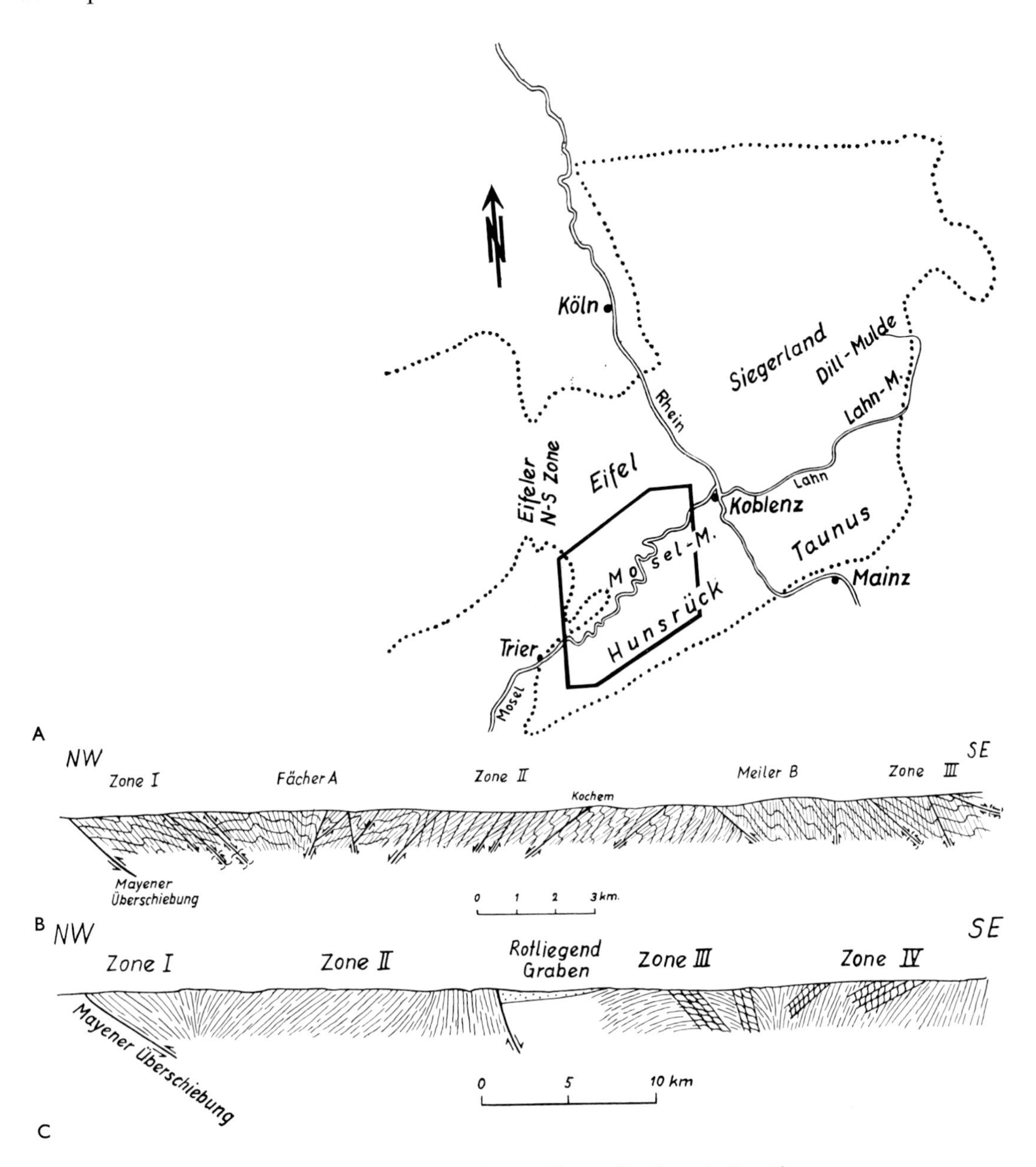

Fig.73. Microtectonics of the Moselle synclinorium. A. Location.
B. Bedding, schistosity and thrust planes of zones *I*, *II* and *III*, separated by the fans *A* and *B*. C. Schematic section, showing arrangement of primary and secondary schistosity and central Permian graben of Wittlich. (After HOEPPENER, 1956.)

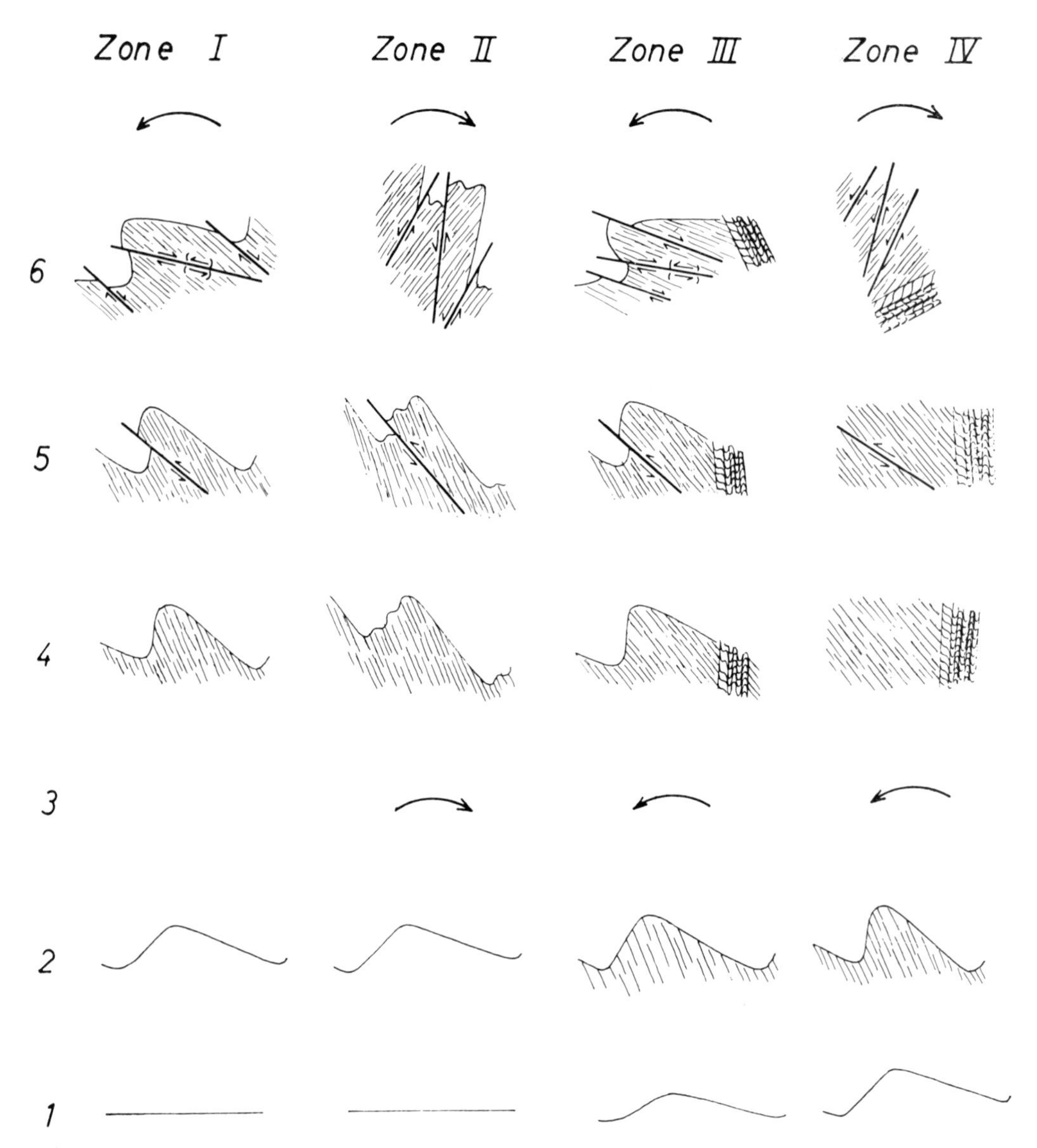

Fig.74. Schematic development of microtectonic features in the zones *I* to *IV* in the Moselle synclinorium. (After HOEPPENER, 1960.)

(time *5*). These were strongly rotated, together with the earlier features, and now in different senses, during the ultimate orogenetic influence of time *6*.

All of these datings, it must be stressed, are utterly relative. The only thing we know is that everything took place between the Lower Devonian (the time of sedimentation of the Hunsrück Slates) and the uppermost Carboniferous (the time of the last Hercynian folding in this region).

In closing this discussion of the tectonics of the R.S. a tribute to the intensive and time-consuming work on microtectonics is warranted. It has by-passed possible pitfalls indicated by the cynical remarks of Hans Cloos, cited above, and it has led to a better understanding of the evolution and the sequence of various tectonic features in several areas in the R.S.

THE SUB-HERCYNIAN BORDER DEPRESSION

North of the R.S. follow the coal measures of the Ruhr Basin. They belong to the paralic facies of the Upper Carboniferous, and are comparable, often in minute detail, with the coal measures of the Namur syncline north of the Ardennes. In regional setting there is, however, this difference between the two: north of the Namur syncline the Brabant Massif has risen since the time of coal measure deposition, whereas north of the R.S. no such younger block has formed. Consequently, in the case of the Namur syncline, stress has always been laid on its tectonic position between the Ardennes and the Brabant Massif, as opposed to the border position cited for the Ruhr Basin. For the history of the deposition of the Upper Carboniferous, and for that of its folding, this distinction is, however, fallacious. We now know that during that time there was no difference between those two areas. So what is said here applies equally well to the Namur syncline and its westerly continuation into the coal fields of northern France.

The concept of the sub-Hercynian border depression (Fig.75), the "subvariscische Saumsenke", in its essence goes back to Stille. But at that time it was called the "subvariscische Vortiefe", the sub-Hercynian fore-deep, in accordance with the "Vortiefe" concept of Suess. For Stille, displacement with time, of both geosynclinal and orogenetic activity, was well established for Hercynian Europe. In the R.S., as elsewhere, three main zones of an orogenetic cycle could be followed. They were all characterized by geosynclinal sedimentation, immediately followed by orogenetic activity. For the three zones mentioned, the latter is thought to have taken place, from south to north, during the Bretonic, the Sudetic and the Asturic phases, respectively.

In this picture the Ruhr Basin forms a fore-deep, although it was situated to the landward or rear side of the geosyncline during the Devonian. This difficulty is of course a general one. Normally there is sediment transport in a geosynclinal belt away from a continent. But the translation of this mobile belt itself with time, and also the "Vergenz" of the orogeny which follows the geosynclinal period, is directed in reverse, i.e., towards the continent. It is debatable whether we must speak of a fore-deep or an after-deep (MacGillavry, 1961). Since, moreover, we are sure that the depositional environment of the coal measures was not that of a deep at all, use of the word "Saumsenke" or border depression is preferable (Paproth and Teichmüller, 1961; Teichmüller, 1962).

Further details of the stratigraphy and the tectonics of the coal measures of the sub-Hercynian border zone can be found in the voluminous literature on coal geology in France (Bouroz et al., 1961), Belgium, The Netherlands and Germany (Hesemann and Dahm, 1962).

We must conclude with a warning against believing geotectonic literature too implicitly, again referring to the alleged evidence for orogenetic phases. As there is this evident shift in time of the thickest and the most clastic sediments, it is postulated that each shift in the geosyncline coincides with an orogenic phase further south.

Paproth and Teichmüller (1961), just to cite one example, assign most of the R.S. to the Bretonic orogenetic phase (Fig.62). As we saw earlier in this chapter, there certainly have been relative variations of considerable size in crustal movements during the Devonian in the R.S. But as far as the evidence goes, these are all vertical movements, dependent upon variation in the rate of sinking of parts of the geosyncline. There is no evidence at

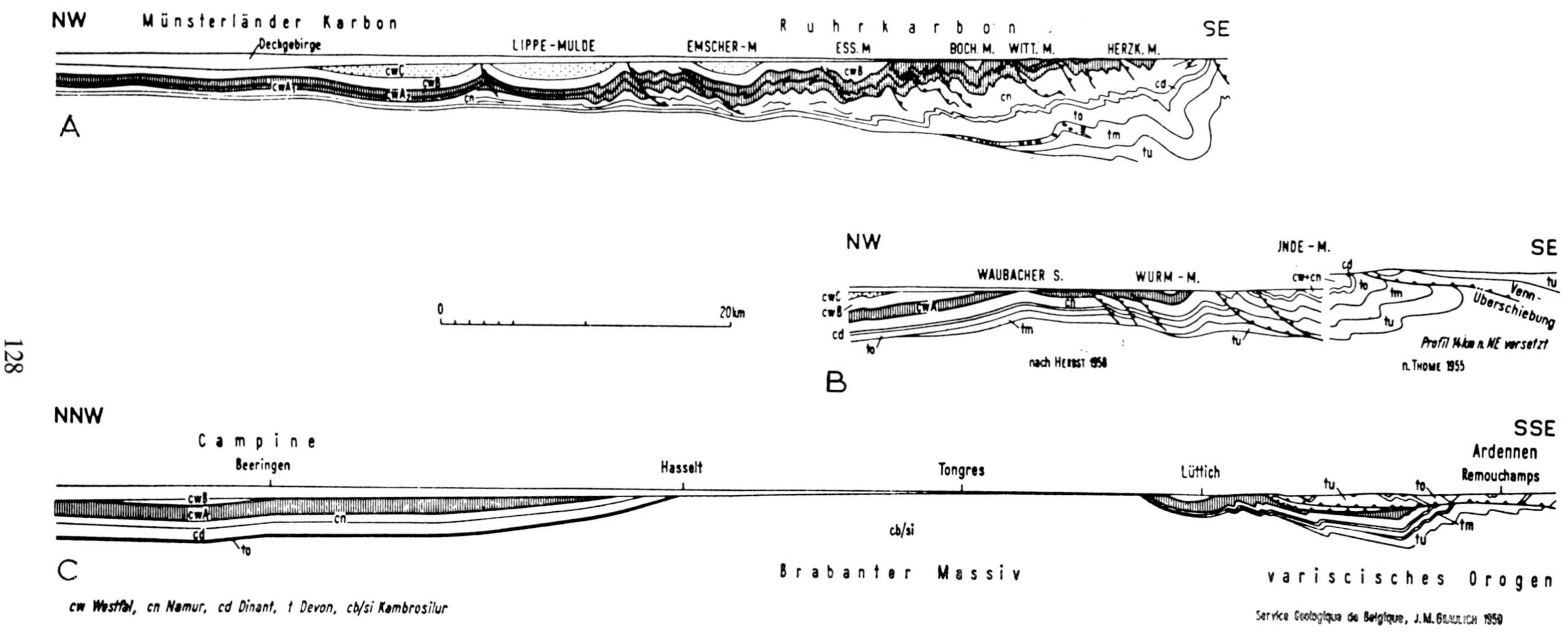

Fig.75. Three sections across the sub-Hercynian border depression.
A. Ruhr Basin. B. Aachen area. C. Campine – Brabant Massif – Ardennes. The equivalent of the Namur syncline is at Liège (Lüttich). (After PAPROTH and TEICHMÜLLER, 1961.)

all for a major orogenetic period. Even the latest research, the microtectonic analysis of the Moselle synclinorium stresses the fact that none of the successive phases of the tectonic evolution can be dated.

KRONBERG et al. (1960) also advocate Bretonic orogenetic movements in the R.S. Their outline sections through two synclines in the Sauerland (KRONBERG et al., 1960, fig. 7, 8) seem convincing indeed, showing Culm Series transgressive and with angular unconformity over Devonian. There is, however, the possibility of a tectonic contact between the highly mobile, thin-bedded Culm and the underlying Devonian. This possibility is treated but lightly. However, in their detailed sections (KRONBERG et al., 1960, fig. 2, 3) only a strongly tectonized zone is indicated at the contact of Culm and Devonian, without any angular unconformity. Emersion, non-sedimentation and erosion may well have occurred. To cite these same authors, "..diastems, conglomerates, smaller disconformities, and rapid changes in facies and thickness abound". But, it seems, nothing more.

Shifts in sedimentation in the geosyncline seem rather poor evidence for postulating orogenetic phases, that is, for major periods of folding in a hinterland, instead of for mere relative vertical crustal movements. It has become still more hazardous at present, because many of the coarser clastic beds are now interpreted as turbidites. These might well have been derived from a hinterland much farther away than formerly supposed, and moreover perhaps unknown, because covered by younger formations.

To conclude, there is no evidence for a Bretonic orogenetic phase in the R.S., though the influence of differential vertical movements in the subsiding geosynclinal belt have been found at many places. But also — this must be added immediately — at many different times. There has been no synchronization of these movements to a definite period or stage.

Moreover, the nearest real evidence for Sudetic orogenetic activity is only found in the Saar Basin. It is, of course, quite probable that the southern part of the R.S., and also of the Ardennes, were already folded at the end of the Lower Carboniferous. This would explain the absence of the coal measures from these areas. But we must keep in mind that the Saar Basin is situated some 50 km south of the southern margin of the R.S.

REFERENCES

BEYER, K., 1941. Zur Kenntniss des Silurs im Rheinischen Schiefergebirge. *Jahrb. Reichsanstalt Bodenforsch.*, 61: 198–266.

BIERTHER, W., 1953. Zur Stratigraphie und Tektonik der metamorphen Zone im südlichen Rheinischen Schiefergebirge. *Geol. Rundschau*, 41: 173–181.

BORN, A., 1929. Über Druckschieferung im variscischen Gebirgskörper. *Fortschr. Geol. Paläontol.*, 7: 22 pp.

BOUROZ, A., CHALARD, J., DALINVAL, A. and STIÉVENARD, M., 1961. La structure du Bassin houiller du Nord, de la région de Douai à la frontière belge. *Ann. Soc. Géol. Nord*, 81: 173–220.

BRINKMANN, R. (Editor), 1955. Tektonik und Lagerstätten im Rheinischen Schiefergebirge. *Geol. Rundschau*, 44: 480 pp.

CLOOS, H., 1940. Über Achsenrampen. *Geol. Rundschau*, 31: 227–229.

CLOOS, H., 1948. Gang und Gehwerk einer Falte. *Z. Deut. Geol. Ges.*, 100: 290–303.

DE SITTER, L. U., 1956. Postscript on schistosity. *Geol. Mijnbouw*, 18: 58–59.

ENGELS, B., 1959. Über neue Ergebnisse kleintektonischer Untersuchungen im Rheinischen Schiefergebirge. *Geol. Rundschau*, 48: 271–280.

ERBEN, H. K., 1962. Zur Analyse und Interpretation der rheinischen und hercynischen Magmafazies der Devons. In: *Symposium Silur–Devon Grenze*. Schweizerbart, Stuttgart, pp. 42–61.

FOURMARIER, P., 1923. Le clivage schisteux dans les terrains paléozoiques de la Belgique. *Congr. Géol. Intern., Compt. Rend., 13e, Bruxelles, 1923*, 1: 517–530.

FOURMARIER, P., 1952. Aperçu sur les déformations intimes des roches en terrains plissés. *Ann. Soc. Géol. Belg.*, *Bull.*, 75: 181–194.

FOURMARIER, P., 1953a. L'allure du front supérieur de schistos dans le Paléozoique de l'Ardenne. *Bull. Acad. Belg.*, *Cl. Sci.*, 5 (39): 838–845.

FOURMARIER, P., 1953b. Schistosité et phénomènes connexes dans les séries plissés. *Congr. Géol. Intern.*, *Compt. Rend.*, *19e*, *Algiers*, *1952*, 3 (3): 117–131.

FOURMARIER, P., 1953c. Schistosité et grande tectonique. *Ann. Soc. Géol. Belg.*, *Bull.*, 76: 275–301.

FOURMARIER, P., 1956. Schistosité et forme des plis. *Ann. Soc. Géol. Belg.*, *Bull.*, 79: 317–346.

HELMBOLD, R., 1952. Beitrag zur Petrographie der Tanner Grauwachen. *Heidelberger Beitr. Mineral. Petrog.*, 3: 253–258.

HELMBOLD, R., 1958. Contribution to the petrography of the Tanner Graywacke. *Bull. Geol. Soc. Am.*, 69: 301–314.

HENTSCHEL, H., 1860. Basischer Magmatismus in der Geosynklinale. *Geol. Rundschau*, 60: 33–45.

HESEMANN, J. and DAHM, H. D. (Editors), 1962. Das Karbon der subvariscischen Saumsenke. *Fortschr. Geol. Rheinland Westfalen*, 3 (1–3): 1282 pp.

HESSISCHES LANDESAMT FÜR BODENFORSCHUNG, 1960. Beitrag zur Geologie der Mittleren Siegener Schichten. *Abhandl. Hess. Landesamtes Bodenforsch.*, 29: 363 pp.

HOEPPENER, R., 1953. Faltung und Klüftung im Nordteil des Rheinischen Schiefergebirges. *Geol. Rundschau*, 41: 128–144.

HOEPPENER, R., 1955. Tektonik im Schiefergebirge. *Geol. Rundschau*, 44: 26–58.

HOEPPENER, R., 1956. Zum Problem der Bruchbildung, Schieferung und Faltung. *Geol. Rundschau*, 45: 247–283.

HOEPPENER, R., 1957. Zur Tektonik des S.W.-Abschnittes der Moselmulde. *Geol. Rundschau*, 46: 318–348.

HOEPPENER, R., 1960. Ein Beispiel für die zeitliche Abfolge tektonischer Bewegungen aus dem Rheinischen Schiefergebirge. *Geol. Mijnbouw*, 39: 181–188.

JAEGER, H., 1962. Das Silur in Thüringen und am Ostrand des Rheinischen Schiefergebirges. In: *Symposium Silur–Devon Grenze*. Schweizerbart, Stuttgart, pp.108–135.

JANKOWSKY, W., 1955. Schichtenfolge, Sedimentation und Tektonik im Unterdevon des Rheintales in der Gegend von Unkel–Remagen. *Geol. Rundschau*, 50: 59–86.

JUX, U., 1960. Die devonischen Riffe im Rheinischen Schiefergebirge. *Neues Jahrb. Geol. Paläontol.*, *Abhandl.*, 110: 186–391.

JUX, U., 1964. Erosionsformen durch Gezeitenströmungen in den unterdevonischen Bensberger Schichten des Bergischen Landes? *Neues Jahrb. Geol. Paläontol.*, *Monatsh.*, 9: 515–530.

KEGEL, W., 1922. Abriss der Geologie der Lahnmulde. *Abhandl. Preus. Geol. Landesamtes*, 86: 81 pp.

KEGEL, W., 1934. Geologie der Dillmulde. *Abhandl. Preus. Geol. Landesamtes*, 160: 48 pp.

KEGEL, W., 1948. Sedimentation und Tektonik in der rheinischen Geosynklinale. *Z. Deut. Geol. Ges.*, 100: 267–289.

KELLER, G., 1948. Die Fortsetzung der Faltung des Ruhrkarbons nach der Tiefe und die Frage der Faltungszeit. *Bergbau-Archiv*, *Essen*, 8: 76 pp.

KNETSCH, G., 1963. *Geologie von Deutschland und einigen Randgebieten.* Enke, Stuttgart, 386 pp.

KOCKEL, C. W., 1958. Schiefergebirge und Hessische Senke um Marburg/Lahn. *Sammlung Geol. Führer*, 37: 248 pp.

KRONBERG, P., PILGER, A., SCHERP, H. and ZIEGLER, W., 1960. Zu den altvariscischen Bewegungen an der Wende Devon–Karbon. *Fortschr. Geol. Rheinland Westfalen*, 3: 1–46.

KRÖMMELBEIN, K., HOTZ, E. E., KRÄUSEL, W. and STRUVE, W., 1955. Zur Geologie der Eifelkalkmulden. *Geol. Jarhb.*, *Beih.*, 17: 204 S.

KUBELLA, K., 1951. Zum tektonischen Werdegang des südlichen Taunus. *Abhandl. Hess. Landesamtes Bodenforsch.*, 3: 1–81.

KUENEN, PH. H. and SANDERS, J. E., 1956. Sedimentation phenomena in Kulm and Flözleeres greywackes, Sauerland and Oberharz, Germany. *Am. J. Sci.*, 254: 649–761.

KUHN, O., 1961. *Die Tierwelt der Bundenbacher Schiefer.* Neue Brehm Bücherei, Wittenberg, 274: 48 pp.

KUTSCHER, F. and SCHMIDT, H., 1958. *Lexique Stratigraphique International. I. Europe*, *5b. Allemagne: Dévonien.* Centr. Natl. Rech. Sci., Paris, 386 pp.

LEHMANN, E., 1952. Beitrag zur Beurteilung der paläozoischen Eruptivgesteine Westdeutschlands. *Z. Deut. Geol. Ges.*, 104: 219–237.

LECOMPTE, M., 1962. Faciès marins et stratigraphie dans le Dévonien de l'Ardenne. *Ann. Soc. Géol. Belg.*, *Bull.*, 85: 17–57.

MACGILLAVRY, H. J., 1961. Deep or not deep, fore-deep or "after-deep"? *Geol. Mijnbouw*, 40: 133–148.

MEISCHNER, K. D., 1962. Rhenaer Kalk und Posidonien Kalke im Kulm des nordöstlichen Rheinischen Schiefergebirges. *Abhandl. Hess. Landesamtes Bodenforsch.*, 39: 47 pp.

PAPROTH, E. and GRAULICH, J. M., 1958. Compte Rendu de la Session extraordinaire dans le Sauerland (Allemagne). *Ann. Soc. Géol. Belg.*, *Bull.*, 67: 329–357.

REFERENCES

PAPROTH, E. and TEICHMÜLLER, R., 1961. Die paläogeographische Entwicklung der Subvariscischen Saumsenke in Nordwestdeutschland im Laufe des Karbons. *Congr. Avan. Études Stratigraph. Géol. Carbonifère, Compte Rendu, 4, Heerlen, 1958*, 2: 471–491.

PILGER, A., 1952. Tektonik, Magmatismus und Vererzung, Zusammenhängen im ostrheinischen Schiefergebirge *Z. Deut. Geol. Ges.*, 104: 198–218.

PILGER, A., 1957. Über den Untergrund des Rheinisches Schiefergebirges und Ruhrgebietes. *Geol. Rundschau*, 46: 197–212.

PLESSMANN, W., 1964. Turbidite in der rechtsrheinischen Geosynklinale. In: A. H. BOUMA and A. BROUWER (Editors), *Turbidites*. Elsevier, Amsterdam, pp.137–141.

RICHTER, D., 1963. Schiefrigkeit und tektonische Achsen im Gebiet des Velberter Sattels. *Forsch. Ber. Nordrhein-land Westfalen*, 1197: 34 pp.

RUTTEN, M. G., 1955. Schistosity in the Rhenic Massif and the Ardennes. *Geol. Mijnbouw*, 17: 104–110.

SCHMIDT, H., 1962. Über die Faziesbereiche im Devon Deutschlands. In: *Symposium Silur–Devon Grenze*. Schweizerbart, Stuttgart, pp.224–230.

SCHMIDT, H. and PLESSMANN, W., 1961. Sauerland. *Sammlung Geol. Führer*, 39: 151 pp.

SCHMIDT, W., 1952. Die palaeogeographische Entwicklung des linksrheinischen Schiefergebirges vom Kambrium bis zum Oberkarbon. *Z. Deut. Geol. Ges.*, 103: 151–178.

SCHMIDT, W., 1955. Die Eruptive in den Kern-Schichten des Hohen Venns. *Geol. Jahrb.*, Hannover, 70: 329–338.

SCHMIDT, W. and SCHRÖDER, E., 1962. *Geologische Erlaüterungen zur Übersichtskarte der nörlichen Eifel, 1:100,000*. Geol. Landesamt Nordrheinland Westfalen, Krefeld, 110 pp.

STENGER, B., 1961. Stratigraphische und gefügetektonische Untersuchungen in der metamorphen Taunus-Südrand-Zone (Rheinisches Schiefergebirge). *Abhandl. Hess. Landesamtes Bodenforsch.*, 36: 68 pp.

TEICHMÜLLER, R., 1962. Die Entwicklung der subvariscischen Saumsenke nach dem derzeitigen Stand unserer Kenntniss. *Fortschr. Geol. Rheinland Westfalen*, 3: 1237–1254.

THOMÉ, K. N., 1955. Die tektonische Prägung des Vennsattels und seiner Umgebung. *Geol. Rundschau*, 44: 266–304.

VAN STRAATEN, L. M. J. U., 1954. Sedimentology of recent tidal flat deposits and the psammites du Condroz (Devonian). *Geol. Mijnbouw*, 16: 25–47.

VAN WIJNEN, J. C., 1953. Étude micro-tectonique dans les Ardennes luxembourgeoises. *Publ. Serv. Géol. Luxemb.*, 10: 60 pp.

VON BUBNOFF, S., 1930. *Geologie von Europa. II. Das ausseralpine Westeuropa. 1. Kaledoniden und Varisciden*. Borntraeger, Berlin, 690 pp.

CHAPTER 7

Hercynian Europe: The Armorican Massif

INTERMEDIATE CHARACTER OF THE ARMORICAN MASSIF

As indicated in Chapter 4, the Armorican Massif forms a large upland, in which, geographically, the main element is formed by Brittany. The upland, however, covers part of the adjoining geographical provinces, i.e., in the north the western and northern part of Normandy, and in the south, across the Loire, the Vendée (Fig.76). Moreover the Channel Islands belong to the same geological unit.

Most of our earlier knowledge of the geology of the Armorican Massif goes back to the work of BARROIS (1930). These earlier investigations have been well summarized by BERTRAND (1946). For Brittany, see also PRUVOST (1958, 1959), for Normandy, DANGEARD (1951). In recent years J. Cogné and M. J. Graindor have been very active in Brittanny and Normandy, and G. Mathieu in the Vendée.

The Armorican Massif has attracted attention, already at an early date, because of the prevalence in this upland of sediments of Early Paleozoic and of Late Precambrian age. Much more of this lower part of the sedimentary column is exposed here, than in either the Ardennes or the Rheinisches Schiefergebirge. Compared with the Massif Central, the Vosges Mountains and the Schwarzwald, much less of these older sediments has become migmatised and more is consequently available for inspection in the Armorican Massif. Only in Bohemia is a comparable situation found.

The Armorican Massif, it follows, occupies an intermediate position in western Europe, both as to the level to which its basement is exposed through later uplift and denudation, and as to the amount of metamorphism.

Moreover, as previously indicated in Chapter 4, it is also intermediate regarding its overall structural directions. In its northern part, from Brest to Cherbourg and Caen, these are southwest–northeast, gradually changing to west–east, south of Caen. South of the Axial syncline, or "syncline of Chateaulin–Laval", or "Median rift zone with Coal Measure basins" (Fig.77), the overall strike is northwest–southeast. Because the southwest–northeast strike in the northern part lines up well with the Ardennes and the Rheinisches Schiefergebirge, this direction is called the Variscan direction in the Armorican Massif. The northwest–southeast trend in the southern part of the Armorican Massif, which can be followed into the western part of the Massif Central, is called the Armorican direction. Both Variscan and Armorican directions are mainly due to the Hercynian orogeny, though earlier fold belts seem to have followed the same general lines.

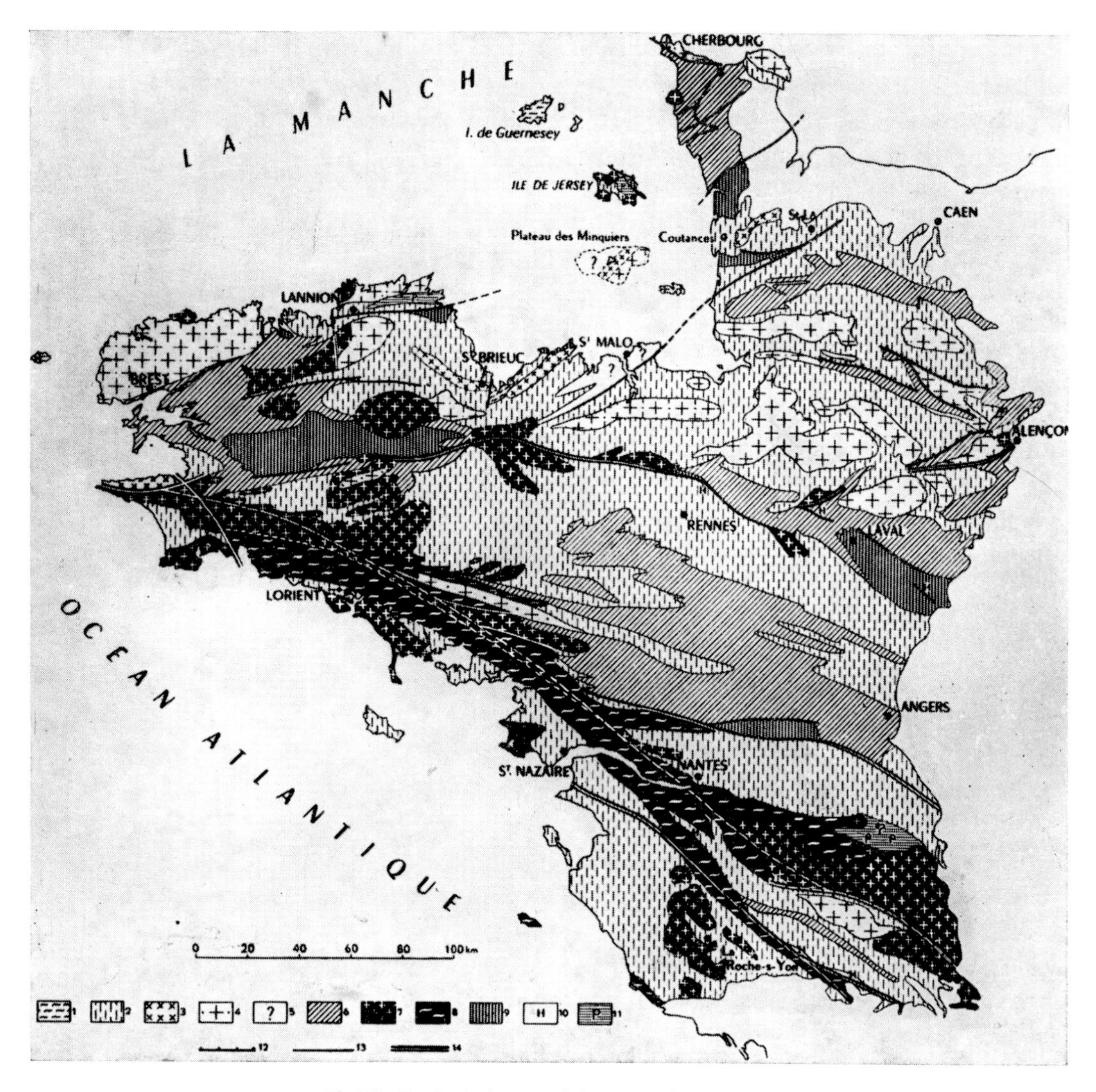

Fig.76. Geological map of the Armorican Massif.
1 = Sarnian basement (pre-Pentevrian orogeny), Precambrian; *2* = Brioverian (partly pre-Cadomian orogeny), Late Precambrian or Eocambrian; *3* = syn-kinematic quartzdiorite; *4* = Precambrian, "Assyntian", granites; *5* = migmatites of unknown age; *6* = zone of pre-Dinantian (Bretonic) folding; *7* = Hercynian granites; *8* = Hercynian migmatites; *9* = zone of post-Dinantian folding; *10* = basins with folded Stephanian; *11* = (?) Hercynian rhyolites. This map offers a good outline of the main structural trends of the Armorican Massif. But many details, in particular the interpretation of the Precambrian elements, are still doubtful. Compare, for instance, the variations in interpretation in a similar map in GRAINDOR (1961). (After GRAINDOR, 1963.)

THE BRIOVERIAN

The most interesting component of the Armorican Massif is probably its thick series of Late Precambrian sediments, assigned to the local stage of the Brioverian (from *Briovera*, the Roman name for the present-day town of Saint Lô). This is a purely clastic series, predominantly formed by shales and sandy shales, but containing coarser members of sandstone and conglomerate.

It has been considered a "*série compréhensive*", or lithostratigraphic unit of unknown, but long time span. Such a designation as "comprehensive series" has, it might be added parenthetically, nothing to do with stratigraphic comprehension. On the contrary, it is often used when the latter fails completely.

Of special importance are the intercalations in the Brioverian of the so-called phtanites (CAYEUX, 1929). These fine-grained carbonaceous quartzites are mainly found in the south-eastern part of the Armorican Massif, in the Vendée. It is from these rocks that CAYEUX (1894a, b), grand old man of sedimentology, described one of the first known well-preserved faunae of the Precambrian.

The fact that fossils of Ordovician and Lower Carboniferous age have been found in beds of similar lithofacies, which hitherto had been considered to belong to the Brioverian, led to the feeling that the Brioverian, although Precambrian, must be Late Precambrian. That is, it was thought to belong to the Eo- or Infracambrian, and consequently to be comparable to the Scandinavian sparagmites. It would thus form the first element of the geosynclinal series of the Caledonian orogenetic cycle.

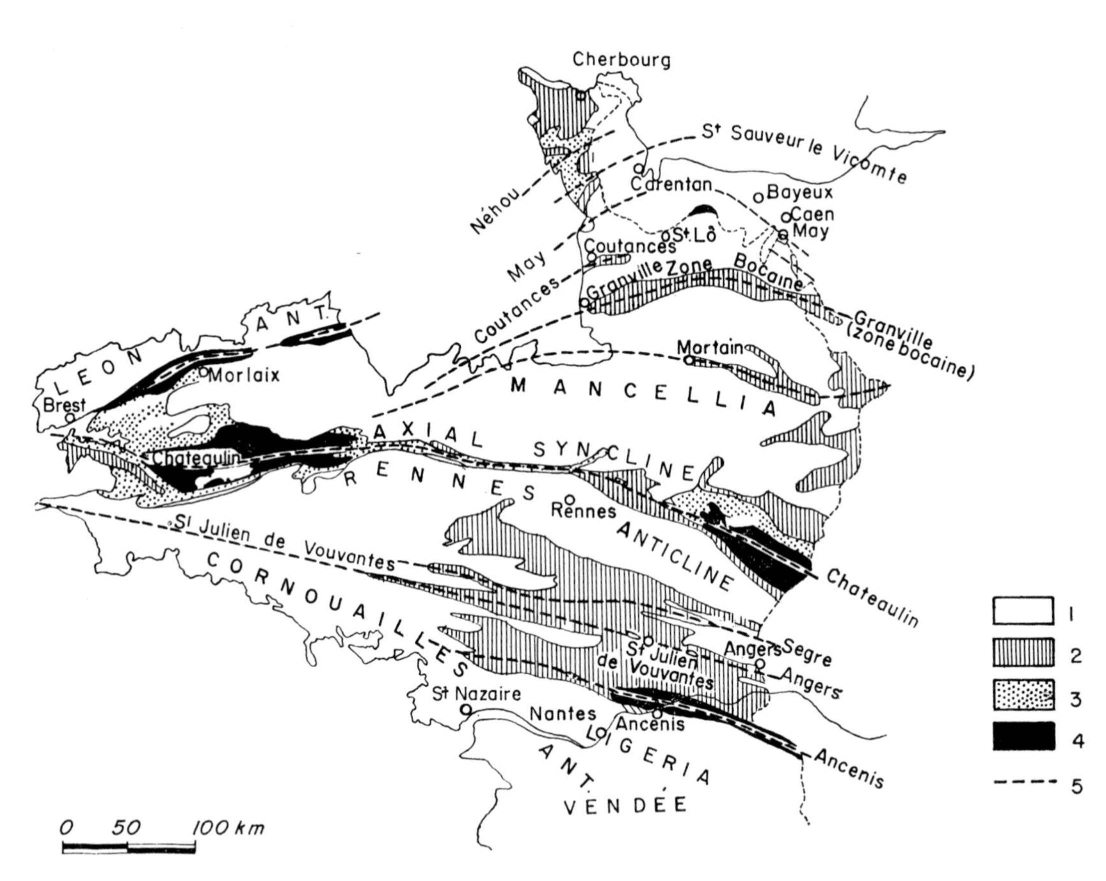

Fig.77. Structural outline of the Armorican Massif, confronting the "anticlinal" areas where the Precambrian crops out (white, nondifferentiated) with the "synclinal" areas filled with Paleozoic sediments.
1 = Precambrian; *2* = Cambrian, Ordovician and Silurian; *3* = Devonian; *4* = Carboniferous; *5* = synclinal axes. (After GIGNOUX, 1950.)

CADOMIAN OROGENY

This idea needs, however, a thorough qualification. In the northern part of the Armorican Massif, in Normandy and northern Brittanny, a major angular unconformity has long been known to exist between the Brioverian and the transgressive base of the Paleozoic, either the Cambrian or a higher stratigraphic unit. This unconformity, well exposed for instance in the classic section along the Orne near Caen (Fig.78), indicates a major orogenetic period, the Cadomian orogeny (from *Cadomus*, the Roman name for Caen)[1]. Because of this unconformity, separating Brioverian from Paleozoic, later authors considered the Brioverian to be really ancient Precambrian, not just Eo- or Infracambrian. According to this view, it cannot be an immediate forerunner of the Early Paleozoic, such as the sparagmites.

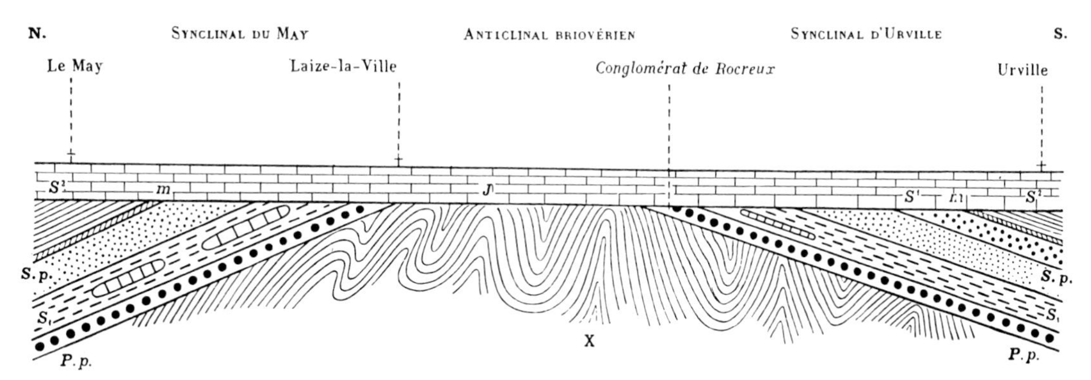

Fig.78. Schematic section of the Cadomian nonconformity south of Caen, Normandy. X = Brioverian; Pp = purple basal conglomerate of the Cambrian; S_1 = shales with lenses with *Archeocyathus*, Acadian–Potsdamian; Sp = "Grès feldspathique", feldspar sandstone, Potsdamian–Tremadocian; S^1 = Armorican Sandstone, Arenigian; S^2 = Anger Shales, Llandeilian; m = oolitic iron ore of Normandy; J = Hettangian, Lower Jurassic. (After Mathieu, 1945.)

Moreover, according to Cogné (1963), the Brioverian in itself forms a full geosynclinal series, complete from initial ophiolites and meso- to catazonal metamorphism to flysch (Dangeard et al., 1961).

Indeed, the definition of the Infracambrian by Pruvost (1951) states that stage as conformable with the Paleozoic, while an angular unconformity characterizes the "real" Precambrian.

As is normally the case in geology, however, things are not nearly as black and white. As is seen in Fig.79, the Cadomian unconformity is only found in the northern part of the Armorican Massif, in Normandy and in northern Brittanny. In its southern part there is no angular unconformity between Brioverian and Paleozoic. Disconformities exist, however, in which all of the Cambrian may be missing.

Accordingly, the Brioverian could be classed as Infracambrian in the south, but should be assigned to the Precambrian in the north of the Armorican Massif.

[1] A proposal for the subdivision of the Cadomian orogeny into a Normandic and an Angevin phase (Klein, 1962) has not been well received (Cogné and Graindor, 1963).

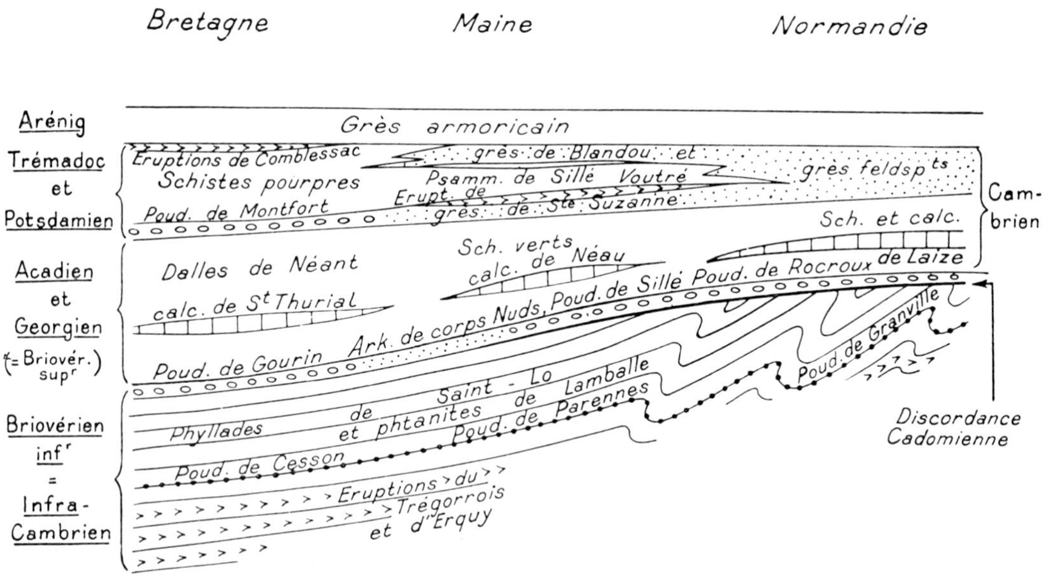

Fig.79. Schematic stratigraphic south–north section showing the relation between Brioverian and Cambrian in the Armorican Massif. The Cambrian covers the Brioverian conformably in the south, unconformably in the north. (After PRUVOST, 1951.)

One must realise that we have as yet no absolute dating on any of these Upper Precambrian series, which may well represent quite longish time spans. It may well be that the apparent difference is greater than the real one, if the Cadomian orogeny is a local orogenetic phenomenon just before the onset of the Cambrian.

This same possibility must be kept in mind, even if the Cadomian orogeny should be found to be represented all over the Armorican Massif. Newer research summarized by COGNÉ (1963) (see also MATHIEU, 1937; CHAUVEL and PHILLIPOT, 1961), seems to indicate such a situation, thereby invalidating Fig.79. For, as long as there are no absolute dates, correlation of these old and non-fossiliferous sediments remains an open question. One may just as well use a correlation based on lithofacies and ice ages (tillites) which would put the Brioverian, together with the sparagmites, in the Eo- or Infracambrian, as one based on orogenetic phases, which would instead relegate the Brioverian, at least in the northern part of the Armorican Massif, to an earlier, "real" Precambrian.

The uncertainty in stratigraphic position of the Brioverian strengthens the opinion of NEUMAN and PALMER (1958) that no valid correlation is possible between these thick clastic series of Late Precambrian age. Classification should remain local, and assignment to either the Eo- or the Infracambrian is no more than pseudo-stratigraphy.

We may, I think, do well in leaving open the question of the age of the Brioverian and of its relation to the Eocambrian elsewhere. It must be noted, however, that the Brioverian rests with another angular unconformity on a still older, strongly metamorphic basement. This basement, only exposed locally near St. Brieuc and St. Malo on the northern shore of Brittanny, consists of a quite varied series of gneisses and accompanying rocks (COGNÉ, 1959). The orogeny separating the Brioverian from this basement is called the Pentevrian (from *Pagus Pentur*, along the bay of St. Brieuc), whereas the basement itself is labelled the ancient continent of Sarnia.

TILLITES

Apart from the questions arising out of the stratigraphic and tectonic position of the Brioverian, this series has evoked general interest by its tillite intercalations. These "tillites", first found in western Normandy near the town of Granville, have since been reported from other localities of the Armorican Massif as well. They form the base of the upper part of the Brioverian, the "X^3" of the geological maps.

The tillites of the Brioverian have always been a strong argument in favour of the correlation with the sparagmites of Norway. Also, exactly as in the case of Norway, doubts have been raised, the latest by WINTERER (1964), about the glacial nature of these conglomerates.

But the pendulum seems to begin to take its swing back. The glacial interpretation of the nature of the intercalations in question has recently been greatly strengthened by the description not only of various tills, but also of fluvioglacial sediments, notably of varved clays (GRAINDOR, 1957; ROBLOT, 1963). This indicates that we do not only have the remnants of a ground moraine, but also of the other types of sediments belonging to a major ice age (Fig.80, 81).

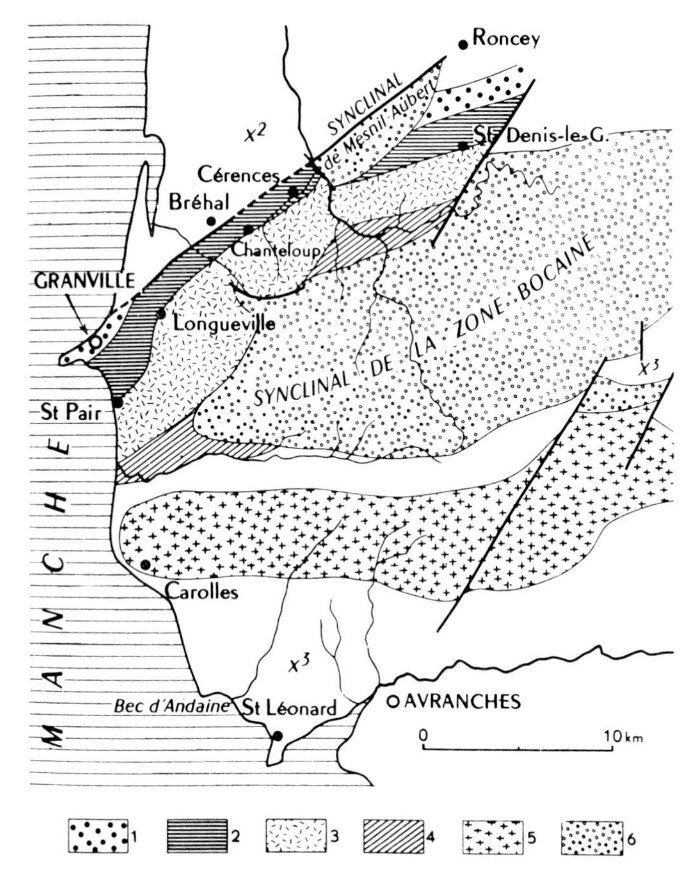

Fig.80. Map of the so-called glacial and periglacial intercalations in the Brioverian near Granville, western Normandy. *1* = tillite; *2* = varved clays and "Gletschermilch" sediments; *3* = graywackes; *4* = "Étage de la Laize" (uppermost Brioverian, post-tillite sediments); *5* = granite; *6* = paleozoic conglomerates; X^2, X^3 = Middle and Upper Brioverian, nondifferentiated. (After ROBLOT, 1962.)

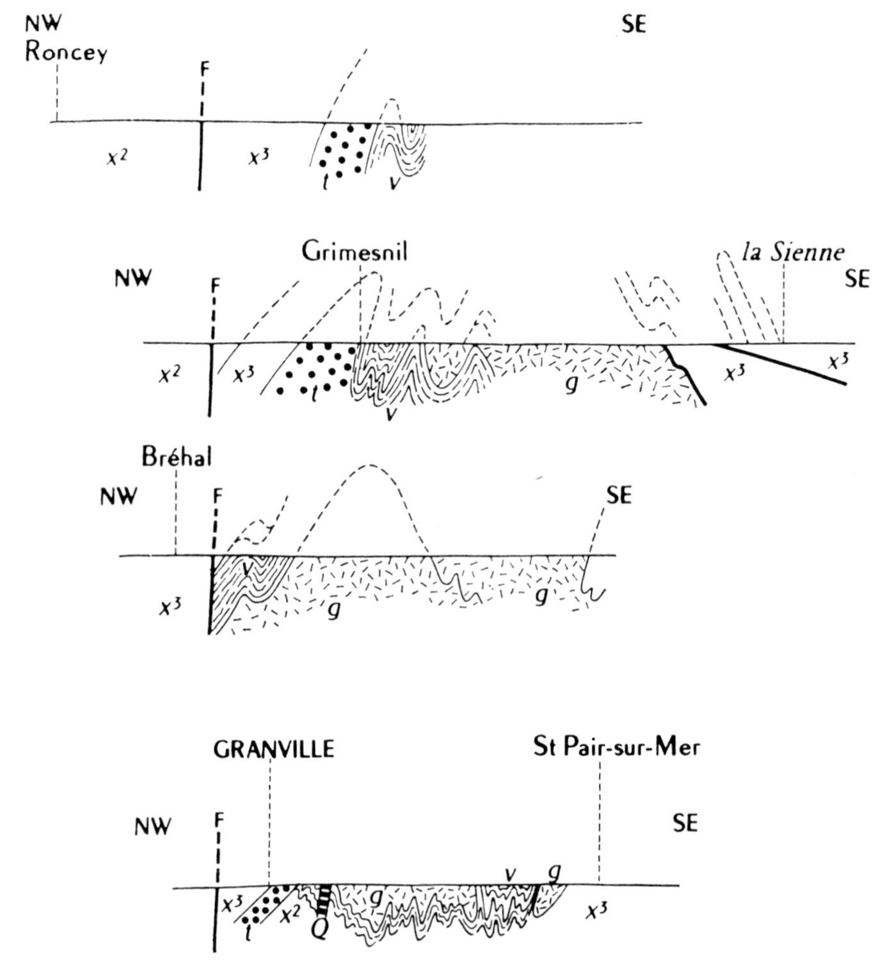

Fig.81. Four sections through the Brioverian with intercalated glacial and periglacial sediments near Granville, western Normandy.
t = tillite; v = varved clays and "Gletschermilch" sediments; g = graywackes; X^2, X^3 = Middle and Upper Brioverian, non differentiated. (After ROBLOT, 1963.)

PALEOZOIC HISTORY

If we now turn to the Paleozoic history it must be kept in mind that this has two quite different aspects. On an upper level, one can follow the sedimentary and tectonic history, preserved in the narrow main synclinal areas of the Armorican Massif. And on a deeper level there is the story of Hercynian metamorphism, migmatization, granitization and intrusion, exposed in the anticlinal areas.

Taking first the sedimentary history, one is left again with the impression that the Armorican massif takes up an intermediate position amidst the various Hercynian uplands of western Europe.

This time it concerns the question of epicontinental versus geosynclinal conditions. The series of the Paleozoic are nowhere complete. Disconformities, at times as large as to involve complete systems, are found almost everywhere and also at almost every level in the stratigraphic sequence (Fig.82; see also PRUVOST, 1958, 1959). On the other hand,

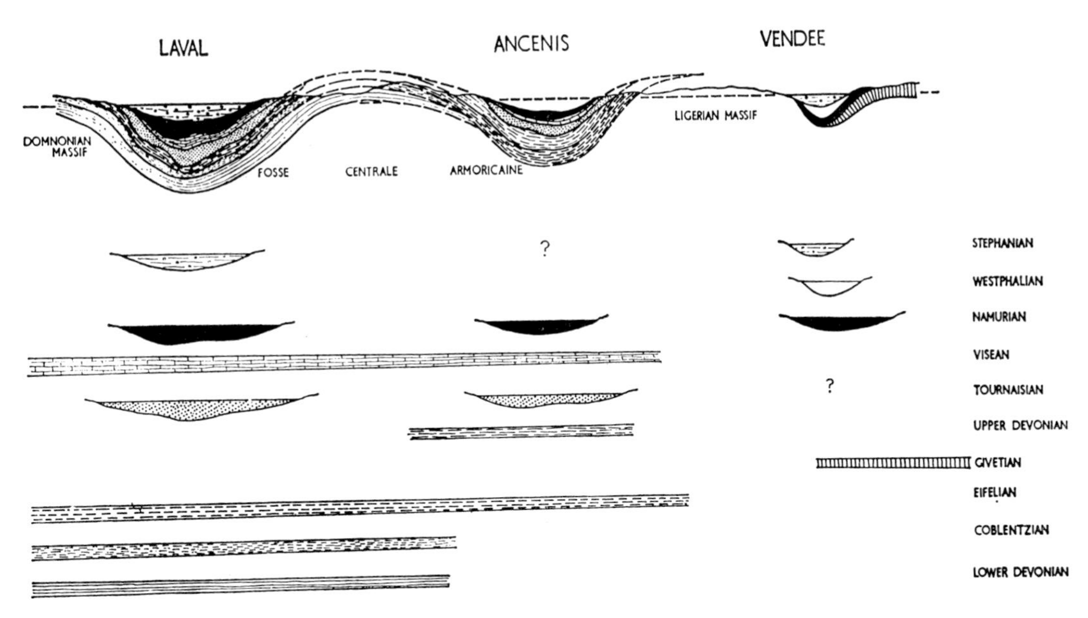

Fig.82. Diagrammatic representation of the Paleozoic sediments preserved in the three main synclinal areas of the Armorican Massif. Upper section: reconstruction of the situation at the end of the Carboniferous. Lower sections: implied Paleozoic sedimentary history of the Armorican Massif. (After GEORGE, 1963.)

thicknesses of a couple of kilometers are quite normal for single systems of the Paleozoic. This is a figure too high for truly epicontinental conditions, but at the same time too low for a truly geosynclinal environment. The best qualification is perhaps: "epicontinental, but very active".

ABSENCE OF CALEDONIAN OROGENY

A second point of general interest in the sedimentary history of the Armorican Massif is the absence of any evidence of the Caledonian orogeny. Of course there have been crustal movements at the time of the Caledonian orogeny, in the Armorican Massif also (PRUVOST, 1949; PÉNEAU, 1962). But these were only vertical movements of an epirogenetic character. We find transgressions, recording differential subsidence, and also local uplift, and maybe tilting. But these are crustal movements similar in character to others which have occurred in this general area both earlier, all through the Cambrian, Ordovician and Silurian, and later, during Devonian and Early Carboniferous. There is nowhere in the Armorican Massif a situation comparable with the Caledonian angular unconformity, such as we know it for instance from Fépin in the Ardennes (p.92).

GRAINDOR (1962) has recently tried to make a case for the Caledonian orogeny in the Armorican massif. He did, however, not succeed in proving more than the existence of crustal movement leading to the disconformities and stratigraphical hiatuses cited.

We will meet with this same feature, the absence of really orogenetic Caledonian movements everywhere further south, both in Hercynian and in Alpine Europe. It has incited Stille and others to draw their boundary between the continent of Paleo-Europe and that of Meso-Europe to the north of the Armorican Massif.

As indicated in Chapter 1, this is not legitimate. To note that a certain orogenetic period, in this case the Caledonian, is restricted to present-day northern Europe, is a statement of geological fact. But the postulate that the continent of Europe did at that time not reach beyond the southern rim of the Caledonian orogeny, is easily disproved. In the Armorican Massif itself, we have seen that the Brioverian preceeded the Paleozoic. And all differences of opinion on the character of the Brioverian apart, after the Cadomian orogeny there is no doubt that this is a truly continental area. It most certainly was something belonging to the domain of the crust, the sial (or whatever name one wishes to use). So it is not true that there has been continental accretion only since the Devonian, just because the influence of the Caledonian orogenetic folding does not reach the Armorican Massif. The continent was already there, long before that, as is witnessed by the Cadomian orogeny which even then already affected Precambrian crustal rocks.

PRE-OROGENETIC CHARACTER OF THE "BRETONIC PHASE"

A much more serious point is the absence too, of a phase of real folding at the transition from Devonian to Carboniferous. I have grown quite dejected by this absence, which I felt as a default: "What, no Bretonic phase in Brittanny?". After all the difficulties encountered in following those of my friends who are Stille disciples in their stories about Bretonic orogenetic movements in the Hercynian uplands of Germany, I always trusted to find beautiful examples of this early phase of the Hercynian orogeny in its type area, in Brittanny.

This has not been the case. The Bretonic phase has been defined by Stille from the maps by Barrois (1930) of the Chateaulin Basin east of Brest, in the Axial Syncline. These early folds are, to use the words of P. Pruvost, "of feeble intensity only". The major proof seems to rest on differences in direction between earlier structures and the later "great synclinal axes" (Pruvost, 1949, p. 357). But as the latter are mainly zones of post-orogenetic, that is post-Sudetic, graben subsidence, in which the Earlier Paleozoic escaped destruction through erosion, they give no indication as to the relative ages of earlier structures. Sudetic folding, as we shall see, has been clearly demonstrated, but there is no reason to assume an earlier, Bretonic orogenetic phase.

Just as was the case at the time of the Caledonian orogeny, there is no doubt that there have been crustal movements during the transition from Devonian to Carboniferous. Even generalizations, such as that after the Famennian there has been general emergence all over the Armorican Massif (Pruvost, 1949), can be accepted, though they rest only on the fragmentary evidence supplied by the present narrow "synclinal" areas within the Armorican Massif. But going further to state that this emergence "accompanied the first Hercynian orogenetic phase" is, I think, stretching the evidence too far. In the newest sections published from the Chateaulin Basin (Pruvost and Lemaitre, 1943) there is overall conformity between Carboniferous and Devonian.

A similar conclusion has recently been arrived at by Cogné (1960a, p.342), who writes: "The Bretonic phase was responsible for the disconformity between Dinantian and Devonian. It was more a phase of uplift and emergence of the Devonian than a case of real folding".

This difficulty, of distinguishing between a period of real folding, on the one hand, and differential vertical crustal movements of a more or less epirogenetic character on the other, is due mainly to two factors. First comes the usage of the word orogeny in the

French literature on the Armorican Massif, where each and every crustal movement, including differential vertical movements and also low-angle regional tilting which only leads to transgressions and regressions, is called orogenetic.

Moreover, and this is the second reason, it has been indicated already that Paleozoic sediments are only preserved in those narrow, parallel "synclinal" structures, separated by much broader anticlinal areas. The direction of these "synclines" is mainly due to post-Sudetic faulting movements and has no relation with the paleogeographic pattern of the Paleozoic. Considerable differences are found in the development of the Paleozoic, when the various "synclines" are compared. But instead of realizing that these differences are due to gradual variations over all of the area, the stratigraphic columns of these widely separated "synclinal" areas are put next to each other, just as if these variations occur at their present boundaries. In this way an image of sudden changes is conjured up. It is a queer sensation to realize that all graphic evidence, offered as proof for a major folding phase, is not a number of sections showing an angular nonconformity, but a set of mismatched stratigraphic columns.

A genetic aproach to the paleogeography of the Paleozoic has been taken by COGNÉ (1966). This author assumes that the development of the Paleozoic is related to that of the migmatite bodies. These are thought to provoke rising tendencies. The Paleozoic consequently is detrital and terrigenous, indicating erosion in a contemporary rising hinterland, near the migmatite areas, and calcareous farther away.

SUDETIC AND LATER OROGENETIC PHASES

The main Hercynian phase in the Armorican Massif is the Sudetic. As is normal the orogenetic movements occurred over a rather extended time interval, from the beginning of the Upper Carboniferous onwards. But Namurian is only rarely found to be transgressive over the earlier Paleozoic, and in most cases the post-orogenetic transgression only begins in the Westphalian or even in the Stephanian. Thus, it is perhaps better to speak of a Sudetic orogenetic phase sensu lato, meaning that it began sometime after the end of the Early Carboniferous.

Several authors use the term Erzgebirgic phase for some of these later movements, but it is doubtful whether there is sufficient proof for the occurrence of a really separate orogenetic phase at that time in the Armorican massif.

The facies of the post-Sudetic Late Carboniferous in the basins of the Armorican massif is the limnic facies typical for the Sudetic part of Hercynian Europe. After the orogeny the sea did not come back until after the Paleozoic. Sedimentation during the Late Carboniferous was restricted to intramontane graben structures. Downfaulting is the primary tectonic movement, but both draping over faults and settling in the graben produced local sharp folding. For the southeastern part of the Armorican massif, the Vendée, this is well represented by MATHIEU (1937), cf. Fig.83.

This younger folding is normally also assigned to specific orogenetic phases, as is seen in Fig.83. Again I do not think it wise to consider these younger movements as regional orogenetic phases. It must be realized that the post-Sudetic history is that of the local development of individual narrow graben basins. The outline of these basins is seen in Fig.76, whereas an idea of their general structure is given in Fig.84. The stratigraphy

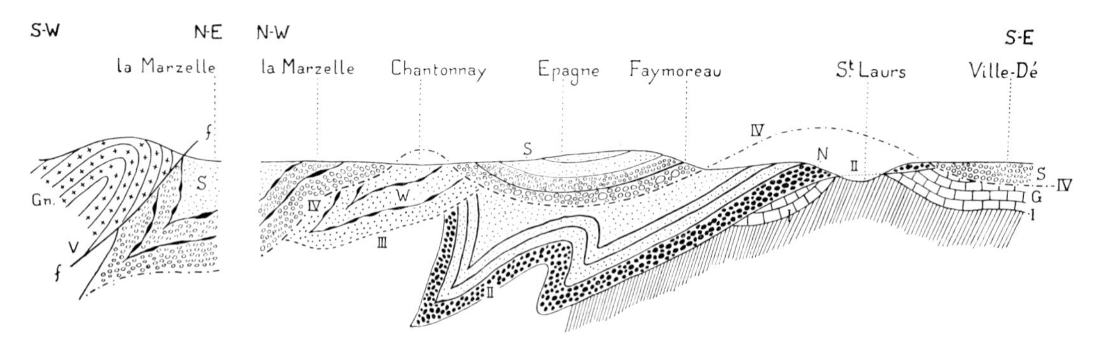

Fig.83. Diagrammatic cross-section through the *Sillon houiller* of the Vendée, showing the main tectonic history. The *Sillon houiller* is formed by narrow basins accompanying a fault zone to the north of the Hercynian migmatites stretching from St. Nazaire to the southeast (compare Fig.76, *H*).

G = Givetian; N = Namurian; W = Westphalian; S = Stephanian; Gn = granulitic gneiss; f = Chantonnay upthrust; I = unconformity between Givetian and Brioverian; II = ditto for Namurian over Brioverian and Givetian, Sudetic phase; III = ditto for Westphalian over Namurian, "Erzgebirgic phase"; IV = ditto for Stephanian over all older series, "Asturic phase"; $V(=f)$ = post-Stephanian, "Saalic" fault. (After MATHIEU, 1937.)

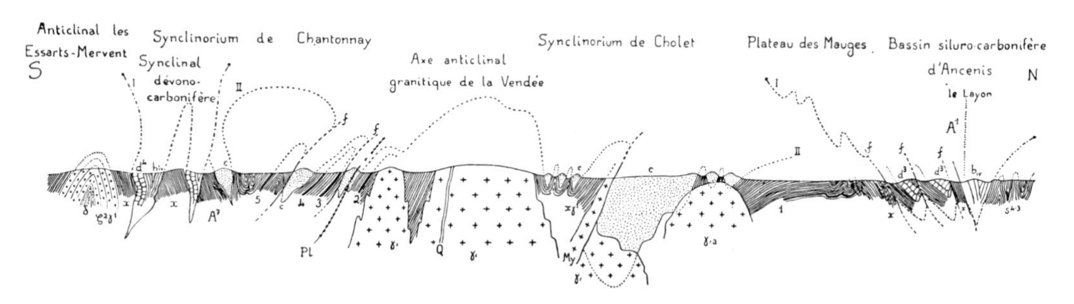

Fig.84. Section through the southeastern part of the Vendée, showing the relation between Paleozoic, Brioverian and plutonic rocks with migmatites.

$\zeta^2\gamma^1$ = gneiss and micaschists; δ = amphibolite- and chloriteschists; γ = various granites; x = Brioverian; c = schists, rhyolites and quartzites; S^{4-3} = Gotlandian; d^3 = Couvinian; d^4 = Givetian; h_{IV} = Namurian. (After MATHIEU, 1937.)

of the limnic facies of the Late Carboniferous differs from basin to basin because sedimentation, as well as the folding within these basins and the accompanying unconformities, are due to local basin subsidence. Such local movements, even if contemporaneous, should not be ranked with regional unconformities outside these basins, for instance in the paralic coal belt.

TECTONIC STYLE

Apart from the acute infolding of the Upper Carboniferous within the narrow graben structures described above, the tectonic style of the earlier Paleozoic is that of a monotonous series of tight anticlinal and synclinal structures. There is normally a regional "Vergenz", but both axial planes and the many upthrusts dip rather steeply. An example of this style in a less strongly folded area of Normandy is presented in Fig.78 (see also the serial sections through the synclinal structures west of Alençon, in the eastern part of the Armorican Massif by GRAINDOR, 1965). Fig.84, on the other hand, from the Vendée depicts a more severely folded region.

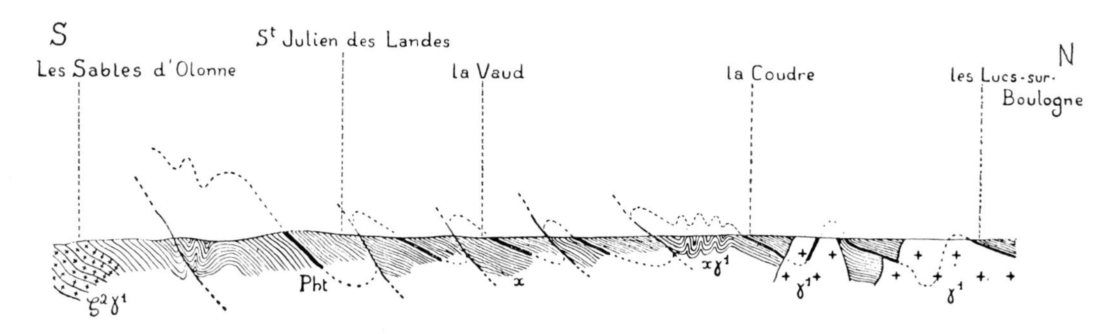

Fig.85. Section through the Brioverian of the western part of the Vendée. $\zeta^2\gamma^1$ = gneiss and micaschists; γ^1 = granulitic granite; $x\gamma^1$ = granitized Brioverian; x = Brioverian. The black bands indicate intercalated phtanites. (After MATHIEU, 1937.)

In those areas where the Brioverian out crops more or less uninterruptedly, that is over large areas in the broad anticlinoria, an imbricate style with a stronger overturn is normally developed. In these regions the upthrusts also have normally acquired the nature of overthrusts (Fig.85).

This difference in style between the Brioverian and the Paleozoic structures might well be due to the fact that the Brioverian, in part at least, has suffered two distinct orogenies, the Cadomian and the Hercynian. However, it must not be forgotten that the well-bedded, slaty series of the Brioverian would have been more mobile than the Paleozoic and might well have reacted more strongly to Hercynian stresses, producing Stockwerk tectonics.

The imbricate structure of the Brioverian is very clearly developed in northern Normandy, especially near Cherbourg. GRAINDOR (1957, 1960, 1961), who mainly worked in this area of the Armorican Massif, postulates such a Stockwerk tectonics due to the Hercynian orogeny. He maintains that the Paleozoic cover slid off the Brioverian basement in large scale "décollement" phenomena. According to Graindor, this would explain the differences in tectonic style between these two elements. His views have, however, been severely criticized on a factual basis by KLEIN (1963).

"ANTICLINAL" AREAS

In the Paleozoic and earlier history of the broad "anticlinal" areas of the Armorican Massif — perhaps better called anticlinoria — the main point of interest lies in their hard-rock geology; in the relations between schists, gneisses and granites.

It has long been known that most of the granites of the Armorican Massif are Hercynian intrusions, but that some are of earlier, Precambrian age. A synthesis of these views has been given by MATHIEU (1944). An additional difficulty in distinguishing the old granites lies in the fact that they have been tectonized during the Hercynian orogeny (GRAINDOR, 1959).

However, according to modern migmatite petrographers, there must have been metamorphic and ultrametamorphic rocks, before one can have granites. It follows that there must have been migmatization and anatexis, not once, but at least during two periods, accompanying the Cadomian and Hercynian orogenies. Studies along these newer lines are much more painstaking and time-consuming than classic petrography. We have at present only one comprehensive modern study at hand, that of COGNÉ (1960a; summary in COGNÉ, 1960b) on the Cornouailles (=Wales) anticlinorium, the most southwesterly structure of the Armorican Massif.

The main points of his results are that he found indeed two successive metamorphic cycles, an earlier Precambrian (Cadomian?), and a younger Hercynian. Each cycle consists of a pre-orogenetic phase of metamorphism, producing ectinites (cf. Chapter 8) and a syn-orogenetic phase of metasomatic granitization.

The Hercynian granites have intruded into predestined localities, that is in tectonic nodes. Over the Armorican Massif a complete series of examples of these granites is exposed, ranging from deep-seated anatexïtic granite to superficial, intrusive microgranites.

COGNÉ (1960a, b) also has distinguished several tectonic Stockwerke in Brittany. That is an upper Stockwerk, comprising the Precambrian Series of migmatites, together with its sedimentary superstructure, and secondly, the Hercynian Series of migmatites. Cogné also admits to large scale "décollement", in this case between his "upper Stockwerk" and the basement. Through this "décollement" it escaped subsequent Hercynian metamorphism. But, it follows that in his view this takes place at the base of the Precambrian migmatites, and not between Paleozoic and Brioverian, such as postulated by Graindor. The interpretation of the tectonics of the Armorican Massif accordingly shows a rather wide variation of opinions.

The results of the research of Cogné are given here in a schematic section through the southern part of the Armorican Massif (Fig.86). We learn from this section that the Hercynian granitization occurs quite independently of the earlier Precambrian granitization. The structures of the latter were in most places destroyed more or less completely by the former. In the section reproduced here, the earlier granitization is still preserved in the northerly St. Julien de Vouvantes synclinorium, where the upper front of the Hercynian migmatization is still deeply buried. In the Cornouailles ticlinorium, on the other hand, the Hercynian zone of granitization has reached much higher, and is cut by the

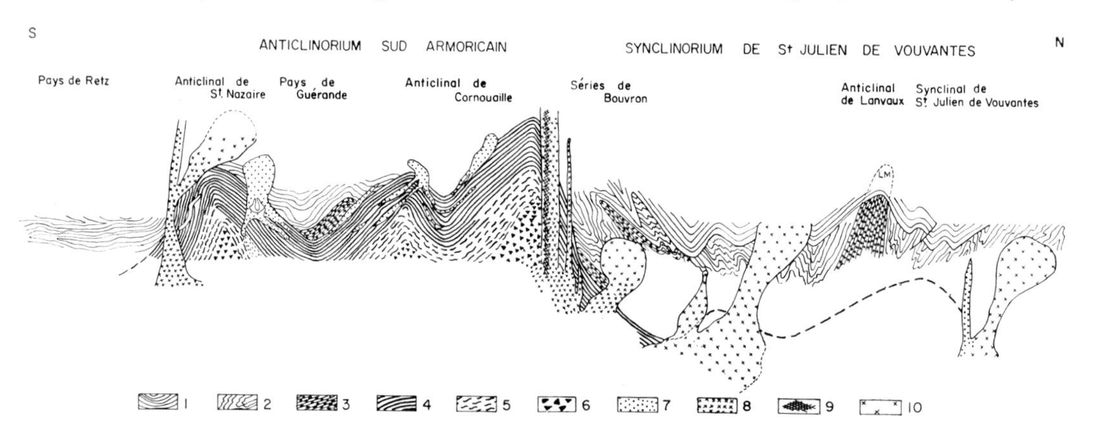

Fig.86. Diagrammatic section through the western part of the Armorican Massif.
1–3: Stockwerk of the superstructure. *1* = Cambro-Silurian; *2* = Brioverian and micaschists; *3* = granite-gneisses, Precambrian granitization.
4–6: Stockwerk of the Hercynian granitization. *4* = layered migmatites, embrechites; *5* = nebulous migmatites, anatexites; *6* = anatexitic granite.
7–10: Intrusive granites. *7* = tardi-migmatitic granites at the limit of the Stockwerk of Hercynian granitization and the sheared off elements of the superstructure; *8* = foliated granite, and *9* = mylonite at the limit of anticlinal and synclinal areas of Hercynian granitization; *10* = Hercynian granite batholiths in the superstructure. (After COGNÉ, 1960b.)

present surface. It has almost completely destroyed all earlier structures. It is thought that the sporadic elements of the older superstructure, which have escaped Hercynian granitization, did so through a "décollement" prior to metamorphism. We will meet a similar problem of two subsequent periods of metamorphism in the Massif Central (p.162).

As indicated before, the situation in the anticlinal areas of the Armorican Massif is extremely complex and with the amount of research proceeding at present, viewpoints may change radically in a few years.

GRANITES OF FLAMANVILLE (NORMANDY), AND OF BARR-ANDLAU (VOSGES)

One of the smaller of the Hercynian granites mentioned in the last section, the Flamanville Granite near Diélette, is one of the two intrusions at the base of a historical controversy between German and French schools of petrography.

Flamanlle, on the west coast of Normandy, south of Cherbourg, is a small oval intrusion in Devonian marly shales and limestones (MARTIN, 1952, 1961; PAVILLON, 1965). It is well exposed in the sea cliffs and in numerous quarries, and shows beautiful contact aureoles. In the inner zones of these aureoles MICHEL-LÉVY (1893) described strong and varied metasomatic processes.

Barr-Andlau is another Hercynian intrusive granite, this time in the Vosges. Its contact aureole had already been described earlier by ROSENBUSCH (1877). That author did not find anything but thermal metamorphism, without a definite change in bulk composition, in the argillaceous slates of the Devonian around the granite intrusion.

This is all early history now, and it has been well realised since then that marls and limestones may suffer strong metasomatism, while argillaceous slates only become thermally metamorphosed, when intruded by granites. But at that time, and for many years to come, there existed two different schools of petrography. The German school, following ROSENBUSCH (1877), insisted that intrusive granites only provoked thermal metamorphism in the country rock. The French school, on the other hand, followed the teaching of MICHEL-LÉVY (1893) and never tired of citing the most beautiful examples of metasomatism to be found in granitic contact aureoles.

REFERENCES

BARROIS, C., 1930. Les grandes lignes de la Bretagne. *Livre Jubilaire Soc. Géol. France*, 1: 83–86.

BERTRAND, L., 1946. *Histoire Géologique du Sol Français. II.* Flammarion, Paris, 365 pp.

CAYEUX, L., 1894a. Les preuves de l'existence d'organismes dans le terrain précambrien. *Bull. Soc. Géol. France*, 22: 197–228.

CAYEUX, L., 1894b. De l'existence de nombreux débris de Spongiaires dans le Précambrien de Bretagne. *Compt. Rend.*, 120: 297–282; also: *Ann. Soc. Géol. Nord*, 23: 52–65.

CAYEUX, L., 1929. Les roches sédimentaires de France. Roches siliceuses. *Mém. Carte Géol. France*, 1929: 774 pp.

CHAUVEL, J. J. and PHILLIPPOT, A., 1961. Sur la discordance de la base du Palaeozoïque dans la région de Rennes. Trois carrières démonstratives. *Bull. Soc. Géol. Mineral. Bretagne*, 1: 1–7.

COGNÉ, J., 1959. Données nouvelles sur l'Antécambrien dans l'ouest de la France: Pentévrien et Briovérien en baie de Saint-Brieux (Côtes-du-Nord). *Bull. Soc. Géol. France*, 7 (1): 112–118.

COGNÉ, J., 1960a. Schistes cristallins et granites en Bretagne méridionale; le domaine de l'anticlinal de Cornouailles. *Mém. Carte. Géol. France*, 1960: 382 pp.

COGNÉ, J., 1960b. Métamorphismes et granitisations en liaison avec l'évolution orogénique en Bretagne méridionale. *Bull. Soc. Géol. France*, 7 (2): 213–228.

COGNÉ, J., 1963. Le Briovérien. Esquisse des caractères stratigraphiques, métamorphiques, structuraux et paléogéographiques de l'Antécambrien récent dans le Massif armoricain. *Bull. Soc. Géol. France*, 4: 413–430.

COGNÉ, J., 1967. Les grand cisaillements Hercyniens dans le Massif armoricain, et les phénomènes de granitisation. In: J. P. SCHAER (Editor), *Colloque Étages Tectoniques*. Baconniere, Neuchâtel, pp.179–192.

COGNÉ, J. and GRAINDOR, M. J., 1963. À propos de prétendues phases "normande" et "angévine" dans l'orogenèse cadomienne (socle antécambrien d'Armorique). *Compt. Rend. Soc. Géol. France*, 1963: 245–247.

DANGEARD, L., 1951. Géologie régionale de la France: La Normandie. *Actes Sci. Ind.*, 1140: 241 pp.

DANGEARD, L., DORÉ, F. and JUIGNET, P., 1961. Le Briovérien supérieur de Basse Normandie (étage de la Laize), série à turbidites à tous les caractères d'un flysch. *Rev. Géograph. Phys. Géol. Dyn., Sér, 2*, 4: 251–259.

GEORGE, T. N., 1963. Devonian and Carboniferous foundations of the Variscides in northwest Europe. In: K. COE (Editor), *Some Aspects of the Variscan Fold Belt*. Manchester Univ. Press, Manchester, pp. 19–47.

GIGNOUX, M., 1950. *Géologie Stratigraphique*, 4e ed. Masson, Paris, 735 pp.

GRAINDOR, M. J., 1957. Le Briovérien dans le nord-est du Massif armoricain. *Mém. Carte Géol. France*, 1957: 211 pp.

GRAINDOR, M. J., 1959. Granites et synclinaux paléozoiques entre Alençon et Mayenne. *Bull. Soc. Géol. France*, 7 (1): 555–566.

GRAINDOR, M. J., 1960. Feuille de Cherbourg au 50,000e. *Bull. Serv. Carte. Géol. France*, 262: 1–82.

GRAINDOR, M. J., 1961a. À propos des mouvements varisques dans le Massif armoricain, *Compt. Rend. Soc. Géol. France.*, 1961: 54–55.

GRAINDOR, M. J., 1961b. Géologie du nordouest du Cotentin. *Bull. Serv. Carte Géol. France*, 272: 81 pp.

GRAINDOR, M. J., 1962. Mouvement orogénique antédevonien en Normandie. *Bull. Soc. Linn. Normandie*, 10: 188–194.

GRAINDOR, M. J., 1963. Le socle armoricain et les contre-coups alpins. In: M. DURAND DELGA (Editor), *Livre à la Mémoire du Professeur Paul Fallot. II.* Soc. Géol. France, Paris, pp. 187–200.

GRAINDOR, M. J., 1965. Géologie de l'extrémité orientale du Massif armoricain (feuille d'Alençon au 80,000e). *Bull. Serv. Carte Géol. France*, 274: 130 pp.

KLEIN, C., 1962. La phase "normande" et la phase "angévine" de l'orogenèse cadomienne. *Compt. Rend.*, 256: 2196–2198.

KLEIN, C., 1963. À propos de la carte géologique de Cherbourg au 50,000e. *Compt. Rend. Soc. Géol. France*, 1963: 309–311.

MARTIN, N. R., 1952. The structure of the granite massif of Flamanville, N.W. France. *Quart. J. Geol. Soc. London*, 108: 311–341.

MARTIN, N. R., 1961. Nouvelle note sur le massif granitique de Flamanville. *Bull. Soc. Linn. Normandie*, 2: 293–301.

MATHIEU, G., 1937. *Recherches Géologiques sur les Terrains Paléozoïques de la Région Vendéenne*. Thesis, Univ. Lille, Lille.

MATHIEU, G., 1944. Essai sur les granites du Massif armoricain. *Rev. Sci.*, 3228: 15–28.

MATHIEU, G., 1945. Le problème du Précambrien dans l'ouest de la France. *Rev. Sci.*, 3241: 75–89.

MATHIEU, G., 1957. Les grandes lignes de la Vendée. *Bull. Serv. Carte Géol. France*, 253: 46 pp.

MICHEL-LÉVY, A., 1893. Contribution à l'étude du granite de Flamanville et des granites françaises en général. *Bull. Serv. Carte Géol. France*, 36: 41 pp.

NEUMAN, R. B. and PALMER, A. R., 1958. Critique of Eocambrian and Infracambrian. *Intern. Geol. Congr., 20th, Mexico, 1956, Rept., El Sistema Cambrico*, 1: 427–435.

PAREYN, C., 1959. Réunion extraordinaire de la Société Géologique de France: Basse Normandie. *Compt. Rend. Soc. Géol. France*, 1959: 245–273.

PAVILLON, M. J., 1965. Paléographie dévonienne et minéralisations ferrugineuses de Diélette et plombo-zincifères de Surtainville (Manche). *Bull. Soc. Géol. France*, 7 (6): 121–126.

PÉNEAU, J., 1962. Silurien supérieur et Dévonien inférieur dans le sud-est du Massif armoricain. In: *Symposium Silur–Devon Grenze*. Schweizerbart, Stuttgart, pp. 191–201.

PRUVOST, P., 1949. La Bretagne aux temps paléozoïques. *Ann. Hébert Haug (Lab. Géol. Fac. Sci. Univ. Paris)*, 7: 345–362.

PRUVOST, P., 1951. L'Infracambrien. *Bull. Soc. Belge Géol. Paléontol. Hydrol.*, 60: 43–65.

PRUVOST P., 1958. Les mers et les terres de Bretagne aux temps paléozoïqus. *Ann. Hébert Haug (Lab. Géol. Fac. Sci. Univ. Paris)*, 7: 345–362.

PRUVOST, P., 1959. Le Cambrien du Massif armoricain. *Ann. Hébert Haug (Lab. Géol. Fac. Sci. Univ. Paris)*, 9: 5–10.

PRUVOST, P. and LEMAITRE, D., 1943. Observations sur la région occidentale du bassin de Chateaulin. *Bull. Serv. Carte Géol. France*, 212: 81–94.

REFERENCES

ROBLOT, M. M., 1963. Le Briovérien supérieur (X^3) aux environs de Granville (Manche). *Bull. Soc. Géol. France*, 7 (4): 565–571.

ROSENBUSCH, H., 1877. Die Steigerschiefer und ihre Contactzone an den Graniten von Barr-Andlau und Howald. *Abhandl. Geol. Spezialkarte Elsass-Lotharingen*, 1: 79–393.

WINTERER, E. L., 1964. Late Precambrian pebbly mudstone in Normandy, France: tillite or tilloid? In: A. NAIRN (Editor), *Problems in Paleoclimatology*. Interscience, New York, N.Y., pp. 159–187.

CHAPTER 8

The Massif Central

INTRODUCTION

"MASSIF CENTRAL" AND "PLATEAU CENTRAL"

The next Hercynian unit of western Europe is the large and complex "Massif central' or Central Massif[1] in south-central France. It more or less coincides with the "Plateau central" or Central Plateau of physical geography. This upland plateau, a former peneplain now uplifted to altitudes between 800 m and 1,200 m, is one of the main topographic barriers of western Europe. Together with the Alps it rather effectively separates the climate of northern Europe from that of southern Europe.

The Massif Central (Fig.87), in the sense of a Hercynian geological unit, does not, however, coincide exactly with this Central Plateau of physical geography. It includes several areas outside the main Plio-Pleistocene uplift, which have either lagged behind, or which form separate orographical units. The most important of these are, first, the Bas Limousin in the west, north of Limoges, which lines up with the Vendée, the southeastern extension of the Armorican Massif. Then, in the northeast is the Morvan, and in the east the Monts du Lyonnais, near Lyon. In the southeast the Hercynian part of the Cevennes, and in the south the Montagne Noire. Geologically all these areas belong to the Central Massif, in spite of strong variations in structure, especially of the Montagne Noire.

The Plio-Pleistocene uplift of the Massif Central shows pronounced differential movement, which resulted in a general tilting towards the northwest. Consequently the highest elevation of the former peneplain is now found in the southeast and in the south, that is in the eastern Velay and in the Lozère. The highest mountains in the Massif Central are, however, not formed by the uplifted peneplain but by Cenozoic volcanic structures sitting on top of it.

As a consequence of this tilting of the former peneplain, the drainage of the Massif Central is predominantly towards the northwest and west, by the Loire and its large tributary the Allier, and by the Dordogne and Tarn rivers. The low-lying Rhône valley, following a graben just beyond the highest eastern rim of the Massif Central, receives only smaller tributaries from that side.

[1] To avoid confusion with the Central Massifs of the Alps, I will use the French term for the Hercynian Massif Central.

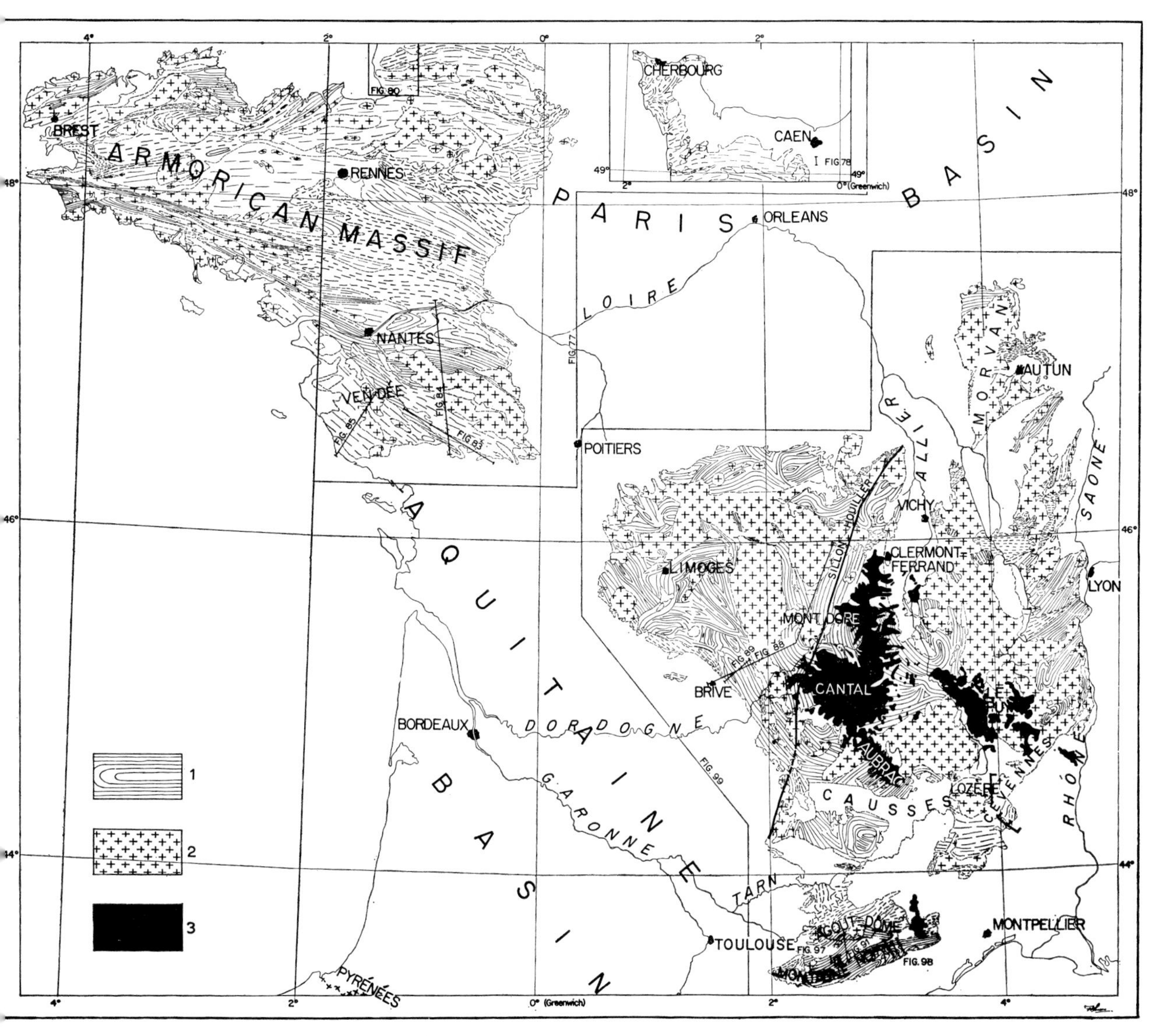

Fig.87. Schematic geological map of the Massif Central.
1 = crystalline schists and gneisses; *2* igneous rocks of the Hercynian uplands; *3* = younger volcanics.

HISTORICAL OUTLINE

The history of the Massif Central is only fragmentarily documented. Therefore it is easy to divide this history into a number of "main events", that is, into periods of which we know something. Although of course the events tath left no historical record might have been just as important. First came the pre-Hercynian, together with the Hercynian history, during which the bulk of unit was formed. Then follows a short post-Hercynian episode, i.e., the formation of the limnic coal basins. This belongs mainly to the Stephanian although it has started locally in the Westphalian, and often continues into the Autunian. This episode not only left its mark in the "Sillon houiller", but also in several other graben structures, scattered widely over the area.

All Mesozoic history was characterized by quiet continental conditions, leading to extensive peneplanation and to intensive erosion.

Block faulting, which had been active during the post-Hercynian episode, again became pronounced during the Cenozoic. It is found reflected in the early history of the big graben structures such as the Limagne Basin. Vertical movements, however, maintained low velocities (RUTTEN, 1962). Contemporaneous erosion still managed to maintain low-level peneplanation over the rising areas. The grabens did not sink very quickly either, and they retained a limnic to brackish facies all through their history.

The ultimate uplift of the Massif Central, during which the vertical movements became much more pronounced, is then usually assigned, on not too secure grounds, to the Plio-Pleistocene.

Volcanism occurred contemporaneously with the block faulting during the Cenozoic. First, the two main central volcanoes, the Cantal and the Mont Dore were formed. Later, plateau basalts followed, and finally linear series of small volcanoes appeared, such as the Chaîne des Puys, west of Clermont Ferrand. These were still active during Late Pleistocene time.

In most regional geology texts of France or of the Auvergne (BERTRAND, 1946; JUNG, 1946; ROQUES and LAPADU-HARGUES, 1958) the Cenozoic volcanism takes up a large part of the description of the Massif Central. As the younger volcanism has, however, no direct relation to the earlier history, it seems advisable to treat it separately, together with the other manifestations of younger volcanism in western Europe. So in this chapter only the earlier history of the Massif Central will be dealt with.

THE CRYSTALLINE BASEMENT

METAMORPHIC SERIES

Most of the more recent work in the Massif Central is of a petrographic nature and has to do with the crystalline basement. The school of Clermont Ferrand University, led first by J. Jung and now by M. Roques (cf. ROQUES, 1941; JUNG, 1946, 1953; JUNG and ROQUES, 1952) has taken the main initiative in this type of research. The mica schists, gneisses and granites, which make up the main body of the crystalline basement, have been interpreted by them from the viewpoint of J. J. Sederholm. They are thought to represent less and more strongly metamorphosed, and in part metasomatized, series of supracrustal rocks.

Although it is not intended to evaluate all of the voluminous, and often controversial, newer petrographic literature on the Massif Central, a survey of the main trends seems in order.

A first point to be retained, when trying to acquire an insight into the ideas of the Clermont Ferrand school, is that in these studies of metamorphism and metasomatism the main classification is always based on rocks of normal clay–sand composition only. Such rocks can be arranged in a beautiful series ranging from shale to granite. The individual steps show a sufficient number of distinctive key minerals for establishing a descriptive "zonéographie" of the successive series of metamorphic or metasomatic rocks.

In the crystalline basement of the Massif Central three main groups are distinguished within the series of normal clay–sand composition. The first group is formed by the non- or slightly metamorphosed rocks. In the second group fall the isochemically metamorphosed rocks, without apport, the so-called ectinites. In the third group fall all rocks in which metamorphism was accompanied by metasomatism, by apport, the so-called migmatites.

The difference between groups one and two is, of course, not fundamental, because the low-grade metamorphics in reality also belong to the ectinites. In mapping, however, the areas occupied by only slightly metamorphosed slates are found to be quite distinct from those occupied by the ectinite–migmatite series, a fact which justifies their separation.

To the first group belong, for instance, quartzites and slates, found in the Bas Limousin, along the western margin of the Massif Central. They form the transition to the Armorican Massif (MATHIEU, 1961), and are not considered in the study of the crystalline basement.

The difference between the second and third groups is also mainly based on field work. The supposition that the migmatites, which are quite different rocks from the ectinites, actually developed from ectinites through apport, through metasomatism, has only rarely been checked by chemical analysis. The genetic basis of this classification therefore rests on but the most slender proof, though, as a descriptive classification, it is extremely useful.

Within the second and third groups further classification is based on two different principles. The isochemically metamorphic ectinites are divided according to ascending order of metamorphism of the rocks. The metasomatic migmatites, on the other hand, according to the structure of the rocks. Or, to be more explicit, according to the gradual loss of their original structure and texture, which is thought to disappear parallel to increasing metasomatism. Such a difference in classification seems at first sight hardly logical. It is, however, eminently geological. It uses those observable features, which, in a given series of rocks, offer the best means for classification.

As we will see, there is no direct relation between the ectinite and the migmatite series. Nor is the upper limit of metasomatism related to a certain level of metamorphism. The "front of migmatization" rises to quite different levels of metamorphism, and presumably to quite different depths of the crust, not only in various parts of the Massif Central, but also in the crystalline basements of other Hercynian units of western Europe (pp.158–161). It is consequently quite permissible to use two different sets of criteria for the classification of these two separate groups of rocks, though they belong to the same basement complex. In the ectinites the degree of metamorphism is the main criterion, as there is no metasomatism. In the migmatites, on the other hand, the evolution of the rocks is hardly dependent on the degree of metamorphism, so that the degree of metasomatism becomes the main criterion.

The difference between the "zonéographie" and the mineral facies system is that in the latter all rocks metamorphosed under similar *pt* conditions, regardless of their composition, are embraced in one mineral facies. In rocks of a composition different from that of the normal clayey-sandy series, such as amphibolites, it is difficult to apply the "zonéographie".

Apart from the amphibolites, two important groups of rocks of such divergent composition are found in the Massif Central: the leptynites and the cipolins. Leptynites are acid, foliated rocks, poor in micas and amphibole. They are comparable, at least in part, with the "Granulite" of the Hercynian uplands of southeastern Germany, as described in the German literature[1]. They show petrographic resemblance to the leptites of the Precambrian of Scandinavia. The latter are thought to represent acid volcanics, but the genesis of the leptynites of the Massif Central has not yet been well established. The cipolins are lens-like bodies of dirty marble, derived from former impure limestone intercalations.

Before going into more details of the "zonéographie", a word about the layered texture of the rocks of the crystalline basement is indicated. This layered texture, which is responsible for the French expression "séries cristallophylliennes", is in general both coarser and more irregular than the schistosity in the slates, for instance, of the Rheinisches Schiefergebirge. It is more of a foliation. There are indications that this layered texture is, in part at least, a primary sedimentary feature. Both metamorphism and metasomatism seem to have followed more or less the anisotropy of the sedimentary alternation of clayey and sandy beds. It is probable that under the influence of these later processes the original layering became more strongly accentuated. But it is thought to have remained parallel to the bedding. If so, the foliation of the crystalline schists is not a tectonic feature, such as the schistosity of the less metamorphic slates, but is of earlier origin. It would be due to the vertical pressure of the sedimentary overburden. It developed parallel to the stratification, at the time when that was still horizontal.

"ZONÉOGRAPHIE"

The various zones which have been distinguished in the crystalline basement are indicated in Table XIII. From the estimated thickness of individual zones in the Massif Central it is evident that together they make up a major part of the thickness of the continental crust. From Table XIII it also follows that in many parts of the Central Massif metasomatism only developed in the deeper levels of the crust. In those areas the "front of migmatization" stayed well down during the time when metasomatism was active. It is only because of later vertical movements and subsequent deep erosion that in so many areas of the Massif Central these lower levels are now available for inspection at the earth's surface.

Within the ectinites the limits between the various zones are normally not clear cut. The parameters of pressure and temperature have of course changed gradually. So, as long as one stays in the same clay–sand facies, the intensity of metamorphism will change gradually too. Only when a change in the original sedimentary facies happens to coincide with the transition between two metamorphic zones, will this contact become more clear-cut.

A gradual transition is often found as well between ectinites and migmatites, although basically their formation is due to two quite different processes. In the literature one even encounters a special transitional category, which actually forms a contradictio interna, i.e., the "metasomatic ectinites". As we will see in the next section, such a gradual transition between ectinites and migmatites is especially found in those areas, where the "front of

[1] In French usage, confusingly, "granulite" denotes a granite or gneiss with two micas, both muscovite and biotite.

TABLE XIII

"ZONÉOGRAPHIE" OF THE CRYSTALLINE SCHISTS[1] COMPARED WITH OTHER CLASSIFICATIONS[2]

	"Zonéographie"	*Notations*	*Key rocks; structure*	*Estimated thickness (m)*	GRUBENMANN and NIGGLI (1924)	P. Eskola[3]	TURNER and VERHOOGEN (1960)
Ectinites	Non- or slightly metamorphic zone	X	shale	4,000			
	Zone of the upper micaschists	Y^2	sericite and chlorite schist	3,000	epizone	greenschist facies	muscovite–chlorite subfacies
	Zone of the lower micaschists	Y^1	micaschist with two micas	3,000	mesozone		biotite–chlorite subfacies
						albite–epidote–amphibolite facies	
	Zone of the upper gneisses	Z^2	gneiss with two micas	4,000		amphibolite facies	sillimanite–almandine subfacies
	Zone of the lower gneisses	Z^1	gneiss with biotite only	6,000	catazone		almandine–diopside–hornblende subfacies
	Zone of the lowermost gneisses[4]	U	gneiss with cordierite and no biotite				
Migmatites	Zone of the embrechites	M^2	regularly layered, amygdalous	1,000[5]			
	Zone of the anatexites	M^1	crumpled, veiny, nebulous	>3,000[5]			
	Anatexiitc granites						

[1] According to JUNG and ROQUES (1952).
[2] Compare PALM (1957).
[3] Compare, e.g., BARTH et al. (1939).
[4] "Gneiss ultra-inférieurs", not known from the Massif Central.
[5] In the Agout dome (cf. SCHUILING, 1961).

migmatization" remained in the deeper levels of the metamorphic zones. Such a gradual transition is normally taken to indicate that regional metamorphism and metasomatism are processes occurring over large areas, both horizontally and vertically, where gradients are small. It is not thought to invalidate the assumption that the formation of ectinites and migmatites is due to two different processes (for an exception to this opinion, see PALM, 1958).

The increase in volume of the rocks in the latter category is not te be disregarded; the specific density is lowered during metasomatosis. Although the actual apport is much smaller, the increase in volume has, for instance, been estimated at 50% of its present volume for the anatexites of the gneiss dome of the Agout in the Montagne Noire (SCHUILING, 1960), a figure arrived at by other authors too.

ANATEXITIC GRANITES

Migmatization will, through metasomatism, ultimately lead to granitization. Originally the transition from the crystalline schists, belonging to the migmatite zones, to the anatexitic granite will have been a gradual one, comparable to the gradual transition from one zone of the crystalline schists to the next. In such a case we will still find mantled domes of crystalline schists and gneisses, with anatexitic granite, the "granite fondamental" of the French authors, in the core.

But the variations in both volume and in specific weight that accompany migmatization tend to become even more accentuated during granitization. This means that both the expansion due to *apport* during metasomatism and the lesser density of the resultant granitized rocks will forcibly lead to tectonic mobilization of these deeper levels of the continental crust.

The result is that, whereas the embrechites and the anatexitic gneisses are normally still more or less in their original position and form simple mantled domes, the granitic material has often become much more mobile. In many, if not in most, cases it has lost its original relation to an enveloping mantled dome, and has evolved into pseudo-magmatic batholiths. The outer form of such batholiths is supposed to depend on when, relative to the orogenetic processes, they began to live a life of their own.

In principle one can distinguish between tectonically molded, or syn-kinematic granites and post-kinematic intrusions of cross-cutting character. But the differentiation between, and the classification of these granites is rather difficult, and often the result of personal interpretation. This distinction is, moreover, not made any easier by the fact that major portions of the Massif Central have suffered the vicissitudes of at least two major orogenies. A discussion of the various ideas on classification and genesis of the granites of the Massif Central can be found in VAN MOORT (1967).

THREE EXAMPLES

In order to avoid the obstruse in these matters of "zonéographie", it seems well to include a short account of three examples. These, the Tulle anticline, the so-called "anticline of the middle Dordogne", and the gneiss dome of the Agout, have been described by ROQUES (1941) in his thesis, and have formed the basis of much of the subsequent interpretations. For the Agout a newer evaluation by SCHUILING (1960) is moreover available. The anticlines of Tulle and the middle Dordogne form part of a west–east cross-section through the western part of the Massif Central, east of Brive. The Agout dome is situated in the Montagne Noire and forms the southernmost unit of the crystalline basement of the Massif Central.

A schematized cross-section through the western part of the Massif Central (Fig.88) shows the relation between the various units of crystalline schists and gneisses and of the granites. The Tulle anticline here forms the western part of the Massif Central, and belongs to the Bas Limousin, which can be correlated with the Vendée of the Armorican Massif.

It is a regularly built, northwest–southeast-trending structure, which is reasonably complete, though cut off by a major fault, the Argentat fault, along its eastern flank. The anticline of the middle Dordogne, separated from the Tulle anticline by the Millevaches syn-kinematic granite (a *granulite*, according to French nomenclature), is more strongly tectonized, and less complete. It now consists only of the western, overturned flank of a

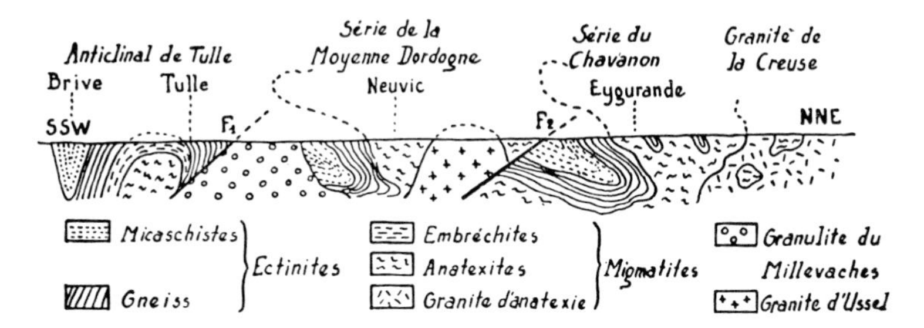

Fig.88. Schematized section through the western part of the Massif Central. Showing the relation between main units of crystalline schists and gneisses, with anatexitic, syn-kinematic, and intrusive granites. (After JUNG, 1946.)

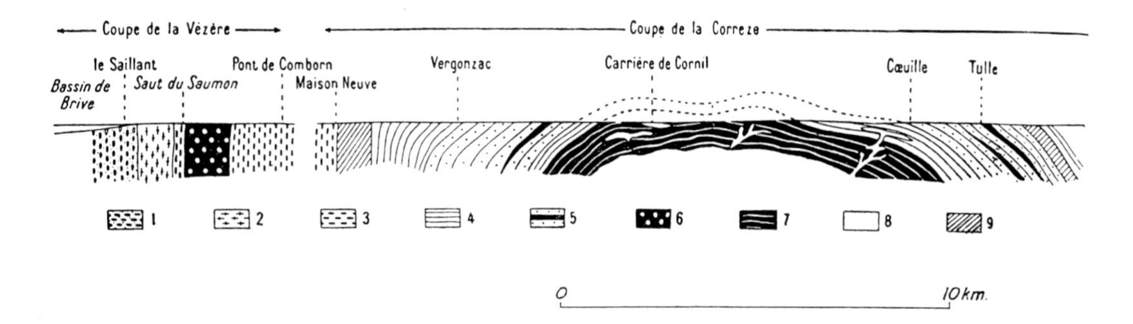

Fig.89. Composite section through the Tulle anticline, taken along the Vézère and Corrèze rivers. Non metamorphic zone: *1* = Allassac slates. Zone of the upper micaschists: *2* = sericite schists of Donzenac. Zone of the lower micaschists: *3* = micaschists with biotite. Zone of the upper gneisses: *4* = two-mica gneiss; *5* = leptynites with intercalated two-mica gneisses. Zone of the migmatites: *7* = embrechites. Plutonic rocks: *6* = syn-kinematic granite; *8* = intrusive granite of Cornil; *9* = quartz diorite. (After JUNG and ROQUES, 1952.)

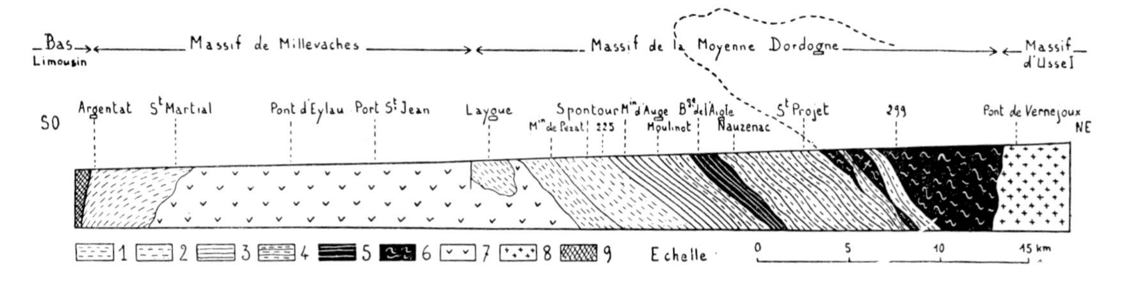

Fig.90. Section through the Millevaches Granite and the anticline of the middle Dordogne. The Argentat fault cuts off the eastern flank of the Tulle anticline (Fig.89). The Millevaches pluton is a syn-kinematic granite.

Zone of the lower micaschists: *1* = micaschists with biotite; *2* = sericite gneiss. Zone of the upper gneisses: *3* = two-mica gneiss. Zone of the lower gneisses: *4* = gneiss with biotite and sillimanite. Migmatites: *5* = augen embrechites; *6* = anatexites with cordierite. Plutonic rocks: *7* = syn-kinematic granite; *8* = intrusive granite. Mylonite: *9* = mylonitic zone of the Argentat fault. (After ROQUES, 1941.)

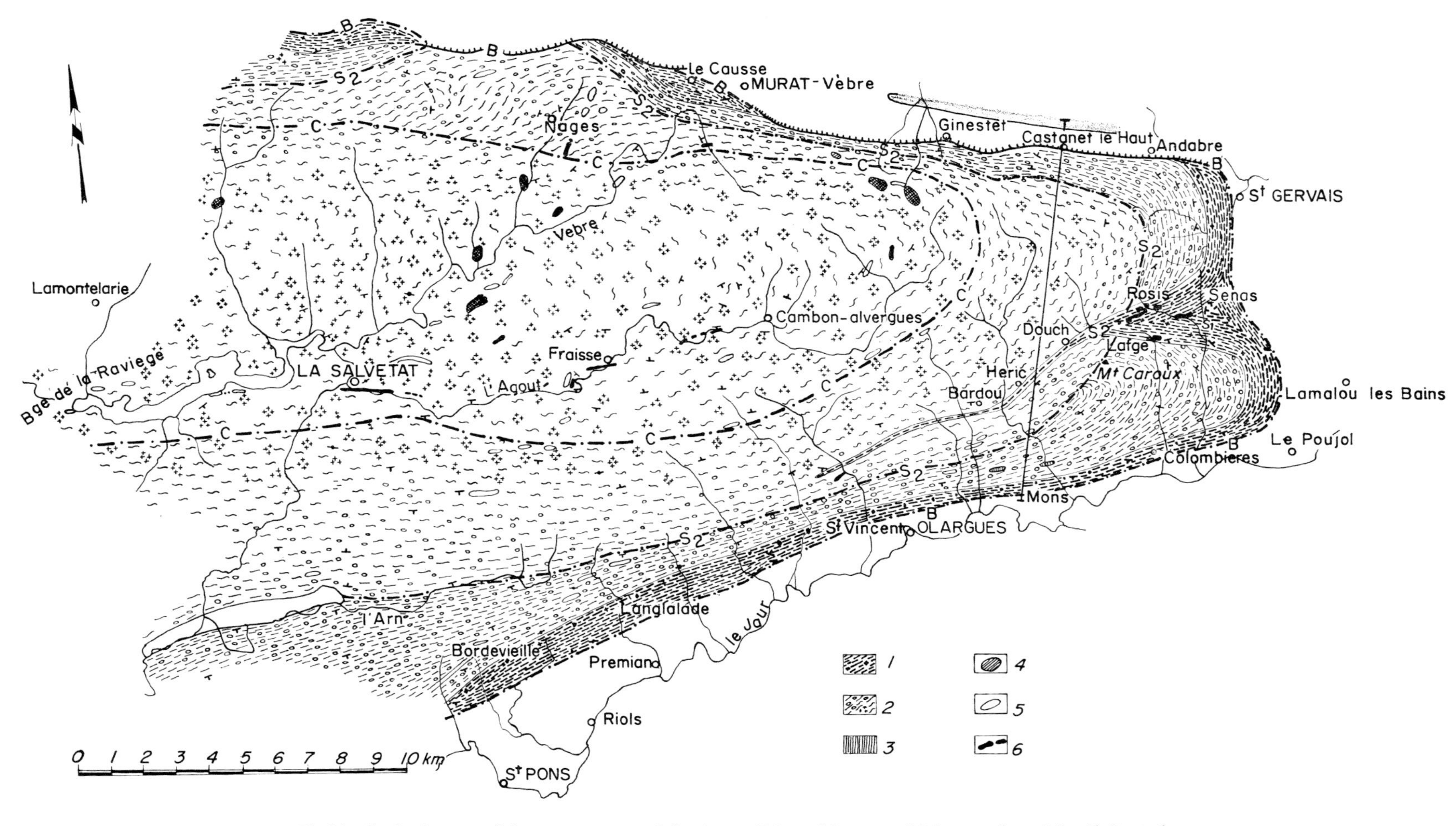

Fig.91. Geologic map of the eastern nose of the Agout Dôme, Montagne Noire, southern Massif Central. White: non metamorphic rocks and zone of the upper micaschists. *1* = Zone of the lower micaschists with andalusite; *2* = migmatites ("Augen Gneis", anatexitic gneiss and anatexitic granite); *3* = muscovite granite ("granulite"); *4* = intrusive granite with andalusite; *5* = ortho-amphibolite; *6* = cipolins. B, S_1, S_2, C = isogrades of biotite, sillimanite 1 and 2, and cordierite (B = upper limit of biotite; S_1 = upper limit of transition of andalusite in sillimanite; S_2 = lower limit of muscovite from which sillimanite and K feldspar form; C = upper limit of cordierite). (After SCHUILING, 1961.)

large anticline. The more easterly, even larger structure of Eygurande, which is also formed by an overturned anticline, is more complete but less well exposed than the middle Dordogne anticline. The latter can be studied in a cross-section along the deeply incised Dordogne valley and has therefore always been used for the elucidation of the "zonéographie".

This schematic section clearly shows the highly fragmentary preservation of the original series of crystalline rocks. It is therefore not surprising that controversies exist in the literature. These refer, however, mainly to the correlation and to the interpretation of the relative histories of the various units. In contrast, the "zonéographie" within each unit, say of the Tulle anticline or of the overturned flank of the middle Dordogne anticline, is rather straightforward. Within each unit the general views on the "zonéographie" seem to be pretty well accepted by everybody.

The composite section of the Tulle anticline of Fig.89 illustrates well the extreme regularity which characterizes the undisturbed mantled domes of the crystalline basement. The "front of migmatization" is here situated at the base of the zone of the upper gneisses, but the transition from ectinites to migmatites is obscured by the intercalation of a series of leptynites. Of the migmatites only the upper part, the zone of the embrechites, is exposed.

The section through the anticline of the middle Dordogne (Fig.90), although in a tectonically overturned position, shows a similar regularity in the development of a series of crystalline schists and gneisses. The series exposed only begins in the zone of the lower micaschists, because all higher elements have either been removed by erosion or cut off by the Millevaches Granite in the west. In the lower horizons it is, however, more complete than the Tulle anticline. For one thing, the "front of migmatization" has in this series only reached the zone of the lower gneisses, so that a more complete section of the ectinites is found. On the other hand the migmatites are represented not only by lenses of embrechites intercalated in the zone of lower gneisses, but also by a major body of anatexites.

Lastly, the Agout dome is, on closer inspection, not so simply built, as was thought earlier. The rocks of the metamorphic series, although they have not lost their overall coherence, have become rather mobile, and have undergone extensive bedding plane slip towards the summit of the structure. Moreover, they have formed two different culminations, separated by a narrow thrust zone, the so-called Rosis syncline (Fig.91, 92).

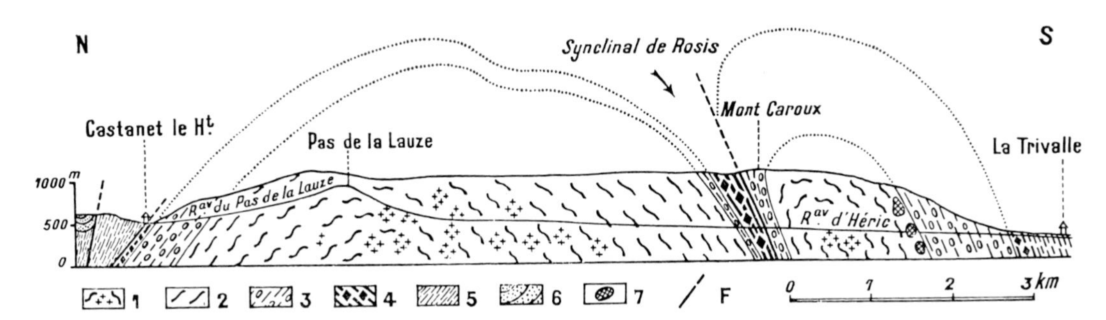

Fig.92. Section through the eastern nose of the Agout Dôme, Montagne Noire, southern Massif Central. *1* = anatexitic gneiss and granite; *2* = anatexitic gneiss with sillimanite; *3* = "Augen Gneis"; *4* = lower micaschists with andalusite; *5* = upper micaschists; *6* = Stephanian; *7* = ortho-amphibolite. (After SCHUILING, 1961.)

VARIATIONS IN THE RELATION BETWEEN METAMORPHISM AND METASOMATISM

As has been stated earlier, and as can be seen in the three examples cited, there is a rather wide, regional variation in the respective relation between metamorphism and metasomatism. In a schematic way this is still well covered by the summary of JUNG and ROQUES (1952), which will be used in this paragraph. They did, however, not yet take into account the possible complications due to recrystallization during two or more orogenetic cycles. This latter topic has been the base of much of the more recent work of the Clermont Ferrand school, and will be reviewed in the next paragraph.

Fig.93 and 94 indicate the scope of the variation in the relation between metamorphism and metasomatism in the Massif Central. In one extreme metasomatism may be limited to the very deepest part of the pile of sediments undergoing regional matamorphism. It consequently does not reach higher than the zone of the lower gneisses. This situation is generally found in the Morvan, in the northernmost part of the Massif Central, and also in the Vosges Mountains. At the other extreme the "front of migmatization" may rise to the very top of the ectinites, and even reach into the zone of the upper micaschists. As shown in Fig.94, this situation is found in the extreme southern and southeastern parts of the Massif Central, that is in the Montagne Noire and in the Cevennes.

PALM (1957) supposed that regional variations in the geothermal gradient at the time of metamorphism have led to variations in the level reached by the Na mobilization, which is held responsible for the metasomatism. Such a simple explanation of the regional variation in the thickness of the ectinite zones and in the level reached by metasomatism at least reduces a little the mystic quality of the "front supérieur migmatitique" in the French literature.

The transition from ectinites to migmatites is found to be very gradual in types *V* and *IV*, where it is situated in the zone of the gneisses. It becomes sharper when migmatization rises higher and develops in less strongly metamorphosed rocks. It may be quite sudden in type *I*, where a direct contact may be found from one stratum to the next, between upper micaschists and embrechites. In the latter situation an unexpected complication may arise out of the fact that the temperature in the embrechites will have been quite different from that in the overlying micaschists, while also, as a result of metasomatic

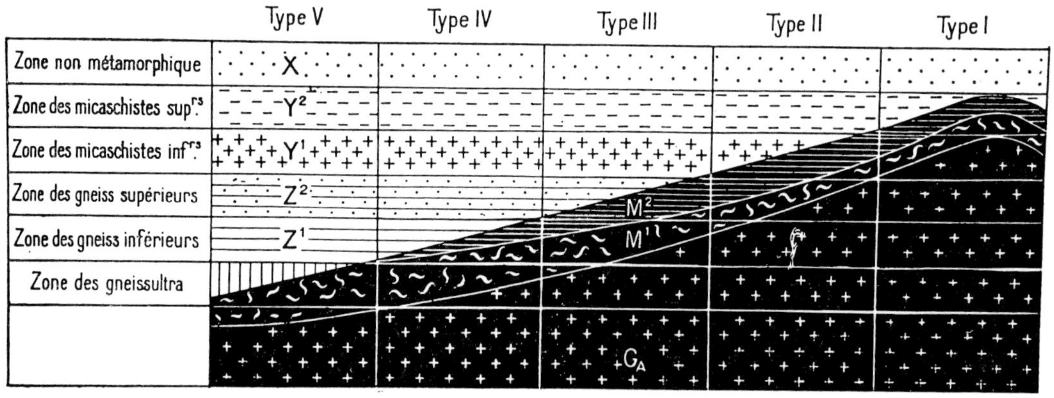

Fig.93. Schematic representation of the different levels within the ectinites, reached by the "front of migmatization" in the Massif Central.
M^2 = embrechites; M^1 = anatexites; G_A = anatexitic granite. For geographical extension of the type *I–V* in the Massif Central, cf. Fig.94. (After JUNG and ROQUES, 1952.)

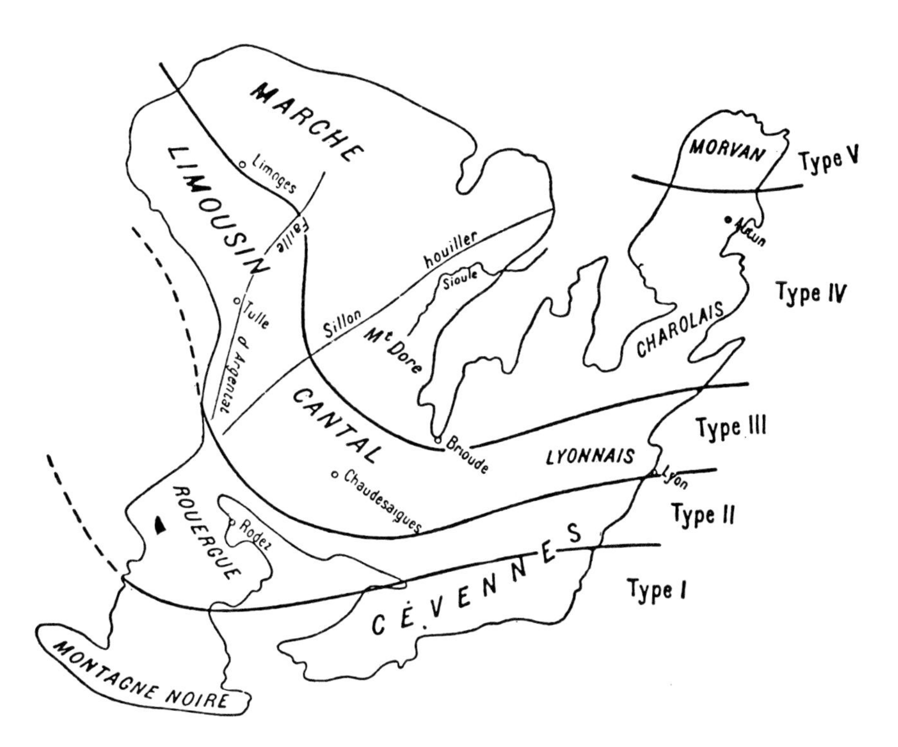

Fig.94. Repartition of the zones in the Massif Central in which metasomatism reached certain levels of the metamorphic series. For levels reached by the "front of migmatization", cf. Fig.93. (After JUNG and ROQUES, 1962.)

apport, the chemical composition of the embrechites was different from the micaschists. Consequently, the latter may show an aureole of contact metamorphism around the migmatites.

JUNG (1953) has further enlarged this notion of geographical variation in the relation between metamorphism and metasomatism. He points out that areas that show different relations of metamorphism and metasomatism of the crystalline basement also show systematic differences in their later, post-metamorphic history during the Paleozoic.

On the one hand, the northern part of the Massif Central, and also the Vosges Mountains, with their thick ectinite zones, have clastic series of the Devonian and of the Lower Carboniferous in Culm facies nonconformably covering the crystalline basement. There is an apparent hiatus, embracing Silurian, Ordovician and Cambrian. On the other hand, in the southern part of the Massif Central, in the Montagne Noire, a thick series of Lower Paleozoic is developed, comprising some 4 km of non-metamorphic Cambrian to Lower Carboniferous overlying a thick series of metamorphic sediments. A similar situation is found in the Pyrenees. Accordingly, as far as concerns its Early and Middle Paleozoic history, the Montagne Noire is much more similar to the Pyrenees than to the more northerly parts of the Massif Central.

JUNG (1953) consequently distinguishes two types of Hercynian units in western Europe, the "type auvergnat", or Auvergne type, and the "type pyrénéen" (Fig.95).

According to Jung, the same distinction can be extended to other parts of the Massif Central, and for other hercynian units of western Europe. The Bas Limousin and the Armorican Massif would belong also to the Pyrenean type (Table XIV), while the reparti-

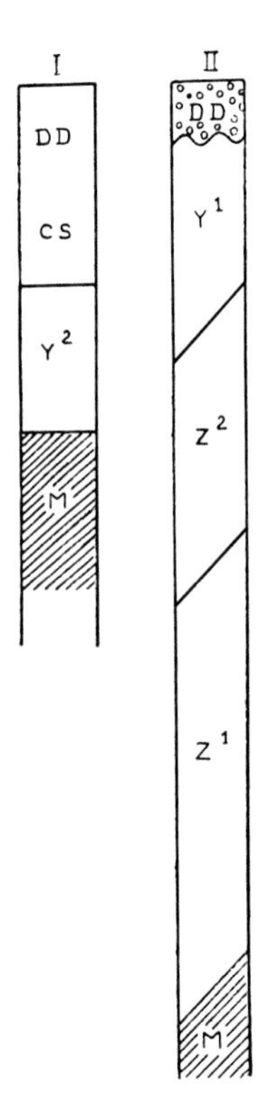

Fig.95. Schematic stratigraphical column of the two types of Hercynian uplands in western Europe (I = Pyrenean type, II = Auvergne and Vosges type).
DD = Lower Carboniferous in culm facies and Devonian; *CS* = non metamorphic Silurian, Ordovician and Cambrian; Y^2 = zone of the upper micaschists; Y^1 = zone of the lower micaschists; Z^2 = zone of the upper gneisses Z^1 = zone of the lower gneisses; *M* = migmatites. (After JUNG, 1953.)

TABLE XIV

RELATION BETWEEN METAMORPHISM AND METASOMATISM AND THE STRATIGRAPHY OF THE LOWER PALEOZOIC IN SEVERAL HERCYNIAN UPLANDS OF WESTERN EUROPE

(After JUNG, 1953)

	Pyrenees	*Armorican Massif*	*Central Massif*		
			southern (Montagne Noire)	*western (Limousin)*	*central and northern (Auvergne)*
Lower Carboniferous	conformable cover	non conformable	conform-able cover	conform-able cover	non conformable
Devonian		non conformable			non conformable
Silurian–Cambrian		non conformable Brioverian			hiatus (?)
Infracambrian	upper micaschists migmatites	upper micaschists lower micaschists migmatites	upper micaschists lower micaschists migmatites	upper micaschists lower micaschists upper gneisses migmatites	eroded lower micaschists upper gneisses lower gneisses migmatites

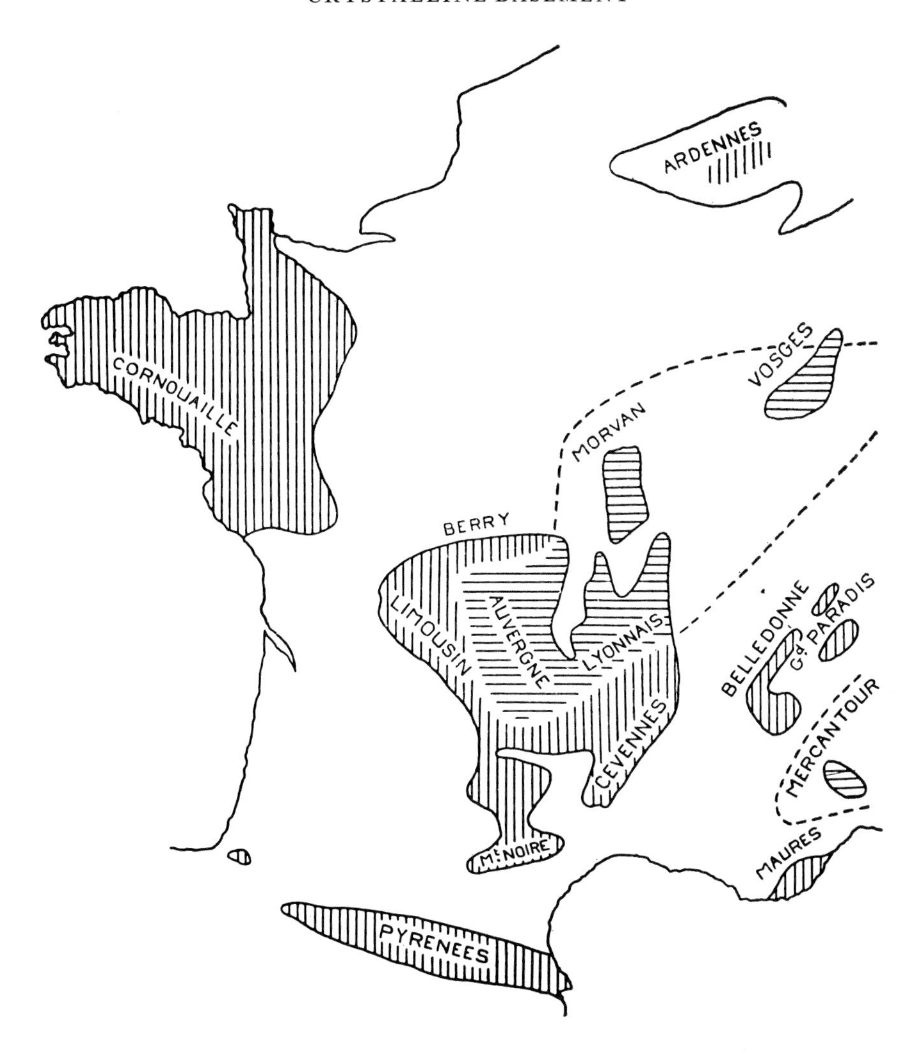

Fig.96. Distribution of the Pyrenean type (vertical shading) and the Auvergne type (horizontal shading) of Hercynian uplands in western Europe. (After JUNG, 1953.)

tion of the two types in the Western Alps is indicated in the simple paleogeographic sketch of Hercynian Europe of Fig.96.

This is too simplified a picture. As regards the crystalline basement, for instance, no account has been taken of the gradual geographical variation in the relation between metamorphism and metasomatism, as expressed in Fig.95, 96. And as regards the later history, during the Early Paleozoic there are quite marked differences between the Pyrenees and the Montagne Noire on one side, and the Bas Limousin and the Armorican Massif (= *Cornouailles* in Fig.96) on the other. Nevertheless, even if things will eventually turn out to be more complicated, Table XIV and Fig.95, 96 confront us with a schematic classification of the Hercynian uplands of western Europe, well worth considering.

NEWER RESULTS

Returning now to the results of more recent research within the confines of the Massif Central, we find that the situation has revealed itself to be much more complicated than supposed from the earlier reconnaissance of Jung and Roques. The newer ideas stem mainly from three test areas, which formed the base, not only for a much more detailed geological analysis, but also for a much more thorough application of the mineral facies concept. These test areas are situated in the northwestern (CHENEVOY, 1958, 1965), the central (Haut-Allier, cf. FORESTIER, 1956) and the eastern (Monts du Lyonnais, cf. PETERLONGO, 1960) parts of the Massif Central.

Apart from the general occurrence of retromorphosis, which poses questions of a technical petrographic character too special to go into here, the main discovery seems to be that of a double sequence of metamorphism and metasomatism; that is, of the influence of two distinct major orogenetic cycles.

In Jung's Auvergne type of Hercynian upland we find the evidence for a Precambrian orogeny, with an old crustal series metamorphosed at an early date, and eroded during the Early Paleozoic. In the Pyrenean type of Hercynian upland, on the other hand, the metamorphosis of the crystalline basement might be younger. Because it forms a conformable sequence with its cover of Early Paleozoic age, it might well have been laid down only in Late Precambrian times. Actual metamorphism of the sedimentary series forming this basement might then have occurred as late as during the Hercynian orogeny.

Now in every one of the three areas of the Massif Central mentioned above, areas that are situated half way between the Auvergne and the Montagne Noire, more detailed mapping and petrographic analysis is thought to have shown the existence of two separate periods of recrystallization. Moreover, these periods of recrystallization show different effects, i.e., have a different style of metamorphism and of metasomatism. According to the modern authors the earlier period has resulted in a type of metamorphism comparable to that found in the Auvergne. A thick series of ectinites developed, and metasomatism did not reach beyond the zone of the lower gneisses. Moreover both the upper limit of the migmatites and also their stratification are conformable with the structure found in the overlying ectinite series. During the younger period of recrystallization, on the other hand, a prevalent influence of metasomatic processes occurred. The younger migmatites are described as stratiform, which seems to be their main characteristic. They occur in distinct beds or zones, which, moreover, cut obliquely through the structures established during the earlier period of recrystallization.

It is tempting to correlate the earlier, Precambrian period of metamorphism with the Cadomian orogeny of the Armorican Massif. The resulting picture then becomes very similar to the situation in the anticlinal areas of the Armorican Massif, particularly so with the southerly, the Cornouailles anticline, as described by Cogné (p. 144). A recent paper by BOYER et al. (1964), on the correlation of the Infracambrian and the Early Paleozoic of the Armorican Massif, the western part of the Massif Central, and the Pyrenees, points in the same direction. It is quite tempting too, to assign the second period of metamorphism to the Hercynian orogeny.

But it is quite possible that this younger period of metamorphism is much older. It might, for instance be related to the Caledonian orogeny, found further north. Or it might be still older, and also belong to Precambrian times. In fact PETERLONGO (1960) claims to have found indications for a Precambrian age of both periods of metamorphism in the Monts des Lyonnais.

There is, of course, no proof whatsoever that the earlier metamorphism should at all be contemporaneous with the Cadomian orogeny in the Armorican Massif. It even remains possible that the earlier period of metamorphism of the Auvergne is not contemporaneous with that found in the other units which now make up the aggregate of the Massif Central. It must be remembered that the delightful simplicity of correlations such as expressed in Table XIV, in the Precambrian rests more often than not on lack of data. As long as absolute dating does not regularly accompany the field work, correlations of the Precambrian normally are really no better than a kind of wishful thinking.

THE MONTAGNE NOIRE

As has been stated before, the southernmost part of the Massif Central, the Montagne Noire, differs from the main part by the presence of non-metamorphic rocks of Early Paleozoic age. It consists of a core, where the crystalline basement is exposed, the Agout dome, bordered both to the north and the south by sediments. An important aspect for Early Paleozoic paleogeography is that their facies closely resembles that of the Pyrenees (pp.349–350).

Two questions of major interest have cropped up in relation to this Early Paleozoic series. The first is, whether there have been found indications for orogenetic movements of Caledonian age. The second is that of the direction from which originated the nappes developed during the Hercynian orogeny.

According to Gèze (1949), the answer to the first question is yes. It is more appropriate, however, to speak of smaller crustal movements of epeirogenetic character, occurring during the Early Paleozoic, than of a major orogeny comparable to the Caledonian movements further north. As to the second question, Gèze has postulated a southern origin for the Hercynian nappes, from a root now covered by the younger sediments of the Aquitaine Basin. Both A. Demay, and L. U. De Sitter and R. Trümpy (in: Gèze et al., 1952), on the other hand, postulate a northerly origin of these nappes.

CALEDONIAN OROGENY?

Gèze (1948, 1949, 1960) is of the opinion that an orogenetic period has occurred during the Early Paleozoic. Although these movements took place a little early, that is, between Lower and Upper Ordovician, or, to be more exact, between Arenigian and Caradocian, they are correlated with the Caledonian cycle.

The main argument is stratigraphical and based on the absence of the Llandeilian. Gèze (1960) also produced a schematic tectonic section through part of the Montagne Noire, as proof for Caledonian folds. In this section the Cambrian is shown as folded in earlier, Jura Mountain type structures, underneath the unfolded and transgressive Caradocian and higher series (Fig.97, lower section).

This would seem to be quite convincing proof for orogenetic movements, though they are a bit early to be assigned to the Caledonian orogeny proper. However, the time available for these supposed orogenetic movements and the subsequent erosion and peneplanation is very short. Only the Llandeilian is lacking between the supposedly folded Arenigian and nonconformable Caradocian. If this leads to some doubts already as to the validity of

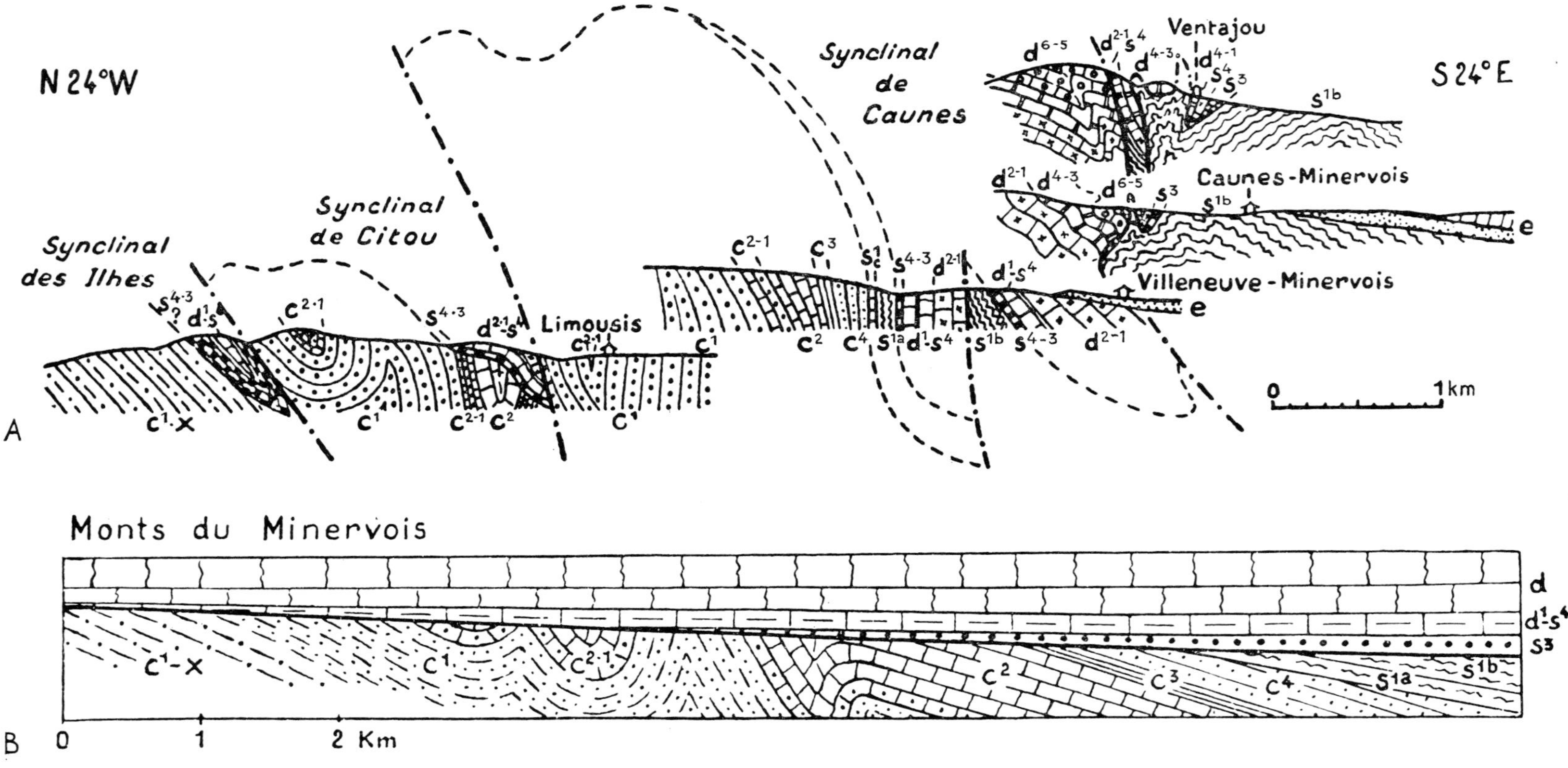

Fig.97. Evidence for Caledonian folding in the Montagne Noire, southern Massif Central, according to Gèze. A. Actual structure. B. Schematic section, with Hercynian structures unfolded.
c^1x = Infracambrian; c^1–c^2 = Cambrian; s^{1a} = Tremadocian; s^{1b-1d} = Arenigian; s^3 = Caradocian and Ashgillian; s^4 = Gotlandian; d = Devonian.
The present Hercynian tectonization of Ordovician and Gotlandian, apparent from the upper sections, throws doubt on the construction of a nonconformity between Lower and Upper Ordovician, as indicated in the lower section. (A: after GÈZE, 1949; B: after GÈZE, 1960.)

Gèze's statements, we might submit there is a flaw in the other part of the argument as well, i.e., in the unfolding of the Hercynian structures. The Hercynian orogeny was rather violent in the Montagne Noire, as shown by the nappe structures. Moreover the shales of the Upper Ordovician, and particularly the graphitic shales of the Gotlandian, form excellent lubricating zones. In the better exposed Pyrenees, these shales are always strongly tectonized and form the boundary between two "Stockwerke" of quite a different tectonic style.

If, with this in mind, we now compare the actual configuration of the Early Paleozoic strata in the Montagne Noire, as figured in the upper part of Fig.97, with the hypothetical pre-Hercynian structure below, this evidence for early Caledonian folding is indeed rather slender. In the Montagne Noire also, it is the Ordovician–Silurian that constitutes the most strongly tectonized areas. This makes it all but impossible to arrive at a proper evaluation of their relative original position.

In the Montagne Noire we therefore find a situation comparable to that in the Armorican Massif. Differential vertical crustal movements of small size, superimposed on the general geosynclinal subsidence, seem to have been active, as is testified, not only by the absence of the Llandeilian, but also by the rather incomplete development of other zones of the Early Paleozoic. But a formation of an actual fold belt at that time, with subsequent peneplanation leading to an unconformable position of Caradocian on Arenigian remains doubtful.

Of course this does not mean that there was no continental crust in the area where the Montagne Noire is now, during the Late Precambrian and Early Paleozoic. A thick series of sialic rocks was laid down in geosynclinal conditions at that time. Absence of true orogenetic movements does not mean absence of sialic rocks. Nor can it be construed in favor of Stile's idea of continental accretion (pp.14–16).

HERCYNIAN NAPPE STRUCTURES

There seems to be no doubt about the existence of Hercynian nappe structures in the Montagne Noire. The fact that older series rest on younger over wide areas seems to be beyond dispute. Nor is there any uncertainty about the age of this tectonic revolution, as it can be dated as post-Dinantian and pre-Stephanian. It consequently took place during the Namurian and possibly the Westphalian, and belongs to the Sudetic phase of the Hercynian orogeny. The Montagne Noire therefore is one Hercynian unit in western Europe, where Hercynian nappe structures of major dimensions seem to have been attested without doubt[1].

There is, however, a controversy about the direction from which these nappes originated. Originally A. Demay postulated a northerly, B. Gèze a southerly provenance (Demay and Gèze, 1950). At that date the actual facts were still under debate. But at a later stage L. U. De Sitter and R. Trümpy, basing themselves on the data collected during the painstaking field work of Gèze, also arrived at the "hypothesis north" (Gèze et al., 1952). Consequently, Fig.98 here reproduces the structure of the Hercynian nappes in

[1] A similar instance can be found in the Cantabrian Mountains of northwestern Spain, not treated in this book. (See Compte, 1959; Rupke, 1965; De Sitter, 1957, 1962, 1965; Van Veen, 1965.)

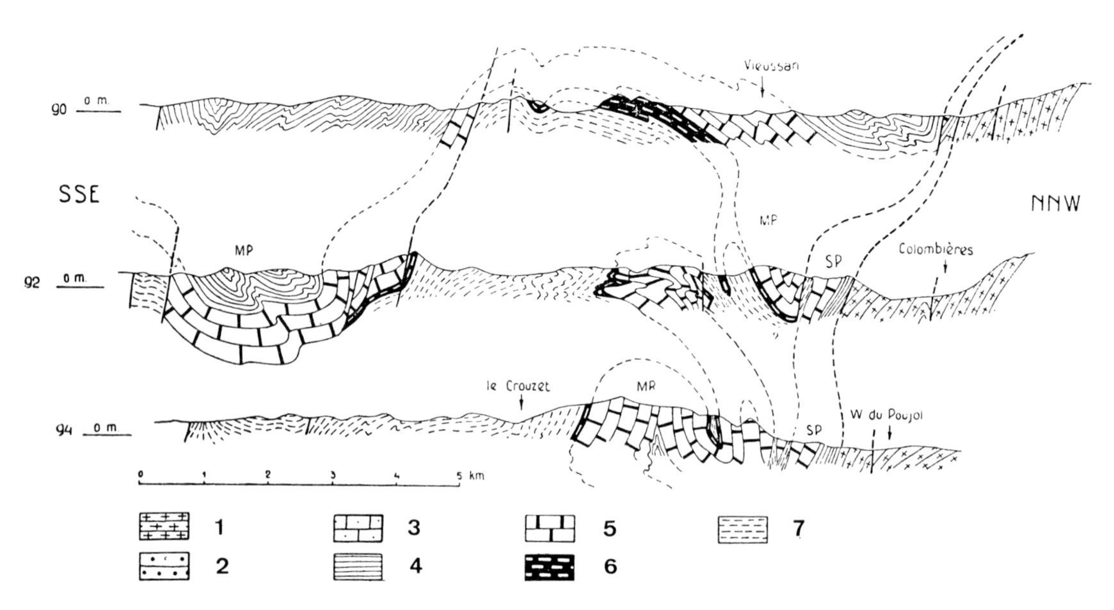

Fig.98. Three sections through the southern sedimentary mantle of the Montagne Noire, southern Massif Central, showing the nappe structure.
1 = Precambrian; *2, 3* = Cambrian; *4* = Ordovician–Silurian; *5* = Devonian; *6* = lydite shales and limestones, lowermost Carboniferous; *7* = Lower Carboniferous in Kulm facies; *MP* = Mt. Peyroux nappe; *SP* = autochthonous mantle of the Agout Dome. (After R. De Sitter and L. U. Trümpy, in: Gèze et al., 1952.)

part of the Montagne Noire, according to the hypothesis of a northern provenance. This seems to be more reasonable than a southern provenance also, because in this case their formation can be explained by gravity tectonics, through "décollement" from the rising mass of the Massif Central.

A group of geologists of Montpellier have recently made microtectonic investigations in the Montagne Noire. They believe their results prove a northerly provenance of the nappes Mattauer, et al., 1967).

THE "SILLON HOUILLER"

Before closing the description of the Massif Central, and with it the description of some selected examples of the Hercynian uplands of western Europe, we must add a short note on the "*Sillon houiller*" or Carboniferous furrow.

The Sillon houiller has two main aspects. On the one hand it is a long, straight, narrow, subvertical fault zone, trending about N 10°E. It separates the western part of the Massif Central, where a northwest–southeast strike prevails, from the eastern part with predominant southwest–northeast strike. That is, it forms the bisectrix in the famous large V, exhibited by the directions of the Hercynian structures in France. As such it is a major tectonic feature of western Europe. On the other hand, along its strike, several disconnected, small, individual coal basins have developed. This situation has been schematically indicated in Fig.99.

The "Sillon houiller" is unique in this combination of a major fault zone studded with localised coal basins. There are other major fault zones in the Massif Central, such as

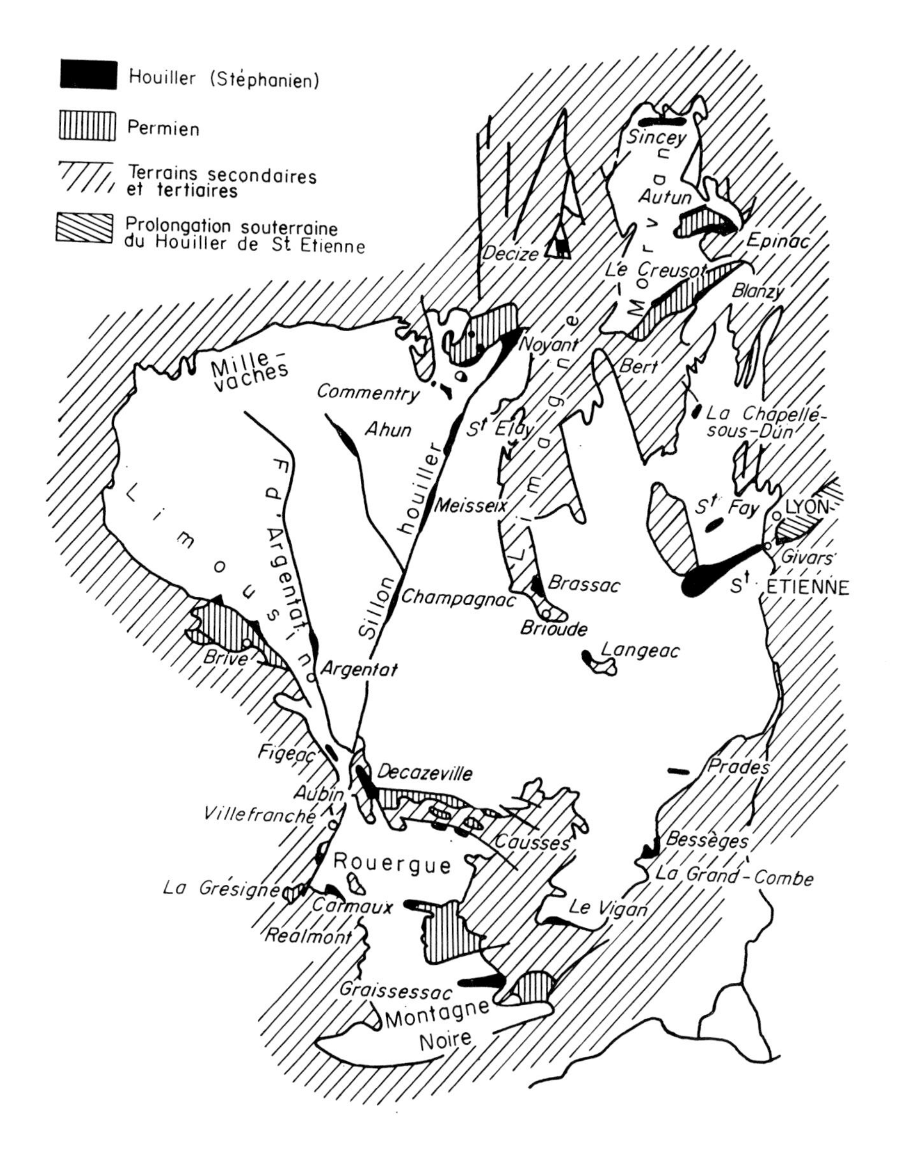

Fig.99. Schematic map of the Massif Central with the *Sillon houiller* and the small post-orogenetic coal basins o the Stephanian and/or the Autunian. (After BERTRAND, 1946.)

the Argentat fault (p. 155). The "Sillon houiller" fault zone is probably much more important than those other major faults, but this question is still open to debate. None of these other fault zones is accompanied by a series of coal basins. Small, individual coal basins, on the other hand, are quite common in the Massif Central, as follows from Fig.99. But all of these, scattered over the Massif Central, are formed by individual graben structures, bordered by ordinary, normal faults.

As stated already in Chapter 4, the "Sillon houiller" has all the earmarks of a major transcurrent fault[1]. But if this is so, this aspect belongs entirely to its earlier history. It

[1] The Villefort fault in the southeastern Massif Central has recently been interpreted as a transcurrent fault (compare VAN MOORT, 1967). Although it is a much smaller feature, with a horizontal displacement of less than 10 km, this at least proves the existence of this type of faulting in the Massif Central.

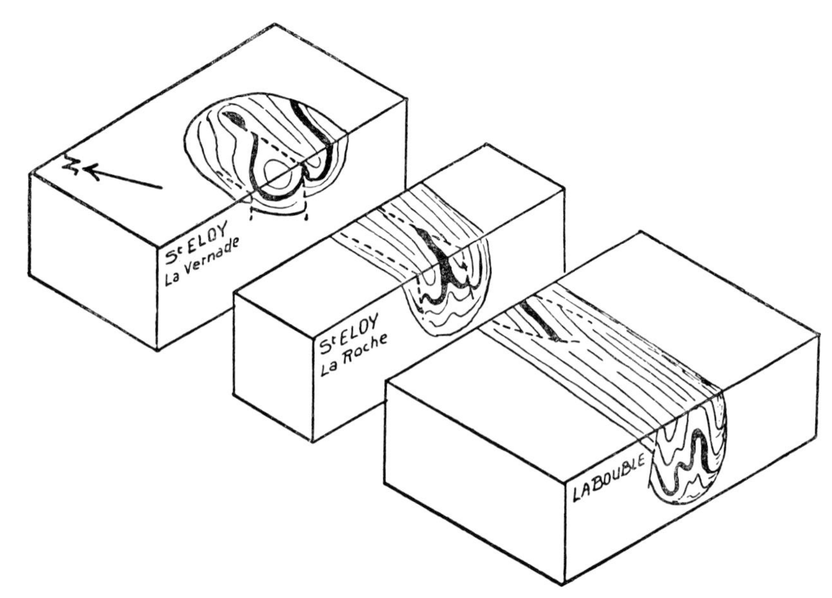

Fig.100. Block diagram of the St. Eloy Coal Basin in the northern part of the *Sillon houiller*, Massif Central. (After JUNG, 1946.)

might have acted as a transcurrent fault during the Hercynian orogeny. Or, in view of the possible older age of structures now exposed in the Massif Central, it might have only acted as such at a more remote period. Since the Hercynian orogeny, however, (and this already holds good for the period directly following the Hercynian orogeny, that is for the post-Hercynian episode of coal formation during the Upper Carboniferous) the "Sillon houiller" has acted as a normal graben structure only.

As is the case with so many graben structures, it has become compressed during some periods of its later history, which has resulted in a rather weird — but always local — tectonic deformation of the Stephanian. In a similar way as in the other little coal basins developed during the Stephanian on the Massif Central, each individual basin along the "Sillon houiller" has its own history. Not only its stratigraphy, and its coal formation, both resulting from variations in basin subsidence, but also the later tectonic deformation, are different in each little individual graben. As an example of such a small coal basin situated on the "Sillon houiller", a block diagram of the St. Eloy Basin is given in Fig.100.

REFERENCES

BARTH, T. F. W., CORRENS, C. W. and ESKOLA, P., 1939. *Die Entstehung der Gesteine*. Springer, Berlin, 422 pp

BERTRAND, L., 1946. *Histoire Géologique du Sol Français. II.* Flammarion, Paris, 365 pp.

BOYER, F., COLLOMB, P. and OVTRACHT, A., 1964. Essai de corrélation stratigraphique dans l'Antécambrien et le Paléozoïque inférieur du sud-ouest et de l'est de la France. *Intern. Geol. Congr., 22nd, New Delhi, 1964, Rept.*, pp. 1–9.

CHENEVOY, M., 1958. Contribution à l'étude des schistes cristallins dans la partie nord-ouest du Massif Central français. *Mém. Carte Géol. France*, 1958: 428 pp.

CHENEVOY, M., 1965. Précisions nouvelles sur les terrains métamorphiques du Mont Pillat (Massif Central) et leur histoire cristallogénétique. *Bull. Soc. Géol. France*, 7 (6): 55–63.

COMPTE, P., 1959. Recherches sur les terrains anciens de la Cordillère Cantabrique. *Mem. Inst. Géol. Minero Espana*, 60: 1–440.

REFERENCES

Demay, A., 1946. *Carte Géologique Schématique des Terrains Précambriens et Paléozoïques du Massif Central, 1:1,000,000 (Avec Notice Explicative).* Imprimerie Nationale, Paris, 10 pp. + map.

Demay, A., 1948. Tectonique antéstéphanienne du Massif Central. *Mém. Carte Géol. France*, 1948: 259 pp.

Demay, A. and Gèze, B., 1950. Réunion extraordinaire dans les Cévennes méridionales et la Montagne Noire. *Compt. Rend. Soc. Géol. France*, 1950: 305–371.

De Sitter, L. U., 1957. The structural history of the southeast corner of the Paleozoic core of the Asturian Moun-

De Sitter, L. U., 1962. The structure of the southern slope of the Cantabrian Mountains: explanation of a geological map with sections, scale 1:100.000. *Leidse Geol. Mededel.*, 26: 255–264.

De Sitter, L. U., 1965. Hercynian and Alpine orogenies in northern Spain. *Geol. Mijnbouw*, 44 (11): 373–383.
tains. *Neues Jahrb. Geol. Paläontol., Abhandl.*, 105: 272–284.

Forestier, F. H., 1956. Unités cristallophylliennes dans le Bassin du Haut-Allier. *Publ. Fac. Sci. Univ. Clermont Ferrand*, fasc. 2: 57–71.

Forestier, F. H., 1963. Métamorphisme hercynien et antéhercynien dans le Bassin du Haut Allier (Massif Central français). *Bull. Serv. Carte Géol. France*, 271: 294 pp.

Gèze, B., 1949a. *Carte Géologique de la Montagne Noire et des Cévennes Méridionales au 1/200.000e (Avec Notice Explicative).* Soc. Géol. France, Paris, 48 pp.

Gèze, B., 1949b. Étude géologique de la Montagne Noire et des Cévennes méridionales. *Mém. Soc. Géol. France*, 29: 215 pp.

Gèze, B., 1956. Les terrains cambriens et antécambriens dans le sud du Massif Central français (Montagne Noire et Cévennes méridionales). *Intern. Geol. Congr., 20th, Mexico, 1956, Rept., El Sistema Cambrico*, 1: 185–234.

Gèze, B., 1960. L'Orogenèse calédonienne dans la Montagne Noire (sud du Massif Central français) et les régions voisines. *Intern. Geol. Congr., 21st, Copenhagen, 1960, Rept. Session, Norden*, 19: 120–125.

Gèze, B., De Sitter, L. U. and Trümpy, R., 1952. Sur le sens de déversement des nappes de la Montagne Noire. *Bull. Soc. Géol. France*, 6 (2): 491–533.

Grubenmann, U. and Niggli, P., 1924. *Die Gesteinsmetamorphose. I.* Borntraeger, Berlin, 539 pp.

Jung, J., 1946. Géologie de l'Auvergne et de ses confins bourbonnais et limousins. *Mém. Carte Géol. France*, 1946: 372 pp.

Jung, J., 1953. Zonéographie et âge des formations crystallophylliennes et des massifs hercyniens français. *Bull. Soc. Géol. France*, 6 (3): 329–339.

Jung, J. and Roques, M., 1952. Introduction à l'étude zonéographique des formations crystallophylliennes. *Bull. Serv. Carte Géol. France*, 235: 62 pp.

Lapadu-Hargues, P., 1947. Les massifs de la Margéride et du Mont Lozère et leurs bordures. *Bull. Serv. Carte Géol. France.*, 222: 397–532.

Mathieu, G., 1961. Un tableau stratigraphique du Primaire sur le bord du Bassin d'Aquitaine. *Bull. Soc. Géol. France*, 7 (2): 11–20.

Mattauer, M., Arthaud, F. and Proust, F., 1967. La structure et la microtectonique des nappes herciennes de la Montagne Noire. In: J. P. Schaer (Editor), *Colloque Étages Tectoniques*. Baconniere, Neuchâtel pp.229–241.

Palm, Q. A., 1957. Les roches cristallines des Cévennes médianes à l'hauteur de Largentière (Ardèche). *Geol. Ultraiectina* 3: 121 pp.

Pavillon, M. J. and Routhier, P., 1963. Prélude à une typologie des massifs granitiques, *Bull. Soc. Géol. France*, 7 (4): 888–897.

Peterlongo, J. M., 1960. Les terrains cristallins des Mont du Lyonnais (Massif Central français). *Ann. Fac. Sci. Univ. Clermont Ferrand*, 4: 187 pp.

Raguin, E., 1957. *Géologie du Granite*, 2e ed. Masson, Paris, 275 pp.

Roques, M., 1941. Les schistes cristallins de la partie sud-ouest du Massif Central. *Mém. Carte Géol. France*, 1941: 527 pp.

Roques, M., 1958. Itinéraires géologiques en Auvergne. *Le Touriste en Auvergne (Éd. Bussac, Clermont-Ferrand)*, 26: 1–42.

Rupke, J., 1965 . The Esla nappe, Cantabrian Mountains, Spain. *Leidse Geol. Mededel.*, 32: 1074

Rutten, M. G., 1962. Cinérites of the Mont Dore, central France. *Geol. Mijnbouw*, 41: 351–355.

Schuiling, R. D., 1961. Le dôme gneissique de l'Agout (Tarn et Hérault). *Mém. Soc. Géol. France*, 91: 1–59.

Turner, F. J. and Verhoogen, J., 1960. *Igneous and Metamorphic Petrology.* 2nd ed. McGraw-Hill, New York, N.Y., 694 pp.

Van Moort, J. C., 1967. Les roches cristallophyllienes des Cevennes et les roches plutoniques du Mont Lozère. *Ann. Fac. Sci. Univ. Clermont Ferrand*, and Thesis Univ. Utrecht, 272 pp.

Van Veen, J., 1965. The tectonic and stratigraphic history of the Cardaño area, Cantabrian Mountains, northwest Spain. *Leidse Geol. Mededel.*, 35 (1966): 45–103.

CHAPTER 9

Alpine Europe: General Introduction

GEOSYNCLINAL AND EPICONTINENTAL ALPINE EUROPE

As already indicated in Chapter 1, Alpine Europe is much more of a unity than the other two main divisions, because structure is related to topography. Structurally it embraces the regions folded during the Alpine orogenetic period. Topographically these fold belts still form mountain regions, because the uplift following the orogeny took place so recently in geologic history.

Nevertheless, controversies still exist as to what ought to be included in the Alpine orogenetic belt, and what ought to be omitted. Although most of the chain developed from a true geosynclinal basin, the Tethys, there are outlying chains that are markedly epicontinental in their sedimentary, pre-orogenetic history. Still others developed in what is thought to be a landlocked basin, and not a true geosyncline, whatever that may be. In still other ranges, of which the Jura Mountains are the type example, epidermis type folding only stripped off and folded the Mesozoic cover, without any apparent relation to its Hercynian basement. Some authors want to restrict the term "Alpine" to the truly geosynclinal part, and consequently use "Alpine Europe" in a narrow sense.

As is so often the case in geology, however, the distinctions cited above are not as sharp as might be thought. For instance, we will see how in the Swiss Alps the northern elements, formed by the Helvetide nappes, are in several points intermediate between the southern elements of the Swiss Alps, the Pennide nappes, and the Jura Mountains. We nowadays characterize the Helvetide nappes as miogeosynclinal, the Pennides as eugeosynclinal. But in the earlier literature, notably in HEIM (1921), one always finds stressed the epicontinental character of the Mesozoic series of the Helvetides. The difference is indeed of a quantitative nature only, the Mesozoic reaching a thickness of the order of 1 km in the Jura Mountains, as against several kilometers in the Helvetides. Moreover, in regard to structure, the Helvetide nappes are of an epidermis type of foldbelt too.

There is thus a rather continuous variation, within the mountain chains folded during the Alpine orogeny, from epicontinental to miogeosynclinal and eugeosynclinal. The same holds true for other characteristics, such as the supposed landlocked situation of a sedimentary basin. It seems therefore better to use "Alpine" in its wider meaning here, and to include within the major division of Alpine Europe all regions folded during the Alpine orogeny.

ENUMERATION OF THE ELEMENTS OF ALPINE EUROPE

In a general way, Alpine Europe stretches across the southern part of the continent, all along the northern shore of the Mediterranean (Fig.101). For completeness, we should add that its African counterpart, along the southern shore, is found only in the western part of the Mediterranean, in the Atlas Mountains in the northern parts of Morocco, Algeria and Tunesia.

Starting in the west, the westernmost element of the main Alpine chain is found in southern Spain, in the Betic Cordillera. This can be followed eastwards from Gibraltar, at first running more or less parallel to the coast. But where the coast turns northerly, the Cordillera runs out into the Mediterranean between the towns of Cartagena and Alicante.

On the other, the northern side of Spain, another element of Alpine Europe, the Pyrenees, stretch right along both sides of the border between Spain and France. As we saw, its supposedly landlocked position, wedged in between the crustal masses of France and the Iberian Peninsula, has posed questions of nomenclature. Although it is a major mountain chain, possessing, locally at least, nappe structures, some geotectonicians have had qualms in assigning true orogenetic status to the Pyrenees. Paleomagnetic evidence, however, has recently indicated large-scale continental drifting for the Pyrenees, comparable to that found for the Southern Alps (pp.343, 345). So the present landlocked position of the Pyrenees might well have nothing to do with its earlier geosynclinal history. Of course, this might ease considerably the difficulties encountered in tectonic systematics.

Between the Betic Cordillera in the south and the Pyrenees in the north, several smaller Alpine chains occur. These are all of the Jura Mountains type. Best known are the Catalanides, found along the coast of the Mediterranean southwestwards of Barcelona. Another is that of the Iberian chain, bordering the Ebro Basin to the south, in a general northwest–

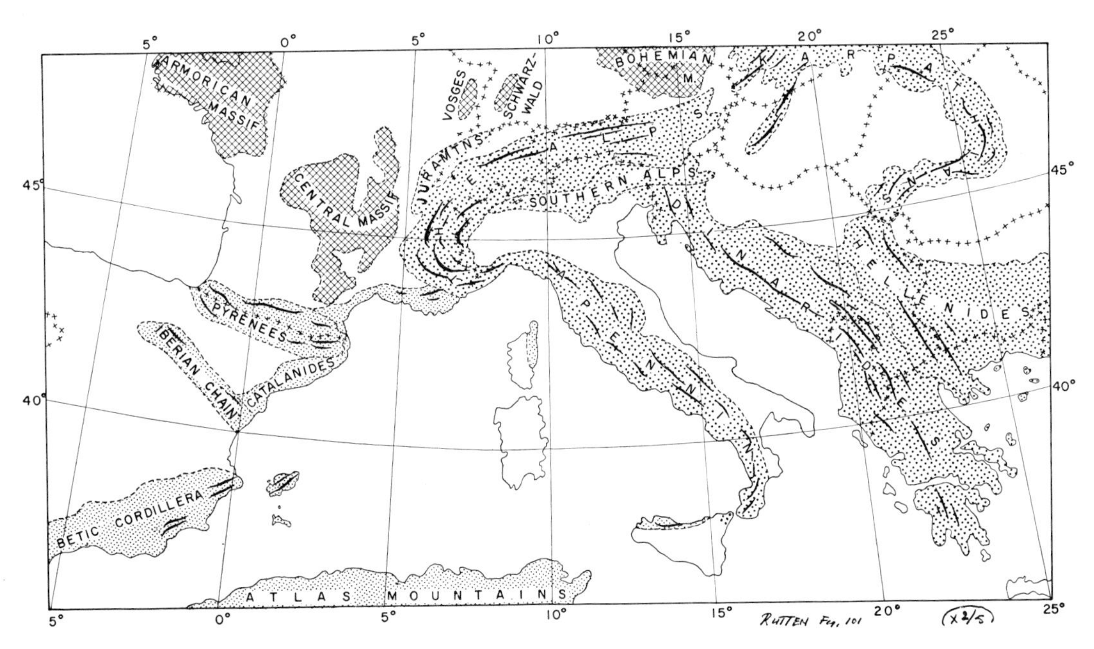

Fig.101. Generalized map of Alpine Europe, showing the main elements.

southeast direction. As is the case with the Jura Mountains, the Triassic is developed here in the Germanic facies, in which thick layers of gypsum in Middle and Upper Triassic have facilitated epidermis type of folding. In contrast to the Jura Mountains, however, the basement reaches the surface in both chains.

From Spain eastward the main Alpine chain of the Betic Cordillera can be correlated with the Balearic Islands, notably with Mallorca. Eastern Corsica also belongs to the Alpine orogeny. But the structural connections, either westward to the Balearic Islands, or northward to the French Alps, are as yet not clear.

Many of the Alpine features of the Pyrenees can also be traced eastward in the west–east trending structures along the Mediterranean in southern France. These continue across the Rhône graben into the Lower Provence, which was also folded mainly during the Pyrenean phase of the Alpine orogeny.

This brings us to the major element of Alpine Europe, to the Alps proper. The Western or French Alps show in their southernmost chains a northwest–southeast strike, against which the west–east trending structures of the Lower Provence abut sharply.

At their southern termination (where, incidentally, the Western Alps extend substantially into Italy), they are cut off indiscriminately by the coast. This part of the Mediterranean, the Ligurian Sea, forms a young subsidence basin, which only developed after the Alpine orogeny.

Topographically, the mountains around Genoa seem to indicate a direct connection between the southern part of the Western Alps and the northwestern part of the Apennines, which form the backbone of the Italian Peninsula. Structurally, however, such a connection is not apparent. Present consensus tends to regard the Western Alps and the Apennines as two quite different elements of Alpine Europe. They are separated by the narrow zone of Sestri-Voltaggio (p.301).

At the southern end of the Apennines, the situation seems to be clearer. There is little doubt that the Apennines connect with the Atlas Mountains of northern Tunesia, via the northern part of the island of Sicily.

The Western Alps curve to a northerly strike already in southern France, in the "Arc of Castellane". They curve further, attaining a northeasterly strike near the French–Swiss boundary. In Switzerland the strike curves still further, until it reaches a true easterly trend in Austria (Fig.102).

The Alps, as the main element of Alpine Europe, have received by far the most attention. Deplorably, this has in many instances led to a semantic babble of differently used regional names. Some of the worst examples will be indicated in the next section, after the present enumeration of the elements of Alpine Europe has been completed.

At the northeastern extremity of the Alps proper, the Eastern Alps can be connected, in a very general way, to the Carpathians. Difficulties in correlation are masked in this area by the intervening lowlands bordering the Vienna Basin.

Along the southeastern border of the Alps another sharp boundary is found, comparable to that between the chains of the Lower Provence and those of the Western Alps. The Alpine chains of Yugoslavia trend northwest–southeast, parallel to the coast of the Adriatic. At their northwestern end, these chains, the Dinarides, abut sharply against the west–east trending Southern Alps. As will be noted further on, however, this sharp natural

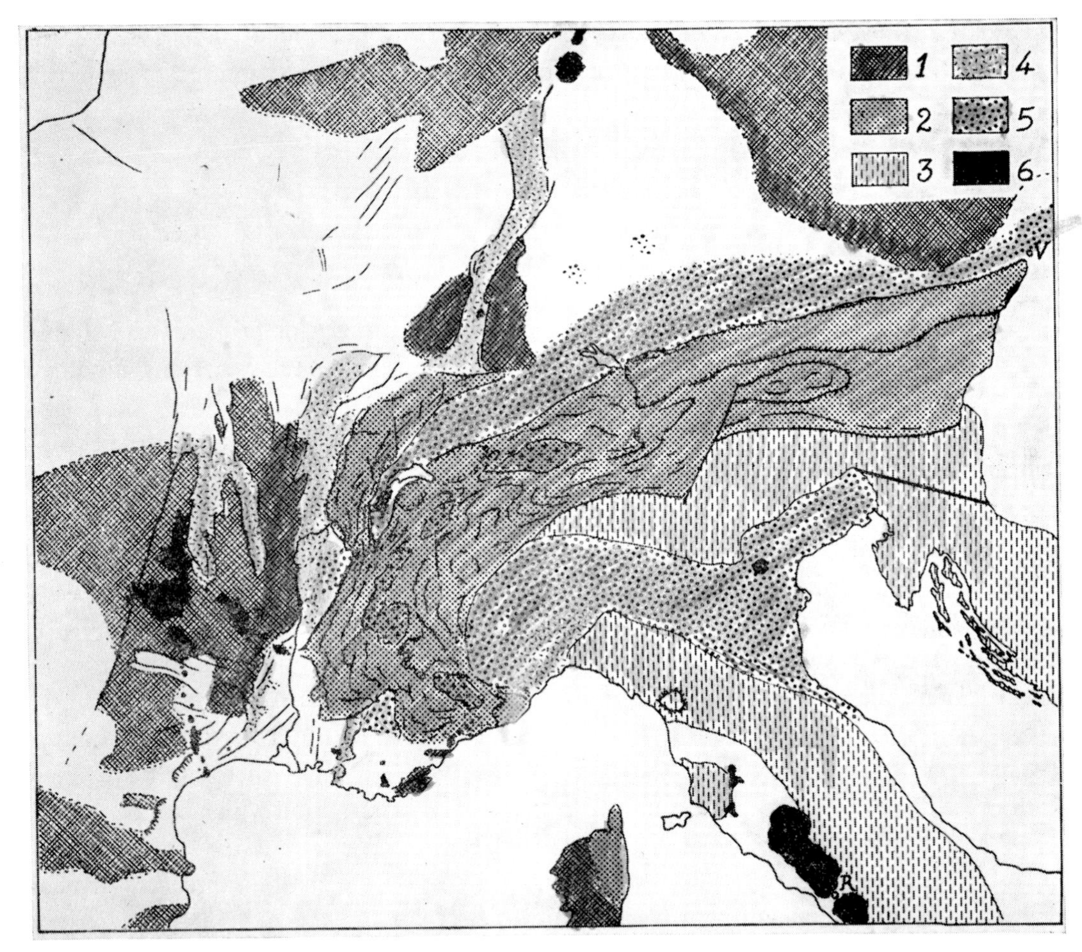

Fig.102. Main structural lines of the French, Swiss and Eastern Alps and the Jura Mountains, placed in their surroundings.
1 = Hercynian massifs (the Pyrenees and Catalanides do not belong to Alpine Europe, according to Goguel); *2* = the Alps, sensu stricto (crosses indicate the main central massifs); *3* = Southern Alps, Dinarides and Apennines; *4* = Oligocene graben; *5* = molasse basins; *6* = Upper Tertiary and Quaternary volcanics.
(After GOGUEL, 1964.)

boundary is masked in the literature by the considerable mix up in the terms "Southern Alps" and "Dinarides".

Eastwards of the Dinarides, the Hellenides follow with the same general strike, to make up most of the territory of Greece and a major part of Bulgaria (cf. AUBOUIN, 1959, 1960; SOCIÉTÉ GÉOLOGIQUE DE FRANCE, 1961, 1963).

Last to be mentioned of the elements of Alpine Europe are the Jura Mountains, which lie to the northwest of the Alps in France and Switzerland. The Jura Mountains have been of great importance in the history of geology for the development of our ideas on folding and overthrusting. Because of their small size, their structure is easily studied. Their tectonics are beautifully exposed along deeply incised transverse river valleys, whereas, already at an early date, several of the principal chains have been pierced by railway tunnels. The structures encountered in these tunnels (Fig.220–222) proved to be quite different from what had been expected from extrapolation of the surface outcrops. This proved the existence, already suspected from field evidence, of a major thrust plane at the base of the Mesozoic sedimentary series, separating the folded epidermis from the basement. The Jura Mountains have since been regarded as the type area of epidermis folding.

THE ALPS AND NOMENCLATURE

Before turning to the main part of Alpine Europe, to the Alps proper, for a slightly more detailed review, a note of caution should be inserted against the variations in nomenclature, as used in the literature.

Historically, the main interest has always centered on the central part of the Alps, on the Swiss Alps. For many a geologist, consequently, "the Alps" is equivalent to "the Swiss Alps". Other regions of the Alps, or of Alpine Europe, are then only seen in their relation to this master picture.

As we shall see, the other parts of the chain deserve better treatment. Far from being exact replicas of the central part, they do not only show important variations upon the theme on which the Swiss Alps are built, but even exhibit features quite different from, or nonexistent in, the central part of the chain. Of course, the Swiss Alps are justly famous for the first brilliant synthesis of the structure of a fold belt built up by a number of nappe systems. In a historical way, this preference for the Swiss Alps may therefore be justified, but at present we should not shut our eyes to the host of information which can be derived from the other parts of the chain.

If one considers the Swiss Alps as the central part of the chain, a threefold division of the Alps is apparent, with the Western or French Alps on one side, and the Eastern or Austrian Alps on the other. But it so happens that the structure of the French and of the Swiss Alps is quite similar in its main lines, and differs rather markedly from that of the Eastern Alps. Many authors therefore prefer a twofold division, into Western and Eastern Alps. In such a twofold division most of the Swiss Alps, that is all of the Helvetides and the Pennide nappe systems, belongs to the Western Alps, whereas the area of the Austride Nappes, both in eastern Switzerland, in southern Germany and in Austria, is equivalent to the Eastern Alps.

The Austride Nappes, this English equivalent for the German "Ostalpine Decken", has, incidentally, nothing to do with the Southern Alps. It is only used because most of the geologic unit of the "Ostalpine Decken" lies in Austria. This name goes back to the reign of Charlemagne, when barbarians had overrun the Mediterranean countries, and civilized Europe lay to the north, in what is now Germany and France. Having conquered Saxony and Bohemia, Charlemagne established a military bufferzone south of the Danube, the so-called *Austrian Mark*, which gave its name to the present state of Austria (PIRENNE, 1956).

The lower part of this Austride nappe system has often been designated separately as the Grisonides, notably by STAUB (1924). According to present usage, however, these are included in the Austrides in this text.

In a threefold division of the Alps, the Swiss Alps quite naturally are designated as the "Central Alps", owing to their position between the French and Austrian Alps. This procedure, though quite sound at first sight, must, however, be avoided, for in most of the German and Austrian literature, the term "Zentralalpin" is used in a quite different sense. In contrast to the transverse threefold division, as used above, the Alps may also be divided lengthwise, because they show a strongly marked longitudinal zonation. In the Eastern Alps the main divisions of this longitudinal zonation are called, from north to south, the Northern Limestone Alps, the "Zentralalpen" or Central Alps, and the Southern Alps.

It seems therefore better to use the term "Swiss Alps" for the central part of the chain in a transverse division, and to retain the term "Central Alpine", or better perhaps "Zentralalpin" for the longitudinal central zone of the Eastern Alps.

As mentioned above, the northerly longitudinal unit of the Eastern Alps, is called the Northern Limestone Alps, or "Nördliche Kalkalpen". This term must be reserved for the Austrides. The Helvetides in Switzerland, also predominantly calcareous, and also occupying a northerly position within the Alpine chain, but belonging to an entirely different tectonic unit, are called, with a nice distinction, the "Hohe Kalkalpen" or High Calcareous Alps.

Another serious confusion exists in the usage of the terms "Southern Alps" and "Dinarides". Realization that these elements are unrelated came only recently. This is mainly due to the fact that for a long time the influence of geotectonic theory prevailed over field evidence.

Theory demanded that an orogeny had to be symmetric, being formed by a geosynclinal sedimentary basin squeezed out between the two jaws of a vice. This is the famous "Doppel-Orogen", which, in many variations has been the basis of most Alpine theories.

Accordingly, as a counterpart to the masses of the Helvetides, the Pennides and the Austrides, all thrust towards the north, a southwards overthrust nappe system had to exist south of the common root zone. As everyone realised, this southward overthrust system was disgracefully narrow, and did not even contain well-developed nappes, in the Southern Alps, mainly found in northern Italy. But, luckily, it seemed to blossom out into the Dinarides of the Balkan Peninsula. The narrow zone of the Southern Alps, although not up to theoretical standard, was consequently integrated with the Dinarides, and the theory of the "Doppel-Orogen" was saved.

Most field workers followed this dictum of geotectonic theory. In all earlier literature one consequently has to distinguish carefully between papers on the "Dinarides" representing what had better be called the Southern Alps, and on the "true" Dinarides of Yugoslavia. Even as late as 1947, for instance, De Sitter's title "Antithesis Alps–Dinarides" was a perfectly good one, although the paper in reality only points out the differences between the Lombardian Alps of northern Italy, forming a part of the Southern Alps, and the Pennide nappes (DE SITTER, 1947). Also most Austrian geologists still at this date lump together Southern Alps and Dinarides.

In the western part of the chain two other terms must be differentiated, viz., the Prealps and the "Subalps" or Subalpine chains. The Prealps are mainly situated in western Switzerland, between Geneva and Bern. There is considerable controversy, both as to their origin and as to their relation to the other main Alpine units (pp.244–255). The French Subalpine chains, on the other hand, are quite generally considered to be the equivalent in France of the Swiss Helvetides. They are much less overthrust, and, if not actually autochthonous, at least parautochthonous. This difference will be discussed later (pp.266–278).

Here, the important thing is that they form much lower chains than the main French Alps. They have consequently been known as "Chaînes péri-alpines", or even as "Chaînes pré-alpines". To prevent possible confusion with the Swiss Prealps, this has in later years been changed to "Chaînes subalpines", or Subalpine chains.

The adjective *pré-alpin*, or "pre-Alpine", is, however, used in yet another sense, i.e., to denote age — for instance, of a sedimentary series, a tectonic movement, a structure or a fabric. In general, anything which belongs to the Alpine orogenetic cycle, not only to the Alpine orogeny, is called "Alpine", everything earlier "pre-Alpine". This implies, of course, that not all things "Alpine" are of the same age. For instance, the Alpine sedimentary series generally starts with the Permian verrucano and terminates with Eocene flysch. Most of the Alpine movements, on the other hand, occur only during the later part of this period.

The nappes of the Prealps can thus not be called a "pre-Alpine" structure, while they also are made up entirely of Alpine sediments. It is only in a geographical sense that they can be called Prealps.

Another difference in nomenclature lies in the designation of the pre-Alpine cores in the French and in the Swiss Alps. In Switzerland a special category is always maintained for the "Zentralmassive" or Central Massifs, thought to be more or less autochthonous. They are found between the Helvetide and Pennide nappe systems. This term incidentally goes back to the theory developed by Leopold von Buch in the 19th century, implying that orogenesis was caused by volcanism. The Central Massifs, in his view, represented the eroded necks of the centrally situated "Erhebungskratere", around which the Alps were folded.

But the Pennide nappes also contain cores of pre-Alpine crystalline basement. These have, however, been mobilized[1] and transported during the Alpine orogeny, and incorporated in the Alpine structures. In Switzerland, the pre-Alpine crystalline cores of individual Pennide nappes are normally not indicated separately.

In the French Alps, on the other hand, a distinction is maintained in the French literature between "massifs externes" and "massifs internes". The first correspond to the Central Massifs of Switzerland, whereas the latter represent the cores, consisting of pre-Alpine rocks, of the Pennide nappes.

As is often the case with such semantic differences, this variation in nomenclature originates from a variation in geology. In the Swiss Alps the bulk of each Pennide nappe is formed by its crystalline pre-Alpine core, surrounded by a relatively thin Mesozoic envelope. From the very beginning geologists have tried to unravel the structure of these nappes as a whole, regarding core and envelope as a single unit. In the French Alps, on the other hand, Alpine sediments are of much greater relative importance, and the crystalline core of individual nappes are situated much farther apart. They appear on the maps as separate crystalline nuclei, and accordingly always have been indicated separately.

[1] This word, incidentally, may be used in quite different ways. Here it means tectonic mobilization, on the scale of the nappe or the fold belt. But it is frequently used also in the petrography of metamorphic rocks, on the scale of the molecule or the crystal. It is normally apparent from the context what type of mobilization or re-mobilization is meant in a particular case, but still it may be quite misleading to a reader who is predominantly occupied with only a single of the different connotations.

This paragraph on nomenclature has been inserted to warn anybody using the detailed literature on the Alps against the frequent differences in meaning of the same terms, both as used by earlier and by later authors, and as used by authors of different nationality. It would be easy to cite more examples of similar possible pitfalls, but let us return to geology proper, to the rocks and to the interpretations derived from them.

NAPPE STRUCTURES

The Alps are of course best known as the first fold belt in which nappe structures were found to play a predominant part. Historically, this concept developed in the Swiss Alps during the latter half of the 19th century. First came the realization, from field evidence, of the existence of large, sub-horizontal overthrusts, where thick series of older rocks rested on younger series. This led later to the theoretical concept of nappes.

The existence of large-scale overthrusting was proved by the Swiss geologist Escher von der Linth during the middle of the 19th century. It was, however, not until 1878, after his death, that it was published by his pupil Albert Heim in his *Untersuchungen über den Mechanismus der Gebirgsbildung* (Heim, 1878).

The most impressive example of this overthrusting lay in the Glarus Mountains of eastern Switzerland, and forms part of what is now called the Helvetide nappe system. In the Glarus overthrust, or "Glarner Überschiebung", sub-horizontal Permian, of a well-known continental red bed-and-conglomerate facies, the verrucano, overrides younger beds, mainly consisting of Eocene, *Nummulites*-bearing flysch. The Glarus overthrust can be followed over many kilometers, and in the section the relatively undisturbed Permian is seen to rest everywhere on top of the folded series of younger rocks.

Overthrusts of such sizes were entirely unknown at that time, while it was also difficult to conceive a mechanism capable of producing such unheard of revolutions of the earth's crust. But the Glarus overthrust was clearly visible, all along the steep walls of the Glarus Mountains. It has been told, how Escher von der Linth, as much an alpinist as he was a geologist, has led quite a number of famous contemporaries high up to the tell-tale outcrops. There, every dissenter was forced to accept the phenomenon.

The Glarus overthrust is still one of the show-pieces of Alpine geology. The difference between the red verrucano lying on top of the whitish or greyish younger series, is quite impressive, and can be seen from afar. Because it is nowadays no longer necessary to convince a possible dissenter, the overthrust concept being an accepted tenet of geology, it is at present mainly studied from afar, through a pair of binoculars. In fact a small outcrop of the overthrust can be reached much more easily in a valley at an accessible place near the town of Schwanden. Although it can be followed here for only a short distance, the privilege of easy access is such, that it is now declared a natural monument.

To minimize the amount of overthrusting, Escher von der Linth, followed by Heim, devised the interpretation as a "Doppelfalte" or double-fronted fold (Fig.103). Using a part situated about midway in the section, where the overlying sheet of verrucano has been interrupted by erosion, he postulated the existence of an original syncline at that point. This syncline was then supposed to have later become overthrust by both flanks. In this way the necessary amount of overthrusting was indeed halved, but the mechanical

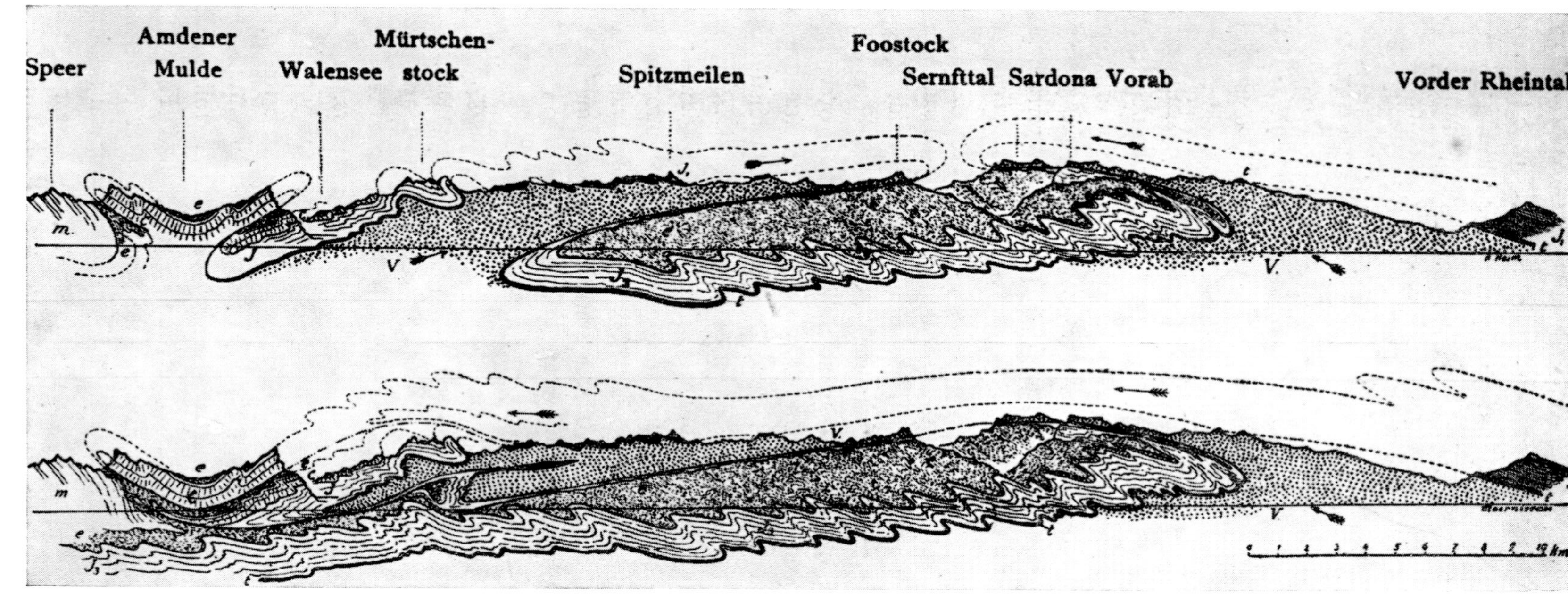

Fig.103. Two sections through the Glarus overthrust, northeastern Switzerland, indicating the different theoretical interpretations as doublefronted overthrust ("Doppelfalte") and as nappe. The Amdener Mulde, to the north of the Glarus overthrust, is interpreted as another structure altogether, as a mushroom fold or "Pilzfalte" in the upper section, as part of the same nappe in the lower section.

m = molasse; e = flysch; c = chalk; J = Jurassic; T = Triassic; V = verrucano (Permian). (After HEIM, 1921.)

difficulties in explaining the origin of such a structure were, of course, not minimized at all.

The "Doppelfalte" concept was applied to other overthrusts as well. Its anticlinal counterpart, the "Pilzfalte" or mushroom fold (Fig.104) was applied in all those cases where the visible lateral extent of the overthrust was small.

The central parts of the sections through both "Doppelfalte" and "Pilzfalte" were always the most difficult to visualize. But it so happened that there was always a part of the section available, where either the overlying thrust sheet had been eroded away, or where the critical area lower down on the slopes was masked by scree or woods. In the first case an interpretation as a "Doppelfalte", such as in the Glarner overthrust, was indicated, whereas in the second case it was relatively easy to construct a "Pilzfalte".

The concept of the nappe was born at a writing desk, not from field work. This should be a warning to all those who now decry all theoretical geology. Although many are the instances, particularly so in the Alps, where pure theorizing has gone too far, we must still strive for a general theory to account for the facts found. At present it seems as if the number of papers, in which a mulish preoccupation with detailed variations of facts prevents the author from acquiring a more general insight, has become at least as large as

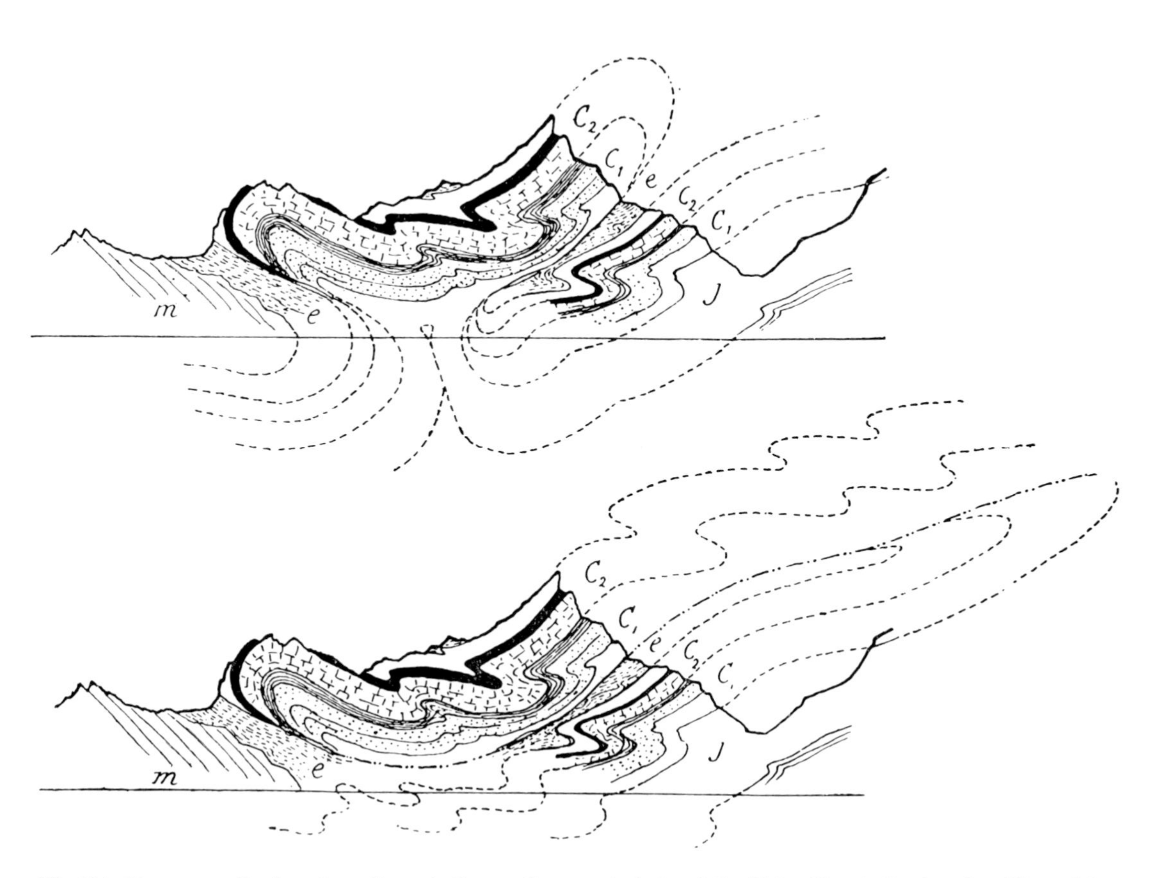

Fig.104. Two generalized sections through the northernmost chain of the Helvetides, indicating the differential theoretical interpretation as a mushroom fold ("Pilzfalte") and as the front of a nappe. A. Theory of the mushroom fold. B. Theory of the fold nappe.
m = molasse; e = flysch (Eocene); c_2 = Seewer limestone, Gault, clints limestone ("Schrattenkalk"); c_1=Neocomian, Valanginian; J = Jurassic. (After HEIM, 1921.)

those papers in which theoretical considerations are extended far beyond their factual basis.

In 1884 Marcel Bertrand, a French geologist, from reading an account of the Glarner Doppelfalte and similar structures conceived a reinterpretation modelled on the Faille du Midi along the northern border of the Ardennes, as described by Gosselet. So complicated is the path of history that BERTRAND (1884) came to know of the Glarner Doppelfalte not from the original publication by HEIM (1878), but through the famous compilation found in SUESS's (1885, 1888, 1901) *Das Antlitz der Erde*. And this again was not even from the original publication but from the French translation by De Margerie. Nevertheless, it is this reinterpretation which has paved the way for the further development of the nappe concept.

Bertrand's interpretation eventually won over Albert Heim, who in his field work in the Swiss Alps was one of the persons most concerned with the difficulties encountered in applying Escher's "Doppelfalte" and "Pilzfalte" concepts to the actual sections. The two figures reproduced here as Fig.103, 104 of the two possible interpretations of the Glarner Doppelfalte and of the Rigi Pilzfalte, are by Heim's hand.

The difference this interpretation has made is revealing, when we compare Heim's earliest major work, his *Untersuchungen über den Mechanismus der Gebirgsbildung* (HEIM, (1878), with his later publications, such as his *Geologie der Schweiz* (HEIM, 1919, 1921, 1922). In the atlas accompanying the first work, which is in fact based on the mapping of the Tödi–Windgällen Group in the Helvetides of northeastern Switzerland, we have already the essentials characterizing all later work: excellent observation and superior draftsmanship. But all sections are interpreted as "Doppelfalten", whereas the same sections appear as nappes in his *Geologie der Schweiz*.

A further major step was taken as a result of the interpretation of the Swiss Prealps based on the detailed field work by SCHARDT (1893). In this work the facies concept was used extensively to delimit a series of superimposed individual nappes. Then, in 1903, after the International Geological Congress in Vienna, another French geologist, Pierre Termier, interpreted the Tauern Mountains in Austria as a tectonic window of the Pennide nappe system, uncovered by erosion underneath the overlying Austride nappes (TERMIER, 1903).

The nappe structural type, in which series of older, or at least different age, are piled on top, either of the autochthonous basement, or of other nappes, has since been recognized not only in many other parts of the Alps, but in a great number of other fold belts as well. I will leave the details of the historical development of this concept, for here we have the delightfully readable essays of an old master who really occupied a ringside seat, viz., BAILEY (1935).

It must be noted though, how, ironically enough, the "Pilzfalte" has recently reappeared in the literature (p.230). Not, however, as a negation of nappe structures, but as a further refinement.

It seems appropriate to conclude this historical section with a remark on the motor, the driving force, postulated for the formation of overthrusts and nappe structures. In 1893, SCHARDT explicitly stated that the field evidence does not permit choosing between tangential push and gravitational sliding. One gets the impression that geologists at that time were not so strongly interested in what lay behind these structures, being fully occupied in proving their existence.

It was only later that the theoretical considerations of the cooling earth, which in fact had already been developed in the middle part of the 19th century, became applied to tectonics. Nappe structures were then claimed as examples of the requisite shrinking and crustal shortening.

In this model, tangential push was of course the appropriate motor. Only through this cross-fertilization between the factual data of overthrusts and nappes and the theory of a cooling and shrinking earth, did tangential push become so universally accepted as the sole motor responsible for the formation of fold belts. Unfortunately it has been generally forgotten that such a contention has never been proved.

Because gravity tectonics is gaining more and more ground at present (pp.202–204) the accidental coincidence of facts and theory, which led to the universal acceptance of tangential push as the driving force in tectonics should be pointed out.

THE CLASSIC PICTURE OF THE ALPS

Based upon the concept of large-scale overthrusting and nappes, and upon the efforts of a number of field geologists, the "classic picture" of the Alps (in particular of the Swiss Alps) arose. This is the picture, as given by HEIM (1919, 1921, 1922); by the combined Swiss geologists in the guide books commemorating the 50th anniversary of the Swiss Geological Society (GAGNEBIN and CHRIST, 1934); and, as recently as 1953, by CADISCH. It is given in English by COLLET (1927), by BAILEY (1935) and UMBGROVE (1950), and it has been applied to the whole chain by STAUB (1924).

The classic picture consists of a number of nappes, which are thought to have originated as recumbent folds formed between the vise of the rigid continents of Europe and Africa. Many variations are found in the literature, variations pertaining, for instance, to the number and limits of individual nappes. There are questions whether both continents were at the same level, or whether one overrode the other, or was, vice versa, underthrust by the other. We find, for instance, speculations as to the cause of the implied shortening, whether this was a shrinking earth, or a mysterious "Polflucht", or continental drift. But apart from such variations, the main substance of the classic picture is the simple model cited above.

Within the domain of the Swiss Alps, which have a general westsouthwest–eastnortheast strike, the classic picture recognizes a number of successive main zones, broadly following the strike. From north to south these are the Helvetide nappes, the Central Massifs, the Pennide nappes, the root zone and the Southern Alps, the latter being cut off by the subsidence of the Po Basin.

The Helvetide nappes are miogeosynclinal and are thought to have originated from the northern border of the Tethys. Their sedimentary facies is predominantly epicontinental. They comprise Alpine sediments only, i.e., Permian (verrucano), Mesozoic and Early Tertiary; the latter developed mainly in flysch facies. These nappes did not travel far, for their roots can be recognized either between, or just south of the Central Massifs.

The latter form part of the pre-Alpine, rigid mass of the European continent. They are thought to be autochthonous, representing crustal material from the Hercynian, or even older, orogenies. The main influence of the Alpine orogeny has been only to lift them up so high that they have become uncovered by erosion.

The Pennide nappes form the main part of the Alpine structure. Their Alpine rocks are typically eugeosynclinal and represent the Tethys geosyncline. They are characterized by Mesozoic in a not too thick, clastic facies, the "Bündner Schiefer" or "Schistes lustrés", containing ophiolites or "roches vertes". But the main bulk of individual nappes is formed by older cores of pre-Alpine elements. Slightly metamorphosed sediments of Carboniferous (?) age, such as the Casanna Schiefer occur, but schists and gneisses predominate. They represent the mobilized ancient basement of the Tethys. The Bündner Schiefer in most cases only forms a relatively thin envelope around the crystalline core of the Pennide nappes.

The root zone of the Pennide nappes is found — in the classic picture — immediately south of this nappe system. So, although the Pennide nappes are thought to have travelled far, the root zone seems to have followed them all the way. The main characteristic of the root zone is its strong vertical tectonization. Moreover, a larger part is obscured by young, post-tectonic, intrusive granite bodies.

Southwards of the root zone the Southern Alps play only a very minor part in the Swiss section of the mountain chain, as has been mentioned already.

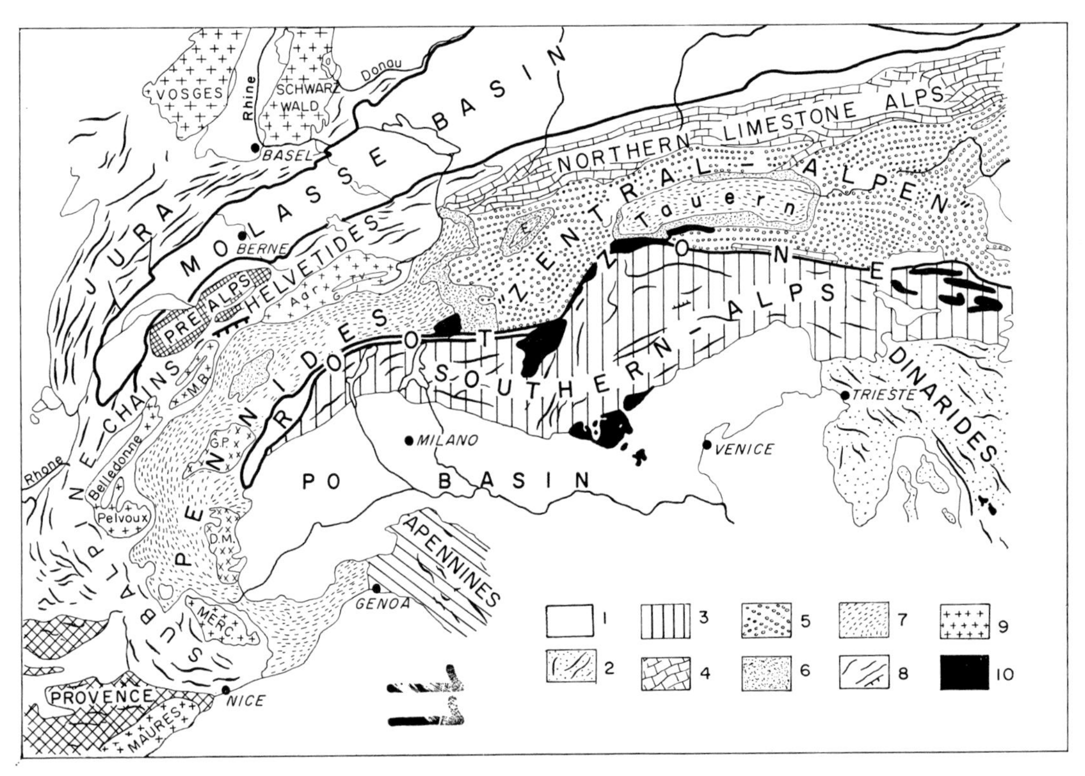

Fig.105. Generalized map of the Alps.
1 = Tertiary and Quaternary of forelands and basins; *2* = Dinarides; *3* = Southern Alps; *4* = Northern Limestone Alps; *5* = "Zentralalpen", predominantly crystalline of the Eastern Alps; *6* = upper Pennides and/or lower Austrides; *7* = Pennide nappes: *8* = Jura Mountains, Subalpine chains and Helvetide nappes; *9* = Central Massifs and Hercynian uplands; *10* = Tertiary and Quaternary igneous rocks. *T.V.* = Tavetscher "Zwischenmassiv"; *E* = Engadin window; *G* = Gotthard Massif; *M.B.* = Mont Blanc Massif; *MERC.* = Mercantour Massif. (After CADISH, 1946.)

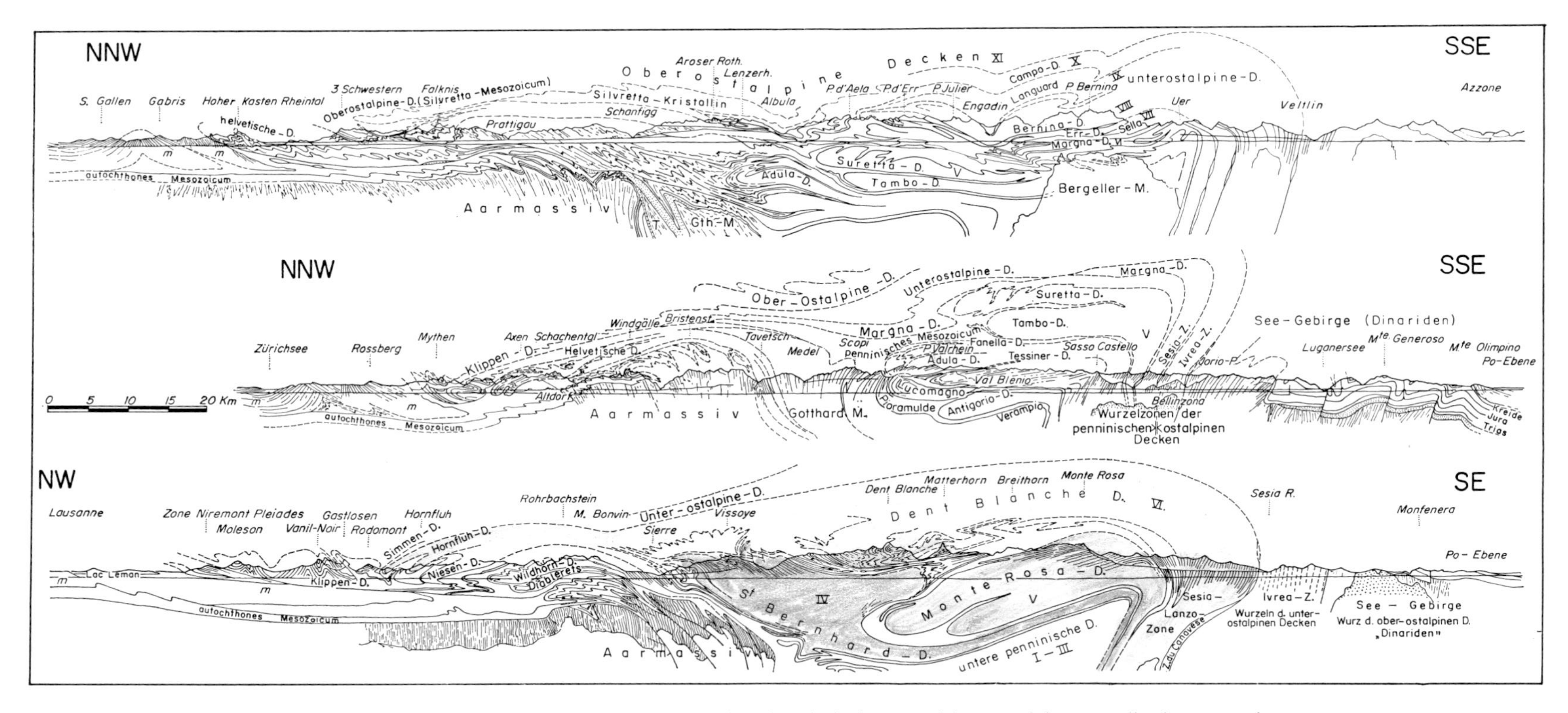

Fig.106. Three generalized sections through the Swiss Alps. As is the case with most of the generalized cross-sections of the Alps, these are purely tectonical sections. They indicate the boundaries of individual nappes, but not their internal stratigraphy. (After HEIM, 1922.)

This classic picture may be completed eastwards, because in that direction a nappe system even higher than the Pennides, the Austrides, is saved from erosion. In western Switzerland the Prealps, which also have already been mentioned, are thought to belong either to the Pennides or to the Austrides, and to have overridden the Helvetides in an area of axial depression of the latter.

The meaning of these terms, too hastily enumerated here, will, I hope, become clearer in the next chapters. Fig.105, a generalized map by CADISCH (1946), and Fig.106 a reproduction of the three generalized sections across the Swiss Alps by HEIM (1922), are included to illustrate the main elements of the classic picture.

To round off this general introduction of the classic picture, we should like to stress the fact that this classic picture of the Swiss Alps in due time became *the* picture of *the* orogen.

It was extrapolated beyond all reason by Alpine geologists, for instance, to other parts of Alpine Europe, when geologists thought to recognise the same Pennide nappes in the Betic Cordillera around Malaga, hundreds of kilometers away. Or when geologists correlated Alpine structures in Celebes in Indonesia with the Pennides of Switzerland. It was not only extrapolated to other parts of the Alpine fold belt, all over the world, but also, in time, to fold belts that have been formed during earlier orogenies. It was, moreover, given to geophysicists, flavoured with a touch of authenticity, as *the* model of an orogen, known beyond any doubt in its genesis and in its form, up to all details, to be used as a basis for numerical calculations.

In short, the classic picture of the Swiss Alps was thought to exemplify, not the evolution of a certain mountain chain and a certain fold belt, but *the* general law of geosynclinal and orogenetic development. If some other fold belt was shown to exhibit a different structure, that was that particular fold belt's defection. Such a local, unorthodox variation could never invalidate the basic laws of the orogen, as embodied in the classic picture of the Swiss Alps.

I do sincerely hope that these remarks about the unwarranted extrapolation of the classic picture of the Swiss Alps sound bitter. That is how they are meant to be. For if ever there is variation, both in time and in space, it is in history, both human and geological. Moreover the presumtuousness of proclaiming the classic picture of the Swiss Alps to be *the* example of *the* general law of tectonics, is, to my mind, as great as the extrapolation used in drawing up this very picture. For let us not forget that it rested on but a most superficial analysis of the rocks themselves.

BASIC ASSUMPTIONS OF THE CLASSIC PICTURE

It has been possible for the early workers to weld the far too scarce field data into the solid structure of the classic picture by the use of geotectonic theory which went back mainly to the Swiss geologist Émile Argand. The basic hypothesis was that the Alps originated out of a series of embryonic cordilleras. These began to develop on the floor of the Tethys already at an early stage, continuing all through the Mesozoic. It was assumed that each embryonic cordillera subsequently became a separate nappe during orogeny.

Before that, the growing cordillera was thought to influence the sediments laid down on its crest and on its flanks, all during the geosynclinal phase.

From this basic assumption two hypotheses follow that have become of utmost importance in the practical application as a guide in mapping and in correlating individual nappes. These deductions are: (*1*) that of continuity of individual nappes all along the fold belt; and (*2*) that of uniformity of facies within a given nappe, and disparity in facies between different nappes.

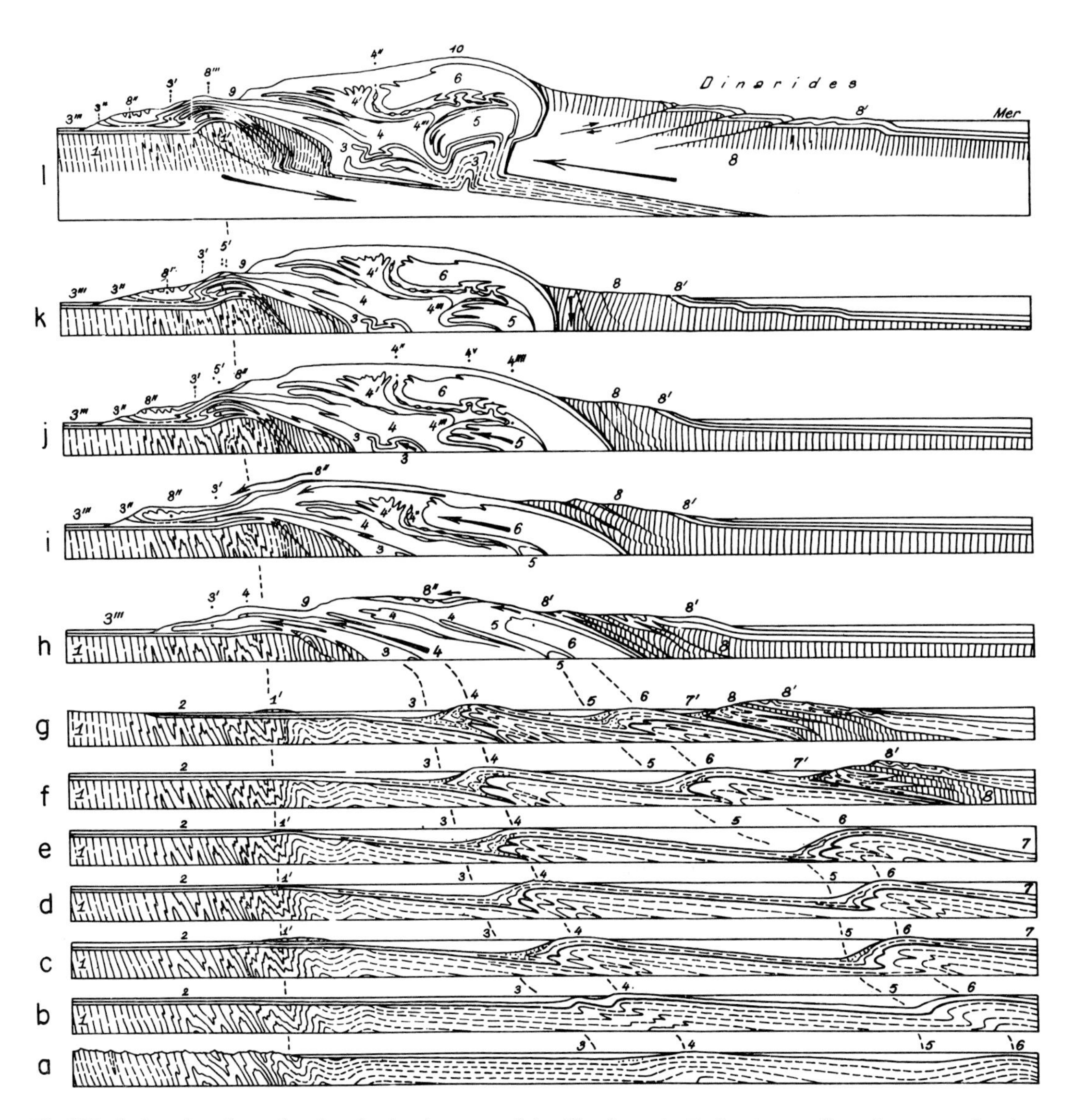

Fig.107. Series of sections, showing the development of the Alps from the Tethys geosyncline, via a stage of embryonic cordilleras.

a = end of Middle Carboniferous; *b* = Middle Triassic; *c* = Liassic; *d* = Middle Jurassic; *e* = Late Jurassic; *f* = Middle and Late Cretaceous; *g* = Middle Eocene and Early Oligocene; *h*–*l* = paroxysmal development, Middle Oligocene. For explanation of notations see Fig.108.

For an identification of the main Pennide nappes (*1*–*6*) in section *l*, compare *I*–*VI* in Fig.109. The "Dinarides" in the upper section would be called today the "Southern Alps" (p. 175). (After ARGAND, 1916.)

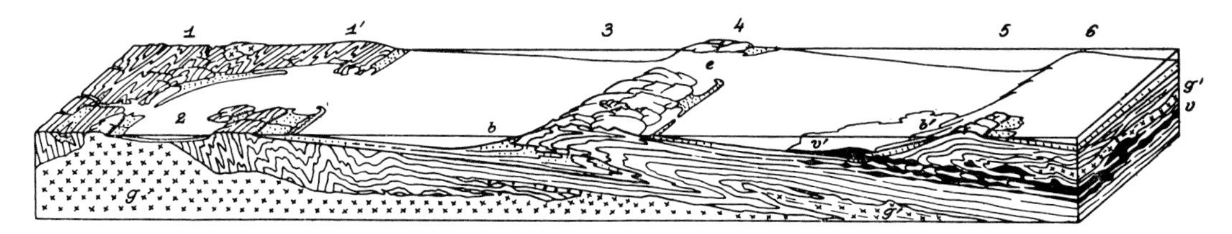

Fig.108. Schematic view of an Alpine mountain chain in its embryonic stage. Although not expressively indicated this corresponds to the situation postulated to have existed from the Jurassic to the Eocene in the series of sections of Fig.107. The numbers and letters are Argand's and, according to his theory, have the following meaning, which is basic to the classic picture.

1 = "foreland", continent rigidified during earlier orogenies; *1'* = slightly mobilized "foreland", at the border of the ancient continent, the present Central Massifs; *2* = epicontinental sea, the present Helvetides; *3,5* = geosynclinal sea, "foredeep" in the nomenclature of Argand; *4,6* = embryonal cordilleras; *b,b'* = clastic series laid down all along the embryonic cordilleras; *e* = lower elevation of embryonic cordillera, related to the lesser resistance of large bays in the "foreland", such as at (*2*); from these regions the later zones of axial depression will develop during the paroxysmal phase; *v,v'* = initial geosynclinal basic volcanites, the ophiolites or *roches vertes*; *g* = granitic crust; *g'* = granitic crust remobilised and metamorphosed in the Pennide nappes. (After ARGAND, 1916.)

The supposed continuity of nappes follows from the assumption that each nappe developed from its own embryonic cordillera, because the latter were thought to extend all along the Tethys. So each nappe had to be continuous too.

The assumption of facies continuity within a given nappe stems from the idea that the embryonic cordilleras had such an influence on geosynclinal sedimentation, that the facies of the sediments developing on and around each of them was continuous and disparate from that found on other cordilleras.

The development of an Alpine mountain chain out of a geosyncline via the stage of embryonic cordilleras is here illustrated by the classic pictures of ARGAND (1916) in Fig.107, 108. The picture is, of course drawn up in accord with a cross-section of the Swiss Alps. Another well-known Argand section, the upper section of Fig.107 is repeated slightly enlarged, and reproduced here in Fig.109.

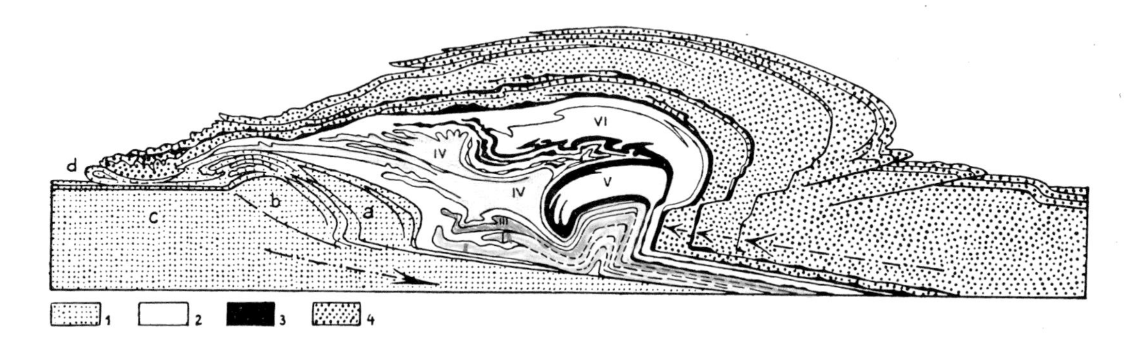

Fig.109. Generalized cross-section, illustrating the classic section of the Alps.

1 = foreland or Eurasia; *a* = crystalline wedges, elsewhere exposed as the Central Massifs; *b* = swelling of the crystalline basement; *c* = non deformed continental crystalline; *d* = Helvetide nappes.

2 = the Pennide nappes, formed in the Alpine geosyncline, which represent the Western Alps; *I–III* = Simplon-Ticino nappes (*I* = Antigorio; *II* = Lebendun; *III* = Monte Leone); *IV* = Great St. Bernard nappe; *V* = Monte Rosa nappe; *VI* = Dent Blanche nappe.

3 = basic rocks; *4* = the Hinterland, or Africa, and the Austrides of the Eastern Alps.

To this text, COLLET (1927) adds: "The Eastern Alps override the Western Alps", and: "Africa overrides Europe". This gives the classic picture in a nutshell indeed! (After COLLET, 1927.)

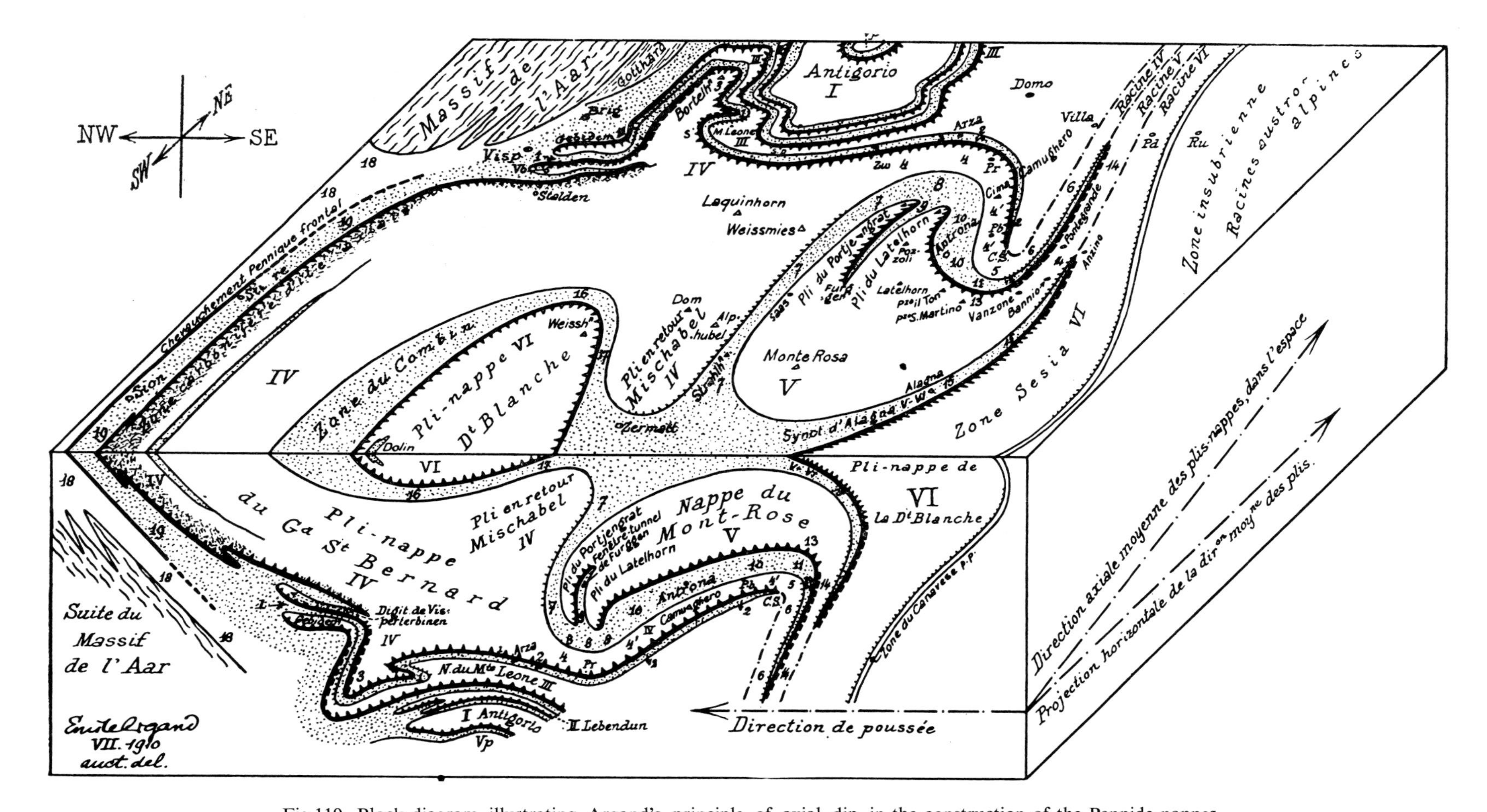

Fig.110. Block diagram illustrating Argand's principle of axial dip in the construction of the Pennide nappes (see also Fig.111). The front of the block crosses the Pennides in a zone of axial depression, in which the highest nappe, the Dent Blanche nappe, is extensively preserved. Towards the northeast follows a zone of axial culmination. In the ideas of the classic picture this is caused by higher resistance from the Aar Massif, coming up towards the upper left hand corner of the block diagram. The lower Pennide nappes, which in the frontal cross-section are supposed to exist deep down in the crust, consequently crop out along the axis of the chain towards the northeast.

The lower Pennide nappes, *I* = Antigoria, *II* = Lebendun and *III* = Monte Leone, are often taken together as the Simplon, or the Simplon–Ticino nappes. The roman numerals for these and higher Pennide nappes, *IV* = Grand St. Bernard, *V* = Monte Rosa, and *VI* = Dent Blanche, are widely used in Swiss nomenclature. The Dent Blanche nappe is nowadays normally classed as lower Austride. (After ARGAND, 1911.)

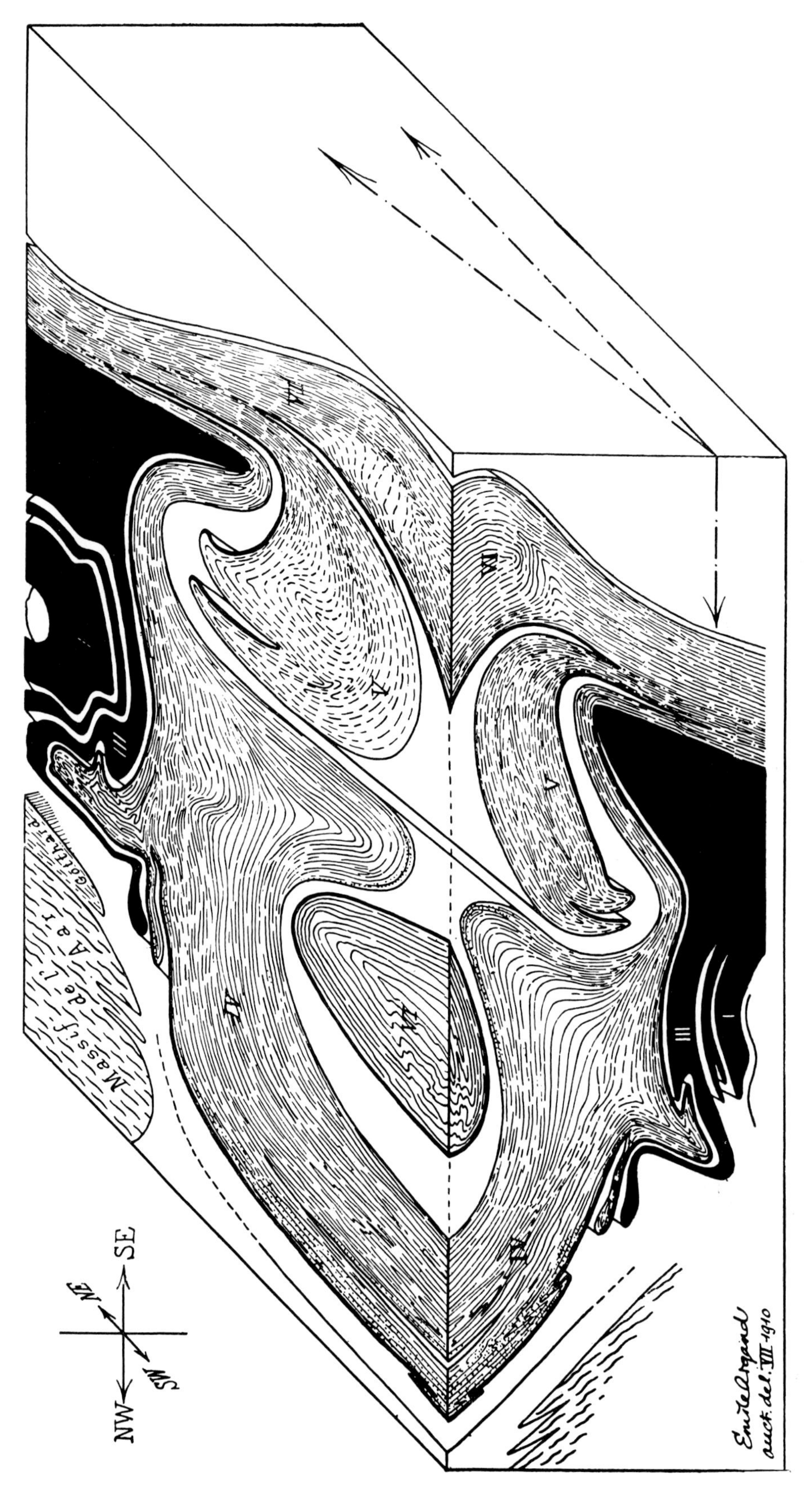

Fig.111. Block diagram of Fig.110, structurally interpreted. Legend see Fig.110. (After ARGAND, 1911.)

The first-mentioned deduction from the basic assumption on the existence of embryonic cordilleras permits the use of axial dip in the construction of serial cross-sections over a fold belt. Every main element present in a given section is supposed to be present, or to have been present, in other sections of the fold belt as well. In cross-sections over axial culminations, this led to the construction of the well known lofty aerial lines for the higher nappe units. In areas of axial depression, on the other hand, the lower nappes were just as rigidly constructed way down, even into depths approaching the base of the continental crust (Fig.110, 111).

The second deduction permits correlation, merely by comparing the facies, of isolated nappe fragments. So well accepted was this tenet of facies uniformity within a given nappe, that correlations of separate nappes, 100 km and more apart, have frequently been based on this assumption alone.

THE "LEITHORIZONT"

A method widely used during the elucidation of the structure of the Alps in drawing up the classic picture, is that of the "Leithorizont", the key or guide horizon.

From a good stratigraphic section, which is accessible and can be studied in detail, a definite, conspicuous horizon, such as a series of whitish limestone, was chosen, to follow the structure of nappes along less accessible valley walls and high mountain peaks. This has made it possible to unravel the general trend of Alpine structures at a much faster rate, than if all of these areas had to be mapped in detail.

The "Leithorizont" method consequently was a legitimate tool during this period of the first general reconnaissance. Its main drawback in the Alps has been that it is often impossible, from a distance, to see whether a "Leithorizont" is right side up or not. If, in a well-exposed, but not so well accessible section, a repetition of a certain "Leithorizont" was observed from afar, the tendency has formerly always been to interpret this through a construction of a series of recumbent folds. Anticlinal bends were thought to have been eroded away, and could easily be visualized in aerial connections, while synclinal bends were just as conveniently located in the lower, not exposed areas.

This interpretation of the repetition of a "Leithorizont" as a series of recumbent folds stems from the ideas on the progressive formation of nappe structures. They were thought to develop gradually from simple folds, by way of asymmetric and recumbent folds. As a historic curiosity Heim's concept in this matter is reproduced here as Fig.112. However, it has become apparent in later times that such gradual transitions are rarely found in nature. Instead, particular fold belts tend to be characterized by quite a narrow variation in tectonic style. They are, for instance, either characterized by simple folds, or by a repetitive imbrication of rather steep upthrusts, or by recumbent folds with flowing contours, or by nappes formed by thrust sheets with subhorizontal thrust planes.

In the Alps, and especially so in the Helvetides, later, more detailed, field work has gradually taught us that most of the strata in the main units are right side up. In those cases the interpretation of sections that show repetition of a "Leithorizont", has consequently changed from a pile of recumbent folds to a series of superimposed thrust sheets. The elusive "middle limbs" of the former recumbent folds, which often escaped detection,

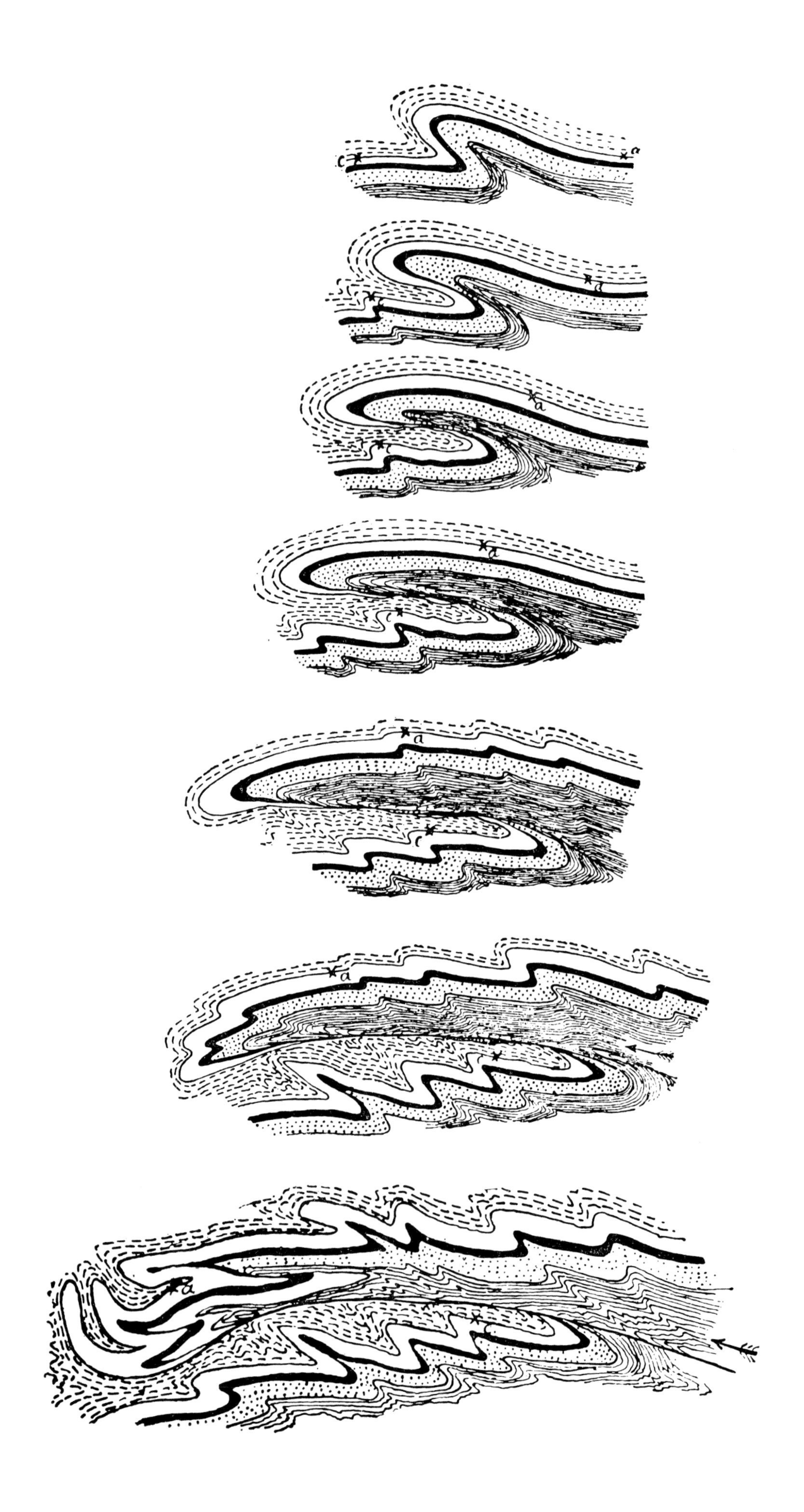

c
a
a
a
c
a
c
a
c
a
a

and then existed only formally as "middle limbs drawn out to zero thickness", have therefore disappeared from most modern interpretations. The same holds true for the large anticlinal bends of these recumbent folds, reconstructed in the air.

Excessive use of the "Leithorizont" method has sometimes been scoffed at by non-Alpine geologists. This has culminated in the widely circulated quip that "A pair of binoculars is the Swiss' geologists hammer". But lest we forget the accomplishments of the early Alpine geologists, we should bear in mind that only through the use of the "Leithorizont" could such a result as the classic picture have been derived.

The classic picture of the Alps was based on a reconnaissance survey only. A relatively small number of field geologists has, over a couple of decades only, been able to supply us with this comprehensive picture of a large and high mountain chain, a chain which moreover was at that date much less accessible than it is now. These early Alpine geologists just had to use any seemingly valid principle in the extrapolation from more detailed analyses of the more accessible to the less accessible parts of the Alps, to be able to present an overall picture in such a remarkably short time.

Perhaps in the coming decades a state of affairs will be reached in the Alps, which comes near to what has been called the "last will" of professor P. Fallot of Paris (in: DURAND DELGA, 1962): "From the other world, in 10 or 20 years, I hope to see you proposing new theories on the basis of tens of well-surveyed 1:50,000 sheets — *essential base of any valid geological theory* —, interpretations that are more adequate than those which now serve us for working hypotheses."

With the recent multiplication of geologists, this might well be the case, but this is no reason to scoff at the methods of the early workers in this field, nor at the classic picture of the Alps.

NEW VIEWS AND OLD

In defence of the classic picture of the Alps, let it be stated that it was, and still is, a thing of beauty. In a sweeping vision the whole history of the mountain chain was understood in a most lucid way. Effect followed cause without fail throughout its development. And all that complicated mass of detail could be forgotten, because of the clear summary presented in the overall picture.

It came up to contemporaneous expectation, fathered by the physics of the time, that a general law in nature had to be simple to be true. But physics, at least the classic physics of those days, is a science governed by statistical laws, in which the singular event plays no role. Geology, on the contrary, and especially so Alpine geology, has almost always to do with singular events, governed by local factors, often strongly varying, both from place to place and with time.

I strongly believe that most of the resistance provoked by assaults on the classic picture of the Alps stems more from a general reluctance to abandon such a beautiful

Fig.112. Series of schematic sections, illustrating the idea how nappe structures were thought to develop continuously from simple folds, by way of recumbent folds with more and more strongly drawn out middle limbs. Nappe structures are now considered, more often than not, as thrust sheets that stand in no direct genetic relation to simple, or even to recumbent folds. (After HEIM, 1921.)

system, than from a proper evaluation of the facts in question. Consequently, it has only slowly filtered through that the classic picture showed points of weakness, even though a case can be made for the statement that newer field work has proved the classic picture to be wrong in all of its main points.

REFERENCES

ARGAND, E., 1911. Les nappes de recouvrement des Alpes pennines et leurs prolongements structuraux. *Matér. Carte Géol. Suisse*, 31: 1–26.
ARGAND, E., 1916. Sur l'arc des Alpes occidentales. *Eclogae Geol. Helv.*, 14: 145–191.
AUBOUIN, J., 1959. I: Contribution à l'étude géologique de la Grèce septentrionale. II: Place des Hellénides parmi les édifices structuraux de la Méditerranée orientale. Thesis (Paris). *Ann. Géol. Helléniques*, 10: 525 pp.
AUBOUIN, J., 1960. Essai sur l'ensemble italo-dinarique et ses rapports avec l'arc alpin. *Bull. Soc. Géol. France*, 7 (2): 487–526.
BAILEY, E. B., 1935. *Tectonic Essays, mainly Alpine*. Clarendon, Oxford, 200 pp.
BERTRAND, M., 1884. Rapports des structures des Alpes de Glaris et du bassin houiller du Nord. *Bull. Soc. Géol. France*, 3 (12): 318–330.
CADISCH, J., 1946. On some problems of Alpine tectonics. *Experientia*, 2: 1–16.
CADISCH, J., 1953. *Geologie der Schweizeralpen*, 2. Aufl. Wepf, Basel, 480 pp.
COLLET, L. W., 1927. *The Structure of the Alps*. Arnold, London, 289 pp.
DE SITTER, L. U., 1947. Anthithesis Alpes–Dinarides. *Geol. Mijnbouw*, 9: 1–47.
DURAND DELGA, M. (Editor), 1962. *Livre à la Mémoire du Professeur Paul Fallot. I.* Soc. Géol. France, Paris, 656 pp.
GAGNEBIN, E. and CHRIST, P. (Editor), 1934. *Geologischer Führer der Schweiz*. 14 vol. Wepf, Basel.
GOGUEL, J., 1964. L'interprétation de l'arc des Alpes occidentales. *Bull. Soc. Géol. France*, 7 (5): 20–33.
HEIM, A., 1878. *Untersuchungen über den Mechanismus der Gebirgsbildung im Anschluss an die geologische Monographie der Tödi–Windgällen Gruppe*. Schwabe, Basel, 246 pp.
HEIM, A., 1919. *Geologie der Schweiz. I. Molasseland und Juragebirge*. Tauchnitz, Leipzig, 704 pp.
HEIM, A., 1921. *Geologie der Schweiz. II (1). Die Schweizer Alpen. I.* Tauchnitz, Leipzig, pp.1–476.
HEIM, A., 1922. *Geologie der Schweiz. II (2). Die Schweizer Alpen. II.* Tauchnitz, Leipzig, pp.477–1118.
LOMBARD, A., BADOUX, H., VUAGNAT, M., WEGMANN, E., GANSSER, A., TRÜMPY, R. and HERB, R., 1962. *Guide Book for the International Field Institute — The Alps 1962*. Am. Geol. Inst., Washington, D.C., 130 pp.
PIRENNE, H., 1956. *A History of Europe*, 1. Doubleday, New York, N.Y., 282 pp.
SCHARDT, H., 1893. Sur l'origine des Préalpes romandes. *Eclogae Geol. Helv.*, 4: 129–142.
SOCIÉTÉ GÉOLOGIQUE DE FRANCE, 1961. Séance sur la géologie des Dinarides. *Bull. Soc. Géol. France*, 7 (2): 363–526.
SOCIÉTÉ GÉOLOGIQUE DE FRANCE, 1963. Séance sur la tectonique de l'Apennin, de la Calabre et de la Sicile. *Bull Soc. Géol. France*, 7 (4): 625–784.
STAUB, R., 1924. Der Bau der Alpen. *Beitr. Geol. Karte Schweiz, N.F.*, 52: 272 pp.
SUESS, E., 1885. *Das Antlitz der Erde. I.* Tempsky, Wien, 778 pp.
SUESS, E., 1888. *Das Antlitz der Erde. II.* Tempsky, Wien, 703 pp.
SUESS, E., 1901. *Das Antlitz der Erde. III.* Tempsky, Wien, 508 pp.
TERMIER, P., 1903. Les nappes des Alpes orientales et la synthèse des Alpes. *Bull. Soc. Géol. France*, 3: 711–765.
UMBGROVE, J. H. F., 1950. *Symphony of the Earth*. Nijhoff, The Hague, 220 pp.

CHAPTER 10

Alpine Europe: Newer Trends in Alpine Geology

INTRODUCTION

Before going into a somewhat more detailed account of several of the main elements of Alpine Europe, we had better summarize shortly along what lines newer Alpine research has resulted in differences from the classic picture.

INVALIDATING THE EMBRYONIC CORDILLERA CONCEPT

The main feature is, to my mind, that newer work tends more and more to invalidate the concept of the embryonic cordillera. This applies to all aspects of the concept, and because the concept was basic to the classic picture, we must look carefully into the various kinds of opposition brought forward in later years. In their main points, the opposition can be classified under three headings. These are: (*1*) data casting doubt on the tenet of facies continuity within a given nappe; (*2*) fresh information about Tethys paleogeography; and (*3*) tectonic views opposed to nappe continuity or "cylindrism".

As we obtain more and more data, it has become increasingly difficult strictly to correlate variations in geosynclinal facies with individual nappes, or, even more so, the position of a given facies with its place in a nappe. Although there are of course nappes characterized by a singular facies, it has become increasingly clear, this does not mean that such a facies remains constant all along that nappe, nor that a similar facies cannot occur in other nappes.

This realisation is, strictly speaking, not entirely due to newer research. The body of stratigraphical facts, on which rested the assumption of the embryonic cordillera with its correlate facies, had already been critically reviewed by HAUG (1925). He arrived at the conclusion that no variation at all in facies, contingent with individual Pennide nappes, could be found for the Jurassic, whereas for the Cretaceous there were only the most meagre of such indications.

One would think that such a warning, coming from France's foremost stratigrapher, would have been heeded. On the contrary, it seems to have passed almost unnoticed. I believe the reason for this is more stylistic than geologic. Haug's critical remarks, 250 pages of them, are no easy reading. It seems to have been more satisfying to leave them alone, and to keep to the dictates of the classic picture.

In detailed descriptions of parts of the Alpine chain, one encounters again and again statements to the effect that correlation between sedimentary facies and tectonics is not as strict as required by theory. Such remarks, however, always remain tentative, and one has never dared to use them as an attack on the theory of the embryonic cordillera. For instance, a passage from ARGAND (1911, p.17) himself is typical: "The existence of transitions of certain facies of the Dauphinois and the Briançonnais zones is interesting, because this shows, to my mind, that the zones of sedimentation are not strictly parallel to the tectonic zones". The weight of the theory of the embryonic cordilleras seems to have been such, that this and similar remarks always were regarded as pertaining to local exceptions, never to be intregrated into a more general picture. In fact, it took forty years, before, as we will see in the next section, the concept of the embryonic cordillera was really attacked in this same Briançonnais zone, cited above.

A sort of compromise can be noted, for instance, in the chapters on Alpine stratigraphy in the well-known text by GIGNOUX (1955). Here the difficulty of reconciling the dictates of the classic picture with the facies actually found is quite apparent, but only between the lines. Although Gignoux adheres to the concept of the embryonic cordilleras, variations of the facies, duly noted, but not commented upon, do not seem to follow the strict rules required by the classic picture. Coarser facies can, in fact, sometimes be correlated with the crests of various nappes, but at other times they have to occupy a seemingly abnormal position on the limbs.

TETHYS PALEOGEOGRAPHY

Research into the paleogeography of the early Tethys, both in the Swiss and in the French Alps, has gone a step further in refuting the overall validity of the concept of the embryonic cordilleras.

In a series of publications on the Swiss Alps R. Trümpy has stressed the fact that Early and Middle Mesozoic times were characterized by blockfaulting of the geosynclinal floor (cf. TRÜMPY, 1960, 1965). Horst and graben structures developed, often with their own sedimentary series. In the early Tethys there was consequently no period of incipient tangential compression. Instead, we find evidence for tension in the geosynclinal floor (Fig.113).

In the French Alps similar observations have been made. Moreover, in this part of the chain a separate main element enters the picture. This is the Briançonnais zone, already cited, which is quite different in structure from an embryonic cordillera, a fact to which LEMOINE (1953, 1961) and DEBELMAS (1957) have first drawn attention.

We will study this zone more in detail in Chapter 12, but let us point out already here that it formed a large, non-subsiding block between two sedimentary troughs, situated between the Helvetides in the west and the Pennides in the east. This situation persisted at least up to the Cretaceous. This centrally situated "Briançonnais geanticline" is a structure quite dissimilar to the concept of the embryonic cordillera, even though the designation as a "geanticline" is perhaps not appropriate. At any rate it formed a "stable" area, that is an area of very minor subsidence and consequently of very minor sedimentation, right in the centre of the Alps.

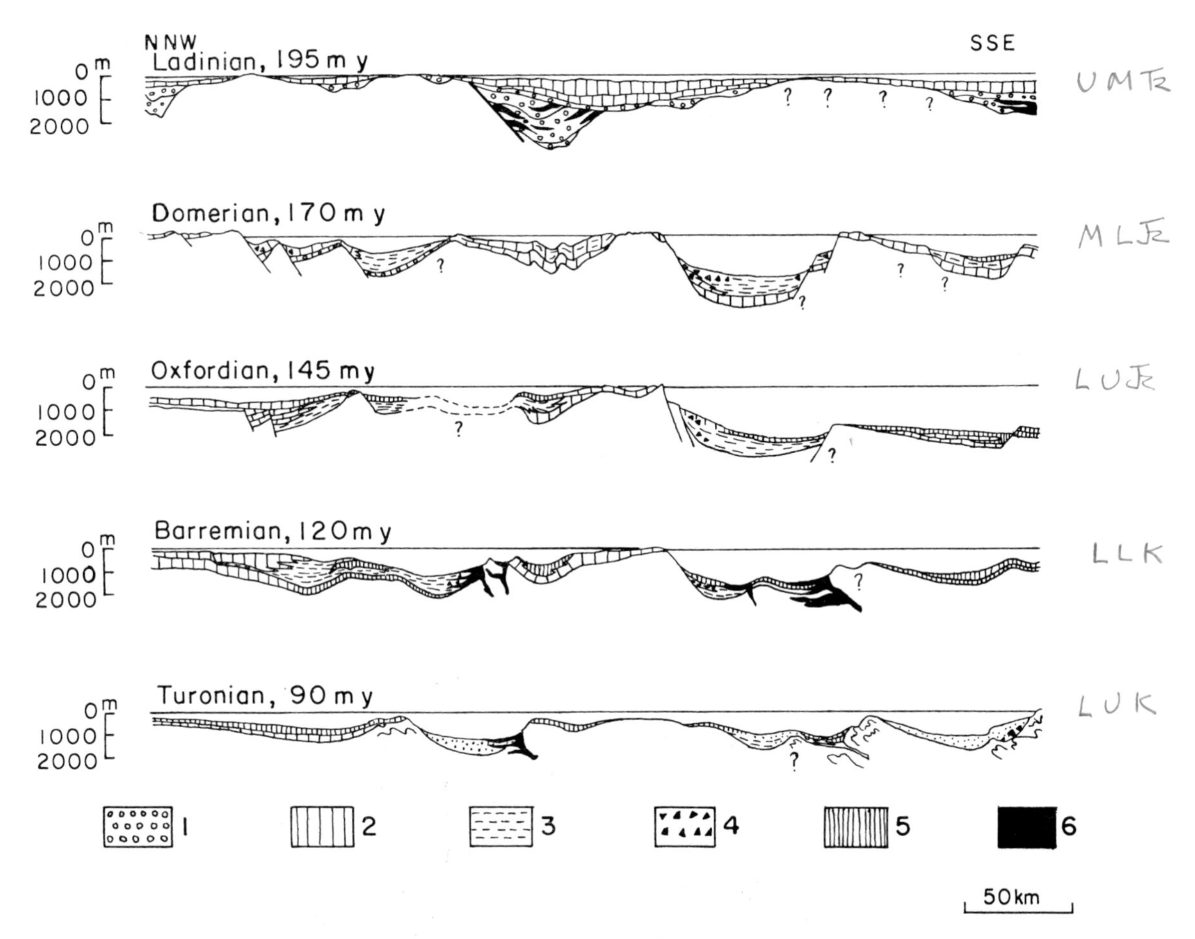

Fig.113. Schematic sections through the Tethys geosyncline, from the Middle Triassic to the Late Cretaceous, illustrating Trümpy's views on Tethys paleogeography.
1 = continental sediments; *2* = neritic sediments; *3* = Bündner Schiefer and related rocks; *4* = marine sedimentary breccias; *5* = pelagic sediments; *6* = geosynclinal basic and ultrabasic igneous rocks. (After TRÜMPY, 1965.)

The newer views are not opposed, therefore, to the existence of differential movements in the Tethys during its geosynclinal stage. They deny, however, both the existence of embryonic cordilleras — together with the causal assumption of tangential compression — and the stringent correlation between geosynclinal paleogeography on the one hand, and nappe facies and tectonics on the other.

VIEWS OPPOSED TO NAPPE CONTINUITY OR "CYLINDRISM"

At a much later date than the initial attack by HAUG (1925), the concept of the embryonic cordilleras was attacked by tectonicians also. This seems to have had far more serious effect on the notion of embryonic cordilleras than the early criticism by stratigrapher Haug. It follows that the two main props upon which this notion rested have now been removed. The paleogeography of the Tethys during Early and Middle Mesozoic is not at all consistent with the picture of a large geosynclinal basin in which, due to incipient compression, folding and even overthrusting was already at work. It has been found, moreover, that individual nappes are not as persistent as was formerly thought. They are even seen to die out occasionally over amazingly short distances.

It has, I believe, never been fully stressed, or even fully realized, how astonishing the fact would be if individual nappes in the Alps should really be continuous over the whole chain. It goes against the grain of comparative tectonics, for it would be a situation which has never been found in any fold belt where it is still possible to really follow individual anticlines and synclines.

In such fold belts one invariably encounters "en échelon" structure, in which individual folds die out and are replaced, along the strike, by others. One could, of course, submit that the Alps, with their nappes, as such are quite different from any other, less folded, orogen. But this contradicts the basic idea that these nappes did in fact develop from simpler folds, that is from the embryonic cordilleras.

To cite only a few authors, FALLOT (1955) has attacked the ideas of nappe continuity for the Eastern Alps, and showed that the published facts are indeed opposed to this notion, whereas ELLENBERGER (1958) has done the same for the French Alps. The latter author concisely summarized these newer views, in stating that detailed mapping left no room for "cylindrism". The model of the continuous nappes was compared to a bundle of thin, long cylinders, or better, half-cylinders, arranged in a continuous sheaf all along the chain. Contrarily, it has now been shown, over and over again, that individual nappes cannot be correlated, that is that there is no cylindrism, sometimes even over so short a distance as the opposite walls of a transverse Alpine valley.

It is, consequently, no longer permitted, in a series of cross-sections over the Alps, to use axial dip in projecting nappes, or even whole nappe systems, far beyond their area of actual outcrop. This eliminates the theoretical basis for the construction of cross sections with the well known piles of nappe systems, such as in Fig.109. Gone is the "proof", both for the lofty aerial connections of nappes across axial culminations, and for their presence deep down in the areas of axial depression.

Nappes, or nappe systems actually are comparable to the folds of more simply built orogens, in that they swell in their zones of maximal development and thin in the direction of replacement. They may well die out altogether, as separate units, along the strike when replaced by other, "higher" or "lower" units, in this respect being in all ways identical with the "en échelon" arrangement of folds in a less tectonized orogen. Such other elements then are only "higher" or "lower" in the area where one unit dies out and is replaced "en échelon" by others. But along the strike they may well take over the total tectonic, volumetric, function of another unit.

This short summary about the success of recent attacks on the concept of nappe continuity or cylindrism, must, as is the case with every notion in geology, be qualified somewhat. For several authors still adhere to it in a general way, whereas other still follow it implicitly. As recently as 1965, TOLLMANN has, for instance, introduced tectonic correlations of nappes, based only on facies, between the French Alps, the Tauern window in Austria, and the Karpathians, leap-frogging distances of about 400 km.

ABSOLUTE DATING OF THE CENTRAL MASSIFS

Another new development is the absolute dating of the crystalline basement that crops out in the Central Massifs.

In the classic picture of the Alps the Central Massifs play the part of the southern mole of pre-Alpine continental Europe. They have become rigid during a former, or during several former orogenies. Their rock material consequently is at least Hercynian, and partly even older. Alpine tectonics were thought to have influenced the Central Massifs only in a superficial way. They have been uplifted and been brought to the present surface. But, according to these views, they only underwent some faulting and shearing in the process, and were not seriously affected.

Absolute dating, mainly carried out by Miss E. Jäger of Bern, has since indicated that the apparent age of at least a part of the mineral content of rocks of the Central Massifs lies in the 25 to 15 million years bracket (cf. JÄGER and FAUL, 1959; JÄGER et al., 1961; JÄGER, 1962; JÄGER and NIGGLI, 1964). That is a truly Alpine age. The geologic interpretation of these radiometric ages is that the older crustal material was seriously affected, and partly regenerated during the Alpine orogeny. For there is no doubt at all of the fact that the original rocks are much older.

According to R. H. Steiger, who studied absolute ages in the Gotthard Massif by the K–Ar method, as against the Rb–Sr method of Jäger, the hornblendes show two definite age groups (STEIGER, 1964; compare also ARMSTRONG et al., 1966). These can be correlated with two distinct tectonic phases, separated by a thermal phase. Together these make up the Alpine metamorphism of the Gotthard rocks.

The earliest Alpine influence resulted in the neoformation, dated as 46 million years old, of hornblendes with a north–south orientation. This is thought to have originated from the drag of the overriding nappes. The subsequent phase of thermal metamorphism, which is also found further south in the so-called Lepontinic region of the lower Pennide nappes in Tessin (compare next section), produced hornblendes with random orientation, from 30 to 16 million years old. The 16 million years Rb–Sr age of the biotites, as reported by Miss Jäger, is thought to represent a final orogenetic event, a period of constriction of the entire region.[1]

Accordingly, as in the classic picture, the Central Massifs are now thought to represent pre-Alpine crustal material. This is confirmed by radiometric dating of the larger constituent minerals of several rocks, which have retained their apparent Hercynian age. The change in viewpoint lies in the metamorphism these rocks are thought to have undergone during the Alpine orogeny. It is quite clear now, that this has been far greater than had been apparent formerly, as deduced from the overall tectonic analysis.

These newer ideas on the relatively strong influence of Alpine tectonics on the Central Massifs confirm the earlier views of Professor B. Sander of Innsbruck. Sander, one of the founders of microtectonics, always maintained that minor structures in the Alps showed Alpine directions everywhere, both in the old cores, and in the younger series. This, according to him was proof of *Überprägung* or "re-stamping". Microtectonic analysis of

[1] A full report on this work has since been published in JÄGER et al. (1967).

the Central Massifs by KVALE (1957) has since corroborated the older views of Sander. But to my mind the Alpine radiometric dates now found for the Central Massifs prove such "Überprägung" more unequivocally than the microtectonic analyses.

We saw, how, according to the classic picture, a fundamental difference exists between the Central Massifs and the crystalline cores of the Pennide nappes; that is, between the "Massifs externes" and the "Massifs internes" of the French nomenclature. As opposed to the Central Massifs, which were formed by earlier, not re-activated crustal material of the European continent, the crystalline cores of the Pennides were thought to have become completely re-activated and built into the nappe system. Moreover, in some interpretations, the crystalline cores of the Pennides belong to the early African continent, to the southern shore of the Tethys. We now know that, although the Central Massifs did not become incorporated into the Alpine macrotectonic elements, they were still so strongly influenced as to undergo more or less complete regeneration of the constituent material, and also a certain degree of microtectonic mobilization.

The difference between "Massifs externes" and "Massifs internes" consequently has been toned down, because the first are now found to have been mobilised during the Alpine orogeny to a much higher degree than formerly was thought possible. In the classic picture this would be difficult to visualize, because in this model we need the rigid crustal blocks of the Central Massifs, to form the southern border of the European continent, against which the sedimentary geosynclinal series of the Tethys had been pushed.

However, according to the newer views on gravity tectonics, where, as we will see, tangential compression no longer plays such a prominent role and can even be dropped out altogether, this mobility of the supposed southern border of the European continent is no longer so important. This newly acquired mobility of the Central Massifs falls in line with the present trend of negation of the existence of the embryonic cordilleras. For, here too, we no longer need a rigid southern border of the European continent, against which the Tethys floor had to be pushed up already in the embryonic phase of the orogeny.

Although the difference between "Massifs externes" and "Massifs internes" consequently now has become more qualified — less black-and-white than in the classic picture — their tectonic position still remains quite different. The Central Massifs still represent part of the basement from which the Alpine sedimentary cover has been stripped off. The crystalline cores of the Pennides, and for that matter of the Austrides, on the other hand, represent parts of the basement that have taken a tectonically active part in the Alpine nappe structures. In the latter elements, however, another new development enters, i. e.,that a part of these "cores" in the deeper nappes is now found to consist of material metamorphosed during the Alpine orogeny. This trend will be reviewed in the next section.

METAMORPHISM AND MICROTECTONICS IN THE PENNIDES

A similar blurring of the classic picture has been going on in the Pennides. The research leading to these newer ideas stems from petrographers and tectonicians, but also from their crossbred offspring, the structural petrologists.

In the classic picture the crystalline cores of the Pennide nappes, which make up the greater part of their bulk, are interpreted as pre-Alpine basement, influenced only through rather light regional metamorphism. This tallies with the fact that their sedimentary envelope, notably the Bündner Schiefer, in general also suffered only light regional metamorphism, which was thought to have developed during the geosynclinal phase. It follows that any massive, granitic rocks within the crystalline cores of the Pennides must represent Alpine magmatic intrusives, according to the classic picture.

E. Niggli drew attention to the fact that within the general area of the Swiss Pennides a certain zone shows relatively stronger metamorphism, both in the nappe cores and in the Bündner Schiefer. This area of stronger metamorphism shows a distinct zonation, with the most strongly altered rocks in the centre (cf. Niggli, 1960).

This zonation is independent of, and cuts across, individual nappes. It must therefore be younger than the formation of the pile of Pennide nappes. For Niggli this extra effect is a static metamorphism, caused by the overburden of the higher nappes, since removed by erosion.

This interpretation has been challenged on several grounds. Chatterjee (1961) has pointed out that Niggli apparently has compared polymetamorphic zones, which would invalidate the whole concept. Wenk (1955, 1967) prefers to think of a deep-seated thermal source which is found reflected in the extra static metamorphism. Wenk's ideas are particularly well illustrated in his 1966 publication, in a composite longitudinal section through the Pennide nappes, in which the distribution of several minerals, some low-, some high-temperature, is indicated. And lastly Plessmann and Wunderlich (1961), and also Chatterjee (1962), stress the importance of Alpine microtectonic structures, and prefer a dynamic over a static interpretation.

E. Niggli and C. R. Niggli (1965) and Niggli and Graeser (1966), in submitting more data, now carefully distinguish between possible influences from polymetamorphism. A number of separate maps for six key minerals confirms the earlier concept that their distribution is not related to individual nappes but to the overall form of the Alps. The interpretation of the phenomenon, which is now left open, is nevertheless thought to be the effect of the multiphase history of the Alpine orogeny.

On a more general scale, it has become clear that Alpine metamorphism, all over the world, has been something special. De Roever and Nijhuis (1963) notably hold that it is characterized by the glaucophane-schist facies. The almost exclusive repartition of glaucophane in the Pennides of the Alps is well illustrated by Van der Plas (1959, pl.I), whereas more descriptive material can be found in Bearth (1958,1962). A possible explanation of this correlation, evidently existing between the occurrence of glaucophane minerals and the Alpine orogeny, is given by De Roever (1965), who supposes that the geothermal gradient has been higher during earlier orogenies, than during the Alpine.

To return to the more local problem of the metamorphic rocks in the Lepontinic gneiss region of the deeper Pennide nappes, we may cite a discussion of the historical development of the interpretation of a single element, i.e., a part of the crystalline core of the Adula nappe (Müller, 1958).

Before the advent of the nappe theory, the crystalline series of this area were thought to represent autochthonous basement. These series are made up by micaschists, amphibolites and gneisses. The latter are characterized by an apple-green mica, called phengite,

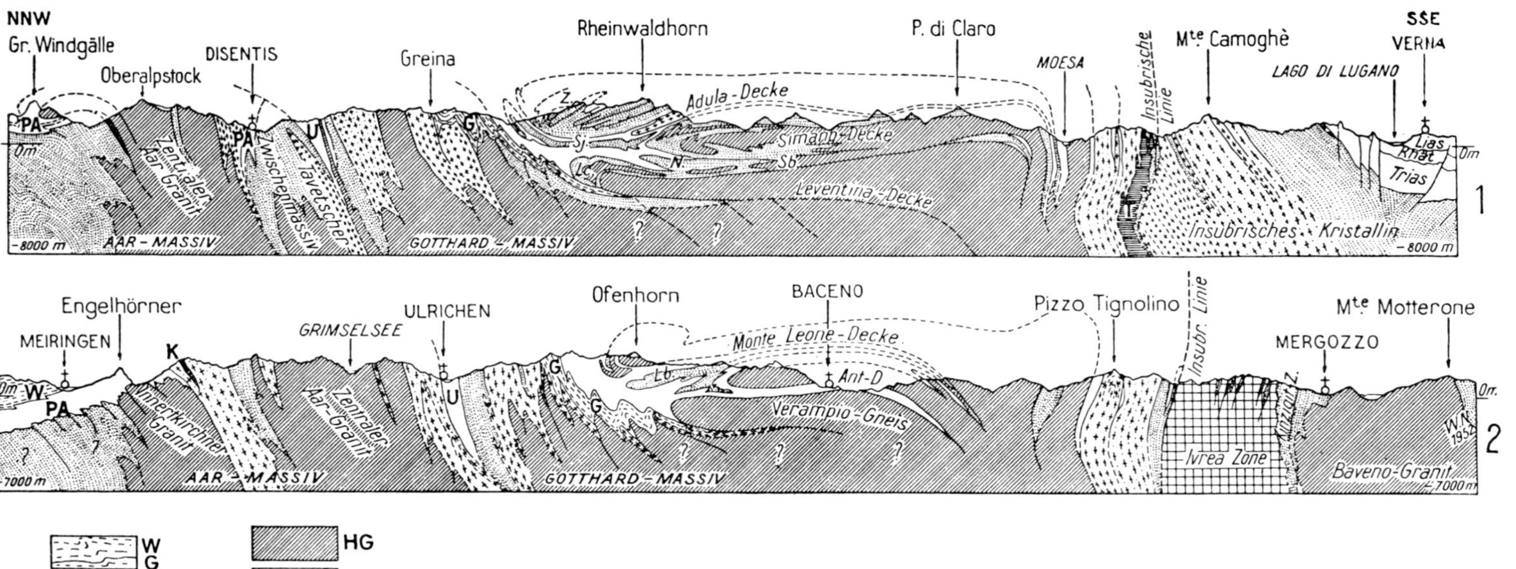

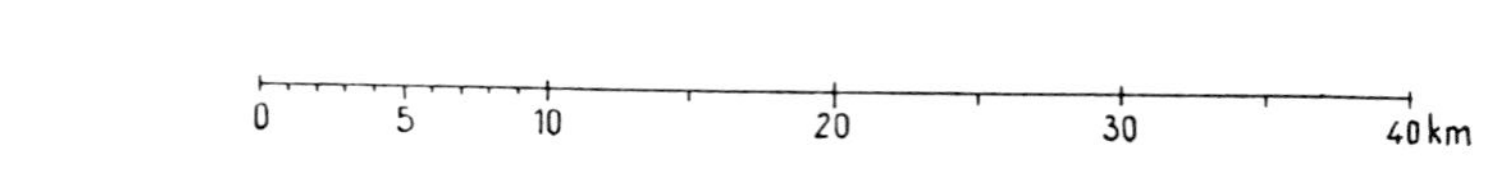

Fig.114. Two sections across central Switzerland, which show the tectonic interpretation of the crystalline cores of the Pennide nappes.

At the left, the Aar Massif with the Helvetide nappes (*PA* = par-autochthonous and autochthonous sedimentary cover; *W* = Wildhorn nappe), the Tavetscher Zwischenmassiv and the Gotthard Massif (*U* = "root zone" of the Helvetides; *G* = sedimentary cover of the Gotthard Massif). Then follow the Pennide nappes, the "root zone" (Ivrea and Insubric zone) and, in the upper right, the Southern Alps.

Apart from their Mesozoic mantle, the Bündner Schiefer, enveloping the Pennide nappes (white on the sections), all of the crystalline cores are interpreted as pre-Hercynian basement (*HG* = Hercynian granites; *O* = orthogneisses; *M* = migmatites; *P* = paragneisses).

According to modern authors, we must interpret the cyrstalline cores of the Pennide nappes to represent older and/or younger material, metamorphosed, metasomatosed, and/or granitized during the Alpine orogeny.

The Baveno Granite, in the lower right hand corner, godmother of the feldspar twins, is here interpreted as the equivalent of the Hercynian Central Massifs. Its age of 270 million years (Jäger and Faul, 1959), placing it at the Carboniferous–Permian boundary, is younger, however. It might possibly be related to other granitic bodies of the Southern Alps, in particular the Cima d'Aste, which has the same age. The latter is combined with the massive ignimbritic rhyolites of the Southern Alps in a late-Hercynian Venetian magmatic province by D' Amico (1964), cf. p.333. (After Nabholz, 1953.)

which plays a prominent part in the regional literature. This interpretation was adhered to, for instance up to 1910, by a first rate authority, such as Grubenmann. Quite apart from the interpretation as autochthonous basement or as nappe cores, a discussion developed as to whether these crystalline rocks were magmatic or migmatic.

The beds of marble and dolomite, known to be intercalated in the gneisses, and noted already for their similarity with elements of the Triassic elsewhere, were, during this first stage, not further considered. These of course came into their own with the subsequent nappist interpretations, in which every intercalation of marble or dolomite was thought to represent a major thrust zone. During this period of predominance of the tectonic interpretation, which persisted up to 1956 (cf. NABHOLZ, 1951, 1953, 1954), the Adula crystalline is thought of only as the tectonically re-activated pre-Alpine basement (Fig.114).

Already since 1936, however, other authors have expressed the opinion that much, if not all, of the present aspect of the Adula crystalline arose during Alpine metamorphism. Müller's own conclusions are that, although the Adula phengite gneisses locally contain remnants of earlier, probably Hercynian, metamorphics, these and younger rocks have become almost totally remobilized during the Alpine metamorphism. The latter, which developed during several stages, has even led to the formation of plastic rocks, which now form the granitic bodies within the migmatite series.

It is consequently no longer necessary to construct a pre-Alpine core for each crystalline body. But the tectonic analysis has become the more difficult, because it now has to be decided first, from case to case, which of these gneisses really contains pre-Alpine elements.

The interpretation based on the microtectonic analyses of H. G. Wunderlich and his fellow workers has shifted almost entirely to a dynamic view. Here it is held that only when the tectonization of the Alps reached the point at which all rocks became mobilized in detail, the possibility for Alpine metamorphism arose.

This research is consequently a sort of revival of the pioneer work of Professor B. Sander of Innsbruck. But Sander was an Alpine geologist to begin with. Although his book (SANDER, 1948, 1950) is quite unreadable, he knew better than to apply microtectonics indiscriminately. His research started from, and was always applied to, separate units of major tectonics. Even then, as we saw in the discussion of the work of Hoeppener in the Rheinisches Schiefergebirge, conclusions about genesis and dating are necessarily vague. Sander's main conclusion, that there exists a strong Alpine influence, an "Überprägung" of all older material, is, as we saw, quite in accord with other, more modern research. But it is hazardous to extend these general conclusions to more specific ones.

It follows that I have strong personal qualms in accepting the value of results such as published by WUNDERLICH (cf. 1964). Microtectonics, to have significant results, must be limited to a statistical treatment of microfeatures within a homogeneous area. It must, moreover, cover, if not all, at least a significant part of such an area. Maps in the scale of 1:2,000,000, or 1:4,000,000, with microtectonic features indicated by thick arrows, scattered as raisins in a fruit cake, are not very convincing. At best they may indicate the possible existence of structures which might have developed, either consistently with macrotectonics, or consistently with metamorphosis, or with both. Structures worthwhile to be studied in more detail in single macrotectonic units, which in this way only might lead to significant results (cf. KVALE, 1957) on the microtectonics of the Gotthard Massif.

GRAVITY TECTONICS

As indicated already on p.181 the question of the motor of nappe formation has originally played no great role in the history of the development of the nappe concept. We find a most poetic description of nappe formation by gravity by the founder of the nappe concept, M. Bertrand. After comparing a local nappe in the Île de Bousset in the Lower Provence (p.362) with the nappes near Glarus and the Faille du Midi in the French–Belgian coal district, BERTRAND (1887, p.702) writes: "..*ces grand plis couchés, qui se déroulent, s'allongent, forment de larges trainées au-dessus des couches plus récentes, et simulent de veritables coulées de terrains sédimentaires, rappelant presque les coulées de basaltes.*" One can hardly think of a more eloquent statement, which I have not dared to translate, of the results of gravity tectonics.

As we saw, SCHARDT (1893) also was mainly interested in the existence of these nappes. This was a reality, apparent from field work, whereas their genesis was another, more theoretical aspect. Because, as we have seen, the model of the cooling and shrinking earth provided the requisite motor for tangential compression, and not (at least not as directly) for gravity tectonics, tangential compression then became universally accepted as the cause of nappe formation.

It follows that tangential compression has never been proved by field evidence. Nevertheless, it has become so well established, that the burden of disproof now lies with the other side, that is with the advocates of gravity tectonics.

From a historical viewpoint, we may note that early heretics, such as E. Haarmann in Germany and O. Ampferer in Austria, were members of technical universities. From their familiarity with the technical properties of matter they were not inclined to believe that tangential compression could move and overthrust such bodies as nappes which are not only extremely thin, but also very extensive in horizontal direction. They supposed instead that gravity, affecting every mass element of a tectonic body, would be the only force able to produce such structures.

These generalized theoretical considerations apparently have come too early for general acceptance. And it remains a curious fact that the main push towards the present growing adherance to gravity tectonics has come, not from a study of the mechanical properties of rocks, but from geological field work. Moreover, it has started, half a century later, from the same area in which H. Schardt originally had stated that field evidence could not decide between tangential compression and gravity — that is from the Swiss Prealps.

The newer ideas have been set off mainly through the effort of Professor E. Gagnebin (LUGEON and GAGNEBIN, 1941; GAGNEBIN, 1945). As we will see, the Prealps are characterized by a stratigraphy and a tectonic style which are both quite different from the other nappe units in Alpine Europe. It is therefore quite understandable that Gagnebin expressly stated that gravity tectonics was proposed only for the Prealps, whereas all other Alpine nappes were still believed to originate through tangential compression.

But although this difference in tectonic style between the Prealps and the other nappe systems exists, this attitude has since changed, slowly and often almost imperceptibly, until at the present time various authors want to explain all nappe formation in Alpine Europe through gravity as the primary motor. GIGNOUX (1950) was one of the first to go all out for gravity tectonics, whereas FALLOT (1950) applied the concept to the Southern Alps in the same year. Later, one finds allusions in the literature stating that the Swiss Helvetides can be explained only by gravity sliding. And more elusively still, it is implied, that if and when the Helvetides should be formed by gravity tectonics, there seems to be no valid reason to suppose that the Pennides were formed by tangential compression.

What evidence there is for gravity tectonics is mostly found in epidermis type nappes. Or, to put it in another, perhaps a better, way: in epidermis type nappes various features occur that are very difficult to explain through tangential compression. Three of these features might be cited here: (*1*) the *Branden der Deckenstirne* and (*2*) the *Voraneilen der jüngeren Schichten* of the Swiss tectonicians; (*3*) the *chevauchements intercutanés* of the French.

The *Branden der Deckenstirne*, the "breakers at the nappe fronts" alludes to the fact that the Helvetide nappes everywhere show an extremely quite, subhorizontal structure, except along their frontal part. Instead the nappe fronts are all crumpled up, one narrow fold following the other. Compare the famous drawing of the serrated folding of the Säntis chain in northeastern Switzerland by A. Heim (Fig.129).

A nappe would most likely get crumpled where the reaction between moving force and resistance was greatest. During gravitational gliding this is indeed at the nappe front, due to frictional resistance. It would be greatest at the rear end of a nappe being pushed from behind.

This feature is not restricted to the Helvetides, though it is perhaps best developed in this nappe system. It is, for instance, well developed in the Adula nappe, already cited above.

Quite often such a frontal crumpled zone shows a type of folding which has been designated as "cascade folds". In these a series of serrated folds does not have the subvertical axial planes, as shown, for instance in the Säntis, cited above, but shows axial planes dipping around 45°, or even less. Such a series of cascade folds can hardly be explained but as a series of beds plunging into a depression through the influence of gravity.

The *Voraneilen der jüngeren Schichten*, the "leading the way by the younger beds", relates to the fact, already alluded to, that of a given unit of epidermis the younger series so often are found to have travelled farthest. Again, this is a feature especially well developed in, and also first described from, the Swiss Helvetides, but found all over the Alpine chain.

One of the most obvious examples is found in the Helvetides of northeastern Switzerland. From north to south, one finds there, first the Kreide-Decke ("Decke" = nappe), the former Drusberg–Säntis-Decke, mainly formed by Cretaceous; then the Jura-Decke, the former Axen-Decke, mainly formed by elements of the Jurassic; and lastly the Verrucano-Decke, the former Mürtschen-Decke, in which the Permian verrucano forms the main element.

This situation follows readily from the concept of gravity tectonics, where the highest elements of a series have the greatest energy potential. It is, on the other hand, difficult to visualize in the tangential compression model. For one then needs differential push, to move the higher parts of a series several tens of kilometers further north. And even if one could arrive at a differential push model, it would be expected that the larger movements were at the base of the series, where the tangential forces are exerted by moving crustal masses.

The classic example of the *chevauchements intercutanés*, the "intercutaneous overthrusts", are those near Roya in the French Alps (FALLOT, 1949; Fig.158, 159). In this type of structure smaller and larger subhorizontal overthrusts are found to exist in a seemingly tranquil, pseudo-autochthonous series of sediments.

Tectonic movements have taken place, not only between the basement and the epidermal sedimentary cover, but also within the epidermis itself. The internal overthrusts have resulted in intercalating foreign elements, of older and of younger age, within the normal series of epidermis, which is itself thrust over the basement. Subhorizontal movements must consequently have taken place both at the lower and at the upper contacts of these foreign elements. And this even notwithstanding the fact that the tranquil upper part of the sedimentary series overlies the internal structure without any apparent break.

To many geologists, the existence of such structures has formed the final argument in favour of gravity sliding and against tangential compression. Intercutaneous overthrusts have later been found in many other parts of the Alpine chain, but the influence of Fallot's description of Roya has been tremendous. I believe it is the main reason why the French geologists have gone out so overwhelmingly in favour of "décollement".

As in the times of H. Schardt, proof for or against either tangential compression or gravity tectonics generally is not found. But more and more local sections have become cited in which gravity tectonics offers an easier explanation. And if we remember how, only a couple of decades ago, tangential compression was the almost universally accepted motor in tectonics, the spreading of the concept of gravity tectonics is remarkable indeed. This is especially true, because the burden of proof still lies with the latter concept.

Of course, there are regional, or even national variations in the rate of spreading of the acceptance of gravity tectonics. In Switzerland, CADISCH (1960) was still violently opposed to the new views, whereas the French literature has become permeated almost completely by the idea of "décollement" in the sense of gravity sliding since about 1950.

TIME OF FOLDING AND OROGENETIC PHASES

There exists a vast literature, to which I will draw only scanty attention, on the time of folding and on the various orogenetic phases. In relation to the concept of gravity tectonics, the exact date of an orogenetic movement (that is, of local or regional folding) has become of far lesser importance than in the tangential compression model. This will permit us to skip most of the controversies existing in this field.

In the tangential compression model, orogenetic movements are a direct result of the primary orogenetic forces. Movement is consequently thought to be simultaneous with an orogenetic phase. In gravity tectonics, on the other hand, orogenetic movements, folding, overthrusting and nappe formation, are secondary phenomena only. They are a consequence of the primary vertical crustal movements, which supply the necessary energy. In this case, folding and overthrusting may either follow immediately upon the primary vertical movements, or may be retarded over a shorter or longer period, depending upon the resistance offered against the gravitational reaction. The latter depends largely upon local circumstances, which may consequently result in strong variation in time of the secondary movements caused by the same set of primary vertical movements.

In the image of the tangential compression model there has been sharp controversy between those who maintained that orogenetic movements were synchronous the world over — the Stille concept — and those holding that they were of local or regional extension in space and time. The different aspect of this problem in the light of gravity tectonics has first been stated expressly by FALLOT (1955), whereas I have once tried to summarize pro and contra in a supposed dialogue between a coal geologist and a university professor (RUTTEN, 1962). The conclusion being that there is indeed a wide variation, both in space and in time, of orogenetic movements, as could be expected from the theory of gravity tectonics.

Another complication in this field stems from the fact that Alpine geomorphologists have often laid inordinate stress on very young movements. The result is that periods of uplift, occurring much later than the folding, but leading to strong erosion, have been indicated as "main phase" of the orogeny. One should shift carefully, when studying Alpine literature, the statements in which such post-orogenetic isostatic adjustment has been confused with orogenetic phases, from those concerning true orogenetic phases in a geological sense.

It must be stressed, however, that there are variations in time of folding which are of major importance and which stand out from the petty bickering alluded to above.

For example, just as in so many other fold belts, the youngest movements seem to have concentrated along the external border. As a result, the Miocene molasse is tectonized along the Alpine borderfault, and has become strongly up- and overthrust, whereas elsewhere it is hardly affected by tectonic movements.

Or, to take another example, two different major phases can be distinguished clearly in the Austrides, but not in the Helvetides and the Pennides. These are an older, pre-Gosau (p.382), which is pre-Late Cretaceous, and a younger, the normal Alpine one, of about Oligocene age.

A third example may be noted in the fact that both Pyrenees and the Provence chains had their main phase in the Eocene "Pyrenean phase", earlier than the Alps proper.

Apart from these larger differences, however, most of the controversies on "age of folding" in the Alps are not of great importance for the overall picture of the development of Alpine Europe, and will consequently be skipped in this text.

THE "ROOT ZONE"

It follows logically that, according to modern views, the root zone concept has lost its importance. Of course, one can maintain that nappes which have glided, through gravity, from a rising portion of the crust, may also be thought to have a "root". But in this case "root" only means the line along which the rear end of a nappe formerly was in contact with that part of the same series which did not take part in gliding. That is the "break-away fault" of PIERCE (1960).

Such a break-away fault is quite a different proposition from the former root concept.

There, the root of a nappe is thought of as a belt of subvertical strata at the crustal cicatrice representing a former wide zone, from which the nappe has been squeezed out.

It might be considered an extra boon if gravity tectonics would be able to solve, or perhaps dissolve, the root zone problem. For amongst all problems of Alpine geology, those pertaining to the root zone were the least agreeable. Such a state of affairs, however, has not yet been generally achieved. It still figures, for instance, as a real root zone according to the classic nappist structure in the synthetic cross-section through the Swiss Alps by TRÜMPY (1965).

Apart from its general interest for the classic picture as the root area of the Pennide nappes, the internal structure of the root zone has been equally important for classic nappist theory. For it should be possible here to distinguish separate, "real" nappes from "secondary" nappe digitations or involutions. According to the theory, only the former can have roots. Nappe systematics, just as any other systematics, has had its lumpers and its splitters, and so attempts were made to solve the controversies by a study of the root zone. Owing to its complicated structure, however, the exegesis of the root zone could be accommodated to whatever theory an investigator adhered to. The lumpers among nappists only found a small number of separate roots but the splitters discovered just as many as needed. The literature on the root zone is seriously cluttered by the existence of these controversies in nappe systematics.

As was the case with the newer ideas on the Central Massifs, it is not so much the existence of a peculiar, separate zone, which has been doubted, as its theoretical interpretation in the root zone concept. It is beyond any doubt that this root zone or (better perhaps, this former root zone, or even "root zone") exists, and forms one of the major structures in the overall plan of the Alps.

A merely descriptive designation seems at present preferable. From west to east the synonymy of the "root zone" then is first the Ivrea zone. That is the Ivrea zone, sensu lato, including such parallel subzones as the Canavese zone, the Sezia zone and the Kinzingite zone, which are sometimes differentiated, sometimes not. It is followed eastwards by the much narrower Tonale zone, often referred to as the "Tonale Linie". To the east of this element, the "root zone" is set off to the north-northeast along the Judicaria fault. From there, it can be followed eastwards again as the "Puster Linie", the "Gail Linie", and finally the "Karawanken Linie".

This profusion of local names is given here as a guide to the local literature, because in this general text it is impossible to go into details. Details are, moreover, less important here, because the main fact is the overall structural similarity of this "root zone". It is a zone of subvertical, metamorphosed and strongly tectonized rocks, 400 km long and 20 km or less wide, in which, moreover, all post-orogenetic intrusions of the Alps have occurred. So it not only separates the Northern Alps — the Pennides in the west and the Austrides in the east — from the Southern Alps, but it also shows its own type of structure.

Regardless of its interpretation, the "root zone" evidently is a major crustal cicatrice. But how and when it was formed is not at all clear at present. Nor even can we say definitely if it was formed during a single, possibly prolonged, tectonic phase, or if it dates back in part to a much older period.

Moreover, it remains possible that the "root zone" is complex. Its character as a strongly tectonized and metamorphosed zone is most typically developed in the western part, south of the Pennide nappes. In its eastern part, after it has been offset to the north by the Judicaria fault, it still is strongly tectonized, but at least the rocks at the surface are only lightly metamorphosed, if at all. They belong to the sedimentary cover (Chapter 13).

The geophysical explorations reported upon in the next paragraph seem to underline the complex build of the "root zone". The gravimetric Ivrean high, which, according to seismic analysis, is formed by a big basic body, seems to have no counterpart farther east along the "root zone".

GEOPHYSICAL INVESTIGATION OF THE ROOT OF THE ALPS

It must be remembered that the "root zone" of the preceding paragraph, is something very different from the geophysical "root of the Alps". The latter is a hypothetical structure, proposed to explain the existence of a belt of negative Bouguer gravity anomalies coinciding with the main body of the Alps.

It is postulated from isostatic considerations to be formed by excess of sialic matter, due to crustal shortening during the formation of the Alps.

The existence of this isostatic root of the Alps, as opposed to the much narrower "root zone", is well known. It is cited in all tectonic textbooks. And though an extra complication has in later years been added by the existence of the narrow positive anomaly of the Ivrean high (Fig.115, 116), it could have been passed over in this text, were it not for some important newer developments.

These are the result of an international cooperation by a group of seismologists. They have been observing from a great many stations the waves propagated from several major artificial blasts up to a maximum charge of 25 tons of explosives in high altitude Alpine lakes. The logistics of these operations were difficult, to say the least. Not only had as many observation posts as possible to be manned at the same time, but the lakes were of difficult access. They had to be situated at such an altitude that no damage occurred from the explosion (CLOSS and LABROUSTE, 1963; CLOSS, 1963, 1965, 1966; FUCHS et al., 1963; PRODEHL, 1965).

Apart from the more local complication offered by the Ivrean high, these seismic explorations revealed that the root, as found gravimetrically, cannot have been formed by a simple downbuckling process of sialic material, due to tangential shortening. It was found that the negative gravity anomalies indeed correspond to a thickening of the crust, but not of the upper, sialic layer. The granite-and-sedimentary layer above the Conrad discontinuity hardly deviates from normal thickness underneath the Alps. Instead, the "basaltic" or "intermediate" layer between the Conrad and Moho discontinuities shows considerable thickening (Fig.117).

To cite CLOSS (1965): "The thickening of the basaltic layer without corresponding change of thickness of the granitic layer may indicate special processes at depths between 20 and 70 km, or deeper, details of which are unknown as yet". And also: "In the main part of the Alps we could not find seismic evidence of a tectonic mixing of granitic and intermediate material".

For the Ivrea body CLOSS (1965) thinks of "a thickening of the basaltic layer due to

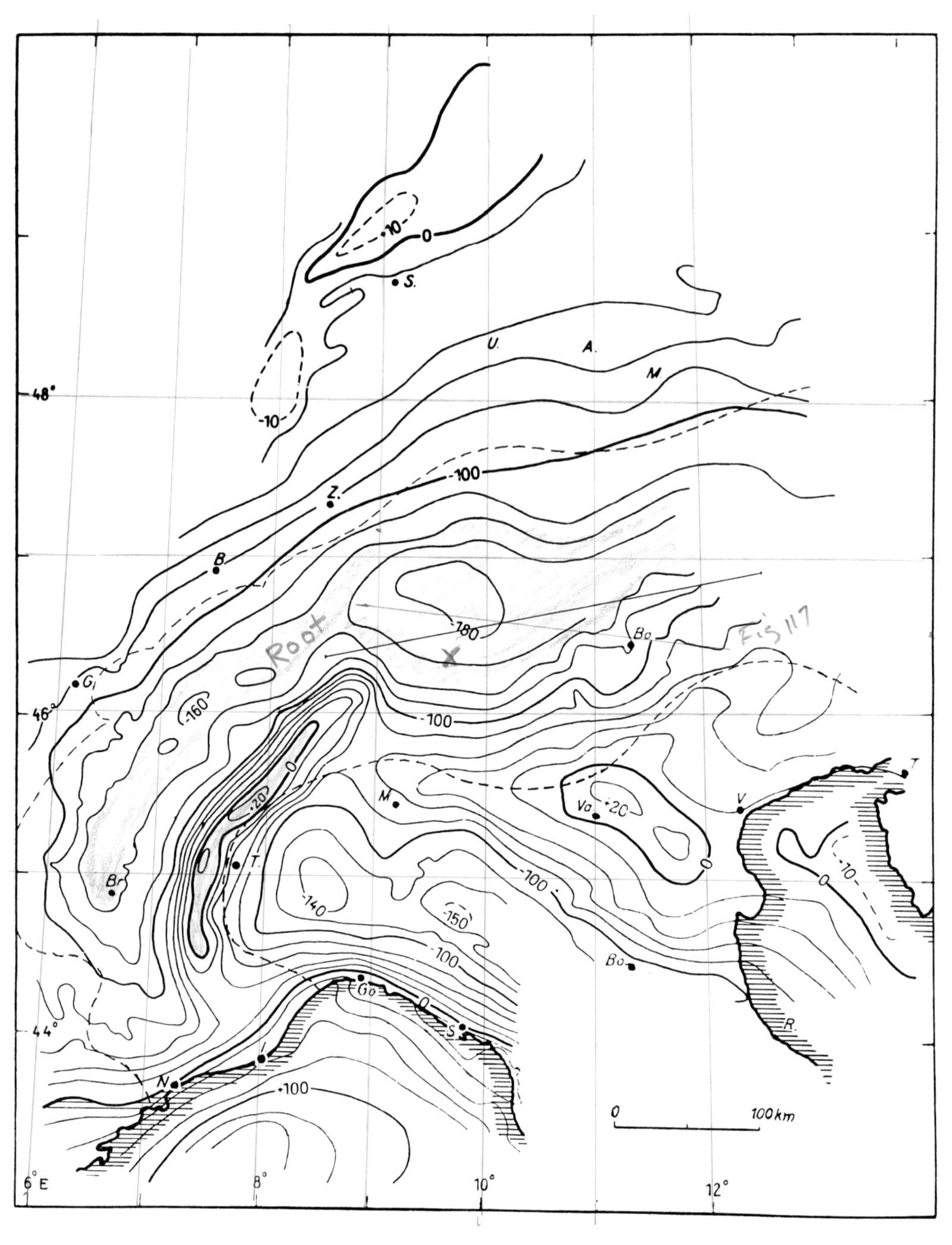

Fig.115. Map of the Bouguer anomalies in the Alps.
A = Augsburg; *B* = Bern; *Bo* = Bolzano and Bologna; *G* = Geneva; *Ga* = Genoa; *I* = Imperia; *M* = Münich and Milano; *N* = Nice; *R* = Rimini; *S* = Stuttgart and Spezia; *T* = Trieste; *U* = Ulm; *V* = Venice; *Z* = Zürich.
The broken line schematically indicates the extent of the Alpine chain.

The root of the Alps is well expressed in the negative anomaly of more than −100 mgal, which follows the curvature of the Alps from north of Nice to south of Münich. A rather unexpected feature is the narrow zone of positive anomalies accompanying the inner flank of the Western Alps. It coincides with the zone of Ivrea and is called Ivrean high in the literature. Further east the "root zone" does not seem to be well expressed geophysically, either in gravimetry, or in seismic data. (After CLOSS, 1965.)

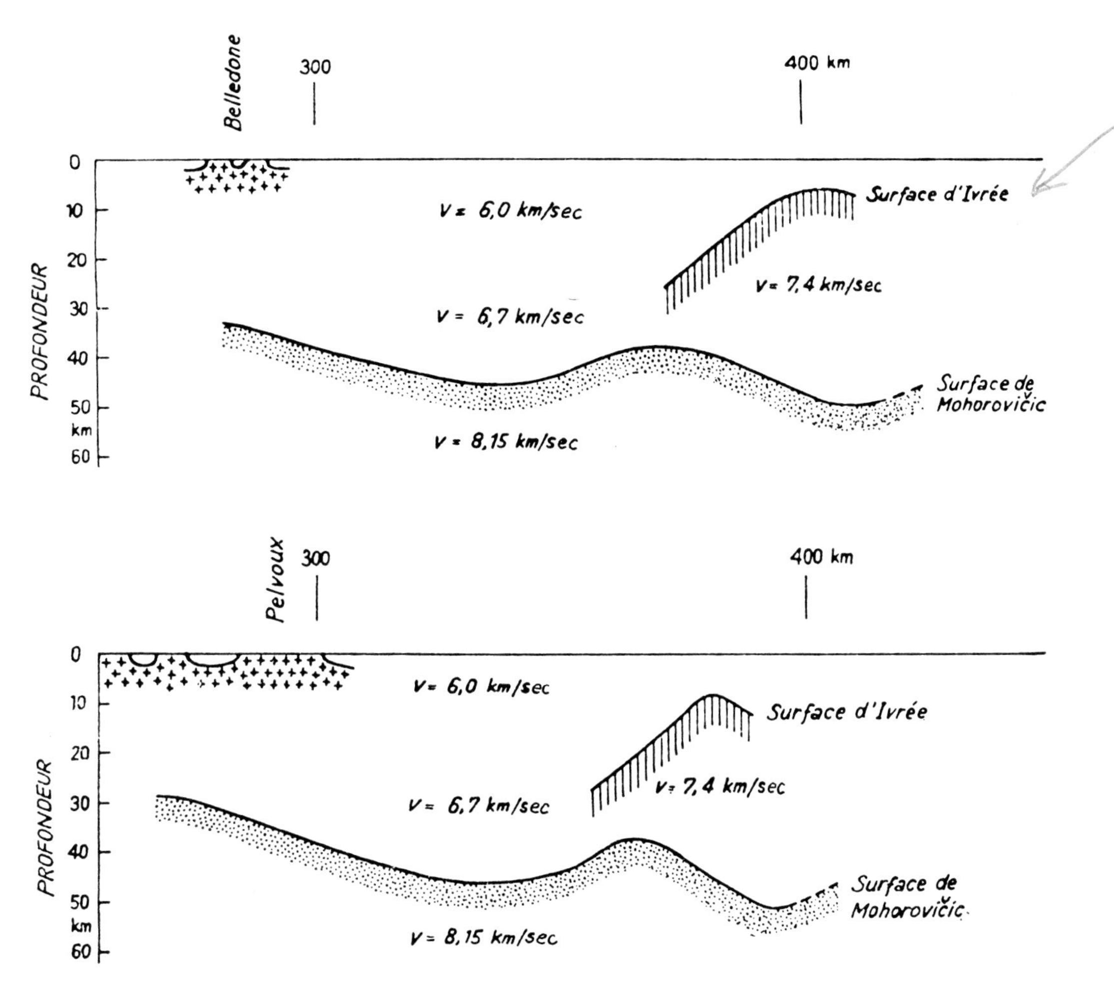

Fig.116. Seismic section across the French Alps. The Ivrean high ("Surface d'Ivrée") of Fig.115 is also found by seismic analysis. Its velocity of 7.4 km/sec indicates a large basic body. (After CLOSS, 1963.)

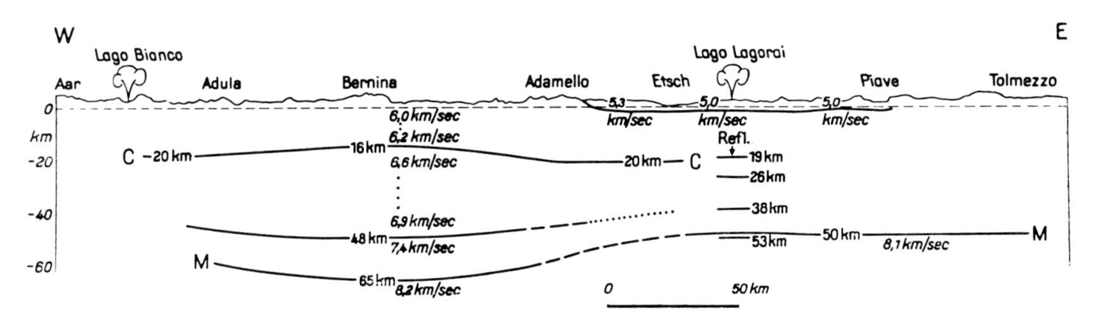

Fig.117. Seismic west–east section across the Alps from Switzerland to northern Italy.
C = Conrad discontinuity; M = Moho discontinuity.
The section obliquely crosses the Pennide nappes and the Southern Alps. The root of the Alps is found at the same location by seismic investigation, as it was found gravimetrically, i.e., underneath the Bernina nappe of the Pennides. The root is, however, not formed by a thickening of the granitic layer above the Conrad discontinuity, but by a thickening of the layers between Conrad and Moho. These were formerly called "basaltic" in the literature, but are now more prudently indicated as "intermediate". (After CLOSS, 1966.)

processes unknown. It might best be compared with similar conditions found in structures considered as "Alps" in the process of formation, such as existing now near Puerto Rico".

As in geology, the conclusion drawn from geophysics seems to be that the classic picture of the Alps is wrong. But, again, we have as yet no new, generally accepted model to replace it with.

To conclude this section, it might be remarked parenthetically that geophysical research in Europe north of the Alps has revealed no trace of a similar root underneath the Hercynian uplands. Any root which would have existed there at the end of the Hercynian orogeny has consequently been obliterated since (PRODEHL, 1964; CLOSS, 1966).

TANGENTIAL SHORTENING

Another wholesome aspect of gravity tectonics is that the amount of shortening, required for the geometrical evolution of the Alps, is much smaller than in the tangential compression model.

If, instead of originating from a recumbent fold, a nappe is postulated to be formed by a thrust plate, this already influences the required amount of shortening. A much greater reduction occurs, when, as for instance in our example of the Helvetides cited in relation to the "Voraneilen der jüngeren Schichten", three or more units, that have been interpreted as separate nappes, can now be re-interpreted as erosional remnants of a single tectonic unit. A similar case is dealt with, in more detail, by FALLOT (1956), in his "Promenade d'hypothèse en hypothèse", for the example of the Swiss Prealps.

The figures actually arrived at for the tangential shortening of the Alps vary from a maximum of the order of 600 km to a minimum of the order of 100 km. The variation depends mainly on the interpretation of the structure of the Helvetides and the Pennides as either thrustplates, broken up by the "Voraneilen", or as recumbent folds, and upon the tectonic position assumed for the Prealps. However, if the most nappist hypothesis for the Prealps, as cited by P. Fallot, would be really applied to unfolding, a much higher figure would be arrived at. The same holds true for the ideas of A. Tollmann for the Austrides (cf. TOLLMANN, 1965).

I will not go further into the figures mentioned in this controversy, because it has lost much of its importance in view of the new results from a new line of research, i.e., paleomagnetism. This will be further reviewed in the next section.

But, curiously enough, just at the time the amount of crustal shortening, necessary for the geometrical evolution of the Alps, could be strongly curtailed, paleomagnetism now seems to indicate the possibility of much greater crustal movements during the Alpine orogenetic cycle.

Of course, the two sets of data are not mutually exclusive. Continental drift over very large distances is not forcibly accompanied by nappe formation. The latter may be restricted to a short period of such drift, or, for all we know, not be related to it at all. It seems therefore the more important to study the amount of crustal shortening which can be deduced from tectonic analysis, and to compare it with the data supplied by paleomagnetism.

PALEOMAGNETISM AND THE ASSEMBLY OF ALPINE EUROPE

In contrast to the figures for crustal shortening derived from tectonic analysis of the Alps, which vary between the order of 100 km and 600 km, evidence from paleomagnetism now points to translations of part of the chain of the order of several thousands of kilometres since the Permian, and relative to the stable block of Meso-Europe. Moreover, it is probable that different parts of Alpine Europe have come from quite different directions. They seem not to have travelled over the same distance, nor to have followed the same path.

In an even more sketchy way, similar results are suggested for older fold belts (RUTTEN, 1965). It seems as if the development of a larger orogen, such as the Alpine chain, is somewhat comparable to the assembly of a mozaic of crustal blocks which originally may have been at great distances from each other (RUTTEN, 1964).

Paleomagnetic measurements from the Permian of the Southern Alps gave the first indication of a pole position deviating strongly from that found for Europe north of the Alpine fold belt, that is for Meso-Europe (VAN HILTEN, 1962). Similar deviations have since been found for other Permian localities elsewhere in the Southern Alps and also for other parts of the chain, i.e., for the Lower Provence, Corsica and the Pyrenees. It is now well established that Permian pole positions from Alpine Europe differ consistently from those found for the Permian of Meso-Europe. There exists, however, no uniformity as to the deviations shown by various elements of the Alpine chain. Large variations are found, both in the declination and in the inclination of the Permian remanent magnetism, indicating large differences, both in the amount of translation, and of rotation, which different elements of the Alpine chain seem to have undergone since the Permian.

Paleomagnetic measurements supply the inclination of the magnetic field at the time of formation of the rocks measured and the declination, as measured against the present polar axis. From this, as a direct result, only the paleolatitude can be calculated, not the paleolongitude. An indirect approach is possible, when, as is the case with the Permian of both the Alps and Meso-Europe, the former declination differs enough from the present one. The paleomagnetic isoclines (and of course also the contemporaneous parallels) derived from the inclinations measured, then form a considerable angle with the present system. Together with the data from regional geology it is in these cases often possible to arrive at more definite conclusions.

For the Southern Alps such a study is supplied by DE BOER (1963), who studied the paleomagnetic history of the Vicentinian Alps in northern Italy. For the Permian he found an average inclination of —20°, instead of the anticipated 0°, inferred from the known equatorial position of the southern border of Meso-Europe for that period. The —20° isocline for Meso-Europe, drawn around the average pole position for the Permian, lies far to the north, in the stable, continental part of Meso-Europe. We are quite sure that no part of the Tethys, southern or northern, can have lain there.

It so happens that the nearest place in which the Permian —20° isocline crosses southward from the stable block of Meso-Eurasia into the Alpine fold belt, is about where now the Himalayas are. This is the position, nearest to their present one, which the Vicentinian Alps can have occupied during the Permian. As we cannot measure their paleolongitude directly, we are not able to exclude the possibility that the Vicentinian Alps had been even farther removed from their present position during the Permian. From paleomagnetic measurements one can only state that they were situated on the contemporaneous —20° isocline. From regional geology a position anywhere on the stable block of Eurasia is excluded, but not a position further out in what is now occupied by the Alpine fold belt.

De Boer (1963), using the minimum value, arrived at above, consequently postulates an extra westward drift of his Vicentinian Alps, in regard to Meso-Europe, of some 5,000 km, since the Permian. This is an extra drift, in so far that it is a translation superposed on top of the curves of polar wandering and continental drift that have been inferred from paleomagnetic research of the continental cores of Europe and North America. The

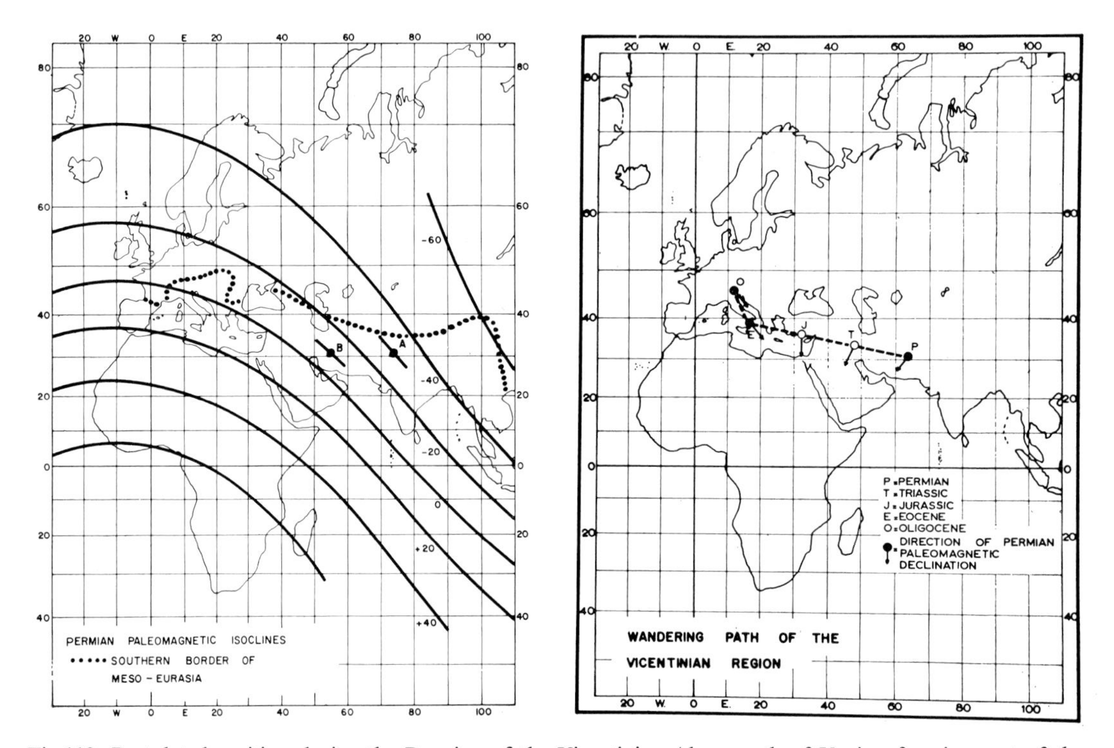

Fig.118. Postulated position during the Permian of the Vicentinian Alps, north of Venice, forming part of the Southern Alps, as derived from the paleomagnetic inclination measured in the Vicentinian Alps, and the position of the Permian isoclines for Meso-Europe.

Position *A* refers to the $-30°$ inclination found by Van Hilten (1960) and De Boer (1963), position *B* refers to the $-10°$ inclination found by Findhammer (unpublished) and Guicherit (1964).

The line of reasoning is that during the Permian the Vicentinian Alps have been on the latitude coindiding with the area between the $-10°$ and the $-30°$ isoclines, which at that time lay to the north in continental Meso-Europe. The position nearest to their present one, and available to the Vicentinian Alps, is where these isoclines cross southwards from Meso-Eurasia into the Tethys.

This deduction is sound, but the actual value arrived at for the drift is rather uncertain, because of the small angle between Permian isoclines and the southern border of Meso-Europe. The isoclines used are those of Van Hilten (1962), based in part on older and not so reliable paleomagnetic analyses of Meso-Europe. A slight shift of the average pole position for Meso-Europe, and of the corresponding isoclines, may well entail a large correction in the amount of drift assumed for the Vicentinian Alps. (After De Boer, 1963.)

Fig.119. Wandering path of the Vicentinian Alps since the Permian, relative to Meso-Europe.

Positions *A* and *B* of Fig.118 have been averaged to *P*. Open circles indicate less reliable measurements. Wandering was apparently completed by Oligocene time. (After De Boer, 1963.)

figure of 5,000 km applies only to the relative drift of this part of the Southern Alps, relative to Meso-Europe (Fig.118, 119).

This figure may be found to be too high, because a better analysis of the Permian poles for Meso-Europe may show some shift in the average pole position, and consequently of the paleo-isoclines (J. D. A. Zijderveld, personal communication, 1965). But even at that, it is a really stupenduous figure, which is most unexpectedly supplied by this entirely new line of research.[1]

We may conclude this section by noting that indication of extra continental drift of such unheard of dimensions does fit in well with the views of some tectonicians on the existence of large scale transcurrent faulting, between several of the major elements of Alpine Europe (cf., for instance, PAVONI, 1961).

NEWER THEORIES

We are now in the uncomfortable position, in which, having destroyed the old, we have not yet erected something new. The theoretical basis for the classic picture of the Alps has been proved wrong on all of its main points, but there is no new theory, which is generally accepted, to replace it.

There are several newer theories, but with the exeption of one of them, all are a couple of decades old, and none of them has become generally accepted in this time. Out of those theories proposed, let me cite those of E. C. Kraus and of R. W. van Bemmelen. The theory of KRAUS (1951, 1960) is the most widely known. It explains the genesis of most of the Alpine structures by a wholesale downward swallowing of the main zones, that is by the well-known "Verschluckung". VAN BEMMELEN (1960, 1964, 1965, 1967), following E. Haarmann, proposes exactly the opposite. That is, a transient rising of the median zones, an undation, which generates sufficient potential energy for the superficial series of the crust to slide outwards and downwards in gravity tectonics.

As the two best known orogenetic theories are diametrically opposed as to the cause underlying the genesis of the Alps, we can only put on record that there is at present as yet no generally accepted substitute for the classic motive force of tangential compression. Nor is there a new synthesis available to replace the tall piles of nappes of the cross-sections of the classic picture.

These might, in due time, be supplied by a new general theory of AUBOUIN (1960, 1961, 1965). In a grand vision this author combines the newer views on gravity tectonics with older knowledge on the differences between miogeosynclinal and eugeosynclinal basins, and on the migration, in time, of the main orogenetic movements, towards the external parts of the fold belts. In a new interpretation of the classic "Doppelorogen", Aubouin stresses the strongly polarized structure of a fold belt, with an overall outward "Vergenz". There has been, as yet, no time to test Aubouin's theory (Fig.120). Personally, I fear that,

[1] Research published after the manuscript was closed has indeed pointed out that this figure is too high. Moreover it seems at the present that the Southern Alps and northern Turkey have moved with the African and not with the Eurasian continent (see VAN HILTEN and ZIJDERVELD, 1966; VAN DER VOO, 1968).

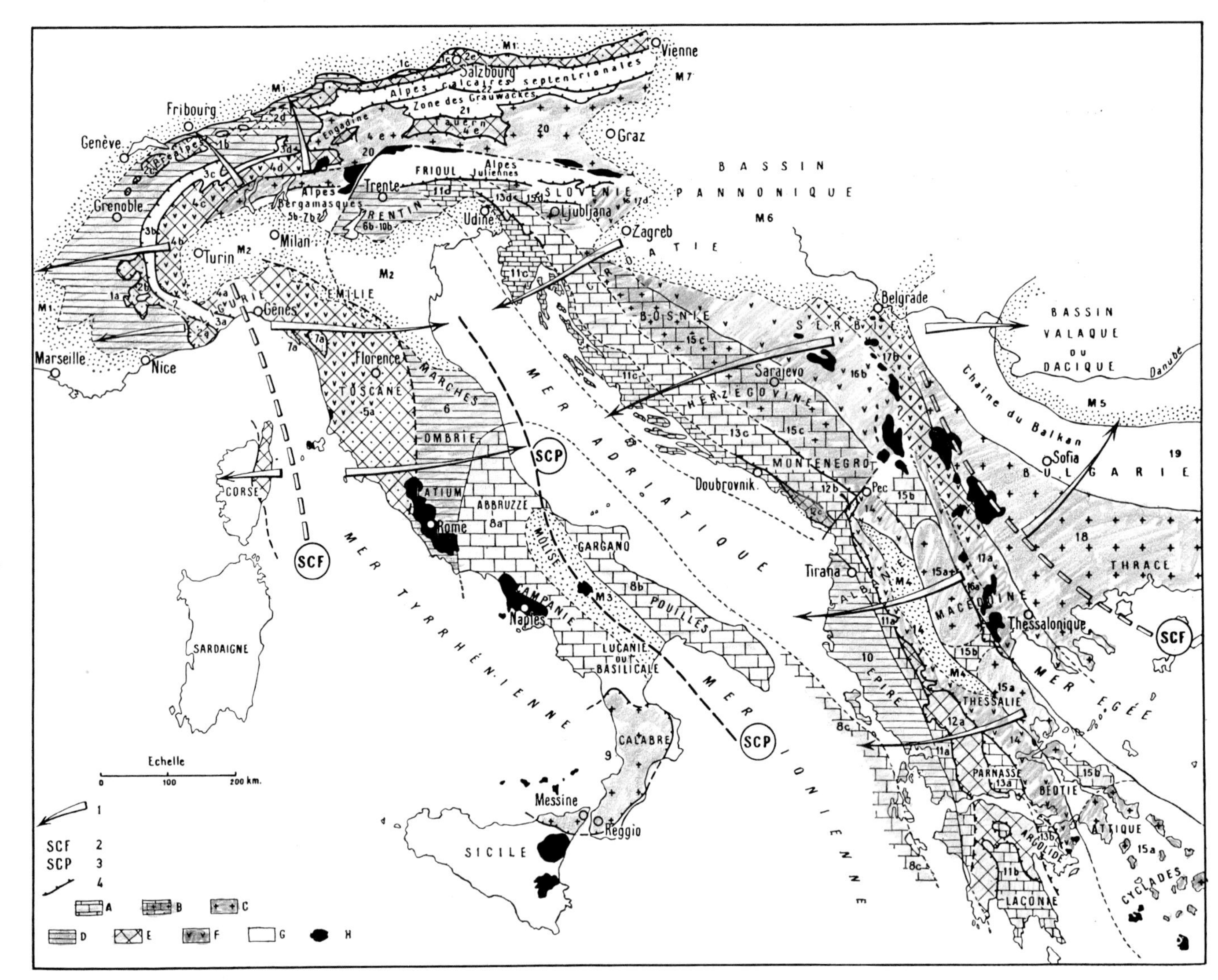

Fig.120. Schematic structural map of the Alps, the Apennines and the Balkan.

The author (Aubouin, 1960, 1961) in certain ways returns to earlier postulates of an orogenetic doublet or "Doppelorogen". Compare the arrows and the notations of *SCF*, centrifugal, and *SCP*, centripetal axes. However, for the Alps, for which such a structure was first proposed, a single outward "Vergenz" is indicated (see arrows), in accordance with modern views. Apart from the tectonic interpretation, for which the reader is referred to the original publications, the figure clearly presents the main elements of the Alpine chain. With the exception of the Northern Limestone Alps of Austria (*Alpes calcaires septentrionales*), French orthography will be recognizable without too much trouble, and has not been changed.

A = zone of neritic or reef sediments; *B* = same with localized outcrops of basement; *C* = basement; *D* = miogeosyncline; *E* = eugeosyncline; *F* = ophiolites; *G* = molasse and younger post-tectonic sedimentary basins; *H* = post-tectonic intrusives and volcanics.

M = post-tectonic basins: *M1* = molasse; *M2* = Po; *M3* = molasse, between Apennines and Dinarides; *M4* = Intra-Hellenic; *M5* = Wallachian or Dacian; *M6* = Pannonian or Hungarian; *M7* = Vienna.

1–4 = Western Alps: *1a* = Dauphinois zone; *1b* = Helvetides; *1c* = same, north of the Austrides; *2a* = *Helminthoïdes* flysch of Liguria; *2b* = same of the Embrunais; *2c* = Prealps; *2d* = "Klippen"; *2e* = flysch zone north of the Austrides; *3a, 3b* = Briançonnais zone. *3c, 3d* = Great St. Bernard and Adula nappes; *4* = Pennides.

5b–7b?, 6b–10b, 11d, 13d, 15d, 16–17d = Southern Alps (in the concept of Aubouin related to the Apennines and the Dinarides).

4e, 20–22 = Eastern Alps: *4e* = Tauern window; *20* = Central Eastern Alps ("Zentralalpen"); *21* = graywacke zone; *22* = Northern Limestone Alps. (After Aubouin, 1960, 1961.)

in a way, it will prove to be too schematized to be true, just as the classic picture. I have become impressed too much by the fact that the various major elements, which together make up the Alpine chain, are strongly divergent in their general structure and history. Such a feeling is strengthened by the first results of paleomagnetism, as reviewed in the last section. This prevents my believing that a single simple scheme could be constructed to encompass the different modes of origin of the many elements out of which the Alpine chain seems to be composed. A critical attitude in this same vein has, for instance, been taken already by DEBELMAS et al. (1966).

Apart from the theory of Aubouin, it even seems, as if Alpine theorizing passes at present through a dormant state. The pendulum has swung far back since the days when the early workers happily extrapolated from meagre data. Today, as F. Ellenberger once remarked, every Alpine geologist is in a way a prisoner of his own valley. Here he has met with so much variable detail, that he dares not extrapolate anymore, even over the slightest distance beyond the divides that separate his valley from the rest of the Alps.

The mass of detail which has now accumulated is so appallingly large, that it seems as if only "people with a theory" can get a grasp on all of the Alps, and are able to arrive at an overall picture. This attitude is, for instance, strongly apparent in the apodictic style of E. C. Kraus. Here one receives the impression that any fact contrary to theory is quickly thrown out and forgotten, clearing the desk for further extrapolation.

Alpine geologists with a more objective approach evidently require more than one lifetime of study. This is, for instance, well illustrated by one of the foremost Alpine geologists, Professor P. Fallot of Paris. Knowing well the western end of Alpine Europe, the Betic Cordillera, and its African counterpart, the Rif of Morocco, he began by an intensive study of the Eastern Alps, but died before arriving at a synthesis of the whole chain.

An inkling of the work this involves can be gleaned from the carefully worded summaries of Fallot's courses at the Collège de France in Paris. Every year for nineteen years, the literature of another part of the chain was digested and discussed critically. As a historical document on such a gradual development of a scientist's knowledge of Alpine Europe, there is no more impressive item than these badly printed "Résumés des leçons", so full of wisdom (FALLOT, 1945–1960).

One more personal aspect of the transition of the classic picture to newer ideas in Alpine geology deserves to be mentioned, because this may help to explain controversies a newcomer may find in the literature. That is, that most authors, who are "modern" in one or two aspects of Alpine geology, are still "classic" on the remainder of the subject. These authors, who just could not stomach the classic way of explaining one familiar feature, still reproduce the classic picture for all other phenomena. They have been bottle-fed on this during their student days, and just dare not to be iconoclasts outside a limited area in which they feel they just have to be critical.

The situation in Alpine geology is in some ways comparable to the period in the nineteen-twenties, when in physics classic views were displaced, first by relativistic, and then by quantum mechanic approach. In geology, however, the transition does take more time than in physics. This is inherent on the fact that in geology the introduction of new views often depends on time-consuming intensive remapping of larger areas, instead of on the acceptance of a new set of formulas.

In these times of transition, one can thus only point out the main elements of the classic picture, and indicate the direction taken by newer developments. For it should be noted, that almost all newer development starts from the classic picture, and in fact is no more than a re-interpretation of the same rocks. These rock units, together with most of their local names, remain the same, although interpretation changes. Whether a certain section is interpreted as a "Doppelfalte", as a recumbent fold nappe, or as a thrust plate nappe, each interpretation still applies to the same section, the same mountain tops, and the same rocks.

This is the reason why the old names never die and why, for most of the newer interpretations, one has to have knowledge of the classic picture as background information, just to know what it is all about.

At present, one already finds indications as to these newer ideas in the books of GIGNOUX (1955) and METZ (1957). Although the former is a stratigrapher, the latter a tectonician, both are Alpine geologists first, and know about their subject. As a general introduction to several of these newer ideas, TRÜMPY (1960) can be cited, whilst special notice must be paid to the excellent summary by DEBELMAS and LEMOINE (1964).

It may be sincerely hoped that, the newest trend in science being compilatory, in the years to come regional volumes on Alpine geology will guide us into its newer and more sophisticated problems. Up to that time, however, we will have to live mainly on a scientific diet of another new concept, that of the symposia. In Alpine geology, for instance, a recent example is the first part of volume 53 of the *Geologische Rundschau* (1964). Further the symposia of the Société Géologique de France on the Betic Cordillera, the Dinarides, and the Apennines, respectively (SOCIÉTÉ GÉOLOGIQUE DE FRANCE, 1961a, b, 1963).

The very best is, however, the two-volume memorial to Professor P. Fallot (DURAND DELGA, 1962, 1963), in which most prominent Alpine geologists have summarized the state of the art in their own area. This, consequently, is a "must" in Alpine geology. An even newer volume, containing much information, is the *Colloque Étages Tectoniques* (SCHAER, 1967).

GERMANIC VERSUS ALPINE FACIES OF THE TRIASSIC

Within this review of newer trends in Alpine geology a section on the difference between, and the occurrence of, the Germanic and the Alpine facies of the Triassic seems in order, because of the misleading connotations of these words. The difference is that between an epicontinental and a geosynclinal sedimentary basin, which does, however, not correlate with the area of the later fold belt.

The Germanic facies is predominantly formed by continental deposits, mostly red beds. These are interrupted occasionally, and only over part of its area, by deposits of an inland sea, such as the Muschelkalk. Evaporites, mainly gypsum, are normally well developed in the higher part, the Keuper, but also may occur elsewhere in the series. Correlation is almost exclusively by way of lithostratigraphy. Total thickness is of the order of several hundreds of meters only.

Although in the Alpine facies the Triassic also begins with continental red beds and an occasional coal, forming the continuation of the facies of the Permian, the Alpine facies is predominantly formed by marine sediments, with a thickness of the order of several kilometers. Correlation is chronostratigraphic, by way of index fossils. Therefore, mutual correlation of stages of the two facies always remains difficult and often is doubtful. It is given here in Table XV.

Although geosynclinal sedimentation in the Alpine facies of the Triassic rarely, if ever, reached deep-sea character, a point further to be elaborated in the next section, Sedimentation was normally in balance with crustal subsidence, and the environment has predominantly been that of an open, shallow sea.

The difference between the sedimentary environment of the Alpine facies and marine intercalations in the Germanic facies, such as the Muschelkalk, lies in the fact that the latter did not form an open shallow sea, but a shallow inland sea. Hypersaline or other extreme conditions normally reigned in the latter sea. In the Muschelkalk, consequently, a fauna is found, which is extremely reduced in the number of species, although locally characterized by the exuberant development of a single species in very large numbers. The well-known *Ceratites*, for instance, is the only ammonite genus which has invaded the Muschelkalk sea. On the other hand, shell beds, formed by massive accumulations of a single species of *Terebratula*, are responsible for the name — "shelly limestone" — of the stage. The fauna of the Alpine facies, on the other hand, is characterized by an abundant development of many different species.

Both subsidence and sedimentation having proceded at a higher rate in the Alpine facies, its limestones reached much greater thickness than did those of the Muschelkalk. Its main elements are the massive reef- and reef-rock limestones and dolomites of the Wetterstein and Dachstein Stages, which normally reach individual thicknesses of the order of 1 km. Diploporids may play an important part in the construction of these reefs of the Alpine facies, but more normal coralline reefs have also developed.

As stated already, there is no correlation between the facies of the Triassic and the area of the Alpine fold belt. The Germanic epicontinental facies also reigns in large parts of the Alps. This is but one more indication that sedimentation and folding are not so strictly related as has often been stated.

The external zones of the Alps, the Swiss Helvetides and the French Dauphinois zone, have the Triassic developed everywhere in the Germanic facies. The Alpine facies in its typical development only occurs in the Austrides and in the Southern Alps. The Pennides

TABLE XV

STAGES OF THE GERMANIC AND THE ALPINE FACIES OF THE EUROPEAN TRIASSIC

Germanic	*Alpine*
Keuper	Norian Carnian
Muschelkalk	Ladinian Anisian or Virglorian
Buntsandstein	Werfenian or Scythian

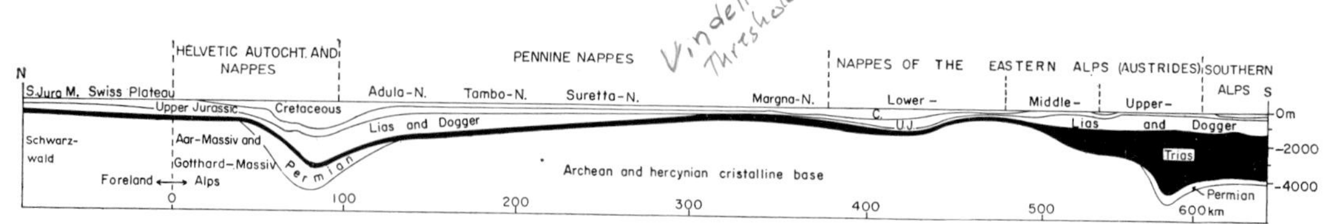

Fig.121. Section across the original sedimentary region of the Swiss Alps. Constructed by unrolling the nappes according to the classic picture and thus restoring the original sequence. Typical development of the Triassic in Alpine facies is restricted to the (upper) Austrides and to the Southern Alps. (After CADISCH, 1946.)

and the Briançonnais zone are intermediate as to the development of the Triassic (Fig.121). Its higher elements are developed predominantly in Alpine facies, whereas the Lower Triassic more often than not is in Germanic facies.

The tectonic implication of this difference in facies will be clear. The evaporites of the Germanic facies form an excellent lubricating horizon, which determines the position of the main zone of "décollement". Together with other lubricating horizons, formed by shales later in the Mesozoic, they are responsible for the tectonic style of the epidermis type nappes of the Helvetides. In the Austrides, on the other hand, the style of the superstructure is mainly determined by the rigid competent blocks of the massive reef limestones of the Alpine facies. "Décollement" is confined to thinner shaly members, intercalated between these massive limestones. This is the reason why the "Nördliche Kalkalpen", the Northern Limestone Alps, though epidermis type nappes just as well, show a tectonic style quite different from the Helvetides.

Professor Gümbel of Munich, in the latter part of the last century, supposed, that this difference between the Germanic and the Alpine facies of the Triassic was due to the existence between the two basins of a high mountain range, the Vindelician chain.

This Vindelician chain has since played a major role in most of the paleogeographic reconstructions of western Europe. Its influence, moreover, has been thought to reach into much younger eras.

It has since transpired, however, that the differences between the two facies realms are not as clear-cut as had originally been thought. Moreover, a real mountain chain in western Europe, during the Triassic seems unlikely for a period, which, by all indications, has been one of extreme peneplanation. So, instead of a Vindelician chain, one nowadays speaks of a Vindelician threshold, a "Seuil vindélicienne" (RICOUR, 1962). Today one rather thinks of an area, which, though submerged, more or less effectively separated the epicontinental sedimentary basin of the Germanic facies from the geosynclinal basin of the Alpine facies of the Triassic.

DEEP OR SHALLOW — OFF-SHORE OR NEAR-SHORE?

We saw how the Alpine Triassic, although geosynclinal, formed as a shallow water deposit. This is confirmed, amongst other things, by its reef limestones. This brings us to a discussion of the evidence cited in the literature for deep- or shallow-water sedimentation and for off-shore or near-shore position of a sedimentary basin.

In most older literature on the Alps, evidence for this type of statement, for paleogeography or paleoecology, is slight or non-existant.

Ideas on paleogeography started in general with the most simple model conceivable, that is that miogeosynclinal sedimentation was on the whole shallow and near-shore, eugeosynclinal sedimentation deep and off-shore. A refinement of this simple model led to the assumption that, irrespective of their position in regard to a mio- or eugeosynclinal basin, all coarse-grained sediments were near-shore, all fine-grained sediments off-shore — hence bathyal, if not abyssal.

The latter statement can be re-interpreted, that most fine-grained, the former "deep", sediments are now called "pelagic". The first statement is more ambiguous, and less easy to translate directly into modern views on paleogeography, because of the existence of coarse-grained sediments both in near shore environments, and in off-shore turbidite series.

Some facies, which formerly were always cited as proof for deep-water sedimentation, have since been re-interpreted. The most important of these is the red ammonite limestone facies such as the Hallstatt Limestone (p.318). Another notable example is that of the *Aptychus* limestones.

The red ammonite limestones, the "ammonitico rosso" of the Italians, was formerly correlated with red deep-sea clay. The voluminous literature on this subject has been reviewed by AUBOUIN (1965).

According to this study, the red ammonite limestones occur in two quite distinct facies, either as red marble, or as red nodular limestone. In the first facies they represent fossil hard grounds, formed on submarine ridges or "haut-fonds", accompanied by large-size sedimentation breaks. The latter may, for instance, extend all over the Dogger or the Malm. The second facies formed in sedimentary basins, within a marly series. It is also accompanied by some telescoping of sedimentation, but of a far lesser extent than in the red marble facies. Neither of the two facies of the "ammonitico rosso" may consequently be interpreted as a deep water sediment.

A similar evolution has taken place in regard to the ideas on *Aptychus* limestones. The quite common limestone series in which ammonite aptychi are the only fossils preserved, invited a comparison with modern deep-sea oozes, in which the deeper oozes only contain siliceous tests. It was assumed that the arragonitic ammonite shells would dissolve at a lesser depth than the calcareous aptychi, and a sedimentation depth of some 4.5 km was arbitrarily postulated.

Such preferential solution may, however, well take place in shallow, well-aerated waters too. It forms no evidence for abyssal sedimentation. Taking into account that many *Aptychus* limestones show lumachella or coquina characters, which points to shallow water sedimentation, the latter environment on second thought even seems the more probable (ZAPFE, 1963).

The only possible deep-water sediments that remain are the clastic series such as the Bündner Schiefer with its radiolarites and its ophiolites (the "Steinmann trinity", p.285), and the flysch with its turbidites. Although, for the latter facies several authors, amongst whom Professor J. Ph. Mangin is the most outspoken, advocate a possible shallow-water origin too. For literature pro and contra, compare MANGIN (1964).

In closing this discussion we may therefore conclude that: (*1*) all statements on paleogeography and paleoecology in the Alps must be viewed with caution; and (*2*) that there are many more shallow-water — although possibly off-shore — sediments, than has formerly been thought.

FLYSCH AND MOLASSE

Finally a word must be added about two terms much used and misused in Alpine geology. Originally the terms flysch and molasse both had a sharply defined meaning. Flysch was the Early Tertiary, thinly bedded series of fine-grained material, derived from the early cordilleras, which at that time did not yet reach high topographic altitude. It became involved in later folding, and thus is not only late geosynclinal, but also synorogenetic. Molasse, on the other hand, was the Middle Tertiary series of coarse material, derived through erosion of the emergent high mountain chain in the process of vertical uplift. It is thus post-orogenetic, and normally it is not folded.

Both terms consequently had a descriptive, as well as a genetic aspect. This was alright for the Swiss Alps, where, minor exceptions apart, the two series are indeed different, both as to their facies and as to their genesis. They have, however, subsequently been used elsewhere, and often been applied in another sense.

With the molasse, it must be noted that it does not always contain the coarse conglomeratic material, the Swiss "Nagelfluh" of its typical development. Between the deltas spread by the rivers draining the rising Alps, finer material, similar to flysch, could be deposited. This does not present serious difficulties, because the deltas could be mapped (cf. FÜCHTBAUER, 1964; MATTER, 1964). Incidentally, this has, already long ago, led CADISCH (1928), to coin one of the nicest titles in geology, i.e., "Das Werden der Alpen im Spiegel der Vorlandsedimentation" (The growth of the Alps in the mirror of its foreland sedimentation).

For the molasse basin to the north of the Alps results from an intensive drilling program have recently enabled FÜCHTBAUER (1967) to produce a considerable refinement of the history of this basin. Variations in composition will not only occur in a geographical sense, due to the position of the river deltas, but also with time, as deeper and deeper levels of the rising mountain chain become available for erosion.

Within the molasse sandstone three types could be distinguished: (*1*) the oldest are calcarenitic sandstones, derived from the erosion of flysch and of the Limestone Alps; (*2*) then follow calcarenitic feldspathic sandstones, the heavy minerals of which indicate added erosion of the Central Massifs; (*3*) the youngest sandstones are still calcarenitic, but contain heavy minerals derived from both igneous and metamorphic rocks, indicating a more general erosion of all zones of the rising Alps.

This relatively simple model is complicated by the fact that the subsidence within the molasse basin was not regular. From time to time it received material directly from the rising Alps, in other periods some part of the basin would sink so fast that it received its sediments lengthwise, not crosswise.

The addition from material eroded from metamorphic rocks may lead to convergences of molasse with flysch. Although it may consequently become difficult to distinguish between finer-grained molasse and coarser-grained flysch sediments, detailed sedimentary petrography will always yield characters to distinguish between these series (WIESENEDER, 1967). It seems, therefore, that those authors who maintain that there really is no difference between flysch and molasse arrived at this conclusion by too superficial a study.

Also, by way of exception, the molasse has become strongly tectonized, tilted and upthrust, along the northern borderfault of the Alps. Along this overthrust the youngest orogenetic movements of the Alpine chain have taken place. This tectonization is most strongly developed in a zone of some 5 km to 10 km wide, bordering the "Alpenrandbruch" north of the Helvetides. Further east, in front of the Austrides, the tectonization of the molasse is much less accentuated (SCHMIDT-THOMÉ, 1962, 1963), whereas further west, in front of the Dauphinois zone, it is all but absent.

In contrast, the term flysch has encountered more vicissitudes in its usage. Several cases will be cited by way of illustration, but this is by no means a complete list:

(*1*) It has been found that "true" flysch already developed during the Upper Cretaceous, so its original definition had to be enlarged.

(*2*) The term flysch has been used for every repetitive series of fine-grained beds, the worst of such abusive uses being the "flysch calcaire" of the French geologists. This applies to Mesozoic, calcareous, marly, repetitive series of various ages of the French Alps. It is neither truly detrital, nor syn-orogenetic. The only resemblance with the Swiss flysch lies in the fact that it is repetitive.

(*3*) The advent of the turbidite concept has led to oversimplifications. On the one hand it has been stated that all turbidite series are equivalent to flysch, on the other that flysch without turbidites is not a "true" flysch. Although neither of these statements is true, as has been well stated by MANGIN (1964) in his attack on the "turbidite dogma", the fact remains that the turbidite concept has greatly clarified our ideas on flysch sedimentation. This applies especially to the facies of the "Wildflysch" (p.239), which formerly was explained along tectonic lines only, and which now is thought to originate, at least in part, through sedimentary slumping and related processes.

(*4*) Finally, as a geological border-line case, in which it is really difficult to trace the limits of the flysch, the situation in Graubünden, in northeastern Switzerland, may be mentioned. In this region the geosynclinal Bündner Schiefer of the Pennide nappes not only reaches well upwards into the Cretaceous, but is also developed in a more coarsely clastic facies (p.259). Hence, Bündner Schiefer and flysch are well-nigh indistinguishable in this region.

REFERENCES

ARGAND, E., 1911. Les nappes de recouvrement des Alpes pennines et leurs prolongements structuraux. *Matér. Carte. Géol. Suisse*, 31 (6): 1–26.

ARMSTRONG, R. L., JÄGER, E. and EBERHARDT, P., 1966. A comparison of K–Ar and Rb–Sr ages on Alpine biotites. *Earth Planetary Sci. Letters*, 1: 13–19.

AUBOUIN, J., 1960. Essai sur l'ensemble italo-dinarique et ses rapports avec l'arc alpin. *Bull. Soc. Géol. France*, 7 (2): 487–526.

AUBOUIN, J., 1961a. Propos sur l'orogénèse. 1. Propos statique. *Bull. Serv. Inform. Géol. Bur. Rech. Géol. Minière*, 52: 1–21.

AUBOUIN, J., 1961b. Propos sur l'orogénèse. 2. Propos dynamique. *Bull. Serv. Inform. Géol. Bur. Rech. Géol. Minière*, 53: 1–24.

AUBOUIN, J., 1964a. Réflexion sur le problème des flyschs et des molasses, son aspect dans les Hellénides (Grèce). *Eclogae Geol. Helv.*, 57: 451–496.

AUBOUIN, J., 1964b. Réflexions sur le faciès "ammonitico rosso". *Bull. Soc. Géol. France*, 7 (6): 475–501.

AUBOUIN, J., 1965. *Geosynclines.* Elsevier, Amsterdam, 335 pp.

BEARTH, P., 1958. Über einen Wechsel der Mineralfazies in der Wurzelzone des Penninikums. *Schweiz. Mineral. Petrog. Mitt.*, 38: 363–373.

BEARTH, P., 1962. Versuch einer Gliederung alpinmetamorpher Serien der Westalpen. *Schweiz. Mineral. Petrog. Mitt.*, 42: 129–137.

BERTRAND, M., 1887. L'îlot triassique du Beausset (Var). Analogie avec le bassin houiller franco-belge et avec les Alpes de Glaris. *Bull. Soc. Géol. France*, 15: 667–702.

CADISCH, J., 1928. Das Werden der Alpen im Spiegel der Vorlandsedimentation. *Geol. Rundschau*, 19: 105–119.

CADISCH, J., 1960. Der Oberbau der Orogene. *Geol. Rundschau*, 50: 53–63.

CHATTERJEE, N. D., 1961. Aspects of Alpine zonal metamorphism in the Swiss Alps. *Nachr. Akad. Wiss. Göttingen, Math. Physik. Kl.*, 5: 59–71.

CHATTERJEE, N. D., 1962. The Alpine metamorphism in the Simplon area, Switzerland and Italy. *Geol. Rundschau*, 51: 1–72.

CLOSS, H., 1963. Der tiefere Untergrund der Alpen nach neuen seismischen Messungen. *Geol. Rundschau*, 53: 630–649.

CLOSS, H., 1965. Results of explosion seismic studies in the Alps and in the German Federal Republic. In: C. H. SMITH and T. SORGENFREI (Editors), *The Upper Mantle Symposium, New Delhi, 1964.* Intern. Union Geol. Sci./ Det Berlingske Boktrykkeri, Copenhagen, pp.94–102.

CLOSS, H., 1966. Der Untergrund der Alpen im Lichte neuerer geophysikalischer Untersuchungen. *Erdöl Kohle*, 19: 81–88.

CLOSS, H. and LABROUSTE, Y. (Editors), 1963. Recherches seismologiques dans les Alpes occidentales au moyen de grandes explosions en 1956, 1957 et 1960. *Année Géophys. Intern., Part. Franç., Centre Natl. Rech. Sci.*, 12 (2): 241 pp.

D'AMICO, C., 1964. Relazioni comagmatiche tra vulcanesimo atesino e plutonismo di Cima d'Asta. La provincia magmatica tardo-ercinica tridentina. *Mineral. Petrog. Acta Bologna*, 10: 157–176.

DEBELMAS, J., 1957. Quelques remarques sur la conception actuelle du terme "cordillère" dans les Alpes internes françaises. *Bull. Soc. Géol. France*, 6 (7): 463–474.

DEBELMAS, J. and LEMOINE, M., 1964. La structure tectonique et l'évolution paléogéographique de la chaîne alpine d'après les travaux récents. *Inform. Sci.*, 1: 1–33.

DEBELMAS, J., LEMOINE, M. and MATTAUER, M., 1966. Quelques remarques sur le concept de "geosynclinal" (à propos d'un récent ouvrage de J. Aubouin). *Rev. Géograph. Phys., Géol. Dyn., Sér. 2*, 8 (2): 133–150.

DE BOER, J., 1963. Geology of the Vicentinian Alps (with special reference to their paleomagnetic history). *Geol. Ultraiectina*, 11: 178 pp.

DE ROEVER, W. P., 1965. On the cause of the preferential distribution of certain metamorphic minerals in orogenic belts of different age. *Geol. Rundschau*, 54: 933–943.

DE ROEVER, W. P. and NIJHUIS, H. J., 1963. Plurifacial alpine metamorphism in the eastern Betic Cordilleras (southeastern Spain) with special reference to the genesis of glaucophane. *Geol. Rundschau*, 53: 324–336.

DURAND DELGA, M. (Editor), 1962. *Livre à la Mémoire du Professeur Paul Fallot. I.* Soc. Géol. France, Paris. 656 pp.

DURAND DELGA, M. (Editor), 1963. *Livre à la Mémoire du Professeur Paul Fallot. II.* Soc. Géol. France, Paris, 712 pp.

ELLENBERGER, F., 1958. Étude géologique du pays de Vanoise. *Mém. Carte Géol. France*, 1958: 545 pp.

FALLOT, P., 1945–1960. Résumé des leçons données dans le chaire de Géologie méditerrannéenne du Collège de France. *Ann. Coll. France*, 42e année–60e année.

FALLOT, P., 1949. Les chevauchements intercutanés de Roya (Alpes maritimes). *Ann. Hébert Haug* (*Lab. Géol. Fac. Sci. Univ. Paris*), 7: 161–169.

FALLOT, P., 1950. Remarques sur la tectonique de couverture dans les Alpes Bergamasques et les Dolomites. *Bull. Soc. Géol. France*, 20 (5): 183–195.

FALLOT, P., 1955. Les dilemmes tectoniques des Alpes Orientales. *Ann. Soc. Géol. Belg., Bull.*, 78: 147–170.

FALLOT, P., 1956. Promenade d'hypothèse en hypothèse. *Gedenkboek H. A. Brouwer—Verhandel. Koninkl. Ned. Geol. Mijnbouwk. Genoot., Geol. Ser.*, 16: 100–113.

FUCHS, K., MÜLLER, S. and PETERSCHMITT, E., 1963. Krustenstruktur der Westalpen nach refraktionsseismischen Messungen. *Gerlands Beitr. Geophys.*, 72: 149–169.

FÜCHTBAUER, H., 1964. Sedimentpetrographische Untersuchungen in der älteren Molasse nördlich der Alpen. *Eclogae Geol. Helv.*, 57: 157–298.

FÜCHTBAUER, H., 1967. Die petrografisch-sedimentologische Stellung der Molassesandsteine. *Geol. Rundschau*, 56: 266–300.

GAGNEBIN, E., 1942. Les idées actuelles sur la formation des Alpes. *Actes Soc. Helv. Sci. Nat.*, 1942: 47–53.

GIGNOUX, M., 1950. Comment les géologues des Alpes françaises conçoivent la tectonique d'écoulement. *Geol. Mijnbouw*, 12: 342–346.

GIGNOUX, M., 1955. *Stratigraphic Geology.* Freeman, San Francisco, Calif., 682 pp.

GUICHERIT, R, 1964. Gravity tectonics, gravity field and palaeomagnetism in N.E. Italy. *Geologica Ultraiectina 14: 125 pp.*

Haug, E., 1925. Contribution à une synthèse des Alpes occidentales. *Bull. Soc. Géol. France*, 25: 97–244.
Jäger, E., 1962. Rb–Sr age determinations on micas and total rocks from the Alps. *J. Geophys. Res.*, 67: 5293–5306.
Jäger, E. and Faul, H., 1959. Age measurements on some granites and gneisses from the Alps. *Bull. Geol. Soc. Am.*, 70: 1553–1558.
Jäger, E. and Niggli, E., 1964. Rubidium Strontium-isotopenanalysen an Mineralien und Gesteinen des Rotondogranites und ihre Geologische Interpretation. *Schweiz. Mineral. Petrog. Mitt.*, 44: 61–81.
Jäger, E., Geiss, J., Niggli, E., Streckeisen, A., Wenk, E. and Wüthrich, H., 1961. Rb–Sr Alter an Gesteinsglimmern der Schweizer Alpen. *Schweiz. Mineral. Petrog. Mitt.*, 41: 255–272.
Jäger, E., Niggli, E. and Wenk, E., 1967. Rb–Sr Altersbestimmungen an Glimmern der Zentralalpen. *Beitr. Geol. Karte Schweiz*, NF, 134: 67 pp.
Kraus, E. C., 1951a. *Baugeschichte der Alpen. 1. Vom Archaikum bis zum Ende der Kreide*. Akademie Verlag, Berlin, 552 pp.
Kraus, E. C., 1951b. *Baugeschichte der Alpen. 2. Neozoikum*. Akademie Verlag, Berlin, 489 pp.
Kraus, E. C., 1960. Das Orogen, Begriff, Bildungsweise und Erscheinungsformen. *Intern. Geol. Congr., 21st, Copenhagen, 1960, Rept. Session, Norden*, 28: 236–248.
Kvale, A., 1957. Gefügestudien im Gotthardmassiv und den angrenzenden Gebieten. *Schweiz. Mineral. Petrog. Mitt.*, 37: 398–434.
Lemoine, M., 1953. Remarques sur les caractères de l'évolution de la paléogéographie de la zone briançonnaise au Secondaire et au Tertiaire. *Bull. Soc. Géol. France*, 6 (3): 105–122.
Lemoine, M., 1961. La marge externe de la fosse piémontaise dans les Alpes occidentales. *Rev. Géograph. Phys. Geol. Dyn., Sér. 2*, 4: 163–180.
Lugeon, M. and Gagnebin, E., 1941. Observations et vues nouvelles sur la géologie des Préalpes romandes. *Bull. Lab. Mineral. Géophys. Musée Géol. Univ. Lausanne*, 72: 90 pp.
Mangin, J. Ph., 1964. Petit historique du dogme des turbidites. *Compt. Rend. Soc. Géol. France*, 1964: 51–52.
Matter, A., 1964. Sedimentologische Untersuchungen im östlichen Napfgebiet (Entlebuch-Tal der Grossen Fontanne, Kt. Luzern). *Eclogae Geol. Helv.*, 57: 315–428.
Metz, K., 1957. *Lehrbuch der tektonische Geologie*. Enke, Stuttgart, 294 pp.
Müller, R. O., 1958. *Petrographische Untersuchungen in der nördlichen Adula*. Thesis. Leeman, Zürich, 173 pp.
Nabholz, W. K., 1951. Beziehungen zwischen Facies und Zeit. *Eclogae Geol. Helv.*, 44: 132–158.
Nabholz, W. K., 1953. Das mechanische Verhalten der granitischen Kernkoerper der tieferen penninischen Decken bei der alpinen Orogenese. *Congr. Géol. Intern., Compt. Rend., 19e, Algiers, 1952*, 3: 9–23.
Nabholz, W. K., 1954. Gesteinsmaterial und Gebirgsbildung im Alpenquerschnitt Aar Massiv–Seegebirge. *Geol. Rundschau*, 42: 155–171.
Niggli, E., 1960. Mineral-Zonen der Alpinen Metamorphose in den Schweizer Alpen. *Intern. Geol. Congr., 21st, Copenhagen, 1960, Rept. Session, Norden*, 13: 132–137.
Niggle, E. and Graeser, S., 1967. Zur Verbreitung der Phengite in den Schweizer Alpen. In: J. P. Schaer (Editor) *Colloque Étages Tectoniques*. Baconniere, Neuchâtel, pp.89–104.
Niggli, E. and Niggli, C. R., 1965. Karten der Verbreitung einiger Mineralien der alpidischen Metamorphose in den Schweizer Alpen (Stilpnomelan, Alkali-Amphibol, Chloritoid, Staurolith, Disthen, Sillimanit). *Eclogae Geol. Helv.*, 58: 335–368.
Pavoni, N., 1961. Faltung durch Horizontalverschiebung. *Eclogae Geol. Helv.*, 54: 515–534.
Pavoni, N., 1962. Rotierende Felder in der Erdkruste. *Abhandl. Deut. Akad. Wiss. Berlin, Kl. Bergbau, Hüttenwesen, Montangeol.*, 1962 (2): 257–270.
Pierce, W. G., 1960. The "break-away" point of the Heart Mountain detachment fault in northwestern Wyoming. *U.S., Geol. Surv., Profess. Papers*, 400B: 236–237.
Plessmann, W. and Wunderlich, H. G., 1961. Eine Achsenkarte des inneren Westalpenbogens, *Neues Jahrb. Paleontol. Mineral.*, 1961: 199–210.
Prodehl, C., 1964. Auswertung von Refraktionsbeobachtungen im bayrischen Alpenvorland im Hinblick auf die Tiefenlage des Grundgebirges. *Z. Geophys.*, 30: 161–181.
Prodehl, C., 1965. Struktur der tieferen Erdkruste in Südbayern und längs eines Querprofiles durch die Ostalpen, abgeleitet aus refraktionsseismischen Messungen bis 1964. *Boll. Geofis. Teorica Appl.*, 7 (N25): 35–88.
Ricour, J., 1962. Contribution à une révision du Trias français. *Mém. Carte Géol. France*, 1962: 471 pp.
Rutten, M. G., 1962. Strata, movement and time: a dialogue. *Congr. Avan. Études Stratigraph. Géol. Carbonifère, Compte Rendu, 4, Heerlen, 1958*, 3: 603–608.
Rutten, M. G., 1964. Paleomagnetism and Tethys. *Geol. Rundschau*, 53: 9–16.
Rutten, M. G., 1965. Recent paleomagnetic work carried out in The Netherlands. *Phil. Trans. Roy. Soc. London, Ser. A*, 258: 159–177.
Sander, B., 1948. *Einführung in die Gefügekunde der geologischen Körper. I.* Springer, Wien, 215 pp.
Sander, B., 1950. *Einführung in die Gefügekunde der geologischen Körper. II.* Springer, Wien, 409 pp.

SCHAER, J. P. (Editor), 1967. *Colloque Étages Tectoniques.* Baconniere, Neuchâtel, 334 pp..
SCHARDT, H., 1893. Sur l'origine des Préalpes romandes. *Eclogae Geol. Helv.*, 4: 129–142.
SCHMIDT-THOMÉ, P., 1962. Paläogeographische und tektonische Strukturen im Alpenrandbereich Südbayerns. *Z. Deut. Geol. Ges.*, 113: 231–260.
SCHMIDT-THOMÉ, P., 1963. Le bassin de la Molasse d'Allemagne du sud. In: M. DURAND DELGA (Editor), *Livre à la Mémoire du Professeur Paul Fallot. II.* Soc. Géol. France, Paris, pp.431–452.
SOCIÉTÉ GÉOLOGIQUE DE FRANCE, 1961a. Séance sur la géologie des Dinarides. *Bull. Soc. Géol. France*, 7 (2): 363–526.
SOCIÉTÉ GÉOLOGIQUE DE FRANCE, 1961b. Séance sur les Cordillères bétiques. *Bull. Soc. Géol. France*, 7 (2): 263–361.
SOC ÉTÉ GÉOLOGIQUE DE FRANCE, 1963. Séance sur la tectonique de l'Apennin, de la Calabre et de la Sicile. *Bull. Soc. Géol. France*, 7 (4): 625–784.
STEIGER, R. H., 1964. Dating of orogenic phases in the Central Alps by K–Ar ages of hornblende. *J. Geophys. Res.*, 69: 5407–5421.
TOLLMANN, A., 1965. Comparaison entre le Pennique des Alpes occidentales et celui des Alpes orientales. *Compt. Rend. Soc. Géol. France*, 1965: 363–365.
TRÜMPY, R., 1957. Quelques problèmes de paléogéographie alpine. Critique de la théorie des plissements précurseurs. *Bull. Soc. Géol. France*, 6 (7): 443–461.
TRÜMPY, R., 1960. Paleotectonic evolution of the Central and Western Alps. *Bull. Geol. Soc. Am.*, 71: 843–908.
TRÜMPY, R., 1965. Zur geosynklinalen Vorgeschichte der Schweizer Alpen. *Umschau*, 18: 573–577.
VAN BEMMELEN, R. W., 1960. Zur Mechanik der ostalpinen Deckenbildung *Geol. Rundschau*, 50: 474–499.
VAN BEMMELEN, R. W., 1964. Les phénomènes géodynamiques à l'échelle de l'orogenèse alpine (la tectonique). *Mém. Soc. Belge. Géol. Paléontol. Hydrol.*, 8: 85–127.
VAN BEMMELEN, R. W., 1965. Der gegenwärtige Stand der Undationstheorie. *Mitt. Geol. Ges., Wien*, 57: 379–400.
VAN BEMMELEN, R. W., 1967. "Stockwerktektonik" sensu lato. In: J. P. SCHAER (Editor), *Colloque Étages Tectoniques.* Baconniere, Neuchâtel, pp.19–40.
VAN DER PLAS, L., 1959. Petrology of the northern Adula region, Switzerland (with particular reference to glaucophane bearing rocks). *Leidse Geol. Mededel.*, 24: 415–498.
VAN DER VOO, R., 1968. Paleomagnetism and the Alpine tectonics of Eurasia, 4. Jurrassic, Cretaceous and Eocene
VAN HILTEN, D., 1960. Geology and Permian paleomagnetism of the Val–di–Non area, W. Dolomites, N. Italy. *Geologica Ultraiectina*, 5: 95 pp.
pole positions from northeastern Turkey. *Tectonophysics*, 6 (3): 251–269.
VAN HILTEN, D., 1962. A deviating Permian pole from rocks in northern Italy. *Geophys. J.*, 6: 377–390.
VAN HILTEN, D. and ZIJDERVELD, J. D. A., 1966. Paleomagnetism and the Alpine tectonics of Eurasia, 2. The magnetism of the Permian porphyries near Lugano (northern Italy, Switzerland). *Tectonophysics*, 3 (5): 429–446.
WENK, E., 1955. Eine Strukturkarte der Tessiner Alpen. *Schweiz. Mineral. Petrog. Mitt.*, 35: 311–319.
WENK, E., 1967. Besonderheiten und Probleme des anatektischen Unterbaues der Alpen. In: J. P. SCHAER (Editor), *Colloque Étages Tectoniques.* Baconniere, Neuchâte pp. 83–88.
WIESENEDER, H., 1967. Zur Petrologie der ostalpinen Flysch. *Geol. Rundschau*, pp.227–241.
WUNDERLICH, H. G., 1964. Zur tektonischen Synthese der Ost- und Westalpen nach 60 Jahren ostalpiner Decken theorie *Geol. Mijnbouw*, 43 :33–51,
ZAPFE, H., 1963. Aptychen-lumachellen. *Ann. Naturhist. Museum Wien*, 66: 261–266.

CHAPTER 11

Alpine Europe: The Swiss Alps

GENERAL REMARKS

The Alps of Switzerland, in our imagination the focal point of the Alpine chain, have formed the basis for the classic picture of the Alps. Their geology has been superbly treated in the classic volumes by Albert Heim. But if we want to find out what is new in this region, we meet with the same difficulty encountered in many other parts of Europe, that there is no modern summary.

Since HEIM (1919, 1921, 1922) and STAUB (1924) and the *Geologischer Führer der Schweiz* of GAGNEBIN and CHRIST (1934)[1], there is of course the admirable summary of CADISCH (1953). This gives an objective account of the literature up to the date of publication. But the wealth of detail, coupled with the fact that there is very little introductory text on the regional tectonics, makes this book no easy reading matter. The more so, because it only contains an index of geographical names. Most of the newer research has been published in only three series, but even so the amount of varied newer information is appalling. These series are: (*1*) the *Beiträge zur geologischen Karte der Schweiz* at Berne; (*2*) the *Eclogae geologicae Helvetiae* at Basel; and (*3*) the *Schweizerische mineralogische und petrographische Mitteilungen* at Zürich.

At present the best outline is given by the new geological map 1:200,000 in eight sheets, and the accompanying explanatory texts. All map sheets have now been published (GEOLOGISCHE GENERALKARTE DER SCHWEIZ, 1942–1965), but of the explanatory texts we have as yet only those belonging to sheets 1–3, 5 and 6. Of course, those geologists who, like the Swiss, like to make a thorough study of their area, will only be able to climb over a limited region during one summer, and will be content with the maps on a larger scale. For these more modern colleagues, who content themselves with "Autobahn geology", excellent summaries of the more scenic routes are given in the "*Routenführer*", *Schweizerische Alpenposten* of the Swiss postal department (P.T.T., Berne). Although touristic, the 22 guides published always contain a short geologic description, normally accompanied by coloured geologic sections and a map.

I will not try here to supply in any great detail factual information about the Swiss Alps, but instead try to give a general presentation of the main features in which as far as possible old facts and new interpretations are interwoven.

[1] Second edition by LOMBARD, (1967).

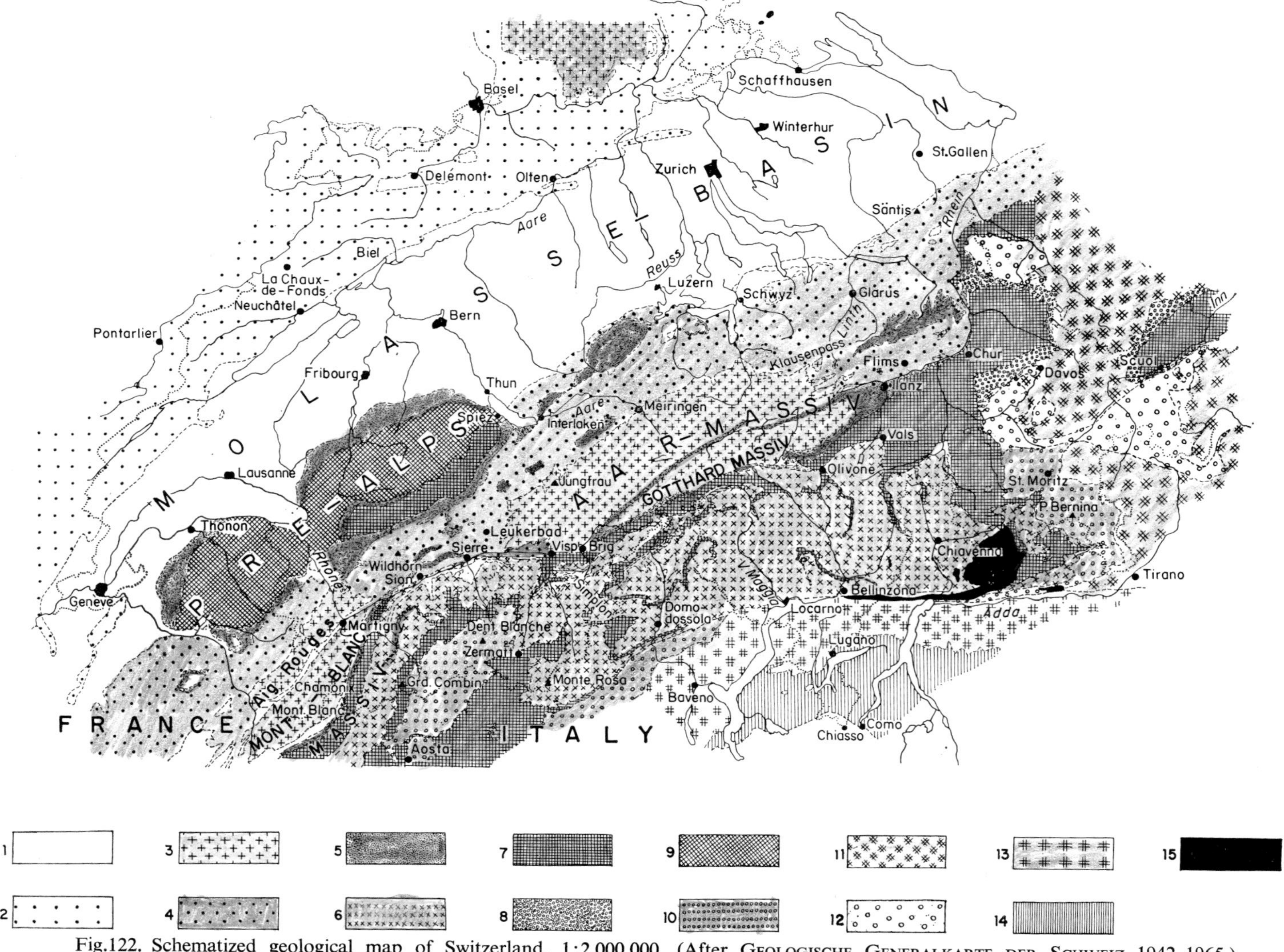

Fig.122. Schematized geological map of Switzerland, 1:2,000,000. (After GEOLOGISCHE GENERALKARTE DER SCHWEIZ, 1942–1965.)
1 = Tertiary basins; *2* = Jura Mountains; *3* = external massifs of the Alps and the Hercynian upland of the Schwarzwald; *4* = autochthonous and nappes of the Helvetides; *5* = Ultrahelvetic; *6* = crystalline cores of the Pennide nappes; *7* = Mesozoic and Early Tertiary of the Pennides; *8* = pre-Alpine nappes in eastern Switzerland; *9* = Prealps; *10* = lower Austride nappes; *11* = crystalline cores of the middle and upper Austrides; *12* = Mesozoic of the middle and upper Austrides; *13* = crystalline of the Southern Alps; *14* = Mesozoic of the Southern Alps; *15* = discordant younger plutons, such as the Bergell Massif east of Chiavenna and younger volcanics of the Hegau in southern Germany. Oligocene to Pliocene.

As stated before, the main elements of the Swiss Alps are the Helvetides, the locally developed Prealps, the Central Massifs, the Pennides, the root zone, the Southern Alps, and towards the northeast, the Austrides (Fig.122). In this chapter the first four units will be treated in some detail.

THE HELVETIDES

The Helvetides (Fig.123) form the northernmost of the continuous zones of the Swiss Alps. Only in the southwest are they locally outdistanced by the Prealps. The Helvetides form a succession of epidermis type of nappes, comprising a miogeosynclinal series of sediments, ranging from the Permian verrucano to the Eocene flysch.

The Swiss name of the Helvetides, the "Hohe Kalkalpen" or High Calcareous Alps, relates to the fact that in their stratigraphy two series occur of massive, whitish limestones, each several hundreds of meters thick. These limestone series design the main topographic outlines of the Helvetides. In their typical development they are the Quintnerkalk or Hochgebirgskalk of the Upper Jurassic and the Schrattenkalk formed by the Lower Cretaceous in the reef- and reef-rock facies of the Urgonian. "Schratten" is a local form for "Karren", which preferentially develop in the massive limestones of the Urgonian.

The Hochgebirgskalk, true to its name, mainly forms the highest peaks in the Berner Oberland, the southern part of the Helvetides, that is in the par-autochthonous cover of the Central Massifs and in the southern nappes. The Schrattenkalk, on the other hand, is best represented in the more northerly nappes.

Both Hochgebirgskalk and Schrattenkalk are, however, predominantly lithostratigraphical units. If one compares detailed stratigraphical columns, such as are found, for instance, in HEIM (1919, 1921, 1922) and in CADISCH (1953), it becomes evident that there is strong lateral variation in chronostratigraphy along the strike of the Helvetides.

In the literature much notice is taken of a deepening of the geosynclinal sea in this or that direction during certain epochs. However, the intercalation of thick, massive reef- and reef-rock limestones and the existence of diastems and larger breaks at various levels of the stratigraphical column, all over the chain, indicate the persistence of a shallow sea environment, even with temporary and locally areas of non-deposition. Sedimentation evidently was able to offset subsidence. It must have been lateral variations in the rate of geosynclinal subsidence, which have led to the variations in thickness and in facies of the various units, and not the existence of deeper seas towards the centre of the geosyncline.

Apart from these lateral variations in facies, the stratigraphy of the Helvetides is, in a general way, quite similar to other external parts of Alpine Europe, particularly to that of the Dauphinois zone in the French Alps, and of the Jura Mountains. And even, though the development is much thinner in the epicontinental facies, to that of the Mesozoic on the platforms of western Europe.

For the tectonic evolution, these lateral variations in stratigraphy are, of course, not important. The main factor is that, apart from the two thick horizons of massive limestones mentioned above, the Mesozoic is formed by a varied series of shales and marls and thin-bedded limestones, while the Early Tertiary flysch is formed by thin-bedded sandstones, shales and an occasional limestone bed. All of these series facilitate differential

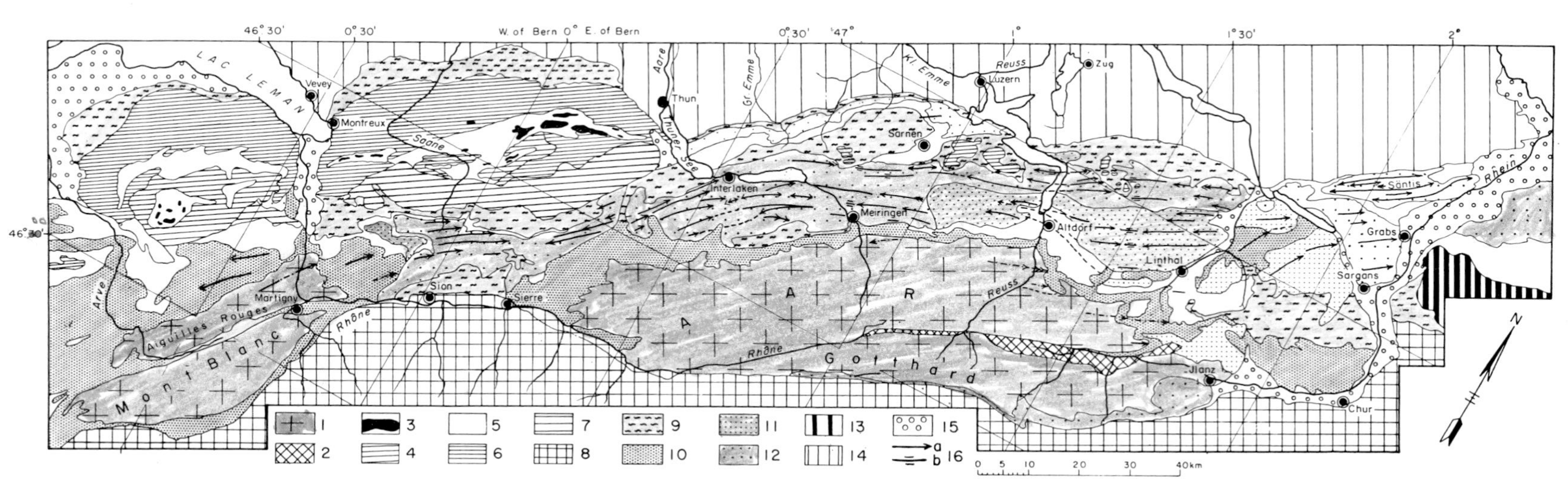

Fig.123. Map of the Helvetides.

1 = Central Massifs; *2* = verrucano; *3* = Simmen nappe; *4* = Breccia nappe; *5* = flysch of the Klippen nappe and Helvetian flysch; *6* = Mesozoic of the Klippen nappe; *7* = Niesen nappe; *8* = Pennide nappes; *9* = Ultrahelvetic flysch and Mesozoic; 10 = authochthonous and par-autochthonous of the Helvetides; *11* = middle Helvetide nappes; *12* = upper Helvetide nappes; *13* = Austride nappes; *14* = molasse; *15* = Quaternary; *16* = main fold axes: *a* = plunging, *b* = horizontal.

Longitude west and east of Berne; 0° Berne is 7° 26′ east of Greenwich. (After ARBENZ, 1934, simplified.)

tectonic movements, that is, "décollement". The basal part of the Alpine series in the Helvetide facies, the Permian verrucano, is formed by red beds, with their thickest development in the northeast. It is not clear what has been the lubricating horizon or horizons, which has made possible the "décollement" of the verrucano from its basement. Presumably this will have been fine-grained silty red beds, the "pelites" of the French nomenclature, so commonly developed in the Lower Permian of western Europe.

The main features of the structure of the Helvetides are still well illustrated by the diagram of ARBENZ (1912; Fig.124). From south to north the following main elements can be distinguished: (*1*) the sedimentary mantle of the Central Massifs, which is only slightly dislocated; (*2*) then follow, in most cases, a number of smaller structures, the par-autochthonous nappes, followed in turn by the main mass of the Helvetide nappes. This can normally be subdivided into three major nappes. In many sections a northernmost border chain, or "Randkette", can be separated from these three nappes.

Owing to a centrally situated zone of axial depression, the development of the Helvetides is best exposed in the southwestern and the northeastern part of the chain. Although a broad correlation of the various nappes in southwestern and in northeastern Switzerland is assumed, the actual names applied to these nappes are different, because they are taken from the main summits in either area. This situation has been indicated diagrammatically by ARBENZ (1912; Fig.125), whereas the newest correlation of Cadisch is given in Table XVI. It must be added, however, that in this correlation the westerly nappes of Diablerets and Wildhorn are placed lower than in most other correlations. For further factual details, I have to refer the reader to the regional literature.

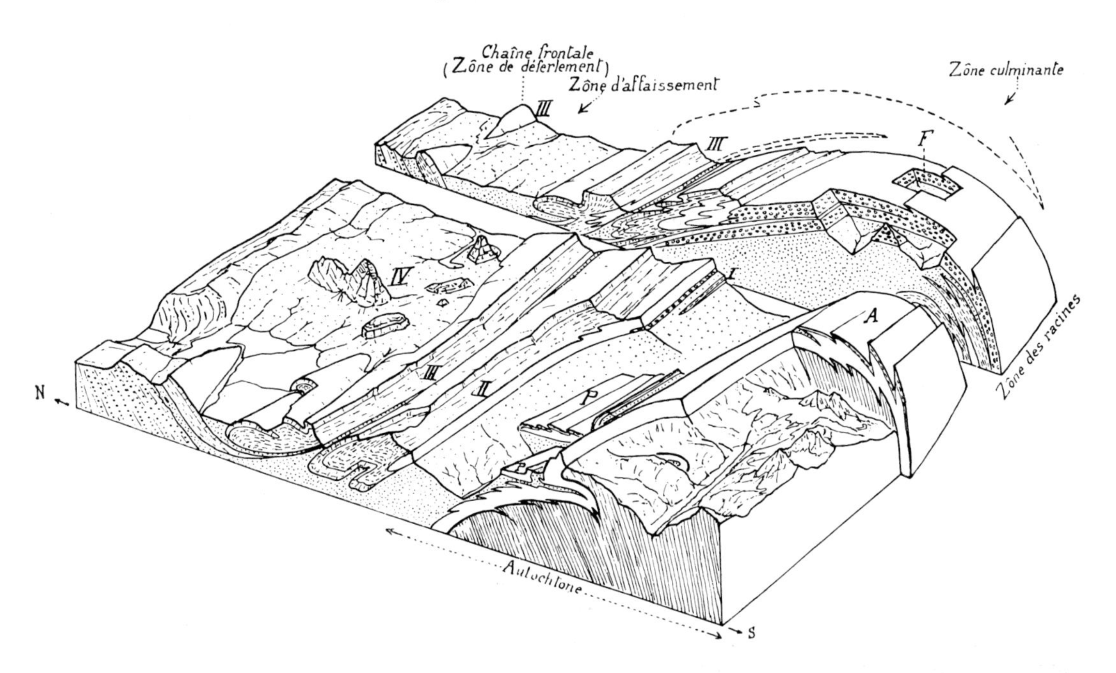

Fig.124. Block diagram of the Helvetides in northeastern Switzerland.
A = autochthonous sedimentary cover of the Aar Massif; *P* = par-autochthonous nappes; *I–III* = main Helvetide nappes; *F* = "Fenster"; window of the crystalline core; *IV* = Klippen nappe, the aequivalent of the Prealps further to the southwest; *zône de déferlement* = surf zone; *zône d'affaissement* = collapse zone; *zône culminante* = culmination; *zône des racines* = root zone. Note the large masses of flysch in which the Mesozoic Klippen swim. (After ARBENZ, 1912.)

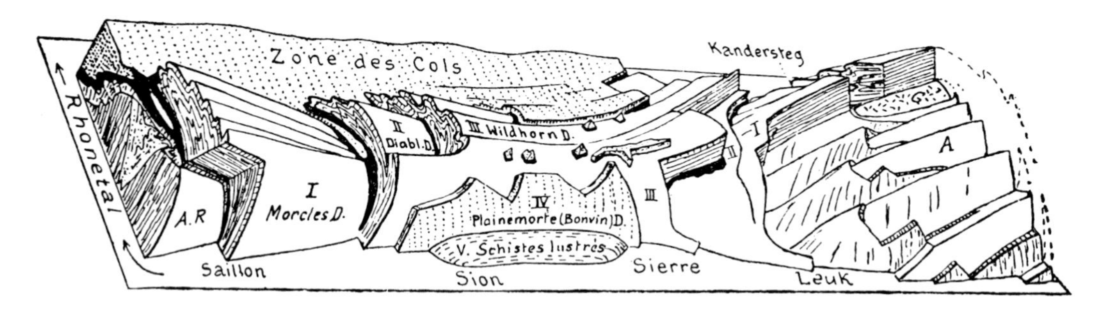

Fig.125. Block diagram of the axial depression of the Helvetides in central Switzerland, showing the correlation between the main nappes in the southwestern and in the northeastern part of the Helvetide chain. *AR* and *A* = Aiguilles Rouges and Aar Central Massif; *I* = Morcles nappe; *II* = Diablerets nappe; *III* = Wildhorn nappe; *zône des Cols* and *nappe de Bonvin* = Ultrahelvetic flysch, belonging to the Prealps. (After ARBENZ, 1912.)

TABLE XVI

CORRELATION OF THE MAIN HELVETIDE NAPPES IN SWITZERLAND

(From CADISCH, 1953. Flysch not considered)

West Switzerland	*Central Switzerland*	*East Switzerland*
Meilleret n. Zone of Bex-Laubhorn n. Tothorn-(or Sex Mort) n.[1] Anzeindaz n. Plaine morte n. } ultrahelvetical	....	Fläscherberg-wedge
Wildhorn n.[2] —	— Schwalmeren-Lobhörner. Scheidegg (Dogger) (Cretaceous-Upper Jurassic)	Drusberg-Fluhbrig n. Räderten n. Wiggis n. } Kreide n.
	—Dreispitz........Harder........ Ober and Niederbauen Seelisberg	
	—Wetterlatte Waldegg Muetterschwanderberg Bürgenstock S.-Rigihochfluh n.	
	«Borderchain», Niederhorn S.-Pilatus n. N.-Pilatus n. } .... N.-Rigihochfluh n.	
	Axen nappe	Upper Jurassic Middle Jurassic } n. Lower Jurassic n.
	←—— Mürtschen n. Elggis wedge	↑
	Saasberg wedges («Glarner n.») | Schilt nappe	
		verrucano n.
Diablerets n. (= Gellihorn-Zwischen n.) Morcles n. (= Doldenhorn n. = Blümlisalp n.)	parautochthonous nappes between Reuss and Linth }	{ parautochthonous fold and nappes in the Ringelspitz and Calanda area
....	autochthonous	....

[1] Former Bonvin nappe.
[2] The connections in the Wildhorn nappe concern the Cretaceous elements.

Of course, once one accepts the possibility of nappe discontinuity, as against the dogma of "cylindrism" of the classic picture, there is no reason to assume physical identity of the various nappes in the northwestern part and in the southeastern part of the Helvetide chain. It is then no longer imperative that the nappes continue, in the crust, underneath a zone of axial depression. On the contrary, such a zone of axial depression might well form, just because the individual nappes thin out in that direction. But even if this is the case, the diagram of Fig.125 and Table XVI have their value, if only in indicating the geometric equivalents in either area.

OLDER AND NEWER TECTONIC INTERPRETATIONS

The differences between the older and the younger interpretations, can perhaps best be estimated from two sections through the same general area, reproduced here as Fig.126, 127. In Fig.126, taken from HEIM (1921), we find the same main elements as in Fig.127, taken from the recent explanatory text to sheet 3 of the geological map 1:200,000.

In the older interpretation, each single overthrust element was interpreted as part of an individual nappe. Such elements were thought to form erosional remnants and consequently had to be completed in the section by aerial or by crustal connections, to make up, for each of them, a separate nappe, complete with normal and reversed limb. In the newer interpretations these elements are considered as parts separated from the same original thrust sheet. Of course, erosion has played its part, but it is no longer necessary to construct the lofty edifices of aerial reconstructions of former nappes, to account for each individual thrusted element.

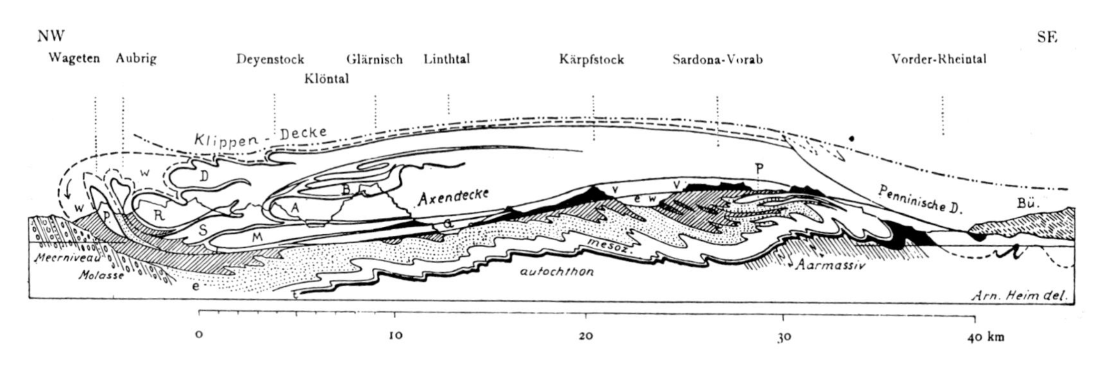

Fig.126. Schematic section through the Alps of Glarus.
G = Glarner nappe; M = Mürtschen nappe; A = Axen nappe; S, R, D = Säntis–Drusberg nappes. (After HEIM, 1921.)

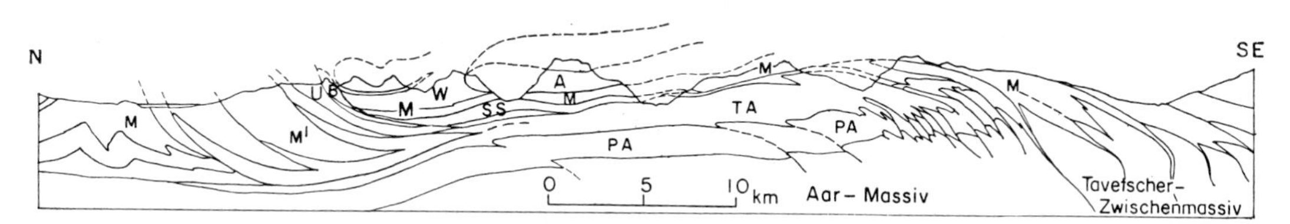

Fig.127. Cross-section through the Alps of Glarus, northeastern Switzerland.
M = molasse; M' = zone of overthrust molasse along the northern border of the Helvetides; U = Ultrahelvetic flysch; W = Kreide-Decke (= Säntis–Drusberg, or Wildhorn nappe); A = Jura-Decke (= Axen nappe); M = Verrucano-Decke (= Mürtschen nappe); SS = Glarner-Decke; TA and PA = par-autochthonous and autochthonous Tertiary flysch and Mesozoic; B = Bündner Schiefer. (After BUXTORF and NABHOLZ, 1957.)

This is not yet quite evident in Fig.127. But this section has been originally compiled by Dr. P. Christ, an old-timer in Alpine geology. It is difficult, apparently, to ignore completely the rounded noses in aerial reconstructions of thrust sheets, once one has become used to drawing them.

Accordingly, it is no longer necessary, in the newer interpretation, to invest each of the thrust sheets with a complete stratigraphic series of Alpine sediments. It is no longer necessary, to complete the Säntis–Drusberg nappe — or nappes, as they were formerly interpreted — which is a thrustsheet formed almost exclusively by Cretaceous, with the rest of the Mesozoic and with the verrucano. Instead it is thought to represent the highest portion of a complete Alpine series, which has travelled farthest north.

Neither is it thought necessary, nowadays, to assume that the Axen nappe formerly enveloped a complete section of Cretaceous, of Triassic and of verrucano. In the same way, the Mürtschen nappe is now considered to form a thrust sheet, mainly composed of verrucano, from which the higher stratigraphical units have been stripped off tectonically, during nappe formation (cf. TRÜMPY and RYF, 1965). A reconstruction of the original sedimentary basin of the Helvetide nappes, based on these newer views is given by DE SITTER (1964, fig.185).

"VORANEILEN DER JÜNGEREN SCHICHTEN"

The fact that the youngest elements of a given stratigraphical series, which originally were, of course, the highest, travel farthest, whereas the oldest rocks, which originally had the lowest position, remain behind, is a general feature in nappe tectonics. It is the "Voraneilen der jüngeren Schichten", already mentioned in Chapter 10 as a proof of gravity tectonics.

In the Helvetides, this phenomenon could develop in such a pronounced form, because of the massive elements of the Schrattenkalk of the Cretaceous, the Hochgebirgskalk of the Jurassic, and the conglomeratic red beds of the verrucano. The intercalated thinly bedded shales and marls could all act as separate lubricating horizons. Once detached, each separated thrustsheet could then moreover override, and eventually be enveloped by flysch. This led to an even greater mobility, and to an even larger separation of elements which originally have been contiguous.

"BRANDEN DER DECKENSTIRNE"

Another conspicuous feature of the Helvetide nappes is the difference between the tranquil, subhorizontal structure found in the greater part of the body of the individual nappes, as compared with the narrow, often serrated folding, encountered in their frontal parts. This "Branden der Deckenstirne" has also been cited already in Chapter 10, as a proof for gravity tectonics.

The tranquil, subhorizontal structure of nappes is, for instance, well illustrated by the famous Glarus overthrust (Fig.103). Here Permian verrucano overlies Lower Tertiary flysch over many kilometers, without any apparent disturbance. The thrustzone may be exceedingly thin, even of the order of 10 cm. This structural type is not restricted to this one nappe, but forms the common feature of the main mass of all the Helvetide nappes.

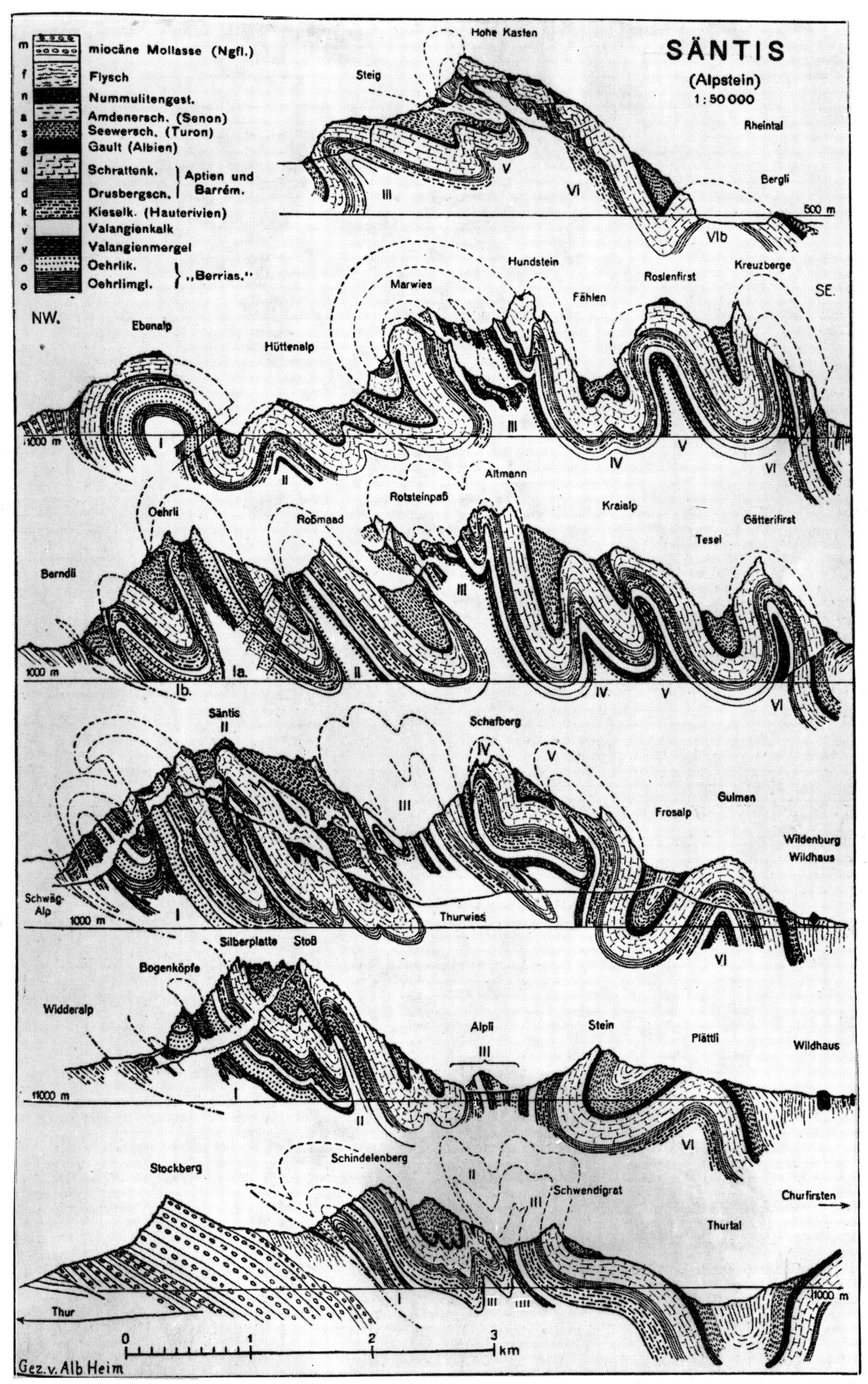

Fig.128. Serial sections through the front of the Säntis nappe in northeastern Switzerland. Showing the detailed structure of the "breakers at the nappe front". Compare Fig.128. (After HEIM, 1921.)

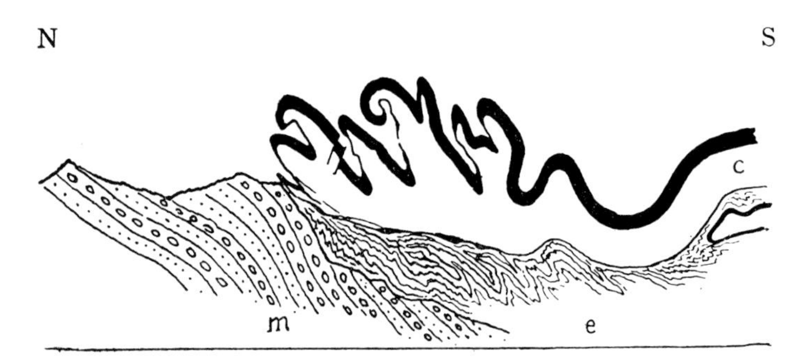

Fig.129. Schematic section through the Säntis nappe in northeastern Switzerland. Showing how at the northern front of the nappe a series of "breakers" develop, in contrast to the quiet overall structure of the nappe. *m* = Miocene; *e* = Eocene; *c* = Cretaceous. The black band is the "Leithorizont" of the Schrattenkalk. (After HEIM, 1921.)

It is, accordingly, met with everywhere, but a single reference will be added, i.e., the beautiful sections through the eastern Helvetides by HELBLING (1938). This, not only because they so impressively show the generally tranquil structures in all main Helvetide nappes, but also because Helbling's sections are based on photogeologic interpretation. Hence, they stand in a class apart, as to their trustworthiness.

A classic picture of the "breakers at the nappe front", is, on the other hand, given by the series of cross-sections through the Säntis chain in northeastern Switzerland, forming the frontal part of the Säntis–Drusberg nappe (Fig.128, 129). In accordance with the theory, requiring that such narrow folds originate through the braking effect of the subsurface on the moving frontal part of the nappe, all folds of Fig.128, 129 are seen to have northward "Vergenz".

The phenomenon of the "Branden der Deckenstirne" is not restricted to the frontal part of the most external nappes, such as the Säntis. It normally occurs in the frontal part of every nappe or nappe digitation. Another well-known locality, is formed by the strongly folded beds of the Lower Cretaceous, along the Urner See north of Fluelen.

Pictures of this section have found their way in many geology textbooks (cf. CLOOS, 1936, fig.137), as one of the most telling examples of folding. It is, however, only found over a zone a couple of hundred meters wide, and forms a limited belt of cascade folds in the frontal part of digitation of the Axen nappe.

ALONG-THE-STRIKE DISCONTINUITIES

Another feature worth while stressing, in relation to the Helvetide nappes, is the discontinuity, along the strike, of individual structural elements. Perhaps this is riding my personal hobby horse too hard. But to one reared in the dogma of nappe continuity, it always comes as a shock, to see how much discontinuity there is, and how very little continuity. Evidence for discontinuity of structural elements in the Helvetides is found on all scales, from that of complete nappes, via structures within an individual nappe, to that of microtectonics.

On the scale of nappes, one finds such discontinuities throughout the chain, with respect to the main nappes. Although we are all familiar with the division of the Helvetides in three main nappes, a glance at the maps of Fig.122, 123 will show how, normally, only

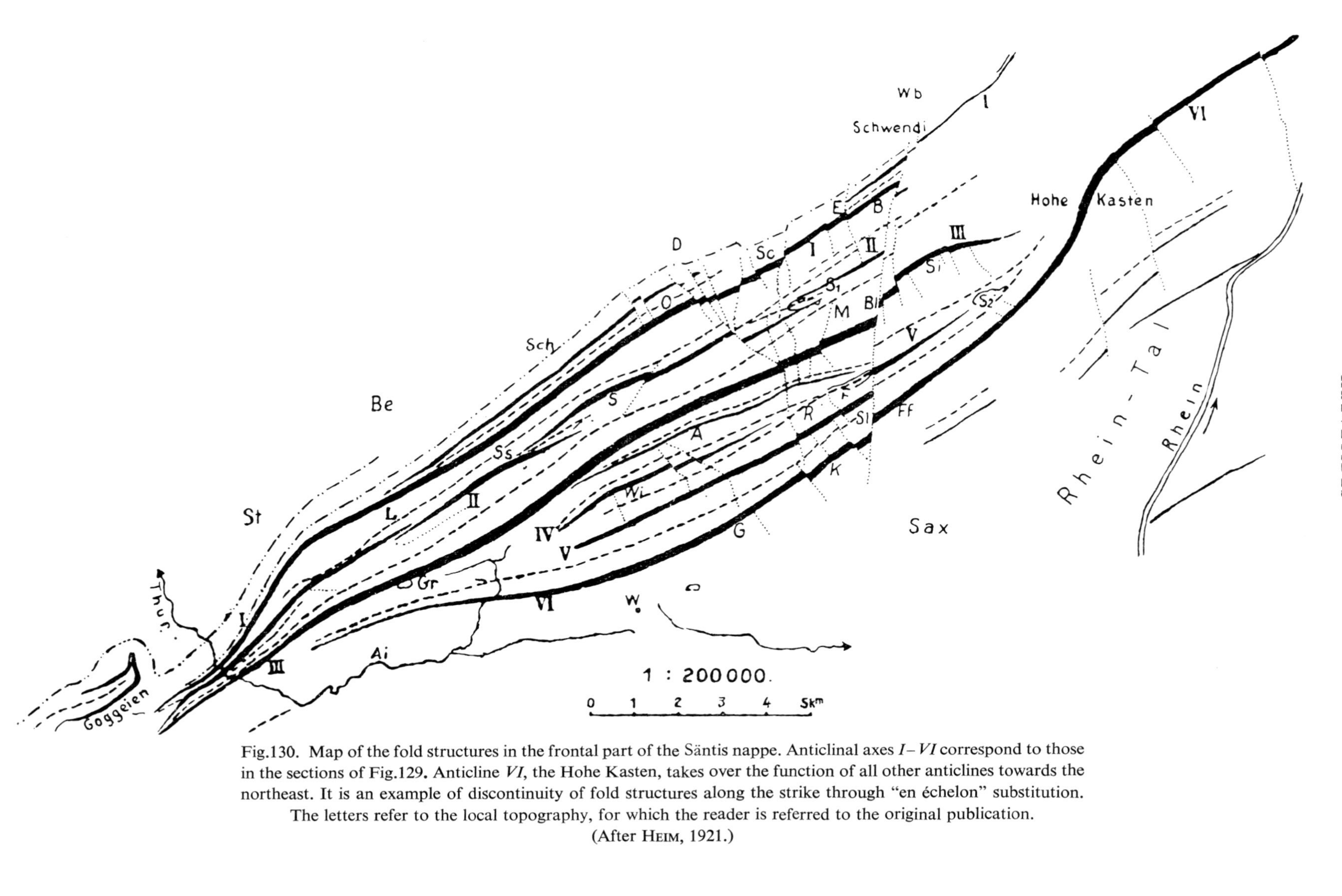

Fig.130. Map of the fold structures in the frontal part of the Säntis nappe. Anticlinal axes *I–VI* correspond to those in the sections of Fig.129. Anticline *VI*, the Hohe Kasten, takes over the function of all other anticlines towards the northeast. It is an example of discontinuity of fold structures along the strike through "en échelon" substitution. The letters refer to the local topography, for which the reader is referred to the original publication. (After HEIM, 1921.)

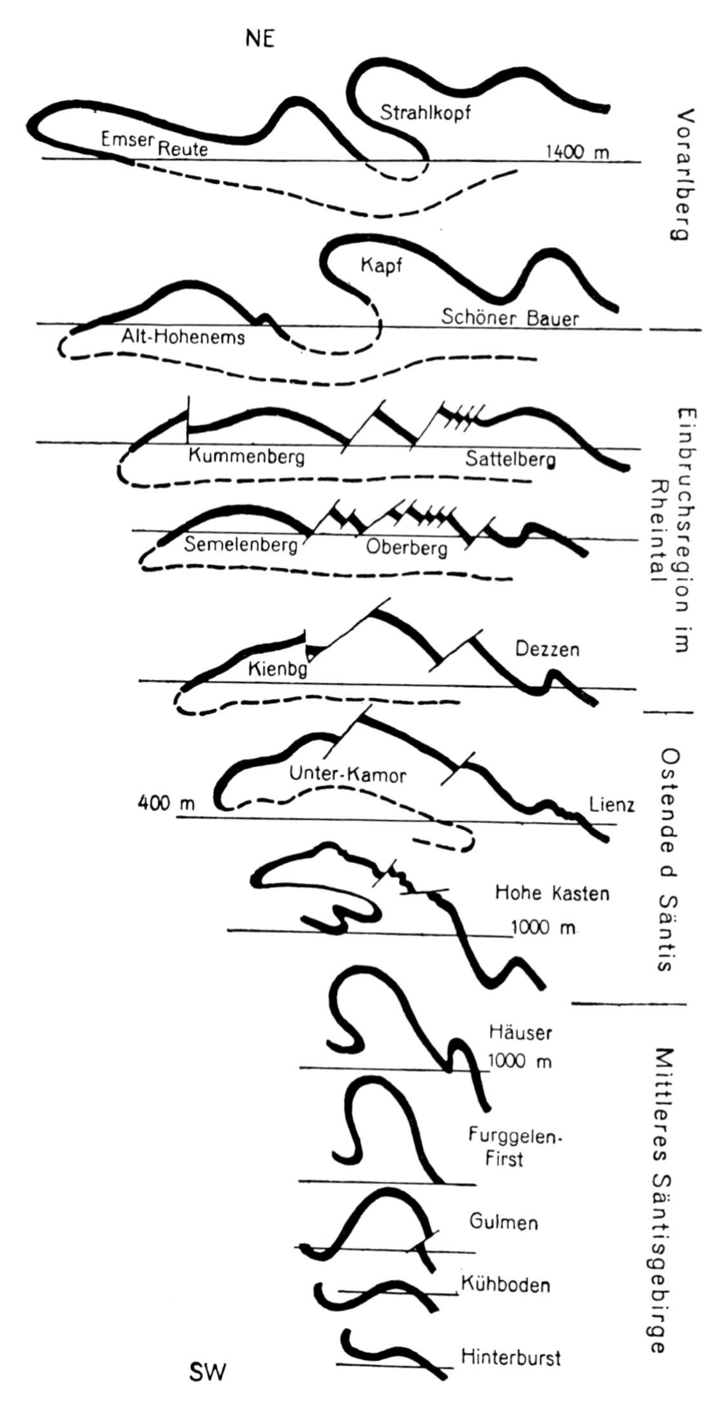

Fig.131. Serial sections through the Hohe Kasten anticline, the southernmost structure of the frontal part of the Säntis nappe (compare Fig.130). The black line represents the (reconstructed) Schrattenkalk of the Lower Cretaceous. The lower sections are the equivalent of the more detailed sections of Fig.129. (After HEIM, 1921.)

two of the main nappes are developed, or how in several cross-sections one only finds a single main nappe represented. Along the strike of the chain there is "en échelon" substitution. Moreover, in a single cross-section one finds all three nappes present only if all of them have subnormal development.

It seems as if there has been a certain amount of overall shortening of the epidermis. This may either be taken up by a single main nappe, in very full development. Or, more normally, by two of the main nappes. Only if this amount of shortening is more or less equally divided over all three main nappes, will we find all three of them, though subnormal and subtypical, in a single cross-section.

An example of similar phenomena of along-the-strike discontinuity on a smaller scale, relating to fold structures within an individual nappe, has been given by HEIM (1921) for the frontal zone of the Säntis. The most rearward of the "breakers at the nappe front", anticline *VI* of Fig.129 takes over the role of all other folds towards the northeast. In doing so, its amount of crustal shortening grows strongly. From a narrow anticline it develops into an overthrust fold, which almost forms a nappe digitation (Fig.130, 131). Here also, we have, in a cross-section, either a number of smaller structures, or a single major structure, which show "en échelon" substitution along the strike, and which have a similar tectonic function.

THE "ROOT ZONE" OF THE HELVETIDES

It has always been understood that the Helvetide nappes originated from a zone situated to the north of the present position of the Pennide nappes. They had a root zone of their own. If, in Alpine geology, "the root zone" is mentioned, this has been understood to mean the root zone of the main nappes, that is of the Pennides, not that of the Helvetides.

There has been general agreement to situate the "root zone" of the Helvetides at the internal side of the most external of the Central Massifs. That is, between Aiguilles Rouges and Mont Blanc in the southwestern part, and between Aar and Gotthard in the northeastern part of the chain. Particularly in the latter area, a narrow, strongly tectonized, crystalline body, intercalated between the main masses of the Aar and the Gotthard, the Tavetscher Zwischenmassiv (Fig.132, 133), has generally been regarded as the Helvetide root zone.

The main reason for this lies in the fact that the par-autochthonous sedimentary cover of the Aar Massif shows a typically Helvetian stratigraphy and facies, whereas the Mesozoic cover on the Gotthard Massif on the whole is more similar to the Mesozoic of the Pennides.

In particular the Lower Jurassic of the sedimentary cover of the Gotthard Massif resembles the Bündner Schiefer of the Pennides. It has, in the literature, even been indicated as "Gotthardmassivischer Bündnerschiefer" or "Gotthardmassivische Jura" (NIGGLI, 1944; NABHOLZ, 1949; MEIER and NABHOLZ, 1950; NABHOLZ and VOLL, 1963). In part, it can be correlated with the Ultrahelvetic facies, which also distinguishes it from the Helvetides, and brings it in line with the Pennide facies (JUNG, 1963).

According to the newer views on gravity tectonics, one does not need a root zone any longer. But even so, one still is in need of an area from which to derive the sedimentary series of the epidermis type of nappes. Concurrently, the original width of the "root zone"

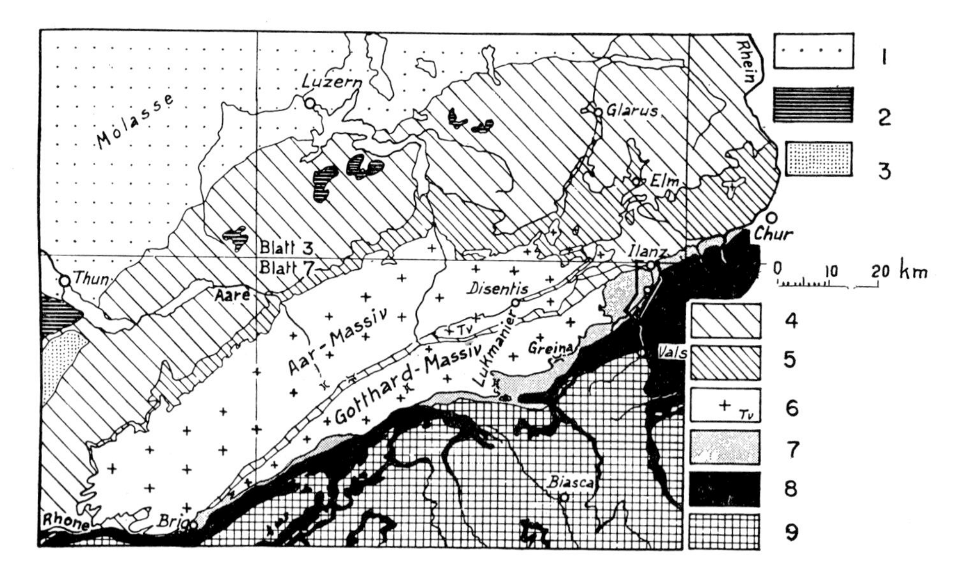

Fig.132. Situation of the Tavetscher Zwischenmassiv, between the Aar and Gotthard Massifs. *1* = molasse; *2* = Klippen and Klippen-Decke of the Prealps; *3* = Breccien-Decke of the Prealps; *4* = Helvetide nappes; *5* = par-autochthonous and autochthonous cover of the Aar Massif in Helvetian facies; *6* = crystalline cores of Central Massifs (*Tv* = Tavetscher Zwischenmassiv); *7* = Mesozoic cover of the Gotthard Massif; *8* = Mesozoic envelopes of Pennide nappes, mainly Bündner Schiefer; *9* = crystalline cores of Pennide nappes. (After NABHOLZ and VOLL, 1963.)

formerly was supposed to be very much larger than that of the "parental region" from which, in the newer views, the thrust sheets of the nappes are thought to have slid. If one really adds up, from the reconstructions of Heim, Staub, and others, all involutions and digitations of nappes, considered as recumbent folds continuous along the strike, the original width of the "root zone" had to be estimated at a couple of hundred kilometers. The width of the "parental region", according to present views, need on the other hand not to be larger than some 20 km.

Now as before, the Tavetscher Zwischenmassiv is thought to represent the area from which the Helvetide nappes originated. But instead of representing a "root zone", from which, through strong compression, the nappes were squeezed out, it is now thought that the nappes were stripped off the area, before it became strongly compressed. The sub-vertical tectonization of the Tavetscher Zwischenmassiv is now thought to be due to the

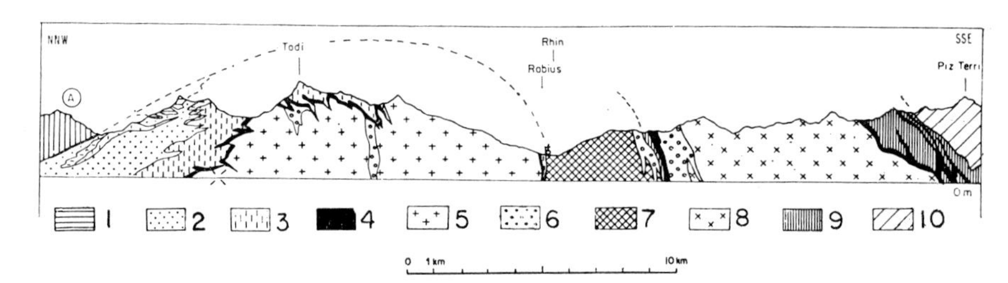

Fig.133. Cross-section over the Aar and Gotthard Massifs, showing the position of the Tavetscher Zwischenmassiv, the Helvetides and the front of the Pennides.
1 = Helvetide nappes; *2* = flysch; *3* = par-autochthonous and autochthonous Mesozoic in Helvetian facies; *4* = Triassic; *5* = Aar Massif; *6* = Permo-Carboniferous; *7* = Tavetscher Zwischenmassiv; *8* = Gotthard Massif; *9* = Mesozoic cover of the Gotthard Massif; *10* = Pennide nappes. (After TRÜMPY, 1963.)

fact that, after nappe formation, the Gotthard Massif approached the Aar Massif over a distance of "at least 20 km" (TRÜMPY, 1963).

ULTRAHELVETIAN, KLIPPEN AND PREALPS

Enveloping the highest Helvetide nappes is a thick, normally strongly tectonized unit, consisting predominantly of flysch. Tectonically this can be interpreted as "higher than the Helvetide nappes", although it is clearly related to this nappe system. Hence it is indicated as *Ultrahelvetian.* As to facies, a considerable part of the flysch which is found in Ultrahelvetian tectonic position is different from the normal flysch adhering to the individual Helvetide nappes, in containing a great amount of erratic blocs. These blocs, varying in size from that of a brick to that of a skyscraper, are moreover incorporated in the finer sediments in a chaotic way. Hence it is called "Wildflysch".

In central Switzerland, floating upon the Ultrahelvetian flysch isolated mountain tops are found, formed by the Mesozoic, the *Klippen.* These are interpreted as isolated remnants of a still higher nappe, which formerly extended as a continuous unit over most of the Ultrahelvetian flysch. The best accessible, and very impressive, of the klippen, are the two Mythen near the town of Schwyz (Fig.134, 135).

Further southwest, these isolated klippen do indeed merge into a continuous nappe, not yet destroyed by erosion. This nappe, the "Klippen-Decke", or "Nappe des Préalpes médianes" as it is called in the French literatures, forms part of the nappe system of the *Prealps* (Fig.123). This is found in southwestern Switzerland, on either side of the Rhone valley, between the Arve and the Lake of Thun, in front of, that is to the northwest of, the Helvetide nappes.

ULTRAHELVETIAN

To return to the Ultrahelvetian flysch, the facies of the Wildflysch, mainly found in the tectonically lowest unit, the Plaine Morte nappe, was formerly interpreted as due purely to tectonization. It is now thought to have developed, at least in its main features, through some sort of sedimentary slumping (cf. SODER, 1949, pp.83–85), the "olistostromes" of the Italian literature. Although it is of course evident that in such mobile material as a flysch series, situated moreover in a frontal zone of a nappe system, tectonization must also have strongly influenced a series which through its sedimentary history had already a chaotic character. In this text, I cannot go into further details on this matter of the Wildflysch, nor on the important finds of small flakes of Mesozoic, found intercalated in the flysch in Ultrahelvetian position, but must refer the reader to the regional literature.

The view that the clastic material of the ordinary, stratified flysch is derived through erosion from embryonic cordilleras, representing nappes in evolution, has been challenged for the Ultrahelvetian flysch by HSU (1960). He noted that paleocurrent directions in turbidites of this series did not show a relation to the nappe system. They even indicated partial derivation from the north. Hsu arrives at a paleogeographic picture consisting of a number of local graben basins and horsts, separated by high angle faulting, the horsts forming local supply areas of terrigenous material. Hsu's interpretation consequently is a sort of extrapolation of the ideas of Trümpy (p. 194) as to the horst-and-graben paleogeography of the Alpine geosyncline during the Mesozoic.

In the strongly tectonized Ultrahelvetian flysch a relatively large number of thin nappes has been differentiated, which I can no more than enumerate here, without entering into any detail. They are, from higher to lower, the Chamossaire, the Meilleret, the Arveyes, the Bex, the Tothorn (the former Bonvin), the Tour d'Anzeinde, and Plaine Morte nappes (BADOUX, 1963; cf. Table XVI). The four highest units are sometimes combined into the Laubhorn nappe, and represent the Upper Ultrahelvetian, whereas the three lowest units together form the Lower Ultrahelvetian.

The Ultrahelvetian nappes are well developed in the zone des Cols, the axial depression of the Helvetides in central Switzerland (p. 229). The position of these nappes in this area, and their supposed mode of origin have recently been summarized by BADOUX (1963), cf. Fig.134, 135.

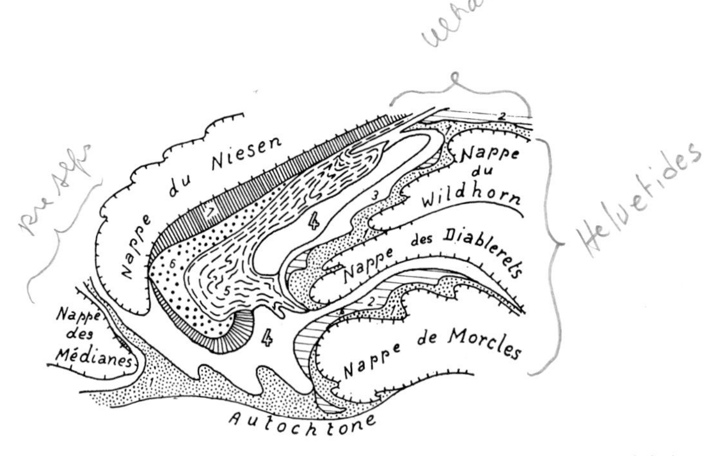

Fig.134. Situation of the Ultrahelvetian nappes (*1–7*) in the "zone des Cols", i.e., in the axial depression of the Helvetides in central Switzerland.
1 = Plaine Morte nappe; *2* = Anzeinde nappe; *3* = Sex Mort nappe; *4* = Bex nappe; *5* = Arveyes nappe; *6* = Meilleret nappe; *7* = Chamossaire nappe.
Helvetides: Morcles, Diablerets and Wildhorn nappes; Prealps: Niesen and Medianes (Klippen) nappes. (After BADOUX, 1963.)

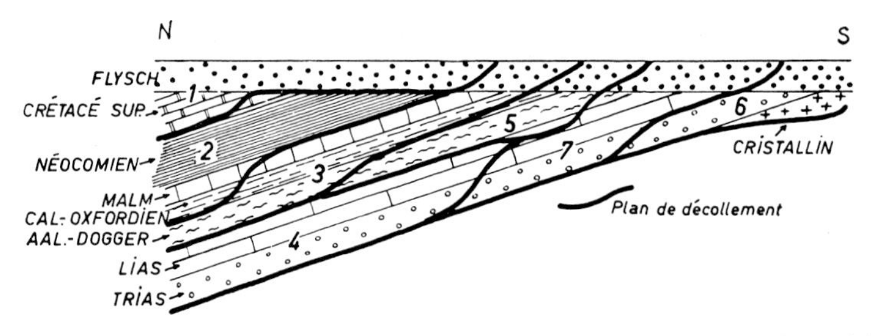

Fig.135. Schematic interpretation of the origin of the Ultrahelvetian nappes in the "zone des Cols" (Fig.134), through differential "décollement" from a single stratigraphic series.
1 = Plaine Morte nappe; *2* = Anzeinde nappe; *3* = Sex Mort nappe; *4* = Bex nappe; *5* = Arveyes nappe; *6* = Meilleret nappe; *7* = Chamossaire nappe. (After BADOUX, 1963.)

KLIPPEN

The klippen, on the other hand, are so fundamental to our notions of nappe tectonics, that a few words on their best publicized example, the Mythen, are in order.

Fig.136 gives a panoramic sketch, the lower section of Fig.137 the geologic interpretation of the Mythen, as seen from the west, from the Vierwaldstätter Lake, both by HEIM (1922). Most of the landscape is formed by soft, rounded hill sides, being the domain of the Ultrahelvetian flysch. On top of this ride the synclinal structures of the Mesozoic of the Mythen (or Mithen, as Heim called them). In the lower righthand corner, on the other hand, the Lower Cretaceous Schrattenkalk of the highest Helvetide nappe, the Wildhorn or Säntis-Drusberg, appears in a typical "breaker at a nappe front" tectonic style. This same situation is met with in all klippen. In a more general way, it is illustrated in Fig.138, a cross-section slightly more westerly, through the Stanserhorn Klippe, south of Luzern.

The geometrics of a series of Mesozoic, riding on top of a thick zone of Tertiary flysch, which in its turn covers the Mesozoic, of course forms one of the best visible instances of nappe tectonics. It has been important, in proving how, even when nappes have been almost eroded away, their former presence can be reconstructed through the erosional remnants of klippen. The main interest in this succession goes, however, far beyond the local geometrics. It lies in the strong difference between the facies of the Mesozoic in the upper and in the lower series. In the Mythen the main development of the Cretaceous is as "couches rouges", in the Wildhorn nappe on the other hand, as Schrattenkalk. The term "couches rouges" in Swiss Alpine geology is not used to indicate continental red beds, but marine, *Globotruncana* bearing, marly limestones. These are interpreted as being formed in relatively deep water. Regardless of their interpretation, they are, however, quite distinct from the massive neritic facies of the Schrattenkalk of the Wildhorn nappe.

It is quite apparent that the sedimentary environment in which the klippen Mesozoic formed, was different from that of the Helvetian Mesozoic. And taking into consideration the persistence of the Schrattenkalk, at least as a lithostratigraphic unit, in the northerly Helvetide nappes, we may even assume that the deposition of the klippen Mesozoic took place in a basin originally situated quite apart from that in which the Helvetian Mesozoic developed.

This conclusion, in such a qualitative way, is quite legitimate, and still stands. It is, however, doubtful if the more daring, specific interpretation of the classic picture, can still be followed. The difference in facies between Mythen Klippe and Wildhorn nappe was interpreted within the concept of uniformity of facies for a given tectonic unit. On the ground of a resemblance of facies between the Cretaceous of the Mythen and that of some of the higher Austride nappes, it was postulated that the klippen had originally belonged to an upper Austride nappe (cf. STAUB, 1958).

Personally I think that this is stretching the information supplied by variations in facies too far. For this would mean that the succession of the klippen on top of the Ultrahelvetian flysch represented an immense structural discordance. The tectonic hiatus in this section then should include all of the Pennide and most of the Austride nappe systems. It seems incomprehensible that only the highest part of this pile of nappes should have been pushed that far, beyond the Helvetides. One should expect to find an occasional

Fig.136. Panoramic view from the Vierwaldstätter See in central Switzerland ("Seefläche 437 m") to the town of Schwyz and the Lesser and Greater Mythen. (After HEIM, 1922.)

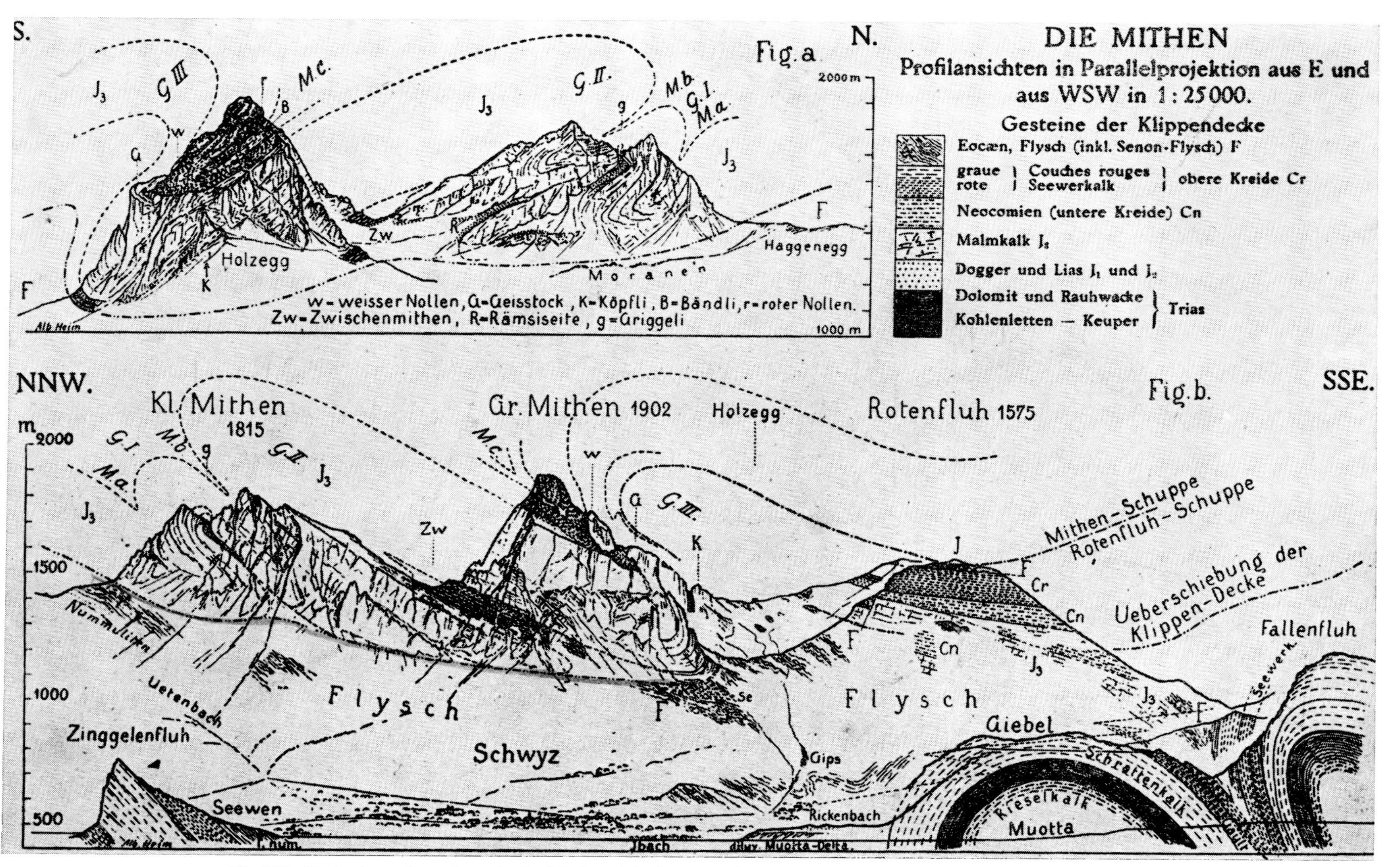

Fig.137. Tectonic sections through the Mythen. The lower section, seen from the west, corresponds approximately to the panoramic view of Fig.136. (After HEIM, 1922.)

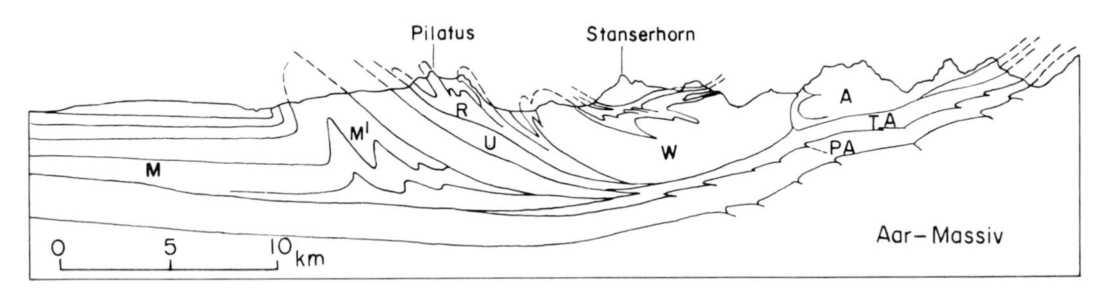

Fig.138. Section through the Helvetides and the Stanserhorn Klippe.
M = molasse; *M'* = zone of folded and overthrust molasse along the northern border of the Helvetides; *U* = Ultrahelvetian flysch; *R* = "Randkette" or border chain of the Helvetides; *W* = Kreide-Decke (= Wildhorn nappe); *A* = Jura-Decke (= Axen nappe); *TA* and *PA* = par-autochthonous and autochthonous flysch and Mesozoic. (After BUXTORF and NABHOLZ, 1957.)

klippe of Pennide, or of Lower or Middle Austride nappes too. To maintain moreover, as has been done by the somewhat more romantic of the defenders of the classic picture, that "in the Mythen one sees the northern part of the African continent overriding Europe", to my ears sounds downright preposterous.

The controversy as to the origin of the klippen is, of course, linked to that of the Klippen-Decke further southwest, and to that of the origin of all of the nappe system of the Prealps. I do not propose to solve this problem, but will nevertheless preceed with some more information on the Prealps.

PREALPS

The Prealps are often called *Préalpes romandes* in the earlier literature, taking their name from the French speaking part of Switzerland, the core of which they form. Topographically, the Prealps consist of two curved mountain chains, on either side of the Rhône valley above the Geneva Lake. The northeastern chain, between the Rhone and the Thuner Lake, the real *Préalpes romandes*, is also designated as *Préalpes fribourgeoises*, the southwestern, between the valleys of the Arve and the Rhône, as the *Préalpes du Chablais*. Or, if only the smaller Swiss part of the latter is meant, the Chablais region overlapping largely into France, the *Préalpes valaisanes*. This morphologic division corresponds to a difference in structure, as the Fribourg Prealps contain a complete set of four nappes, whereas the Chablais Prealps are almost entirely formed by the two middle ones only, the Breccia- and the Klippen nappe.

In a way, the existence of the Prealps is a nuisance. For they do not fit into the nice and simple schematization of the Alpine chain into the Helvetides, the Central Massifs, the Pennides, Austrides, root zone and Southern Alps. They seem to be redundant, so what to do about them? They are higher than Helvetide, tectonically, but how much higher really? Should one consistently make use of the concept of uniformity of facies, though the facies of the various nappes of the Prealps is already quite different inter se, and although this would mean that their tectonic position has to be interpreted as very extreme, that is as upper Austride? Or should one try for a more modest fit?

The influence of such exceptional situations, as formed by the Prealps in Switzerland, invariably becomes less of a nuisance, when, paradoxically, there is more of them. In regard to the Prealps, one finds an example of this state of affairs, when comparing the tone of the Swiss and the French publications. The reason is, that in the French Alps two more structures exist, in some ways comparable to the Prealps, i.e., part of the Alpes Maritimes near Nice, and the *Helminthoïdes* flysch nappe of the Embrunais (Fig.139).

DEBELMAS and LEMOINE (1964) in all tranquility describe all three areas as "bavures", as slobber from the internal nappes over the external units.

A glance at Fig.139 shows, however, that the Prealps differ in two ways from the more southerly areas. In the first place, much of the Prealps is formed by "normal" Mesozoic series, quite distinct from the *Helminthoïdes* flysch. A flysch series comparable to the latter only occupies a relatively small area in the higher of the nappes of the Prealps, that

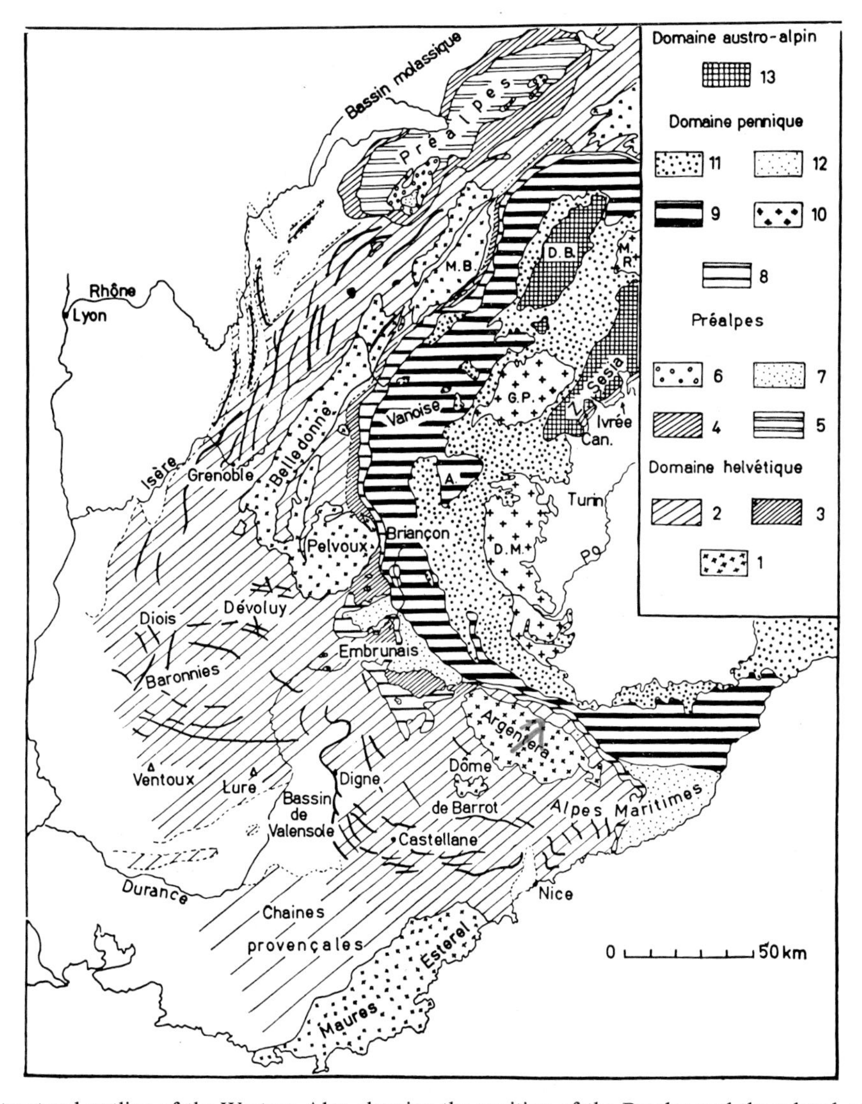

Fig.139. Structural outline of the Western Alps, showing the position of the Prealps and the related areas of the Embrunais and part of the Maritime Alps northeast of Nice.
1 = Central Massifs, ("Massifs externes"); *2* and *3* = Dauphinois zone and Helvetides; *4–7* = Prealps: *4* = Ultrahelvetian and Niesen nappe; *5* = Klippen nappe; *6* = Breccien nappe; *7* = exotic flysch on *6*; *8–12* = Pennides: *8* = Subbriançonnais zone; *9* = Briançonnais zone; *10* = crystalline cores ("Massifs internes") of the Pennide nappes; *11* = Bündner Schiefer of the Pennide nappes; *12* = nappe of the *Helminthoïdes flysch*; *13* = Austrides (Dent Blanche nappe and Sezia–Lanzo zone); *D.B.* = Dent Blanche; *D.M.* = Dora Maira; *G.P.* = Gran Paradis; *M.B.* = Mont Blanc. (After DEBELMAS and LEMOINE, 1964.)

is in the Simme nappe (BADOUX and MERCANTON, 1962; KLAUS, 1953). Secondly, the two southern areas still have retained physical connection with the Pennides, whereas the Prealps has lost its rearward connection through erosion. This offers of course a wider choice of possible correlations.

In a general way the localization of these "bavures" is closely related to zones of axial depression of the external nappe chain. These are, from south to north: (*1*) the easterly plunge of the Argentera Massif for the *Helminthoïdes* flysch nappe of the Maritime Alps and the adjoining area in the Italian Alps; (*2*) for the Embrunais, the depression between Argentera and Pelvoux Massifs; and (*3*) for the Prealps, the depression between Aiguilles Rouges and Mont Blanc Massifs in the southwest and the Aar and Gotthard Massifs in the northeast.

The Prealps have been correlated, on the basis of stratigraphic similarity, of parts of their Jurassic and Cretaceous, with the Austrides (cf. CADISCH, 1953; Table XVIII); and with the Briançonnais zone of the French Alps (pp.289–294; cf., for instance, GIGNOUX, 1955; TRÜMPY, 1965), and with the Embrunais nappe of the French Alps (cf. DEBELMAS and LEMOINE, 1964), in account of the *Helminthoïdes* flysch of the Upper Cretaceous.

Professor P. Fallot has commented on the various hypotheses as to the origin of the nappes of the Prealps in a charming paper, entitled "Promenade d'hypothèse en hypothèse" (FALLOT, 1956). He stressed the fact that the difficulties of such a correlation grow stronger, the higher, tectonically, the units with which the Prealps are correlated. Anything higher than the Pennides would, according to Fallot, require an amount of overthrusting beyond imagination. Fallot's faintly bantering tone, has, however, not convinced his opponents. The question of the correlation of the nappes of the Prealps, and of their origin, is still open.

From my personal reactions to the dogma of nappe continuity, it will be clear that I am very much in favour of a more moderate solution to this problem, as advocated mainly by the French authors. It does not matter so much, whether the Prealps have to be correlated with either the Briançonnais zone or the *Helminthoïdes* flysch nappe, or with (parts of) both units. For it is quite possible that more or less separate basins have developed outwards of the more persistent basins from which the Pennide nappes originated. From one of these the Briançonnais zone might stem, whilst another basin with partly comparable, but partly diverging, stratigraphy, was at the origin of the Prealps. In the same way one or more *Helminthoïdes* flysch basins, extending even to the present Apennines, might have developed without a proper relation to the later main structural zones.

Such a situation would become even more probable, if, according to the evidence supplied by paleomagnetic research, a fold belt can be seen as a mosaic of crustal elements which have drifted together, either before or during orogenesis, instead of as the result of a single geosynclinal fold belt.

To close this discussion, it must be noted that another difficulty of semantics resides in the word "high", as referring to the *Helminthoïdes* flysch. In the "Swiss" sense of the word, "high" means "tectonically high", which also means that such a tectonic unit stems from a relatively internal position. In the "French" sense, "high" in relation to the *Helminthoïdes* flysch means "stratigraphically high". It originally formed the top of the sedimentary pile, when, through a rise in the country of origin, the Pennide nappes began to form through gravity gliding. In this model, the *Helminthoïdes* flysch nappe is no more than another example of the "Voraneilen der jüngeren Schichten", and no tectonic correlation with any of the other nappes of the Pennides is meant, when it is indicated as "high".

Tectonic structure of the Prealps

Apart from their geotectonic interpretation, from the various ways in which the Prealps can be fitted into the general picture of the Alps, two features are important in a more descriptive way. These two features, which we presume to be interrelated, are their aberrant stratigraphy and tectonic style.

The facies of the Mesozoic is strongly variable. The variations in facies, not only between the several nappes of the Prealps, but also within single nappes, are so large that it is difficult to give a simple summary. Again, the reader is referred to the regional literature, in which the stratigraphy of the Prealps is treated in great detail (cf. CADISCH, 1953). Here we need only to retain that the Mesozoic consists mainly of marls and marly limestones and thinly bedded limestones.

It lacks the massive limestones of the Helvetides, such as the Hochgebirgskalk or the Schrattenkalk. The tectonic style exhibited by the nappes with a major Mesozoic core consequently is quite different from that of the Helvetides.

The main structure of the nappe system of the Prealps follows from the generalized cross section of Fig.140. It comprises, from the highest downwards, the Simme, Breccien (or Brèche), Klippen (or Médianes), and Niesen nappes.

Of these the Niesen nappe is formed almost exclusively by Upper Cretaceous flysch. Although diligent search succeeds in turning up more and more intercalations of Mesozoic and of crystalline rocks, partly at least with lower Austride affinities, the Niesen nappe is currently regarded as either Ultrahelvetian or Pennide.

In contrast to the Niesen nappe, the three higher nappes of the Prealps are largely formed by Mesozoic. The tectonic style, peculiar to these nappes was already well illustrated in the sections in the standard work of HEIM (1922). It is here given from a newer cross-section by GAGNEBIN (1942) in Fig.141. The tectonic style of the Mesozoic in the

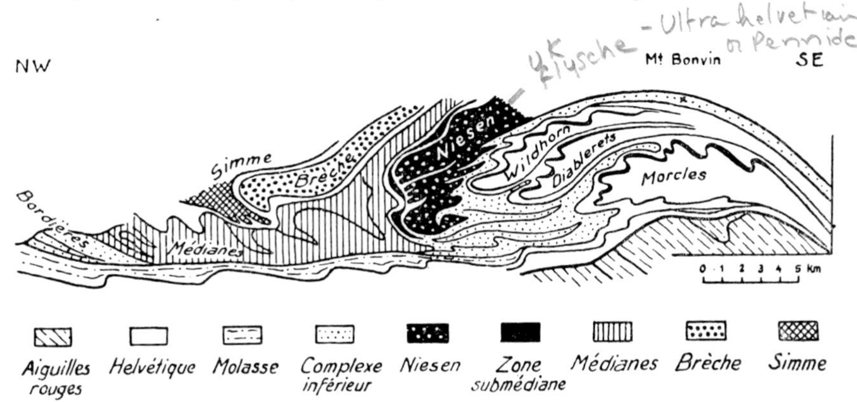

Fig.140. Schematic cross-section through the Prealps, to show the general succession within the nappe system of the Prealps.

Médianes = Klippen-Decke; *Brèche* = Breccien-Decke. Mt. Bonvin, to the right of the Prealps, indicates the zone of axial depression in the Helvetide nappe system, the "zone des Cols", over which the Prealps could move forward (Fig.125). (After GAGNEBIN, 1934.)

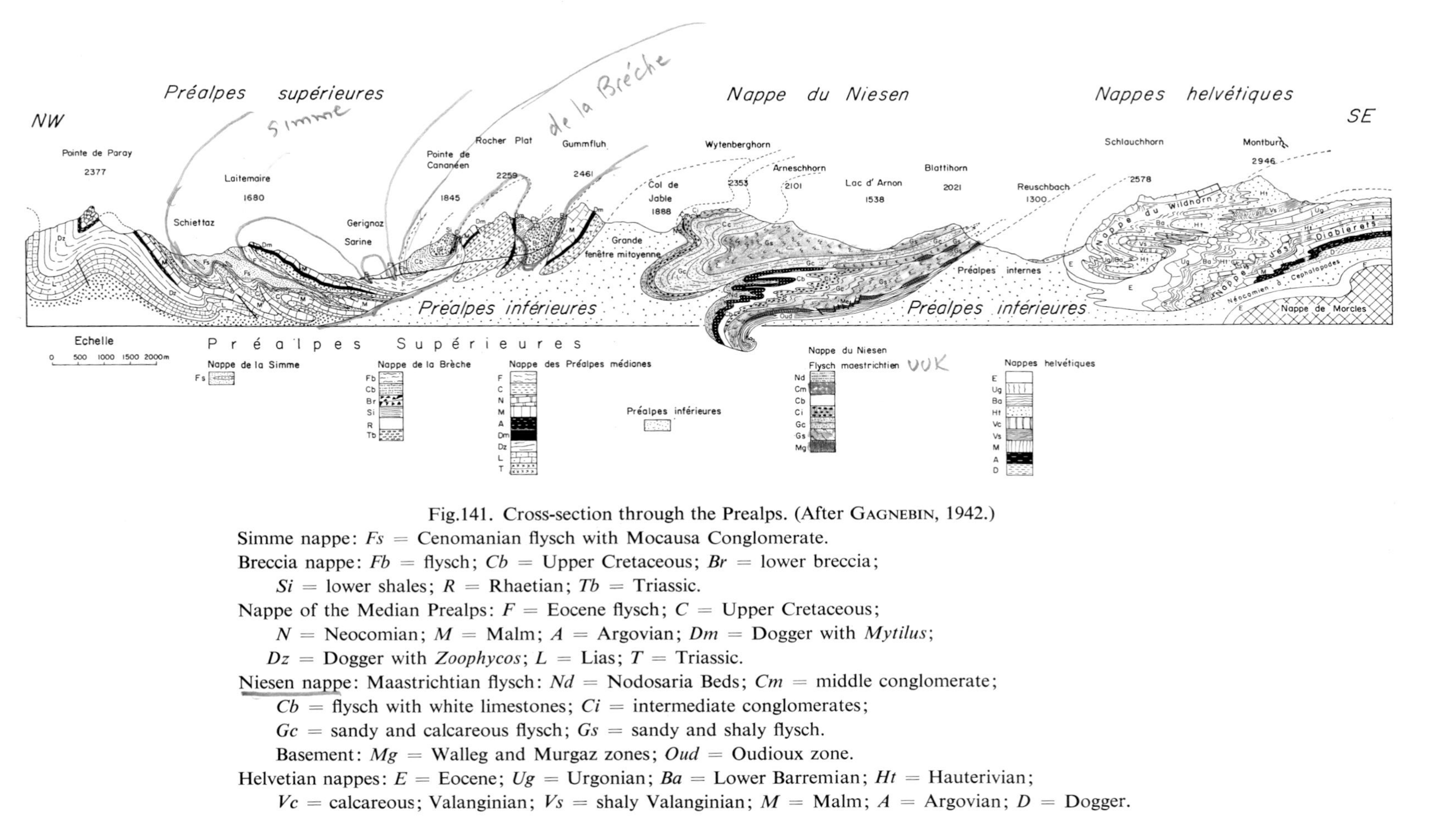

Fig.141. Cross-section through the Prealps. (After GAGNEBIN, 1942.)
Simme nappe: *Fs* = Cenomanian flysch with Mocausa Conglomerate.
Breccia nappe: *Fb* = flysch; *Cb* = Upper Cretaceous; *Br* = lower breccia;
Si = lower shales; *R* = Rhaetian; *Tb* = Triassic.
Nappe of the Median Prealps: *F* = Eocene flysch; *C* = Upper Cretaceous;
N = Neocomian; *M* = Malm; *A* = Argovian; *Dm* = Dogger with *Mytilus*;
Dz = Dogger with *Zoophycos*; *L* = Lias; *T* = Triassic.
Niesen nappe: Maastrichtian flysch: *Nd* = Nodosaria Beds; *Cm* = middle conglomerate;
Cb = flysch with white limestones; *Ci* = intermediate conglomerates;
Gc = sandy and calcareous flysch; *Gs* = sandy and shaly flysch.
Basement: *Mg* = Walleg and Murgaz zones; *Oud* = Oudioux zone.
Helvetian nappes: *E* = Eocene; *Ug* = Urgonian; *Ba* = Lower Barremian; *Ht* = Hauterivian;
Vc = calcareous; Valanginian; *Vs* = shaly Valanginian; *M* = Malm; *A* = Argovian; *D* = Dogger.

nappes of the Prealps is much more fluid than in the Helvetides. This phenomenon can be related both to the absence of thick, massive limestone series, and to the presence of much thinner, well-bedded limestone members, intercalated in mobile marly horizons.

This difference in tectonic style is quite apparent from a comparison of Fig.141 with sections through the Helvetide nappes. I only wanted to stress this difference, because it provides a beautiful example of variations in tectonic style within the same mountain chain, due to variations in stratigraphy.

THE CENTRAL MASSIFS

I have to be very short on the Central Massifs of Switzerland, again being obliged to refer the reader to the regional literature. The main newer development, i.e., realization of the strong influence Alpine metamorphism has had on the older crystalline rocks of the basement, as revealed by absolute dating, has already been indicated on pp.197–198.

A recent summary of the structure of the Aar Massif, combining the results of petrological investigations with the absolute datings, is given by WÜTRICH (1963), cf. Fig.142, Table XVII. From a comparison of the ages determined with the rock series from which the samples have been taken, it follows that the influence of the Alpine metamorphism has varied widely, both as to different samples from one rock series, and as to different minerals from one sample.

The relation of the Central Massifs to the "root zone", or "parental region", of the Helvetides, has been cited on p.237.

TABLE XVII

Rb–Sr DATES OF AAR MASSIF ROCKS[1]

(From WÜTRICH, 1963)

Sample	*Rock series*	*Rock or mineral tested*	*Age (million years)*
1	Gastern Granite	biotite	271 ± 11
2	Mittagfluh Granite	whole rock	256 ± 22
		biotite	54 ± 3
		microcline	102 ± 8
		albite	2,200 ± 250
3	Central Aar Granite	biotite	18.5 ± 2
4	Central Aar Granite, aplitic border	whole rock	249 ± 40
5	Erstfeld Gneiss	biotite	298 ± 12
			305 ± 12
6	Erstfeld Gneiss	biotite (bleached)	170 ± 27
7	Pegmatite in Tödi Granite	muscovite	312 ± 12
8	Pegmatite in tectonized zone in Aar Granite	muscovite	287 ± 12
9	Pegmatite in northern mantle of crystalline schists	muscovite	309 ± 30

[1] For location and geologic horizon of samples, compare Fig.142.

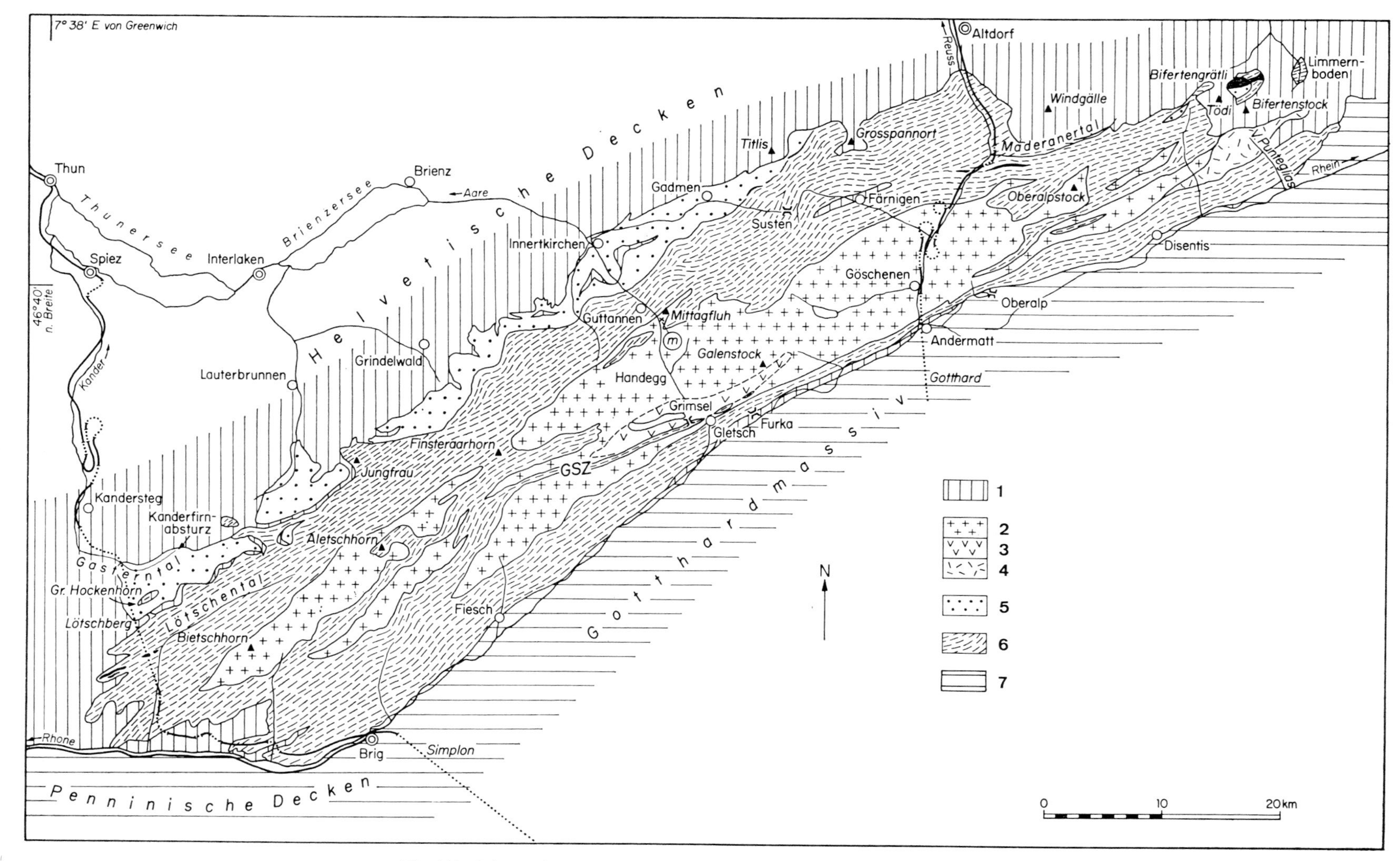

Fig.142. Schematic map of the Aar Massif, with location of absolute datings.

1 = autochthonous and par-autochthonous sedimentary cover; *2* = central Aar Granite; *3* = Grimsel Granite; *4* = Punteglias Granite; *5* = Gastern, Lauterbrunnen Innertkirchner, and Tödi Granites; *6* = mantle of crystalline schists of the Aar Massif; *7* = Gotthard Massif and Pennide nappes, non differentiated. (After WÜTRICH, 1963.)

THE PENNIDES

As was the case with the Helvetides, I cannot go into too much factual detail about individual nappes or about nappe correlations within the system of the Pennide nappes. This would require a wealth of detail far to voluminous for this text. As an outline the summary by CADISCH (1953) is reproduced here as Table XVIII. The general position of the main nappes follows from the map of Fig.122; their tectonic features from cross sections in Fig.106, 109, 110.

In abstracting ourselves from the details of the development of individual nappes, for which the regional literature must be consulted, our interest centers quite naturally on the two main features that characterize the Pennide nappes and distinguish them from other nappe systems. These are: (*1*) the stratigraphy of their Mesozoic; (*2*) the fact, already

TABLE XVIII

CORRELATION OF THE PENNIDE AND AUSTRIDE NAPPES IN SWITZERLAND

(From CADISCH, 1953)

The correlation of the Prealps (Präalpen) follows the Swiss interpretation

West Switzerland			*East Switzerland, Voralberg and Tirol*			
		Prealps (Klippen n.)	Middlebünden	Inntal n. (s)*	Oetztal) n. (Cr)	upper Austride
				Lechtal n. (s)	Silvretta n.	
				Allgäu n. (s)	(Cr)	
				possibly Middle East alpine		
			Tschirpen n.?	Scarl n. (s)		
				Unterengadiner Dolom. n.		
				Umbrail n.		
			Aroser Dolomite n.	Ortler n. (s) Languard n. (Cr)**	Campo n. (Cr)	middle Austride
		Simmen n.				
VI Dent Blanche n. s.l.	Dent Blanche n. s.s.	Breccien n. Klippen n.	Sulzfluh n. Falknis n. Aroser wedgezone	Bernina n. Err n.	Engadiner ns.	lower Austride
	Mt Mary n. Mt Emilius n.		Schamser ns.---	Sella n. Margna n.	upper-	Pennide
Mischabel n.	V Bernhard mass IV Monte Rosa mass			Suretta n. Tambo n.	middle-	
Simplon	III Monte Leone n. II Lebendun n. (according to H. Preiswerk to be derived from IV) I Antigorio n. Verampio Granite (Crodo)		Maggia-, Sambuco-, Campo Tencia festoons; Verzasca Gneiss	Adula n. Soja n. Simano ns. Lucomagno n. Tessiner Gneiss (Leventina n.) Tessin	lower-	

* (s) means: mostly sedimentary complex.
** (Cr) means: mostly old crystalline complex

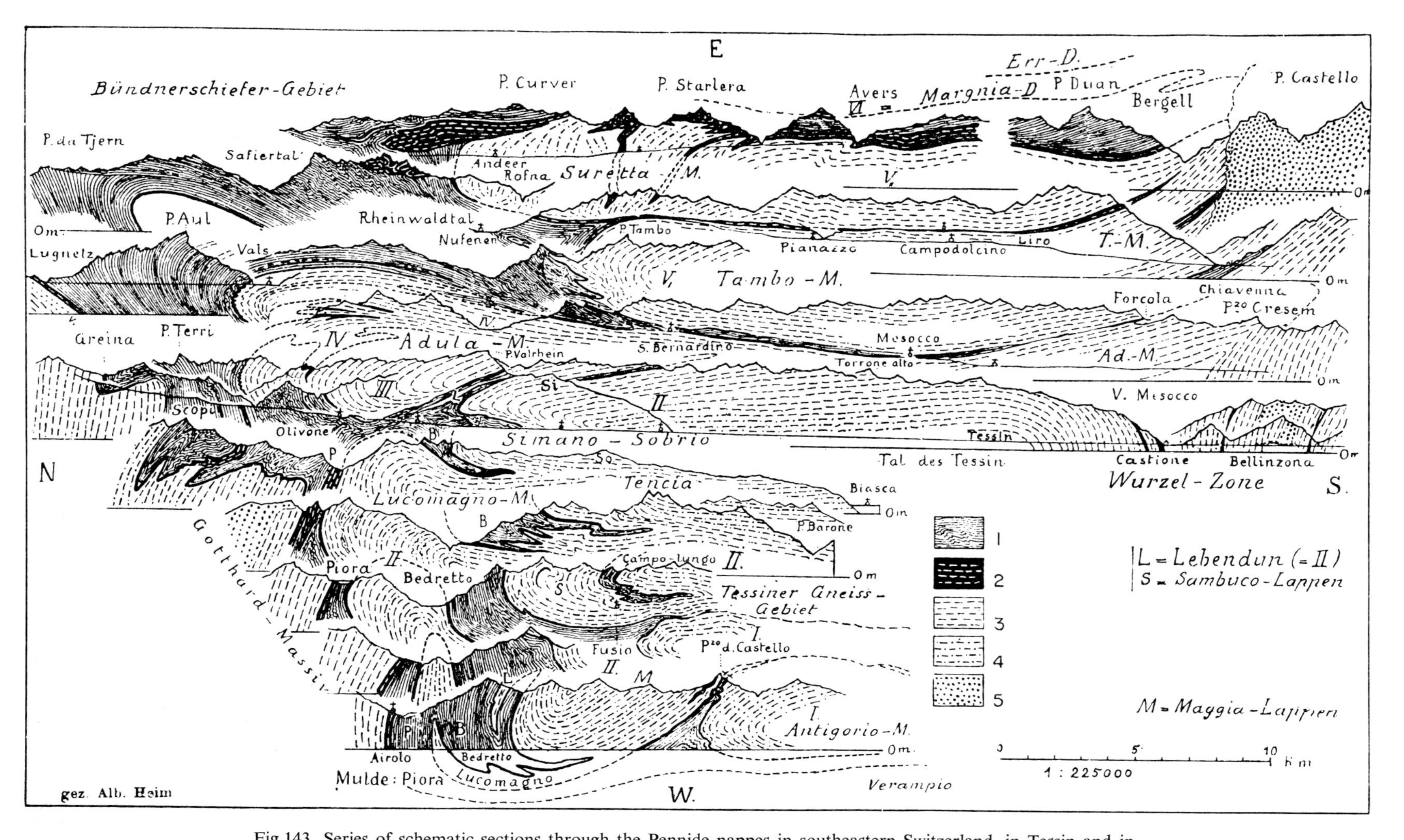

Fig.143. Series of schematic sections through the Pennide nappes in southeastern Switzerland, in Tessin and in Graubünden. Showing the relative importance, within individual nappes, of the crystalline cores and the envelopes of Bündner Schiefer.

1 = Bündner Schiefer; *2* = Triassic; *3* = crystalline schists; *4* = "ortho" gneiss; *5* = granite, tonalite. (After HEIM, 1922.)

touched upon, that tectonically most nappes are formed predominantly by relatively large cores of crystalline rocks, enveloped by a rather thin, in some cases even an extremely thin mantle of Mesozoic. This situation follows readily from Fig.143–146.

We may classify the Pennides as derma-type nappes. The pre-Alpine basement has been largely remobilized tectonically, quite apart from any younger metamorphism, during the Alpine orogeny. This is the main contrast with the epidermis type of nappes of the Helvetides. Accordingly, one normally finds in the Pennides a much more fluid style of tectonics than in the Helvetides. The Pennide style belongs to what in French tectonic theory is always indicated as "tectonique de fond", that is deeper-layer tectonics.

The stratigraphy of their Mesozoic will also have played its part in the development of this tectonic style. The Pennide Mesozoic normally consists of a relatively thin series of highly mobile schists, shales and sandstones, the Bündner Schiefer, whilst thick, massive resistant limestone series, so characteristic for the Helvetide Mesozoic stratigraphy, are absent. But the difference in behaviour of the basement of the Helvetides and the Pennides must also be noted. That is, between the brittle reaction, in a typical sort of "tectonique cassante" of the basement in the general area of the Helvetides, that is of the Central Massifs, as against the supple remobilization of the basement in the cores of the Pennide nappes.

The Pennides form the major element, the backbone of the Alpine chain in Switzerland. And of this backbone, the crystalline cores of the nappes, are, as we saw, the most important part. The reason why these nappes are so often called after major Alpine peaks, such as the Dent Blanche, the Monte Rosa, or the Grand St. Bernard, is because of the resistance to erosion of their crystalline cores (Fig.147). River valleys and cols, on the other hand, have normally developed in the sedimentary envelope.

Although thus forming the major element of the Alps, I must of necessity bypass the crystalline cores of the Pennide nappes in this text, for lack of space. The reader is referred to the regional literature, such as CADISCH (1953), or BADOUX (1967). The reason lies in the fact that it is hard to give a survey of the crystalline cores of the Pennide nappes on a sort of intermediate level. Even the short summary by Badoux indicates the enumeration of a wealth of detailed information, necessary for an outline review of this part of the Alps.

In a very general way, it can be stated that, in the western part of the Pennides, thick, slightly metamorphic series of Upper Carboniferous age are found. These can be dis-

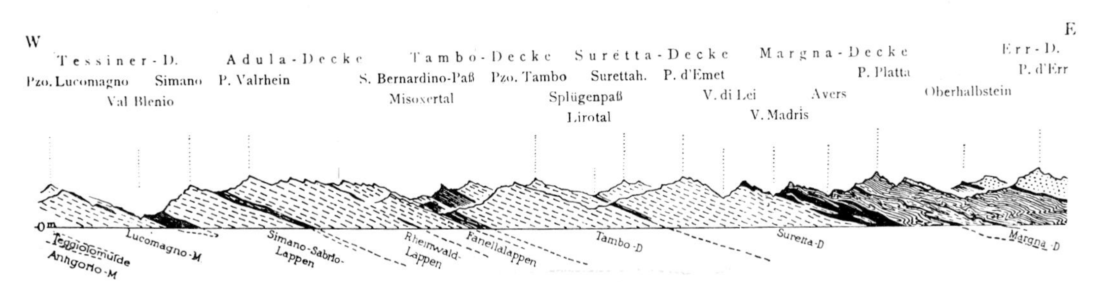

Fig.144. Schematic longitudinal section along the Pennide nappes in southeastern Switzerland, in Tessin and in Graubünden. More clearly than Fig.140, this schematic longitudinal section shows the relative importance of the envelopes of Bündner Schiefer around individual nappes. Very thin in Tessin, it grows thicker in eastern Switzerland, in the upper Pennide nappes, near the boundary with the Austrides. Particularly the Margna nappe is characterized by a predominance of Bündner Schiefer.

Dashes: crystalline of the nappe cores; lines: Bündner Schiefer; black: Triassic. (After HEIM, 1922.)

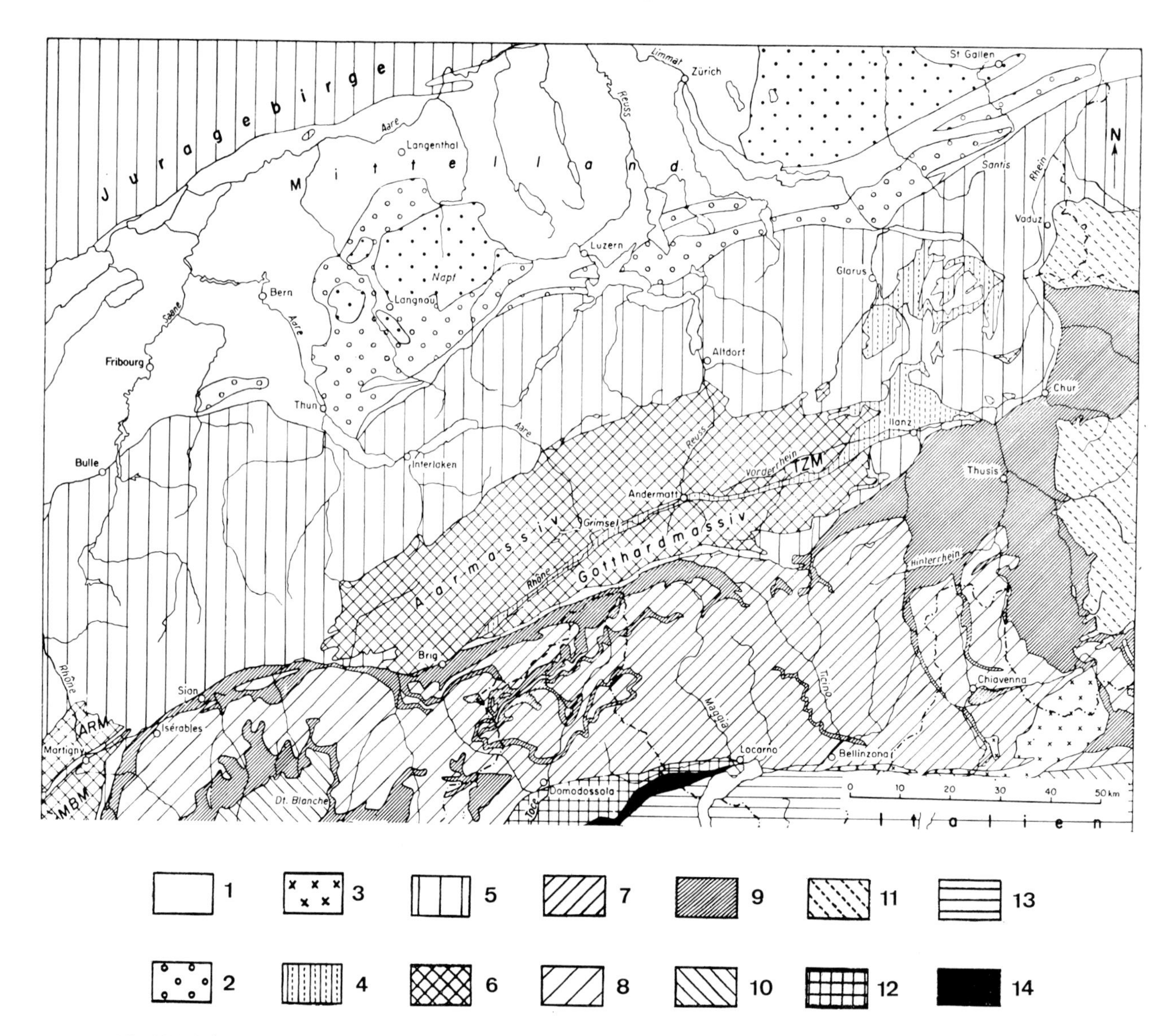

Fig.145. Schematic tectonic map of southern Switzerland. Showing the relative importance of the crystalline cores and the Mesozoic envelopes of the Pennide nappes. The importance of the Bündner Schiefer increases eastwards, towards Graubünden, where it is even predominant around Chur. It is in this general area that the Bündner Schiefer lithofacies persists into the Upper Cretaceous.

1 = molasse; *2* = molasse conglomerates ("Nagelfluh"); *3* = Alpine Bergeller Granite; *4* = verrucano, Permian; *5* = Helvetides; *6* = Central Massifs; *7* = lower Pennide nappes; *8* = middle and upper Pennide nappes; *9* = Mesozoic of the Pennide nappes, mainly Bündner Schiefer; *10* = crystalline cores of the Austride nappes; *11* = verrucano of the Austrides; *12* = root zone crystalline; *13* = crystalline of the Southern Alps; *14* = basic rocks of the root zone. (After HÜGI, 1963.)

tinguished readily from the stronger metamorphic older rocks. A regional unit of this series, famous in the literature, is the Casanna Schiefer.

These series can be compared to the Carboniferous of the Briançonnais zone in the French Alps, and perhaps also to slightly metamorphic clastic series around the crystalline cores, the "Massifs internes", of the Piemonte zone. Because the structure of the Alps becomes less complicated towards the southwest, the relation between these series of the

Fig.146. Outline map of the various Pennide nappes in the Simplon-Tessin area in southeastern Switzerland *AD* = Adula nappe; *M* = Maggia nappe; *MR* = Monte Rosa nappe; *GS* = Grand St. Bernard nappe; *ML* = Monte Leone nappe; *S* = Simano nappe; *A* = Antigorio nappe; *L* = Lebendun and Soja nappes; *V* = Vergeletto; *CL* = Cima Lunga; *C* = Camuguera; *Lu* = Lucomagno; *Le* = Leventina; black: sedimentary envelopes of the crystalline cores of the nappes: Trias, Bündner Schiefer and ophiolites. Scale: 1/70,000 (After BADOUX, 1967.)

Upper Carboniferous with the main tectonic zones can better be studied there (pp.198–201).

Towards the east the crystallinity of the cores of the Pennides nappes becomes more pronounced, a fact, at least in part, due to Alpine metamorphism. It consequently becomes more difficult to distinguish between the older series. The development of our ideas in the influence of Alpine metamorphism on the cores of the Pennide nappes has already been indicated in the preceding chapter.

BÜNDNER SCHIEFER

Returning now to the first-mentioned characteristic feature of the Pennides, the stratigraphy of its Mesozoic, we find that the most important member is the Bündner Schiefer (French: schistes lustrés; Italian: calcescisti). Apart from rather thin and incomplete Triassic, all of the Pennide Mesozoic belongs to this series.

It is a clastic series, originally consisting mainly of thinly bedded clays and sandstones. It shows a variable amount of calcite (the "calce" of the Italian designation). It has, moreover, undergone regional epizonal metamorphism, and become transferred into schists (the "scisti" of the Italian). The coarser members especially are shiny (the "lustré" of the French term) from micas.

It is a monotonous series, fossils being extremely rare. In its typical development, the

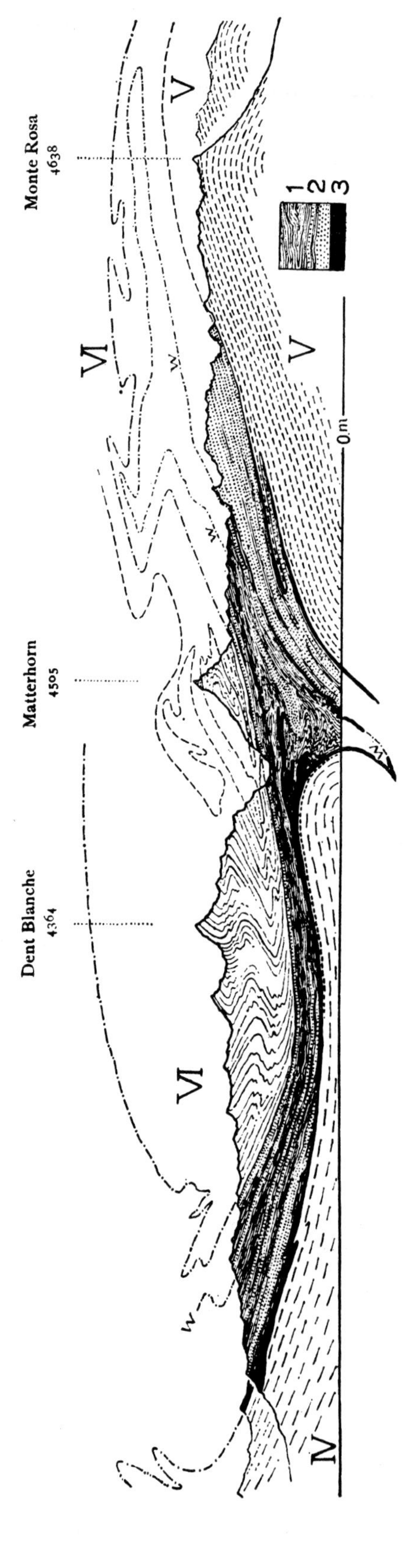

Fig.147. Cross-section over the Dent Blanche, the Matterhorn and the Monte Rosa, showing how the major mountain peaks in the Pennides are formed by the crystalline nappe cores.
1 = Bündner Schiefer; *2* = ophiolites; *3* = Triassic; *IV* = Great St. Bernard nappe; *V* = Monte Rosa nappe; *VI* = Dent Blanche nappe. (After HEIM, 1922.)

Bündner Schiefer carries fossils of Liassic age and is supposed to cover all of the Jurassic. In eastern Switzerland, and in the French–Italian border district, it is, however, known to reach into the Upper Cretaceous. The Bündner Schiefer thus forms an excellent example of what in the French literature is indicated by the term "série compréhensive", a monotonous series, forming a lithostratigraphic unit of unknown, but "comprehensive" duration.

In all probability the Bündner Schiefer represents a deep-water sediment, formed in the geosynclinal trough, or in several contiguous geosynclinal troughs, through turbidite activity. In its normal development, the series is rather thin, its thickness being of the order of several hundred meters only. It represents the partial filling of an undernourished geosyncline.

Characteristic for the Mesozoic of the Pennides is the association of the terrigenous clastic sediments of the Bündner Schiefer series with basic and ultrabasic igneous rocks and with radiolarites.

OPHIOLITES

The basic and ultrabasic rocks, called ophiolites (German: Ophiolithe) in the Alps, are normally equalized with the "roches vertes", the greenstones of the French literature. There is, however, a difference. For the "Ophiolithe" comprise both basic rocks, such as gabbro and/or diabase, and ultrabasic rocks, mostly serpentinized peridotite. A "roche verte" series may also be built up of both basic and ultrabasic rocks, but it may, on the other hand, also consist of basic rocks only. Whereas in the ophiolites of the Pennides both basic and ultrabasic rocks are found, there are other eugeosynclinal series which only carry basic rocks. "Roches vertes" series consisting of basic rocks only, without the ultrabasic members of the "Ophiolithe", may moreover form also outside geosynclinal basins.

In the past, "Ophiolithe" and "roches vertes" have always been lumped together in the literature, regardless of the fact whether the series in question contained ultrabasic rocks or not. This state of affairs rests on the postulation that the ultrabasic and the basic rocks of the "Ophiolithe" are consanguineous, and therefore must have a similar history. We will see, however, how in recent years doubt has been cast, both on the geometric and on the genetic relationship of the ultrabasic and the basic rocks classed together under the term ophiolite. One must, consequently, become more careful, and distinguish properly between the occurrence of ultrabasic and of basic rocks in geosynclinal series.

The concept of a narrow relation between ultrabasic and basic rocks in the ophiolites of the Pennides mainly goes back to STEINMANN (1906, 1927). The rather extensive literature which has since cropped up on the subject has been ably reviewed recently by VUAGNAT (1964), whereas a magmatic-tectonic interpretation of the regional setting of the ophiolites by BRUNN (1961) has also to be mentioned.

According to Steinmann the ophiolites occur as intrusives in the Bündner Schiefer in the stratigraphic succession serpentinite–gabbro–diabase (that is diabase German style, meaning a strongly altered basaltic rock). This succession was thought to represent a chronological succession, the consecutive formation of the three different members being due to magmatic differentiation.

The ophiolites were, moreover, thought to be always related to radiolarites intercalated in the abyssal type of clastic sediments of Bündner Schiefer facies, and the whole occurred only in the special types of geosynclines we now call eugeosynclines. Hence, there is a general relation, a law, governing the occurrence of what since has often been called the "Steinmann-trinity", comprising (*1*) ophiolites, (*2*) radiolarites, (*3*) abyssal sediments.

Newer research has since proved that the age relationship of the three different members of the ophiolites is not as strict as was supposed by Steinmann. There have been found older diabases and younger serpentinites. Moreover, there have frequently been found only one or two of the three members of the ophiolites occurring together, which also tended to unloosen the strict petrographic theory of Steinmann in relation to the formation of the ophiolites.

Considering for a moment only the basic rocks of the ophiolite series, their interpretation as submarine lavas and submarine shallow depth sills has in the meantime acquired almost general approval. RITTMANN (1960) has, amongst others, concisely summarized the differences that must develop between basaltic lavas erupting subaerially on continents and islands, and basaltic lavas erupting subaquatically, beneath a couple of kilometers of water, on the ocean floor. Apart from the pillow structures, generally found, the fact that many of the basic members of the ophiolite series are spilitic tallies well with this picture. This is generally ascribed to some sort of reaction with the sodium of the ocean water.

The relation between the basic rocks and the ultrabasics has, on the other hand, become less clear. It must of course be remembered that consanguinity of any ultrabasic rock with basic rocks is difficult to prove. A more serious objection stems from a group of petrologists like ERNST (1935) and DE ROEVER (1957, 1961) who suppose that most ultrabasic rocks are not related at all to basic rocks, because they represent samples of the mantle. According to these views, the ultrabasic rocks have arrived at the surface of the earth through two different modes of transportation, either through volcanic, or through tectonic action. The first mode of transport results in the olivine nodules, so profusely scattered in many basalt flows, whereas through the second mode of transport peridotite slices are wedged in between orogenetically disturbed deep-sea sediments.

This difference of opinion as to the nature of the ultrabasic members of the ophiolites will be difficult to solve. For it so happens that all ultrabasic inclusions in the Bündner Schiefer, although now serpentinized, originally formed massive elements in relation to the highly mobile series of clastic sediments. The result being that all intercalations of ultrabasic rocks in the Bündner Schiefer show tectonic contacts (W. K. Nabholz, personal communication, 1964). This would plead for the ideas of De Roever, were it not that such contacts could well be of a secondary nature, due to disharmonic folding.

RADIOLARITES

The other member of the "Steinmann-trinity", the radiolarites, also have a more general importance, because such rocks are known from many eugeosynclinal series outside the Alps. In a similar way as the ophiolites, the distribution and the genesis of radiolarites has given rise to a voluminous literature.

In this context it is proper to distinguish the state of affairs before and after the introduction of the concept of turbidites. "Before the turbidites" a major difficulty resided in how to combine the near shore characteristics of the coarser clastic elements of the Bündner Schiefer, with the deep water character of radiolarites. It was supposed that during the submarine extrusions of the ophiolites extra silica was supplied, which led to a sort of Radiolaria bloom. How it came about that basic rocks, poor in silica, were able to supply this commodity in quantity to the ocean water, then of course presented a "petrogenetic problem" (WENK, 1949).

It must be stressed that never any proof has been supplied for a silica supply through the ophiolites. In general the close association of ophiolites and radiolarites was accepted as such. But there has never been any proper geochemical investigation. Moreover, when studied in detail, even the close association of radiolarites and ophiolites becomes chimeric. GRUNAU (1947) maintains that such a close association, that is a primary stratigraphic contact of radiolarite and basic extrusive, could only be established in the Breccien-Decke of the Prealps, and nowhere in the Bündner Schiefer.

This difficulty can now be overcome, because we now know that turbidites may distribute coarse terrigenic clastic material over the deeper ocean floor. In this environment basaltic extrusions will acquire "roches vertes" and spilitic character, while also Radiolaria may be deposited between turbidite sedimentation. The Radiolaria are present in this environment of the deeper ocean floor, not because of a local supply of silica, but because, just as at present, in many ocean basins siliceous shells are the only remnants of the plankton that are not dissolved during sedimentation.

THE BÜNDNER SCHIEFER IN GRAUBÜNDEN

We will end up this short survey of the Bündner Schiefer, by touching upon the situation in eastern Switzerland, in Graubünden, where the series took its name. As stated above, it is in this region that the Bündner Schiefer forms the predominant element of the higher Pennide nappes. This situation leads, geologically, to a strong tectonization, and morphologically to gentler slopes and less good outcrops.

The classic description of a part of this region is by NABHOLZ (1945), who has since pursued his investigations in the same general area. Although the reason why the Bündner Schiefer makes up such a large part of the higher Pennide nappes, lies in part in tectonic repetition, there really is much more of it, stratigraphically. The Bündner Schiefer is in this eastern part of Switzerland developed in a much thicker, and also a coarser, series, than further west. Although the strong tectonization makes it difficult to give exact figures, one might estimate the total thickness as at least in the order of 1 km, and probably as considerably higher.

This greater thickness is, at least partly, the result of the fact that the lithofacies of the Bündner Schiefer in Graubünden persists into the Cretaceous, as has now been proved by microfossils (BOLLI and NABHOLZ, 1959). Apart from that it seems as if in this region the Pennide nappes represent a sedimentary series from a geosynclinal basin that was not undernourished, such as is the situation further to the southwest.

As stated already on p. 220, a consequence is, that the differentiation between the clastic series of the Cretaceous elements of the flysch and the Cretaceous elements of the Bündner Schiefer will be difficult. A case in point is the often cited Prätigau flysch, named after a major tributary valley of the Rhine river in northeastern Graubünden. From the literature one sometimes receives the impression that there is, nowadays, hardly any difference to distinguish Bündner Schiefer from Prätigau fllysch. This is definitely not the case. It is due to the fact that research into and controversies about what must be called flysch and what Bündner Schiefer tend to concentrate upon this development in eastern Switzerland. Here both series contain members in a typical development, which, moreover, show a sort of convergent evolution of both facies and stratigraphy. In their typical development, the two series are still easily told apart.

THE AUSTRIDES, THE ROOT ZONE AND THE SOUTHERN ALPS

The three remaining main elements of the Alps of Switzerland will not be further described in this chapter. The Austrides of Switzerland will be treated together with their Austrian equivalents in Chapter 13, in which also the description of the Swiss southern Alps is incorporated.

As to the root zone, the more general newer ideas have already been referred to on pp.205–206. A more detailed idea of the structure of this zone, and of the rocks which are found in it, may be distilled from the many local investigations. For instance from the description of a small area northwest of Lago Maggiore by SCHILLING (1957). An example of the post-orogenetic, young granitic intrusions that accompany the root zone is found in its largest representative, the Bergell Massif, as described by STAUB (1918).

REFERENCES

ARBENZ, P., 1912a. Der Gebirgsbau der Zentralschweiz. *Verhandl. Schweiz. Naturforsch. Ges., II*, 2: 95–123.

ARBENZ, P., 1912b. La structure des Alpes de la Suisse centrale. *Arch. Sci. Phys. Naturforsch., Genève*, 34: 401–425.

ARBENZ, P., 1913. Die Faltenbogen der Zentral- u. Ostschweiz. *Vierteljahresschr. Naturforsch. Ges. Zürich*, 38: 15–34.

ARBENZ, P., 1934. Die helvetische Region. In: E. GAGNEBIN and P. CHRIST (Editors), *Geologischer Führer der Schweiz II*. Wepf, Basel, pp.96–120.

AUBERT, D. and BADOUX, H., 1956. *Notice explicative, feuille 1: Neuchâtel. Carte Géologique Générale de la Suisse*. Kümmerly Frey, Berne, 27 pp.

BADOUX, H., 1963. Les unités ultrahelvétiques de la zone des Cols. *Eclogae Geol. Helv.*, 56: 1–14.

BADOUX, H., 1967. Géologie abrégée de la Suisse. In: A. LOMBARD, W. K. NABHOLZ and R. TRÜMPY (Editors), *Geologischer Führer der Schweiz*, 1. Wepf, Basel, pp.1–44.

BADOUX, H. and MERCANTON, C. H., 1962. Essai sur l'évolution des Préalpes médianes du Chablais. *Eclogae Geol. Helv.*, 55: 137–188.

BOLLI, H. and NABHOLZ, W. K., 1959. Bündnerschiefer, ähnliche fossilarme Serien und ihr Gehalt an Mikrofossilien. *Eclogae Geol. Helv.*, 52: 237–270.

BRUNN, J. H., 1961. Les sutures ophiolitiques. Contribution à l'étude des relations entre phénomènes magmatiques et orogéniques. *Rev. Géograph. Phys. Géol. Dynamique Sér. 2*, 4: 89–96; 181–202.

BUXTORF, A., 1951. *Erläuterungen zu Blatt 2: Basel–Bern. Geologische Generalkarte der Schweiz*. Kümmerly Frey, Berne, 39 pp.

BUXTORF, A. and NABHOLZ, W., 1957. *Erläuterungen zu Blatt 3: Zürich–Glarus. Geologische Generalkarte der Schweiz*. Kümmerly Frey, Berne, 81 pp.

CADISCH, J., 1953. *Geologie der Schweizeralpen*. 2nd ed. Wepf, Basel, 480 pp.

CLOOS, H., 1936. *Einführung in die Geologie*. Borntraeger, Berlin, 305 pp.

COLLET, L. W., 1927. *The structure of the Alps*. Arnold, London, 289 pp.

COLLET, L. W., 1955. *Notice explicative, feuille 5: Genève–Lausanne. Carte Géologique Générale de la Suisse*. Kümmerly Frey, Berne, 47 pp.

DEBELMAS, J. and LEMOINE, M., 1964. La structure tectonique et l'évolution paléogéographique de la chaîne alpine d'après les travaux récents. *Inform. Sci.*, 1: 1–33.

DE ROEVER, W. P., 1957. Sind die alpinen Peridotitmassen vielleicht tectonisch verfrachtete Bruchstücke der Peridotitschale? *Geol. Rundschau*, 46: 137–146.

DE SITTER, L. U., 1969. *Structural Geology*, 2nd ed. McGraw-Hill, New York, N.Y., 551 pp.

DE ROEVER, W. P., 1961. Mantelgesteine und Magmen tiefer Herkunft. *Fortschr. Mineral.*, 39: 96–107.

ELLENBERGER, F., 1952. Sur l'extension des faciès briançonnais en Suisse, dans les Préalpes médianes et les Pennides. *Eclogae Geol. Helv.*, 45: 285–286.

ERNST, T., 1935. Olivinknollen der Basalte als Bruchstücke alter Olivinfelse. *Nachr. Ges. Wiss. Göttingen, Math. Phys. Kl., Fachr. IV*, 1: 147–154.

FALLOT, P., 1956. Promenade d'hypothèse en hypothèse. *Gedenkboek H. A. Brouwer — Verhandel. Koninkl. Ned. Geol. Mijnbouwk. Genoot., Geol. Ser.*, 16: 100–113.

GAGNEBIN, E., 1934. Les Préalpes et les "klippes". In: E. GAGNEBIN and P. CHRIST (Editors), *Geologischer Führer der Schweiz. II.* Wepf, Basel, pp.79–95.

GAGNEBIN, E., 1942. Les idées actuelles sur la formation des Alpes. *Acta Soc. Helv. Sci. Nat.*, 1942: 47–53.

GAGNEBIN, E. and CHRIST, P. (Editors), 1934. *Geologischer Führer der Schweiz.* Wepf, Basel, 14 vol.

GEOLOGISCHE GENERALKARTE DER SCHWEIZ, 1942–1965. *Blatt 1: Neuchâtel, 2: Basel–Bern, 3: Zürich–Glarus, 4: St. Gallen–Chur, 5: Genève–Lausanne, 6: Sion, 7: Ticino, 8: Engadin. 1:200,000.* Kümmerly Frey, Bern.

GIGNOUX, M., 1955. *Stratigraphic Geology.* Freeman, San Francisco, Calif., 682 pp.

GRUNAU, H., 1947a. *Geologie von Arosa (Graubünden), mit besonderer Berücksichtigung des Radiolarit-problems.* Thesis, Univ. Bern, Bern, 109 pp.

GRUNAU, H., 1947b. Die Vergesellschaftung von Radiolariten und Ophioliten in den Schweizer Alpen. *Eclogae Geol. Helv.*, 39: 256–260.

HEIM, A., 1919. *Geologie der Schweiz. I. Molasseland und Juragebirge.* Tauchnitz, Leipzig, 704 pp.

HEIM, A., 1921. *Geologie der Schweiz. II. (1). Die Schweizer Alpen I.* Tauchnitz, Leipzig, pp.1–476.

HEIM, A., 1922. *Geologie der Schweiz. II (2). Die Schweizer Alpen II.* Tauchnitz, Leipzig, pp.477–1118.

HELBLING, R., 1938. Zur Tektonik des St. Gallener Oberlandes und der Glarneralpen. (Die Anwendung der Photogrammetrie bei geologischen Kartierungen.) *Beitr. Geol. Karte Schweiz*, 106: 71–133.

HSU, K. J., 1960. Paleocurrent structures and paleogeography of the Ultrahelvetic flysch basins, Switzerland. *Bull. Geol. Soc. Am.*, 71: 577–610.

HÜGI, T., 1963. Uranvorkommen in der Schweiz. *Atomwirtschaft*, 8: 524–529.

JUNG, W., 1963. Sedimente am Südostrand des Gotthardmassivs. *Eclogae Geol. Helv.*, 56: 653–754.

KLAUS, J., 1953. Les couches rouges et le flysch au sud-est de Gastlosen (Préalpes romandes). *Bull. Soc. Fribourg. Sci. Nat.*, 42: 8–128.

LOMBARD, A., BADOUX, H., VUAGNAT, M., WEGMANN, E., GANSSER, A., TRÜMPY, R. and HERB, R., 1962. *Guide book for the International Field Institute — The Alps* 1962. Am. Geol. Inst., Washington, D.C., 130 pp.

LOMBARD, A., NABHOLZ, W. and TRÜMPY, R. (Editors), 1967. *Geologischer Führer der Schweiz*, 2. ed. Wepf, Basel, 915 pp.

LUGEON, M. and GAGNEBIN, E., 1941. Observations et vues nouvelles sur la géologie des Préalpes romandes. *Bull. Lab. Géol. Mineral. Géophys. Musée Géol. Univ. Lausanne*, 72: 90 pp.

MEIER, P. und NABHOLZ, W. K., 1950. Die mesozoische Hülle des westlichen Gotthard-Massivs im Wallis. *Eclogae Geol. Helv.*, 42: 197–214.

NABHOLZ, W. K., 1945. Geologie der Bündnerschiefergebirge zwischen Rheinwald, Valser- und Safiental. *Eclogae Geol. Helv.*, 38: 1–121.

NABHOLZ, W. K., 1949. Das Ostende des mesozoischen Schieferhülle des Gotthard-Massivs im Vorderrheintal. *Eclogae Geol. Helv.*, 41: 247–268.

NABHOLZ, W. K. and VOLL, G., 1963. Bau und Bewegung im gotthardmassivischen Mesozoikum bei Ilanz (Graubünden). *Eclogae Geol. Helv.*, 56: 755–808.

NIGGLI, E., 1944. Das westliche Tavetscher Zwischenmassiv und der angrenzende Nordrand des Gotthardmassivs. *Schweiz. Mineral. Petrog. Mitt.*, 24: 58–301.

RITTMANN, A., 1960. *Vulkane und ihre Tätigkeit*, 2. Aufl. Enke, Stuttgart, 336 pp.

SCHILLING, S., 1957. Petrografisch-geologische Untersuchungen in der unteren Val d'Ossola. Ein Beitrag zur Kenntniss der Ivrea zone. *Schweiz. Mineral. Petrog. Mitt.*, 57: 435–544.

SODER, P. A., 1949. Geologische Untersuchung der Schrattenfluh und des südlich anschliessenden Teiles der Habkern-Mulde (Kt. Luzern). *Eclogae Geol. Helv.*, 42: 39–109.

STAUB, R., 1918. Geologische Beobachtungen am Bergellermassiv. *Vierteljahresschr. Naturforsch. Ges. Zürich*, 63: 1–18.

STAUB, R., 1924. Der Bau der Alpen. *Beitr. Geol. Karte Schweiz*, 52: 272 pp.

STAUB, R., 1958. Klippendecke und Zentralalpenbau. *Beitr. Geol. Karte Schweiz*, 103: 184 pp.

STEINMANN, G., 1906. Geologische Beachtungen in den Alpen. II: Die Schardt'sche Überfaltungstheorie und die geologische Bedeutung der Tiefseeabsätze und der ophiolitischen Massengesteine. *Ber. Naturforsch. Ges. Freiburg*, 16: 18–67.

STEINMANN, G., 1927. Die ophiolithischen Zonen in den Mediterranen Kettengebirgen. *Congr. Géol. Intern., Compt. Rend., 14e, Madrid, 1926*, 2: 637–668.

TRÜMPY, R., 1955. Remarques sur la corrélation des unités penniques externes entre la Savoie et la Valais et sur l'origine des nappes préalpines. *Bull. Soc. Géol. France*, 5 (6): 217–231.

TRÜMPY, R., 1963. Sur les racines des nappes helvétiques. In: M. DURAND DELGA (Editor), *Livre à la Mémoire du Professeur Paul Fallot. II.* Soc. Géol. France, Paris, pp.419–428.

TRÜMPY, R. and RYF, W., 1965. *Symposium sul "Verrucano". Erläuterungen zur Exkursion in die Glarner Alpen.* Inst. Geol. Paleontol. Univ. Pisa, Pisa, 22 pp.

VUAGNAT, M., 1964. Remarques sur la trilogie serpentines–gabbros–diabases dans le bassin de la Méditerranée occidentale. *Geol. Rundschau*, 53: 336–357.

WENK, E., 1949. Die Assoziation von Radiolarienhornsteinen mit ophiolithischen Erstarrungsgesteinen als petrogenetisches Problem. *Experientia*, 5: 226–232.

WÜTRICH, H., 1963. Rb–Sr-Altersbestimmungen an Gesteinen aus den Aarmassiv. *Eclogae Geol. Helv.*, 56: 103–112.

CHAPTER 12

Alpine Europe: The French Alps

INTRODUCTION

As stated already in Chapter 9, the French Alps are usually taken together with the main part of the Alps of Switzerland (with the exception of those parts of the Austrides and the Southern Alps that extend on Swiss territory), and designated as the "Western Alps". With some authors, however, "Western Alps" is more or less synonymous with "French Alps", whilst the Swiss part is then indicated as "Central Alps" (for instance, TRÜMPY, 1960). Use of the term "Central Alps" leads, however, to ambiguities. For in Austria, in the Austrides "Zentralalpin" indicates an east–west striking, "central" portion of the Austrides, in a north–south division of this part of the Alps. For another, we have met already the "Central Massifs" as a unit within the Western Alps.

A case can be made for a main northwest–southeast division line within the "Western Alps", because there are marked differences in structure between the more westerly and the more easterly parts of this major element of the Alpine chain. As these differences moreover coincide rather closely with the state border, I have preferred the terms Swiss and French Alps in this text, and to give a separate account of these two main elements of the "Western Alps".

There is another reason for this, which lies in a difference in the state of the art, regarding Swiss and French geologic literature. In France a number of modern regional descriptions make a survey of Alpine geology easier. Moreover, quite a few of the newer notions on Alpine geology have originated in the French literature.

As to the first point, the factual differences between the French and Swiss Alps, one is tempted by the idea that they are all more or less related, geometrically, if not causally, to the fact that the French Alps are much broader than the Swiss Alps. Any cross-section over the Swiss Alps, north of the root zone, is not longer than 100 km, whereas the French Alps widen to almost 200 km in the section across Gap.

This results in a situation, where various phenomena, that are either completely masked, or difficult to reconstruct by the pronounced nappe tectonics in the Swiss Alps, are still recognizable in the French Alps. This difference in tectonic style is most apparent in the external zone of the chain. As we will see, nappes are almost absent in the Dauphinois zone, which is regarded as the equivalent of the Helvetides. Moreover, an entirely new unit, the Briançonnais zone, can be distinguished in the French Alps between the Dauphinois zone and the Pennides. Elements equivalent to the Briançonnais zone can be traced

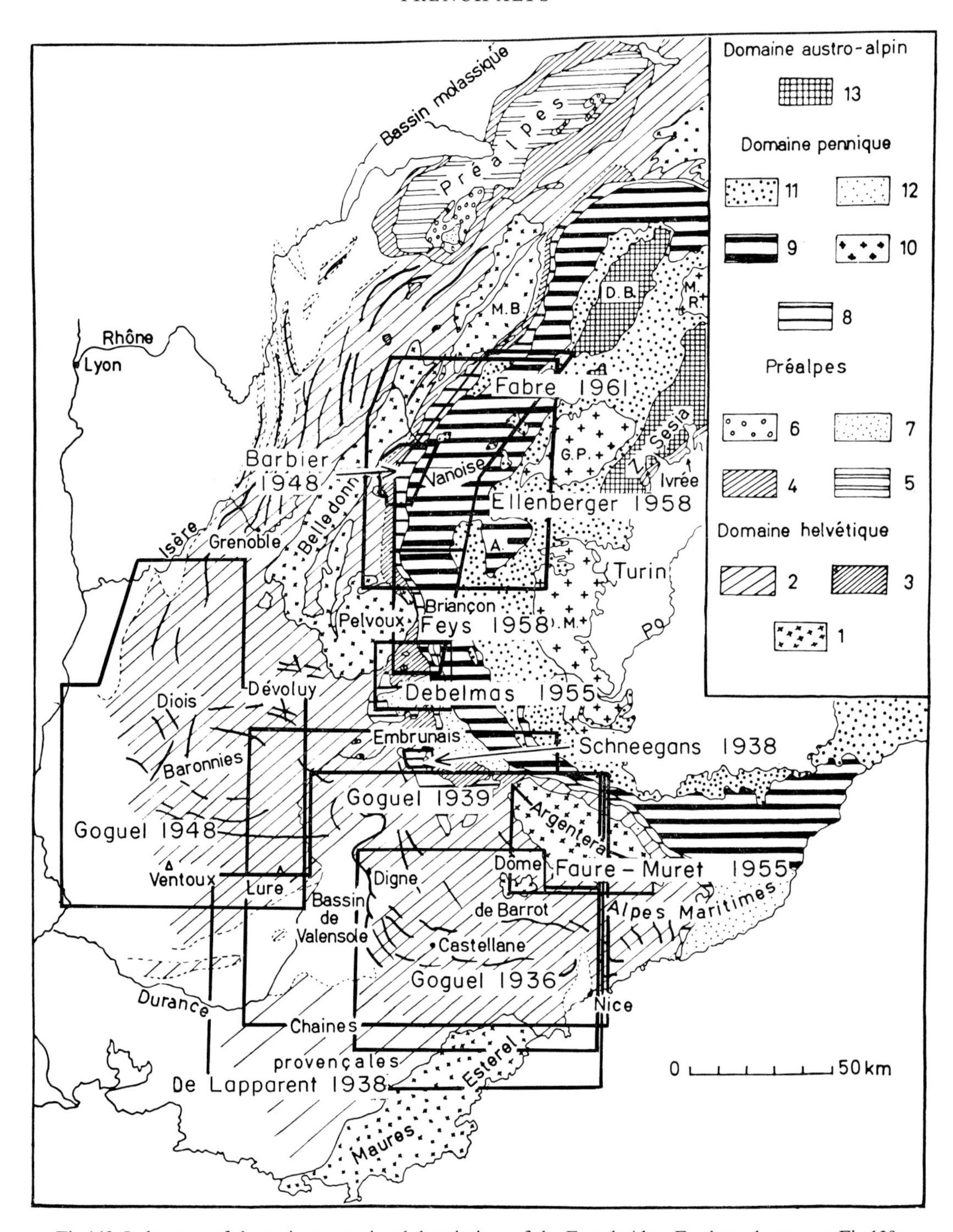

Fig.148. Index map of the main new regional descriptions of the French Alps. For legend compare Fig.139. (Feys, 1958, has been superseded by FEYS, 1963.)

into the western part of the Pennides of the Swiss Alps, but it is not known if these elements really die out further eastwards, or if they only become unrecognizable within the crystalline nappe cores, through stronger Alpine metamorphism and tectonization.

As to the second item, the difference of interpretation, the modern regional descriptions of the various parts of the French Alps are mainly the result of theses, published to acquire a "doctorat d'état". This kind of D.Sc. is of an extremely high level, as compared with a mere university doctoral thesis, even of the Paris Sorbonne. As a prerequisite to obtain a French university professorship, normally some ten years of hard work are incorporated in any of these volumes. In geology, they are mostly published by the *Service de la Carte géologique*. Those most important for Alpine geology are indicated in Fig.148. Throughout this volume of research one feels in its quality the influence of that master of Alpine geology, Professor M. Gignoux of Grenoble.

The ideas on the development of the French Alps are admirably presented in *Du Pelvoux au Viso*, a colour film (VALÉRIEN et al., 1964), which can be loaned from any French Embassy. The paleogeography of the different zones of the Alps, at various stages of their development, and the tectonic structure, are rendered in cartoons, which are blended with aerial views of the mountains as they are at present.

I must, at this stage, warn the reader that the jump from the very general text presented here, to the detailed factual representation found in the regional descriptions, is a formidable one. Each of them has, however, a short survey of the area studied, either as an introductory chapter, or as a summary of results and these may well serve as stepping stones.

THE MAJOR DIVISIONS

Just as in the Swiss Alps, the French Alps can be divided into an outer (external), and an inner (internal) part. Because of the arcuate trend of the Alpine chain, the boundaries between the main elements of the French Alps strike generally north–south (Fig.139).

The external part comprises, again as in the Swiss Alps, a number of older massifs, together with a sedimentary cover. The main Central Massifs are, from north to south, the Belledonne, the Pelvoux and the Argentera (or Mercantour). Contrasting with the usage in the Swiss Alps, both the sedimentary cover and the crystalline massifs (the "massifs externes" of the French), are classed together in the Dauphinois zone. The sedimentary cover on the external side of the Central Massifs is designated as the Subalpine chains, which would be the exact equivalent of the Helvetide nappes. On the internal side of the Central Massifs the sedimentary cover locally shows a facies different from that in the Subalpine chains. It is called the Ultradauphinois facies, or Ultradauphinois zone, and is correlated with the Ultrahelvetian nappes.

Stratigraphically, the main difference between the Subalpine chains and the Ultradauphinois zone is that the latter reflects the influence of pre-Tertiary tectonic movements, resulting in the formation of a thick series of Lower Tertiary flysch, the Aiguilles d'Arves Flysch (BARBIER, 1956). Flysch is absent in the Subalpine chains, a first indication as to differences existing between this part of the Dauphinois zone and the Helvetide nappes, with which we will come into contact later in this chapter. The Ultradauphinois facies consequently is more closely comparable to that of the Helvetides. The stratigraphy of this Ultradauphinois zone is, however, rather variable along the strike (BARBIER, 1963). Moreover, as it is a narrow zone, we will not go into details here.

There is here a slight mix-up in terminology. The term "Dauphinois facies" is reserved for the sedimentary cover of the external part of the French Alps, that is for the Subalpine

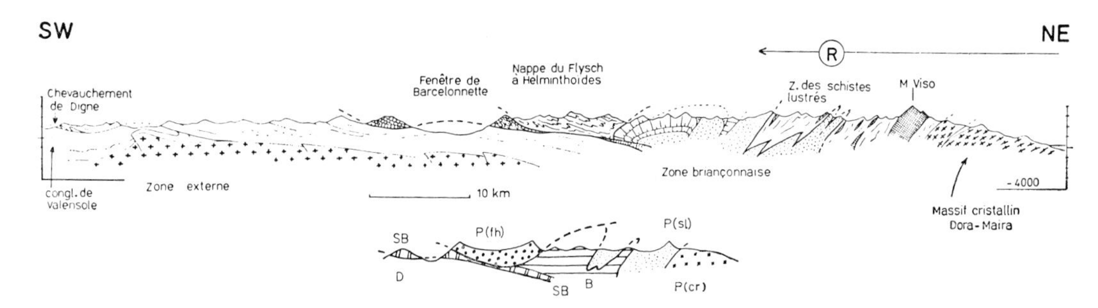

Fig.149. Schematic section across the French Alps. Upper figure is actual section, lower figure indicates the main elements distinguished within the French Alps.

The external part, the Dauphinois zone (*D*), comprises the crystalline basement (which on this section does not crop out as a central massif), and its sedimentary cover, from which the Ultradauphinois zone has not been distinguished in this section. It is overthrust by the Subbriançonnais (*SB*) and Briançonnais (*B*) zones, the latter showing a fan-like arrangement, the much discussed "éventail briançonnais" (p. 294). It is overridden by the *Helminthoïdes* flysch nappe — *P(fh)*. To the east follow the Pennides, with the "schistes lustrés" — *P(sl)* — and their crystalline cores, such as the Dora Maira — *P(cr)*. The Mount Viso is formed by an exeptionally thick intercalation of ophiolites. (After DEBELMAS and LEMOINE, 1964.)

chains and for some parautochthonous sediments at the internal side of the Central massifs. The term "Dauphinois zone", on the other hand, embraces both the sedimentary cover and the Central Massifs, and sometimes even includes the Ultradauphinois zone.

The internal part of the French Alps comprises, from west to east, the Briançonnais zone and the Pennides. Along the western, that is the external, side of the Briançonnais zone, a subsidiary zone is separated, the Subbriançonnais zone. It forms a transition from the Dauphinois zone, mainly from its internal element, the Ultradauphinois zone, to the Briançonnais zone. As the facies of the Subbriançonnais resembles that of the Briançonnais zone, and because we are here occupied with the main divisions rather than with details, it will not be treated further, but will be regarded as a subzone of the Briançonnais zone.

The equivalents of the two major elements in the Swiss Alps, more internal than the Pennides, that is the root zone and the Southern Alps, are not observed in the French Alps. Although probably developed in a similar way, the innermost part of the French Alps has since subsided in the post-orogenetic Po Basin, and became covered by a thick series of younger sediments. The boundaries of this younger subsidence of the Po Basin cut obliquely across the earlier structural trends of the Alpine chain (Fig.139).

This younger subsidence affects not only the equivalents of the Southern Alps and "root zone", but further south also causes the eastern parts of the Pennides, to disappear, until, in the extreme south, only a narrow zone of Pennides remains.

From the Pennides, which overlap largely across the state border into Italy, and consequently are called the Piemonte zone, from the Piemonte province of northwestern Italy, another subzone or element is always separated, i.e., the Embrunais flysch, or *Helminthoïdes* flysch nappe. We have already mentioned this unit in Chapter 11, in the discussion of the Prealps.

A very generalized section through the French Alps, with a schematic indication of its main elements, is given in Fig.149; an indication of the normal stratigraphical development within the main units, in Fig.150, and a generalized map of the central region in Fig.176.

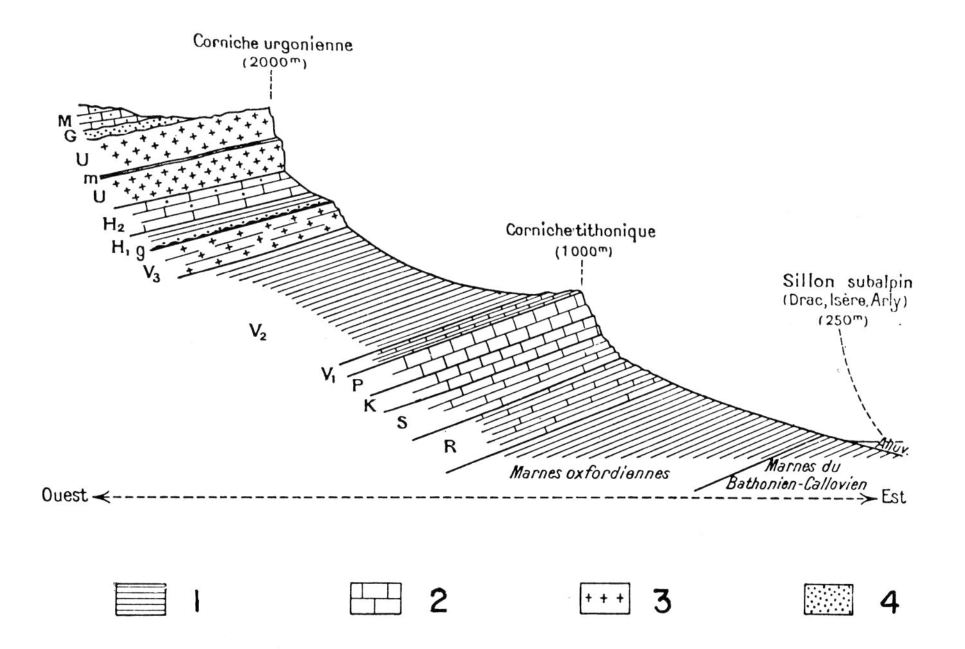

Fig.151. Schematic section, showing the two step morphology along river valleys in the Subalpine chains. *1* = marls; *2* = limestones; *3* = organo-detrital limestones; *4* = sandstones; *M* = Maestrichtian: limestones, organo-detrital, sandy or marly; *G* = Albian: glauconitic sand and crinoid limestone; *U* = Urgonian (reef-rock limestones; *m* = *Orbitolina* limestones); H_2 = Upper Hauterivian (sandy marls with *Spatangus*); H_1 = Lower Hauterivian (marls; *g* = glauconite); V_3 = Upper Valanginian (Fontanil limestones); V_2 = Middle Valanginian (marls); V_1 = Lower Valanginian (marly limestones, cementstones; = Berriasian); *P* = Portlandian (massive limestones); *K* = Kimmeridgian (massive limestones; *P* + *K* = "Tithonique"); *S* = Sequanian (marly limestones); *R* = Rauracian (marly limestones). (After GIGNOUX, 1950.)

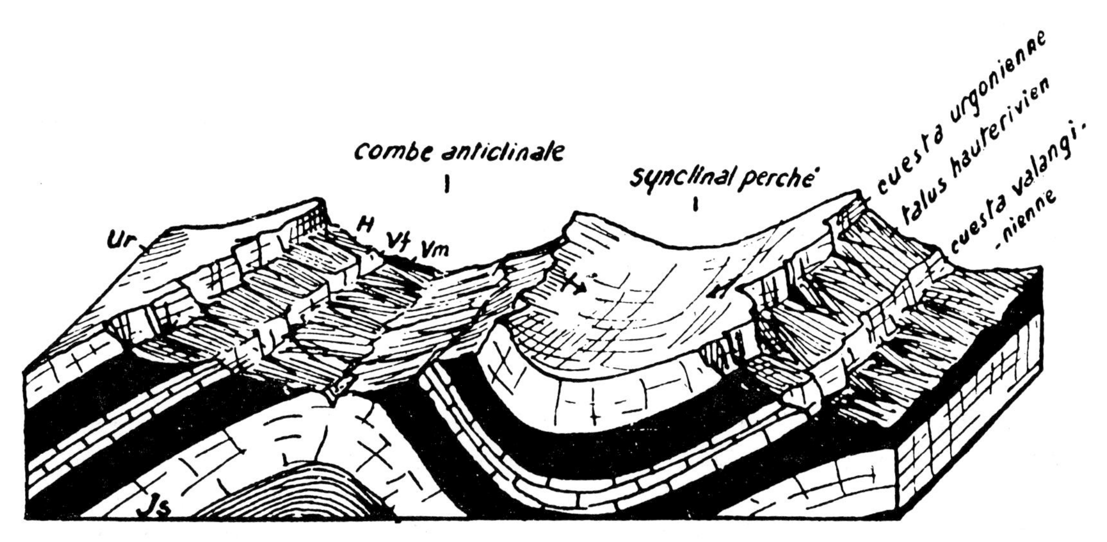

Fig.152. Schematic representation of the the reversal of relief, common in the Subalpine chains around Grenoble. This causes the bastion-like sheer walls of the Urgonian around the high plateaus formed by synclinal areas. (After GIGNOUX and MORET, 1944.)

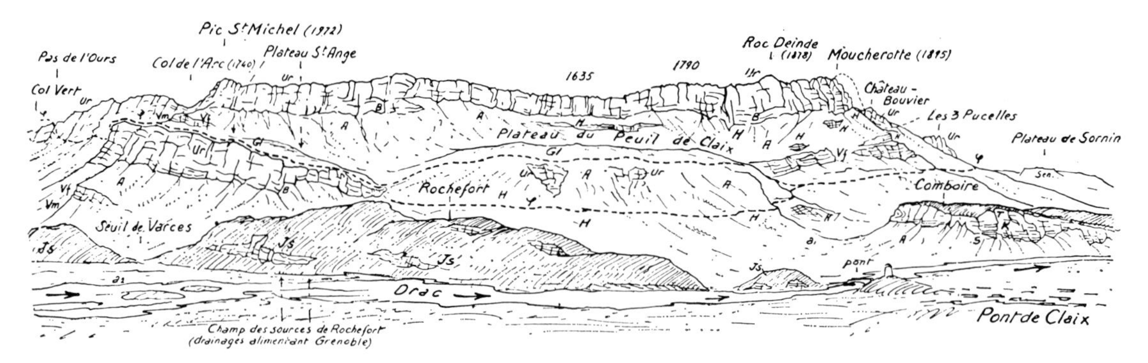

Fig.153. Panoramic view of the eastern front of the Vercors high plateau, south of Grenoble. Illustrating the bastion-like mode of outcrop of the Urgonian.
Js = Jurassic; *S* = Sequanian; *K* = Kimmeridgian; *T* = Tithonique; *Vm* = Berriasian and Valanginian marls; *Vf* = Fontanil limestones, Upper Valanginian; *H* = Hauterivian; *Ur* = Urgonian; *a* = detritus; *Φ* = overthrust. (After GIGNOUX and MORET, 1944.)

The former, which may attain a thickness of up to 1 km, are of the type of the black marts, indicating hypertrophic environment during their deposition. The Pleistocene erosive surfaces, which developed along their outcrops during a periglacial climate, are unstable in the present climatic conditions. As a consequence deforestation has led to the formation of extensive badlands along many of the larger valleys. Their forceful erosion is mainly responsible for the extremely muddy character of all rivers in the Subalpine chains during flash floods.

In the areas, where this alternation of major marl series with thick limestone series is best developed, for instance in the classic region around Grenoble, one always finds a definite two step morphology of the Subalpine chains. A lower step is formed by the Tithonique, and upper step by the Urgonian (Fig.151). The latter is most impressive in the larger, flat lying synclinal areas, such as the Vercors and the Grande Chartreuse south and north of Grenoble. Here a reversal of relief has led to the formation of extensive high plateaus exposing the Urgonian bastion-like in a vertical drop of several hundreds of meters towards the surrounding valleys (Fig.152, 153).

In other areas of the Subalpine chains, however, this difference is not so pronounced. The stratigraphy there resembles more that represented in the left-hand column of Fig.150. It is in these regions that an intermediate facies may develop, that of the marl-and-limestone series. Such thin-bedded formations, in which marls alternate with limestones, either dense, marly or nodular, may reach considerable thickness. In the French literature they have been called most inaptly "flysch calcaire" (p. 221).

STRUCTURE

Structurally, the difference between Helvetides and Subalpine chains appears both in ground plan and in section. Looking at the ground plan first, we get a good idea of the overall structure of the Subalpine chains from Fig.154. They are formed by a large number of well-separated, more or less parallel and relatively short folds. In cross-section, these are quite variable, from broad anticlines and wide shallow synclines, to pinched, asymmetric forms. They show not only overall "en échelon" replacement of fold axes, but also quite strong variations in the intensity of folding and/or local overthrusting along the strike of the same structural unit. Compared to the Helvetides, the Subalpine chains are made up by a far greater number of individual units; each unit being of far lesser dimensions. The tectonic style of the Subalpine chains accordingly is much more similar to that of the Jura mountains, although of a magnitude about one order larger, than to that of the Helvetides.

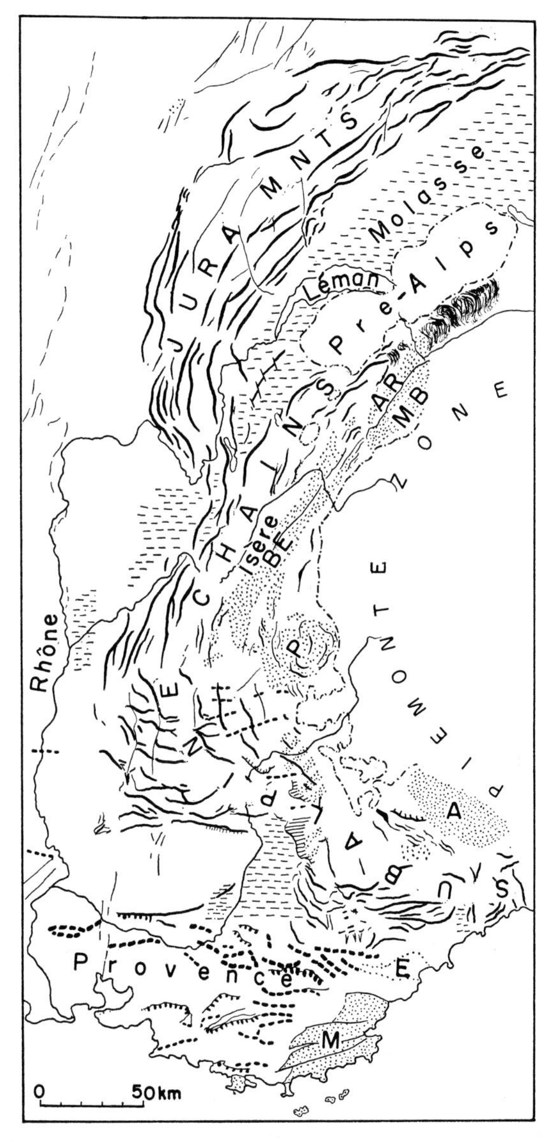

Fig.154. Schematic tectonic map showing the main fold axes of the Jura Mountains, the Subalpine chains (full lines) and the Provence (blocks).

According to Goguel, the influence of the earlier east–west folding found in the Provence, formed during the Pyrenean phase, can still be recognized in the southern half of the Subalpine chains, though overlain by the younger Alpine folding.

Stippled: Central Massifs. *M* = Maures; *E* = Estérel; *A* = Argentera–Mercantour; *P* = Pelvoux; *BE* = Belledonne; *AR* = Aiguilles Rouges; *MB* = Mont Blanc; *B* = Permian Barrot Dome; *Piémonte zone* = Pennides (After GOGUEL, 1963.)

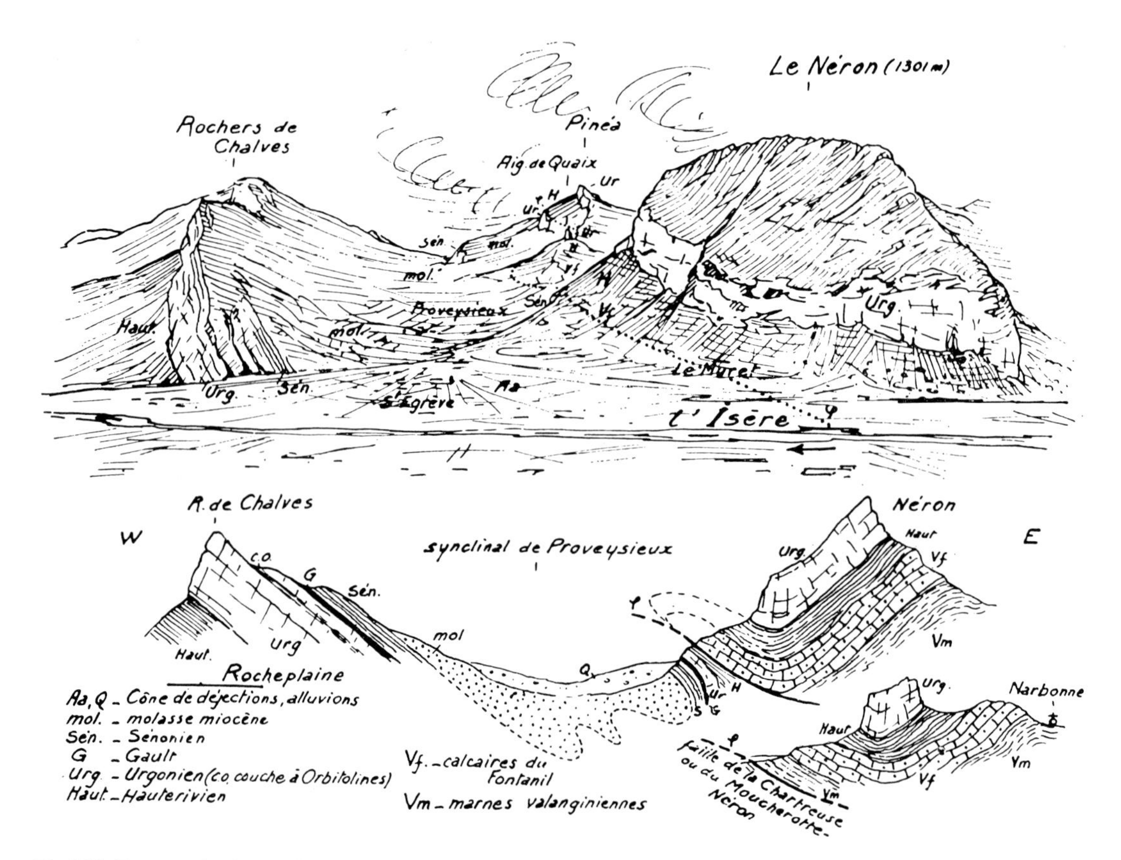

Fig.155. Panoramic view and cross-section of the Proveysieux syncline and the Néron structure, north of Grenoble For an explanation of the stratigraphy, compare Fig.151 and 153.
(After GIGNOUX and MORET, 1944.)

Most individual structures are asymmetric, but there is no general "Vergenz". The typical development is that of the asymmetric syncline, and the overturned anticline. Both may develop along the strike into a localized overthrust. This is the "pli-faille" type of structure, of which the pli-faille de Sassenage, well exposed in the Isère valley west of Grenoble (HAUG, 1921, pl.VII) is the locus typicus.

This style of folding is of course largely determined by the stratigraphy of the Subalpine chains, with its marked difference between competent limestone series and incompetent marls, which may serve as secondary planes of "décollement". I will not go into any detailed account here of the various major structures forming the Subalpine chains, but refer the reader to the general literature. An idea of the tectonic style characteristic for this fold belt can be gleaned from Fig.155, 156, 157, chosen amongst many other more or less similar ones.

Within the northern half of the Subalpine chains a more or less broadly north–south arrangement is apparent. In the southern half the strike pattern becomes more complicated, and a number of arches can be seen in the ground plan. The most important of these are the "Arc de Castellane" (GOGUEL, 1936, 1953; MENNESSIER, 1964) and the "Arc de Nice" (GÈZE, 1963).

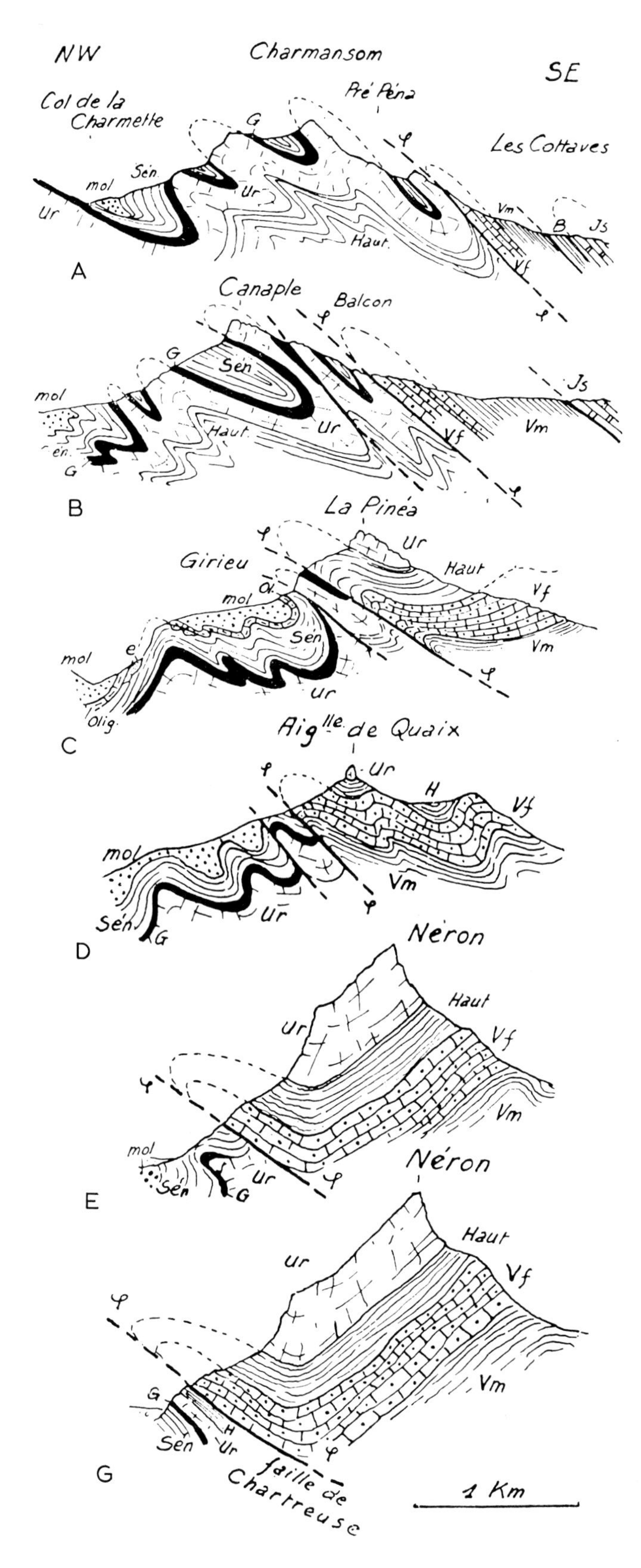

Fig.156. Series of sections to show the development from north to south of the Néron structure north of Grenoble. "En échelon" replacement of the asymmetric, complicated anticlinal structure of the Charmansom by the asymmetric "pli-faille" of the Néron "synclinal" structure. (After GIGNOUX and MORET, 1944.)

Fig.157. Serial sections through the Teillon, east of Castellane.
Stratigraphic section in left upper corner: t = Triassic; l = Lias; J_{I-IV} = Dogger; J^{8-1} = Malm; c_{III-V} = Lower Cretaceous; c^{5-8} = Upper Cretaceous. (After GOGUEL, 1936.)

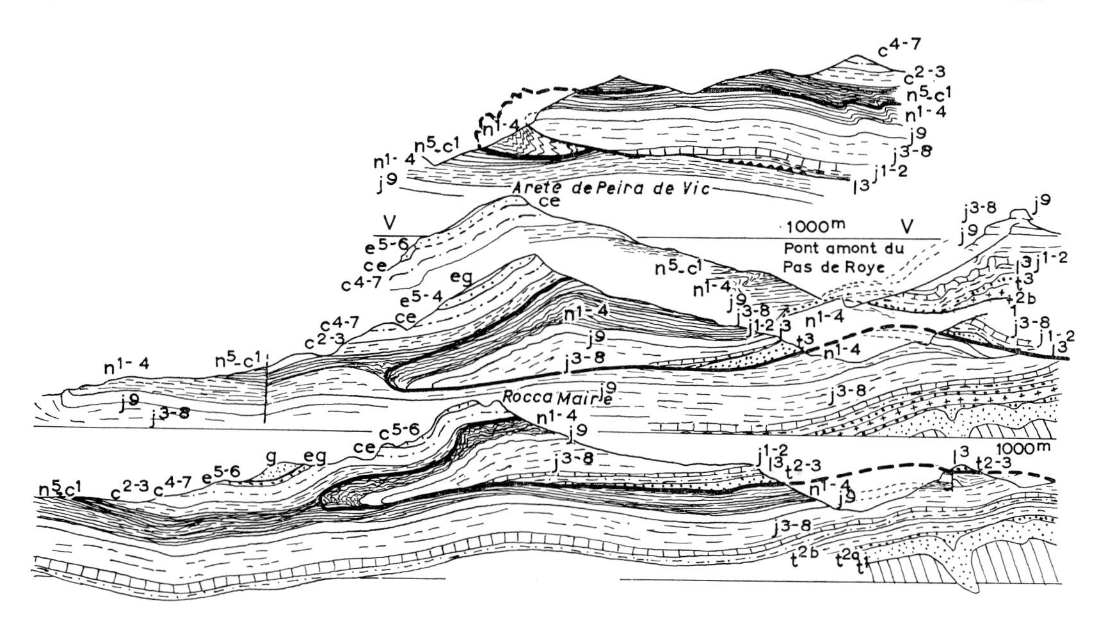

Fig.158. Sections through the intercutaneous structures of Roya (Alpes Maritimes, France), 1:40,000. g = Grès d'Annot; eg = Priabonian; e^{5-6} = Auversian; ce = conglomerates at the Cretaceous–Eocene border; c^{1-7} = Senonian limestones; c^{2-3} = Turonian limestones; n^5-c^1 = Aptian–Albian black marls; n^{1-4} = Neocomian marls; J^9 = Malm limestones; J^{3-8} = Oxfordian marls; J^{1-2} = Dogger limestones; l^3 = Lias; t^3 = Triassic, upper "cargneules"; t^{2b} = Middle Triassic dolomites; t^{2a} = lower "cargneules"; t^1 = Werfenian; (After FALLOT, 1949.

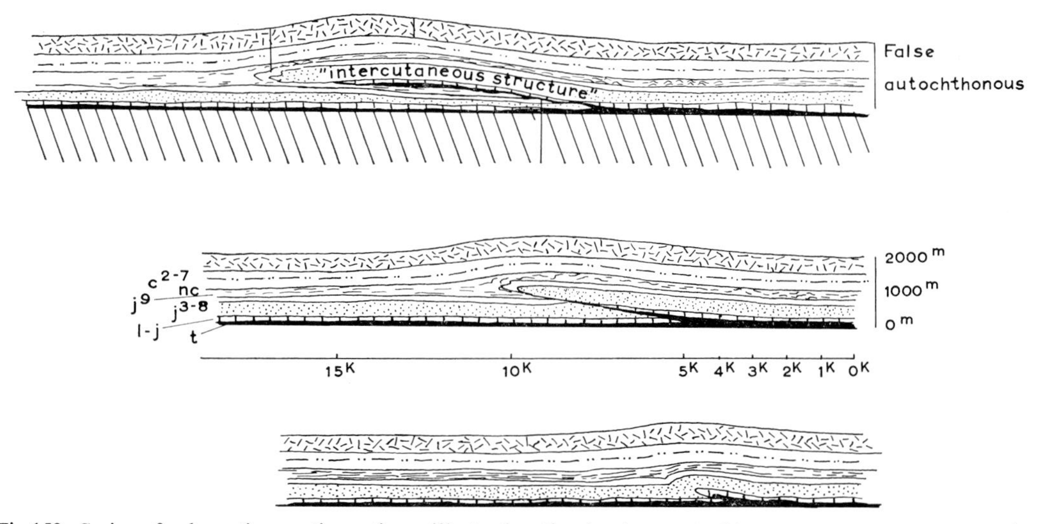

Fig.159. Series of schematic genetic sections, illustrating the development of intercutaneous structures, such as found near Roya. For legend, see Fig.158. (After FALLOT, 1949.)

THE INTRACUTANEOUS STRUCTURES OF ROYA

Locally more flat-lying thrustplanes developed, such as in the intracutaneous folds and overthrusts of Roya, which have played a major part in the development of the ideas on gravity as the motor for nappe tectonics (p.221).

These are found in the southwestern mantle of the Argentera Massif, due south of

St. Étienne de Tinée (cf. Fig.169; the area is situated in the headwaters of the Vallon de Roya). The structures consist of a lower series of Jurassic, including part of the marls of the Lower Cretaceous, and an upper series comprising Cretaceous and Lower Tertiary, which starts at the same marls of the Lower Cretaceous. In many sections one gets the impression of a normal, concordant sedimentary complex. Only where the shearing movements in the marls of the Lower Cretaceous were oblique to the bedding, and hence are accompanied by recumbent folds and by overthrusts, is the existence of such movements, and also their order of magnitude realized (Fig.158).

The upper series, above the shear zone in the marls, is indicated as "false autochthonous" (Fig.159) because outside the area of the actual folds and overthrusts it appears to be autochthonous. This term indicates, however, only its position with respect to the lower sedimentary series. The Jurassic itself is of course not autochthonous, being separated from the basement by the "décollement" in the underlying Triassic strata.

THE "SILLON SUBALPIN"

The internal side of the Subalpine chains is marked, for most of their length by a broad longitudinal depression, the "Sillon subalpin", or Subalpine groove (Fig.151, 174). Erosion produces this feature, wherever the marly series of the Lower Jurassic overlie the crystalline basement of the Central Massifs.

This belt of Lower Jurassic in many places shows some extra tectonization into a number of smaller, par-autochthonous overthrust or nappe structures. These, however, do not warrant a separation of the "Sillon subalpin" as a distinct zone of the Subalpine chains, and are no more than a local complication.

THE "FOSSE VOCONTIENNE"

As stated already on p. 268, there is one major element of aberrant facies in the Subalpine chains, which merits separate mention. This is the so-called "Fosse vocontienne", or Vocontian deep (Fig.160). It is a roughly east–west striking zone, in which the thick limestone series of the normal Dauphinois facies of the Cretaceous are replaced by marls. Moreover, the series is much thinner.

This aberrant facies, called the Diois facies from the town of Die, differs strikingly from the normal Dauphinois facies. Moreover, it is of course very much in evidence in the morphology of the Subalpine chains.

Terrigenous, fine-grained sedimentation has in the past been equated with sedimentation far from the continents, and hence in deep water, whereas the so-called reef facies of the Urgonian was held to indicate shallow-water sedimentation. The paleogeographic interpretation of the Diois facies consequently resulted in the picture of a narrow east–west deep, the "Fosse vocontienne", bordered by shallow platforms, on which reefs could grow. This interpretation, conceived at an early date in French geology, has since persisted in all later textbooks.

Its justification is slender, as we have no indication whatsoever as to actual depth of sedimentation. One might just as well conceive a diametrically opposed paleogeography, in which the shaly Diois facies represents a deltaic area, a sort of Wadden Sea, surrounded

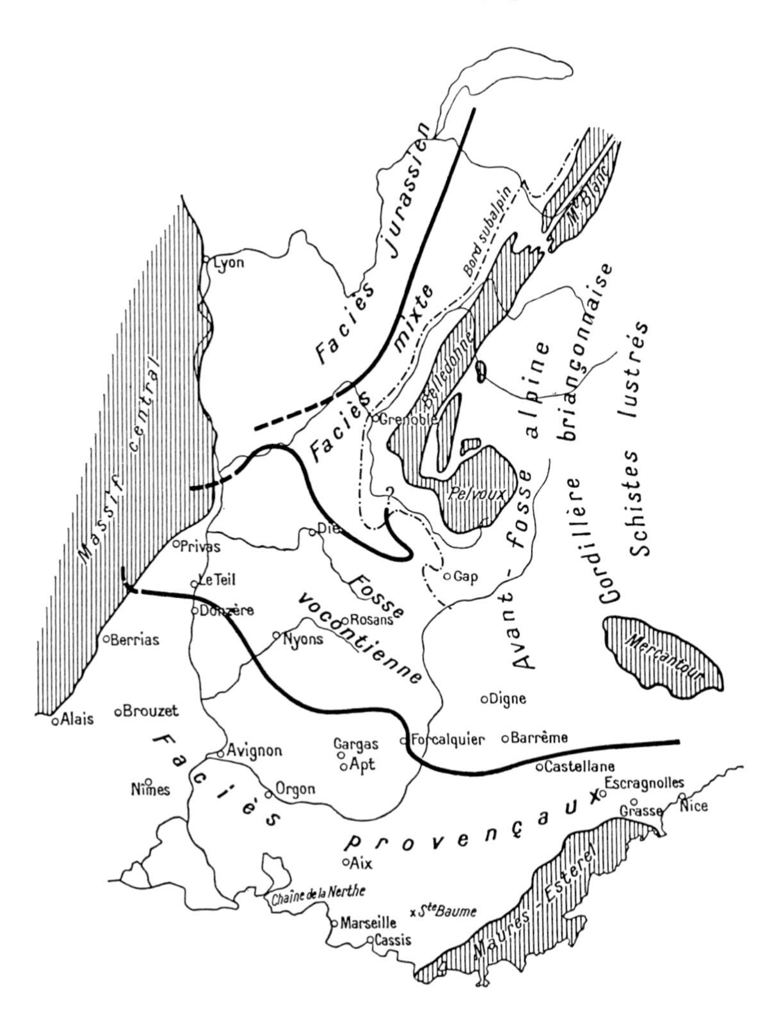

Fig.160. The extent of the so-called "Fosse vocontienne" or Vocontian deep during the Early Cretaceous. The Jurassic facies in the north and the Provençal facies in the south are similar, and characterized by a thick development of the Urgonian in reef-rock facies. The Diois facies within the Vocontian deep carries a relatively thin series of marls instead. (After GIGNOUX, 1950.)

by clear water areas in which reef-rock was deposited. As long as no reliable sedimentological criteria have been marshalled, both interpretations are equally probable. Moreover, differential crustal subsidence during sedimentation has played its role because the thickness of the normal Dauphinois facies of the Cretaceous is about double that of the Diois facies.

As is so often the case in geology, the rocks are there! No doubt remains that the Diois facies is different from the normal Dauphinois facies. It is only the interpretation — in this case the paleogeographic interpretation — that is doubtful.

THE BARROT DOME

The exact relation between epidermis and basement is rarely visible in the Subalpine chains. It is, however, very well exposed around the Barrot Dome, some 45 km north-

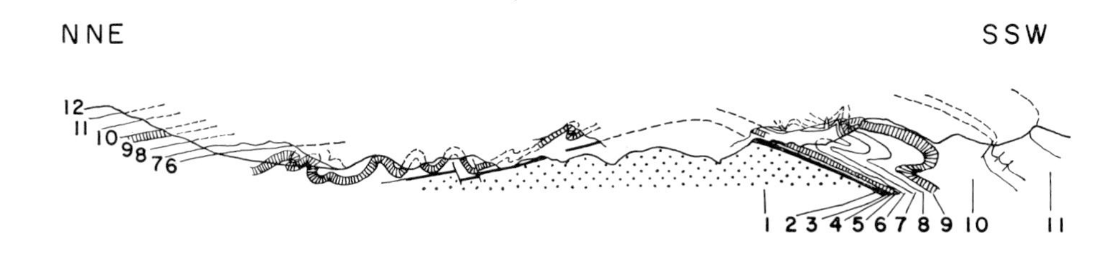

Fig.161. Section through the Barrot Dome.
1 = Permian, conformable with Triassic; *2* = Werfenian (= Buntsandstein); *3* = lower "cargneules" and evaporites; *4* = Muschelkalk; *5* = upper "cargneules" and evaporites; *6* = Lias; *7* = Dogger; *8* = Oxfordian; *9* = Malm; *10* = Lower Cretaceous; *11* = Aptian; *12* = Upper Cretaceous.
The "décollement" may take place either below the Muschelkalk, such as on the northern flank, or above the Muschelkalk, such as on the southern flank of the Permian dome. (After P. BORDET, 1950.)

noıthwest of Nice. This is an exposure of Lower Triassic, some 15 km in radius, and of Permian in the southern half.

Both the Permian and the Lower Triassic have an extremely regular structure, with low, radial, outward dip. The "décollement" occurs in the gypsum of the Middle Triassic, so that the Upper Triassic belongs to the epidermis of higher Mesozoic (BORDET, 1950; SCHUILING, 1957; Fig.161, 162).

There is a general feeling that "décollement" in the Subalpine chains has never reached larger dimensions, the epidermal series is thought to be nearly autochthonous. The representation of the folding and faulting by the deformation of a net of squares, supposedly drawn on a guide horizon before tectonization (GOGUEL, 1939, pl.II, 1965, pp.115, 116; MENNESSIER, 1964), is indicative of this interpretation. In this way only the deformations of the epidermis can be taken into account, not a wholesale "décollement" of a sedimentary epidermis. This feeling rests, however, merely on an absence of data. Since large-scale overthrusting has in later years been proved to exist in the similarly built Jura Mountains, a drilling program in the Subalpine chains might well come up with some unsuspected facts too.

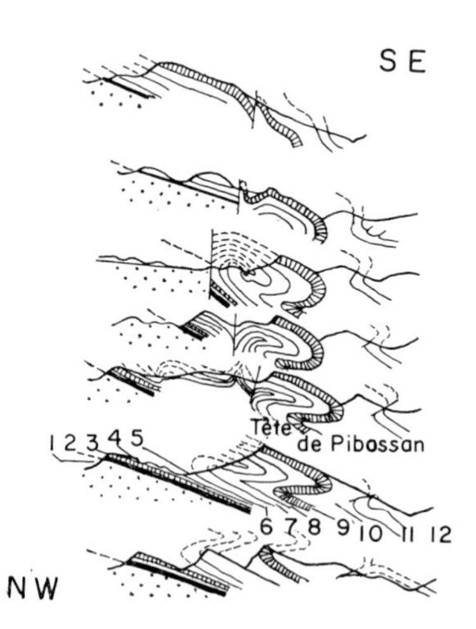

Fig.162. Serial sections through the Tête de Pibossan structure southeast of the Barrot Dome. Showing along the strike variations and the influence of later faults. For explanation of numbers, see Fig.161. (After P. BORDET, 1950.)

THE CENTRAL MASSIFS

The Central Massifs of the French Alps are, from north to south, the Belledonne, with its easterly outlier the Grandes Rousses, the Pelvoux and the Argentera or Mercantour.

In the French Alps one does not find a double set of Central Massifs, such as the Aar–Gotthard and Aiguilles Rouges–Mont Blanc in the Swiss Alps. The small Grandes Rousses Massif, which lies behind the southern tip of the Belledonne, is not a separate structure, but part of the Belledonne, being separated only by an inlier of sediments.

In a way, the absence of doublets of Central Massifs in the French Alps is a nice fact. For we have seen how in the Swiss Alps the "root zone" of the Helvetide nappes is located between this double set of Central Massifs. It might therefore be more than a coincidence, when in the French Alps, where the Subalpine chains do not possess nappe structures, and hence have no "root zone", we also find only a single set of Central Massifs, and no indication of something comparable to the Tavetscher Zwischenmassiv (pp.237–238).

The central massifs of the French Alps show a great dissimilarity in their structure. Moreover, as the geological research has progressed along quite different lines in these various massifs, short separate notes on their structure will follow.

THE BELLEDONNE MASSIF

The most striking feature of the Belledonne Massif is its long, narrow form, of over 150 km by less than 15 km. We have a general map by P. BORDET and C. BORDET (1963; Fig.163), and some more detailed petrographical analyses, notably that of TOBI (1959).

The long, narrow outline of the Belledonne Massif is reflected in a part only of its inner structure. We find a narrow "median syncline", or synclines, which is more in the nature of an overthrust zone (Fig.164), which runs the entire length of the massif. It separates an external zone, the "rameau externe", from an internal one, the "rameau interne".

In the external zone are found only epizonal metamorphics, albite–sericite–chlorite schists, the "série satinée" of authors. The internal zone, on the other hand, contains amphibolites, micaschists, gneisses and granites, some of which have suffered retrograde metamorphism. In this rather varied complex P. BORDET and C. BORDET (1963) distinguish a green series; a brown series; a group of cordierite gneisses—the fully series; and granites. When these are taken as stratigraphical units, it follows that the structure of the internal zone is not parallel to, and therefore older than, the present trend of the massif (Fig.163).

The stratigraphy used is, of course, no more than a lithostratigraphy. In polymetamorphic rock series, part of which have undergone retrograde metamorphism, the application of lithostratigraphic criteria can be just as dangerous as in sedimentary series. It must therefore be feared that more thorough analysis of the mineral facies will show the present picture to be oversimplified.

THE PELVOUX MASSIF

There is, as yet, no new description of the Pelvoux Massif, so TERMIER's note of 1896 still remains the basis from which to start. A later memoir by BELLAIR (1948) has not

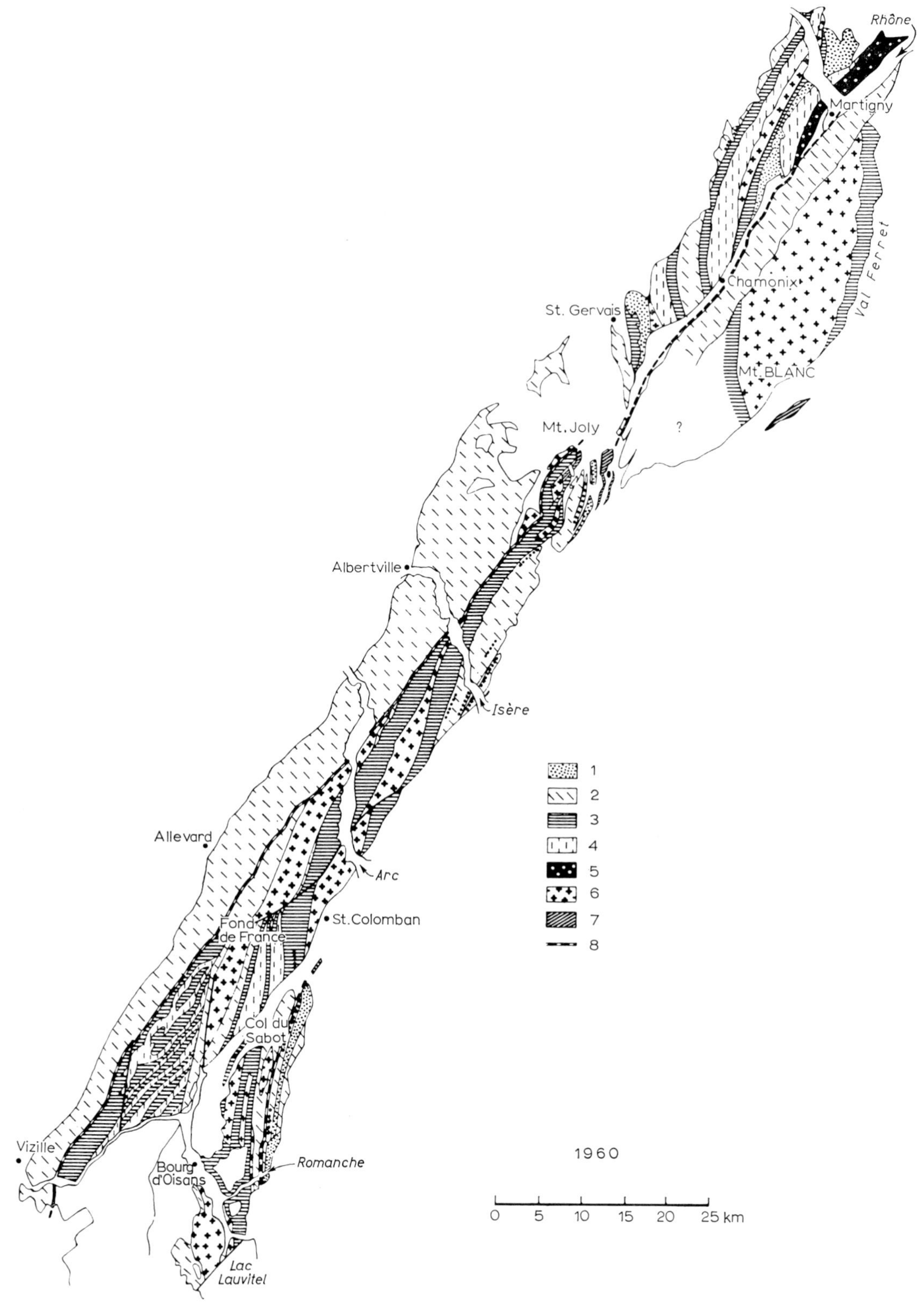

Fig.163. General structural map of the Belledonne, Aiguilles Rouges and Mont Blanc Massifs, 1:1,000,000. *1* = post-Hercynian Upper Carboniferous and Permian; *2* = "série satinée", albite–sericite–chlorite schists; *3* = "série verte", varied schists; *4* = "série brune", micaschists and gneisses; *5* = fully series, cordierite gneisses; *6* = granites; *7* = allochthonous series; *8* = median syncline(s). (After P. BORDET and C. BORDET, 1963.)

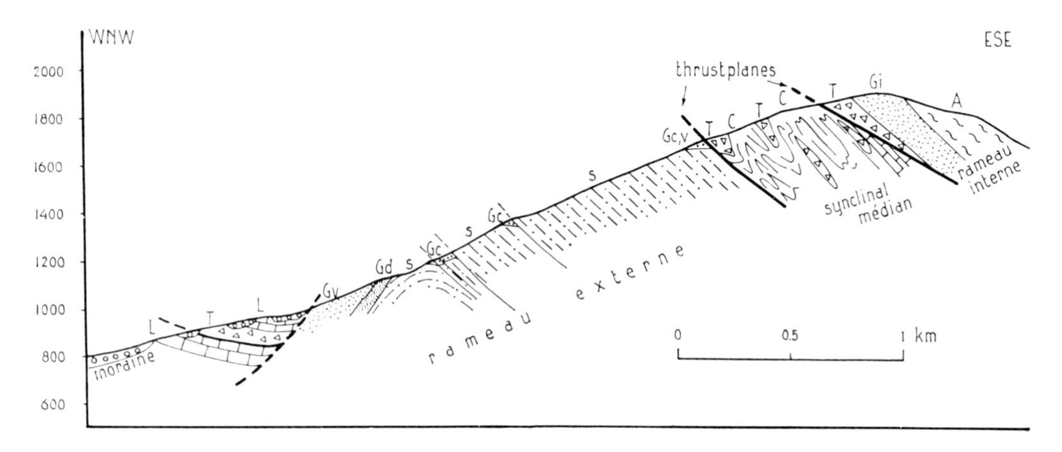

Fig.164. Cross-section through part of the Belledonne Massif, showing the relation between the external and internal zones and the median syncline.
L = Lias; *T* = Triassic; *Gv*, *Gd*, *Gc*, *Gi* = Allevard Sandstone, Permian; *S* = albite–sericite–chlorite schists; *A* = amphibolites. (After TOBI, 1959.)

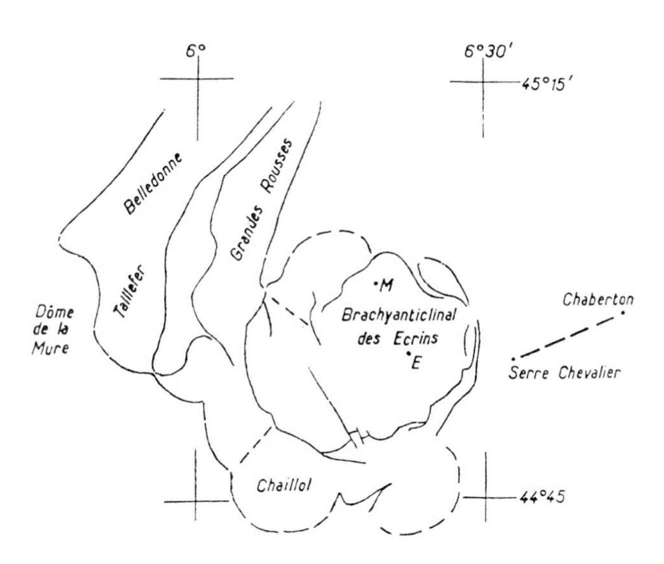

Fig.165. Outline sketch of the Central Massifs of the central French Alps.
M = La Meije; *E* = Les Écrins. Longitudes east of Paris. (After VERNET, 1952.)

resulted in essentially new insight. For this we have to turn to the scattered notes that form a prelude to a thesis by VERNET (1951, 1952, 1954, 1956) the latest worker in the Pelvoux Massif[1].

If we compare Pelvoux with Belledonne (Fig.165) we find at once a striking difference in outline, the Pelvoux being nearly circular. This outline conforms to a number of concentric structures which are found in the western half of the massif. In the eastern half, however, from the La Meije–Les Écrins (*M* and *E* in Fig.165) anticline, a north-northwest–south-southeast trend dominates.

[1] See VERNET (1965), published after this text was written.

As in other Central Massifs, we find again in the Pelvoux that it is not just a part of the Hercynian basement, updomed more or less uniformily during the Alpine orogeny. On the contrary, it is of a composite nature, and not only its present outline, but much of its internal structure is due also to Alpine movements. E. Haarmann, the German geologist who was one of the first to stress the importance of gravity tectonics, and who was therefore opposed to the buttress role ascribed to the Central Massifs in the classic picture, is reputed to have said: "The Pelvoux is not at all a buttress formed by the Hercynian basement. It was highly mobile during Alpine orogeny, and as a consequence acquired a structure as composite as an artichoke".

The structure of the Pelvoux is interesting in two different aspects: (*1*) The complex system of peculiar anticlines and synclines formed during the Alpine orogeny; and (*2*) the completely different mechanism of deformation of the Hercynian basement and of its sedimentary cover due to the Alpine movements.

As to the first point, the Alpine structural elements of the Pelvoux, as elucidated from the folding of the partly preserved post-Hercynian sedimentary cover, show an interplay of broad anticlines and narrow, pinched synclines (Fig.166). The axes of the anticlinal structures are short, they form in fact a sort of brachyanticline or dome. Where several of these brachyanticlines meet, synclinal pits may develop (Fig.167).

An intensification of this tectonic style may even lead to a complete pinching-out of the synclines, leaving only local amygdales of sediments along the axial planes (Fig.168). These indicate that there was no general, overall direction in the tectonic movements leading to this type of deformation.

Taking up Haarmann's simile, this tectonic style is indeed reminiscent of an inverted artichoke. VERNET (1954) thinks tangentional compression must nevertheless be invoked as the cause of this peculiar tectonic pattern. It might, however, also be related to the rising of a number of localized migmatitic domes, within the confines of the present massif.

Another special feature of the Pelvoux is that its eastern half shows strong westerly "Vergenz" and overturned folds, in contrast to the western half, which shows more or less vertical axial planes. This westerly "Vergenz" in the eastern part of the massif might of course be correlated with an influence exerted by the nappes of the Briançonnais zone, sliding down towards the eastern boundary of the Pelvoux. It is difficult, however, to

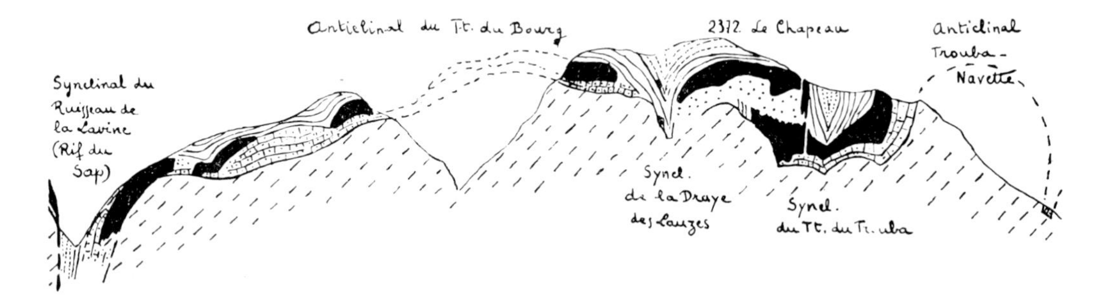

Fig.166. Cross-section over part of the western half of the Pelvoux Massif. For notations compare Fig.167. Showing broad dome-like anticlines and pinched synclines of the sedimentary cover, whereas the basement has reacted by differential movement along planes of schistosity or along minor fault planes, that have a quite constant attitude. (After VERNET, 1954.)

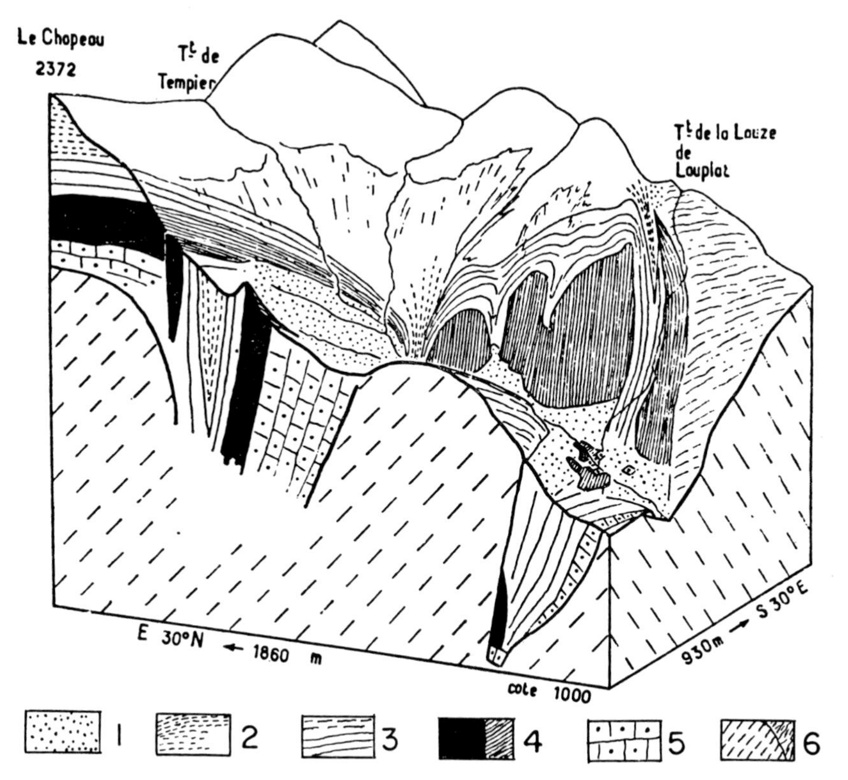

Fig.167. Block diagram of a part of the Pelvoux Massif, showing relation between a number of anticlinal domes and pinched synclines.
1 = alluvium; *2* = Lias shales; *3* = Lias limestones; *4* = melaphyres; *5* = Triassic; *6* = crystalline basement. (After VERNET, 1951.)

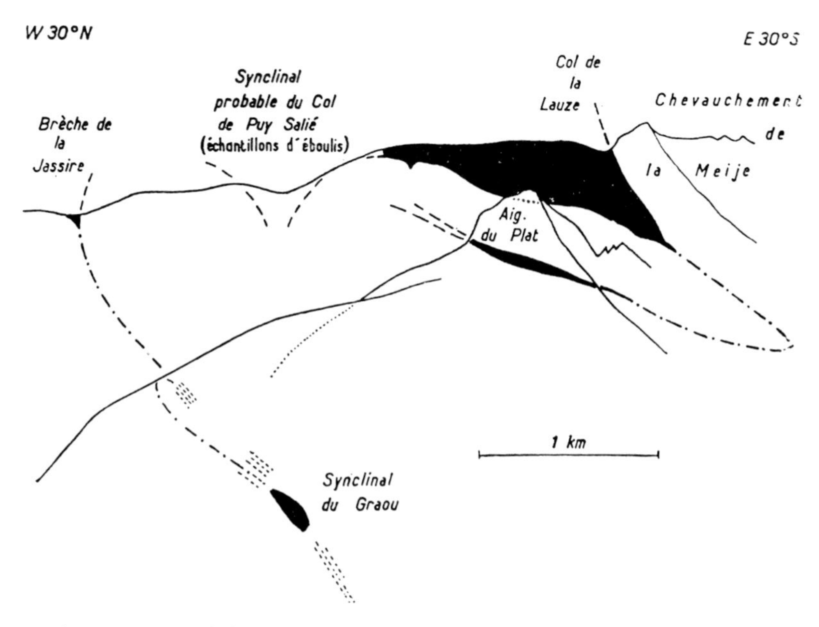

Fig.168. Cross-section over part of the Pelvoux Massif, west of La Meije, showing completely pinched syncline with local amygdales of sediment and westward "Vergenz" in the eastern part of the massif. (After VERNET, 1952.)

visualize such an effect exerted by a system of nappes which originated by gravity tectonics, as is normally supposed to be the case for the Briançonnais nappes. This relationship is therefore not definitely established. It might be no more than another example of the jumping at conclusions which has so largely prevailed in Alpine geology.

As to the second point, the difference in reaction to Alpine orogenetic influences by the crystalline basement and the sedimentary cover, we see that the sedimentary cover is everywhere deformed in normal, concentric type folding, with bedding plane slip. The underlying basement, however, shows uniform, steep dips all the way, regardless of anticlinal and synclinal structures. It has been deformed by differential shearing along sets of parallel planes, provided either by the already existing schistosity, or by sets of closely spaced fractures, of several centimeters apart. We meet here with a similar phenomenon, as in the Ardennes, where an earlier, anisotropic basement formed shearfolds by differential gliding, whereas its sedimentary cover reacted by concentric folding to the same orogenetic influences (pp.89–92).

According to Vernet, this shear folding of the basement explains the complete antithesis in the opinions expressed earlier as to the tectonic style of the Pelvoux. In fact TERMIER (1896) wrote: "There are no faults. Under the influence of orogenetic forces of extraordinary intensity all rock series, granites, granulites and gneisses included, behaved as plastic materials during the folding". Whereas BELLAIR (1948) writes of the same massif: "Very fractured, broken in multiple splinters by successive compression and decompression, the massif shows a veritable mosaic of fractures".

One should distinguish between the style of deformation on different scales. On a larger scale, the main elements of the Pelvoux do indeed show striking mobility and plasticity. Whereas on a smaller scale the movements that produced this large-scale flowage in the basement may either be of a plastic nature too, of flowage along schistosity planes, or of a more discontinuous nature, along sets of smaller parallel fractures crosscutting the earlier structures. In this way the apparent antithesis of the tectonics of the Pelvoux is resolved into a single coherent picture.

Quite apart from its internal structure, the Pelvoux Massif must have been elevated considerably during the youngest geologic history. As is the case with the other Central Massifs too, it is only in this way that the pre-Alpine basement is now exposed at more than 3,000 m above sea level, higher in fact than the youngest members of the geosynclinal series surrounding it. In the Pelvoux Massif it is thought that this vertical rise must have taken place after the main nappes of the internal zones had been emplaced. Because of this late rising of the Pelvoux Massif and part of its surroundings, a seesaw movement of the internal zones would have been produced, strong enough to cause a limited amount of eastward gravitational reaction of parts of the internal zones. This would account for the phenomenon of back sliding or "rétrocharriage", encountered in the internal zones of the French Alps (pp.294–298).

THE ARGENTERA (MERCANTOUR) MASSIF

In the French literature the southernmost of the Central Massifs of the French Alps was formerly indicated mostly by the name of Mercantour, whereas the Italian literature spoke of the Argentera Massif. The latter name has in later years been used more and more

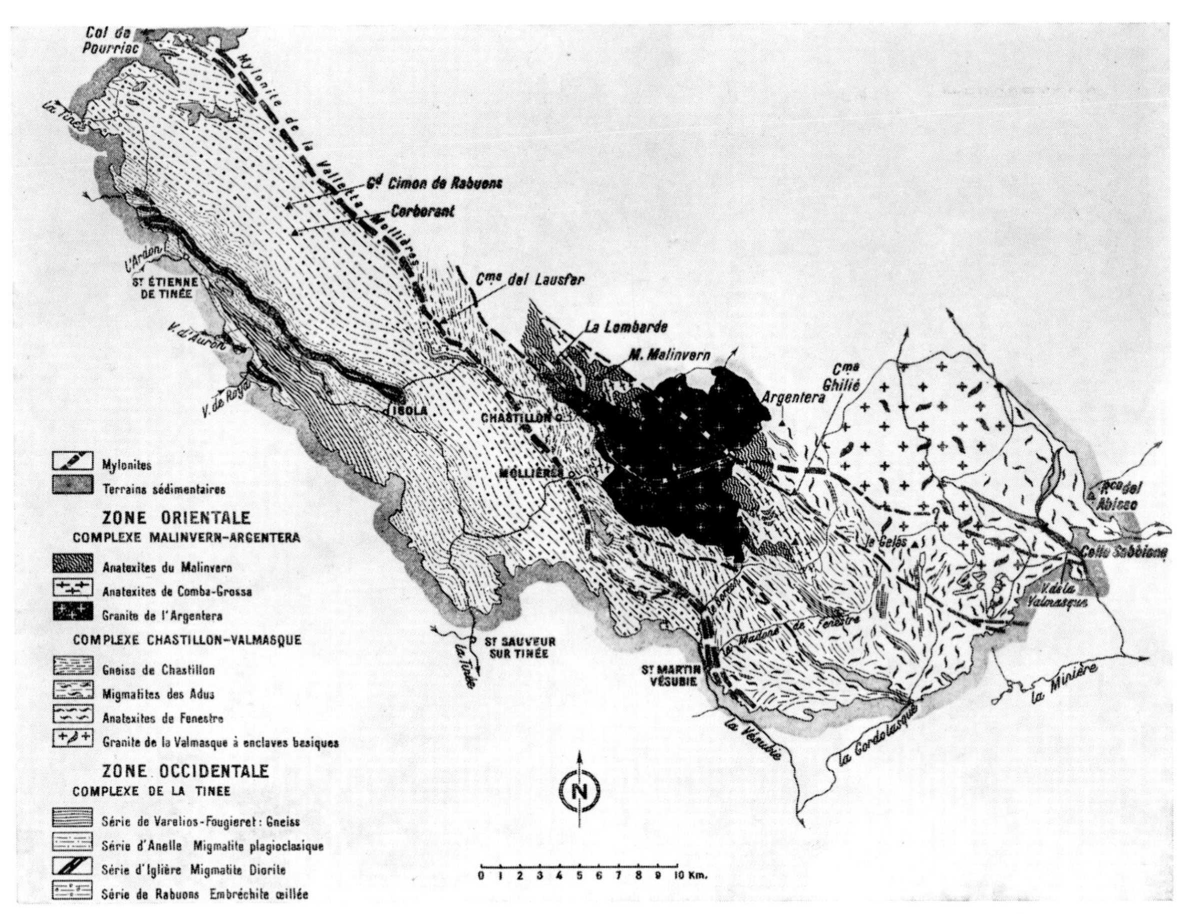

Fig.169. General map of the Argentera (or Mercantour) Massif. (After FAURE-MURET, 1955.)

in the French literature too. The French part has recently been ably studied by FAURE-MURET (1955), and a short summary of its main features, based on her monograph, follows.

Just as in the more northerly Central Massifs, the structure of the Argentera Massif is composite. The various zones which together make up the present massif follow more or less its outline. Through the change in direction of the Alpine chain, they strike northwest–southeast (Fig.169).

Starting from the external side, from the southwest, we first meet with a complex of crystalline schists, showing progressive migmatization, the Tinée Complex. It is separated from the internal zone by a narrow band of mylonite, the Valetta–Mollières Mylonite. This runs along the entire length of the massif, and must represent a major tectonic dislocation zone. The internal zone of the massif is formed by two very different elements, the Malinvern–Argentera Complex and the Chastillon–Valmasque Complex. Both complexes contain migmatite series, comprising notably anatexites and granites. They differ, however, in that the series of the Malinvern–Argentera Complex show only progressive metamorphism, whereas the Chastillon–Valmasque rocks have undergone extensive retrograde metamorphism.

FAURE-MURET (1955) points out that the Argentera, also, is not just a part of a

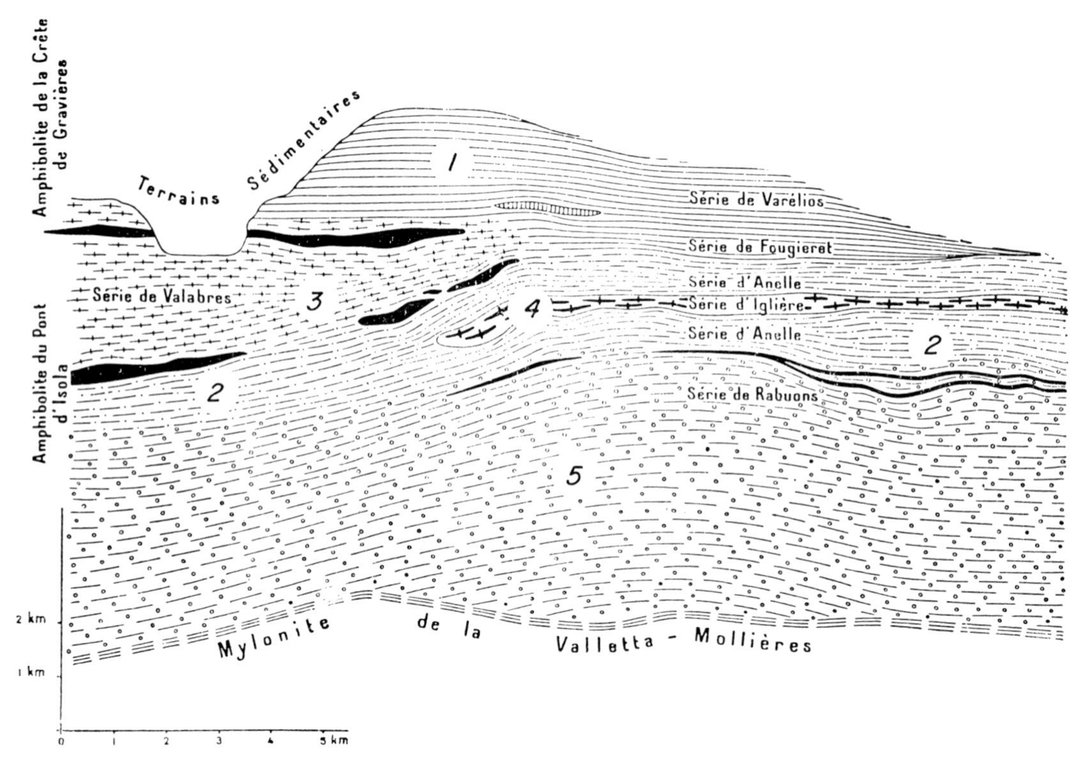

Fig.170. Hypothetical stratigraphic section through the Tinée Complex, according to the most probable assumption of present synclinal structure.
1 = paragneisses; *2, 3* = Ca + Na migmatites; *4* = pseudo-dioritic migmatite; *5* = K + Na migmatites. (After Faure-Muret, 1955.)

Hercynian orogenetic fold belt pushed upwards during the Alpine revolution. Part of the rocks forming the crystalline basement in the massif have already been affected by earlier orogenies. The pre-Alpine basement consists of a mosaic of elements which had already quite different earlier histories. It seems to be more or less due to chance which of these elements later became incorporated into the Central Massifs. This composite structure of a central massif has since been found to exist in several other Central Massifs, a fact duly noted in this text. Faure-Muret was, however, the first to stress this important aspect, and to map such elements with a different earlier history within the confines of a single central massif.

In the case of the Argentera Massif, it is thought most probable that both the external Tinée Complex and that part of the internal zone taken up by the Malinvern–Argentera Complex, which suffered only one cycle of metamorphism and migmatization, acquired this deformation during the Hercynian orogenetic cycle. The Chastillon–Valmasque Complex, on the other hand, must have gone through two metamorphic cycles. Migmatization took place during the first, whereas during the second retrograde metamorphism developed. In view of the absence of any real orogenetic diastrophism during Caledonian times in southern Europe, it is probable that the first period of metamorphism and migmatization is due to a Precambrian orogenetic cycle, whereas the second period probably was Hercynian.

An aspect which must be noted, even in this short summary, is that the progressive migmatization in the external Tinée Complex seems to be very similar to that found in the Massif Central and elsewhere in Hercynian Europe. Of course, Alpine tectonics have strongly influenced the earlier Hercynian succession. The interpretation of the succession of migmatite zones therefore remains somewhat hypothetical. In Fig.170, following the reconstruction thought most plausible by Faure-Muret, we see, however, a series of progressive migmatization of exactly the same order of magnitude as found in those parts of Hercynian Europe which have not subsequently been disturbed by the Alpine orogeny.

The Annot Sandstone

Around the Argentera Massif we find more indications of the vertical rise of a central massif during the youngest geological history. Of course, just as the other Central Massifs, the Argentera must have been elevated considerably during recent times to attain its present topographical and morphological situation. Sedimentological investigations of Tertiary sediments from its southern mantle have recently substantiated the youthfulness of these vertical movements.

The story concerns a thick series of clastics, conglomerates, sandstones and flysch-type deposits of Eo–Oligocene age in the Maritime Alps north of Nice. The Annot Sandstones, as this series is called from its best known member, were thought to be composed of

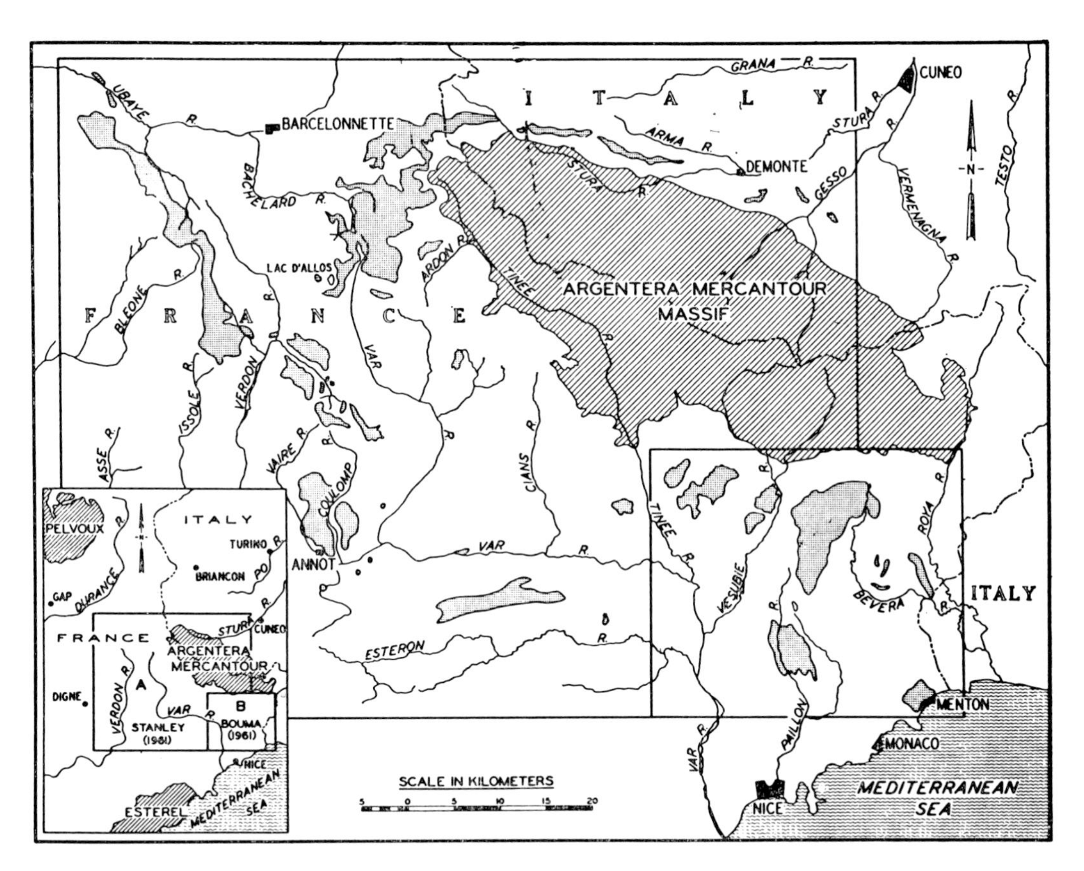

Fig.171. Map of southeastern France, showing the Argentera Massif (hatched) and the areas of the Annot Sandstone (stippled). (After STANLEY and BOUMA, 1964.)

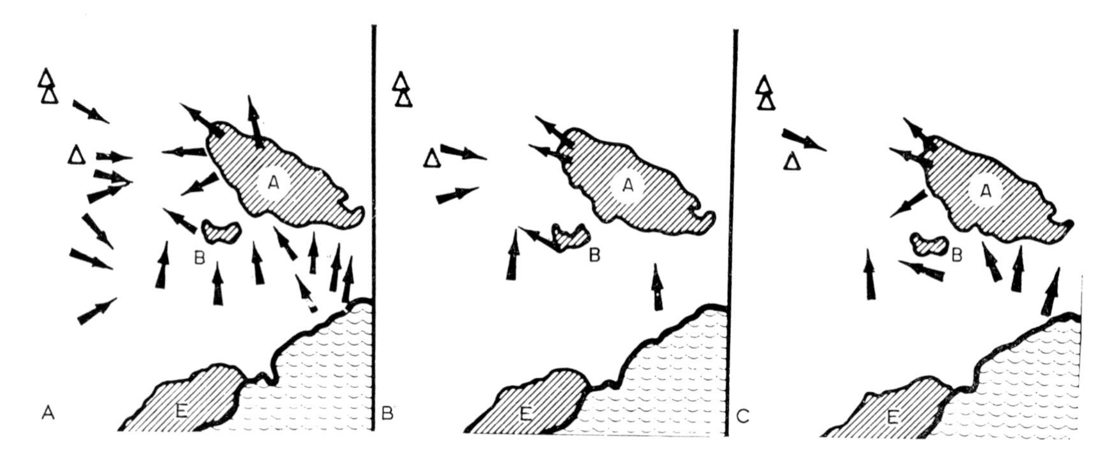

Fig.172. Results of sedimentological analysis of the Eo-Oligocene Annot Sandstone in the Maritime Alps south of the Argentera Central Massif.
A. Direction of sediment transport, as inferred from sedimentary structures, mainly sole markings. B. Sand/shale ratio, arrows point to decreasing ratio. C. Thinning of sandstone beds, arrows point towards thinner beds. *A* = Argentera Massif; *B* = Barrot dome; *E* = Estérel (Provence). (After STANLEY and BOUMA, 1964.)

material eroded from the slowly rising Argentera Massif, a view, for instance, expressly stated by FAURE-MURET (1955).

A cursory examination of the region by Kuenen, accompanied, amongst others, by Faure-Muret and by Fallot (KUENEN et al., 1957) revealed that sedimentary structures, such as sole markings, consistently indicated northward transport during the sedimentation of the Annot Sandstones. This excluded the Argentera Massif as a source area for this clastic series.

This has since been substantiated (STANLEY and BOUMA, 1964). Except for its northwestern edge, the Argentera Massif does not seem to have supplied any clastic material to the Eo–Oligocene seas (Fig.171, 172).

Instead, the quartzose, terrigenous detrital material that forms the bulk of the Annot Sandstone complex arrived mainly from the south. This means that during Eo–Oligocene times a relatively high crystalline massif must have occupied a position somewhere south of Nice, whereas the present Argentera Massif was still at so low an elevation that it could not supply any detrital material to the south. Since that time, the hypothetical crystalline massif south of Nice has subsided into the Mediterranean, to a depth of —4,000 m, whereas the Argentera Massif rose to an altitude of over 3,000 m.

This is one of the best substantiated cases proving the existence of very strong post-orogenetic vertical movements, in which both positive, rising movements, and negative, subsiding movements were involved. Due to the obviousness of horizontal movements in folding and in nappe tectonics, the importance of such differential vertical crustal movements in the Alpine chain has always been more or less overlooked. The Argentera Massif and its surroundings offer an example of the importance such vertical movements, however difficult to detect normally, may in fact have had. Because we need such vertical movements as the motor for postulated gravity tectonics, it is reassuring to find that movements of such a large order of magnitude can indeed be proven to have taken place.

THE INTERNAL ZONES

As for the internal zones (cf. BARBIER et al., 1963), we already saw in the introduction to this chapter that there are two main internal zones, the Briançonnais and the Piemonte; the latter being the equivalent of the Pennides of the Swiss Alps. For brevity we have not gone into the properties of the two minor zones, normally distinguished between the Dauphinois and Briançonnais zones, i.e., the Ultradauphinois and Subbriançonnais zones (Fig.173, 174). An idea of their relative importance can be gained through the panoramic view from the Lautaret Pass, reproduced here as Fig.175.

A recent review of our knowledge of the internal zones of the French Alps, which pays special attention to the many problems as yet unsolved, is by BARBIER et al. (1963). A field description of these zones, mainly based on three cross-sections in the central region, is by the SOCIÉTÉ GÉOLOGIQUE DE FRANCE (1965).

THE BRIANÇONNAIS ZONE

This unit stands out amongst the other subdivisions of the French Alps both by its stratigraphy and by its structure. The stratigraphy is unique in two ways, first in the pre-Hercynian Upper Carboniferous, and second in its post-Hercynian, Alpine development of the Mesozoic. Structurally, it is characterized by the fact that within this zone the "Vergenz" changes from westerly in the western part to easterly in the eastern part. In cross-section, the Briançonnais zone consequently shows a fan-like structure, the much advertised "éventail briançonnais" (Fig.173).

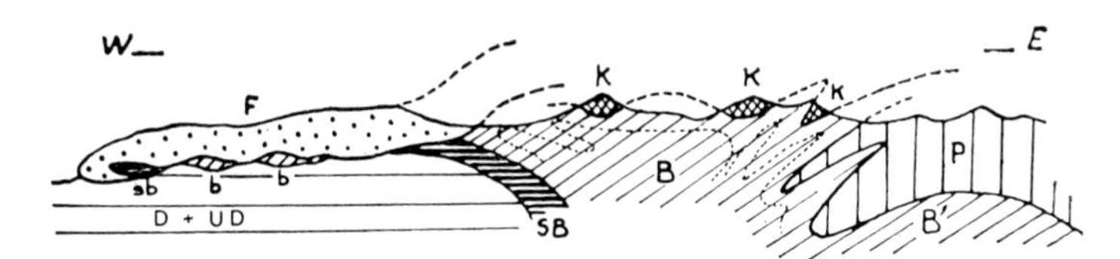

Fig.173. Schematic section through the internal zones of the French Alps.
$D + UD$ = Dauphinois and Ultradauphinois zones; SB = Subbriançonnais zone (sb = elements of this zone, transported by the *Helminthoïdes* flysch); B = Briançonnais zone (b = elements of this zone, transported by the *Helminthoïdes* flysch); P = Piemonte zone or Pennides; F = *Helminthoïdes* flysch nappe, belonging to P. K = various "Klippen", not treated in this text. (After BARBIER et al., 1963.)

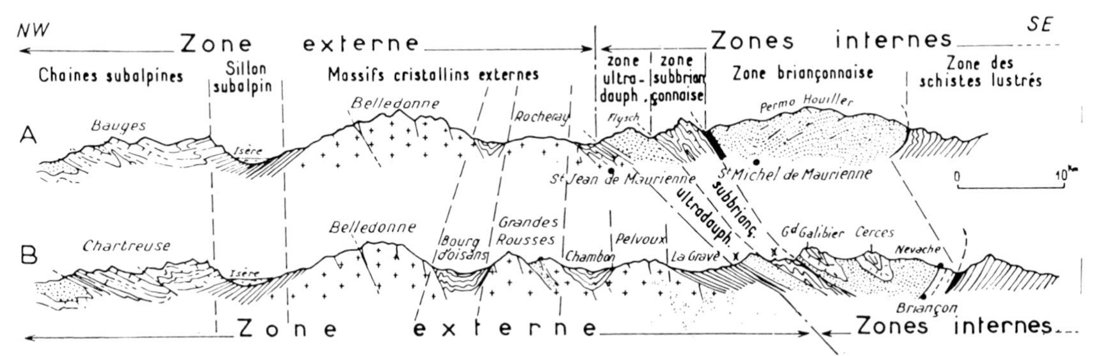

Fig.174. Two cross-sections over the central part of the French Alps, showing the structure of the main zones. The upper section is drawn over the river Arc and the Maurienne district, the lower over the Romanche river (compare Fig.173). (After FEYS et al., 1964.)

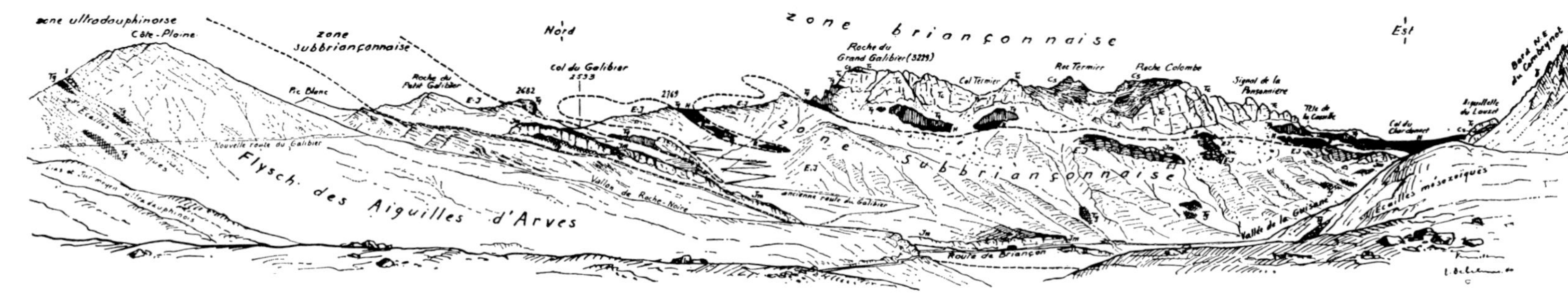

Fig.175. Panoramic view from the Col du Lautaret, in the central part of the French Alps, showing the development of the Ultradauphinois, the Subbriançonnais and the Briançonnais zones. This panorama corresponds to part of the lower section in Fig.174, east of La Grave. (After J. Debelmas, in: Feys et al., 1964.)

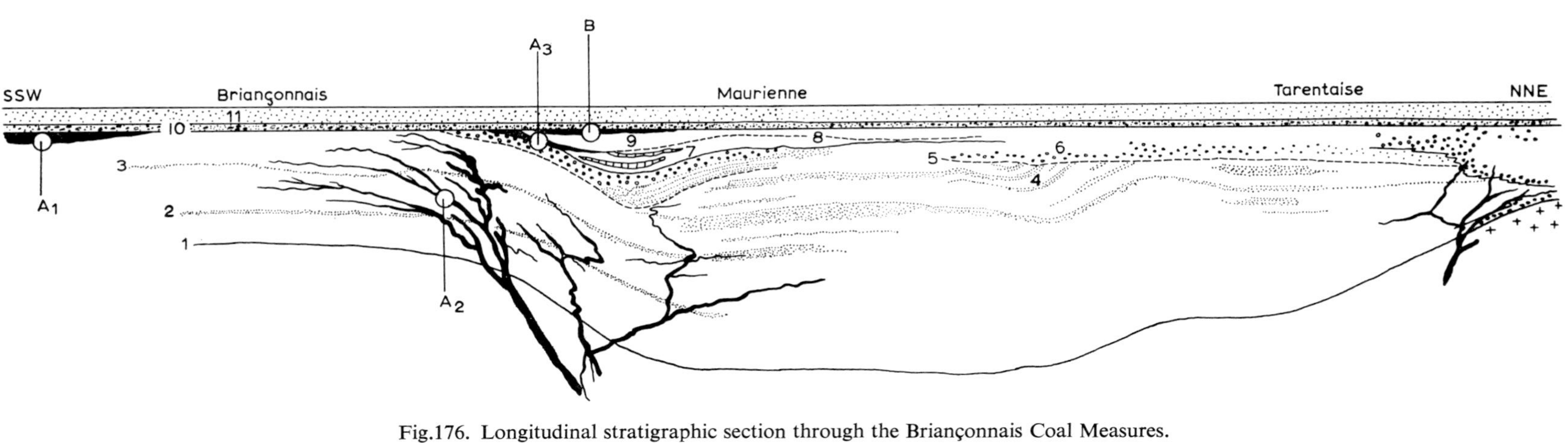

Fig.176. Longitudinal stratigraphic section through the Briançonnais Coal Measures.
1 = Pasquier coal seam, Namurian; *2* = Benoîte coal seam, Westphalian A; *3* = Madeleine coal seam, Westphalian C; *4* = Tarentaise series, Upper Westphalian and Lower Stephanian; *5* = "Asturian" disconformity; *6* = Courchevel series, Middle and Upper Stephanian; *7* = Lower Permian; *8* = "Saalian" disconformity; *9* = Poisonniere series, Upper Permian; *10* = Permo-Triassic in verrucano facies; *11* = Lower Triassic.
A_1 = ignimbrite of the Guil river; A_2 = microdioritic dikes and sills of the Briançonnais; A_3 = dacitic lava of Poisonniere; B = rhyolitic ignimbrite of Poisonniere. (After J. Fabre and R. Feys, in: BARBIER et al., 1963.)

Briançonnais Coal Measures

Starting with the stratigraphy of the Upper Carboniferous, we find in the Briançonnais zone a coal basin which begins in the Namurian, persists up to the Upper Westphalian, and with a slight discontinuity, into the Stephanian (Fig.176). For the paleogeography of Europe during the Late Carboniferous, this is very important, because it indicates a marked difference between the coal basins in the external zones and in the Briançonnais zone of the French Alps. Whereas the former, such as the La Mure Basin in the Subalpine chains (HAUDOUR and SARROT-REYNAULD, 1964; cf. Fig.177), are post-orogenetic, and only contain sediments of Stéphanian age, the Briançonnais Coal Measures are pre-orogenetic (FEYS et al., 1964). Although there are, as yet, no marine transgressions known from the Briançonnais Coal Measures, they correspond to the paralic coal basins elsewhere in Europe. Whereas the coal basins in the external zones of the French Alps correspond to the limnic basins (p.77).

Another difference with the stratigraphy of the Coal Measures in the normal paralic coal basins of the Upper Carboniferous, is the volcanic activity encountered in the Briançonnais. This is expressed in the large number of intrusive dikes, connected to sills that quite often follow the base of a coal seam. For various reasons this volcanicity is thought to be of Permian age.

The Briançonnais Coal Measures have never been exploited by wholesale mining operations. Instead a great number of peasant exploitations have pockmarked the deep glacial valleys. Also a modern reconnaissance, to which we owe the monographies of FEYS (1963) and FABRE (1961), did not lead to any major exploitation.

Two more points must still be mentioned in relation to the Briançonnais Coal Measures. First the weakness of the Hercynian orogeny, and second the weakness of the Alpine metamorphosis. As to the first point, we learn from Fig.176 that a slight discontinuity only exists between the Westphalian and the post-Hercynian Stephanian and Permian. This situation is comparable with that found for instance in the northern parts of The Netherland and Germany, where also the geosynclinal sedimentation of the Coal Measures was not followed by any orogenetic disturbances and only a slight discontinuity separates pre-Hercynian Coal Measures from post-Hercynian Permian.

Briançonnais "geanticline"

The second point, the slight metamorphism of the Coal Measures, is directly related to the peculiar situation of the Briançonnais zone during Alpine sedimentation. This follows, for instance, from the ideal sections of Fig.150. We see there that the Triassic, although beginning with evaporites (of course a very important factor in the tectonics of the Briançonnais zone), then is characterized by a series of carbonates in Alpine facies. This is, however, a reduced form of the truly Alpine facies, as it reaches a thickness of only 300 m. After that, a period of non-deposition follows for most of the Mesozoic, with a total column of sediments only between 100 m and 200 m thick.

This is the evidence for the "Briançonnais geanticline", famous in the literature, which has served as the main starting point in the opposition towards the embryonic cordillera concept (pp.194–195). In French such an area of non-deposition is normally designated by

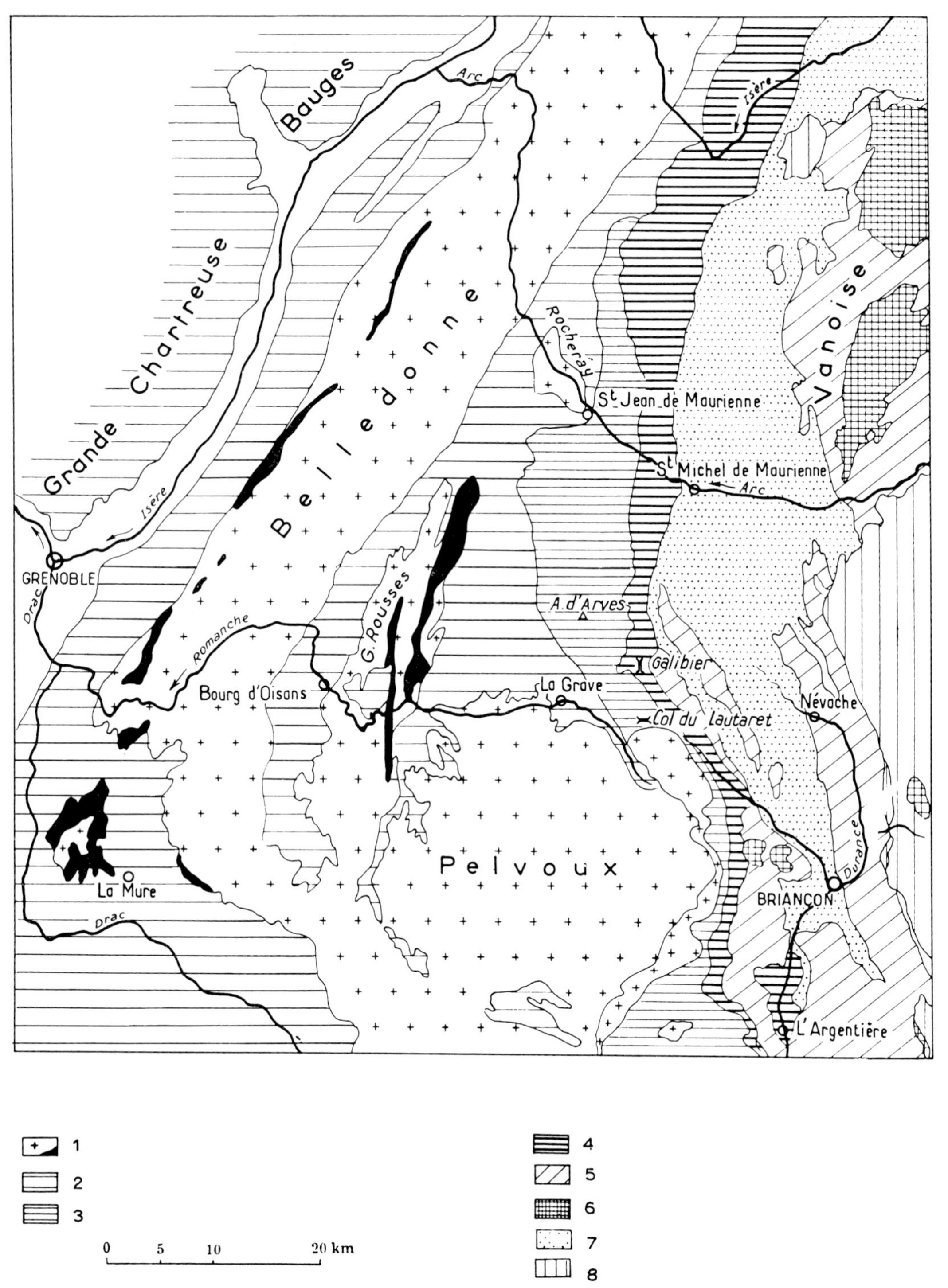

Fig.177. Map of the central part of the French Alps showing the various zones of the French Alps and the relation of the limnic coal basins in the external zone (black) to the coal basin of the Briançonnais zone.

1 = Central Massifs with post-Hercynian coal basin (black); *2* = sediments of the Dauphinois zone; *3* = Ultradauphinois zone; *4* = Subbriançonnais zone; *5* = non metamorphic Permo-Carboniferous of the Briançonnais zone; *6* = metamorphic idem; *7* = younger sediments of the Briançonnais zone; *8* = Pennide nappes.

(After FEYS et al., 1964.)

the term "haut-fond", meaning a shoal area, which did not recieve sediments at a time when sedimentation in its surroundings went on at full pace.

In my opinion the term "haut-fond" — that is a "shoal", but is perhaps best translated by "high-bottom" — describes best the situation in the Briançonnais during Jurassic and Cretaceous times. At this time a zone existed, some 50 km wide (LEMOINE, 1953, 1961), in-between the miogeosynclinal Dauphinois zone to the west and the eugeosynclinal Piemonte zone to the east, which subsided only slightly. The newer term of Briançonnais "geanticline" is a misnomer, because it is used to describe a situation during the geosynclinal phase, and not during the orogenetic phase of the Alpine cycle.

The Briançonnais fan

Apart from its stratigraphy, the Briançonnais zone, as we saw already, is known for its structure. This shows, generally, west "Vergenz" in its western part, sub-vertical position in its median zone, and east "Vergenz" in its eastern limits. This structure is well illustrated in Fig.173, 174, 178, 179.

As stated, this fan-like arrangement is thought to be the result of a secondary phenomenon, the "rétrocharriage" or backthrust. Originally, up to the Oligocene, nappes were formed by overthrusting towards the west. Following the formation of these nappes, an area to the west in the Briançon traverse (notably the Pelvoux Massif) began to rise vertically and part of the pile of nappes slid back towards the east. The eastern part of the Briançonnais zone now consequently overrides the more internal Piemonte zone over a distance of at least 10 km.

This backthrust was elucidated only in later years. It occurs in the Swiss Alps too, but it has not there been cited as explicitly as in the French Alps. As an illustration of the thesis that the rocks do not change, but only the tectonic interpretation, a series of historical sections through a part of the Briançonnais zone is given in Fig.180.

From a theoretical point of view the newer interpretation of the genesis of the Briançonnais fan is important, because it stresses the variety of successive movements which succeeded in time, each contributing to the present structure. It is probable that the picture is even much more complicated, because a nappe was such a superficial phenomenon that erosion took place not only directly after, but even during nappe formation. Several newer publications, too specialized, to be related here, point in this way (cf., for instance, BLOCH, 1965).

The composite sequence of events leading to the present structure might also have had important effects elsewhere in the French Alps. It is, for instance quite possible that the late rise of the Central Massifs caused the tectonization of the "Sillon subalpin" (p.276), to the west. But in the Dauphinois zone such later tectonization will have the same "Vergenz" as earlier thrusts, and will consequently be difficult to detect. It is through the situation of the Briançonnais zone, internal to the Central Massifs, that "rétrocharriage" with a "Vergenz" opposite to the earlier nappes can be detected more easily.

The phenomenon of backthrust in the Briançonnais therefore tells us not only that tectonic movements may follow each other in a more or less composite sequence, but also that they may vary in character and direction. All three earlier interpretations (Fig.180) only consider the possibility of outward movements, from east to west. This tallies well

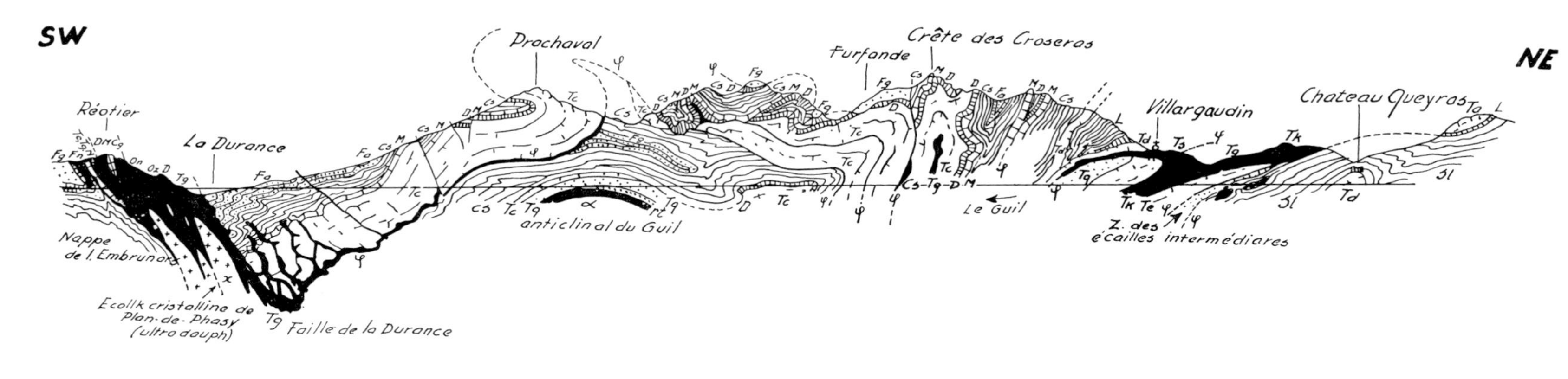

Fig.178. Section across the Briançonnais zone, showing its fanlike structure, the "éventail briançonnais".

The western edge of the Briançonnais zone, west of the Durance river, is affected by young subvertical faulting, along which the Pelvoux, north of this section, has risen. In the eastern part of the section, with eastward "Vergenz", the inner part of the Briançonnais zone is thrust over the the "schistes lustrés" (*sl*) of the Piemonte zone. The extent of this backthrust or "rétrocharriage" is of the order of several tens of kilometers, as indicated by the erosional remnant of Triassic dolomites (*Td*), in Briançonnais facies, on top of the "schistes lustrés", east of Chateau Queiras (cf. Fig.179).

Ultradauphinois zone: *O* = Callovian; *x* = crystalline basement. Brianconnais zone: *Fn* = basal flysch; *Cs* = Upper Cretaceous; *M* = Malm; *S* = Dogger; *T* = Triassic; *Tk* = cargneules; *Tq* = quartzite; *rt* = Permo-triassic; *r* = Permian; α = Permian andesite.

Piemonte zone: *Gg* = sandy flysch of the Embrunais; *sl* = "schistes lustrés"; *rv* = "roches vertes", ophiolites; *L* = Lias of the pre-piemont; φ = overthrust.

(After J. Debelmas, in: Barbier et al., 1963.)

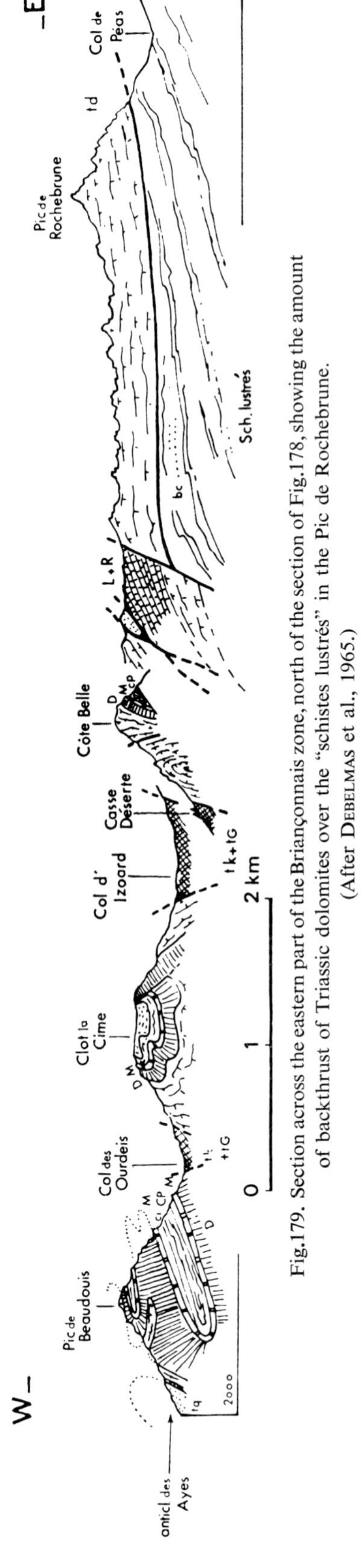

Fig.179. Section across the eastern part of the Briançonnais zone, north of the section of Fig.178, showing the amount of backthrust of Triassic dolomites over the "schistes lustrés" in the Pic de Rochebrune. (After DEBELMAS et al., 1965.)

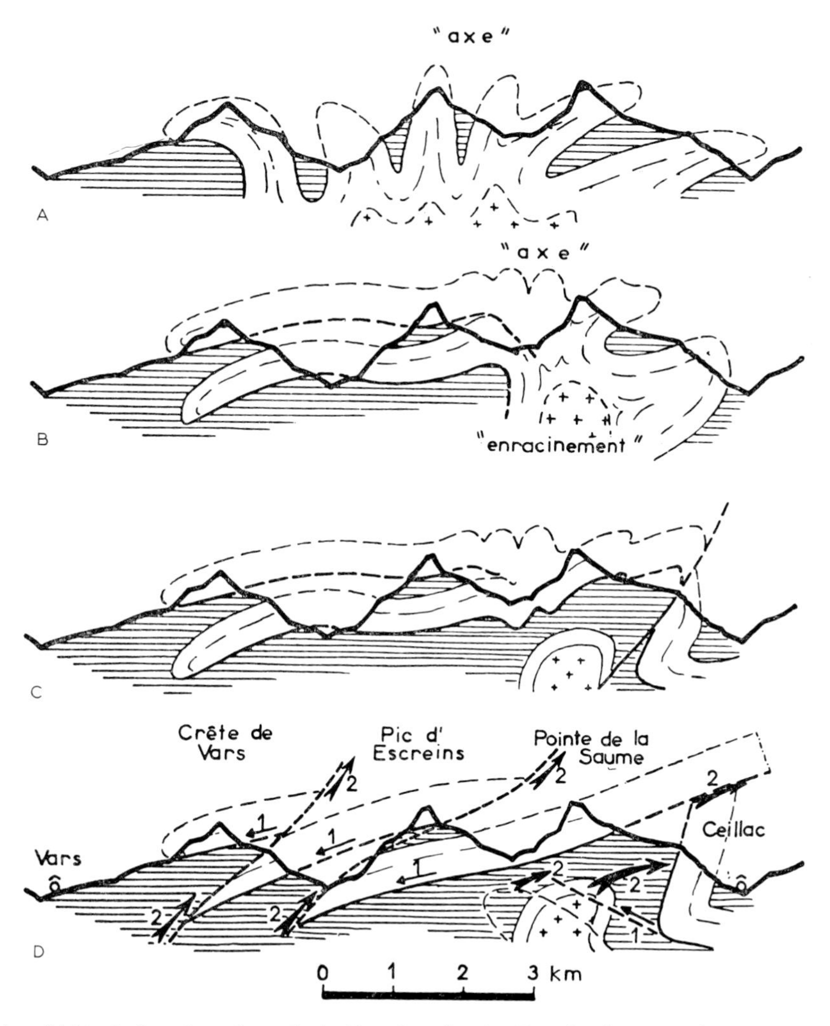

Fig.180. Series of historical sections through the Escreins, showing Permian basement (crosses), "schistes lustrés" of the Piemonte zone (hatched), and Mesozoic cover of the Briançonnais zone (white).
A. Interpretation of W. Kilian in 1900.
B. Interpretation of F. Blanchet in 1935.
C. Interpretation of J. Goguel in 1950.
D. Interpretation of M. Gidon in 1958.
In the last section the number *1* arrows indicate tectonic movements during the earlier period of nappe formation, the number *2* arrows movements during the younger period of backthrust or "rétrocharriage". (After GIDON, 1962.)

of course, with the notion of tangential compression, of a solid block moving in on an abutment, which was the philosophy underlying the two earliest interpretations. But it also tallies with an interpretation as one-step gravity tectonics, such as advocated by J. Goguel in section *C* of Fig.180. The backthrust phenomenon does, however, indicate two-step gravity tectonics. This must have been caused by an early rise of an internal area, bringing about nappe formation by "décollement", and followed by a rise of the Pelvoux Massif and related areas, causing the backthrust.

Just as in the case of the Annot Sandstone (pp.287–288) south of the Argentera Massif, the composite sequence of tectonic events, as analyzed now by detailed investigations, points to a sort of oscillation of crustal blocks. No single impetus causes all of the tectonic events in a given fold belt. Tectonics is not something of one cast. The final structure is not implicit from the beginning, but determined by a succession of factors. In the case of the Briançonnais fan, and also for the Annot Sandstone, we must postulate a sort of giant seesaw motion of the basement as the motor for the variously directed gravitative reactions of the sedimentary cover.

THE PENNIDES OR PIEMONTE ZONE

After having gone into some detail about the Dauphinois and Briançonnais zones, I must of necessity be very short about the Pennides of the French and Italian Alps. Geographically they mainly constitute the eastern flank of the Alps, that is the upper slopes of the Italian Piemonte region, from which the zone takes its name. Its most important development is in the Cottian Alps, between Briançon and the Dora Maira Massif.

Its bulk is quite comparable, geologically, to the Pennides of Switzerland, being built up by a monotonous clastic series, barren, or almost barren of fossils, in an epimetamorphic facies, with a relatively important number of ophiolitic intercalations. It is a Bündner Schiefer, or "schistes lustrés" or "calcescisti" complex.

The age of this complex in the Piemonte zone is not known. Conti (1953, 1955) speaks of a comprehensive series, ranging from Triassic to Lias. Others (cf. Barbier et al., 1963) think it restricted to the Lias. It is quite possible that further analysis will show the presence of two series of different age and provenance (Michard and Sturani, 1964).

The single important Pennide element not belonging to the "schistes lustrés" complex, the *Helminthoïdes* flysch, will be treated in the next section.

Just as in the Pennides of the Swiss Alps, pre-Alpine crystalline rocks crop out, in the Piemonte zone, in two larger "internal massifs", those of the Grand Paradis and the Dora Maira (Fig.139). It is not clear, however, whether these massifs are tectonically autochthonous, or form the older cores of nappes of the "schistes lustrés". That is, whether they form the basement of, or belong to, the Alpine fold belt (Michard, 1963).

Between the Briançonnais and the Piemonte zones a number of subzones can be distinguished, both as to their present position, and as to their supposed area of sedimentation. There seems to be, consequently, a more or less gradual transition between these two main zones. But it must be noted that any correlation remains difficult, due to the strong Alpine tectonization. Again I can do no more than mention the presence of these subzones of intermediate character, referring the reader to the literature for more details (cf. Barbier et al., 1963; Lemoine, 1961, 1964).

The Helminthoïdes flysch

Only a short summary of this important series is possible here. It is also a monotonous clastic series, but it is non-metamorphic, and is characterized by the occurrence, on some bedding planes, of animal tracks of intricate pattern, that have received the name of *Helminthoïdes* (Fig.181).

Originally considered to form part of the flysch of the external zones, it has since been dated as Cretaceous, mainly by its microfauna, particularly by species of the genus *Globo-*

Fig.181. Bedding plane covered by *Helminthoïdes*.

truncana. It is now thought to represent a deep-water, turbidite sequence, and to form the highest Pennide unit (BARBIER et al., 1963).

As stated earlier (p. 246), *Helminthoïdes* flysch is present in the Alps in three nappes that more or less override the external zones. These are situated in the Swiss Prealps and in the Embrun and Nice nappes in France (Fig.139). It moreover occurs in the northern Apennines (Fig.182).

It is thought probable that the *Helminthoïdes* flysch originated in a vast sedimentary basin situated between those of the "schistes lustrés" and of the Apennines (LANTEAUME, 1963; LANTEAUME et al., 1963). This paleogeographic interpretation is complicated by the difficult question of where to locate the rising crustal blocks which provided the clastic debris for this series.

It is still an open question, however, whether the *Helminthoïdes* flysch is only a facies of the Upper Cretaceous developed in various basins, or whether the present *Helminthoïdes* flysch nappes are really related as to their area of sedimentation and/or as to their tectonic history.

THE JUNCTION OF ALPS AND APENNINES

Although we find *Helminthoïdes* flysch in both Alps and Apennines, just as we find ophiolites in either area, and although there are more similarities, there exist, on the other hand, marked differences between these two members of the Alpine fold belt. The most conspicuous being that the French Alps have westerly, the Apennines on the contrary, easterly "Vergenz".

For want of space, I will not go further into these differences here, but only note

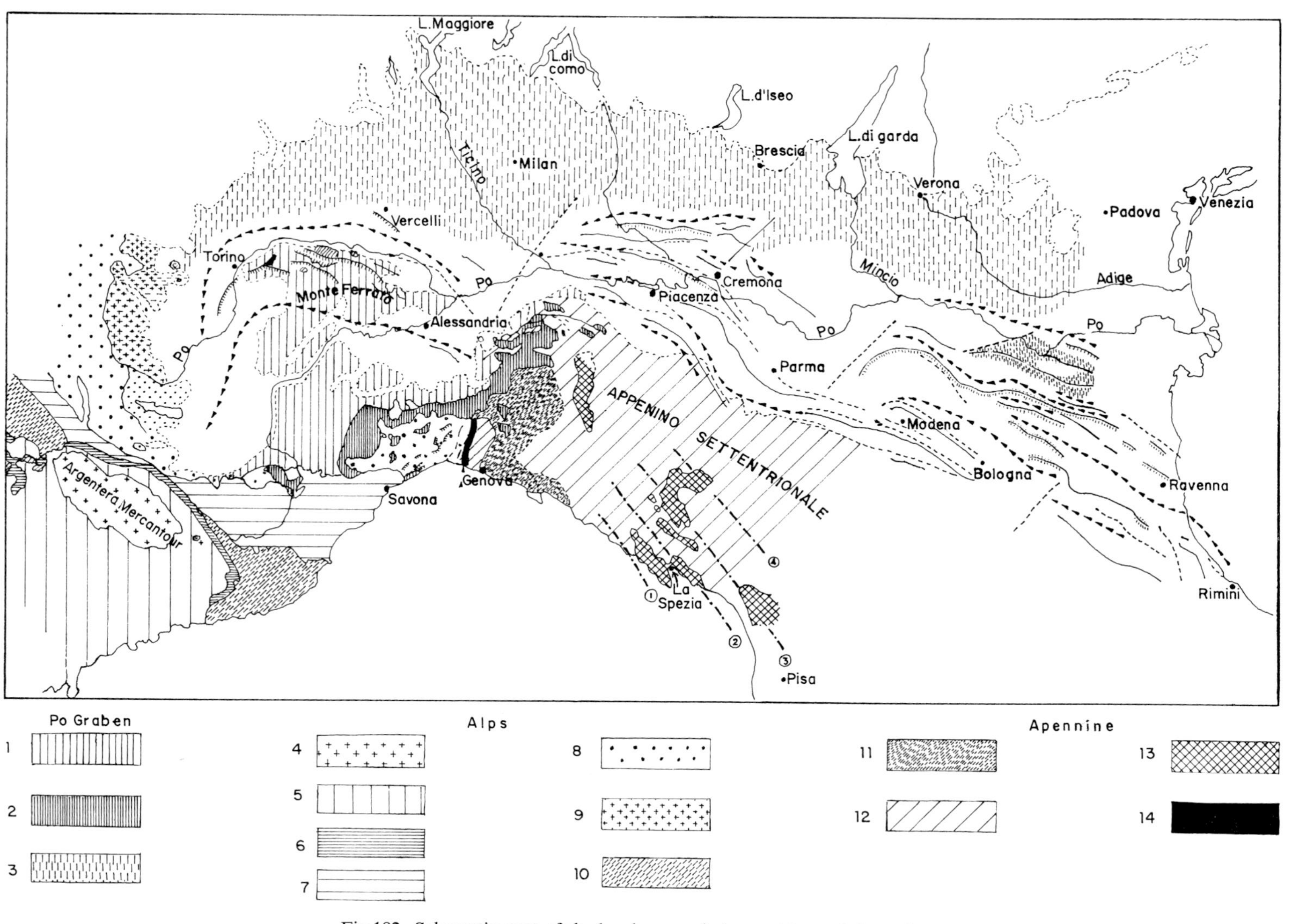

Fig.182. Schematic map of the border zone between Alps and Apennines.

1–3 = Po Basin; *1* = Neogene; *2* = Oligocene; *3* = "Pede Alpine" zone; *4* = Central Massifs; *5* = sedimentary cover of the external zones of the French Alps; *6* = Subbriançonnais zone; *7* = Briançonnais zone; *8* = "schistes lustrés"; *9* = crystalline (Dora Maira Massif); *10* = *Helminthoïdes* flysch nappes of the Piemonte zone; *11* = *Helminthoïdes* flysch; *12, 13* = other units of the Apennines; *14* = zone of Sestri–Voltaggio. (After LANTEAUME et al., 1963.)

that in recent years the zone of Sestri–Voltaggio west of Genoa has become quite generally accepted as the dividing line between Alps and Apennines (cf. BONI, 1963; Fig.182). This zone is formed by a number of strongly tectonized, subvertical units (GÖRLER and IBBEKEN, 1964). It forms a major crustal cicatrice, in a way comparable to the "root zone" of the Alps.

SCHEMATIC HISTORY OF THE FRENCH ALPS

The history of the French Alps, starting with the Lias, has been summarized by DEBELMAS and LEMOINE (1964; cf. Fig.183, which is self-explanatory). A much more detailed account can be found in DEBELMAS (1964).

In the series of schematic genetic sections of Fig.183 the external border of the Briançonnais zone is kept at the same position. It seems that both the external and the internal zones have moved in on this line, due to shortening of the epidermis of sediments through folding and nappe formation. This is, of course, only an artifice used in drawing this series of genetic sections. DEBELMAS and LEMOINE (1964) express no opinion as to how far, and as to which zones of the present structure of the Alps have moved in relation to, say, the stable continent of Meso-Europe.

Such opinions have been repeatedly given, the latest being the quite elegant solution offered by GOGUEL (1964). But as long as we have no paleomagnetic analysis of at least some parts of the French Alps, such reconstructions are premature.

CORRELATIONS

We now have to study the homologies between the various units distinguished in the Swiss and French Alps.

Starting from the internal side, we saw how in the French — and Italian — Alps, neither the equivalent of the Southern Alps, nor that of the "root zone" is present. This is due to the later, post-orogenetic subsidence of the Po Basin, which has affected these zones from the area of Ivrea southwards. They may well be present at depth, but this remains conjectural. Moreover, with figures of between 10 km and 20 km subsidence given for the Po Basin, one cannot but wonder at the amount of metamorphism these units will have undergone at any rate.

For the next zone, the Pennides and the Piemonte zone, there is general acceptance of homology, both as to stratigraphy and structure.

The Briançonnais zone, on the other hand, cannot be properly followed into Switzerland. Very schematically it seems that this separate unit of the French Alps becomes either involved in, or overriden by the Pennide nappe system.

For the Dauphinois zone, on the other hand, one commonly finds stated that it is the equivalent of the Swiss Helvetides. This comparison breaks down, however, on a number of counts. The stratigraphy of the Mesozoic of both units is quite similar, but the Dauphinois zone has no flysch to speak of. Moreover, it became folded at an earlier date, during the Paleocene, and did not develop nappe structures.

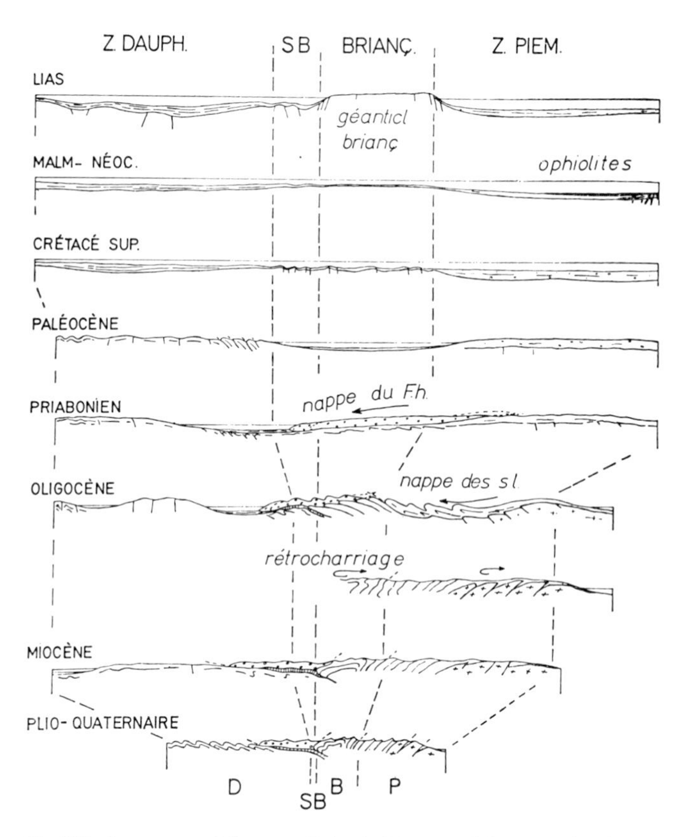

Fig.183. Summary of the post-Triassic history of the French Alps.
During the Lias a marked difference existed between the external basin of sedimentation, in which the Dauphinois facies developed, and the internal basin with the "schistes lustrés" of the Piemonte zone. These basins were separated by the so-called Briançonnais geanticline.
This distinction persisted all through the Jurassic and the Cretaceous. In the Piemonte zone ophiolites were intercalated in the "schistes lustrés" during the Cretaceous, and the *Helminthoïdes* flysch was formed during the Late Cretaceous.
During the Paleocene the Dauphinois zone was folded, while the Briançonnais zone, and of course also the Ultra-dauphinois zone, subsided, receiving flysch type sediments.
During the Late Eocene the Pennides rose far enough to cause the higher part of the sedimentary pile to slide down and form the *Helminthoïdes* flysch nappe (*F.h.*). This was followed by the "schistes lustrés" (*s.l.*) nappe(s) and accompanied by the tectonization of the Briançonnais zone. Rising of the area between the Dauphinois zone and the Briançonnais zone then led to backthrusting in the eastern part of the Briançonnais zone.
Further rise during the Neogene and the Quaternary led to intensive erosion to form the present picture.
(After DEBELMAS and LEMOINE, 1964.)

The similarity of the stratigraphy of the Mesozoic has led to a certain mixing of stratigraphical and structural criteria. This is a typical result of the Argand maxim that in an orogeny of Alpine type structure must always be related to facies.

The fallacy of this tenet becomes immediately evident when we bear in mind that Mesozoic stratigraphy is comparable not only in the Helvetides and the Dauphinois zone, but also in the Jura Mountains, and even, to some extent, in the younger basins, such as the eastern rime of the Paris Basin and the Basin of Swabia in southern Germany.

Broad areas of similar stratigraphic development of the Mesozoic consequently have either remained practically undisturbed, or were folded without nappe formation, such as in the Jura Mountains and in the Dauphinois zone, or did develop nappe systems, such as in the Helvetides.

Structurally speaking, both as to time of folding, the absence of flysch and the tectonic style produced, the Dauphinois zone shows much better correlations with the Jura Mountains, than with the Helvetides. A case can be made for a distinction of all three units as separate entities, within the major context of the Alpine orogeny.

This way of thinking incidentally disposes of an old controversy, namely where to draw the boundary between the Jura Mountains and the Dauphinois zone. As is evident from Fig.154, a couple of the westerly structures of the Dauphinois zone can be followed north of the Isère river into the Jura Mountains. Structurally and stratigraphically there are no sharp limits. The importance of the Cretaceous gradually diminishes northwards, until in the Jura mountains the structures are mostly formed by the Jurassic.

Towards the northeast, however, there is a sharp tectonic limit between Dauphinois zone and the southernmost nappes of the Helvetides.

REFERENCES

Barbier, R., 1948. Les zones ultradauphinoises et subbriançonnaises entre l'Arc et l'Isère. *Mém. Carte Géol. France*, 1948: 291 pp.

Barbier, R., 1956. L'importance de la tectonique "anténummulitique" dans la zone ultradauphinoise au nord du Pelvoux: La Chaîne Arvinche. *Bull. Soc. Géol. France*, 6 (6): 355–370.

Barbier, R., 1963. Réflexions sur la zone dauphinoise et la zone ultradauphinoise. In: M. Durand Delga (Editor), *Livre à la Mémoire du Professeur Paul Fallot. II.* Soc. Géol. France, Paris, pp.321–329.

Barbier, R., Bloch, J. P., Debelmas, J., Ellenberger, F., Fabre, J., Feys, R., Gidon, M., Goguel, J., Gubler, Y., Lanteaume, M., Latreille, M. and Lemoine, M., 1963. Problèmes paléogéographiques et structuraux dans les zones internes des Alpes occidentales entre Savoie et Méditerranée. In: M. Durand Delga (Editor), *Livre à la Mémoire du Professeur Paul Fallot. II.* Soc. Géol. France, Paris, pp. 331–377.

Bellair, P., 1948. Pétrographie et tectonique des Massifs Centraux de la France. 1. Le Haut Massif. *Mém. Carte Géol. France*, 1948: 345 pp.

Bloch, J. P., 1965. Nappe de cisaillement et érosion précoce dans les Alpes ligures. *Compt. Rend. Acad. Sci., Paris*, 260: 4016–4019.

Boni, A., 1963. Lignes et problèmes tectoniques du secteur nord-ouest de l'Apennin septentrional. *Bull. Soc. Géol. France*, 7 (4): 644–656.

Bordet, P., 1950. Le Dôme permien de Barrot (A. M.) et son aureole de terrains secondaires. *Bull. Serv. Carte Géol. France*, 228: 89 pp.

Bordet, P. and Bordet, C., 1963. Belledonne–Grandes Rousses et Aiguilles Rouges–Mont Blanc. Quelques données nouvelles sur leurs rapports structuraux. In: M. Durand Delga (Editor), *Livre à la Mémoire du ProfesseurPaul Fallot. II.* Soc. Géol. France, Paris, pp. 309–316.

Conti, S., 1953. Studi geologici sulle Alpi occidentali. 1, 2. *Bol. Serv. Geol. Italia*, 75: 491–525.

Conti, S., 1955. Studi geologici sulle Alpi occidentali. 3. *Bol. Serv. Geol. Italia*, 77: 275–326.

Debelmas, J., 1955. Les zones subbriançonnaises et briançonnaises occidentale entre Vallouise et Guilestre (Hautes Alpes). *Mém. Carte Géol. France*, 1955: 164 pp.

Debelmas, J., 1964. Essai sur le déroulement du paroxysme alpin dans les Alpes franco-italiennes. *Geol. Rundschau*, 53: 133–152.

Debelmas, J. and Lemoine, M., 1964. La structure tectonique et l'évolution paléogéographique de la chaîne alpine d'après les travaux récents. *Inform. Sci.*, Paris, 1: 1–33.

Debelmas, J., Gidon, M. and Haccard, D., 1965. *Réunion extraordinaire de la Société géologique de France, Briançonnais (Hautes Alpes). Compt. Rend. Soc. Géol. France*, 1964, pp.433–471.

De Lapparent, A., 1938. Études géologiques dans les régions provençales et alpines entre le Var et la Durance. *Bull. Serv. Carte Géol. France*, 198: 31 pp.

Ellenberger, F., 1958. Étude géologique du pays de Vanoise. *Mém. Carte Géol. France*, 1958: 545 pp.

FABRE, J., 1961. Contribution à l'étude de la zone houillère en Maurienne et en Tarentaise (Alpes de Savoie). *Mém. Bur. Rech. Géol. Minières*, 2: 315 pp.

FALLOT, P., 1949. Les chevauchements intercutanés de Roya (A. M.). *Ann. Hébert Haug (Lab. Géol. Fac. Sci. Univ. Paris)*, 7: 161–170.

FAURE-MURET, A., 1955. Études géologiques sur le massif de l'Argentera–Mercantour et ses enveloppes sédimentaires. *Mém. Carte Géol. France*, 1955: 336 pp.

FEYS, R., 1963. Étude géologique du Carbonifère briançonnais (Hautes Alpes). *Mém. Bur. Rech. Géol. Minière*, 6: 388 pp.

FEYS, R., GREBER, CH., DEBELMAS, J., LEMOINE, M. and FABRE, J., 1964. Bassin houiller briançonnais. *Congr. Intern. Stratigraph. Géol. Carbonifère, Compte Rendu, 5, Paris, 1963*, 1: 93–116.

GÈZE, B., 1963. Caractères structuraux de l'Arc de Nice. In: M. DURAND DELGA (Editor), *Livre à la Mémoire du Professeur Paul Fallot. II.* Soc. Géol. France, Paris, pp. 289–300.

GIDON, M., 1962. A propos de l'éventail briançonnais. *Compt. Rend. Soc. Géol. France*, 1962: 12–13.

GIDON, M., 1966. La zone briançonnaise en Haute-Ubaye (B. A.) et son longement au Sud-Est. *Mém. Carte Géol. France*, 1966.

GIGNOUX, M., 1950. *Géologie Stratigraphique*, 4th ed. Masson, Paris, 735 pp.

GIGNOUX, M. and MORET, L., 1944. *Géologie Dauphinoise ou Initiation à la Géologie par l'Étude des Environs de Grenoble.* Masson, Paris, 391 pp.

GOGUEL, J., 1936. Description tectonique de la bordure des Alpes de la Bléone au Var. *Mém. Carte Géol. France*, 1936: 360 pp.

GOGUEL, J., 1939. Tectonique des chaînes subalpines entre la Bléone et la Durance. *Bull. Carte Géol. France*, 202: 48 pp.

GOGUEL, J., 1948. Recherches sur la tectonique des chaînes subalpines entre le Ventoux et le Vercors. *Bull. Serv. Carte Géol. France*, 223: 46 pp.

GOGUEL, J., 1953. *Géologie Régionale de la France, 8. Les Alpes de Provence.* Hermann, Paris, 124 pp.

GOGUEL, J., 1963. Les problèmes des chaînes subalpines. In: M. DURAND DELGA (Editor), *Livre à la Mémoire du Professeur Paul Fallot. II.* Soc. Géol. France, Paris, pp. 301–307.

GOGUEL, J., 1964. L'interprétation de l'arc des Alpes occidentales. *Bull. Soc. Géol. France*, 7 (5): 20–33.

GOGUEL, J., 1965. *Traité de Tectonique*, 2nd ed. Masson, Paris, 457 pp.

GÖRLER, K. and IBBEKEN, H., 1964. Die Bedeutung der Zone Sestri–Voltaggio als Grenze zwischen Alpen und Apennin. *Geol. Rundschau*, 53: 73–83.

HAUDOUR, J. and SARROT-REYNAULD, J., 1964. Le Carbonifère des zones externes des Alpes françaises. *Congr. Intern. Stratigraph., Géol. Carbonifère, Compte Rend., 5, Paris, 1963*, 1: 119–173.

HAUG, E., 1921. *Traité de Géologie.* Colin, Paris, 4 vol., 2024 pp.

KUENEN, PH. H., FAURE-MURET, A., LANTEAUME, M. and FALLOT, P., 1957. Observations sur les flysch des Alpes maritimes françaises et italiennes. *Bull. Soc. Géol. France*, 6 (7): 11–26.

LANTEAUME, M., 1963. Considérations paléogéographiques sur la patrie supposée des nappes de Flysch à Helminthoïdes des Alpes et des Apennins. *Bull. Soc. Géol. France*, 7 (4): 625–643.

LANTEAUME, M., HACCARD, D., LABESSE, B. and LORENZ, C., 1963. L'origine de la nappe du Flysch à Helminthoïdes et la liaison Alpes–Apennins. In: M. DURAND DELGA (Editor), *Livre à la Mémoire du Professeur Paul Fallot. II.* Soc. Géol. France, Paris, pp. 257–272.

LEMOINE, M., 1953. Remarques sur les caractères de l'évolution de la paléogéographie de la zone briançonnaise au Secondaire et au Tertiaire. *Bull. Soc. Géol. France*, 6 (3): 105–122.

LEMOINE, M., 1961. La marge externe de la fosse piémontaise dans les Alpes Occidentales. *Rev. Géograph. Phys. Géol. Dyn., Sér. 2*, 4: 163–180.

LEMOINE, M., 1964. Le problème des relations des schistes lustrés piémontais avec la zone briançonnaise dans les Alpes cottiennes. *Geol. Rundschau*, 53: 113–132.

MENNESSIER, G., 1964. Sur l'évolution tectonique et morphologique des chainons externes de l'Arc de Castellane entre le Verdon et la Siagne (Haute Provence). *Rev. Géograph. Phys. Géol. Dyn., Ser. 2*, 6: 91–113.

MICHARD, A., 1963. Sur quelques aspects de la zonéographie alpine dans les Alpes cottiennes méridionales. *Bull. Soc. Géol. France*, 7 (4): 477–491.

MICHARD, A. and STURANI, C., 1964. La zone piémontaise dans les Alpes cottiennes du Cuneese: nouveaux résultats et nouvelles questions. *Compt. Rend. Soc. Géol. France*, 1964: 382–385.

RUTTEN, M. G., 1949. Actualism in epeirogenetic oceans. *Geol. Mijnbouw*, 11: 222–228.

SCHNEEGANS, D., 1938. La géologie des nappes de l'Ubaye-Embrunais entre la Durance et l'Ubaye. *Mém. Carte Géol. France*, 1938: 389 pp.

SCHUILING, R. D., 1957. Jointing in the Permian Dome de Barrot, S. France. *Geol. Mijnbouw*, 18: 227–234.

SOCIÉTÉ GÉOLOGIQUE DE FRANCE, 1965. Réunion extraordinaire. Briançonnais (Hautes Alpes). *Compt. Rend. Soc. Géol. France*, 1965: 433–471.

REFERENCES

Stanley, D. J. and Bouma, A. H., 1964. Methodology and paleogeographic interpretation of flysch formations: a summary of study in the Maritime Alps. In: A. H. Bouma and A. Brouwer (Editors), *Turbidites*. Elsevier, Amsterdam, pp. 34-63.

Termier, P., 1896. Sur la tectonique du massif du Pelvoux. *Bull. Soc. Géol. France*, 3, 24: 734–757.

Tobi, A. C., 1959. Petrographical and geological investigations in the Merdaret–Lac Crop region (Belledonne Massif, France). *Leidse Geol. Mededel.*, 24: 181–281.

Trümpy, R., 1960. Paleotectonic evolution of the Central and the Western Alps. *Bull. Geol. Soc. Am.*, 71: 843–908.

Valérien, J., Debelmas, J., Lemoine, M. and Ministère des Armées, 1964. *Du Pelvoux au Viso — Un Survol Géologique des Alpes*. Serv. Film Rech. Sci., Off. Natl. Univ. Écol. Franç., Paris, 16 mm, couleurs, sonore optique, 351 m, 32 min. (English version available.)

Vernet, J., 1951. Le synclinorium de l'Aiguille de Morges et le style de déformations alpines du Cristallin de Pelvoux. *Bull. Soc. Géol. France*, 6 (1): 169–183.

Vernet, J., 1952. Les déformations d'age alpin du cristallin du Pelvoux à la lumière d'observations nouvelles. *Bull. Soc. Géol. France*, 6 (2): 175–190.

Vernet, J., 1954. Idées nouvelles sur la tectonique du Massif du Pelvoux et de ses annexes (Dome Haut-Dauphinois). *Bull. Soc. Études Hautes-Alpes*, 46: 1–7.

Vernet, J., 1965. La zone "Pelvoux–Argentera". Études sur la tectonique alpine du socle dans la zone des massifs cristallins externes du Sud des Alpes occidentales. *Bull. Serv. Carte Géol. France*, 275: 294 pp.

CHAPTER 13

Alpine Europe: Eastern and Southern Alps

INTRODUCTION

It is only for the sake of a convenient subdivision, that the Eastern and Southern Alps have been combined into one chapter. For, although there are some resemblances, notably in the development of their Alpine Triassic, they are quite different tectonically, and separated by the "root zone". Moreover, though the broadest development of the Southern Alps is indeed situated south of the Eastern Alps, the former can, as we saw, be followed much further westward, to the Lugano area, south of the Swiss Pennides (Fig.184).

EASTERN ALPS

The Eastern Alps are largely composed of the Austride nappes and by the two tectonic windows of the Engadin and the Tauern, in which the underlying Pennide nappes crop out. But in addition, a narrow zone of foothills is found, all along their northern front, which formerly was correlated with the Swiss Helvetide nappes. The reasons being that, on the one hand it consists mainly of flysch, and on the other that a nappe system present at one cross-section just had to be present anywhere else along the fold chain. It will be clear from the following that this correlation is no longer valid, but it is a thing to remember, because it is indicated as such on all older maps.

In the Eastern Alps we are in the exceptional position of having both a modern geological map 1:1,000,000 (BECK-MANNAGETTA, 1964) with explanatory text (BECK-MANNAGETTA et al., 1966) and an introduction to the geology of Austria, with a translation into both English and French (EXNER, 1966). Moreover there are several detailed guide books of field trips. The best and most recent of these has been published at the convention of the Deutsche Geologische Gesellschaft in Vienna in September 1964 (BRAUMÜLLER et al., 1964).

Moreover, a new insight into the structure of the northern border of the Alps, and notably on the amount of overthrusting of the Alps over the molasse basins, has in recent years been gained from the exploration for oil (BRIX and GÖTZINGER, 1964; KAPOUNEK et al., 1965). After having started in the molasse basins, exploratory drilling has been extended to the flysch zone, and will in future even englobe the Northern Limestone Alps.

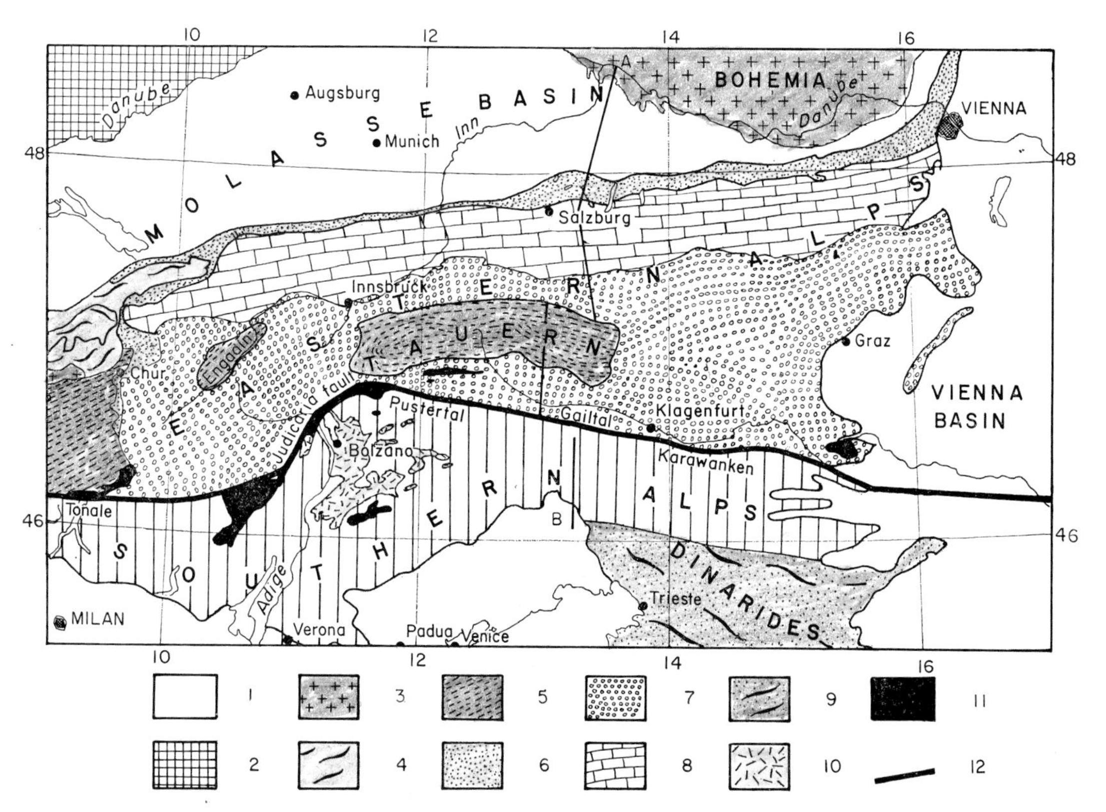

Fig.184. Outline map of the Eastern and Southern Alps. 1:5.000.000.
1 = molasse basins; *2* = Jurassic of Swabia; *3* = Hercynian uplands of Bohemia and eastern tip of the Aar Massif; *4* = eastern end of the Helvetide nappes; *5* = Pennide nappes; *6* = northern border zone; flysch with elements of the Helvetides in the west, of the Pienide "Klippen" in the east; *7* = "Zentralalpen" of the Eastern Alps; *8* = Northern Limestone Alps; *9* = Dinarides; *10* = Bolzano Permian ignimbrites; *11* = Tertiary and Quaternary igneous rocks; *12* = "root zone". *A–B* = location of section of Fig.186.

Moreover, apart from the review of the stratigraphy of the Eastern Alps in the *Lexique Stratigraphique International* (KÜHN, 1962), we have at hand the modern synopsis of the Tertiary by JANOSCHEK (1964), of the Mesozoic by ZAPFE (1964) and the Paleozoic by FLÜGEL (1964), and in addition the micropaleontological summary by OBERHAUSER (1963). For a more thorough treatment of the subject, the reader is referred to SCHAFFER (1951).

The Eastern Alps being well served with introductory texts, I will consequently confine myself to submitting a general outline, completed by a discussion of several controversial points.

At present the main controversy in the Eastern Alps is not so much a question of the formation of the Alps, or a question of tangential compression versus gravity tectonics, but is the interpretation and subdivision of the Austrides.

Geotectonic theories, such as those of Kraus or Van Bemmelen, cited in Chapter 10, do not at the moment, it seems, interest the minds of geologists very much. FALLOT's (1960) considered objections to the theory of Kraus have not convinced the latter author, nor his followers, but on the other hand the newer publications by KRAUS (cf. 1966) have also not made much impact. For Van Bemmelen's theory we find a thoughtful review by KÜPPER (1966), who accepts most of his gravity tectonics, but not his geotumors. But

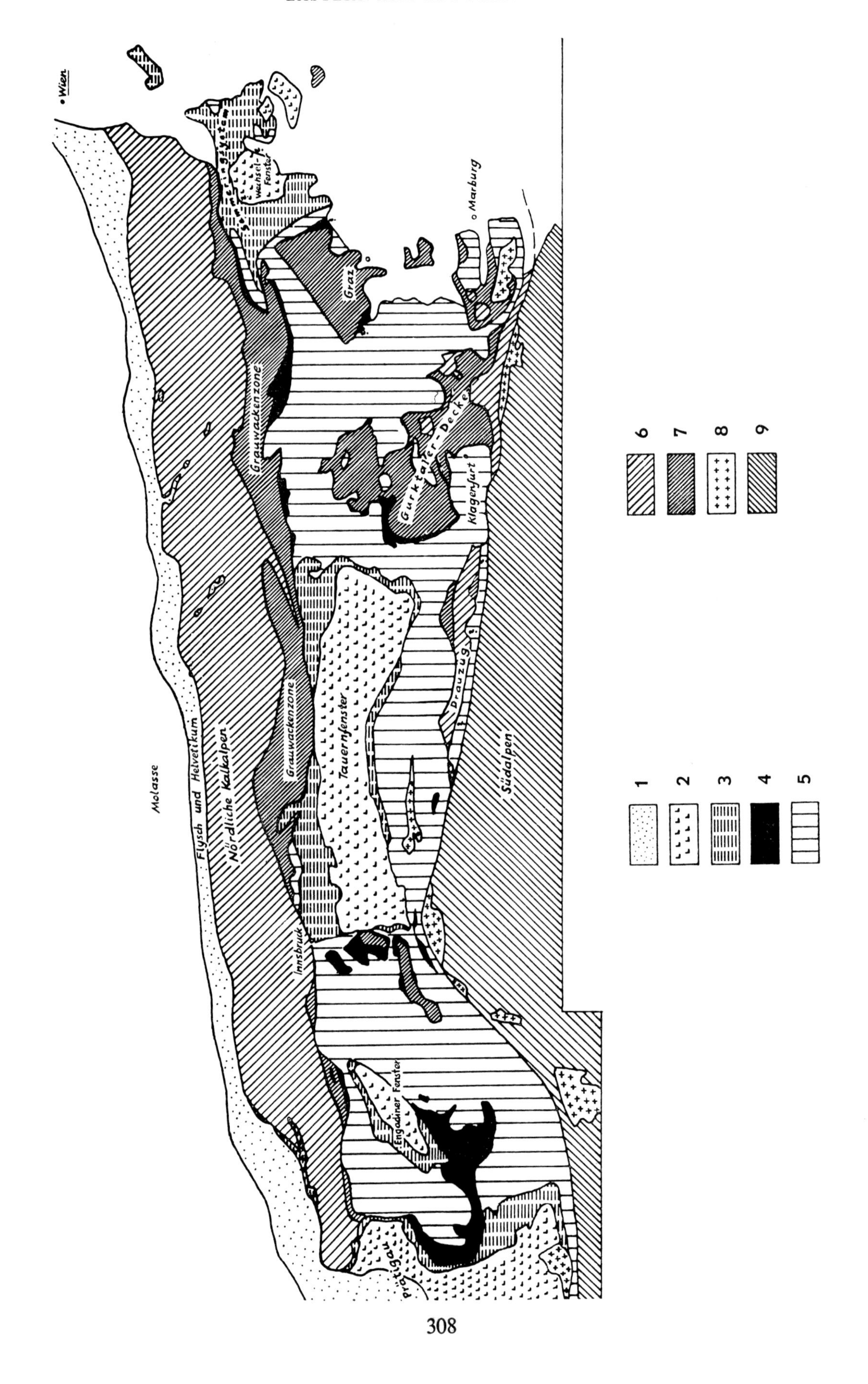

Wien
Semmeringsystem
Wechsel-Fenster
Marburg
Graz
Grauwackenzone
Gurktaler-Decke
Klagenfurt
Drauzug
Südalpen
Tauernfenster
Grauwackenzone
Molasse
Flysch und Helvetikum
Nördliche Kalkalpen
Innsbruck
Engadiner Fenster
Prätigau
1
2
3
4
5
6
7
8
9

Fig.185. Map of the Eastern Alps, with the main units according to Tollmann.

1 = Helvetide nappes, continuing as the northern border zone, comprising Helvetide elements, flysch and Pienide "Klippen"; *2* = Pennides; *3* = lower Austrides; *4* = Permo-Mesozoic of the middle Austrides; *5* = crystalline basement of the middle Austrides; *6* = Mesozoic of the upper Austrides; *7* = Paleozoic of the upper Austrides; *8* = peri-Adriatic younger intrusives; *9* = Southern Alps.

In this interpretation the Prätigau flysch of northeastern Switzerland (pp. 259–260), is considered to belong to the Pennides. In other interpretations, such as CLAR (1965) the flysch of the northern border zone is considered to belong to the Pennides too.

In the Austrides Tollmann's interpretation of three nappe systems is opposed to that of most older geologists, who only accept two divisions. In their view the Mesozoic of the "Zentralalpen" (*4*) is the (par) autochthonous equivalent of that of the Northern Limestone Alps (*6*), whereas the Paleozoic of the graywacke zone (*7*) represents the non-metamorphic part of their basement (*5*). (After TOLLMANN, 1965.)

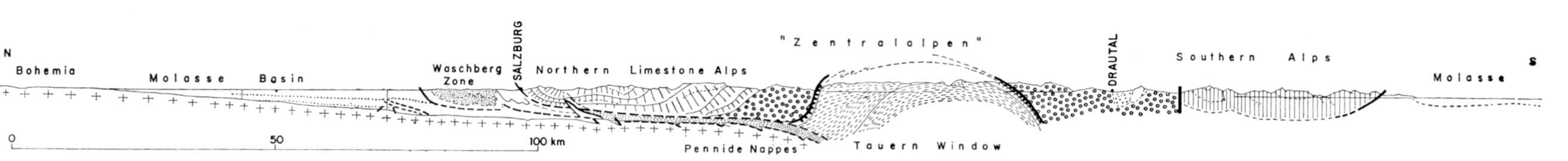

Fig.186. Schematic section across Eastern and Southern Alps. (After H. Küpper, in: BECK-MANNAGETTA, 1964; EXNER, 1966.)

in the main we are under the impression that most geologists at present tend to leave these theories alone.

The question of tangential push versus gravity tectonics has been much less discussed for the Eastern Alps, than in the more westerly parts of the Alpine chain. The reason might well lie in the fact that the tectonic style of the Austrides offers less oppoıtunity for an evaluation of this controversy.

But the subdivision of the Austrides is a hot question indeed. Two groups are formed by, on the one side the more conservative, and often also the older, Austrian geologists, and on the other by Dr. A. Tollmann of Vienna and his school. The difference narrows down the question to whether the Austrides are built up by two, or by three nappe systems. That is, if they comprise only lower and upper units ("Unterostalpin" and "Oberostalpin") or lower, middle and upper units ("Unterostalpin", "Mittelostalpin" and "Oberostalpin") (pp.315–317; Fig.185, 186).

Other controversies are: (*1*) the correlation of the various elements found in the northern border zone; (*2*) the question of autochthony or nappe structure of the Northern Limestone Alps; and (*3*) the history of the "root zone", here represented by a linear succession of more or less individual elements, the Lienzer Dolomiten and the Gailtaler Alpen, which together form the Drauzug and the Karawanken.

THE NORTHERN BORDER ZONE

As stated already, this narrow zone along the northern border of the Northern Limestone Alps, also called flysch zone, and formerly interpreted as the easterly continuation of the Helvetides, is of complex structure.

Starting in the west, we find that in northeastern Switzerland the Helvetides, together with the Aar and Gotthard Massifs, show a pronounced easterly plunge in approaching the Rhine valley north of Chur. Only the highest unit, the Säntis nappe, is not affected by this axial plunge, and continues 40 km further eastwards, in front of the Northern Limestone Alps, before terminating as a tectonic unit.

Up to the region of Salzburg one still finds elements of limestone, comparable to the Helvetides. These are surrounded by marly sediments, often richly fossiliferous, the "Buntmergel" or variegated marls, which represent a more pelagic facies. But their structure is quite different. Subhorizontal nappe units are no longer found, but instead a succession of long, thin, imbricate slices in subvertical position (Prey, 1962; Janoschek, 1964).

East of Salzburg the northern border zone is mainly formed by clastic, detrital sediments of flysch type. Overridden by the flysch, and in tectonic windows, the variegated marls are, however, still found.

This flysch series of the eastern part of the northern border zone has no great affinities with the Helvetides to the west. Instead, it can be correlated directly eastward with the broad zone of flysch nappes comprising the external part of the Carpathians.

Within the flysch mentioned above, other Carpathian elements occur, the so-called "Klippen" zone of Gersten and St. Veith, which can be correlated with the Pienide "Klippen" of the Carpathians (Fig.187). It must be stressed that the name of this zone is completely misleading. The Pienide "Klippen" are not erosional remnants of a higher, subhorizontal nappe, such as the Swiss "Klippen". Instead, they are a narrow, strongly tectonized

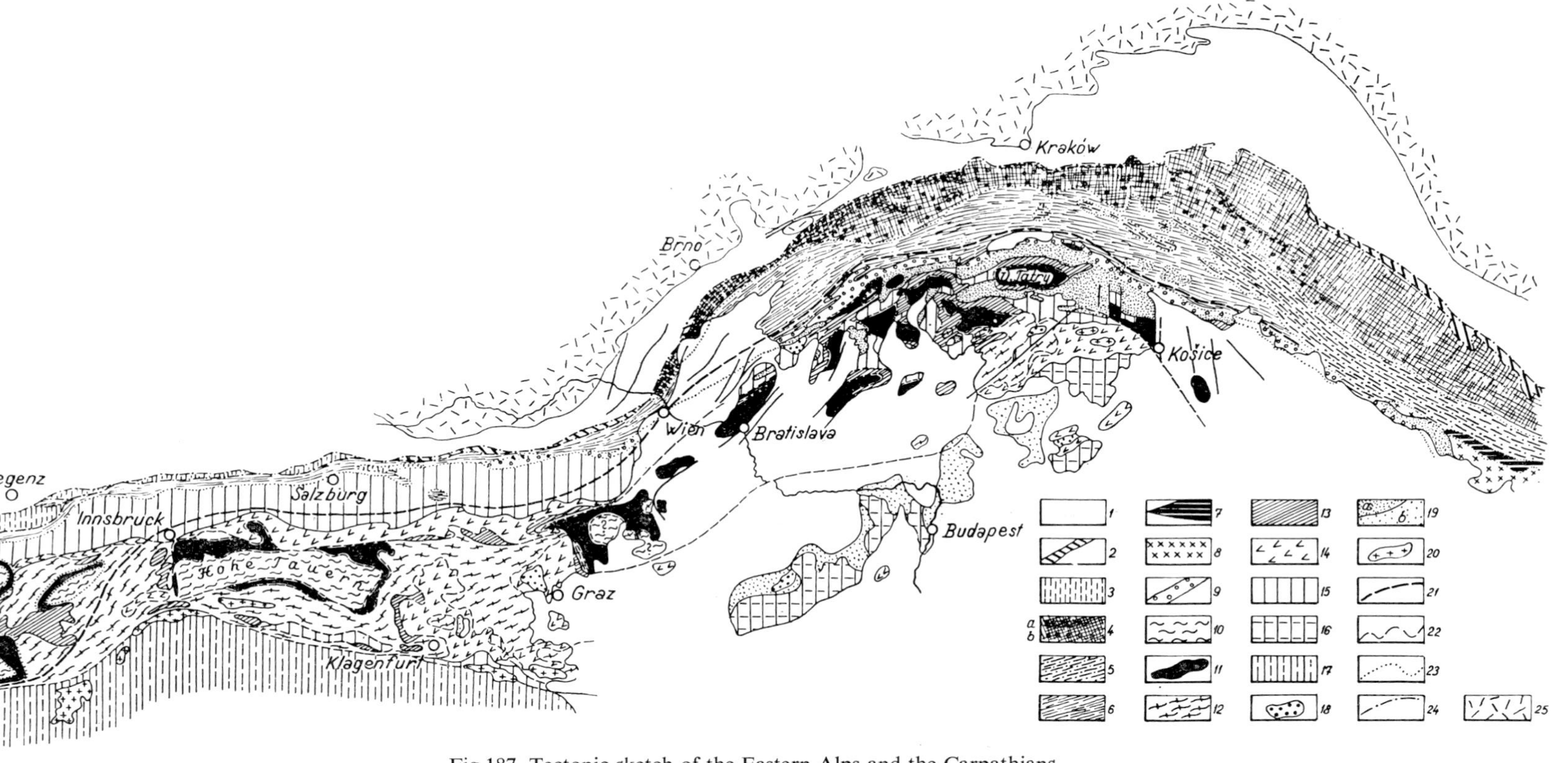

Fig.187. Tectonic sketch of the Eastern Alps and the Carpathians.

1 = molasse along the Alpine front; *2* = tectonized molasse along the Alpine front (e.g., Waschberg zone); *3* = Helvetides; *4–8* = external flysch nappes of the Carpathians (*4a* = Frydek nappe; *4b* = Silesian nappe; *5* = Dukla zone; *6* = Magura zone; *7* = Rachow–Sinaja zone of the eastern Carpathians; *8* = Marmaros zone); *9* = Pennide Klippen zone; *10* = Pennide nappes; *11* = lower Austrides of the Eastern Alps, Tatrides of the Carpathians; *12–13* = middle Austrides and Veporides of the Carpathians; (*12* = pre-Carboniferous crystalline basement; *13* = Mesozoic) *14–15* = upper Austrides, and Choc nappe and northern Gemeride nappes of the Carpathians (*14* = Paleozoic; *15* = Mesozoic); *16* = Mesozoic of the southern Gemerides; *17* = Southern Alps; *18* = Senonian; *19* = Paleogene; *20* = younger intrusives; *21* = deep-seated overthrust; *22* = pre-Gosau overthrusts; *23* = post-Eocene overthrusts; *24* = inter-Miocene overthrusts; *25* = Hercynian uplands north of the Alps and the Carpathians. (After ANDRUSOV, 1963.)

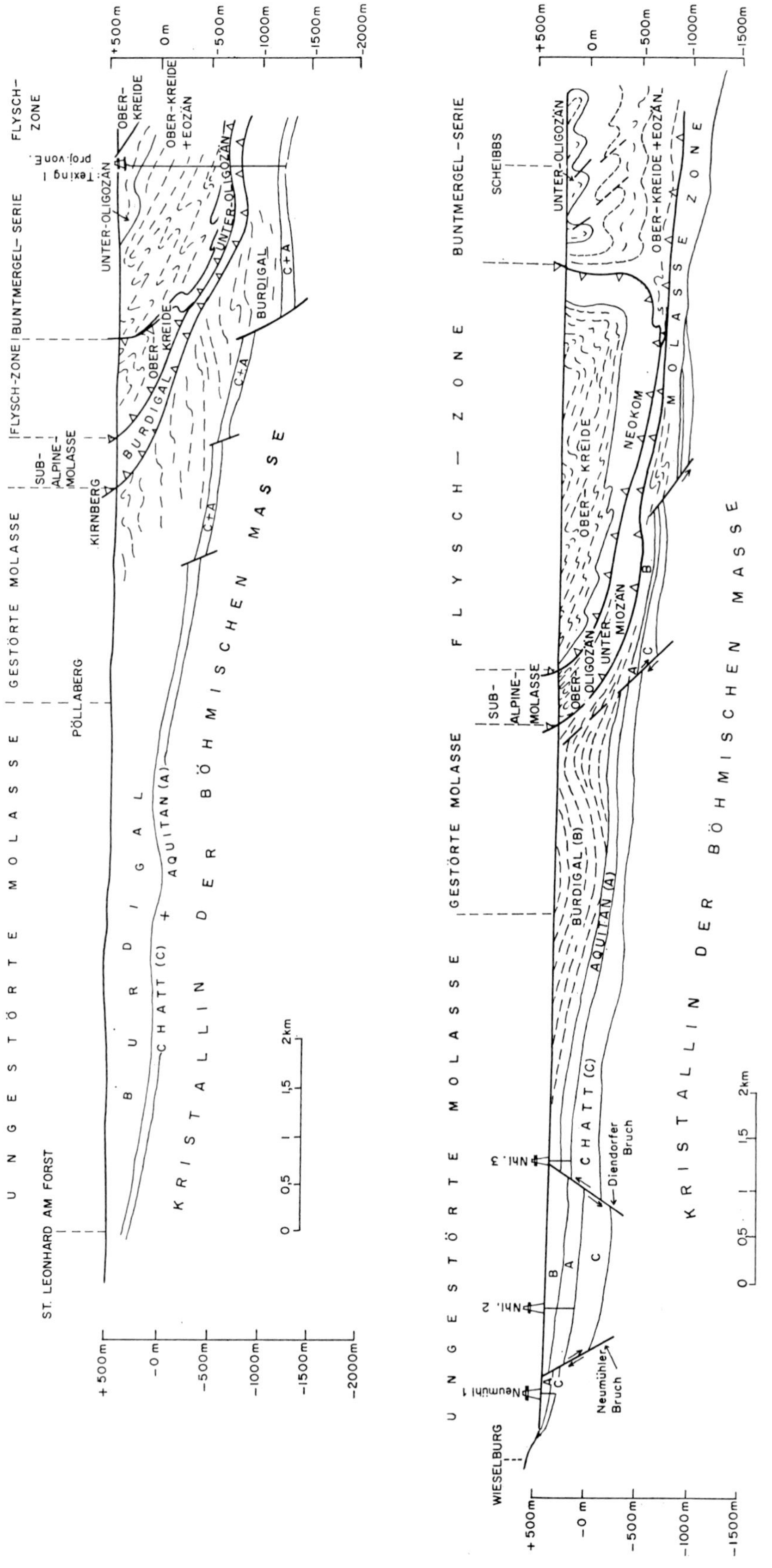

Fig.188. Two sections across the molasse basin, the tectonized Subalpine molasse and the northern border zone, west of Vienna.

The crystalline basement, which crops out in the Hercynian upland of Bohemia, can be followed all the way beneath the molasse basin and the northern border zone (After BRIX and GÖTZINGER, 1964.)

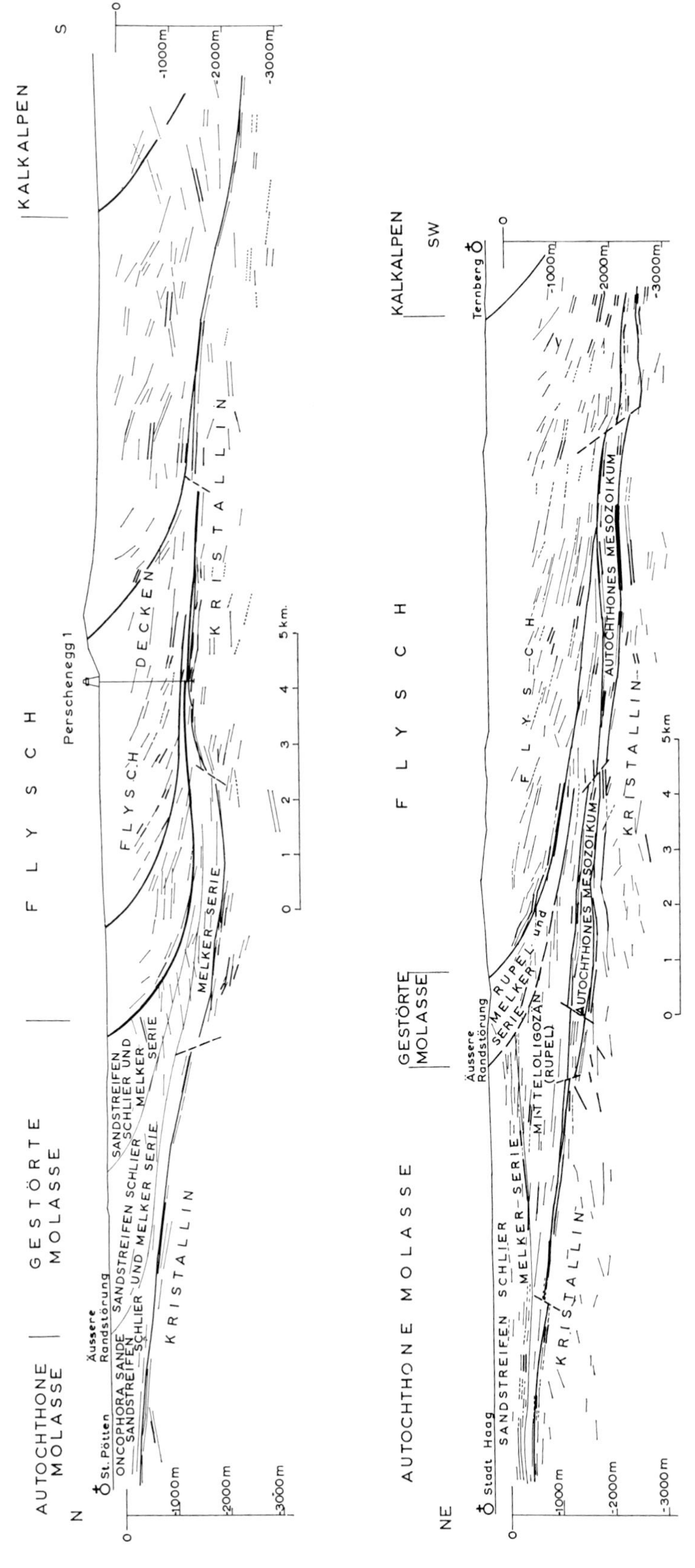

Fig.189. Two sections across the southern border of the molasse basin and the northern border zone of the Alps, showing actual reflections. The upper section is about 50 km, the lower section about 130 km west of Vienna. (After KAPOUNEK et al., 1965.)

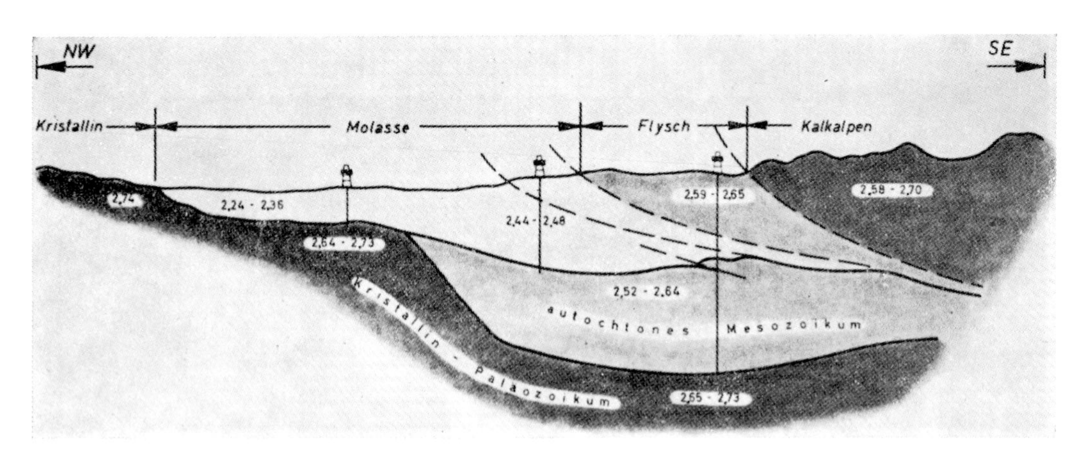

Fig.190. Schematic section from the Hercynian upland of Bohemia, across the molasse basin to the Northern Limestone Alps, showing velocity distribution. (After KRÖLL, 1964.)

zone, in which all kinds of elements of various facies are aligned in a subvertical position. In the Carpathians they form a dividing line between the external flysch nappes and the internal nappes which can be correlated with the Austrides. They have all the earmarks of a major transcurrent fault.

If, as is proposed by several authors, one wishes to correlate the flysch of the northern border zone with the Pennides, via the Tauern and Engadin windows, this is an example of cylindrism well worth noting. But in any case, it is not permissible to correlate all of the northern border zone with either the Helvetides (cf. TOLLMANN, 1965), or with the Pennides (cf. CLAR, 1965). The complex character of this zone should be duly taken note of. D. Andrusov's sketch of Fig.187 is already much better.

Turning now to the structure of the northern border zone, we learn from Fig.188, 189, based on the recent drilling programs for oil exploration, that it is strongly tectonized, and in the main overthrust towards the north.

Starting from the north, we find the southward dipping post-Hercynian peneplain on top of the Bohemian Massif covered by autochthonous Mesozoic and molasse. Near the front of the Alps the molasse has become tectonized and overthrust northwards. This southern zone of the molasse basin is called either the Subalpine molasse, or the Waschberg zone. On top of the Waschberg zone follows the system of imbricate slices of flysch and Pienide "Klippen", and on top of that, in about the same position, the overthrust masses of the Northern Limestone Alps.

Similar sections for the Vienna region can be found in KÜPPER (1965), and a discussion of the velocity distribution is given by KRÖLL (1964, see Fig.190).

THE AUSTRIDE NAPPES

The Austride nappes, the "Ostalpine Decken", consist of three elements very different in stratigraphy and in structure. First, there are the Lower Austrides, the "Unterostalpin" formerly called the Grisonides, which occupy a sort of transition zone between the Pennides and the bulk of the Austrides, situated on the border between Switzerland and Austria. The bulk of the Austrides may then be divided, on purely lithological features, into a southern and a northern zone. The southern zone is formed by the predominantly crystalline "Zentralalpen", as opposed to the entirely sedimentary northern zone of the Northern Limestone Alps.

Taking the lower Austrides first, we find that this system is formed by a large number of strongly imbricate minor nappe structures. They consist mainly of schists which resemble the Bündner Schiefer, enveloping elements of the crystalline basement, and of Triassic in Alpine facies. It is difficult to draw an exact limit between the upper Pennide nappes, which, as we saw are also characterized by a strong development of Bündner Schiefer, and the lower Austride nappes. For further details the reader is referred to CADISCH (1953) and OBERHAUSER (1965). Of the various nappes (cf. Table XVIII, XIX) the Aroser "Schuppenzone", the imbricate Arosa zone, is perhaps the most typical element of the lower Austrides.

According to many authors the lower Austrides are not met with further east in the chain, not even around the Tauern window. FALLOT (1955) has used this interpretation to stress the point that even in the reconstructions of the most dogmatic of nappists, cylindrism evidently did not exist here. For it seems as if the lower Austrides had died out between eastern Switzerland and the Tauern window. A. Tollmann (cf. Fig.185) has in fact in later years correlated the zones surrounding the Tauern window, and also those surrounding the newly found window further east, the Wechsel window, with the lower Austrides. But I am not sure whether it is rather the dogma of nappe continuity than actual analogies, which has led him to this correlation.

TABLE XIX

SCHEMATIC PRESENTATION OF THE MAIN FACIES ELEMENTS OF THE TRIASSIC IN THE NORTHERN LIMESTONE ALPS AND IN THE SOUTHERN ALPS

(After GIGNOUX, 1955)

	Bavarian nappe	*Hallstatt nappe*	*Dachstein nappe*	*Southern Alps (Dolomites, Dinarides)*		
Norian	Main Dolomite	Hallstatt Limestone (deep-water facies)	Dachstein coral limestones or dolomites	Main Dolomite		
Carnian	beds with *Cardita* (littoral)		Raibl Beds with *Cardita*	Raibl Beds		
Ladinian	coral limestones of the Wetterstein Partnach Shales	dolomites or limestones with cephalopods	Ramsau Dolomite	marly facies (with tuffs): St. Cassian Wengen Buchenstein	coral facies: Limestone of Marmolata	Schlern Dolomite
Virglorian	Virgloria Limestones sandy marls with *Myophoria*			dolomites with diploporids sandy limestones with cephalopods		
Werfenian	sandstone conglomerate (verrucano facies)	Werfen sandstone and shale (Salzkammergut salt)		sandstones and shales (salts)		

Turning now to the higher part of the Austrides, the summary indication of the "Zentralalpen" as predominantly crystalline must be qualified. For not only is some Permian and Mesozoic — mostly Triassic — found as the autochthonous cover of the crystalline basement, but the basement itself consists in part of "Alt-Paleozoikum", of pre-Hercynian Paleozoic which is only slightly metamorphosed, and even carries fossils, though very scarcely so. Moreover there is a distinct lithological difference between the graywacke zone, found all along the border between the "Zentralalpen" and the Northern Limestone Alps, and the other occurrences of Paleozoic found in the "Zentralalpen".

Recent literature on the Paleozoic of the Zentralalpen is by METZ (1954, 1958, 1962), FLÜGEL (1960, 1964), FRITSCH (1962) and CLAR et al. (1963). As an introduction to the local names of the Paleozoic in question, the tables in TOLLMANN (1963) are useful.

We may now go into a schematic review of the differences of opinion on the structure of the Austrides. The conservative school estimates that the Paleozoic of the graywacke zone belongs to the Mesozoic of the Northern Limestone Alps and that both were sheared off from their crystalline basement, as still found in the "Zentralalpen". Both the Northern Limestone Alps, and the graywacke zone, together with the crystalline rocks of the "Zentralalpen" consequently belong to one single unit, the "Oberostalpin" or upper Austrides. The Permo-Triassic found scattered on top of the crystalline basement of the "Zentralalpen" is thought to represent the (par)autochthonous equivalent of the nappes of the Northern Limestone Alps.

In TOLLMANN's (cf. 1963, 1965) view, the Permo-Triassic of the "Zentralalpen" is so different in facies from the Triassic of the Northern Limestone Alps — a fact denied by the conservative school — that it just has to belong to a different nappe system. Once this is accepted, the easiest solution is to combine it with the underlying crystalline basement to a middle Austride nappe system. In that case one can combine the Paleozoic of the graywacke zone with the Mesozoic of the Northern Limestone Alps, to form the upper Austride nappe system.

In the estimate of the conservative school nappe formation would have resulted in crustal shortening of some 200 km, whereas in Tollmann's view this figure would rise to 300 km. This is what all the present hullabaloo is about (cf. BECK-MANNAGETTA, 1960).

I am of the opinion that one might at least give Dr. Tollmann the benefit of the doubt.

Arguments in favour of his view are, first that, according to a number of geologists, the Triassic of the Northern Limestone Alps resembles more that of the Drauzug than that of the "Zentralalpen". While, moreover, the graywacke zone is rather similar to the Paleozoic basement of the Southern Alps. An area of origin — or "root" — of the Northern Limestone Alps seems therefore indicated south of the "Zentralalpen", possibly more or less in the area where the Drauzug and its lateral equivalents are now.

A second argument in favour of a southerly origin of the Northern Limestone Alps lies in the fact that nonconformable Gosau Basins are found not only on the Northern Limestone Alps, but also on the "Zentralalpen" (pp.320–321). This would point to an early, pre-Cenomanian, "décollement" of the Northern Limestone Alps, from a southerly area of origin, clear across the "Zentralalpen".

On the other hand, it must be remembered that Tollmann is a dogmatist par excellence, who, in a most pure application of the classic theory of the Alps, is convinced that a resemblance of facies must mean consanguinity of nappes. Such a disposition may tend to overlook facts which are contrary to theory. Moreover, the series involved are not excessively fossiliferous, so that much depends on the interpretation of a single fossil find, and on the extrapolation which it is thought might be based on it.

CLAR (1965), who is perhaps the most cautious of the conservative school, maintains that various tectonic lineaments in the "Zentralalpen" can be retraced in the Northern Limestone Alps, which should prove their consanguinity. I do not think, however, that this line of argument has a higher heuristic value than Tollmann's arguments of facies resemblance.

The situation has, by the way, become so ambiguous that in the Vienna Geological Institute, where Clar teaches the conservative standpoint, the corridor walls curiously enough are hung with exhibits illustrating Tollmann's viewpoint.

A further discussion of this question, with an extensive bibliography, is reported by WUNDERLICH (1964). I can, however, not follow that author's contention that a solution to the riddle of the Austrides can be found by comparing them with the nappe systems of the French Alps. This is stretching too far the trust in the validity of cylindrism or nappe continuity.

Space does not permit me to go further into the problems of the "Zentralalpen" here, the reader is referred to the literature cited above. I will limit this text to some further remarks on the Northern Limestone Alps and on the Tauern window.

NORTHERN LIMESTONE ALPS

As stated before, regardless of theoretical considerations, the Northern Limestone Alps form a quite distinct unit. It is an epidermis nappe system of massive, subhorizontal thrust plates with relatively minor folding. It is composed mainly of Alpine Triassic and of equally calcareous Jurassic. These series are nonconformably overlain by paralic and largely detrital sediments of Late Cretaceous and Early Tertiary age, the famous Gosau Beds.

The main members of the Northern Limestone Alps are the massive carbonate series of the Middle and Upper Triassic, which may reach thicknesses of over 1 km. Well-known members are, for instance, the Wetterstein Limestone and its lateral equivalent, the Wetterstein Dolomite of Ladinian age. And the Hauptdolomit or Main Dolomite of the Norian, grading into its lateral equivalent the Dachstein Limestone, which, however, extends into the Rhaetian.

Despite their geosynclinal thickness, these carbonate series are shallow water deposits. They contain reefs, but are mostly formed by reef-rock sediments (cf. ZAPFE, 1962, 1963; ZANKL, 1966; and earlier papers for relevant literature). Even red cephalopod limestone, such as typically found in the Hallstätter Limestone, is now considered to represent shallow water deposits, as we saw already in the discussion of the "ammonitico rosso" facies on p.150. A series of about 200 consecutive cyclothems, each of them of paralic facies and formed by a succession of intertidal and subtidal deposits, has been described from the Dachstein Limestone near Salzburg by FISCHER (1964). Evidently organo-detrital sedimentation was able to offset geosynclinal crustal subsidence.

To understand the orogenetic movements it is important to note that below and between the massive carbonate members thinly-bedded series are found with shales and evaporites. The two main representatives of these predestined "décollement" horizons are found at the base of the Triassic, in the Werfenian — or Skyth(ian) as the Austrians like to call this Stage — and in the middle part, in the Carnian.

The Werfenian contains "Buntsandstein", and further shales and evaporites. Amongst the latter the rock-salt deposits, locally called "Haselgebirge" in the Hallstatt area are well known from the literature. Exploited since prehistoric days along the steep and high lake walls, they gave rise to an intensive salt trade, on which the Hallstatt Culture was founded. Recently, however, these evaporites have been dated palynologically as Late Permian (KLAUS, 1963, 1965).

The best-known elements of the second thin-bedded series, that of the Carnian, are the Raibler series and the paralic Lunzer series, which may even contain coal.

The Werfenian forms the basal thrust plane of the epidermis thrust blocks, whereas both Werfenian and Carnian participate in the internal thrust planes of the Northern Limestone Alps.

Table XIX gives a very schematic representation of the main facies relationships of the Triassic of the Northern Limestone Alps. A much more detailed table is given by E. Spengler (in: SCHAFFER, 1951). The main point to retain from the more detailed tables is that the various facies areas do not coincide with the later nappes (SPENGLER, 1953, 1956, 1959). It is only on superficial inspection that such coincidences seem to be present. We meet here again with the old story of gradual facies changes within a given tectonic unit, which only appear to be sudden when we go from one tectonic unit to another, because the gradual transition zone is now missing.

This is an important point, because the schematization present in all review articles and in all general maps tends to stress the differences and to obscure the gradual transitions. The erroneous conclusion that sedimentary paleogeography already followed the same lines as later tectonics, is thus all too easily arrived at.

In the case under discussion, I would insist that the reader turns to the map of SPENGLER (1963), unfortunately too detailed to reproduce here, in which it is clear that facies boundaries only rarely do follow nappe boundaries. And if they do, notably in the Hallstatt nappe, it is only incidentally so, because the intermediate areas are lost to observation. Moreover, it should always be kept in mind that the differences in facies of the Lower Mesozoic of the Northern Limestone Alps, often so stringently described, are of minor importance. They do not involve more than quantitative changes in the proportion of, say, limestone and dolomite, or even only of reef limestone and bedded limestone, in an otherwise remarkably uniform carbonate series (Fig.194).

In the interpretation of the many west–east faults and river valleys as transcurrent faults (PAVONI, 1961; FISCHER, 1965) facies differences should also, to my mind, be used with extreme caution.

Leaving the stratigraphy of the Northern Limestone Alps, we may point out that their structure reveals two main points of interest: (*1*) the formation of nappes and thrust blocks formed by the carbonate series of Triassic and Jurassic; (*2*) the tectonization of these units, as attested by the overlying Gosau Beds. I will consider the first aspect here, and return to an overall appraisal of the tectonic history of the Northern Limestone Alps after a discussion of the Gosau Beds.

The massive carbonate members of the Northern Limestone Alps normally show a tranquil, subhorizontal attitude. They are separated by more strongly tectonized, often imbricate zones, formed by the more thinly bedded, or by the slaty members. Strict application of the nappe theory has led to the distinguishing of a great many individual nappes (see, for instance, the map in SCHAFFER, 1951).

Following HAHN (1913), these individual nappes are often grouped together into three units. These are, from north to south: (*1*) the Bajuvaric (= Bavarian nappe of Table XIX), (*2*) the Tirolic and (*3*) the Juvaric. The Bajuvaric Group, forming the northernmost border of the Limestone Alps, includes the Allgäu and the Frankenfels nappes. The main mass of the Limestone Alps belongs to the Tirolic, with, amongst others, the Lechtal, Inntal, Totengebirgs, Örtscher and Lunzer nappes. The southerly Juvaric Group embraces the Halstätter and the Dachstein nappes.

Recognition of the fact that there is no recumbent-fold nappe style in the Limestone Alps, but instead a series of subhorizontal thrust plates, has led to the insight that all or most of these nappes are no more than local developments (German: "Teildecke") of one or several major nappes.

A serious objection against the nappe character of the Northern Limestone Alps has, in Bavaria, been recently raised by a group of German geologists (cf. KOCKEL, 1956). In the area of the Allgäu, the Lechtal and the Inntal nappes a number of tectonic windows have been known of old (Fig.191). These were now re-interpreted as pressed-out former synclines which originally were contiguous with what now seems to be the surrounding higher nappe.

C. W. Kockel and his co-workers (HAMANN and KOCKEL, 1956; JACOBSHAGEN and

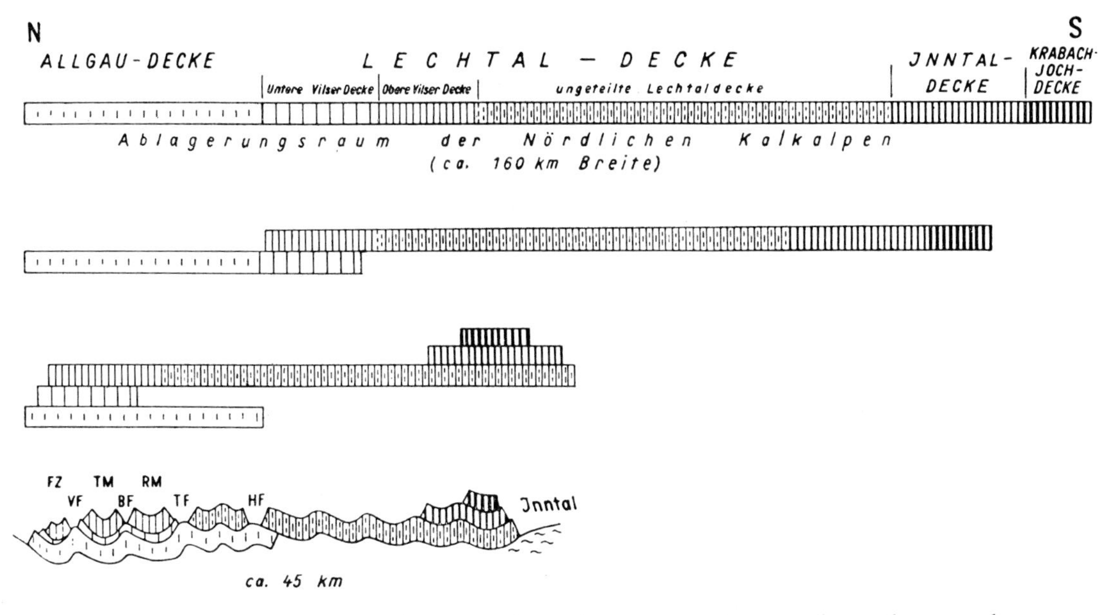

Fig.191. Schematic reconstruction of the formation of the Bavarian Alps according to the nappe theory. *FZ* = Falkenstein chain; *VF*: Vilstal window; *TM* = Tegelberg syncline; *BF* = Benna window; *RM* = Reintal syncline; *TF* = Tannheimer halfwindow; *HF* = Hornback halfwindow.
According to some German authors (e.g., KOCKEL, 1956), the "windows" are pressed-out former synclines, which originally were in lateral connection with what has erroneously been interpreted as a "higher" nappe.
(After ZACHER, 1961.)

KOCKEL, 1960) have, for instance, interpreted such structures as the Benna window as "Pilzfalten" and have thereby returned completely to the pre-nappe ideas of autochthony.

The apodictic style used in this controversy may be evident from the following translation of part of KOCKEL's (1956) introduction: "The classic interpretation of the structure of the Northern Limestone Alps has broke down. It was still praised in 1948 by RICHTER, in 1953 by SPENGLER, and sung of in 1953 by KOCKEL. But in the meantime a complete change over to the "Gebundene Tektonik", to autochthonous tectonics, has taken place. BEURLEN (1944), and RICHTER and SCHÖNENBERG (1954) have brought down the Inntal nappe. KOCKEL (1954) the Krabjoch nappe, and HAMANN and KOCKEL (1956) the Lechtal nappe. It is anticipated that this evolution will develop in easterly direction, as far as the Vienna area". One can but wonder about the feelings of Austrian geologists on hearing their northern neighbours bragging in this way!

Although it is doubtful if the nappes really exist as beautifully outlined as indicated in the earlier literature (ZACHER, 1961), the interpretation of the narrow, tectonized structures as "Pilzfalten" seems equally doubtful at least. For one thing, top and bottom features, not studied by Kockel and his co-workers, seem to contradict their interpretation of reversed flanks, pertinent to the "Pilzfalte" model. A further discussion of these ideas can be found in SARNTHEIM (1962).

Quite apart from the interpretation of such local, more tectonized zones, as windows or as pressed-out former synclines, the nappe structure of the Northern Limestone Alps as a whole has since been proved on two different lines of evidence. One resides with the geophysical exploration of the Alps, the other with the existence of windows of flysch of the northern border zone, underneath the Northern Limestone Alps.

The geophysical explorations cited in Chapter 10 have shown that the Northern Limestone Alps form but a thin sheet of sediments lying directly on top of the crystalline basement, without an intervening layer of Paleozoic. This proves their allochthonous, epidermal, nappe structure.

The windows of flysch of the northern border zone are now known to exist halfway across the breadth of the Northern Limestone Alps, proving an overthrust of the latter of some 30 km (PLÖCHINGER, 1964).

GOSAU BEDS

The Gosau Beds, of Late Cretaceous to Early Tertiary age, are found in small disjunct graben on top of the Northern Limestone Alps, and also on top of the crystalline basement of the "Zentralalpen".

They have been faulted and folded, but they always overlie unconformably the earlier Mesozoic of the Northern Limestone Alps, and they even transgress over nappe limits. It is evident that the main features of the nappes of the Northern Limestone Alps existed already, before they locally broke into graben, into which the Gosau Beds were deposited.

The Gosau Beds are famous for their paleontology. They were formed either in shallow marine, or in paralic facies. Strongly fossiliferous strata were deposited in several basins, of which the ammonites and the rudistids are perhaps the best known. Moreover, in several of the paralic basins, coal has developed.

The stratigraphy of the various basins is independent of that in other basins, indicating a local subsidence history for each basin. As a result, the Gosau Beds are difficult to correlate from one basin to another. Micropaleontological studies lead, however, nowadays to a better understanding (OBERHAUSER, 1963).

For further details I have to refer the reader to the local literature. For the Gosau graben on top of the Northern Limestone Alps we may cite PLÖCHINGER (1961), HERM (1961), VON HILLEBRANDT (1961), and KOLLMANN (1964) and for the southern basins the two publications by VAN HINTE (1963, 1965).

The Gosau Beds are important for our understanding of the genesis of the Eastern Alps, in particular for the formation of the Northern Limestone Alps. They underline the fundamental difference existing between the two carbonate nappe systems of the Alps: the "Hohe Kalkalpen" of the Helvetides and the "Nördliche Kalkalpen" of the Austrides. These nappe systems show so many similarities, that at first sight one might be tempted to assume that they were comparable.

But in the Helvetides sedimentation was more or less continuous during the whole of the Jurassic and the Cretaceous, whereas in the Northern Limestone Alps it stopped at the end of the Jurassic, to be renewed again during the Late Cretaceous. Correlated with this difference in stratigraphy, is the difference in tectonic history. The main orogenetic phase in the Helvetides was post-Cretaceous, that in the Northern Limestone Alps most probably was of Early Cretaceous age.

This follows from the fact that the tectonization of the Gosau Beds is rather light, and related to the blockfaulting of the graben. The main phase of the Alpine orogeney in the Northern Limestone Alps is consequently much earlier than in the other elements of the chain. The Tertiary movements have done no more than accentuate the structures formed earlier.

We have no indication how far the nappes of the Northern Limestone Alps, with the Gosau Beds sitting on their backs, have travelled. Or, to use the terms coined by TOLLMANN (1955), how extensive the *mise en marche*, the "putting in move", or "entrainment", of pre-Gosau times has been, and what was the influence of the post-Gosau *mise en place*, the "putting in place" or "emplacement", in reaching their present position.

FALLOT (1955) thought that the Gosau Beds had developed south of the "Zentralalpen" on top of the sedimentary basin of the present Northern Limestone Alps. He assumed that they travelled together for more than 250 km northward to occupy their present position. The existence of Gosau graben on top of the easterly "Zentralalpen" would, on the other hand, induce me to postulate that the Northern Limestone Alps had arrived already north of the present "Zentralalpen" during the pre-Gosau orogenetic phase.

The fact that the Northern Limestone Alps must indeed have moved northwards in post-Gosau times is attested by the windows of underlying flysch in this nappe system. A post-Gosau thrust of some 30 km is, however, but a minor distance compared with the total displacement, assumed to be of the order of 200 or 300 km.

A post-Gosau "mise en place" of the order of 30 km may well be related to the young uplift of the high mountain chain of the present "Zentralalpen". It would in that case be a phenomenon comparable to, and of the same order of magnitude as, the "rétrocharriage" in the French Alps. Only, in the case of the Northern Limestone Alps, the recently uplifted parts of the mountain chain do lie to the rear of the nappe system. The earlier orogenetic nappe formation, and the later thrusting consequently have the same direction. So for the Austrides it is difficult to say, what part of the distance travelled by the nappes was "mise en marche", and what was "mise en place".

ENGADIN AND TAUERN WINDOWS

In all this controversy, it is a comforting fact that the basic argument for the nappe character of the Austrides, i.e., the existence of tectonic windows, has been very generally accepted. Ever since TERMIER in 1903 recognized the Tauern as a tectonic window of Pennide nappes cropping out underneath the Austrides, dissident views have been heard less and less. Moreover, next to the second window, the Engadin window, the existence of a third window, the Wechsel window south of Vienna has recently been proposed (Fig.185).

The Engadin window is much smaller than the Tauern window. It has an oval shape, its longer axis measuring about 50 km. It is rather monotonous, consisting almost entirely of strongly tectonized Bündner Schiefer, which contains ophiolites (CADISCH, 1951, 1953).

The more easterly Tauern window is not only bigger, but also more complicated in structure. Its longer axis measures about 170 km. Within the window a number of more or less concentric zones can be distinguished. Schematically, we find an outer schist mantle, surrounding a crystalline core, which is in itself of complex build (Fig.192).

Within the schist mantle a less metamorphic outer zone can again be distinguished from a stronger metamorphic inner zone. The former, the "Obere Schieferhülle", is correlated with the Bündner Schiefer, the latter, the "Untere Schieferhülle", with Paleozoic schists of the Pennide nappes. The crystalline core is mainly composed of gneisses, which are, at least in part, polymetamorphic (EXNER, 1963). It is therefore comparable to the crystalline cores of the Pennide nappes farther west.

On superficial inspection the Tauern window, with its concentric structure, would thus seem to represent a single, large Pennide nappe. But the presence of gneiss lamina in the schist mantle, and also the complex pattern of the gneisses of the core, point to a certain repetition and duplication. This indicates that we have at present exposed, in the Tauern window, a certain number of individual Pennide nappes. Just how many distinct nappes can be separated, is, however, still a matter of discussion. It will perhaps eventually be solved by a more detailed cartography. But in such a region of high Alpine peaks this is a study that progresses but very slowly.

THE "ROOT ZONE" OF THE EASTERN ALPS

The "root zone" of the Eastern Alps, when seen at large, forms a continuation of the "root zone" of the Swiss Alps, as discussed already in Chapter 10.

In detail, however, there are differences worth noting. From Fig.184 we learn how this zone, which has a west–east direction in the eastern part of the Swiss Alps, is set off to the northeast over 80 km along the so-called Judicaria fault, before continuing again in an easterly direction. The relation, if any, of this Judicaria fault to the "root zone" is doubtful. Secondly, after continuing for some 70 km in a direction of south by east as a narrow fault zone, separating the crystalline of the Austrides from the metamorphosed

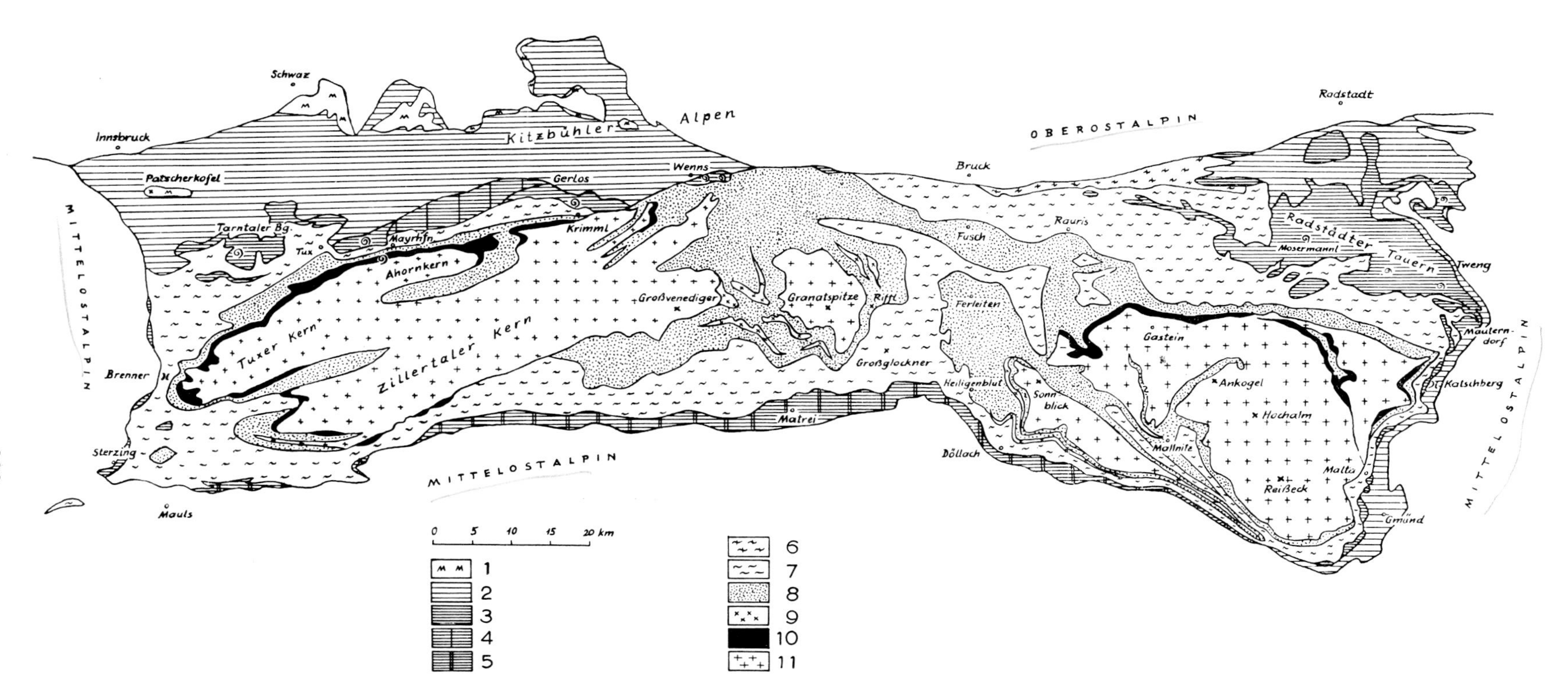

Fig.192. Map of the Tauern window. The actual window is, in Tollmann's opinion, surrounded by a narrow zone of lower Austride nappes. Around these follow, still in Tollmann's conception, the middle Austrides to the west, south and east, and the upper Austride graywacke zone (white) to the north.

1 = middle Austride elements on lower Austrides; *2* = quartzphyllite nappe; *3* = other nappes of lower Austrides; *4* = verrucano of the lower Austrides; *5* = Matrei zone (lower Austride or Pennide); *6* = Klammkalk; *7* = upper schists of the schist mantle; *8* = lower schists of the schist mantle; *9* = gneiss lenses in the schist mantle; *10* = Mesozoic cover of the crystalline cores (the so-called Hochstegen facies); *11* = gneiss cores.

(After TOLLMANN, 1963.)

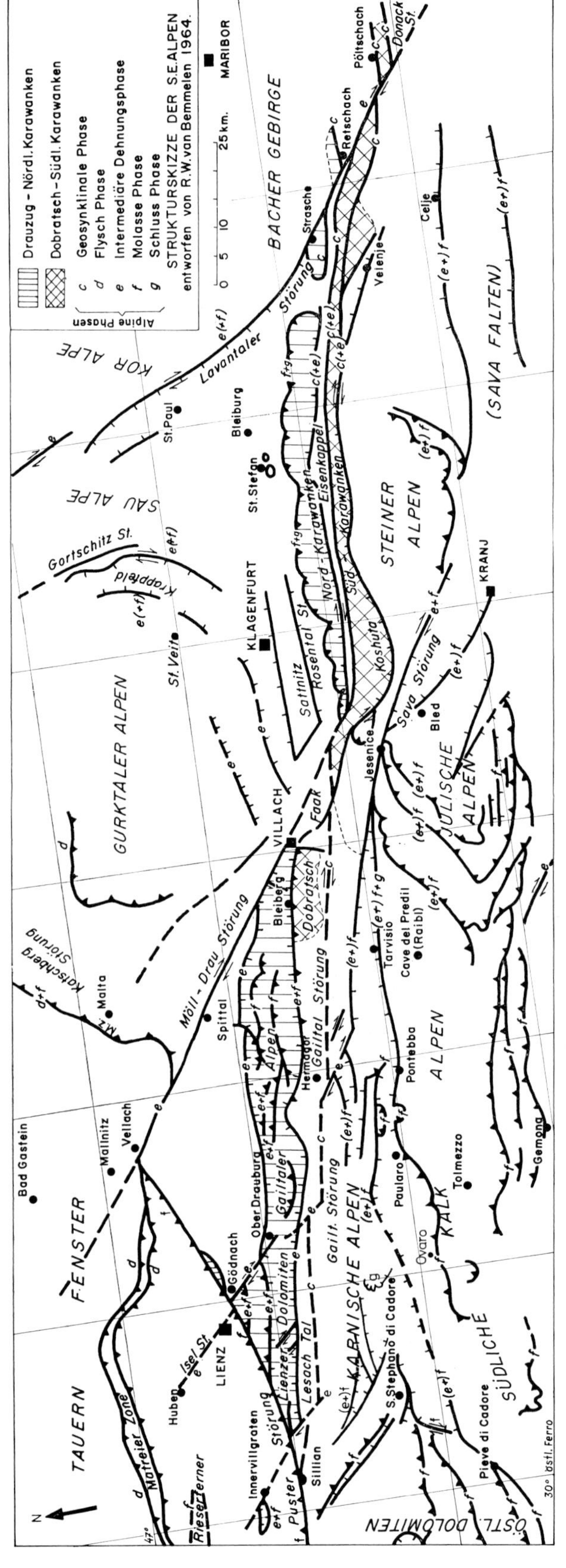

Fig.193. Structural map of the Drauzug and the Karawanken and their surroundings. (After VAN BEMMELEN and MEULENKAMP, 1965.)

Early Paleozoic (?) of the Southern Alps, the "root zone" opens up, to form a zone mainly occupied by Alpine Triassic, the Drauzug and the Karawanken (Fig.193).

The Drauzug has in recent years been studied by R. W. van Bemmelen and his students (VAN BEMMELEN, 1957, 1961; VAN BEMMELEN and MEULENKAMP, 1965). What follows is based on their results. I have confined myself, however, to a descriptive summary. Those readers, interested in how the geology of the Drauzug can be incorporated into Van Bemmelen's geotectonic views, are referred to the original publications (see also VAN BEMMELEN, 1965, 1966a).

The Drauzug is formed by a crystalline basement, only rarely exposed, nonconformably covered by Upper Carboniferous and Permian, which is followed by thick, geosynclinal Triassic in Alpine facies, and by Jurassic. The Upper Carboniferous and the Permian are

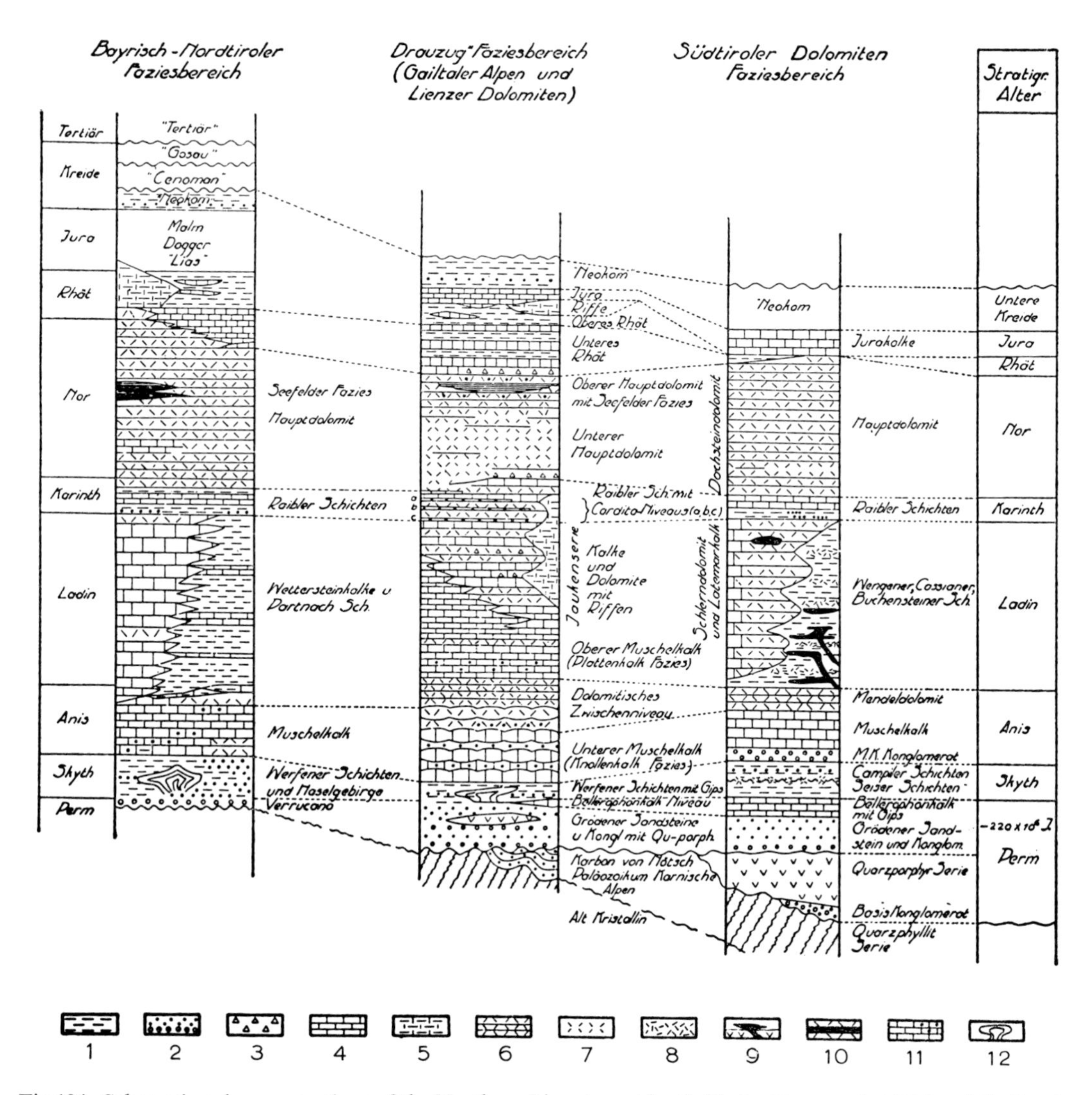

Fig.194. Schematic columnar sections of the Northern Limestone Alps (left), the Drauzug (middle) and the Southern Alps (right).
1 = clay and marl; *2* = sandstone and conglomerate; *3* = sedimentary breccia; *4* = bedded limestone; *5* = reef limestone; *6* = bedded dolomite; *7* = massive dolomite; *8* = tuff; *9* = volcanics; *10* = bituminous shale; *11* = chert in limestone; *12* = evaporites ("Haselgebirge"). (After VAN BEMMELEN, 1961.)

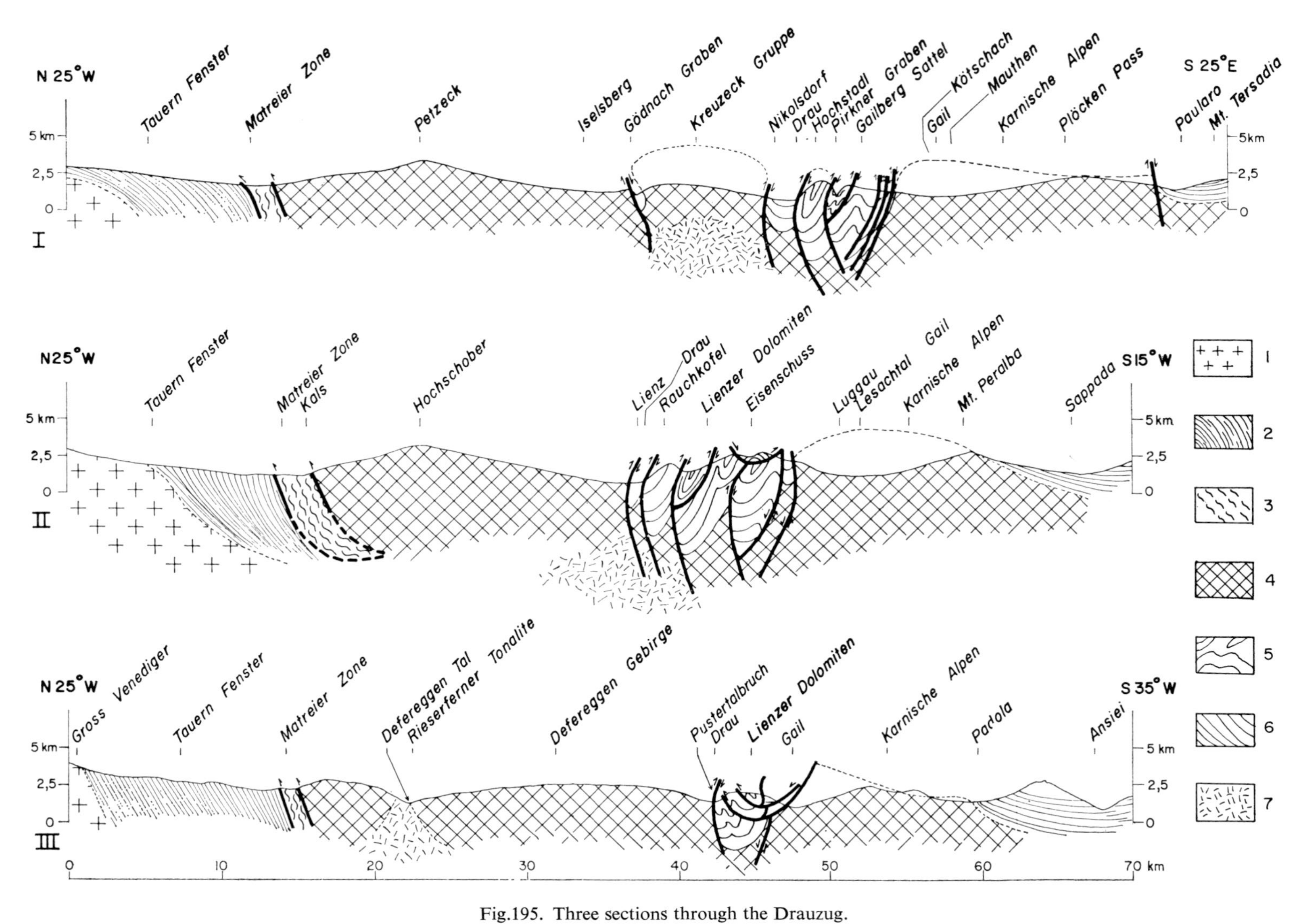

Fig.195. Three sections through the Drauzug.
1 = tonalite of the Tauern window; *2* = schists of the schist mantle of the Tauern window; *3* = schists of the Matrei zone; *4* = crystalline of the Austrides and of the Southern Alps; *5* = Mesozoic of the Drauzug; *6* = Mesozoic of the Southern Alps;*7* = peri-Adriatic younger intrusives.
(After VAN BEMMELEN and MEULENKAMP, 1965.)

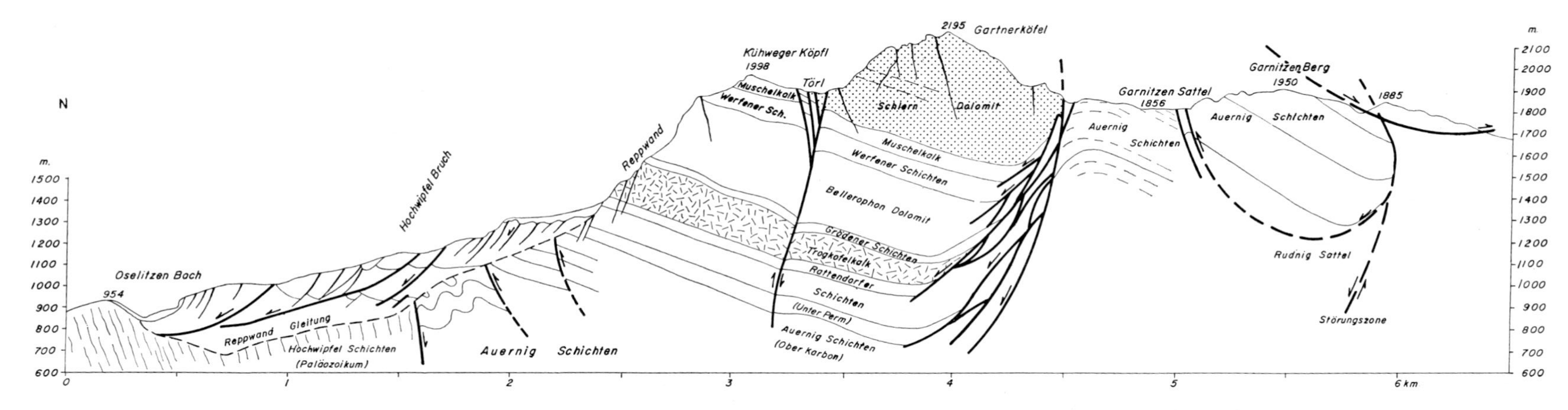

Fig.196. Section through the Drauzug, showing normal graben tectonics in the middle and at right, and post-erosional slumping, transgressing the graben border, at left. (After VAN BEMMELEN and MEULENKAMP, 1965.)

similar in development to those of the Southern Alps, whereas the Triassic and the Jurassic are similar to both those of the Northern Limestone Alps and the Southern Alps.

For the Triassic this overall similarity of the stratigraphy of the Northern Limestone Alps, the Drauzug and the Southern Alps follows from Fig.194. It once again stresses the amazing crustal quietness which must have reigned in Europe during this period. The continent must then have been base-levelled to an extent, which we, in our post-orogenetic environment, can at present hardly imagine.

We already took note of this fact in Chapter 10, in our discussion of the "Vindelician chain", but now are again confronted with this situation. If crustal movements were slight, the Germanic facies developed. But in the case of stronger, geosynclinal subsidence, sedimentation was predominantly organic or organo-detrital, because there was hardly any hinterland available for the supply of enough detrital material through erosion. In this carbonate environment smaller variations in subsidence and/or in organic production of skeletal remains, led to differences in the facies of the carbonate series. Whereas during the rare periods of apport of terrigenous clay, marls were formed.

As to structure, the Drauzug forms a graben, when seen in large. The basement of the sedimentary series exposed on the surface at either side, has subsided in the Drauzug to a depth of several kilometers. But when seen in detail, the sedimentary cover is found to be strongly tectonized with a predominantly subvertical, imbricate style (Fig.195).

According to Van Bemmelen, this tectonization of the sedimentary cover is due to gravitational reaction to the formation of the graben. The epidermis slid into the graben, which is, at most, only four times wider than high. Moreover, a certain constriction of the original graben, steepening and accentuating the imbricate structure, is thought to have also taken place.

The thick series of Triassic carbonates, often of massive character, separated by marly horizons, are of course pre-destined for gravitational tectonics. These same properties have also led to recent or subrecent slumping of mountain sides (German: "Freigleitung"), when erosion had sculptured out to steeply the resistant Triassic sediments. This recent slumping may produce results very similar to the earlier gravity tectonics. In some cases, however, it clearly transgresses the border of the Drauzug graben, and can easily be recognized as such (Fig.196). We will meet again with such recent post-erosional tectonics in the Dolomites, where it has also been confused with normal tectonics until a short while ago.

Basing himself on the results of paleomagnetic studies in the Southern Alps (cf. VAN HILTEN, 1962, 1964), and on the structure of the Eastern and the Southern Alps, VAN BEMMELEN (cf. 1966a) sees in the "root zone" a major transcurrent fault, along which the Southern Alps have moved westward, relative to the Eastern Alps. If this is so, then the structure of the Drauzug at present seems to indicate that such translation must have taken place in the early history of the Tethys, during Permian and Early Triassic.

THE SOUTHERN ALPS

The Southern Alps are ambiguous. They are Alps in the sense of a high mountain chain. They are Alps also in the geosynclinal development of their Mesozoic, notably that of the Triassic. But in structure they hardly present the typical picture of an Alpine chain,

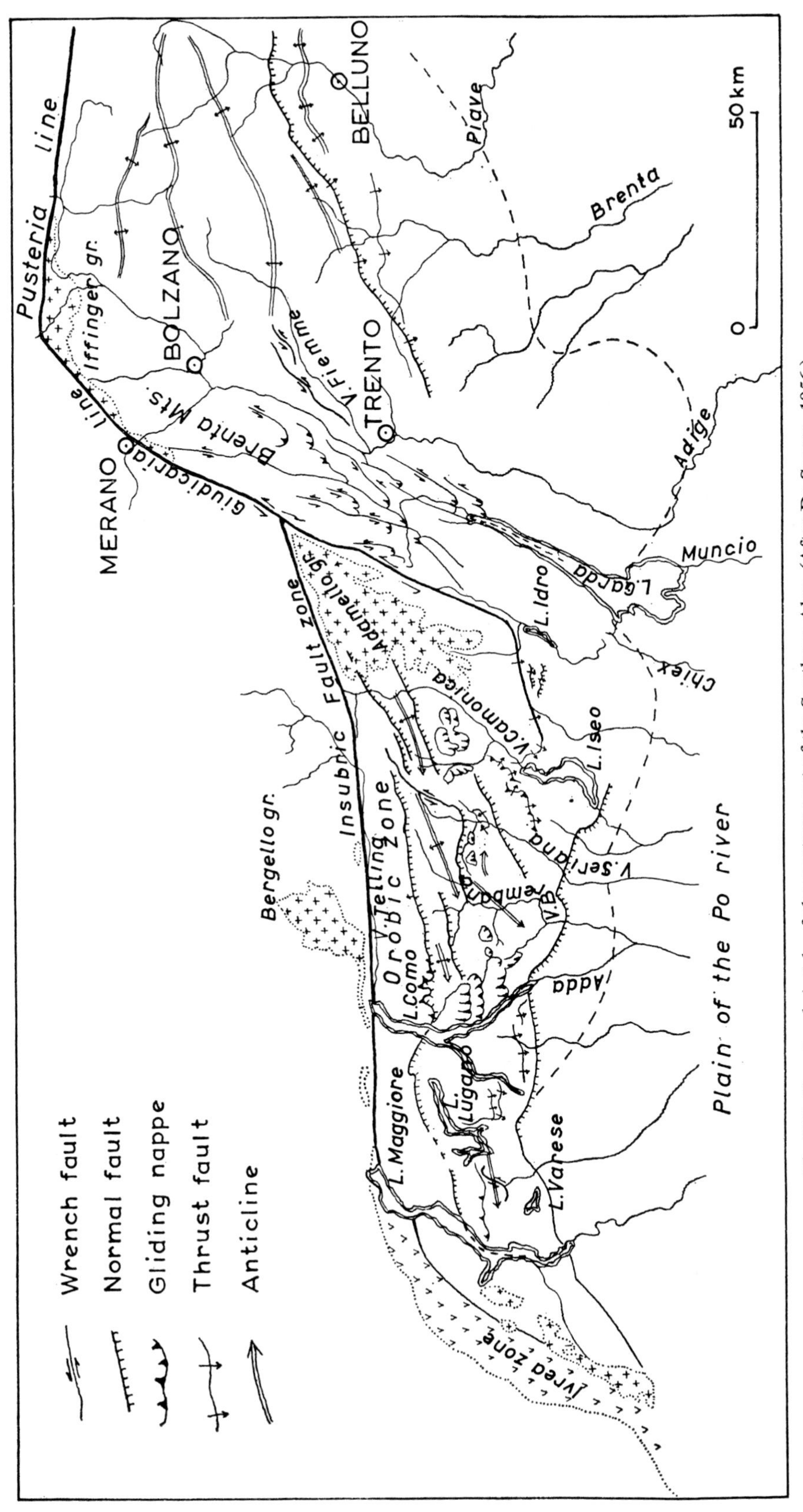

Fig.197. Structural sketch of the western part of the Southern Alps. (After DE SITTER, 1956.)

that is of a fold belt formed by a series of major nappes. Instead, they are on the whole autochthonous, grading into parautochthonous, whereas smaller, localized nappes are only found in their western part (DE SITTER, 1947; Fig.197).

We must strongly keep in mind that not every element of an alpine chain always posesses all the characteristics that, according to theory, together should belong to such a chain. Efforts to force the Southern Alps into the straight jacket of theory have, in the past, resulted in rather weird descriptions of their structure. A case in point can be found in the monograph on the Southern Alps, or Dinarides, as he called the chain of STAUB (1950). The map and local descriptions are still basic, but the accompanying sections are somehow unrealistic.

The Southern Alps do not contain such large, more or less homogeneous subunits such as the Pennide or Austride nappe systems. They are therefore normally subdivided into geographical regions. Of these, the most westerly is formed by the Alps of Lombardy, south of, and overlapping into, southern Switzerland. Their most important element is the Bergamo (or Bergamask) Alps (DE SITTER and DE SITTER-KOOMANS, 1949; DE SITTER, 1963; and, for the Swiss part, BERNOULLI, 1964; REINHARDT, 1964).

The following main element is the Dolomites (LEONARDI, 1955; and students of Van Bemmelen, notably VAN HILTEN, 1960; AGTERBERG, 1961; DE BOER, 1963; ENGELEN, 1963; GUICHERIT, 1964). In the Dolomites geological difficulties are aggravated by language difficulties. Because much of this area belongs to Southern Tirol, German papers still persist in using the old German names, instead of the Italian names, which have been the official ones for almost forty years now.

To the east of the Dolomites follow the Carnic and the Venetian Alps, the former lying to the north of the latter. Then follow the Julian Alps (SELLI, 1963), which overlap into Yugoslavia.

STRATIGRAPHY OF THE SOUTHERN ALPS

The rock series of the Southern Alps is, as elsewhere in the Alps, divided into a pre-Alpine and an Alpine series[1].

The pre-Alpine basement consists of metamorphics ranging from phyllites to gneisses, and of igneous rocks. Apart from the paper by REINHARDT (1964), no new petrographic analyses of the basement of the Southern Alps are available.

The Alpine series differs from that of the Eastern Alps mainly on the following two points: (*1*) the amount of Permian present in the Southern Alps; and (*2*) the volcanic influences found during the Triassic.

The Permian is not only developed as continental red beds in verrucano facies, but is also characterized by the large amount of continental, acid volcanics. The existence of verrucano, similar to that of the Helvetides, in the Southern Alps, indicates a Permian paleogeography that is entirely different from that of the Triassic.

The volcanics are present in the form of massive, thick effusions of porphyries, mainly found in two areas. The westerly, the Lugano Porphyries (DE SITTER, 1939), lies around

[1] The reader is especially referred to the comprehensive publication on the Southern Alps by VAN BEMMELEN (1966), which only appeared after the closing of the manuscript.

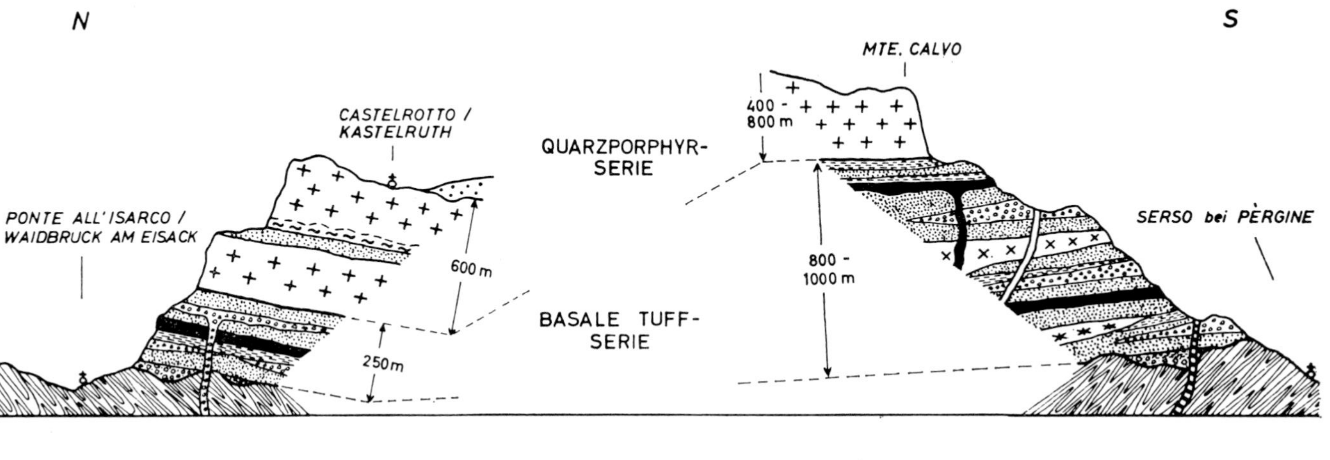

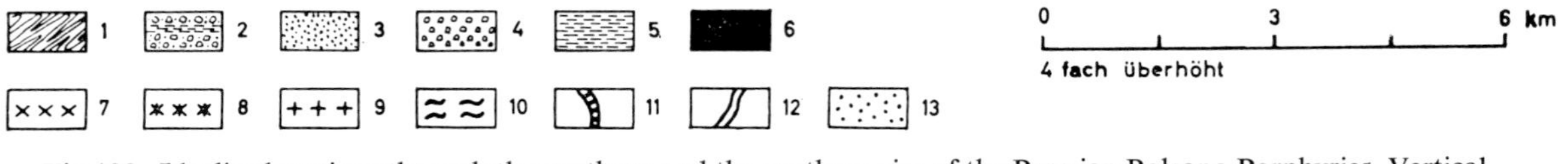

Fig.198. Idealized sections through the northern and the southern rim of the Permian Bolzano Porphyries. Vertical scale four times horizontal scale.

1 = basement; *2* = Permo-Carboniferous sediments; *3* = tuff; *4* = agglomeratic tuffs; *5* = tephra; *6* = melaphyres and basic porphyrites; *7* = porphyrites; *8* = quarzporphyrites; *9* = quarzporphyries; *10* = vitrophyric quarzporphyry; *11* = augiteporphyrite; *12* = quarz dikes; *13* = Upper Permian Grödener Sandstone. (After H. Pichler in: MAUCHER and PICHLER, 1959.)

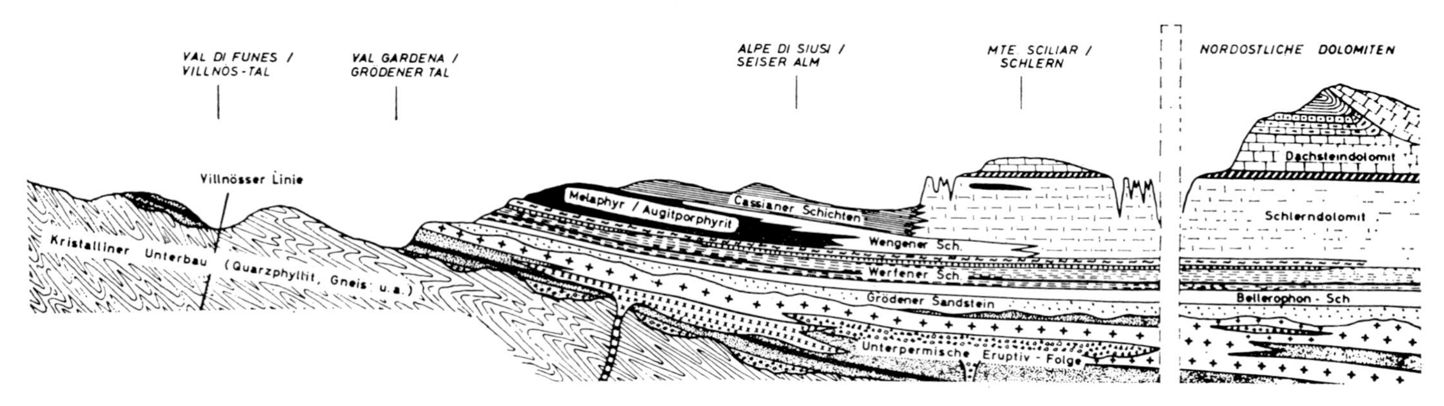

Fig.199. Idealized section through the Dolomites, showing the abrupt lateral facies changes between the reef- and reef-rock carbonate series and the volcanoclastic series of the Triassic. For details of the Permian volcanic series (*Unterpermische Eruptiv-Folge*), see Fig.198. (After H. Pichler in: MAUCHER and PICHLER, 1959.)

the southern part of the Lago Maggiore, the easterly, the Bolzano Porphyries, north of Venice (MITTEMPERCHER, 1958).

These volcanic areas in the Southern Alps are yet another occurrence of the voluminous Permian volcanism, already met with in the Oslo area and in the Estérel in southeastern France. The pile of porphyries (Fig.198), over 1 km thick, and forming a plate of at least 65 km by 60 km in the Bolzano area, shows all characteristics of ash flow eruptions (MAUCHER, 1960; VAN BEMMELEN, 1961; MITTEMPERCHER, 1963).

A further parallel with the Oslo province is that subvolcanic rocks of the same age also occur in the Southern Alps. This relates to granites of the Cima d'Aste, which have been dated as 271–280 million years old. D'AMICO (1964) has combined the plutonic rocks of the Cima d'Aste and related intrusions with the volcanics of Lugano and Bolzano, which are very similar petrographically, into the *provincio magmatica tardo-ercinica tridentina*. That is the "late-Hercynian magmatic province of Trento". It is named after the Roman name *Tridentum* for the present Trento, a name also commonly used for this whole region, in "Veneto tridentino".

Of the normal Alpine geosynclinal rock series of the Southern Alps the massive Triassic carbonates are widely known. They form the backbone of the Dolomites. They are characterized by the occurrence of reefs formed either by Algae or by corals. The first kind of reefs has been of old described by VON PIA (cf. 1941), whereas a good review article is given in MÄGDEFRAU (1946, pp.223–229). They are formed by *Diplopora* and related genera of the Dasycladaceae, a group of algae, characterized, among other things, by the fact that they secrete external, species-specific, carbonate skeletons. Already in 1882 the more normal coral reefs have been described excellently by H. E. von Mojsisovics, while newer descriptions are given by LEONARDI (1955, 1966).

In lateral alternation with these carbonate reef series submarine volcanic deposits occur, which are a special development of the Southern Alps. In the Dolomites this contemporaneous volcanism, in the Strati di Valle or Wengener Schichten of the Ladinian and in the Cassianer Schichten of the Carnian, has seriously limited the extension of the reefs. As a result one finds abrupt facies changes from the massive reef- and reef-rock carbonate series to the thinly bedded volcanoclastic facies (Fig.199). Although some interfingering occurs, the reefs probably towered high above the volcanoclastic deposits in the Triassic sea, and, schematically, the border of the reef facies is now subvertical, often over hundreds of meters. The volcanoclastic facies is, of course, much more mobile than the massive carbonate facies, a point we will return to in the last paragraph of this chapter.

When seen as a whole, the thickness and the facies of the Southern Alps show rapid changes in east–west direction. Apparently there have been alternating "haut-fonds" and subsiding basins. This is thought to be due to blockfaulting in the basement along a set of north–south faults by BOSSELINI (1965).

The higher Mesozoic, the Jurassic and Cretaceous, is also developed in a predominately carbonatic facies, but well-bedded limestones are more common than in the massive limestone and dolomite facies of the underlying Triassic. For an excellent review article the reader is referred to AUBOUIN (1964, 1965).

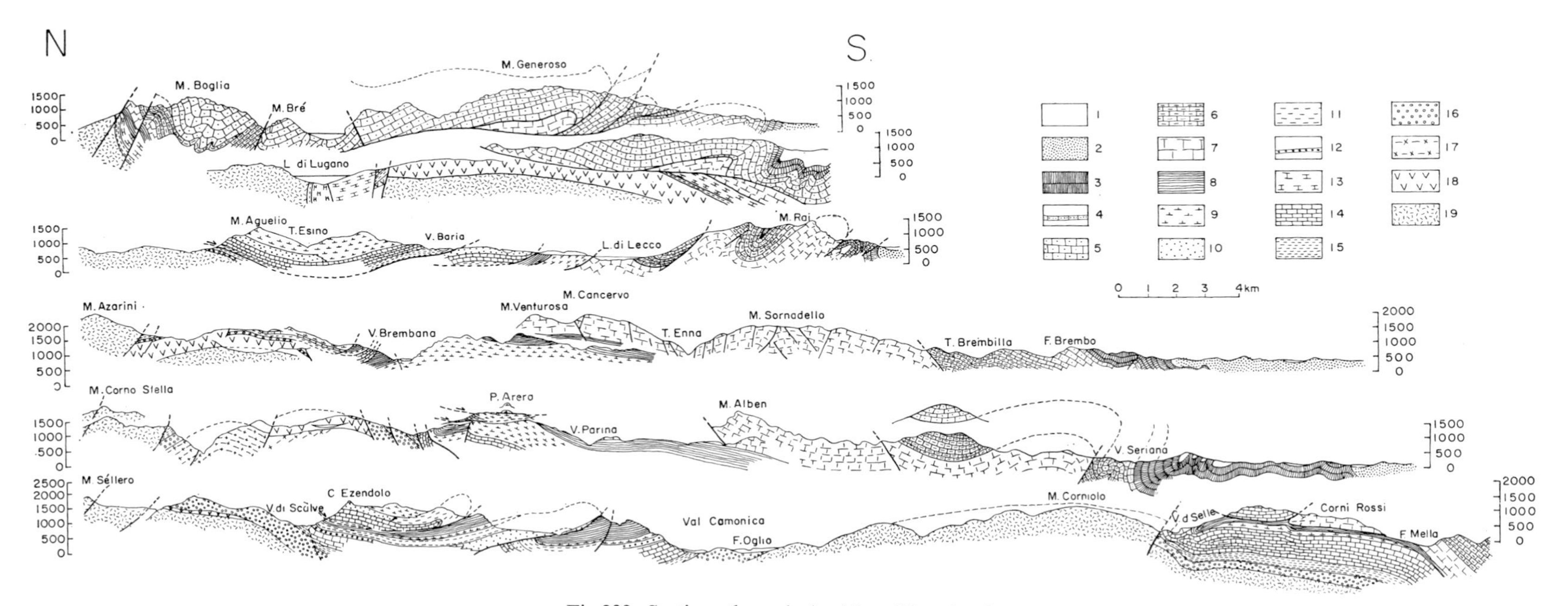

Fig.200. Sections through the Alps of Lombardy.
1 = Cenozoic; *2–3* = Cretaceous (*2* = Scaglia flysch; *3* = Majolica); *4–5* = Jurassic (*4* = ammonitico rosso; *5* = limestone with chert); *6–15* = Triassic (*6* = Raehtian; *7* = Norian "Hauptdolomit"; *8* = Carnian; *9–12* = Ladinian; *13* = Ladinian–Anisian; *14* = Anisian; *15* = Werfenian); *16–18* = Permian; *19* = pre-Alpine basement. (After DE SITTER, 1963.)

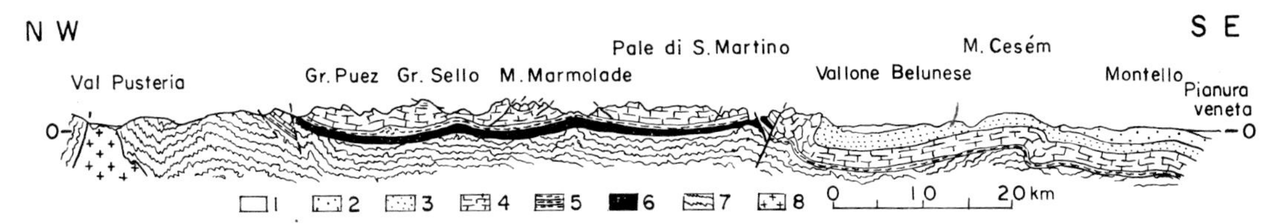

Fig.201. Schematic section through the Dolomites.
1 = Quaternary; *2* = Tertiary; *3* = Cretaceous and Jurassic; *4* = Triassic; *5* = Permian; *6* = Permian volcanics; *7* = pre-Alpine basement; *8* = peri-Adriatic younger intrusives. (After LEONARDI, 1963.)

STRUCTURE OF THE SOUTHERN ALPS

As stated above the tectonic style of the Southern Alps is that of (par)autochthony, with only minor nappes. This relatively quiet style is illustrated in Fig.200, 201.

As DE SITTER (1956) remarked, tectonization is more vigorous in the Alps of Lombardy than in the Dolomites. The nappe structures are only found in the western region (Fig.184, 200), whereas the Dolomites are pretty well autochthonous (Fig.201).

De Sitter attributes this difference in tectonization to the effect of the Judicaria fault (Italian: "Giudicaria"; cf. Fig.197), which he interprets as a wrench fault. Neither VAN HILTEN (1963), nor BONI (1964) could find, however, indications in the field for the horizontal translations belonging to a wrench fault. The difference in tectonization west and east of the fault could also be arrived at in assuming the Judicaria fault to be a hinge fault, whose throw would increase northwards. The much broader block east of the fault would in this case have become tilted much less than the narrower, and consequently steeper, block to the west. It stands to reason that any gravitative reaction of the epidermis will in this case also have been stronger in the Alps of Lombardy, than in the Dolomites.

"EROSION TECTONICS"

Not only DE SITTER (1956), but already FALLOT (1950), and later Van Bemmelen and his school, have stressed the importance of gravity tectonics for the Southern Alps. Most of the movement of the epidermis of Alpine rocks is due, in their view, to gravitational reactions resulting from rise and descent, and also from tilting, of basement blocks.

One of the most outspoken of the latter group is ENGELEN (1963), who has, however, contaminated his excellent factual descriptions with not quite digested theory. The reason for mentioning this author lies, however, in another part of his thesis, in which he has tried to distinguish between "real" and "pseudo" tectonics in the Dolomites.

"Real" tectonics is the sliding and draping of the epidermis over the mobilized basement blocks. "Pseudo" tectonics relates to movements over a stable basement, due to selective erosion. This phenomenon, which, for want of a better word, I will call "erosion tectonics" here, is strongly developed in the central parts of the Dolomites, where, moreover, "real" tectonics is quite unobtrusive.

When, after an initial uplift of the Southern Alps during (?) Plio-Pleistocene time, erosion had cut through the carapace of limestones of Cretaceous and Jurassic age, a great difference was encountered in further erosion, depending upon the chance whether the underlying Triassic was found to be developed in the reef limestone or in the volcanoclastic facies.

Further uplift now led to strong erosion in the latter areas, which resulted in the modelling out of the massive reefs of the Dolomites. Present-day topography follows rather closely the paleogeography of the Triassic. Putting it schematically, the mountain peaks of the Dolomites represent former reef areas, the valleys those of the former submarine volcanic activity.

The valley pattern thus developed has no correlation to the main structural lines of the Alps. Much of the charm of the Dolomites is even due to these deep and sinuous valleys, curving more or less capriciously around the massive Dolomite Mountains.

The stage was now set for the next development. Because of the mobility of the basal sands and shales of the Triassic, these were squeezed out from underneath the heavy massive reef mountains, and thrust up in anticlines in the intervening valleys. This explains the fact why in the Dolomites the anticlines follow the valleys so closely, even though the direction of these valleys shows an irregular pattern, and does not follow the main Alpine trend.

The mobility of the beds underlying the massive reef mountains was so great that they were squeezed out in many cases in horizontal diapirs at the foot of the steep valley walls. Careful mapping has moreover revealed that the life expectancy of such a diapir is related to the process of valley incision. The diapirs develop, when the valley floor is stable for some time, which gives the diapir time to develop. If erosion proceeds and the valley floor is then stabilized at a new, lower level, a new diapir will arise, and the older diapir dies. Diapirs of different ages have been found to be strictly correlated with the river terraces indicating temporary valley stability.

Although these reactions to the selective erosion of the reef facies and the volcano-clastic facies of the Triassic thus first affects the valleys, this phenomenon will eventually lead to a tilting, or even to a breaking up, of the reef blocks, which gradually loose their support.

Tilting of the massive reef blocks will thereupon have a new, secondary effect. This relates to the cover of limestones of the younger Mesozoic, when and where still existent on top of the reef blocks. These well-bedded limestones, alternating with a minor amount of marly layers, will tend to slide down independently from the underlying rigid reef blocks.

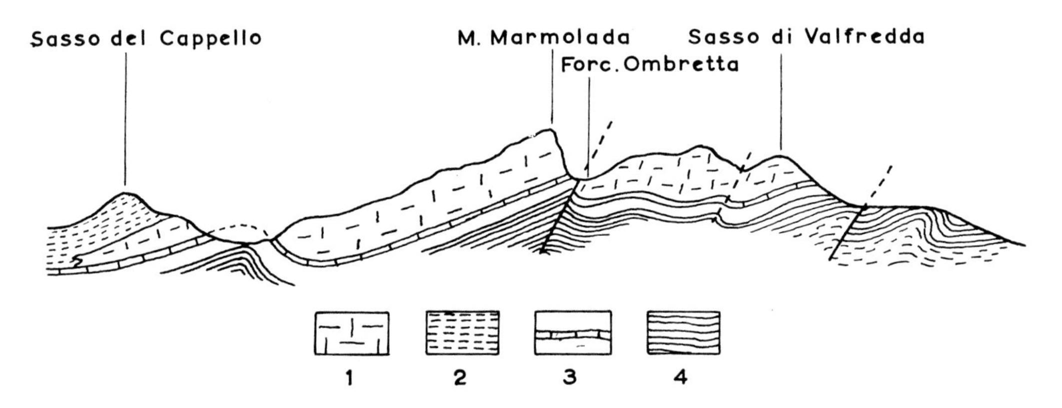

Fig.202. Section through the Marmolada Group of the Dolomites. showing the normal intrepretation of its tectonization, through conformable folding.
1 = reef limestones of the Ladinian and Carnian; *2* = La Valle Beds; *3* = Raibl Beds; *4* = Werfenian and Lower Anisian. (After LEONARDI, 1955.)

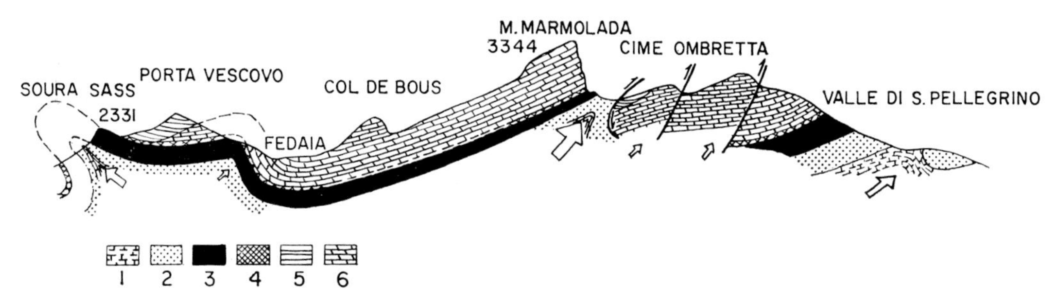

Fig.203. Section through the Marmolada Group of the Dolomites.
Instead of the tectonic interpretation of Fig.202, ENGELEN (1963) thinks both the valley tectonics and the tilting of the Triassic reef blocks as a reaction to selective erosion. The areas of volcanoclastic facies were strongly eroded, mobile material was squeezed out into valley anticlines and valley wall diapirs from underneath the reef blocks, which subsequently became tilted.
1 = *Bellerophon* limestone; *2* = Werfenian; *3* = Anisian; *4* = Livinalongo Beds; *5* = La Valle Beds; *6* = Marmolada Limestone. (After ENGELEN, 1963.)

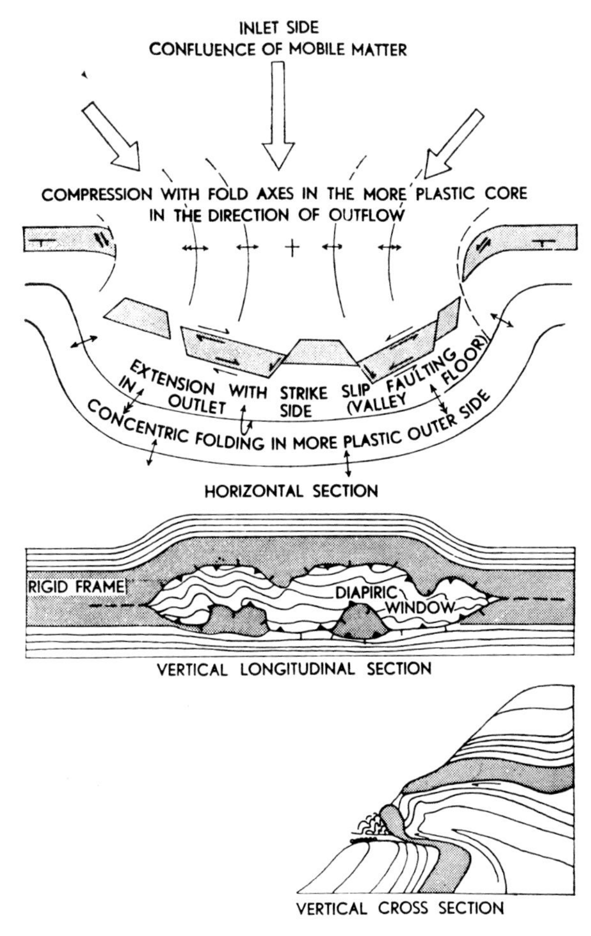

Fig.204. Schematic picture of a valley wall diapir, originating through erosion tectonics in the Dolomites.
The cross-section in the lower picture explains the genesis of such diapirs. The middle figure gives a view of the base of a valley wall, the upper picture a schematized ground plan. The length of such foot-of-a-valley-wall diapirs varies from less than 100 m to several kilometers. (After ENGELEN, 1963.)

They will accumulate into the enigmatic "summit structures", known for a long time to occur on top of, or alongside from, the massive Triassic which was only tilted.

All of the local, highly diversified, structures of the Dolomites can thus be explained as the result of erosion tectonics. It is tried in Fig.202–204 to give an idea of the dimensions of this phenomenon.

Erosion tectonics is, of course, so forcibly expressed in the Dolomites, because of the abrupt facies changes between the massive reef areas and the intervening volcanoclastic beds of the Triassic. It is, moreover, easy to follow in the Dolomites, because the influence of "real" tectonics has been very slight in this region.

I am convinced that in the light of this analysis of the structure of the Dolomites, it will become apparent that erosion tectonics may be spread far wider than is known by now. It will be less well expressed in other areas, where, on the one hand, facies changes are not so abrupt as in the Dolomites, and/or where, on the other, "real" tectonics has had a stronger influence. Nevertheless, it is necessary in any deeply incised mountain valley, to assess carefully, what might be "real", and what might be "erosion" tectonics.

REFERENCES

AGTERBERG, F. P., 1961. Tectonics of the crystalline basement of the Dolomites in North Italy. *Geol. Ultraiectina*, 8: 232 pp.

ANDRUSOV, D., 1963. Les principaux plissements alpins dans le domaine des Carpathes occidentales. In: M. DURAND DELGA (Editor), *Livre à la Mémoire du Professeur Paul Fallot. II.* Soc. Géol. France, Paris, pp. 519–528.

AUBOUIN, J., 1964. Essai sur la paléogéographie posttriassique et l'évolution secondaire et tertiaire du versant sud des Alpes orientales. *Bull. Soc. Géol. France*, 7 (5): 730–766.

BECK-MANNAGETTA, P., 1960. Bemerkungen zu A. Tollmann's tektonische Synthese der Ostalpen. *Geol. Rundschau*, 50: 517–524.

BECK-MANNAGETTA, P., 1964. *Geologische Übersichtskarte der Republik Österreich mit tektonischer Gliederung, 1: 1.000.000.* Geol. Bundesanstalt, Wien.

BECK-MANNAGETTA, P., 1966. Erläuterungen zur Geologischen und zur Lagerstätten-Karte 1:1.000.000 von Österreich. *Verhandl. Geol. Bundesanstalt (Austria)*, 1966: 7–64.

BECK-MANNAGETTA, P., GRILL, R., HOLZER, H. and PREY, S., 1966. Erläuterungen zur Geologisch und zur Lagerstatten-Karte 1: 1.000.000 von Österreich. *Verhandl. Geol. Bundesanstalt*, pp.7–64.

BERNOULI, D., 1964. Zur Geologie des Monte Generoso (Lombardische Alpen). *Beitr. Geol. Karte Schweiz*, 148: 118 pp.

BEURLEN, K., 1944. Zum Problem der Inntaldecke. *Sitz. Ber. Bayr. Akad. Wiss. München, Math. Naturw. Abt.*

BONI, A., 1964. La ligne judicarienne et la limite nordouest de l'Apennin septentrional. *Geol. Rundschau*, 53: 84–100.

BOSSELINI, A., 1965. Lineamenti strutturali delle Alpi Meridionali durante il Permo-Trias. *Mem. Museo Storia Nat. Venezia Tridentina*, 28: 68 pp.

BRAUMÜLLER, E., GRILL, R., JANOSCHEK, R., KÜPPER and SALZER, H., 1964. Geologischer Führer zu Exkursionen durch die Ostalpen. *Mitt. Geol. Ges. Wien*, 57 (1): 377 pp.

BRIX, F. and GÖTZINGER, K., 1964. Zur Geologie der Beckenfüllung, des Rahmens und des Untergrundes. *Erdöl Z.*, 80: 57–76.

CADISCH, J., 1951. Prätigauer Halbfenster und Unterengadines Fenster, ein Vergleich. *Eclogae Geol. Helv.*, 43: 172–180.

CADISCH, J., 1953. *Geologie der Schweizeralpen*, 2. Aufl. Wepf, Basel, 480 pp.

CLAR, E., 1953. Zur Einfügung der Hohen Tauern in den Ostalpenbau. *Verhandl. Geol. Bundesanstalt (Austria)*, 1953: 93–104.

CLAR, E., 1965. Zum Bewegungsbild des Gebirgsbaues der Ostalpen. *Verhandl. Geol. Bundesanstalt (Austria), Sonderh.*, G: 11–35.

CLAR, E., FRITSCH, W., MEIXNER, H., PILGER, A. and SCHÖNENBERG, R., 1953. Die geologische Neaufnahme des Saualpen-Kristallins, Kärnten, VI. *Carinthia II*, 73: 23–51.

REFERENCES

D'Amico, C., 1964. Relazioni comagmatiche tra vulcanismo atesino e plutonismo di Cima d'Asta. La provincia magmatica tardo-ercinica tridentina. *Mineral. Petrogr. Acta Bologna*, 10: 157–176.

De Boer, J., 1963. The geology of the Vicentinian Alps (NE-Italy) with special reference to their paleomagnetic history. *Geol. Ultraiectina*, 11: 178 pp.

De Sitter, L. U., 1939. Les porphyres Luganois et leurs enveloppes. *Leidse Geol. Mededel.*, 11: 1–61.

De Sitter, L. U., 1947. Anthithesis Alpes–Dinarides. *Geol. Mijnbouw*, 9: 1–47.

De Sitter, L. U., 1956. A comparison between the Lombardy Alps and the Dolomites. *Geol. Mijnbouw*, 18: 70–77.

De Sitter, L. U., 1963. La structure des Alpes Lombardes. In: M. Durand Delga (Editor), *Livre à la Mémoire du Professeur Paul Fallot. II.* Soc. Géol. France, Paris, pp. 245–256.

De Sitter, L. U. and De Sitter-Koomans, C. M., 1949. The geology of the Bergamasc Alps, Lombardy, Italy. *Leidse Geol. Mededel.*, 14B: 1–257.

Engelen, G. B., 1963. Gravity tectonics in the northwest Dolomites (N. Italy). *Geol. Ultraiectina*, 13: 92 pp.

Exner, C., 1963. Structures anciennes et récentes dans les gneiss polymétamorphiques de la zone pennique des Hohe Tauern. In: M. Durand Delga (Editor), *Livre à la Mémoire du Professeur Paul Fallot. II.* Soc. Géol. France, Paris, pp. 503–515.

Exner, C., 1966. Einführung in die Geologie von Österreich. *Verhandl. Geol. Bundesanstalt (Austria)*, 1966: 65–91 (with English translation; avec traduction française).

Fallot, P., 1950. Remarques sur la tectonique de couverture dans les Alpes Bergamasques et les Dolomites. *Bull. Soc. Géol. France*, 5 (20): 183–195.

Fallot, P., 1955a. Les dilemnes tectoniques des Alpes orientales. *Ann. Soc. Géol. Belg., Bull.*, 78: 147–170.

Fallot, P., 1955b. *Schéma Structural des Alpes Orientales.* Monsanglant, Paris.

Fallot, P., 1960. Le problème de l'espace en tectonique. *Festschr. Ernst Kraus — Abhandl. Deut. Akad. Wiss. Berlin*, 3 (1): 48–58.

Fischer, A. G., 1964. The Lofer cyclothems in the Alpine Triassic. *Bull. Kansas Geol. Surv.*, 169: 107–149.

Fischer, A. G., 1965. Eine lateralverschiebung in den Salzburger Kalkalpen. *Verhandl. Geol. Bundesanstalt (Austria)*, 1965: 20–33.

Flügel, H., 1960a. Geologie des Grazer Berglandes. *Mitt. Museum Joanneum Graz*, 23: 212 pp.

Flügel, H., 1960b. Die tektonische Stellung des "Alt-Kristallins" östlich der Hohen Tauern. *Neues Jahrb. Geol. Paläontol., Monatsh.*, 1960: 202–220.

Flügel, H., 1963. Das Paläozoikum in Österreich. *Mitt. Geol. Ges. Wien.*, 56: 400–444.

Flügel, H., 1964. Versuch einer geologischen Interpretation einiger absoluter Alterbestimmungen aus ostalpinen Kristallin. *Neues Jahrb. Geol. Paläontol., Montash.*, 1964: 613–625.

Fritsch, W., 1962. Von der "Anchi" zur Katazone im kristallinen Grundgebirge Ostkärntes. *Geol. Rundschau*, 52: 202–210.

Gignoux, M., 1950. *Stratigraphic Geology.* Freeman, San Francisco, Calif., 682 pp.

Guicherit, R., 1964. Gravity tectonics, gravity field and palaeomagnetism in northeast Italy. *Geol. Ultraiectina*, 14: 125 pp.

Hahn, F. F., 1913. Grundzüge des Baues der nördlichen Kalkalpen zwischen Inn und Ems. *Mitt. Geol. Ges. Wien*, 6: 238–357; 374–501.

Hamann, P. and Kockel, C. W., 1956. Luitpoldzone, Bärgündele und das Ende der Lechtaldecke. *Geol. Rundschau*, 45: 204–213.

Herm, D., 1961. Die Schichten der Oberkreide (Untere, Mittlere und Obere Gosau) im Becken von Reichenhall (Bayrische/Salzburger Alpen). *Z. Deut. Geol. Ges.*, 113: 320–338.

Jacobshagen, V. and Kockel, C. W., 1960. Überprüfung des "Benna-Deckensattels" in den Hohen Schwangauer Alpen. *Neues Jahrb. Geol. Paläontol., Monatsh.*, 99–110.

Janoschek, R., 1964a. Das Tertiär in Österreich. *Mitt. Geol. Ges. Wien.*, 56: 319–360.

Janoschek, R., 1964b. Geologie der Flyschzone und der Helvetischen Zone zwischen Atersee und Traunsee. *Jahrb. Geol. Bundesanstalt (Austria)*, 107: 161–214.

Kapounek, J., Kröll, A., Papp, A. and Turnovsky, K., 1965. Die Verbreitung von Oligozän, Unter- und Mittelmiozän in Niederösterreich. *Erdoel Z.*, 81: 109–116.

Klaus, W., 1963. Sporen aus dem südalpinen Perm. *Jahrb. Geol. Bundesanstalt (Austria)*, 106: 229–363.

Klaus, W., 1965. Zur Einstufung alpiner Saltztone mittels Sporen. *Verhandl. Geol. Bundesanstalt (Austria), Sonderh.*, G: 544–548.

Kockel, C. W., 1953. Beobachtungen im Hornbachfenster (Lechtaler Alpen). *Neues Jahrb. Geol. Paläontol., Abhandl.*, 96: 339–356.

Kockel, C. W., 1954. Die Larsen-Scholle bei Imst (Tirol). *Neues Jahrb. Geol. Paläontol., Monatsh.*, 1953: 520–533.

Kockel, C. W., 1956. Der Umbau der nördlichen Kalkalpen und seine Schwierigkeiten. *Verhandl. Geol. Bundesanstalt (Austria)*, 1956: 205–211.

KOLLMANN, H., 1964. Zur stratigraphischen Gliederung der Gosauschichten von Gams. *Mitt. Ges. Geol. Bergbau Stud. Wien*, 13: 189–212.

KRAUS, E. C., 1960. Das Orogen, Begriff, Bildungsweise und Erscheinungsformen. *Intern. Geol. Congr., 21st, Copenhagen, 1960, Rept. Session Norden*, 18: 236–248.

KRÖLL, A., 1964. Ergebnisse der geophysikalischen Untersuchungen. *Erdöl Z.*, 80: 221–227.

KÜHN, O., 1947. Zur Stratigraphie und Tektonik der Gosauschichten. *Sitz. Ber. Akad. Wiss. Wien, Math. Naturw. Kl., Abt. 1*, 156: 181–200.

KÜHN, O., 1962. *Lexique Stratigraphique International. I. Europe. 8. Autriche.* Centre Natl. Rech. Sci., Paris, 646 pp.

KÜPPER, H., 1965a. *Geologie von Wien.* Hollinek, Wien, 194 pp.

KÜPPER, H., 1965b. Quasicraton und Orthogeosynklinale (Ostalpen und Böhmische Masse im Kenntnisbild der heutigen Geologie). *Eclogae Geol. Helv.*, 58: 73–85.

KÜPPER, H., 1966. Elemente eines Profils von der Böhmischen Masse zum Bakony. *Verhandl. Geol. Bundesanstalt (Austria), Sonderh.*, G: 52–55.

LEONARDI, P., 1955. *Breve Sintesi Geologica delle Dolomiti Occidentali.* Failli, Rome, 80 pp.

LEONARDI, P., 1963. Die Tektonik der Dolomiten im Rahmen des südalpinen Baues. *Geol. Rundschau*, 53: 101–112.

LEONARDI, P., 1965. Tettonica e tettogenesi delle Dolomiti. *Mem. Accad. Lincei Roma*, 8 (7): 85–212.

LEONARDI, P., 1967. *Le Dolomiti. Geologica dei Monti tra Isarco e Piave. Consiglio Nazionale delle Richerche e della Giunta Provinciale di Trento.* Manfrini, Rovertero.

MÄGDEFRAU, K., 1946. *Paläobiologie der Pflanzen*, 3. Aufl. Fischer, Jena, 451 pp.

MAUCHER, A., 1960. Der Permische Vulkanismus im Raum von Trient. *Geol. Rundschau*, 49: 487–498.

MAUCHER, A. und PICHLER, H., 1959. *Führer zur Pfingstexcursion (1959) der Geologischen Vereinigung nach Südtirol zum Studium des permischen Vulkanismus.* Inst. Angew. Geol. Mineral., Munich, 23 pp.

METZ, K., 1954. *Geologische Karte der Steiermark, 1:300,000.* Akad. Verlagsanstalt, Graz.

METZ, K., 1958. Gedanken zu baugeschichtlichen Fragen der steirisch-kärntnerischen Zentralalpen. *Mitt. Geol. Ges. Wien*, 50: 201–250.

METZ, K., 1962. Das ostalpine Kristallin der Niedern Tauern im Bauplan der NE-Alpen. *Geol. Rundschau*, 52: 210–225.

MITTEMPERGHER, M., 1958. La serie effusiva paleozoica del Trentino, Alto Adige. *Com. Naz. Rich. Nucl. Rome, Div. Geomineral.*, 1: 1–87.

MITTEMPERGHER, M., 1963. Rilevamento e studio petrografico delle vulcaniti paleozoiche della Val Garbena. *Atti Soc. Tosc. Sci. Nat., A*, 2: 1–41.

OBERHAUSER, R., 1963. Die Kreide im Ostalpenraum österreichs in mikropaleontologischer Sicht. *Jahrb. Geol. Bundesanstalt (Austria)*, 106: 1–88.

OBERHAUSER, R., 1965. Zur Geologie der West-Ostalpen-Grenzzone in Voralberg und im Prätigau unter besonderer Berücksichtigung der tektonischen Lagebeziehungen. *Verhandl. Geol. Bundesanstalt (Austria), Sonderh.*, G: 184–190.

PAVONI, N., 1961. Faltung durch Horizontalverschiebung. *Eclogae Geol. Helv.*, 54: 515–534.

PLÖCHINGER, B., 1961. Die Gosaumulde von Grünbach und der Neuen Welt (N.Ö.). *Jahrb. Geol. Bundesanstalt (Austria)*, 104: 359–441.

PLÖCHINGER, B., 1964. Die tektonischen Fenster von St. Gilgen und Strobl am Wolfgangsee (Salzburg, Österreich). *Jahrb. Geol. Bundesanstalt (Austria)*, 107: 11–70.

PREY, S., 1962. Flysch und Helvetikum in Salzburg und Oberösterreich. *Z. Deut. Geol. Ges.*, 113: 282–292.

REINHARD, M., 1964. Über das Grundgebirge des Sottoceneri im Süd-Tessin und die darin auftretenden Ganggesteine. *Beitr. Geol. Karte Schweiz*, 117: 89 pp.

RICHTER, M., 1948. Die Entwicklung der Anschauung über den Bau der deutschen Alpen. *Z. Deut. Geol. Ges.*, 100: 338–347.

RICHTER, M. and SCHÖNENBERG, R., 1954. Über den Bau der Lechtaler Alpen. *Z. Deut. Geol. Ges.*, 105: 55–79.

SARNTHEIM, M., 1962. Beiträge zur Tektonik der Berge zwischen Memminger und Würtemberger Hütte (Lechtaler Alpen). *Jahrb. Geol. Bundesanstalt (Austria)*, 105: 141–172.

SCHAFFER, F. X. (Editor), 1951. *Geologie von Osterreich*, 2. Aufl. Deuticke, Wien, 810 pp.

SCHÖNENBERG, R., 1959. *Die Tektonik im Gebiet der Memminger Hütte und ihre Bedeutung für den Bau der Lechtaler Alpen.* Deut. Alpenverein, Memmingen.

SELLI, R., 1963. Schema geologico delle Alpi Carniche et Giulie Occidentali. *Ann. Museo Geol. Bologna*, 30: 1–121.

SPENGLER, E., 1953. Versuch einer Rekonstruktion des Ablagerungsraumes der Decken der Nördlichen Kalkalpen. I. Der Westabschnitt. *Jahrb. Geol. Bundesanstalt (Austria)*, 96: 1–94.

SPENGLER, E., 1956. Versuch einer Rekonstruktion des Ablagerungsraumes der Decken der Nördlichen Kalkalpen. II. Der Mittelabschnitt. *Jahrb. Geol. Bundesanstalt (Austria)*, 99: 1–74.

SPENGLER, E., 1959. Versuch einer Rekonstruktion des Ablagerungsraumes der Decken der Nördlichen Kalkalpen. III. Der Ostabschnitt. *Jahrb. Geol. Bundesanstalt (Austria)*, 102: 193–312.

SPENGLER, E., 1963. Les zones de faciès du Trais des Alpes calcaires septentrionales et leurs rapports avec la structure des nappes. In: M. DURAND DELGA (Editor), *Livre à la Mémoire du Professeur P. Fallot. II.* Soc. Géol. France, Paris, pp. 465–475.

STAUB, R., 1950. Betrachtungen über den Bau der Südalpen. *Eclogae Geol. Helv.*, 42: 220–408.

TERMIER, P., 1903. Les nappes des Alpes orientales et la synthèse des Alpes. *Bull. Soc. Géol. France*, 4 (3): 711–765.

TOLLMANN, A., 1963a. Der Baustil der tieferen tektonischen Einheiten der Ostalpen im Tauernfenster und in seinem Rahmen. *Geol. Rundschau*, 52: 226–237.

TOLLMANN, A., 1963b. *Ostalpensynthese.* Deuticke, Wien, 256 pp.

TOLLMANN, A. 1963c. Résultats nouveaux sur la position, la subdivision et le style structural des zones helvétiques, penniques et austro-alpines des Alpes orientales. In: M. DURAND DELGA (Editor), *Livre à la Mémoire du Professeur Paul Fallot. II.* Soc. Géol. France, Paris, pp. 477–490.

TOLLMANN, A., 1963d. Tabelle des Paläozoikums der Ostalpen. *Mitt. Ges. Geol. Bergbau Stud. Wien*, 13: 213–228.

TOLLMANN, A., 1964. Die Fortsetzung des Briançonnais in den Ostalpen. *Mitt. Geol. Ges. Wien*, 57: 469–478.

TOLLMANN, A., 1965. Die Neuergebnisse der geologischen Forschung in Österreich. *Naturhistorikertagung 1965, Beiblätter*, pp.3–57.

VAN BEMMELEN, R. W., 1957. Beitrag zur Geologie der westlichen Gailtaler Alpen (Kärnten, Österreich) *Jahrb. Geol. Bundesanstalt (Austria)*, 100: 179–212.

VAN BEMMELEN, R. W., 1961a. Volcanology and geology of ignimbrites in Indonesia, North Italy and the U.S.A. *Geol. Mijnbouw*, 40: 399–411.

VAN BEMMELEN, R. W., 1961b. Beitrag zur Geologie der Gailtaler Alpen (Kärnten, Österreich). II. Die zentralen Gailtaler Alpen. *Jahrb. Geol. Bundesanstalt (Austria)*, 104: 213–237.

VAN BEMMELEN, R. W., 1965. Der gegenwärtige Stand der Undationstheorie. *Mitt. Geol. Ges. Wien.*, 57: 379–400.

VAN BEMMELEN, R. W., 1966. The structural evolution of the Southern Alps. *Geol. Mijnbouw*, 45: 405–444.

VAN BEMMELEN, R. W., 1967. "Stockwerktektonik" sensu lato. In: J. P. SCHAER (Editor), *Colloque Étages Tectoniques.* Baconniere, Neuchâtel, pp.19–40.

VAN BEMMELEN, R. W. and MEULENKAMP, J. E., 1965. Beiträge zur Geologie des Drauzuges (Kärnten, Österreich). III. Die Lienzer Dolomiten und ihre geodynamische Bedeutung für die Ostalpen. *Jahrb. Geol. Bundesanstalt (Austria)*, 108: 213–268.

VAN HILTEN, D., 1960. Geology and Permian paleomagnetism of the Val-di-Non area, W. Dolomites, N. Italy. *Geol. Ultraiectina*, 5: 95 pp.

VAN HILTEN, D., 1962. Presentation of paleomagnetic data, polar wandering, and continental drift. *Am. J. Sci.*, 260: 401–426.

VAN HILTEN, D., 1964. Evaluation of some geotectonic hypotheses by paleomagnetism. *Tectonophysics*, 1: 3–71.

VAN HINTE, J. E., 1963. Zur Stratigraphie und Mikropaläontologie der Oberkreide und des Eozäns des Krappfeldes (Kärnten). *Jahrb. Geol. Bundesanstalt (Austria), Sonderb.*, 8: 147 S.

VAN HINTE, J. E., 1965. Remarks on the Kainach Gosau (Styria, Austria). *Kon. Ned. Akad. Wetenschap., Proc., Ser. B*, 68: 72–92.

VON HILLEBRANDT, A., 1961. Das Alttertiär im Becken von Reichenhall und Salzburg. *Z. Deut. Geol. Ges.*, 113: 339–358.

VON PIA, J., 1941. Kalkalgen der Adria und ihre fossilen Verwandten. *Natur Volk*, 71: 39–49; 84–90.

WUNDERLICH, H. G., 1964. Zur tektonischen Synthese der Ost- und Westalpen nach 60 Jahren ostalpiner Deckentheorie. *Geol. Mijnbouw.*, 43: 33–51.

ZACHER, W., 1961. Zur tektonischen Stellung der Vilser Alpen. *Z. Deut. Geol. Ges.*, 113: 390–408.

ZANKL, H., 1968. Die Karbonat sedimente der Obertrias in den Nördlichen Kalkalpen. *Geol. Rundschau*, in press.

ZAPFE, H., 1962. Untersuchungen im obertriadischen Riff des Gosaukammes (Dachsteingebiet, Oberösterreich). *Verhandl. Geol. Bundesanstalt (Austria)*, 1962: 346–360.

ZAPFE, H., 1963a. Das Mesozoikum in Österreich. *Mitt. Geol. Ges. Wien*, 56: 360–400.

ZAPFE, H., 1963b. Beiträge zur Paläontologie der nordalpinen Riffe. *Ann. Naturhist. Museum Wien*, 66: 207–259.

CHAPTER 14

Alpine Europe: Pyrenees and Lower Provence

INTRODUCTION

The title of this chapter needs some justification. Pyrenees and Lower Provence are treated together, because they show a certain similarity in their Alpine history. But, as will be seen, this does not mean overall identity, neither during Alpine nor during earlier history.

Moreover, the exact content of these names, as used in this context, needs some elaboration. "Pyrenees" in this text only indicates the mountain range in its geographical sense, not its westward, geological, continuation. And of the Provence, only the southern half of this southeastern province of France, the Lower Provence, will be considered in this chapter.

As for the Pyrenees, these end, geographically speaking, as a separate mountain chain in the west at the coast of the Atlantic. Their structure, however, can be followed westwards, along the northern coast of Spain. West of the Pyrenees proper follows an area of axial depression, in which only Mesozoic and Tertiary strata crop out (cf. LLOPIS LLADO, 1948; MANGIN and RAT, 1962). This region has recently been called the Vascogothian chain by WIEDMANN (1962). But further west the fold axes rise again, and consequently the Hercynian basement crops out in the Asturian and Cantabric Mountains, which culminate topographically in the Picos de Europa (DE SITTER, 1949, 1962). These ranges will, however, not be considered in this text.

As for the Provence province, that is for southeastern France east of the Rhône, this belongs geologically to two major elements. Its northern part, the Higher Provence, comprises the southern slopes of the French Alps. It is formed by elements of the Subalpine chains, and goes by the name of Provençal Alps. It hence belongs to the Alps proper. Its southern part, the Lower Provence, on the other hand, is taken up by west–east trending structures, which were folded during the Pyreneean phase of the Alpine orogeny. Both in their general strike, in their time of folding and in their stratigraphy, they show more resemblances to the Pyrenees, than to the Alps towards the north. This is the reason why the southern part of the Provence, geologically the most important part of this province, is included in the same chapter with the Pyrenees here.

Notwithstanding these similarities, there are, however, also marked differences between Pyrenees and Lower Provence. The Pyrenees, for instance, can very schematically be presented as a symmetrical fold belt, formed by the longitudinal Axial zone of Paleozoic rocks, bordered on both sides by Mesozoic and Tertiary. The Provence, on the other hand is formed by Alpine structures of much shorter dimensions, amongst which the Hercynian basement only crops out in two areas which are reminiscent of the Central Massifs of the Alps. Or, to cite only one other difference, the Pyrenees have often been cited as a good example of a land-locked orogen, situated between the main continental part of Europe and the Iberian subcontinent. Such a situation has never been proposed for the Lower Provence, for the very good reason that it adjoins the Mediterranean, and consequently can hardly be called land-locked.

Although similar in various ways, it follows that there are also marked differences in structure between Pyrenees and Lower Provence.

THE PYRENEES

The Pyrenees form the long (400 km), straight and narrow (40–80 km) mountain chain that separates France from Spain. It runs all the way from the Atlantic to the Mediterranean in the same WNW–ESE direction (Fig.205).

Geologically, the most important characteristic of the Pyrenees is that we have here an example of a mountain chain built up by crustal elements formed by two consecutive major orogenetic periods, i.e., the Hercynian and the Alpine, in which the older rocks have not been more or less completely metamorphosed.

This is due, mainly, to a feature often brought into a causal relationship with the supposed land-locked origin of the mountain chain, that is that Mesozoic and Tertiary are not of truly geosynclinal, but of epicontinental character. This has resulted in a relatively thin cover of Alpine sediments, and a correspondingly mild metamorphism of the earlier rocks. In fact, practically all of the metamorphism of the Paleozoic and earlier rocks dates from the Hercynian orogenetic cycle. In the Pyrenees this embraces rocks of Late Precambrian and Early Paleozoic age, due to the absence of the Caledonian orogeny. In contrast to the Alpine orogeny, it is characterized by a truly geosynclinal phase.

PYRENEES AND ALPS

I should like, at this point, to go into somewhat more detail about this supposed land-locked condition of the Pyrenees. This was thought to prove that, whereas the Alps were a real first class orogen, the Pyrenees were not. For were not the Alps derived from the Tethys geosyncline, and the Pyrenees from some intercontinental basin? In this way it was also easy to understand, why the Alps were asymmetrical and the Pyrenees symmetrical (cf., for instance, DE SITTER, 1956; or more recently, MATTAUER, 1968). According to several authors the Pyrenees cannot even be reckoned to the Alpine fold belt, as we saw already in Chapter 9.

At present the Pyrenees are, of course, land-locked, wedged in between Meso-Europe and the Spanish Meseta. But what was their position, we may ask, before the Alpine orogeny. CAREY (1958), for instance, has come up with a supposed anti-clockwise rotation of Spain over 40°, which would already make room for a much larger basin of sedimentation for the Pyrenees. And paleomagnetic measurements for the Permo-Triassic of the southern part of the axial zone (VAN DER LINGEN, 1960; SCHWARZ, 1962) found

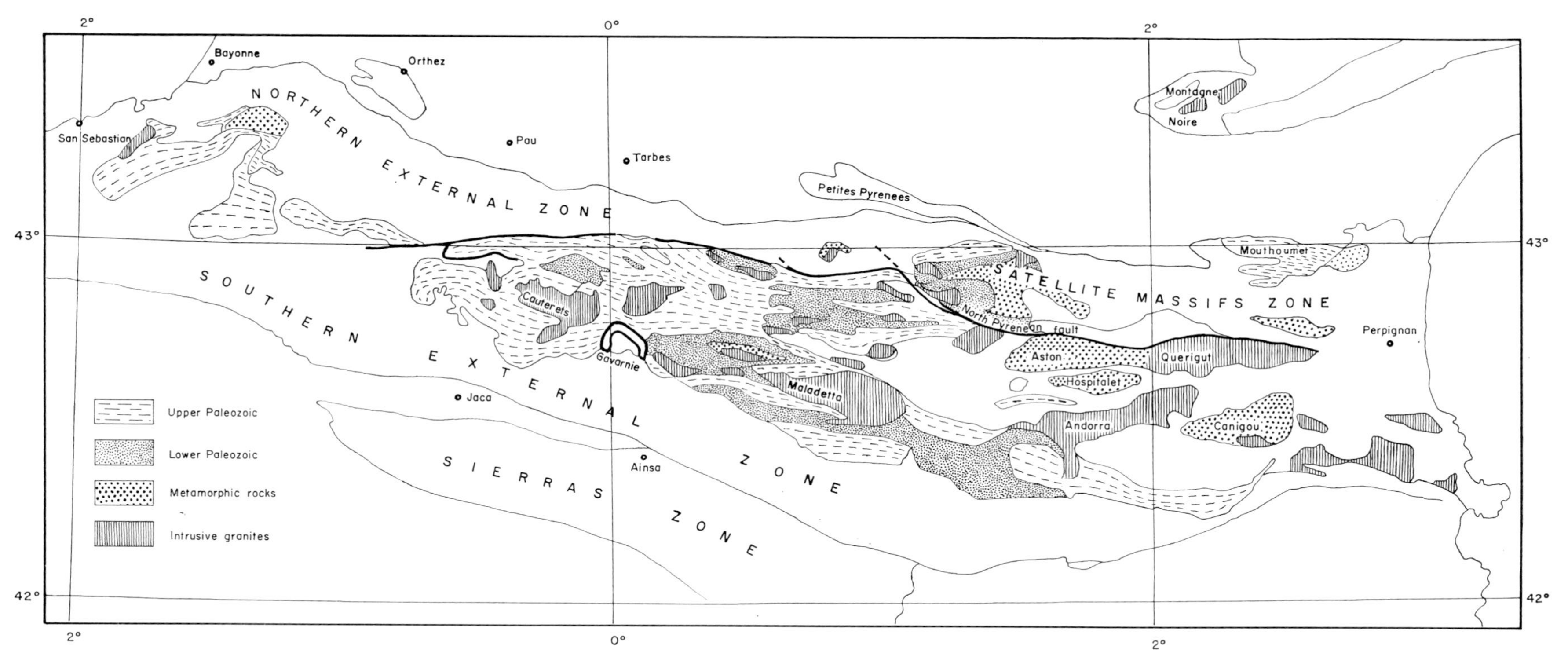

Fig.205. Schematic geologic map of the Pyrenees.

inclinations deviating considerably from those of Meso-Europe. This would even indicate a considerably larger extra continental drift of both Spain and the Pyrenees relative to Meso-Europe.

It is well possible that further investigations will considerably diminish the differences supposed to exist between Alps and Pyrenees.

It must moreover be stressed that these differences were drawn up at a time when an orogen was supposed to be something of a single, simple unit. At that time, differences found to exist between "the" Alps and "the" Pyrenees, had all to be explained by some major factor, by which all differences thought to exist between two orogenes could be reduced to a single, and possibly simple, cause. Since then, it has become obvious that the structure of orogenes is much more complicated. And the existence of along-the-strike differences, found to occur within what originally seemed to be one single, simple orogen, has of course diminished the emphasis formerly laid upon differences between two or more orogenes.

MAIN DIVISIONS OF THE PYRENEES

For the Pyrenees also, the structure is more complex than has originally been supposed. The simplest schematization, cited already is that of a symmetrical orogen, consisting of an axial zone accompanied, both to the north and to the south, by external zones (JACOB, 1930). The axial zone, containing pre-Alpine, Hercynian, and even older, rocks, is thought to be the oldest. The external zones interpreted as marginal troughs, are younger, and filled with Mesozoic and Tertiary strata. Moreover it is assumed that Alpine orogenesis was no more than a rejuvenation of the "Hercynian chain".

Still further outward follow smaller folds of Jura Mountains type, amongst them the "Petites" or Lesser Pyrenees, in the north (CUVILLIER et al., 1962), and the Sierra zone (SELZER, 1934) in the south. These structures are, however, always distinctly separate from the Pyrenees proper, and consequently do not enter into this discussion.

This simple, longitudinal division into a symmetric set of zones holds good only for the western and part of the central Pyrenees[1]. Towards the east the axial zone not only widens and shows stronger differentiation, but the structure of the Pyrenees is complicated by the addition of a peculiar zone to the north of the axial zone. This element consists of strongly folded Mesozoic, with Lower Cretaceous in an aberrantly thick development, surrounding the so-called satellite massifs. The latter resemble the axial zone, in being built up by granites, gneiss, schists, and less metamorphosed Paleozoic sediments.

This extra unit of a "northeastern Pyrenean zone with satellite massifs", or "satellite massifs zone" for short, is separated from the axial zone by a major fault, the north Pyrenean fault. This has all the earmarks of a transcurrent fault. It is quite probable that because of this fault crustal blocks which were once widely separated have now been brought into near contact.

DE SITTER (cf. 1956, 1964), who, with his school, mapped the eastern central Pyrenees,

[1] In the Pyrenees, quite sensibly, "central" indicates the central part of the mountain chain, between its western and eastern parts. There is none of the ambiguity found in the Alps, where "Zentralalpin" indicates a longitudinal zone of the Austrian Alps.

in which not only the axial zone is at its widest, but also the satellite massifs zone, has always tried to extrapolate the latter zone as forming a major unit, not only all along the northern flank, but also along the southern flank of the Pyrenees. This is contrary to the factual data, because both to the northwest and to the south of the axial zone no satellite massifs are found. To me this is but one other example of a tectonician trying to squirm a foldbelt into a preconceived model.

This difference of opinion notwithstanding, we may be thankful to De Sitter, for he is the only modern author who has dared to present a general picture of the Pyrenees. In view of the importance of his generalizations for the general reader, a certain mix-up in his terminology must, however, be noted. The satellite massifs zone and its supposed southern counterpart are called "north Pyrenean zone with satellite massifs" and "south Pyrenean zone with gliding tectonics" by DE SITTER (1956, fig.1). They are, however, indicated as "internal zones" in DE SITTER'S (1956) text, and in his fig.2, which is reproduced here as Fig.206, also because of their position between the axial zone and the marginal troughs (cf. also DE SITTER and ZWART, 1959). Usage of the word "internal" is rather misleading, because in the Pyrenees it is seen to mean something quite different from the internal zones of the nappes in the Alps. In DE SITTER (1964) these same zones are indicated as "external zones", in the text, although in DE SITTER'S (1964) fig.282, copied from the 1956 publication, they still retain their old names. The southern external zone has been rechristened "Nogueras zone" (cf. DALLONI, 1910) in DE SITTER (1965), cf. Fig.207.

But apart from the complication of an added "northeastern Pyrenean zone with satellite massifs", things are more complicated than transpires from the simple scheme of an older axial zone, flanked by younger external zones.

In the western Pyrenees, where the axial zone is much narrower, so that its relation to the marginal zones becomes more apparent, no indication is found of a wedging-out of either the Cretaceous or the Eocene towards the axial zone (VAN ELSBERG, 1966). This is moreover borne out along the westward plunging nose of the axial zone. The existence of an erosional remnant of Cenomanian on the top of the Balaïtous in the centre of the axial zone of the western Pyrenees (MIROUSE and SOUQUET, 1964) points in the same direction.

Differences certainly exist in the stratigraphy of the northern external zone and the southern external zone. However, variations in stratigraphy also exist along the strike within the same zone and it is hard to tell whether the variations across the chain are really more important than those along the chain.

So, in a general way, bot Mesozoic and Eocene extended over the present axial zone before the Alpine orogeny. Uplift of the axial zone, leading to erosion of the younger sediments has since uncovered the Paleozoic rocks. It follows that the present trend of the

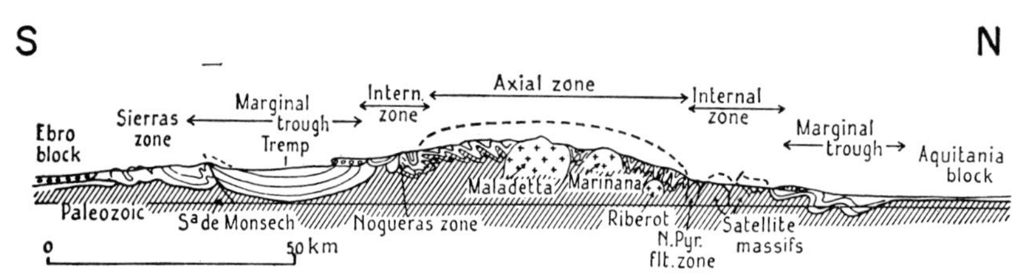

Fig.206. Schematic cross-section through the central Pyrenees. (After DE SITTER, 1956.)

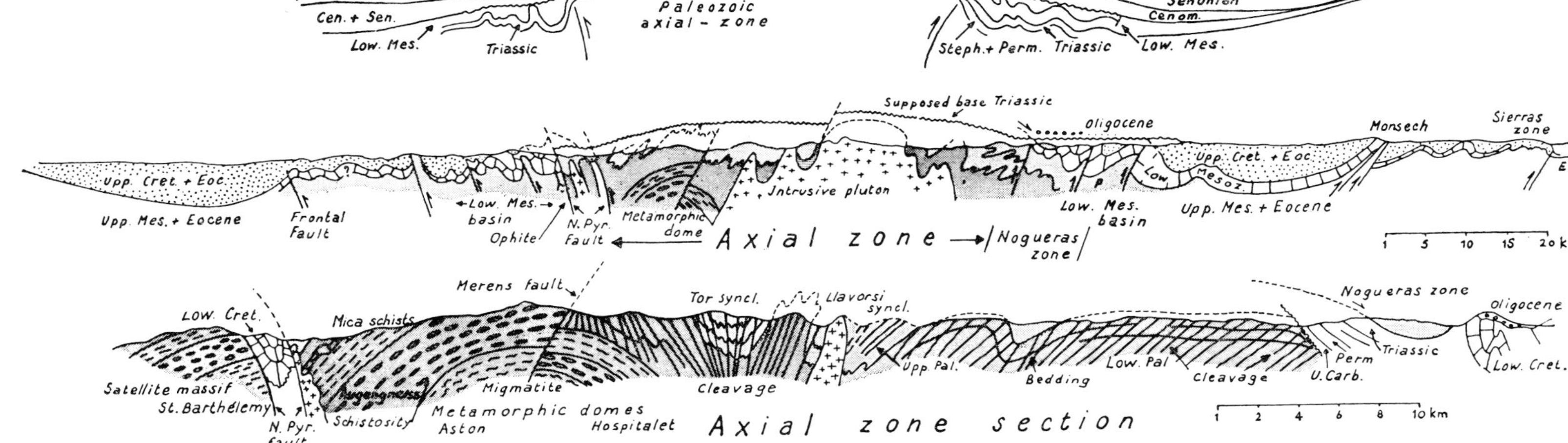

Fig.207. Cross-sections through the eastern part of the central Pyrenees.

The upper section is a reconstruction of the situation after the pre-Cenomanian folding. It is supposed that individualization of the axial zone, the external zones and the marginal basins had begun at that time. The middle section is a generalized actual cross-section. The lower section is a section through the axial zone only.

North of the north Pyrenean fault (*N.Pyr. fault* in the middle section), lies the satellite massifs zone. The Nogueras zone south of the axial zone was formerly called southern internal (or external) zone. (After De Sitter, 1965.)

Pyrenees is largely due to the Alpine orogeny. It is quite true that in the axial zone the oldest rocks are exposed, but it is not true that the zone as such is older. Nor are the marginal zones younger, only the rocks exposed are younger further outwards.

Nor is it true that the Alpine structure is no more than a rejuvenation of the earlier Hercynian fold belt. Detailed analyses of Hercynian structures, both in the axial zone and in the satellite massifs zone have revealed structural trends quire distinct from the pronounced west-northwest–east-southeast Alpine direction (pp.351–352). Moreover, the stratigraphy of the Lower Paleozoic is in many ways similar to that of the Montagne Noire (pp.163–166). It seems consequently more realistic to consider the present axial zone of the Pyrenees as part of an earlier, much more extensive, Hercynian orogen, which has been more or less forfuituously uplifted and uncovered in the centre of a cross cutting Alpine orogen.

OROGENETIC PHASES

The two orogenies which have been active in the formation of the Pyrenees, the Hercynian and the Alpine, can be defined somewhat more closely as to time of folding.

The main Hercynian folding took place sometime during the Early Westphalian, although earlier nonconformities within the Lower Carboniferous have also been cited. Upper Westphalian and Stephanian, although folded themselves, are clearly transgressive at several points along the axis of the Pyrenees. This relegates the Hercynian folding in the Pyrenees to the Sudetic phase, quite in accordance with the position of the Pyrenees in their relation to the areas of limnic and paralic coal basins in western Europe (pp.127–129).

As to the Alpine orogeny, a distinction is always maintained between a precursory and a main phase. The earlier phase is correlated with the stratigraphic gap of varying width preceding the "Cenomanian transgression", which also itself varies in age. It is consequently dated as "pre-Cenomanian". It is, however, not a true folding phase, its influence being limited almost everywhere to vertical movements and tilting of crustal blocks.

The main phase of the Alpine orogeny in the Pyrenees, the Pyrenean phase, derived its name from this mountain range. Stille placed it in the Late Eocene between Bartonian and Ludian. It has since aged, and is now generally held to have occurred at the end of the Eocene, at the top of the Lutetian (GIGNOUX, 1955, pp.506, 509).

This dating has lately been contested by MANGIN (1959, 1960), on the basis of micropaleontological studies of the Lower Tertiary around the western Pyrenees (see also SELZER, 1934). This forms a renewed application of Professor J. Cadisch's tenet of using foreland sedimentation in deciphering mountain building (p. 220). Mangin derives the Eocene flysch type sediments from the south, from the area of the present Ebro Basin. In his ideas, at least part of this basin formed a high during the Eocene.

A "Pyrenean" main phase has been defended for the Pyrenees by DE SITTER (1961), and the issue is still open (see also VAN DE VELDE, 1967). It might of course well be possible that folding in the axial zone — De Sitter's area — preceded that of Mangin's area in the peripheral sedimentary basins.

LITERATURE

To close this short introduction on the Pyrenees, a note on the literature seems in order. Skipping all earlier work, one might start with Léon Bertrand, who is, incidentally, no relation of Marcel Bertrand, the French geologist who had earlier convinced Alpine geologists of the existence of major nappe structures. In his work in the Pyrenees, in the beginning of this century, L. BERTRAND (1908, 1911), postulated major nappe structures for the Alpine tectonics of the Pyrenees. With some exceptions, notably the Gavarnie nappe (p. 356), these have, however, not stood up to further investigation. Most of the Alpine overthrusting in the Pyrenees is of the order of magnitude which in the Alps is at most regarded as par-autochthonous (cf. CASTERAS, 1933).

Several earlier and younger French authors have been mainly interested in stratigraphy, whereas a group of German investigators of the school of Stille were so heavily ingrained with preconceived structural ideas that, with the exception of the thesis of MISCH (1934), their publications lack sufficient stratigraphic backing.

The Pyrenees have suffered from the fact that they form a border chain. Most investigators have studied only one flank, either the French, or the Spanish side of the mountains. This of course precludes a sound general view. De Sitter with his students have, from the onset, been opposed to this. Their work, mainly published in the *Leidse Geologische Mededelingen*, now forms the only modern mapping of a cross-section of the axial zone of the Pyrenees (cf. DE SITTER, 1959; DE SITTER and ZWART, 1959, 1962; KLEINSMIEDE, 1960; ZANDVLIET, 1960).

For the eastern part of the chain we have moreover a modern description in the guide book to the 1958 "réunion extraordinaire" of the SOCIÉTÉ GÉOLOGIQUE DE FRANCE (1959).

Apart from this, several detailed studies from Utrecht University, published in *Estudios Geologicos*, and from various French Universities, mainly published or to be published by the *Service de la Carte Géologique*, form the fragmentary basis of our knowledge of the Pyrenees.

THE AXIAL ZONE OF THE PYRENEES

Within the axial zone three major elements can be recognized. These are not arranged on subzones, but form a more or less irregular mosaic within the axial zone. These three distinct elements are: (*1*) large areas of concordant sedimentary series; (*2*) large gneiss domes; and (*3*) granite batholiths.

Hercynian sedimentary series

The sedimentary series range all the way from (?) Upper Precambrian — the oldest fossils known from the Pyrenees belong to the Caradocian — to the Lower Carboniferous. Within the lower part of this series, beneath the Gothlandian, large, structurally simple domes are found, in which metamorphism and metasomatism have produced mantled gneiss domes. A well-known example is that of the Canigou in the eastern Pyrenees (GUITARD, 1954). The intrusive, larger and smaller granite batholiths belong to the group of the "circumscribed granites" of Professor E. Raguin of Paris, that have lost their connection with the areas of granitization, from which they presumably derive.

Within the sedimentary series a lower and an upper part are quite different in facies (cf. VIENNOT, 1927; SCHMIDT, 1941; CAVET, 1959; MIROUSE, 1962; WENSINK, 1962; CLIN, 1964). The lower part of the series, from the Caradocian downwards, consists of a monotonous clastic series of sands and clays, several kilometers thick. It is presumed that it reaches down into the Upper Precambrian. The upper part of the series is formed by an alternation of shales, marls and limestones, not over 1 km thick. Several of the limestones are formed by massive reefs, sometimes over 100 m thick, which play a prominent part in the tectonics of the axial zone.

Between the lower part and the upper part of the series of the axial zone the Silurian forms a relatively thin — up to 100 m thick — series. Apart from the local occurrence of conglomerates and limestones, it is characterized by black, fissile, carbonaceous graptolite shales. This zone is strongly tectonized almost everywhere in the Pyrenees. It normally forms the boundary between a lower tectonic "Stockwerk", characterized by large, simple dome-like structures, and a higher "Stockwerk", characterized by strong folding, overthrusting, and, may be, by minor nappes.

The situation described above is most clearly developed in the western Pyrenees. Towards the east, the Silurian becomes more sandy, and it is probably due to this difference in facies that there all of the Paleozoic sediments can be taken together as forming a single "Stockwerk". This has recently been done by RAGUIN (1963, 1965), who, beneath this "Stockwerk" of more or less metamorphosed Paleozoic sediments, distinguishes two major "Stockwerke", i.e., a middle one formed by migmatitic domes, and a lower one formed by the Precambrian basement which is less re-activated by metamorphism and migmatization.

It follows that crustal mobility was at its highest during the sedimentation of the lower part of the sedimentary series of the axial zone. It thus led to the filling of a basin with clastic sediments several kilometers thick, an event which was, of course, accompanied by the appropriate erosion of a rising hinterland somewhere. The Silurian was a period of tranquility, whereas in the Devonian crustal unrest started anew. In the upper part of the sedimentary series, in the Devonian and the Lower Carboniferous, fossiliferous sequences permit the distinction of a number of hiatuses of varying importance, and, moreover, of lateral changes in facies, which has, as yet, not been possible in the earlier sedimentary series. This does, however, not indicate stronger crustal movements during the later period. Instead, the thick clastic sedimentary series of the earlier part indicates the strongest crustal movement during the earlier geosynclinal stage in the Pyrenees.

Just as is the case in the Hercynian uplands of France, there is no indication of orogenetic movements during Caledonian times in the Pyrenees. Nor is there found any indication for an orogenetic Bretonic phase of the Hercynian orogeny. The supposed absence of Upper Devonian sediments is generally cited in favour of this phase, but this could well have indicated somewhat stronger movements of epeirogenetical character, and no more.

Moreover, the supposed gap between the Middle Devonian and the Lower Carboniferous has in later years been narrowed through conodont studies. It must be stated as a general fact that the Paleozoic of the Pyrenees is poor in fossils. Correlations are often based only on very few lucky fossil finds, if not on lithostratigraphy or on enigmatic remains such as *Pleurodictyon*. The more sweeping correlations, not only across the whole of the Pyrenees, but also over adjoining orogenes, must be regarded with scepsis.

Gneisses and granites

The influence of metamorphism and metasomatism within the axial zone shows strong lateral variation. It is, at present, not known which factors have contributed to this variation (RAGUIN, 1962).

Metamorphism is normally restricted to the lower part of the sedimentary series, but it does locally extend further upwards, reaching into the Devonian. On the other hand, although metamorphism and metasomatism has been very active in the lower part of the sedimentary series locally, the major part of the clastic series below the Gotlandian is found to be only very slightly metamorphosed in other areas of the Pyrenees.

The strong variation in the gradient of metamorphism and metasomatism in adjacent domes in the eastern Pyrenees has, for instance, been cited by GUITARD (1959), and is here illustrated in Fig.208.

Moreover, the metasomatism of the mantled gneiss domes seems to be not as strict, as, for instance, in the Massif Central. Instead of a single sequence from lightly metamorphic towards strongly migmatized rocks, repetitions are found to occur. Two-mica schists and gneisses are, for instance, repeated in the same section, such as that of the Canigou in the eastern Pyrenees (Fig.209).

Such repetition has been explained to be due to an alternation in the original composition of the sedimentary series out of which these anatectites developed, or to the incorporation of a pre-Hercynian basement in the Hercynian structure (RAGUIN, 1965). Such explanations form the easiest way out, when, as has generally been the case, these large mantled gneiss domes of the axial zone, are interpreted as forming part of the original autochthonous basement. There is, however, another possible explanation, recently put forward by GUITARD (1964). This author has postulated that the Canigou is not an autochthonous mantled dome, but that its core of gneiss, intercalated between two-mica schists, can be interpreted as the crystalline core of a major nappe. This then would be a nappe resembling the Pennide nappes of the Alps in structure, but being formed during the Hercynian orogeny.

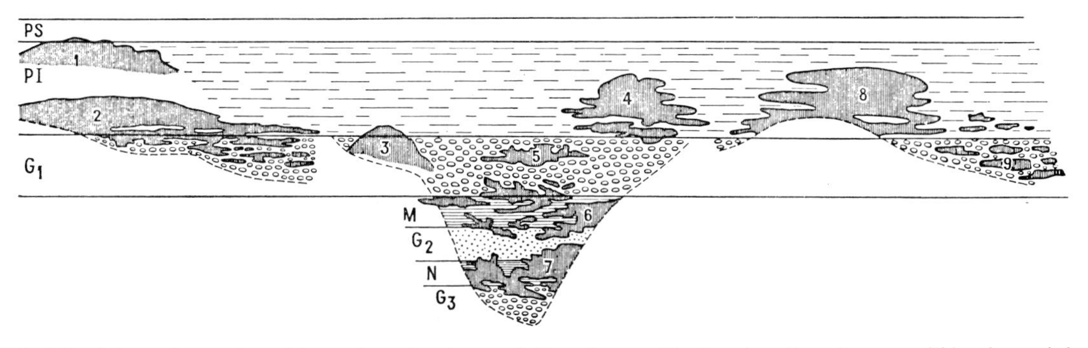

Fig.208. Schematic stratigraphic section, showing variations in granitization in adjacent areas within the axial zone of the eastern Pyrenees.

Stratigraphy: *PS* = Upper Paleozoic; *PI* = Lower Paleozoic; G_1 = stratified gneiss of the Canigou and the Carança; *M* = Balatg Micaschists; G_2 = Casemi Gneiss; *N* = deeper micaschists of the Canigou; G_3 = deeper gneiss of the Canigou.

Regional distribution: *1* = Querigut Granite; *2* = Mont-Louis Granite; *3* = Costabonne Granite; *4* = Batere Granite; *5* = Valmanaya Granite; *6* = Py Granite; *7* = Deeper Canigou Granite; *8* = Saint Laurent Granite. (After GUITARD, 1959.)

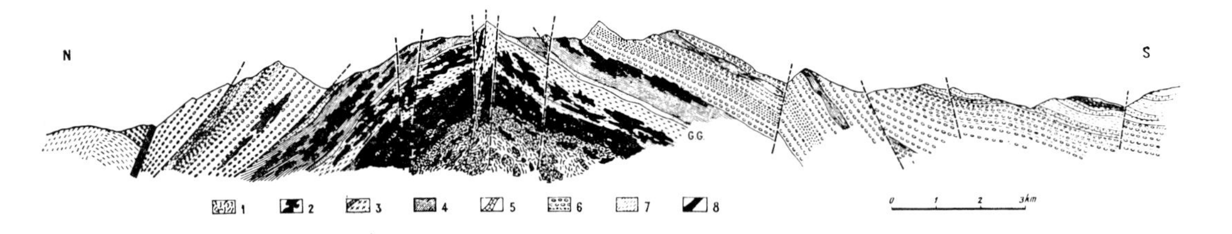

Fig.209. Cross-section through the Canigou mantled gneiss dome in the eastern Pyrenees. *1* = Sierra del Cadi Gneiss; *2* = leucocratic granite; *3* = Casemi Gneiss; *4* = Balatg Micaschists; *5* = "cipolin" marble; *6* = stratified gneiss; *7* = Canaveilles Micaschists; *8* = granular gneiss. (After GUITARD, 1959.)

Another aspect of the metamorphic areas within the axial zone is that they have led to detailed studies on microtectonics and/or petrofabrics by GUITARD (1959, 1961) and ZWART (1960, 1963). Both authors distinguish between several cinematic and metamorphic phases during the Hercynian orogeny. But though it is possible to date the sequence of these events, it is, as yet, not possible to assign them to a general chronology, be it relative or absolute. We meet here with a difficulty similar to that encountered by R. Hoeppener in his analysis of the microtectonics of the Rheinisches Schiefergebirge (p. 125).

There is, however, one major result of these studies which transcends their more localized importance. That is, that they have proved the existence of Hercynian orogenetic directions (folding lineations and such), strongly at variance with, and in part even perpendicular to, the WNW–ESE Alpine direction. It is these findings, which led me to postulate on p. 348 that the present Pyrenees, that is the Alpine orogen, show in their axial zone a basement consisting of part of an earlier, entirely unrelated, Hercynian orogen.

The intrusive granite batholiths, the "circumscribed granites", finally, are quite normal granite batholiths, without special features of their own. Interested readers will have to turn to the regional literature for further information.

It has, however, been seriously questioned whether the classification used in the Pyrenees, distinguishing between deep-seated migmatite gneisses and granite batholiths intruded at a high level into the crust, is at all sensible. GLANGEAUD (1959) has, for instance, stressed the fact that there must be transitions between the two. His general ideas are expressed in a well-documented genetical section. But it is much harder to indicate the presence of these transitional elements in the field. A glance at the map of Fig.205 shows that there are more migmatite domes in the eastern Pyrenees and more intrusive batholiths in the western and part of the central Pyrenees. If one interprets this with Glangeaud to indicate a generally stronger gradient of metamorphism and metasomatism in the eastern Pyrenees, a zone of transition ought to be found somewhere at the meridian of Andorra. In this area, matters are complicated by semantic variations between the systematics of the French and the Leyden schools. De Sitter and his school (cf. DE SITTER, 1956) distinguish between syntectonic leucocratic granites related to the migmatites, and post-tectonic granodiorites, which makes an understanding rather difficult.

One of these "transitional" elements, however, seems recently to be found in the Lys-Caillaouas Granite of the central Pyrenees (CLIN et al., 1963).

It is, of course, probable that transitions between the deep-seated migmatite domes and the much higher intrusive granites exist. Such transitions have been noted, in fact, elsewhere, such as in the Massif Central. It is, however, not necessary that such transitions must be cut by the present topography, so they might consequently escape detection.

It is quite possible that the fact why in the Pyrenees we only find either deep-seated migmatite domes or the superficially intruded end members of what ought to be a more or less continuous series of granitization is due to the pronounced "Stockwerk" tectonics within the axial zone.

As a result of the "décollement" over the Silurian shales, the upper part of the Paleozoic, the Devonian and the Lower Carboniferous, together with the granites intruded to a high level in the crust, forming the higher "Stockwerk", have virtually lost contact with the lower "Stockwerk", consisting of the Lower Paleozoic containing the migmatite domes.

THE EXTERNAL ZONES OF THE PYRENEES

The northern and southern external zones form the flanks of the Pyrenees, in which the sedimentary series of the Alpine orogenetic cycle have not yet been removed by erosion. Although there are marked variations in facies, both between the northern and the southern zone, and along the strike of each zone, there is one overall property, i.e., that Alpine sedimentation has been of epicontinental character.

Alpine sedimentary series

Sedimentation has been intermittent, and also the Alpine sediments form only a relatively thin cover. All during Early and Middle Mesozoic sedimentation has been interrupted repeatedly. It is only from the Late Cretaceous onward that sedimentation was more or less continuous over the larger part of the Pyrenees. The thickness of the Mesozoic generally is of the order of 1 km, a figure rarely doubled.

Post-Hercynian sedimentation starts within localized basins in Late Westphalian times, which state of affairs continues in a haphazard way into the Permian. A first more generalized sedimentation is formed by continental red beds, as yet undated, and generally referred to as "Permo-Triassic". Most of the Triassic is developed in Germanic facies. Its most important feature is the occurrence of evaporites, which have been a major factor determining the development of the thrust sheets during the Alpine orogenesis.

A "Spanish" element of the Pyrenean Triassic is formed by its volcanics. Some of these occur as flows and intrusions of andesitic and related composition in the red beds (Van der Lingen, 1960; Schwarz, 1962; Morre and Thiébault, 1963; Mirouse, 1966), such as the famous Pic du Midi d'Ossau. But the greater part is formed by what in Spain is called "ophites". These are formed by basic greenstones, without a clearly defined structure. A difference with the ophiolites of the Alps is that they occur in epicontinental, and even in continental series, and consequently have nothing to do with an eugeosynclinal development. The ophites are always strongly tectonized, which has obliterated all intrusive contacts, and which has, up to now, prevented a genetic interpretation.

The Jurassic and Lower Cretaceous are sporadically represented in the northern external zone and from both the western and eastern ends of the southern external zone (Casteras, 1933; Casteras et al., 1957; Société Géologique de France, 1959; Cuvillier et al., 1962).

The Upper Cretaceous is represented in both external zones, but differences in facies occur. Very schematically, the Upper Cretaceous in the north generally starts with a transgression during the Campanian with conglomerates and clastic series. During the Maastrichtian, flysch type sediments have been deposited. Moreover, the upper part of the Maastrichtian is developed in the northeastern part of the mountain chain in the continental Garumnian facies, which stretches all the way to the Provence. In the south, Upper Cretaceous sedimentation in some districts starts also with the Cenomanian, but elsewhere it may only begin with the Campanian. Moreover, the series is largely carbonatic, formed by marls and limestones and subordinated dolomites (MISCH, 1934; VAN DE VELDE, 1966; VAN ELSBERG, 1966).

The difference in facies between the northern and southern external zones, cited above, has always been stressed. It should, however, not be forgotten, that within each of these zones, in most formations, there exist along the strike variations in facies of the same order of magnitude.

The differences in stratigraphy between the northern and southern external zones have perhaps been overstressed owing to the fact that these differences appear suddenly, because of the absence of data from the intervening area of the axial zone. In contrast, the variations along the strike in either of the external zones can be followed step by step and hence give a much more gradual impression.

A word must be added about the supposed "marginal troughs" which according to many writers are thought to accompany the external zones along their outer borders. A marginal trough is a requisite feature of a tectonic model for a symmetrical orogen composed of an older axial zone, bordered by sedimentation troughs which spread outwards with time.

In the Pyrenees, however, the axial zone, as a tectonic element, is not older than the external zones. Nor has the sedimentation in the latter areas been influenced by the later tectonic lines. A theoretical necessity for two marginal troughs therefore does not exist.

Nevertheless it is true that in various parts along the Pyrenean flanks, thicker sedimentary series have accumulated at various times. But to connect such basins, that are localized both in area and in time, and to represent them as marginal troughs, is an oversimplification, based, in most cases more on lack of data than on anything else.

When more data become available, such as from the western part of the Aquitanian Basin to the northwest of the Pyrenees (CUVILLIER et al., 1962), both the limited nature of the sedimentary basins at a given time, and the rather irregular variations of the thicknesses reached, at once become apparent (Fig.210).

The only real trough in this area is that of the Late Cretaceous and Early Eocene flysch, which parallels the axis of the Pyrenees, and reaches a thickness of several kilometers. But during the Late Cretaceous this trough was not situated marginally, and covered the major part of the present northern external zone. Only during the Early Eocene did it shift northward, and only during that short time there existed a sort of "marginal trough" to the northwest of the Pyrenees.

Similar situations seem to occur in the much less known southern external zone, where TEN HAAF (1968) described an apparently localized accumulation of flysch of over 4 km thick during the relatively short span of Eocene time represented by the Landenian and Bartonian Ages.

With MANGIN (1960) we are, moreover, of the opinion that the latter flysch series did

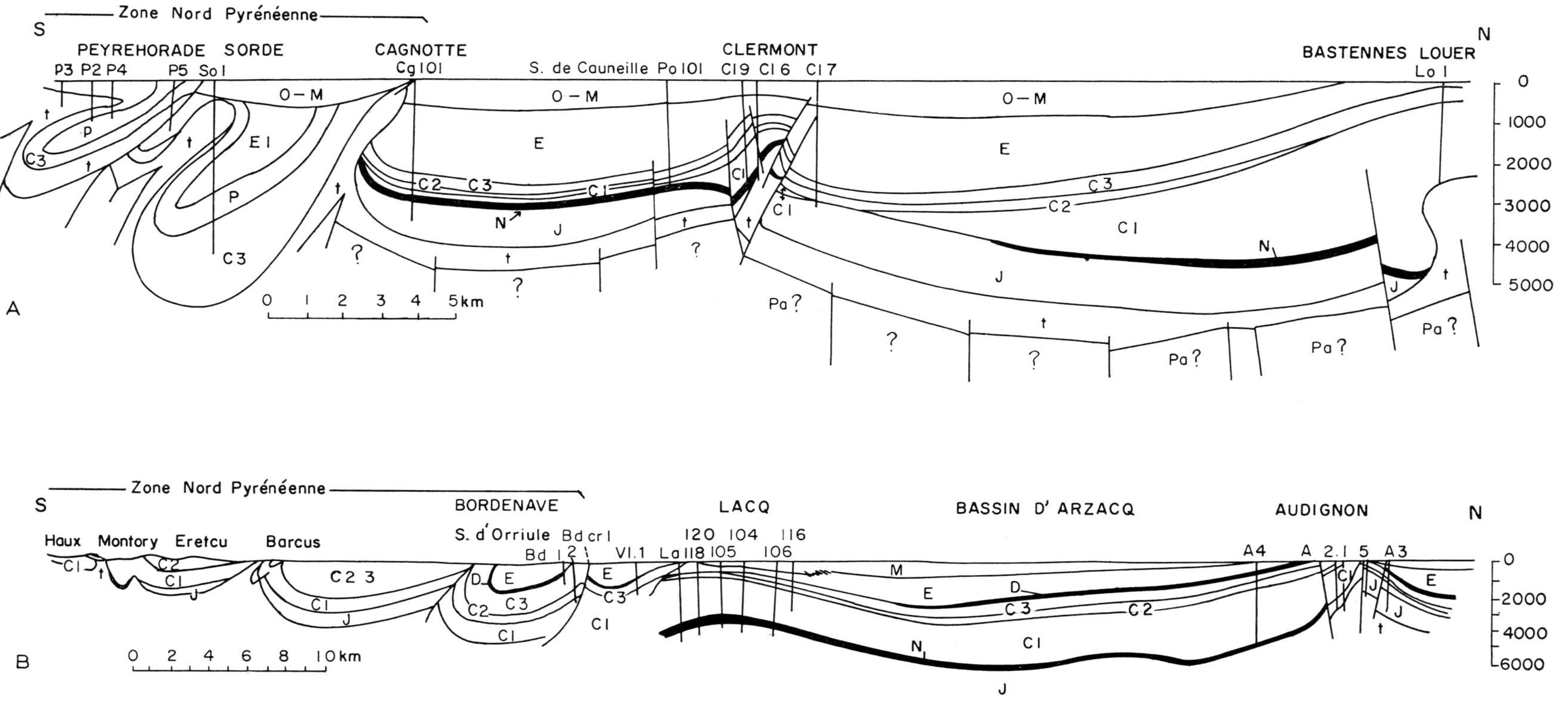

Fig.210. Two sections through the western part of the northern external zone and the southern border of the Aquitaine Basin.

The sections show the irregular distribution of variations in thickness of the Upper Cretaceous. Moreover, these have no relation to the post-sedimentary Pyrenean orogeny.

O–M = Oligocene–Pliocene; *E* = Eocene; E_1 = Lower Eocene; *Y* = Ypresian; *P* = Paleocene; *D* = Danian; C^3 = Senonian; C^2 = Cenomanian–Turonian; C^1 = Lower Cretaceous; *N* = Neocomian; *J* = Jurassic; *t* = Triassic; *Pa* = Paleozoic. (After CUVILLIER et al., 1962.)

not derive from the Pyrenees, but from some massif that has since subsided in the Ebro Basin to the south. Although in the margin of the Pyrenees, it did consequently not extend far along its axis, whilst its genesis probably was not directly related to that of the Pyrenees.

For the north Pyrenean flysch too, a provenance from the axial zone seems improbable. This zone simply never had the required amount of the sort of rocks whose detritus now forms the flysch.

If we want to look at this problem of the marginal troughs with an eye unbiassed by tectonic theory, we can do no better than quote CUVILLIER et al. (1962), who state that, apart from the flysch trough of the Late Cretaceous and the Early Eocene just mentioned: "Here and there troughs developed, rather capriciously localized. Their formation is without doubt related to small scale isostatic readjustment, (for) the Tertiary sediments are normally thinner where the Early Cretaceous sediments are thickest. During Middle Lutetian times, the Pyrenean chain arose in a brutal way; this is the real Pyrenean orogenesis".

Structure

Structurally, there is no great difference between the northern and southern external zones. Both are characterized by the appearance of subhorizontal thrust sheets, and by smaller, often acute, folds. The latter may take the form of cascade folds (Fig.211).

In the earlier nappist interpretations of the Alpine structures of the Pyrenees large recumbent folds and nappes have been the "Leitmotiv". One by one these have been replaced by series of superimposed "thrust" or gliding sheets, each formed by a right side up slice of the sedimentary series. The famous Gavarnie nappe in the southern external zone of the western Pyrenees, with subhorizontal overthrusting of the order of 10 km, forms a case in point (MENGAUD, 1939; CLIN, 1964; VAN LITH, 1966, who also gave a historical review of the various interpretations). Even the overlying Ordessa thrust mass, which for MENGAUD (1939) still formed a recumbent fold, has been re-analyzed and for its main part is found to consist of normal thrust sheets. Only in the thinly bedded Lower Tertiary have local recumbent folds been piled up on top of the main Ordessa thrust sheet, to form the Monte Perdido (MISCH, 1934; VAN DE VELDE, 1966).

The only larger example of a recumbent fold with a real overturned limb of the order of 10 km in diameter seems to have been preserved in the peculiar graben formed by the Coustouges Basin in the eastern Pyrenees (SOCIÉTÉ GÉOLOGIQUE DE FRANCE, 1959).

On the other hand, HENRY (1966) has recently returned to the interpretation of large recumbent folds at the northern side of the axial zone in the western Pyrenees, in the Les Eaux Chaudes or Lacoura thrust sheet of the western Pyrenees (Fig.212).

DE SITTER (1964) maintained a distinction between the "north Pyrenean zone with satellite massifs", and the "south Pyrenean zone with gliding tectonics". Such a distinction is, however, unrealistic. As we have already seen, the so-called "north-Pyrenean zone with satellite massifs" occurs only in the northeastern part of the mountain chain, and forms an extra element. Further west the northern external zone is structurally similar to the southern external zone.

De Sitter's schematic view of the Pyrenees was drawn up in the fifties, when our notions of gravitational tectonics were in their early stage. At that time, any area where

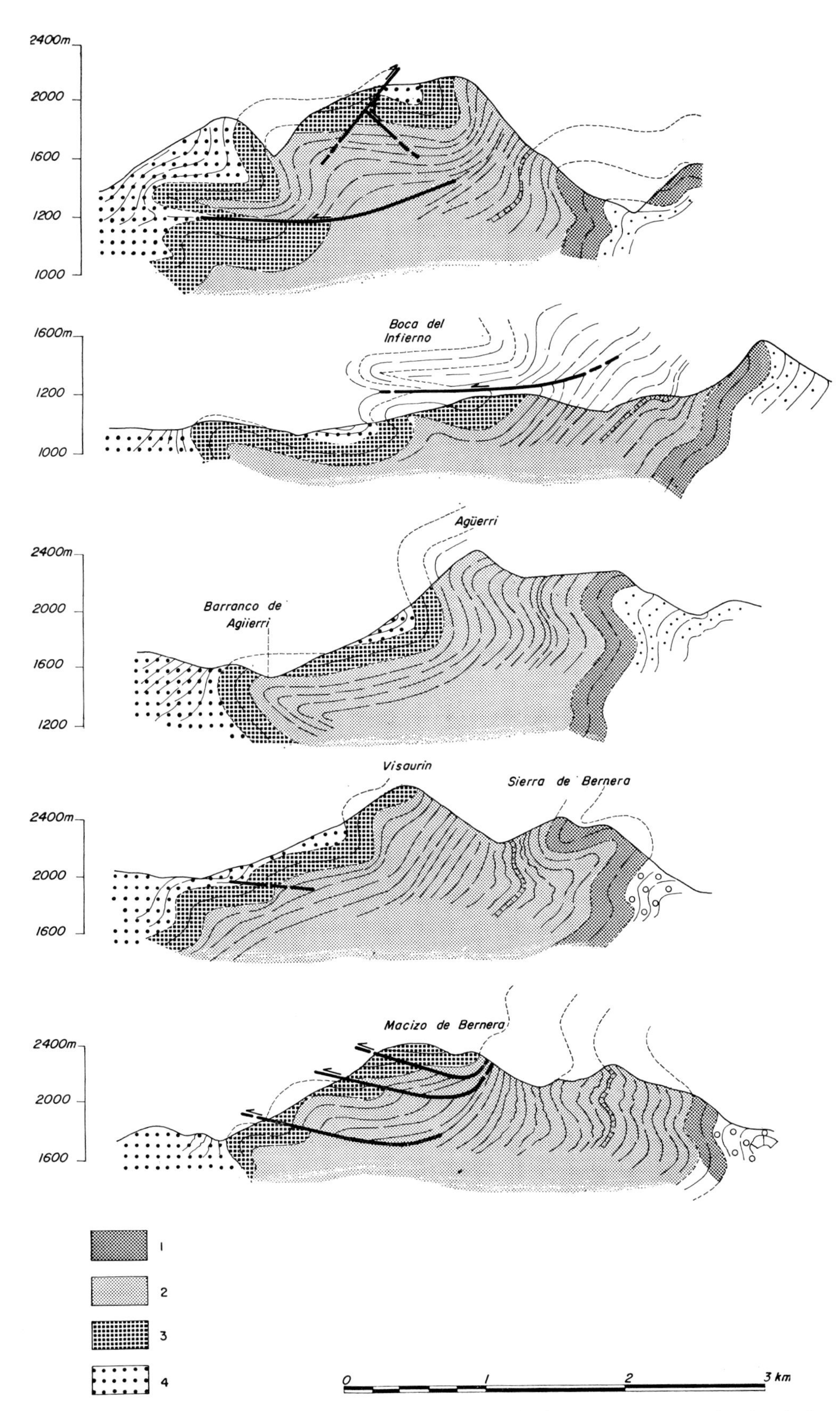

Fig.211. Serial sections through part of the southern external zone in the western Pyrenees, showing the form of a set of cascade folds and its along-the-strike variations. The distances between the various sections are of the order of 3 km.

1 = Coniacian–Santonian–Campanian; *2* = Maastrichtian; *3* = "Dano-Montian"–Lower Ilerdian; *4* = flysch, Upper Ilerdian (Ilerdian being a local stage name, not quite the equivalent of the Landenian). The distances between the various sections are of the order of 3 km. (After VAN ELSBERG, 1966.)

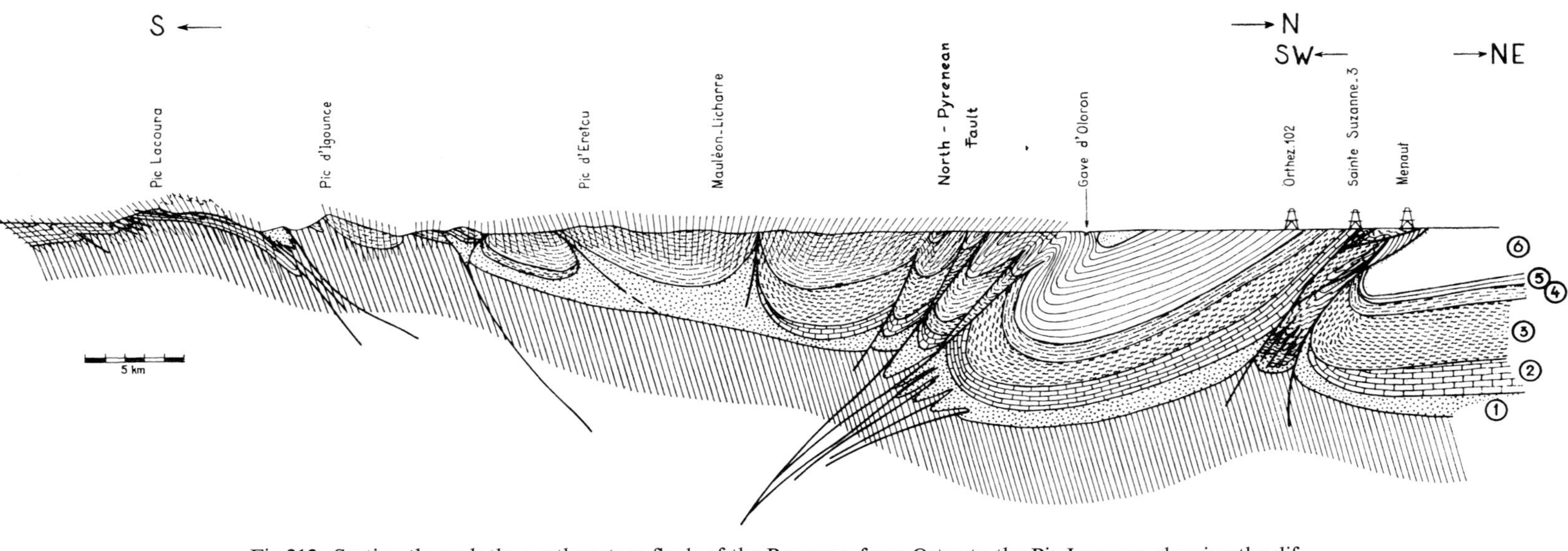

Fig.212. Section through the northwestern flank of the Pyrenees, from Ortez to the Pic Lacoura, showing the difference in tectonic style to the north and to the south of the north-Pyrenean fault, and the fact that schistosity only develops south of this structure.

1 = Permo-Triassic; *2* = Jurassic; *3* = Lower Cretaceous; *4* = Upper Albian–Coniacian; *5* = Senonian; *6* = Tertiary. (After HENRY, 1966.)

sliding and gravitational tectonics was thought provable, received extra indication as such. For the southern external zone the beautiful cascade folds, such as reproduced in Fig.211, present such a decisive indication for gravitational tectonics.

But to maintain this distinction at this date is no more than a heritage from those times. In explaining the tectonics of both the southern and the northern external zones, most geologists nowadays adhere, I think, to a "décollement", a gravitational reaction of the sedimentary cover, due to the rising of at least part of the axial zone.

Instead, it seems that the north Pyrenean fault, cited already as a major tectonic element on p. 345, forms a much more important boundary between a southern and a northern part of the Pyrenees, then does the axial zone.

For one thing, it has a marked correlation with the Alpine metamorphism in the eastern Pyrenees. Using crystallinity of illite as an index of metamorphism, KUBLER (1967) has shown that to the north of this fault almost only diagenesis has occurred. Southwards of this fault, on the other hand, metamorphism develops, first of the anchi- and then of the epizone.

In the western Pyrenees, south of the Aquitaine Basin, drilling has shown the north Pyrenean fault to be a very complex fracture zone, over which the Mesozoic has been draped (HENRY, 1966). In this region its character as a major tectonic cicatrice follows from the fact that a different tectonic style has developed to the north and to the south of this fault. To the north, a series of open folds is found, which all have northward "Vergenz". To the south, on the other hand, recumbent folds of the order of 10 km of overthrust are found, which all have southward "Vergenz".

So, when we look to the Alpine folding, it is not the axial zone, which forms the boundary between a northern and a southern flank, but the north-Pyrenean fault. Once again it transpires that the rise and erosion of the axial zone is not necessarily related to the earlier folding. Also, it is well possible that these younger vertical movements, which led to the rise of the present axial zone, have reversed the dip of earlier thrust planes. So what may have glided down south of the north Pyrenean fault at present looks as if it had been pushed up against the axial zone (Fig.212).

That this rising of the axial zone, which was to form the actual mountain of the Pyrenees, is indeed a quite recent event only, has been attested for a part of the eastern Pyrenees by BOISSEVAIN (1934). It has been dated as post-Miocene (post-Pontian) on the basis of the Pontian sediments found in the Cerdaña (French: Cerdagne) Basin. These, now situated at 1,600–1,700 m of altitude, contain lignites which were evidently deposited in marshy lowlands.

Thus the post-Pontian rising of the axial zone elevated the former peneplain and lowland basin to a present general "Gipfelfluhr" at 2,800–2,900 m of altitude, whereas the Cerdaña Basin remained in a graben position at an altitude of about 1 km less elevation.

In the Pyrenees the Alpine orogenesis has normally only slightly influenced the earlier Hercynian tectonics of the axial zone (WENSINK, 1962). Only in those cases where the basal thrust planes of the Alpine structures have been installed within the upper part of the Paleozoic (as for instance, in the Gavarnie nappe, and further west (see VAN ELSBERG, 1966), has this upper part of the Paleozoic been more strongly influenced by the Alpine movements. Similar conclusions have been reached by MATTAUER and SEGURET (1966) and by SEGURET (1966) for the eastern Pyrenees.

THE LOWER PROVENCE

The Lower Provence, that is the southern part of the province, is not only morphologically distinct from the northern part, the Higher Provence, but it is also markedly different in structure. The Higher Provence, the Provençal Alps (GOGUEL, 1943), belong to the Subalpine chains of the Alps, whereas the Lower Provence in its geology shows more similarities with the Pyrenees (Fig.213).

The extent of the Provençal chains in relation to the Subalpine chains follows from Fig.214. For the Lower Provence a newer map by AUBOUIN and MENNESSIER (1963) gives more information, but it is less legible for a first reconnaissance, so the older map by GOGUEL (1936) is preferred here.

Between the Castellane arc of the Subalpine chains and the Lower Provence a zone of transition is found, formed by the *Anticlinaux coupés*, the "Broken Anticlines", of Comps, south of Castellane. This area, in which compressive forces forming anticlines and tensional forces resulting in blockfaulting, have alternated in space and time, is important for our ideas on general tectonics. It has recently been described in great detail by MENNESSIER (1964). To this work the reader must be referred, because a further discussion of this area, interesting though it may be, is beyond the scope of this text.

The geology of the Lower Provence has been in the limelight since Marcel Bertrand recognised major nappe structures both in the Alps and in the Provence. One of his

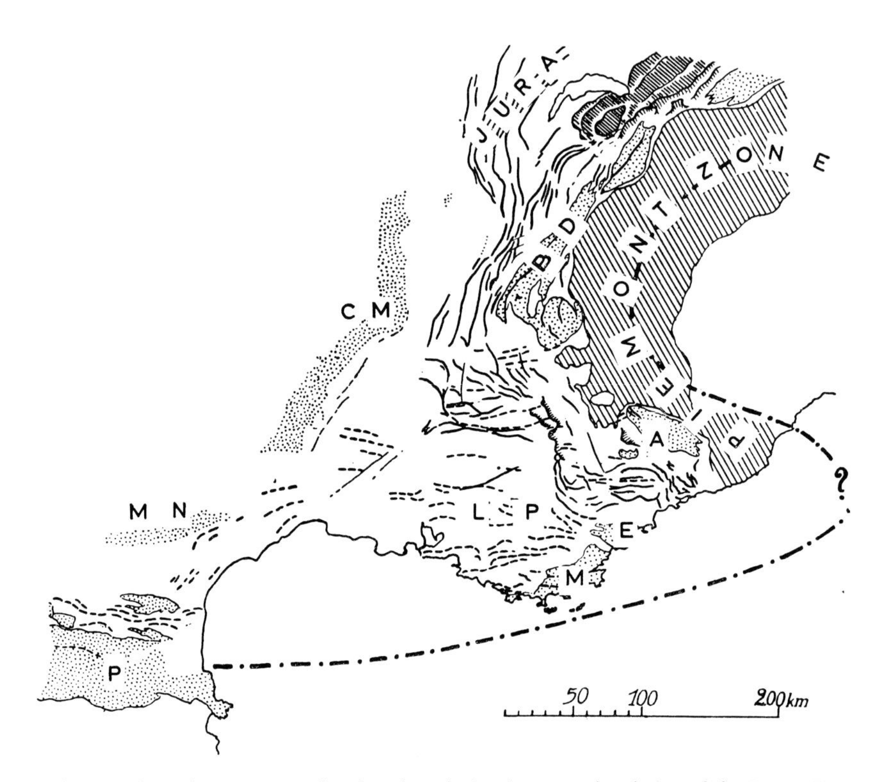

Fig.213. Sketch map of southern France, showing the relation between the chains of the Lower Provence (dashes), with those of the Subalpine chain (full lines).
A = Argentera Massif; *BD* = Belledonne Massif; *CM* = Massif Central; *E* = Estérel; *LP* = Lower Provence; *M* = Maures; *MN* = Montagne Noire; *P* = Pyrenees. (After GOGUEL, 1965.)

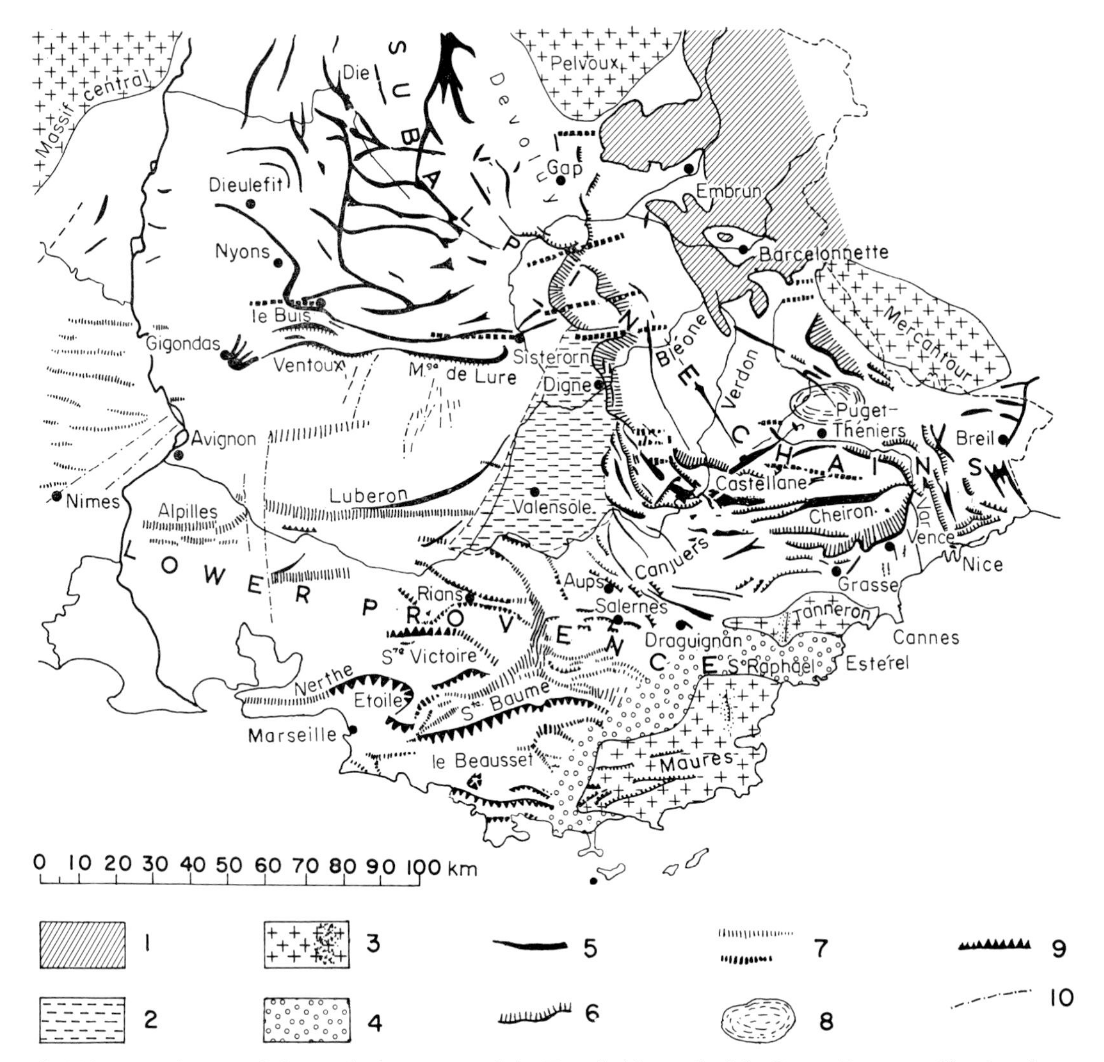

Fig.214. Tectonic map of the southeastern part of the French Alps and of the Lower Provence. Showing the relation between the Subalpine Chains and the structures of the Lower Provence.

1 = internal nappes of the French Alps; *2* = Valensole conglomerates, a Miocene molasse basin; *3* = crystalline massifs; *4* = Permian cover of the crystalline massifs; *5* = Alpine anticlinal structures; Miocene (rather Upper Oligocene, or Oligo-Miocene).; *6* = Alpine overthrusts; *7* = Provençal anticlines, Eocene; *8* = Barrot dome; *9* = Provençal overthrusts; *10* = faults. (After GOGUEL, 1936.)

descriptions of a Lower Provençal structure, that of the Beausset structure, has in fact already been cited on p. 202 (M. BERTRAND, 1887).

For the Lower Provence the existence of major nappes has since been largely refuted, just as has been the case for the Pyrenees. This does not mean that overthrusting and recumbent folding is non existant, but it is now generally held that this pertains to local structures. In the earlier nappist interpretations, on the other hand, these local structures were thought to represent erosional remnants of, originally continuous, much larger nappes of the magnitude of several tens of kilometers. In the newer interpretations, such horizontal displacements are estimated to be of the order of 1 km. Fig.215 gives a historical development of the various interpretations of a cross-section through one of the most

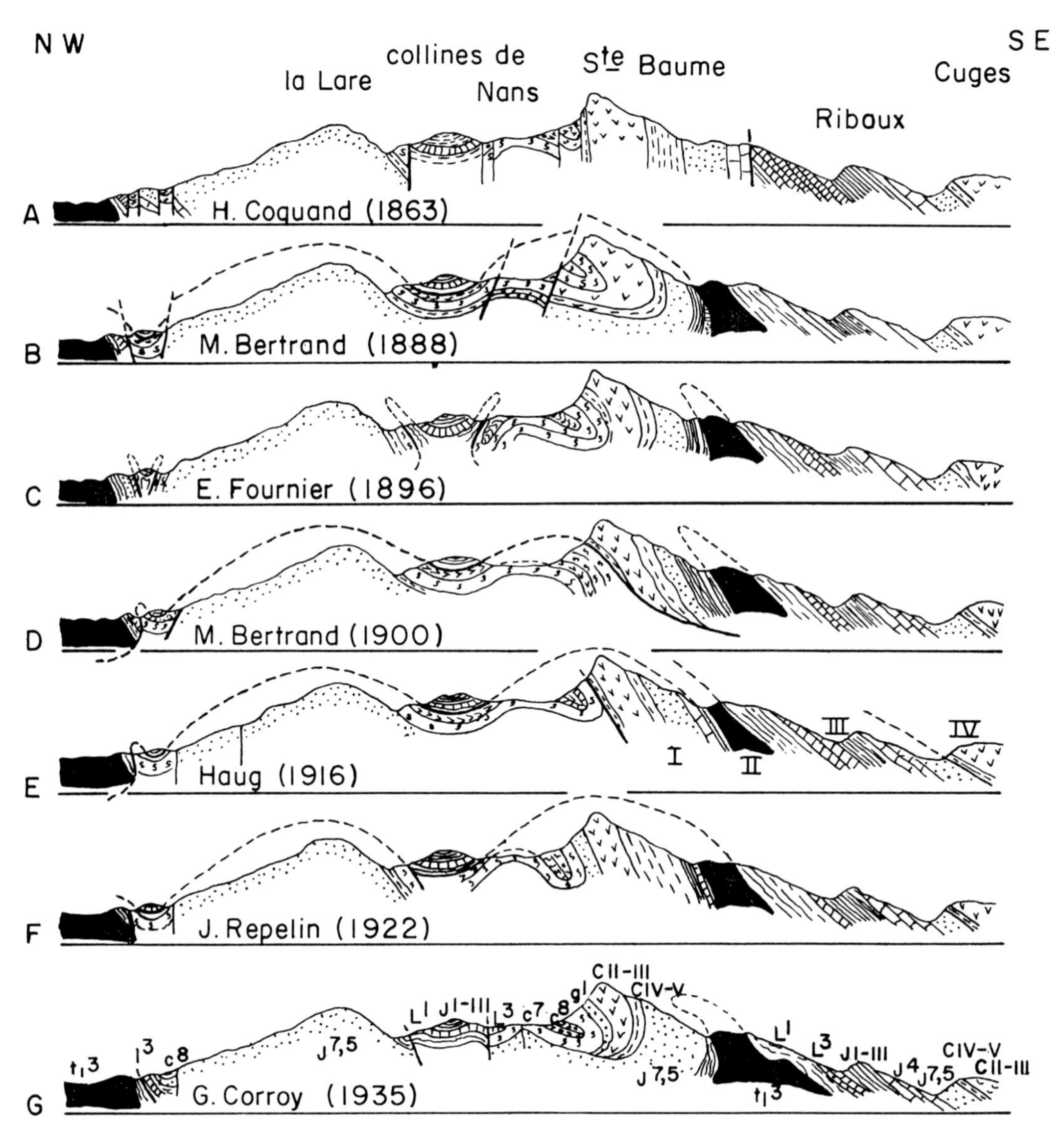

Fig.215. Sections through the Sainte Baume anticline, east of Marseille (cf. Fig.214), showing the history of the interpretation of its structure. For stratigraphic notations see Fig.218.

The nappist interpretation of Marcel Bertrand (*B*, 1888) has been attacked, without result, by a "Pilzfalte" interpretation by Fournier (*C*, 1896), but, apart from that, held true until 1922 (*F*). The newer interpretation by Corroy (*G*, 1935) shows an interplay of subvertical crustal movements with much more localized horizontal reactions of the sedimentary cover.

In a still more recent interpretation (cf. Fig.218) the nappist view has, however, again gained some ground. The *Collines de Nans* belong to the same structure as the *Île de Bousset* of Marcel Bertrand. (After CORROY, 1939.)

interesting areas of the Lower Provence. The Sainte Beaume structure on this section has since been interpreted as a recumbent fold (cf. Fig.217).

A further step in the interpretation of the structure of the Lower Provence was taken by LUTAUD (1924), who in his study of the crystalline massifs of the Provence noted that these were strongly influenced by Tertiary faulting. This indicated Alpine tectonization of these Hercynian massifs. Lutaud postulated that the reason why these massifs exist as they do today, lies in the fact that their sedimentary cover had been stripped off during the Alpine orogeny. As elsewhere, the "décollement" has taken place over the evaporite horizons of the Triassic.

The Lower Provence has since been a type area for epidermal tectonics, just as the Jura Mountains. Later studies have stressed the influence of sub-vertical dislocations, not only within the area of the crystalline basement, but also in the sedimentary cover. This led to the assumption that such vertical movements have been responsible for the horizontal tectonics. That is that the "décollement" has been due to gravity gliding. Noting moreover that such subvertical dislocations in many cases show reversal of directions with time, it is also thought that the fact that many of the overthrust planes now dip upwards in the direction of thrust, is accounted for by a later reversal of the rising movements of the area from which the "décollement" took place (Fig.216).

Such seesaw movements of crustal blocks are found more and more through detailed tectonic analyses. We have seen a major example of such an alternation of the direction of vertical crustal movements in the nearby Maritime Alps (pp.187–188). It must be noted that such movements seem to belong to a general rule rather than be an exception.

Turning now to a bird's eye view of the Lower Provence, a major division becomes at once apparent, i.e., between the crystalline massifs in the east, and the sedimentary region. Although post-Hercynian, the Permian cover of the crystalline massifs can best be assigned to the area of the massifs, because it lies beneath the evaporite horizons of the Triassic and therefore structurally belongs to the high area of the massifs. Topographically, on the other hand, it forms a depression, the "dépression perméenne" of the literature, because the softer sediments of the Permian weather much more easily than both the crystalline basement and the calcareous younger sediments.

The two massifs, the Maures (LUTAUD, 1924; GUEIRARD, 1959) and the Tanneron (BORDET, 1951), belong to the Hercynian basement, and are comparable to the external massifs of the Alps. Apart from a local graben filled with some Upper Carboniferous,

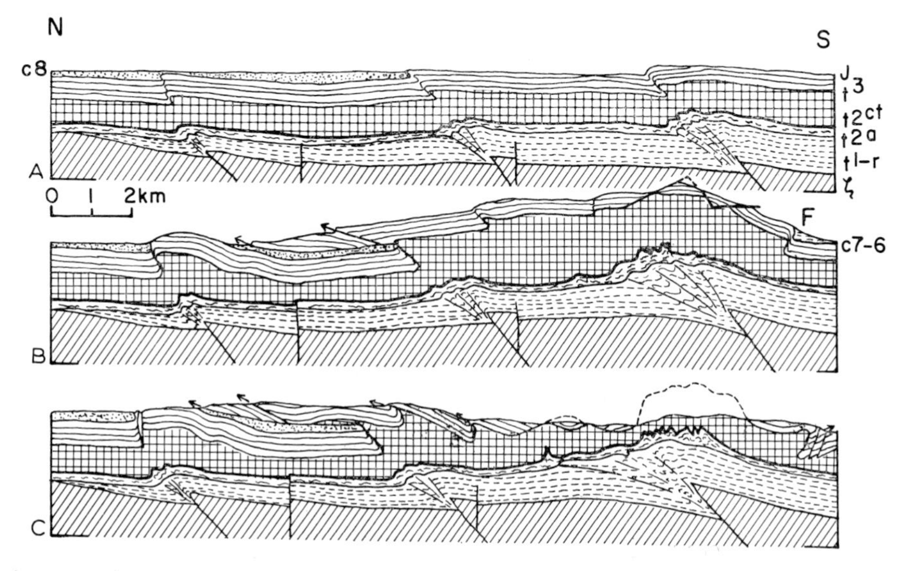

Fig.216. Three genetic sections, illustrating the main points of the development of the Provençal chains. (A. Danian phase. B. Provençal phase, Eocene. C. End of the Provençal phase, Late Eocene.
ζ = basement; t^{l-r} = Lower Triassic–Permian; t^{2a} = Middle Triassic; t^{3} = Keuper; J = Jurassic; c^{8} = Danian; c^{7-6} = "Valdo–Fuvelian". (After MENNESSIER, 1960.)

they are covered by the Permian. This is mainly developed in the normal facies of red beds (BOUILLET and LUTAUD, 1958). In its eastern part, between the Maures and the Tanneron, extensive Permian volcanism has occurred. The volcanic series is formed in its major part by thick, reddish, massive ignimbritic flows, which form the scenic coast line of the Côte d'Azur between St. Raphael and Cannes. More inland, however, a Permian lava and agglomerates also occur (RUTTEN, 1959, 1963).

The younger sedimentary history of the remaining, major, part of the Lower Provence is rather complicated. It starts all over the area with Triassic in Germanic facies. Its lower red beds remain glued to the underlying Permian. Then follows a series of evaporites, a carbonate series and again a reddish series. The latter, normally by far the thicker, is correlated with the Keuper. The carbonate series is always called Muschelkalk, a correlation resting on pure lithology; it may well be younger (RICOUR, 1962).

Cretaceous and Jurassic are well developed only in the southwestern part of the Lower Provence. The many variations in stratigraphy will not be gone into here, and the reader is referred to CORROY (1963) and AUBOUIN and MENNESSIER (1963). Towards the north of the area the Lower Cretaceous wedges out towards an old paleogeographic high, the "Isthme durancienne" or Durance Isthmus. It runs from Nice westwards, following more or less the course of the present lower Durance river. In the eastern part of the area the Upper Cretaceous wedges out too.

This brings us to the uppermost Cretaceous and Lower Eocene, which are developed in a continental facies, comparable to the Garumnian, but indicated by a number of local stage names.

The main tectonic phase of the Lower Provence lies in the higher Eocene. It has been locally dated as between Ludian and Late Lutetian, that is between Late and Middle Eocene. There is, however, no reason why the gravitational reactions to the vertical movements of underlying crustal blocks should be contemporaneous all over the area. There are, moreover, some indications that the main phase did not occur exactly at the date given above in other spots. A more elastic dating of the main orogenetic phase of the Lower Provence, i.e., as pre-Early Oligocene (pre-Sannoisian), seems more realistic. As stated before, this correlates rather well with the Pyrenean phase of the Pyrenees. Notwithstanding this, one always finds it indicated as "provençal" in the literature on the Lower Provence.

Structurally the tectonic units of the Lower Provence are characterized first by their small dimensions, and second by their irregularity, expressed in extremely strong variation of the same structure along its strike.

Both features distinguish the Lower Provence from that other epidermal orogen, the Jura Mountains. In fact, the latter may really be called a fold belt, whereas the Lower Provence forms more of a mosaic of folded structures. These same features resemble, on the other hand, the structure of the Subalpine chains. They are, however, much stronger expressed in the former area. Within the Lower Provence both the small size and the irregularity of individual structures are more evident in the eastern than in the western part. This has been correlated with the fact that in the eastern part the Cretaceous is missing, resulting in a much thinner sedimentary cover.

The tectonics of the Lower Provence have been described in recent years by CORROY (1939), CORROY and DENIZOT (1943), GOGUEL (1943, 1944), and by AUBOUIN and MENNES-

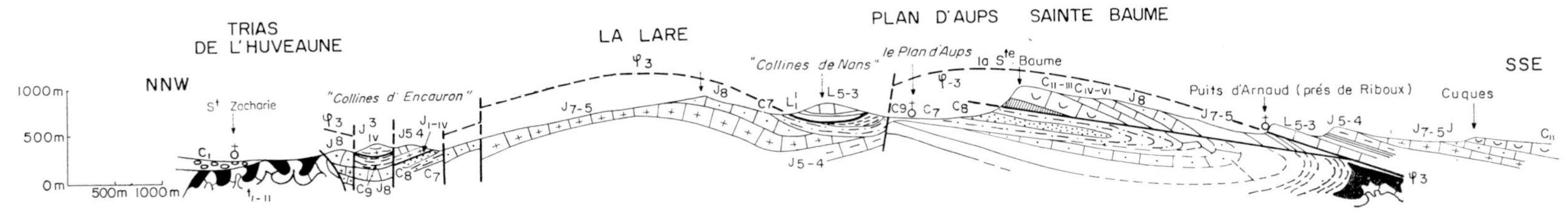

Fig.217. Section through the Sainte Baume structure, to show an example of exeptional recumbent folds in the Lower Provence. C_9 = Maastrichtian; C_8 = Fuvelian–Valdonian; C_7 = Santonian; C_{II-III} = Urgonian; C_{IV} = Hauterivian; C_V = Valanginian; C_{VI} = Berriasian; J_8 = Upper Portlandian; J_{7-5} = Lower Portlandian–Kimmeridgian; J_4 = Kimmeridgian–Sequanian; J_{I-IV} = Bathonian–Bajocian; L_{5-3} = Upper and Middle Lias; t_{1-2} = Muschelkalk; φ = overthrust. (After AUBOUIN and MENNESSIER, 1963.)

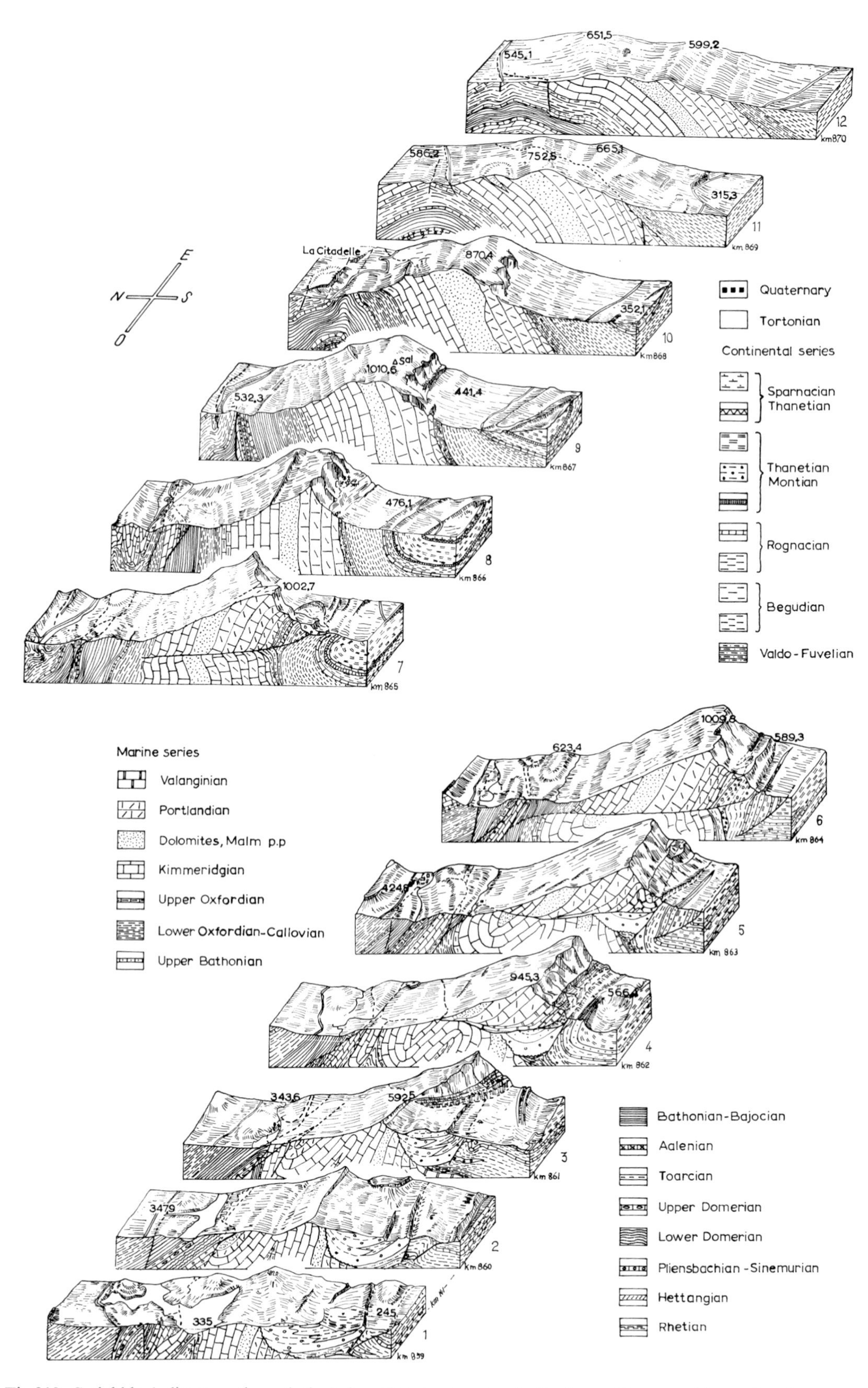

Fig.218. Serial block diagrams through the Sainte Victoire structure in the Lower Provence, illustrating the extreme rapidity of along-the-strike structural variations. (After CORROY et al., 1965.)

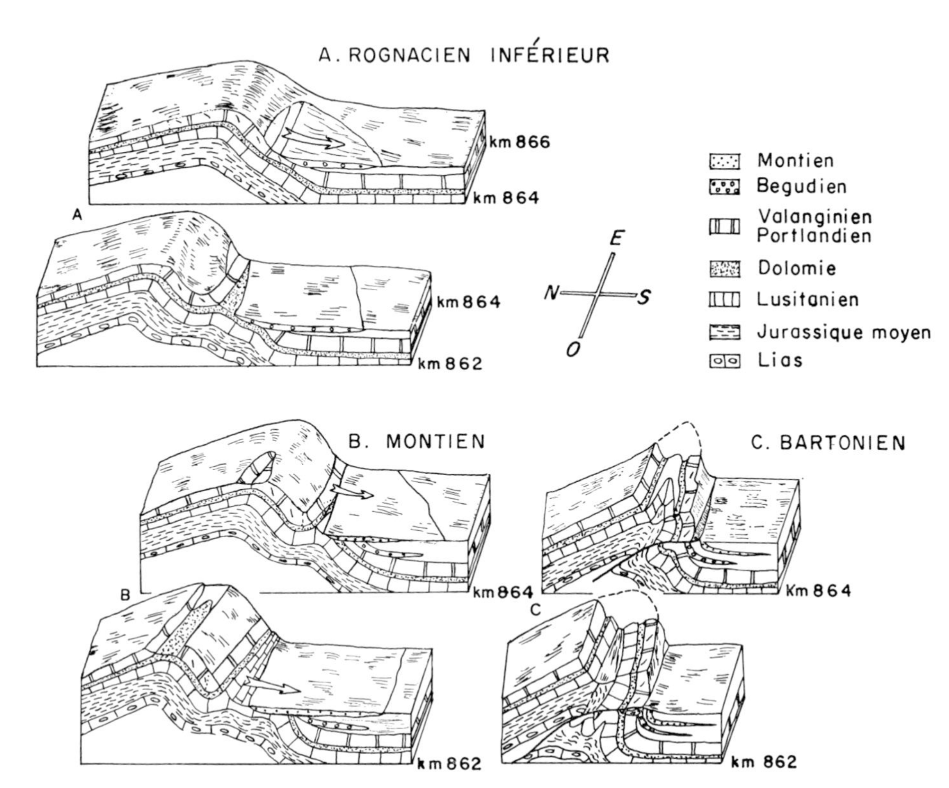

Fig.219. Genetic blockdiagrams, illustrating the development of the Sainte Victoire structure in the Lower Provence. A. Uplift during the Upper Cretaceous, leading to superficial "décollement". B. Further uplift during the Paleocene. C. Main phase, overthrusting during the Eocene. (After CORROY et al., 1965.)

SIER (1963). The junction between the Lower Provence and the Subalpine chains has been more expressly investigated by DE LAPPARENT (1938), GINSBURG (1960) and MENNESSIER (1960).

The most common forms found are narrow, asymmetric anticlines, overthrust folds ("pli–failles"), and overthrusts, alternating with much broader synclines. Recumbent folds (Fig.217) have developed only rarely. A good example of the Provençal structures is the Sainte Victoire anticline. Its modern, detailed analysis by CORROY et al. (1965) also offers a good example of the order of magnitude of the along-the-strike variations in structure (Fig.218, 219).

It is with these illustrations that our short summary of our knowledge of the Lower Provence has to be closed.

REFERENCES

AUBOUIN, J. and MENNESSIER, G., 1963. Essai sur la structure de la Provence. In: M. DURAND DELGA (Editor), *Livre à la Mémoire du Professeur P. Fallot. II.* Soc. Géol. France, Paris, pp. 45–98.

BERTRAND, L., 1908. Contribution à l'histoire stratigraphique et tectonique des Pyrénées orientales et centrales. *Bull. Serv. Carte Géol. France*, 118: 1–183.

BERTRAND, L., 1911. Sur la structure géologique des Pyrénées occidentales et leur relation avec les Pyrénées orientales et centrales. *Bull. Soc. Géol. France*, 4 (11): 122–153.

BERTRAND, M., 1887. Îlot triassique du Beausset (Var). Analogie avec le bassin houiller franco-belge et avec les Alpes de Glaris. *Bull. Soc. Géol. France*, 3 (15): 667–702.

BERTRAND, M., 1888. Nouvelles études sur la chaîne de la Sainte-Beaume. Allure sinueuse des plis de la Provence. *Bull. Soc. Géol. France*, 3(16) : 748–778.

BOISSEVAIN, H., 1934. Étude géologique et géomorphologique d'une partie de la vallée de la haute Sègre (Pyrénées catalanes). *Bull. Soc. Hist. Nat. Toulouse*, 66: 33–70.

BORDET, P., 1951. Étude géologique et pétrographique de l'Estérel. *Mém. Carte Géol. France*, 1951: 207 pp.

BOUILLET, G. and LUTAUD, L., 1958. Contribution à l'étude paléogéographique de la période permienne dans la Provence cristalline. *Bull. Soc. Géol. France*, 6 (8): 447–462.

CAREY, S. W., 1958. A tectonic approach to continental drift. *Symp. Continental Drift, Hobart, Tasmania*, pp.177–355.

CASTERAS, M., 1933. Recherches sur la structure du versant nord des Pyrénées centrales et orientales. *Bull. Serv. Carte Géol. France*, 189: 524 pp.

CASTERAS, M., CUVILLIER, J., ARNOULD, M., BUROLLET, P. F., CLAVIER, B. and DUFAURE, P., 1957. Sur la présence du Jurassique supérieur et du Néocomien dans les Pyrénées orientales et centrales françaises. *Bull. Soc. Hist. Nat. Toulouse*, 92: 297–347.

CAVET, P., 1959. Stratigraphie du Paléozoïque de la zone axial pyrénéenne à l'est de l'Ariège. *Bull. Soc. Géol. France*, 6 (8): 853–867.

CAVET, P., 1959. Le Paléozoïque de la zone axiale des Pyrénées Orientales Françaises entre le Rousillon et l'Andorre. *Bull. Serv. Carte Géol. France*, 254: 305–518.

CLIN, M., 1964. Étude géologique de la haute chaîne des Pyrénées centrales entre le Cirque de Troumouse et le Cirque de Lys. *Mém. Bur. Rech. Géol. Minières*, 27: 379 pp.

CLIN, M. and PERRIAUX, J., 1968. Contribution à la géologie des feuilles de Morains-en-Montagne et de Nantua au 50.000e. *Bull. Serv. Carte Géol. France*, 276, in press.

CLIN, M., DE LA ROCHE, H. and LELONG, F., 1963. Nouvelles observations sur le massif granitique du Lys-Caillaouas (Pyrénées centrales). *Sci. Terre*, 9: 149–174.

COQUAND, H., 1863. Description géologique du massif montaigneux de la Sainte-Beaume. *Soc. Émulation Provence*, 3: 73.

CORROY, G., 1939. Le Massif de la Sainte-Beaume. *Bull. Serv. Carte Géol. France*, 201: 124 pp.

CORROY, G., 1963. L'évolution paléogéographique post-hercynienne de la Provence. In: M. DURAND DELGA (Editor), *Livre à la Mémoire du Professeur P. Fallot. II.* Soc. Géol. France, Paris, pp.19–43.

CORROY, G. and DENIZOT, G., 1943. *La Provence occidentale.* Hermann, Paris, 184 pp.

CORROY, G., DURAND, J. P. and TEMPIER, C., 1965. Évolution tectonique de la montagne Sainte-Victoire en Provence. *Bull. Soc. Géol. France*, 7 (6): 91–106.

CUVILLIER, J., FOURMENTRAUX, J., HENRY, J., JENNER, P., PONTALIER, Y. and SCHOEFFLER, J., 1962. État actuel des connaissances géologiques sur le Bassin de l'Aquitaine au sud de l'Adour. In: M. DURAND DELGA (Editor), *Livre à la Mémoire du Professeur P. Fallot. I.* Soc. Géol. France, Paris, pp. 367–382.

DALLONI, M., 1910. Étude géologique des Pyrénées de l'Aragon. *Ann. Fac. Sci. Marseille*, 19: 444 pp.

DE LAPPARENT, A. F., 1938. Études géologiques dans les régions provençales et alpines entre le Var et le Durance. *Bull. Serv. Carte Géol. France*, 198: 301 pp.

DE SITTER, L. U., 1949. The development of the Paleozoic in northwest Spain. *Geol. Mijnbouw*, 11: 312–319; 325–340.

DE SITTER, L. U., 1956. A cross-section through the Pyrenees. *Geol. Rundschau*, 45: 214–234.

DE SITTER, L. U., 1959. The structure of the axial zone of the Pyrenees in the province of Lerida. *Estud. Geol. Inst. Invest. Geol. "Lucas Mallada" (Madrid)*, 15: 349–360.

DE SITTER, L. U., 1961. La phase tectogénique pyrénéenne dans les Pyrénées méridionales. *Compt. Rend. Soc. Géol. France*, 1961: 224–225.

DE SITTER, L. U., 1962. The structure of the southern slope of the Cantabrian Mountains: Explanation of a geological map with sections scale 1:100,000. *Leidse Geol. Mededel.*, 26: 255–264.

DE SITTER, L. U., 1964. *Structural Geology*, 2nd ed. McGraw-Hill, New York, N.Y., 551 pp.

DE SITTER, L. U., 1965. Hercynian and Alpine orogenies in northern Spain. *Geol. Mijnbouw*, 44: 373–383.

DE SITTER, L. U. and ZWART, H. J., 1959. Explanatory text to the geological map of the Paleozoic of the Central Pyrenees. Sheet 3, Ariège, France, 1:50,000. *Leidse Geol. Mededel.*, 22: 351–418.

DE SITTER, L. U. and ZWART, H. J., 1962. Geological map of the Central Pyrenees. Sheets 1: Garonne, 2: Salat, France, 1:50,000. *Leidse Geol. Mededel.*, 27: 191–236.

FOURNIER, E., 1896. Le pli de la Sainte-Beaume et son raccord avec le pli périphérique d'Allauch. *Bull. Soc. Géol. France*, 3 (24): 663.

GIGNOUX, M., 1955. *Stratigraphic Geology.* Freeman, San Francisco, Calif., 682 pp.

GINSBURG, L., 1960. Étude géologique de la bordure subalpine à l'ouest de la basse vallée du Var. *Bull. Serv. Carte Géol. France*, 259: 38 pp.

GLANGEAUD, L., 1959. Le plutonisme sialique, ses relations avec le métamorphisme dans les Pyrénées orientales et centrales. *Bull. Soc. Géol. France*, 6 (8): 961–978.

GOGUEL, J., 1936. Description tectonique de la bordure des Alpes de la Bléone au Var. *Mém. Carte Géol. France*, 1936: 360 pp.

GOGUEL, J., 1943. Essai d'une synthèse tectonique de la Provence. *Bull. Soc. Géol. France*, 5 (13): 367–382.

GOGUEL, J., 1944. Description géologique des Alpilles. *Bull. Serv. Carte Géol. France*, 214: 28 pp.

GOGUEL, J., 1965. *Traité de Tectonique*, 2nd ed. Masson, Paris, 457 pp.

GUEIRARD, S., 1959. Description pétrographique et zonéographique des schistes cristallins des Maures (Var). *Trav. Lab. Géol. Fac. Sci. Univ. Aix–Marseille*, 6: 71–264.

GUITARD, G., 1959. La Structure du Massif du Canigou. Aperçu sur la métamorphisme régional dans la zone axiale des Pyrénées orientales. *Bull. Soc. Géol. France*, 6 (3): 907–924.

GUITARD, G., 1961. Linéations, schistosité et phases de plissement durant l'orogénèse hercynienne dans les terrains anciens des Pyrénées orientales, leur relations avec le métamorphisme et la granitisation. *Bull. Soc. Géol. France*, 7 (2): 862–887.

GUITARD, G., 1964. Un exemple de structure en nappe de style pennique dans la chaîne hercynienne. Les gneiss stratoïdes du Canigou (Pyrénées orientales). *Compt. Rend. Acad. Sci. Paris*, 258: 4597–4599.

HAUG, C., 1916. La tectonique du Massif de la Sainte-Beaume. *Bull. Soc. Géol. France*, 4 (15): 113.

HENRY, J., 1967. Le problème des étages tectoniques dans les Pyrénées occidentales. Comparaison entre les accidents nord-pyrénéennes et l'accident de Larrau-Gourette. In: J. P. SCHAER (Editor), *Colloque Étages Tectoniques*. Baconniere, Neuchâtel, pp.253–267.

JACOB, C., 1930. Zone axiale, versant sud et versant nord des Pyrénées. *Livre Jubilaire Soc. Géol. France*, 1: 389–410.

KLEINSMIEDE, W. F. J., 1960. Geology of the Valle de Arán. Geological map of the Central Pyrenees, sheet 4, Valle de Arán. *Leidse Geol. Mededel.*, 25: 131–244.

KUBLER, B., 1967. La cristallinité de l'illite et les zones tout à fait supérieures du métamorphisme. In: J. P. SCHAER (Editor), *Colloque Étages Tectoniques*. Baconniere, Neuchâtel, pp.105–122.

LAVARDIÈRE, J. W., 1930. Contribution à l'étude des terrains paléozoiques dans les Pyrénées occidentales. *Mém. Soc. Géol. Nord*, 10: 131 pp.

LLOPIS LLADO, N., 1948. Sobre la estructura de Navarra y los enlaces occidentales del Pirineo. *Miscellanea Almera (Barcelona)*, 7 (1): 159–186.

LUTAUD, L., 1924. Étude tectonique et morphologique de la Provence cristalline. *Rev. Géograph.*, 12: 1–270.

MANGIN, J. PH., 1959. Données nouvelles sur le Nummulitique pyrénéen. *Bull. Soc. Géol. France*, 7 (1): 16–30.

MANGIN, J. PH., 1960. Le Nummulitique sud-pyrénéen à l'Ouest de l'Aragon. Thesis. *Pirineos*, 51–58: 631 pp.

MANGIN, J. PH. and RAT, P., 1962. L'évolution post-hercynienne entre Asturies et Aragon (Espagne). In: M. DURAND DELGA (Editor), *Livre à la Mémoire du Professeur P. Fallot. I.* Soc. Géol. France, Paris, pp.333–350.

MATTAUER, M., 1968. Les traits structuraux essentiels de la chaine Pyrénéenne. *Rev. Géograph. Phys. Géol. Dyn., Sér. 2*, 10 (3): 3–12.

MATTAUER, M. and SEGURET, M., 1966. Sur le style des déformations tertiaires de la zone axiale hercynienne des Pyrénées. *Compt. Rend. Soc. Géol. France*, 1966: 10–12.

MENGAUD, L., 1939. Études géologiques dans la région de Gavarnie et du Mont Perdu. *Bull. Serv. Carte Géol. France*, 204: 197–218.

MENNESSIER, G., 1960. Le style tectonique de la chaîne provençale dans la région de Draguignan. *Rev. Géograph. Phys. Géol. Dyn., Sér. 2*, 3: 3–14.

MENNESSIER, G., 1964. Sur l'évolution tectonique et morphologique des chaînons externes de l'arc de Castellane entre le Verdon et la Siagne (Haute Provence). *Rev. Géograph. Phys. Géol. Dyn., Sér. 2*, 6: 91–113.

MIROUSE, R., 1958. Extension et relation des séries permiennes sur la feuille d'Urdos au 80.000e. *Bull. Serv. Carte Géol. France*, 257: 209–218.

MIROUSE, R., 1962. Observations sur le Dévonien inférieur de la partie occidentale de la zone axiale dans les Pyrénées françaises. In: *Symposium Silur–Devon Grenze*. Schweizerbach, Stuttgart, pp.165–174.

MIROUSE, R., 1968. Recherches géologiques dans la partie occidentale de la zone primaire axiale des Pyrénées. *Bull. Serv. Carte Géol. France*, in press.

MIROUSE, R. and SOUQUET, P., 1964. Présence du Cénomanien au sommet du Pic Balaïtous (Hautes-Pyrénées). *Compt. Rend. Soc. Géol. France*, 1964: 308–309.

MISCH, P., 1934. Der Bau der mittleren Südpyrenäen. *Abhandl. Ges. Wiss. Göttingen, Math. Phys. Kl.*, 3 (12): 168 pp.

MORRE, N. and THIÉBAULT, J., 1963. Les roches volcaniques du Trias inférieur du versant nord des Pyrénées. *Bull. Soc. Géol. France*, 7 (4): 539–546.

MORRE, N. and THIÉBAULT, J., 1965. Constitution de quelques roches volcaniques permiennes de la Sierra del Cadi (Pyrénées catalanese). *Bull. Soc. Géol. France*, 7 (6): 389–396.

RAGUIN, E., 1963. Sur la structure en grand des massifs de gneiss des Pyrénées. *Geol. Rundschau*, 52: 246–249.

RAGUIN, E., 1965. Les problèmes du massif d'Aston dans les Pyrénées de l'Ariège. *Bull. Soc. Géol. France*, 7(6): 64–86.

RÉPELIN, J., 1922. Monographie géologique du Massif de la Sainte-Beaume, *Ann. Fac. Sci. Marseilles*, 25, fasc. 1.

RICOUR, J., 1962. Contribution à une révision du Trias français. *Mém, Carte Géol. France*, 1962 : 471 pp.

RUTTEN, M. G., 1959. Ignimbrites permiennes de l'Estérel. *Compt. Rend. Soc. Géol. France*, 1959: 168.

RUTTEN, M. G., 1963. Acid lava flow structure, as compared to ignimbrites. *Bull. Volcanol.*, 25: 111–121.

SCHMIDT, H., 1941. Das Paläozoikum der spanischen Pyrenäen. *Abhandl. Ges. Wiss. Göttingen, Math. Phys. Kl.*, 3 (5): 85 pp.

SCHWARZ, E. J., 1962. Geology and paleomagnetism of the valley of the Rio Aragon Subordan N and E of Oza, Spanish Western Pyrenees, province of Huesca. *Estud. Geol. Inst. Invest. Geol. "Lucas Mallada" (Madrid)*, 18: 193–240.

SEGURET, M., 1966. Sur les charriages de la zone des Nogueras (Versant sud des Pyrénées). *Compt. Rend. Soc. Géol. France*, 1966: 17–18.

SELZER, G., 1934. Geologie der Südpyrenäischen Sierren. *Neues Jahrb. Mineral., Geol. Paläontol.*, 71B: 370–407.

SOCIÉTÉ GÉOLOGIQUE DE FRANCE, 1959. Réunion extraordinaire, Pyrénées orientales. *Bull. Soc. Géol. France*, 6 (8): 807–1005.

TEN HAAF, E., 1967. Le flysch sud-pyrénéen le long du Rio Ara. *Actes Congr. Intern. Études Pyrenaiques, 5e, Inst. Estud. Pyrenaicas (Zaragoza)*, in press.

VAN DER LINGEN, G. J., 1960. Geology of the Spanish Pyrenees, North of Canfranc, Huesca Province. *Estud. Geol., Inst. Invest. Geol. "Lucas Mallada" (Madrid)*, 16: 205–242.

VAN DE VELDE, E. J., 1967. Geology of the Ordessa overthrust mass, Spanish Pyrenees, Province of Huesca. *Estud. Geol., Inst. Invest. Geol. "Lucas Mallada" (Madrid)*, 13: 163–201.

VAN ELSBERG, J. N., 1966. Geology of the Upper Cretaceous and part of the Lower Tertiary north of Hecho and Arragues del Puerto. *Estud. Geol., Inst. Invest. Geol. "Lucas Mallada" (Madrid)*, in press.

VAN LITH, J. G. J., 1966. Geology of the Spanish part of the Gavarnie Nappe (Pyrenees) and its underlying sediments near Bielsa (Province of Huesca). *Estud. Geol., Inst. Invest. Geol. "Lucas Mallada" (Madrid)*, in press.

VIENNOT, P., 1927. Recherches structurales dans les Pyrénées occidentales françaises. *Bull. Serv. Carte Géol. France*, 163: 267 pp.

WENSINK, H., 1962. Paleozoic of the Upper Gállego and Ara Valleys, Huesca Province, Spanish Pyrenees. *Estud. Geol. Inst. Invest. Geol. "Lucas Mallada" (Madrid)*, 18: 1–74.

WIEDMANN, J., 1962. Contribution à la paléogéographie du Crétacé vascogothique et celtibérique septentrional (Espagne). In: M. DURAND DELGA (Editor), *Livre à la Mémoire du Professeur P. Fallot. I.* Soc. Géol. France, Paris, pp. 351–366.

ZANDVLIET, J., 1960. The geology of the Upper Salat and Pallaresa valleys. Geological map of the Central Pyrenees, sheet 5, Pallaresa. *Leidse Geol. Mededel.*, 25: 1–127.

ZWART, H. J., 1960. The chronological succession of folding and metamorphism in the central Pyrenees. *Geol. Rundschau*, 50: 203–218.

ZWART, H. J., 1960. Relations between folding and metamorphism in the Central Pyrenees and their chronological succession. *Geol. Mijnbouw*, 39: 163–180.

ZWART, H. J., 1963a. Some examples of the relations between deformation and metamorphism from the Central Pyrenees. *Geol. Mijnbouw*, 42: 143–154.

ZWART, H. J., 1963b. Metamorphic history of the Central Pyrenees. 2. Valle d'Aran. Sheet 4. *Leidse Geol. Mededel.*, 28: 321–376.

CHAPTER 15

Alpine Europe: The Jura Mountains

INTRODUCTION

As stated already in Chapter 9, the Jura Mountains have, amongst the Alpine chains of western Europe, always attracted a fair share of attention. This arose from several factors. Not only are found here the beautiful sections exposed in the various deeply incised transverse river canyons (French: "cluses"; German: "Kluse"). But, moreover, because of their small size, they formed an ideal region to demonstrate the wealth of forms encountered in structural geology. Here also could be found a clear demonstration of the correlation, either positive or negative, dependant on the particular case, between structure and morphology. Moreover, it has of course been the first fold belt where railway tunnels proved the existence of a strikingly different folding style at the surface and deeper down in the crust. The Jura Mountains have since served as the type area of epidermis folding.

Perhaps it is not even necessary, one could ask, to include a chapter on the Jura Mountains in this text, because so much attention has already been given to them in various text-books? But as long as a geologist who recently visited the area with an A.G.I. Field Institute, thinks it feasible to sum up the geology as follows: "The concave innerhalf was strongly folded in the Tertiary into *symmetrical* northeast–southwest anticlines and synclines. Deformation of the Folded Jura was *to recent for inversion of relief to have occurred*" (FERNOW, 1965), it seems just as well to present a more qualified description.

When speaking about the Jura Mountains, it is well to remember that, though they form a type area for structural geology, this does not apply to stratigraphy. The Jurassic, to which they gave their name, occurs widely also outside the Jura Mountains, in the low plateaus in the post-Hercynian basins, of western Europe. In fact, part of the Jurassic aureole of the Paris Basin forms the type area for the stratigraphy of the Jurassic (GIGNOUX, 1955).

Although not the type area of Jurassic stratigraphy, most of the visible part of the Jura Mountains is, nevertheless, built up by rocks of that period. They belong to the same epicontinental domain of the surrounding low plateaus and the external chains of the Alps, and are developed predominantly in a carbonate facies. For a modern stratigraphic analysis of a smaller area in the southern part of the chain, see SIGAL (1962). The thickness of the Jurassic varies between 500 m and 1,000 m.

Underneath the Jurassic only Upper and Middle Triassic in Germanic facies are exposed in the more deeply incised of the more steeper anticlinal cores. The gypsum, mainly belonging to the Muschelkalk, is the main factor in the development of the epidermis type of fold belt, in which the basement is nowhere exposed.

On top of the Jurassic follows some Cretaceous, mainly found in the broad synclinal basins. It is thought that the Cretaceous formerly was developed more completely, comparable to the situation found in the surrounding areas, but not much proof of this contention can now be offered. The Early Tertiary, at any rate, represents a period of regression. Continental conglomeratic iron ores developed during this period, the "Bohnerze", which have been of economic importance.

A first folding period was during the Oligocene. It was followed by the deposition of coarsely conglomeratic post-orogenetic molasse deposits during the Miocene, which in turn have become tectonized, but only more locally, mostly along major faults and overthrusts.

As to the structure, historically the recognition of the epidermis character of the Jura Mountains mainly goes back to the descriptions by A. Buxtorf of the Weissenstein tunnel (BUXTORF, 1908) and of the Hauensteinbasis and Grenchenberg tunnels (BUXTORF, 1916). This may now seem to be rather old history, but we might do well to remember the immense impetus to geologic understanding, derived from the confrontation of the original mistaken prognoses and the structures actually found. As a practical result, this resulted in much better forecasts for the newer tunnels. The original prognoses and actual structure for these tunnels are given in Fig.220–222. Methodologically, one might note that the prognoses were wrong mainly where diapiric anticlinal cores were cut, whereas they compared much better with the reality found in the more tranquil structures. It was not the actual mapping which was so much at fault, but the construction of the sections, in which the epidermal character of the fold belt had not yet been taken into account.

As to the relation between structure and morphology, the important fact is that the Jurassic of the Jura Mountains, just as in the Subalpine chains, is built up by an alter-

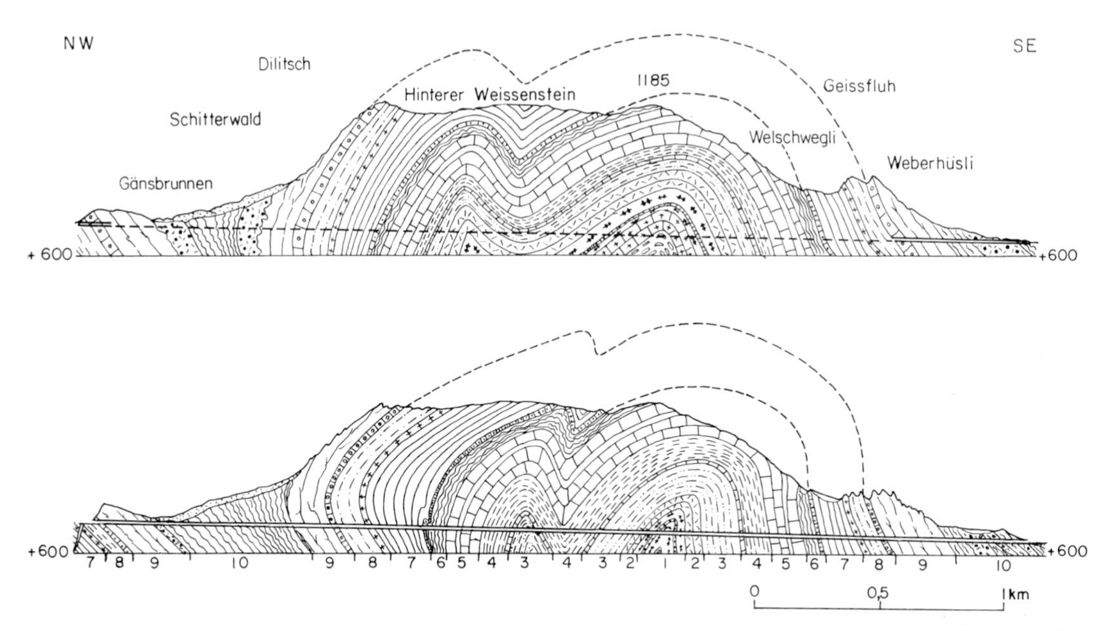

Fig.220. Prognosis and structure of the Weissenstein tunnel. Upper figure: prognosis by C. Schmidt in 1904. Lower figure: section by Buxtorf after completion of the tunnel. The difference between prognosis and actual structure in a rather quiet type of anticlines is relatively slight.

1 = Triassic; *2* = Lias; *3* = Lower Dogger; *4* = Middle Dogger; *5* = "Hauptrogenstein", Dogger oolite; *6* = Oxfordian–Callovian; *7–9* = Malm; *10* = Lower Tertiary. (After BUXTORF, 1908.)

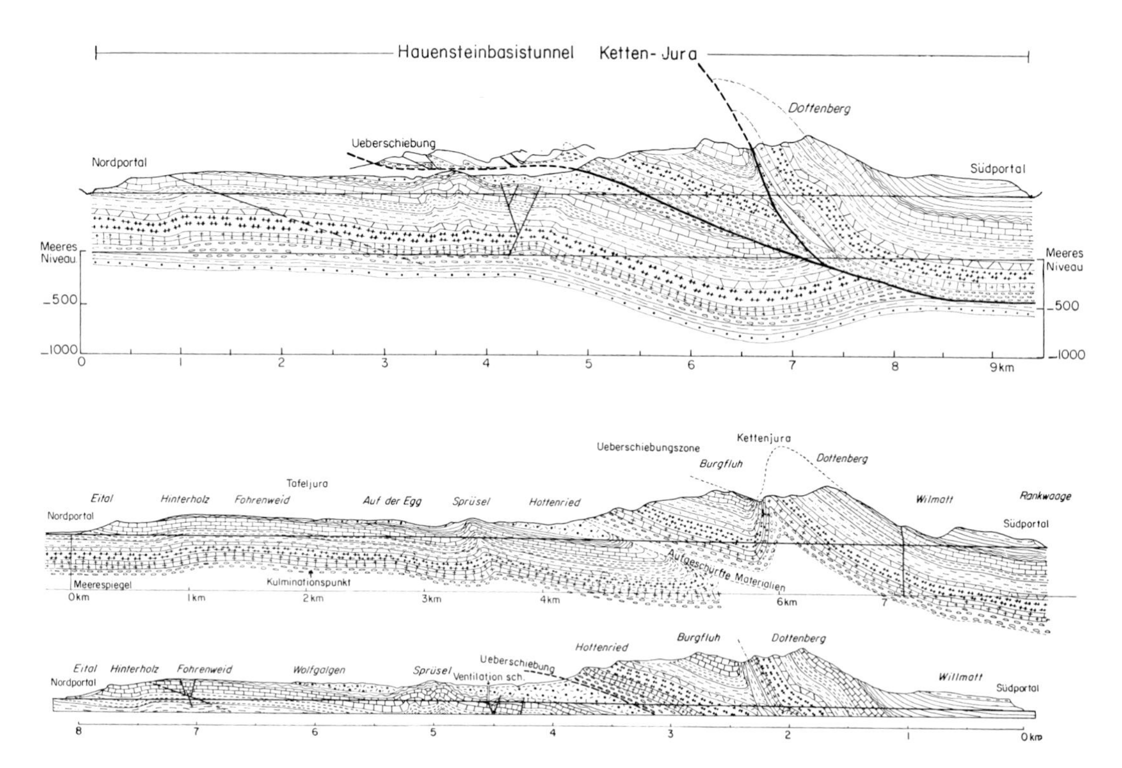

Fig.221. Prognosis, structure and interpretation of the Hauensteinbasis tunnel. Upper figure: prognosis by F. Mühlberg in 1910. Middle figure: actual structure, as found in 1914. Lower figure: interpretation in 1916, both by Buxtorf. For stratigraphy, compare Fig.220.

The upthrust over Triassic between Burgfluh and Dottenberg was completely unsuspected in the prognosis. (After BUXTORF, 1916.)

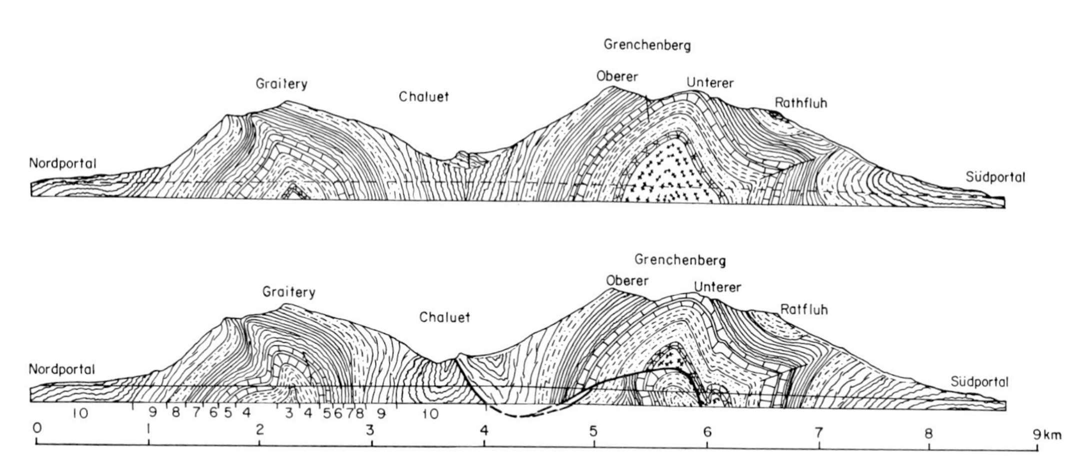

Fig.222. Prognosis and real structure of the Grenchenberg tunnel. Upper figure: prognosis in 1912. Lower figure: actual structure, as found in 1914. For stratigraphy, cf. Fig.220.

The more simply built Graitery anticline has presented no difficulties. In the Grenchenberg structure, on the other hand, a completely unsuspected undulating overthrust was found. (After BUXTORF, 1916.)

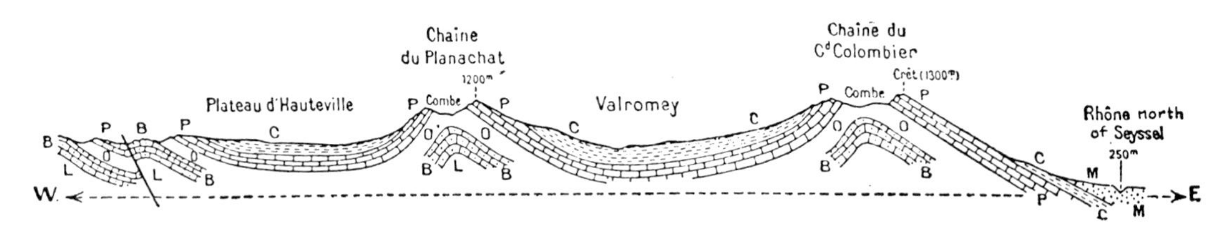

Fig.223. Schematic section through two chains in the southwestern part of the Jura Mountains, to show the relation between structure and morphology. (Scale: horizontal 1:225,000; vertical 1:112,500.) *M* = Tertiary molasse; *C* = Lower Cretaceous; *P* = Upper Jurassic limestones; *O* = Oxfordian marls; *B* = oolitic limestones of the Bajocian–Bathonian = "Hauptrogenstein"; *L* = Lias marls. (After GIGNOUX, 1955.)

nation of more marly series alternating with more resistant limestones. The main horizons of the latter are formed by the thick series of coarse oolitic limestones, the "Hauptrogenstein" of the higher Dogger, and by the limestones of the Upper Malm.

In the more shallow anticlines, a carapace of one of these limestone series will protect the marls. Consequently, morphology will correlate with structure, and anticlines will form ridges. In the steeper structures, however, the higher elements of the limestone carapace are destroyed by erosion, and a reversal of relief develops. Valleys of varying width and depth (French: "combes") will be cut into the anticlinal crests (Fig.223). This situation is more fully treated, for instance already by HEIM (1919), who distinguishes between synclinal, isoclinal and anticlinal valleys.

THE CLASSIC PICTURE

Again, just as with the Swiss Alps, the basic description of the Swiss part of the Jura Mountains is by HEIM (1919). Further information can be found in the explanatory texts to the Swiss 1:200,000 geological map (BUXTORF, 1951; COLLET, 1955; AUBERT and BADOUX, 1956). For the French part of the chain the treatise of DE MARGERIE (1936) is basic. Most of the newer literature on all of the Jura Mountains is cited by CAIRE (1963).

We may use the same procedure for the Jura Mountains as for the Alps, and consequently describe first the classic picture, followed by a short survey of some of the more important newer ideas.

The Jura Mountains may be divided into three parts, i.e., the Table Jura ("Tafeljura"), the Plateau Jura and the Folded Jura or High Chain ("Kettenjura", "Faltenjura", "Faisceau helvétique", "Haute chaîne").

As follows from Fig.224, the Plateau Jura occupies a wide area in the western and northwestern part of the chain, but it wedges out both southwards and eastwards. The High Chain, moreover, narrows northeastwards, and finally is represented by a single anticlinal structure (cf. Fig.221), thereby accentuating the difference between Table Jura and Folded Jura. For the most articulate of the writers on Jura Mountains geology, that is for Professor Buxtorf of Basel, the main antithesis therefore lay in the difference between the Table Jura and the High Chain, whereas the Plateau Jura was thought to represent a feature of more secondary interest. Not only French geologists, but also their Swiss colleagues from Geneva, Lausanne and Neuchatel, have, on the other hand, often been more preoccupied with the Plateau Jura.

The Table Jura is formed by the normal post-Hercynian sedimentary cover of the Vosges and the Schwarzwald. It is more strongly faulted north of the High Chain, whereas individual blocks may show some tilting. This somewhat stronger tectonization remains, however, of small importance.

The High Chain is the narrow concave foldbelt at the inner side of the Jura Mountains. It is normally formed by a number of parallel, strongly tectonized structures, but towards its northeastern end it is represented by a single chain only. The High Chain is for many geologists the most representative element of all of the Jura Mountains.

The Plateau Jura, is, on the other hand, formed by a rather large number of more or

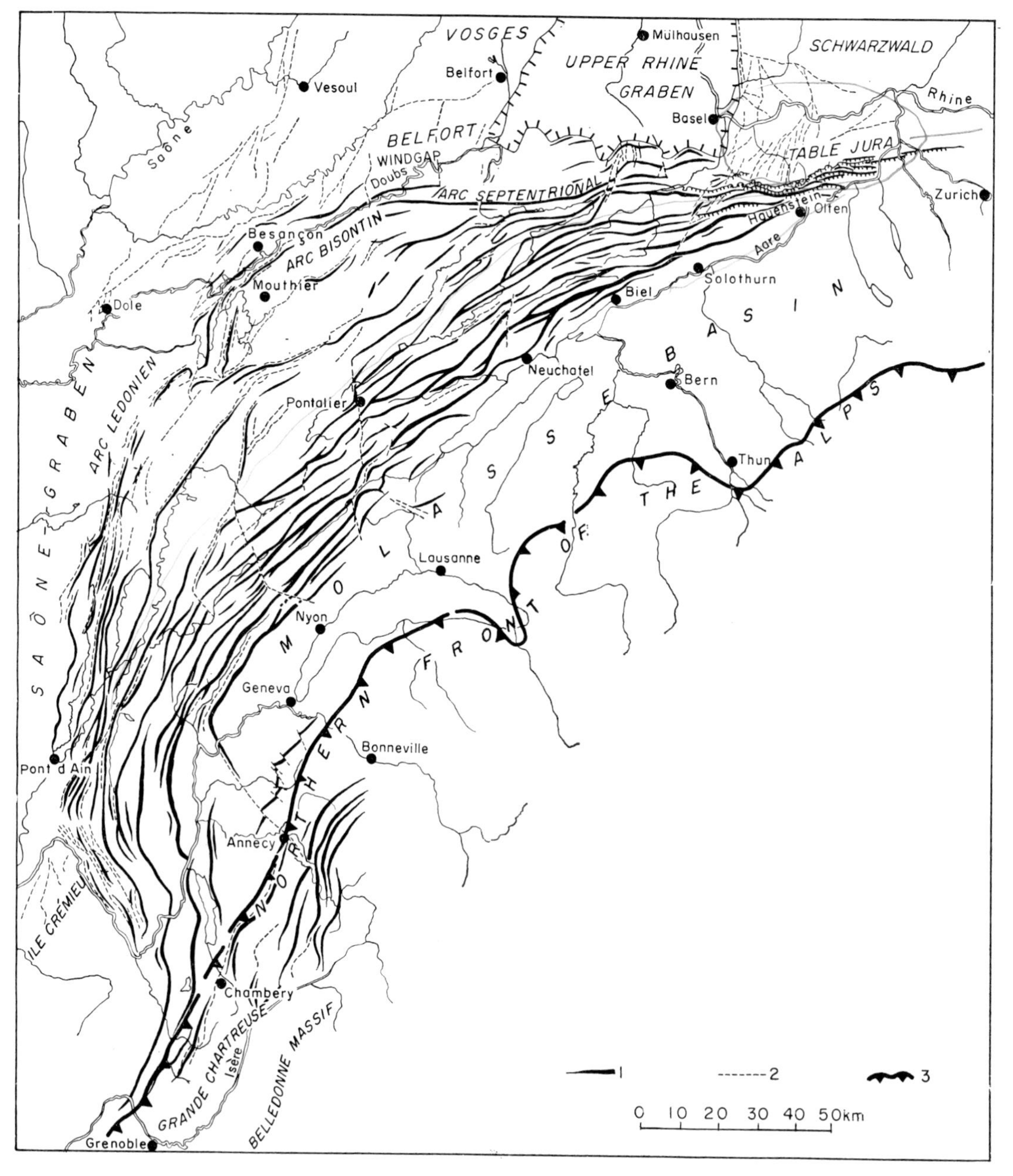

Fig.224. Schematic map of the Jura Mountains. (After BERSIER, 1942.)

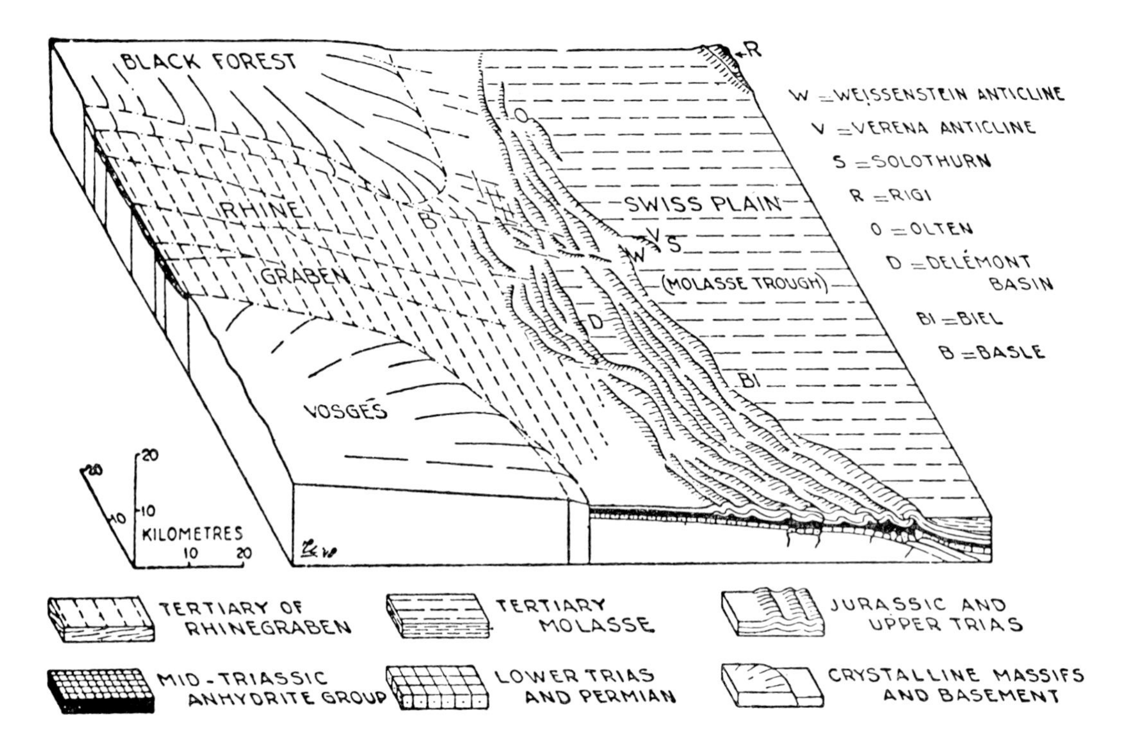

Fig.225. Schematic block diagram of the eastern end of the Jura Mountains, showing the High Chain, thought to be pushed up against the Vosges Mountains and the Schwarzwald or Black Forest. In front of the latter the Table Jura developed, whereas in front of the Upper Rhine graben north of the Delémont Basin folded structures could develop, owing to the lower resistance offered by the graben. The latter structures, the "Vorland Falten" or foreland folds of the Swiss literature, form the easterly continuation of the Plateau Jura. In front of the Vosges Mountains no Table Jura developed, due to the existance of the graben of the Burgundian gate or Belfort windgap. This graben connects the Upper Rhine graben with that of the Saône–Rhône further south. (After UMBGROVE, 1950.)

less stable blocks of various size and form, forming synclinal areas, which are separated by narrow fold belts. Of the latter, the more external ones, the "Arc lédonien" named after Lons-le-Saunier, the "Arc bisontin" named after Besançon, and the "Arc septentrional", south of the Belfort windgap or Burgundian gate, are most commonly cited in the literature.

The boundary between Table Jura and Folded Jura is quite sharp, that between Folded Jura and Plateau Jura, on the other hand, is gradual. Along the first boundary one normally finds the northernmost anticline of the High Chain thrust over the Table Jura. In the second case, the synclines of the High Chain gradually widen northwestwards, and more or less imperceptibly pass into the plateaus of the Plateau Jura.

According to the classic picture, the Jura Mountains developed as a fold belt as a result of tangentional thrust exerted by the nappes of the Alps through the intermediary of the molasse basin. This interpretation is, for instance, illustrated by UMBGROVE (1950), in a block diagram, reproduced here in Fig.225.

As a result of the curvature of the foldbelt, and also of the obliqueness of the thrust from the Alpine nappes, the Jura Mountains were thought to have undergone a relative lengthening during the folding process. This was effected, according to the theory, by a set of wrench faults, as illustrated in Fig.224.

It must be stressed expressly that the horizontal translations along these wrench faults, the "décrochements horizontaux", is small. It is normally of the order of 100 m, only exeptionally reaching 1,000 m or more. Moreover, it is found to be strongly variable, and may even be of opposite sign, along the strike of one and the same wrench fault. Also, it must be remembered that they are known only from the superficial, folded, sedimentary epidermis. This has been widely speculated upon, but they are not definitely known to have any connection with the basement.

The Jura Mountains wrench faults, already schematized by HEIM (1919) in the picture reproduced here as Fig.226, have since become still more schematized in textbooks on structural geology. I will return to this subject in the next paragraph, to see how modern views on the genesis of the Jura Mountains are related to the wrench faults, but it seems well to present already at this moment a somewhat more qualified picture.

Fig.227 shows part of the Jura Mountains with two of the most prominent wrench faults on a larger scale. It follows immediately that there is not one transverse fault, nor even a zone of parallel faults. The larger of the "wrench faults" had better be described as zones separating parts of the fold belt, in which the folding is different. This follows from the many structures which die out near the "wrench fault", and are replaced "en

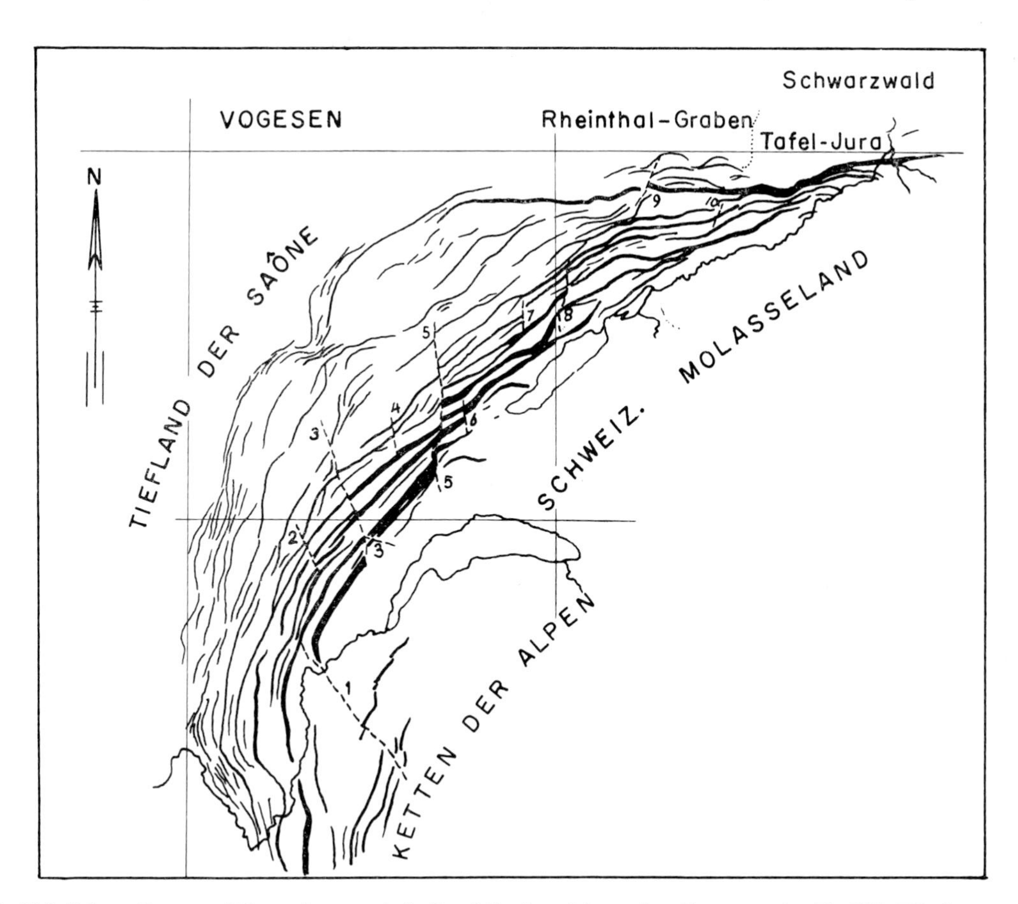

Fig.226. Schematic map of the main wrench faults of the Jura Mountains. Compare also Fig.227, 228, for a more detailed representation of the wrench faults numbers *3* and *5*.
1 = a sheaf of smaller faults along the Salève; *2* = the St. Claude fault; *3* = the Dôle–Champagnole fault, passing through Morez; *4* = the Mouthe fault; *5* = the Vallorbe–Pontarlier fault; *6* = the Mont Suchet fault; *7* = the Les Brenets fault; *8* = the Montruz fault; *9* = a sheaf of smaller faults near Montmelon; *10* = smaller faults (about 20), forming the fault zone of Gänsbrunnen. (After HEIM, 1919.)

échelon" by another structure at the other side. In trying to connect these "en échelon" structures, and by assuming that these were originally continuous, one arrives at the most variable figures for the horizontal translation along the supposed fault. But this is only the result of an improper reconstruction.

This image of two parts of a fold belt, each folded on its own, is even more clearly demonstrated by a modern analysis of part of the main "wrench fault", that of Pontarlier, reproduced in Fig.228.

The Jura Mountain wrench faults consequently belong to a group of features entirely different from the larger wrench faults of the San Andreas type, which play such a large part in more modern geotectonic analyses.

To round off the classic picture of the genesis of the Jura Mountains it must be noted that, according to this picture, the actual shortening of the cover must have been the same as that of the basement. We do not know this basement exposed, due to the "décollement", but it was through shortening of the "entire" crust that the Jura Mountains were thought to have originated, not through some transportation on its own of the sedimentary cover. Though the basement consequently was thought to have been shortened over the same distance, it was not required that this shortening should take place in a similar way. The most common solution proposed was that the basement had been affected by low angle overthrusts (cf., for instance, AUBERT, 1945).

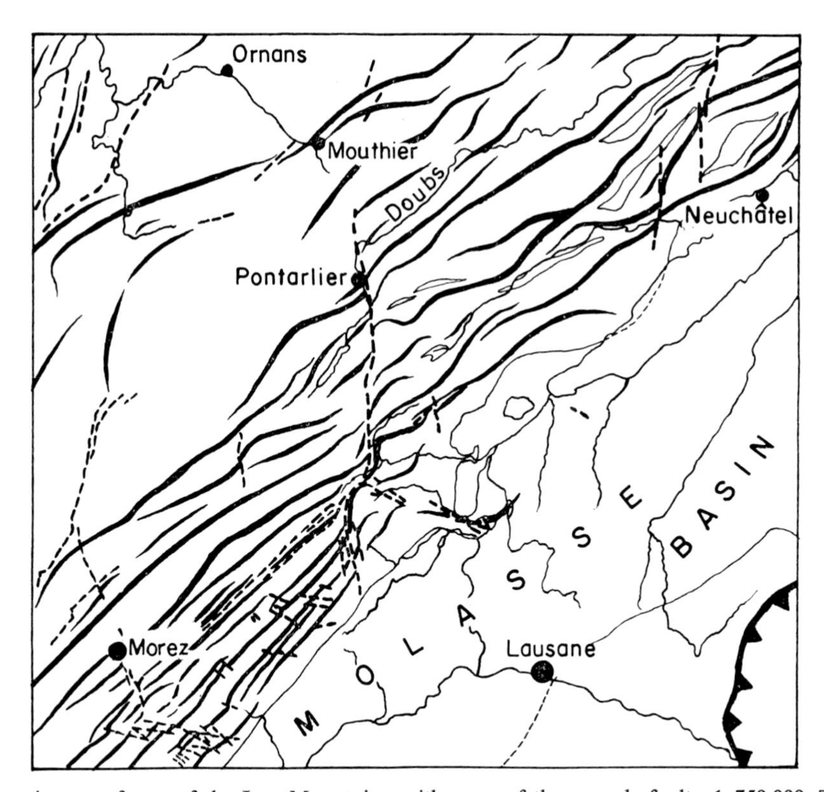

Fig.227. Tectonic map of part of the Jura Mountains, with some of the wrench faults, 1:750,000. The so-called wrench faults apparently are much less continuous, than normally presented in schematized figures such as Fig. 226. (After AUBERT and BADOUX, 1956 and COLLET, 1955.)

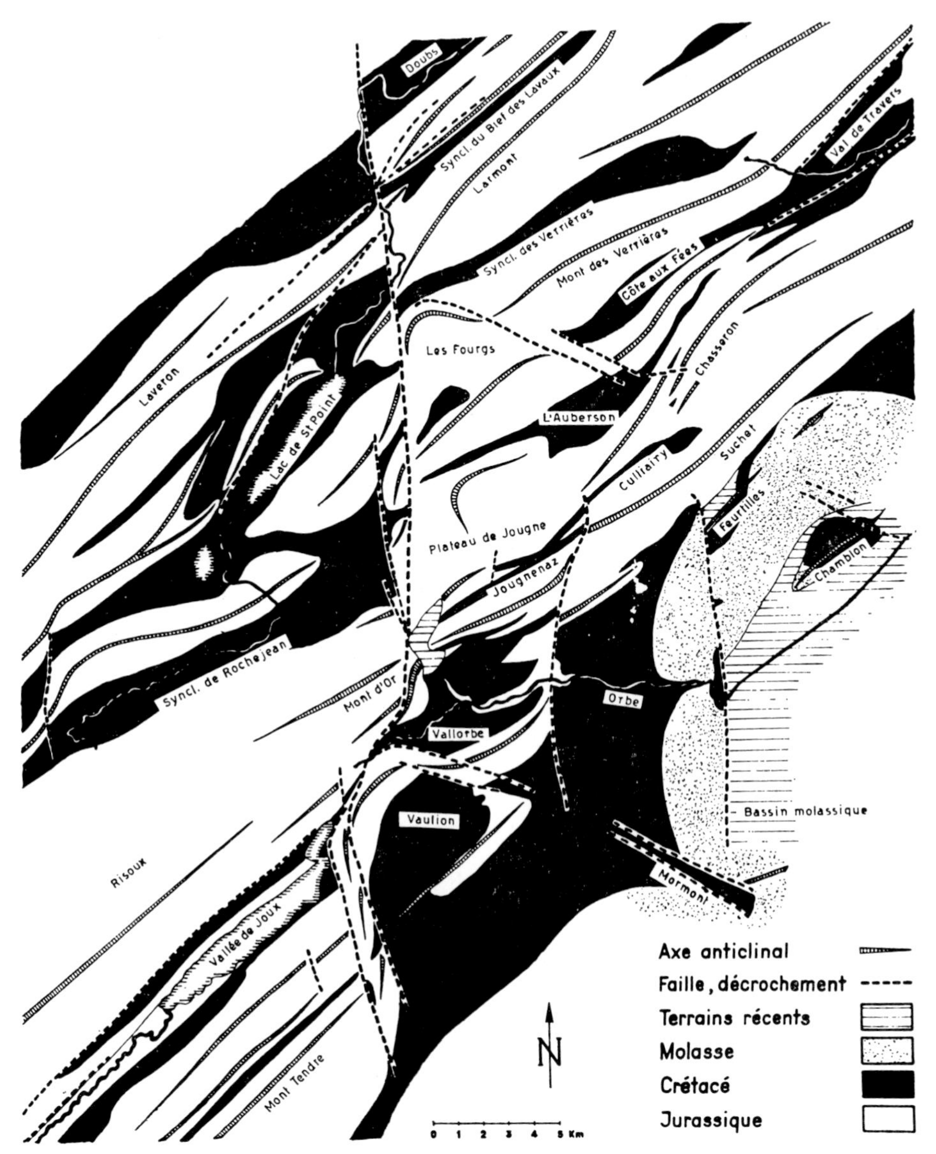

Fig.228. Tectonic map of the Jura Mountains along part of the major wrench fault of Pontarlier, north of Vallorbe. There is no horizontal translation apparent along the so-called wrench fault, but it separates two parts of the Jura Mountains which were folded on their own. (After AUBERT, 1959.)

NEWER IDEAS

GRAVITATIONAL TECTONICS

Fresh ideas as to the genesis of the Jura Mountains start with LUGEON (1941). He noted that in all earlier papers on the Jura Mountains the intimate relationship between Jura and Alps had been stressed; the Jura branching off from the Subalpine chains at the latitude of Chambéry. On the basis of this relationship everyone had assumed that if the Alps had been formed by tangentional compression, a similar origin had to be accepted for the Jura also. If, however, we assume an origin through gravitational tectonics, if not for all of the Alps, at least for the Prealps, which lie nearest to the Jura, such as had been proposed by E. Gagnebin (p. 202), we might well assume that the Jura Mountains were also formed by this mechanism. That is, that they originated through gravitational "décollement" into the Saône–Rhône graben, in a manner comparable to the Prealps sliding into the molasse basin.

LUGEON's (1941) ideas were, however, not acceptable to GAGNEBIN (1942), who thought gravitational tectonics was restricted to the Prealps, and, as we will see, the question is still open.

A factor which must be taken into consideration in this discussion is, that the overall outward "Vergenz", required by the theory of tangentional compression, does not exist. In most schematic pictures of the Jura Mountains, such as in the blockdiagram of Fig.225, this outward "Vergenz" is prominent, but this is an artificial impression only. Long before Lugeon, other geologists had insisted on this fact, a point later taken up by WEGMANN (1961, 1963) of Neuchatel.

Apart from the extreme eastern ending of the High Chain, which is here reduced to a single anticline, indeed with a northward "Vergenz", individual structures may either "face" southeast or northwest. In my opinion this indicates that the Jura Mountains do not form a single homogeneous fold belt. They are not of one casting, not "aus einem Guss", but the result of a number of phenomena, probably interrelated, but affecting different parts of the area in quite varied ways.

COMPLEX HISTORY OF THE "ARC BISONTIN"

At the other side of the chain, in the neighbourhood of Besançon, GLANGEAUD (1949, 1950) stressed the composite history of the Plateau Jura. Its genesis seems not in any way related to that of the far-away Alps, but instead to a succession of vertical movements of the local basement.

At the beginning of the Jura Mountains formation, larger crustal blocks, the later plateaus, were separated by narrow graben structures. Within the latter at first tension and

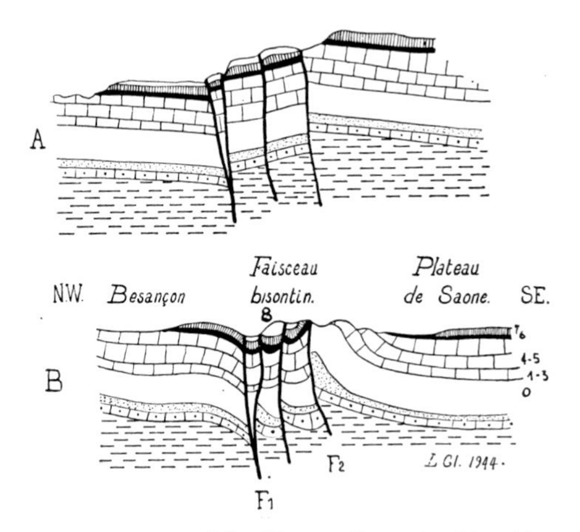

Fig.229. Genesis of one of the outer structures of the Plateau Jura, the "Arc bisontin", near Besançon, according to the theory of GLANGEAUD (1950). Upper section: final Oligocene. Lower section: present situation. F_1 = Doubs fault; F_2 = Argel–Beure fault; *O* = Lias; *1–5* = Dogger; *6* = Oxfordian; *7* = Rauracian; *8* = Sequanian–Kimmeridgian. (After GLANGEAUD, 1950.)

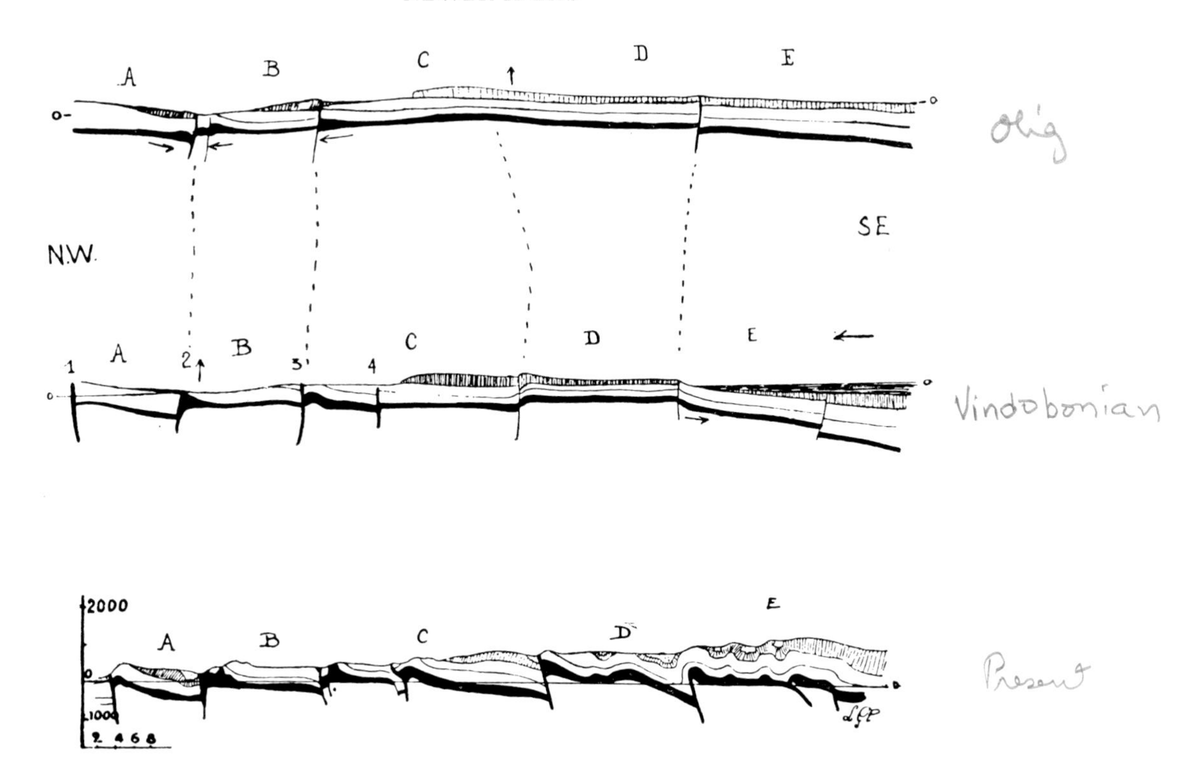

Fig.230. Genetic sections through the Jura Mountains between Besançon and Pontarlier. Vertical scale 4 × horizontal scale. Upper section: Oligocene. Middle section: Vindobonian. Lower section: Present.
A = plateau and marginal fault–fold north of Ognon; *B* = same for the Avants–Monts plateau and the Bisontin chain; *C* = Ornans plateau with Mamirolle chain; *D* = Laveron plateau with Salinois chain; *E* = High Chain. (After GLANGEAUD, 1950.)

subsidence occurred, but in a later stage, owing to a "décollement" of the sedimentary cover of the plateaus, which became slightly tilted, the former graben were compressed and their sedimentary series folded more or less strongly by a kind of pinching. The narrow anticlinal structures of the Plateau Jura consequently at present form a sort of draping over these original graben structures (Fig.229).

Originally postulated for the "Arc bisontin" near Besançon, it quickly became apparent that a similar genesis can be postulated for all of the Plateau Jura, if not for all of the folded Jura Mountains, inclusive of the High Chain (Fig.230).

A new elemental form of descriptive structural geology was introduced, that of the *faille–pli* or "fault–fold". In contrast to the well known *pli–faille*, first described by E.

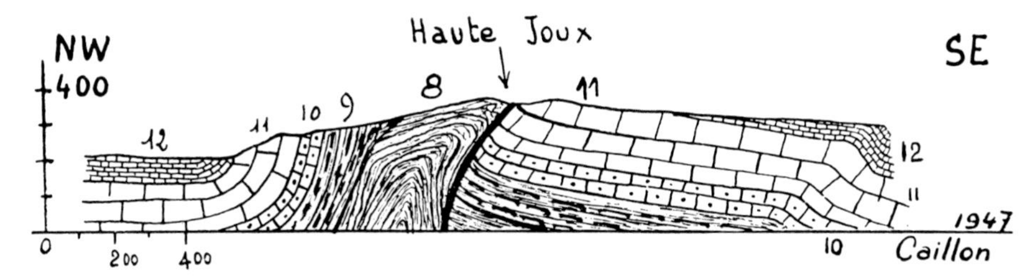

Fig.231. Type of a *faille–pli* or "fault–fold". Anticline of Haute Joux, southeast of Pontarlier. The plastic core is formed by the marls of the Oxfordian. For location of this anticline, compare Fig.237.
8 = Oxfordian marls; *9* = Rauracian; *10* = Sequanian; *11* = Kimmeridgian; *12* = Portlandian. (After GLANGEAUD and SCHNEEGANS, 1950.)

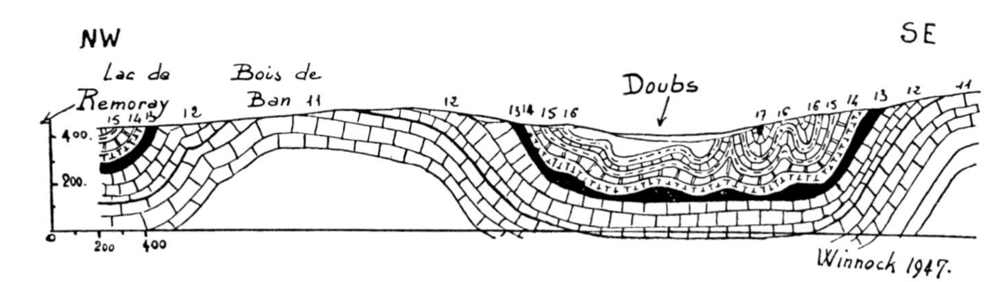

Fig.232. Section through part of the Plateau Jura, showing collapse structures and "décollement" in the Cretaceous, at a level well above the normal "décollement" in the gypsiferous Triassic.
11 = Kimmeridgian; *12* = Portlandian; *13* = Purbeckian; *14* = Valanginian; *15* = Hauterivian; *16* = Barremian; *17* = Albian–Aptian; *18* = Cenomanian. (After GLANGEAUD and SCHNEEGANS, 1950.)

Haug from the Subalpine chains (p. 272), this is formed by a fault which subsequently developed into a fold (Fig.231).

Moreover, it was found by detailed field work that "décollement" was not restricted to the gypsiferous Triassic, but can occur in any reasonably mobile horizon (Fig.232). This finding fits, of course, exactly with what has been found in the Helvetides and in the Subalpine chains.

The difference in view between LUGEON (1941) and GLANGEAUD (1949, 1950) rests mainly in the fact that the first author thinks of one major plane of "décollement", affecting all of the Jura Mountains, and extending all under it, whereas the latter author thinks more of partial "décollement" on top of the separate blocks formed by the basement underlying the plateau areas. These might have moved up and down, and/or become tilted more or less independantly through forces as yet unknown, operative on the basement.

The mosaic resulting from the movements envisaged by Glangeaud is, to my opinion, compatable with the irregularity of the "Vergenz" of the individual structures, as cited above.

LARGE-SCALE OVERTHRUSTING

The main trend of the ideas presented by Glangeaud has since been corroborated in a remarkable way through the research carried out by the B.R.G.M. (Bureau de Recherches Géologiques et Minières) in the Lons-le-Saunier region, as reported by LIENHARDT (1962). The B.R.G.M. began prospecting for potassium salt, then for gas, and ultimately found coal. But, ingrained by the tenet of Professor Raguin of Paris that to become a good economic geologist, one had first to be a good geologist "tout court", the B.R.G.M.'s most important result might well be the new light shed on the structure of part of the outer chain of the Jura Mountains.

This research by the B.R.G.M. deep drilling has revealed the structure of the basement in the Lons-le-Saunier area, this being the first factual information about the basement, anywhere, in the Jura Mountains. The basement is formed by pre-Hercynian rocks, mainly micaschists, gneisses and granites, in which a post-orogenetic limnic coal basin of Stephanian age developed, some 700 m thick, with several coal beds.

The Stephanian has suffered only from blockfaulting tectonics (Fig.233). After a period of peneplanation it was conformably covered by the Permian and Triassic. Although the later "décollement" makes this difficult to prove, the younger Mesozoic probably also was deposited conformably (Fig.234, *5*).

The region was then tilted during the Eocene (Fig.234, *6*), which was followed by strong blockfaulting during the Oligocene. The major structure which developed during this period was the border fault between the Saône graben and the Jura Mountains, which has a throw of at least 3 km. Many other faults developed during this period too (Fig.234, *7*).

The present outer chain of the folded Jura, the "Arc lédonien", was then thrust over this border fault along an extremely level, subhorizontal thrust plane. This thrust plane is known over a distance of 13 km, and is consequently a feature rather distinct from the smaller overthrusts imagined by Glangeaud (Fig.234, 9).

The front of the overthrust sheet spread out even beyond the main border fault of

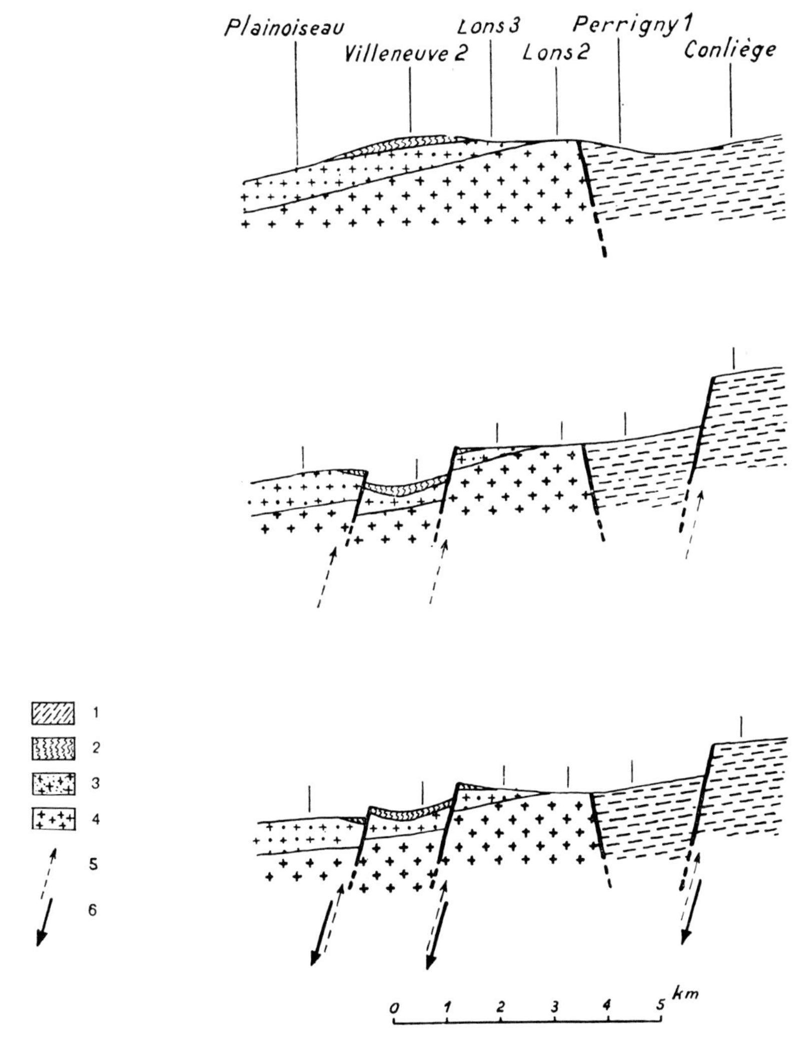

Fig.233. Schematic section through the pre-Stephanian basement of the Jura Mountains near Lons-le-Saunier. Vertical scale 5 × horizontal scale. Upper section: pre-Stephanian peneplain. Middle section: influence of post-Stephanian, pre-Permian blockfaulting. Lower section: present situation. The Tertiary blockfaulting in part reversed the movements of the older blockfaulting period.
1 = micaschists; *2* = gneisses; *3* = migmatites; *4* = granite; *5* = Hercynian movements; *6* = Alpine movements. (After LIENHARDT, 1962.)

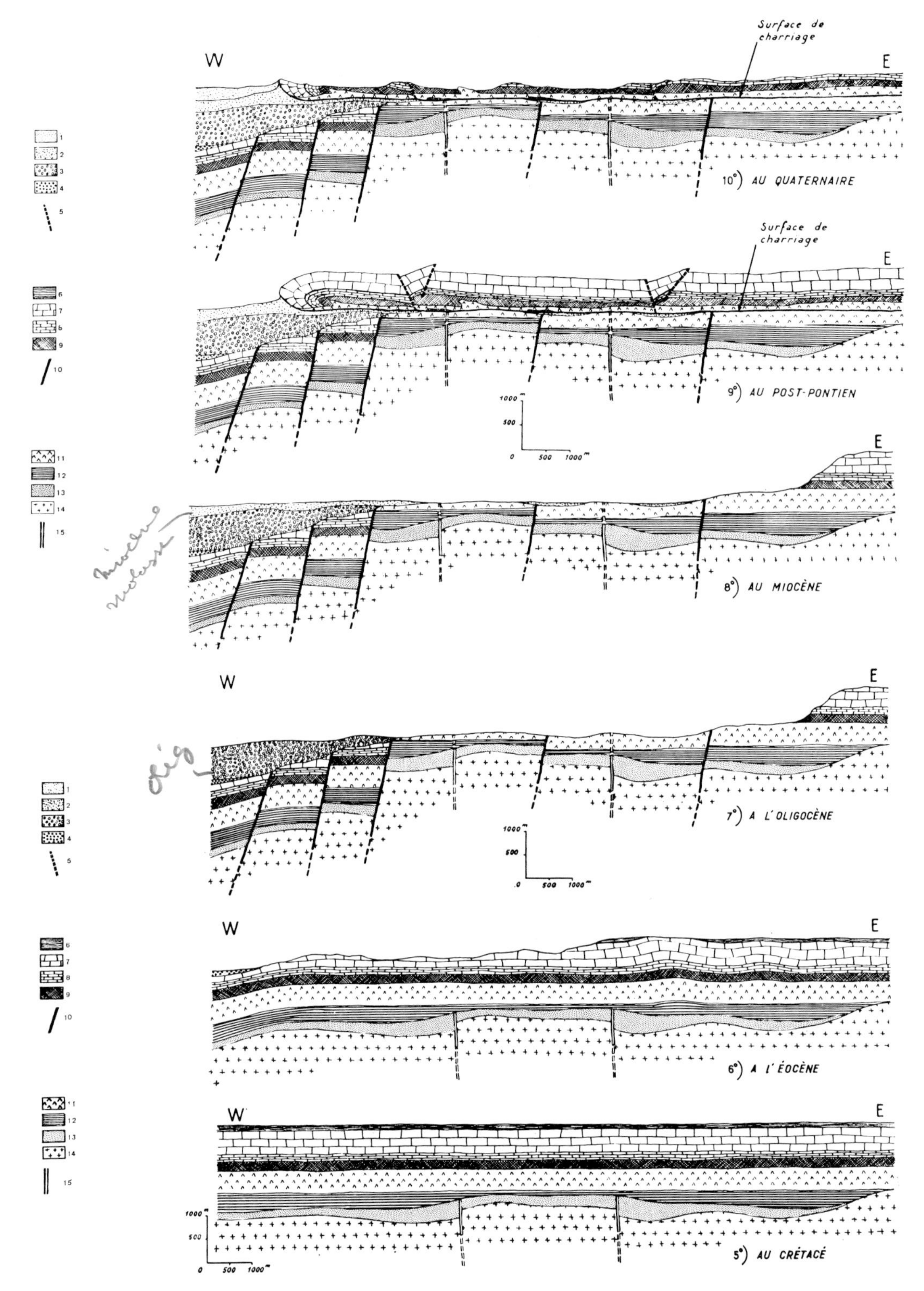

Fig.234. Series of genetic sections through the outer structure of the Jura Mountains, the "Arc lédonien", near Lons-le-Saunier.

1 = Pliocene; *2* = Miocene molasse; *3* = Oligocene; *5* = Eocene; *4* = faults accompanying the overthrusting; *6* = Cretaceous; *7* = Upper Jurassic; *8* = Lower Jurassic; *9* = Lias; *10* = Oligocene blockfaults; *11* = Triassic; *12* = Permian; *13* = Stephanian; *14* = pre-Hercynian basement; *15* = post-Stephanian, pre-Permian blockfaults.

After the post-orogenetic deposition of the Stephanian, and the ensuing blockfaulting, Permian and Mesozoic were deposited conformably (lower section). During the Eocene tilting and erosion took place, followed by strong blockfaulting during the Oligocene. Renewed erosion led to renewed peneplanation, over which surface the thrust-sheet spread, which ultimately was itself attacked by erosion (upper section). (After LIENHARDT, 1962.)

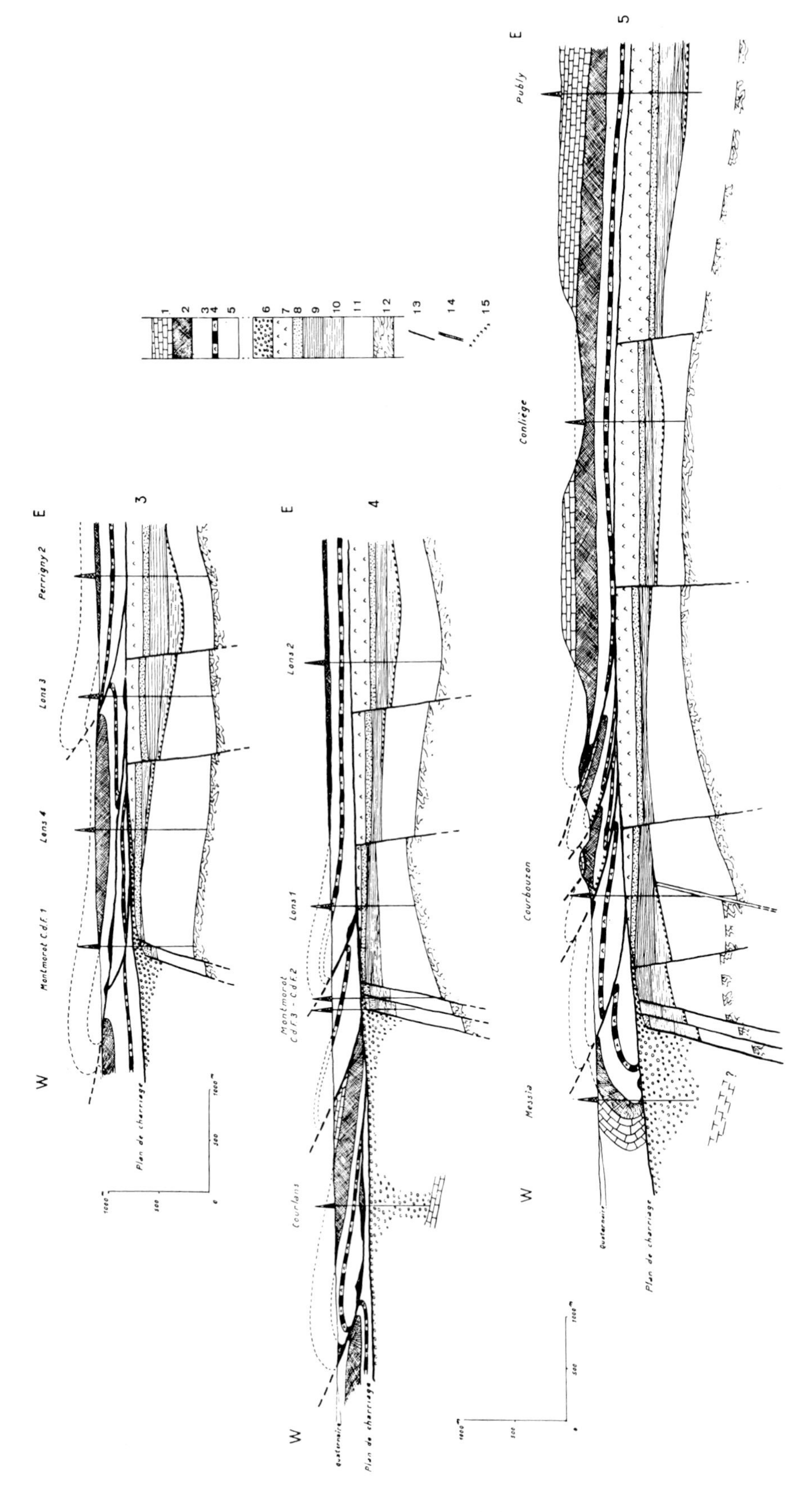

Fig.235. Three sections through the outer chain of the Jura Mountains, the "Arc lédonien", near Lons-le-Saunier. Elements of the overthrust sheet: *1–5*. Autochthonous elements: *6–12*.

1 = Middle and Upper Jurassic; *2* = Lias; *3* = Upper Keuper; *4* = thinly bedded dolomite; *5* = Lower Keuper; *6* = Tertiary molasse; *7* = Triassic; *8* = Lower Triassic Buntsandstein; *9* = red Saxonian; 10 = violet Saxonian; *11* = Stephanian; *12* = crystalline basement; *13* = Oligocene blockfault; *14* = post-Stephanian–pre-Permian blockfault; *15* = Keuper squeezed into overthrust planes. (After Lienhardt, 1962.)

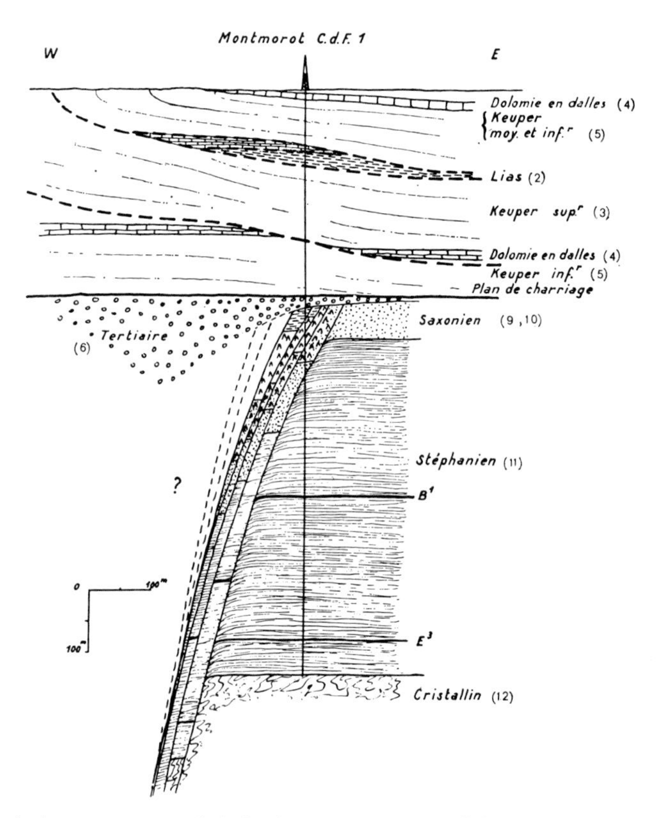

Fig.236. Section through the Montmoret 1 hole in the outer structure of the Jura Mountains, the "Arc lédonien" near Lons-le-Saunier. For location, compare the upper section of Fig.235. This section shows the composite natur of the main thrust plane. For notations, see Fig.235. (After LIENHARDT, 1962.)

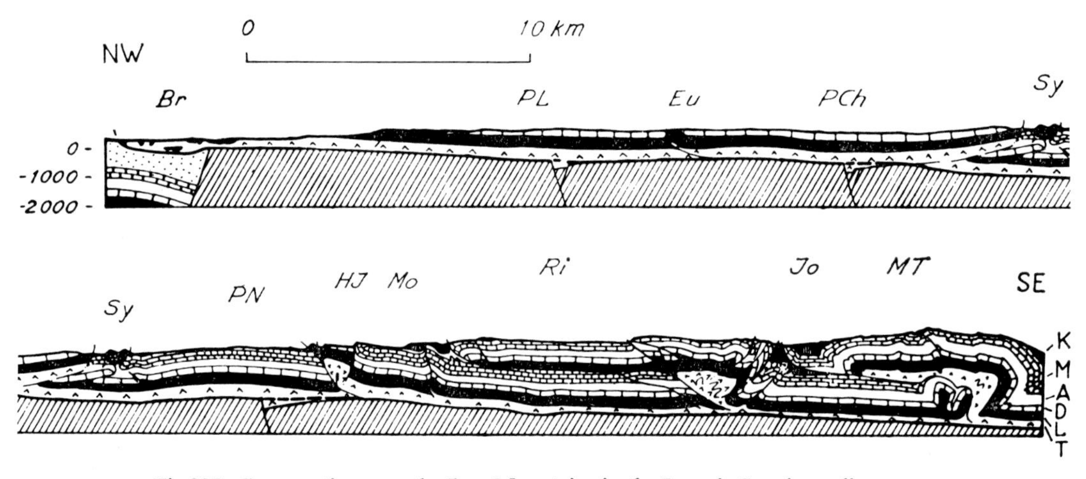

Fig.237. Cross-section over the Jura Mountains in the Lons-le-Saunier radius.
Topography: *Br* = Bresse (Saône) graben, overthrust by the "Arc lédonien"; *Pl* = Lons-le-Saunier plateau; *Eu* = Euthe anticline; *PCh* = Champagnole plateau; *Sy* = Syam structures; *PN* = Nozeroy plateau; *HJ* = Haut Joux anticline; *Mo* = Mouthe syncline; *Ri* = Risoux anticline, with an overthrust of 10 km; *Jo* = Joux–Tals anticline; *MT* = Mont Tendre anticline, the innermost structure of the High Chain.
Stratigraphy: *K* = Cretaceous; *M* = Malm; *A* = Argovian; *D* = Upper Dogger; *L* = Lias and Lower Dogger; *T* = Triassic.
In this section no account has been taken of the thick development of the Triassic underneath the folds of the Jura Mountains, as found by exploratory drilling and geophysical exploration (p.390). Underneath the Risoux (*Ri*) composite thrust block, the Triassic is over 1 km thick. (After LAUBSCHER, 1965.)

the Saône graben and its adjoining steps, and here overrides autochthonous molasse conglomerates of Oligocene and Miocene age. These derive from elsewhere in the Jura Mountains, where either through Eocene tilting, or through Oligocene blockfaulting, or due to both processes, parts of the foldbelt were already rising and could be attacked by contemporaneous erosion.

The overthrust, at least in its outer part, where it overrides the molasse sediments mentioned, consequently took place at a very late date, i.e., post-Miocene. Such a young age for this particular overthrust is, of course, compatible with its position at the external side of the chain, a position in which in other foldbelts too, the youngest movements tend to occur. On the other hand, we see in Fig.235, how the overthrust is composite. The distance over which the molasse is overridden is only between 2 km and 3 km long. It might consequently well be possible that it was formed already in the main at an earlier date, whilst the post-Miocene thrust is more of the order of a posthumous movement. We do not have, however, any means, of dating exactly the various thrust planes which together make up this spectacular feature.

The facts recorded by LIENHARDT (1962) are to my view of great importance for our comprehension of the genesis of the Jura Mountains. A rather complete collection of his more general sections has therefore been reproduced in Fig.233–236.

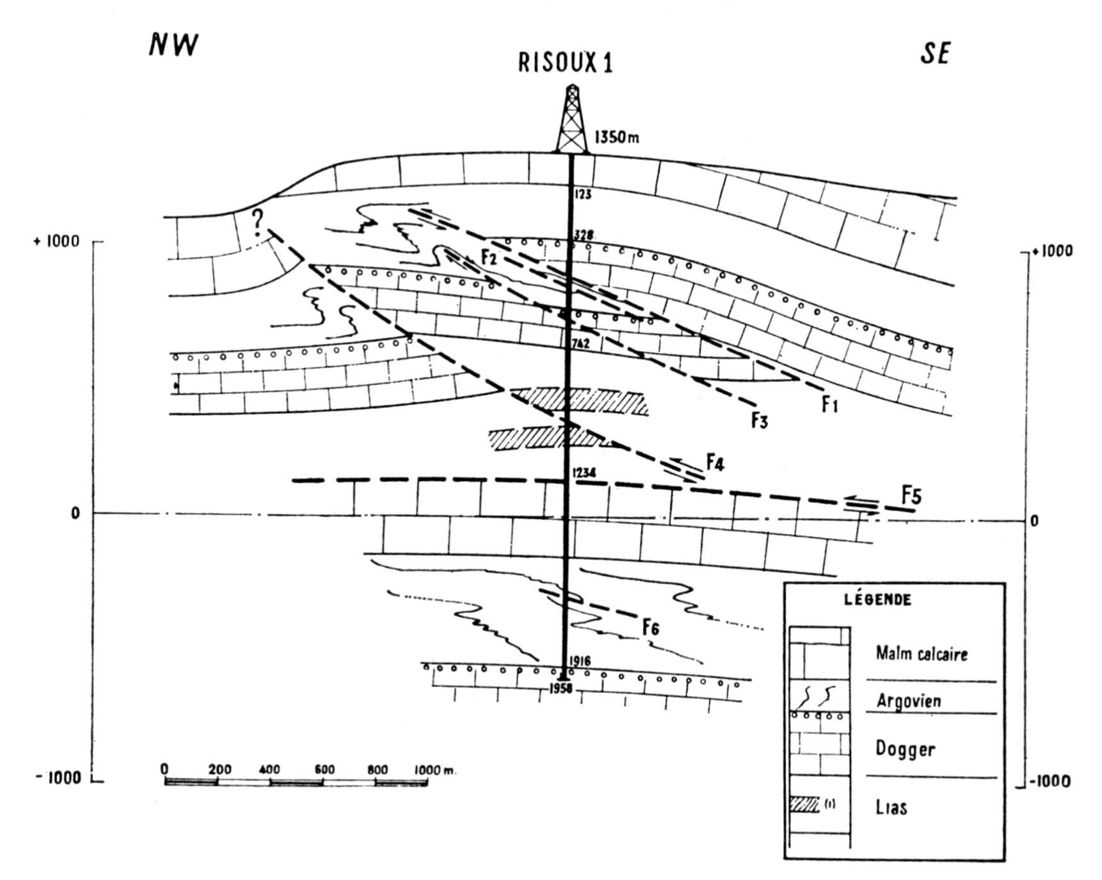

Fig.238. Section through the Risoux anticline in the Jura Mountains.
The structure, in reality more a broad plateau than an anticline, shows some disharmonic folding and faulting in relation to the shaly Argovian. Below a depth of 1,234 m, it does, however, encounter a complete duplicate series, starting with the Malm. Neither the Triassic, nor the basement have as yet been reached. (After WINNOCK, 1961.)

Such large scale overthrusting is not restricted to the outer chains of the Jura Mountains. It has also been found in the High Chain, such as, for instance, in the Risoux anticline. This structure is situated on the same radius as the "Arc lédonien" (Fig.237, 238). Similar overthrusts will probably be found in many more of the Jura Mountains structures, when exploration of the basement will proceed further.

The conspicuous elements of the modernised section through the Jura Mountains of Fig.237 are: (*1*) the smoothness of the plane of "décollement"; (*2*) the relative importance of overthrusting; and (*3*) the rather small difference existing between the Plateau Jura and the High Chain. As to the first point, it must be mentioned that the position of the basement is still only known from a small number of bore holes. The two interpretations given simultaneously in Fig.237 (that of blocks separated by faults, and that of overall smoothness), are equally possible.

WRENCH FAULTS OF THE BASEMENT?

Turning now from facts to fancies, we have to record how WEGMANN (1961, 1963) and PAVONI (1961) have independently proposed a solution for the folding of the Jura Mountains as a result of horizontal translation and wrench faulting of underlying basement blocks. Pavoni's map, which is better drawn than Wegmann's more schematic figures, well indicates the difficulties encountered in this solution. Only in the case of the "wrench faults" of long standing, found in the sedimentary epidermis, and described by HEIM (1919), which according to the views mentioned above, would correspond to wrench faults in the basement, is the requisite angle between wrench fault and fold axis found. In other areas of the Jura Mountains this relation is not apparent.

Personally I still think L. Glangeaud's idea of tilted basement blocks with local gravitational reaction of the epidermis gives the best model for the origin of Jura Mountain folding. This feeling rests, however, not so much on the Jura Mountains, where we do not know enough of the position of the basement, but on the structure of the comparable, but more deeply eroded Catalanides, where the tilted basement blocks are well visible (pp.409–414).

This model does, however, not explain the large westward overthrust of the western structure, the "Arc lédonien", as described by LIENHARDT (1962). But every single explanation has, up to now, failed to explain both the mosaic of smaller structures with opposed "Vergenz", and the single, large outward overthrust of the external anticline. We must conclude that no single phenomenon can have been at the base of all the features developed by the Jura Mountains.

In the model of a gravitational reaction to crustal subvertical movements and tilting, this can be found by assuming a two step "décollement". A first, slight, but general, outward tilting of the basement would in this case have produced the large outward overthrusts. Further movements would then have resulted in a breaking up of the Jura Mountains basement into smaller blocks along faults, some of them would be antithetical. The smaller structures could then have developed on top of these smaller blocks, and could, according to the direction of tilting of their own basement block, develop opposite "Vergenz".

A solution of the genesis of the Jura Mountains which shows a superficial resemblance to that of Glangeaud has recently been proposed by LAUBSCHER (1965). For the (north) eastern part of the Jura he assumes the existence of a number of blocks in which the folding has been autonomous. At the border between such blocks, each with their own folding, differing from that in the adjacent block, "wrench faults" — in the epidermis — may form.

Laubscher is in the advantageous position of having re-mapped a larger part of this section of the Jura Mountains, and his models are consequently adapted in detail to the present situation. He thinks it is possible to distinguish between an earlier tectonic phase of the epidermis, followed by denudation, and then by a "morpho-tectonic" phase. The latter phase is thought to imply gravitational settling of surface elements, due to differences in surface level, produced by erosion. It is not a primary tectonic phase. LAUBSCHER's (1962) "Zweiphasenhypothese" therefore is quite a different proposition than the "two phase tectonics" often assumed by earlier writers.

The folding within the blocks is thought to have resulted from a primary clockwise rotation of the entire Jura chain over some 8°, accompanied by local variations in reaction of individual blocks. This modern analysis of the Jura Mountains is completed, rather unexpectedly, by the assumption of an oldfashioned motor, i.e., tangential push. This is thought to have been exerted, as advanced in older theories, far from the south (the "Fernschubhypothese"; LAUBSCHER, 1961), by the intermediary of the molasse basin, which is thought to have behaved as a competent block.

Just to show how one can never hope to really prove something in geology, we must note the curious coincidence that Laubscher bases this oldfashioned explanation on the very notion gravity tectonicians have lately used to contradict tangential-push tectonicians, which is the role of fluid pressure in overthrust tectonics, as proposed by RUBEY and HUBBERT (1959). If, Laubscher says, this phenomenon develops under a certain overburden, it will not only have existed underneath the Jura Mountains, but also underneath the molasse basin. This will have reduced friction so considerably that it will have been easy for even the smallest of tangentional forces exerted by the Alps to have pushed forward the competent molasse basin, which in turn rotated and folded the Jura Mountains.

To conclude this exposé on the Jura Mountains, we see how they still present us with a great number of problems. One thing is certain, however. The older notions about the Jura being a miniature fold belt which could serve as a model for the Alps, can no longer be maintained.

THE JURA MOUNTAINS AND THE EXTENSION OF THE SALT OF THE TRIASSIC

This has been underlined by the results of the recent drilling campaign in the Jura Mountains. It was found that the Folded Jura is limited both by the thickness of the underlying Triassic, and by the fact whether it contains salt.

These results are added here as an afterthought. They have, as yet, not been incorporated into any theory of the genesis of the Jura Mountains, and consequently also not in any of the cross-sections used in this text (E. Winnock, communication at the *Colloque Étages Tectoniques*, Neuchâtel, April, 1966).

The Triassic underneath the Jura Mountains, is thicker than 500 m, reaching 1,300 m in the Lavéron hole in the central part of the High Chain. In the molasse basins, both in the Bresse Basin to the west, and the Swiss Basin to the east, the Triassic is thinner.

Moreover, the area of the development of salt or other evaporites in both the Muschelkalk and the Keuper is almost exactly the same as that of the Jura Mountains, although the Muschelkalk evaporites show a slightly larger area. The thickness of the evaporite series increases towards the centre of the chain, till it reaches 700 m in the Lavéron hole cited above.

One conclusion therefore is: no salt, no folding. A second conclusion is, that there has been inversion of relief since the Triassic. What was at that time a minor sedimentation basin, is now a minor mountain chain.

REFERENCES

AUBERT, D., 1945. Le Jura et la tectonique d'écoulement. *Bull. Lab. Géol. Mineral. Géophys. Musée Géol. Univ. Lausanne*, 83: 20 pp.

AUBERT, D., 1959. Le décrochement de Pontarlier et l'orogénèse du Jura. *Mém. Soc. Vaudoise Sci. Nat.*, 12: 93–152.

AUBERT, D. and BADOUX, H., 1956. *Notice explicative, feuille 1: Neuchâtel. Carte Géologique de la Suisse, 1:200.000.* Kümmerley und Frey, Bern, 27 pp.

BERSIER, A., 1942. Carte tectonique du Jura. In: E. GAGNEBIN and P. CHRIST (Editors), *Geologischer Führer der Schweiz. I.* Wepf, Basel,

BUXTORF, A., 1908. Geologische Beschreibung des Weissensteintunnels und seiner Umgebung. *Mém. Carte Géol. Suisse*, 21: 125 pp.

BUXTORF, A., 1916. Prognosen und Befunden beim Hauensteinbasis und Grenchenbergtunnel und die Bedeutung der letzteren für die Geologie des Juragebirges. *Verhandl. Naturforsch. Ges. Basel*, 27: 184–254.

BUXTORF, A., 1951. *Erläuterungen zu Blatt 2: Basel–Bern. Geologische General Karte der Schweiz, 1:200.000.* Kümmerley und Frey, Bern, 39 pp.

CAIRE, A., 1963. Problèmes de tectonique et de morphologie jurassiennes. In: M. DURAND DELGA (Editor) *Livre à la Mémoire du Professeur P. Fallot. II.* Soc. Géol. France, Paris, pp. 105–158.

COLLET, L. W., 1955. *Notice Explicative Feuille 5 : Genève – Lausanne. Carte Géologique de la Suisse, 1:200.000.* Kümmerley und Frey, Bern, 47 pp.

DE MARGERIE, E., 1922. Le Jura. I. Bibliographie sommaire du Jura franco-suisse. *Mém. Carte Géol. France*, 1922: 642 pp.

DE MARGERIE, E., 1936. Le Jura. II. Commentaire de la carte structurale du Jura. *Mém. Carte Géol. France*, 1936: 900 pp.

FERNOW, L. R., 1965. Paris Basin. *Geotimes*, 10 (4): 9–13.

GAGNEBIN, E., 1942. Vues nouvelles sur la géologie des Alpes et du Jura. *Bull. Soc. Neuchâteloise Sci. Nat.*, 67: 1–4.

GIGNOUX, M., 1955. *Stratigraphic Geology.* Freeman, San Francisco, Calif., 682 pp.

GLANGEAUD, L., 1949. Les caractères structuraux du Jura. *Bull. Soc. Géol. France*, 19: 670–688.

GLANGEAUD, L., 1950. Le rôle du socle dans la Tectonique du Jura. *Ann. Soc. Géol. Belg.*, 73: 57–94.

GLANGEAUD, L. and SCHNEEGANS, D., 1950. Les caractères généraux du style jurassien. *Ann. Soc. Géol. Belg., Bull.*, 73: 131–146.

HEIM, A., 1919. *Geologie der Schweiz. I. Molasseland und Juragebirge.* Tauchnitz, Leipzig, 704 pp.

LAUBSCHER, H. P., 1961. Die Fernschubhypothese der Jurafaltung. *Eclogae Geol. Helv.*, 54: 221–282.

LAUBSCHER, H. P., 1962. Die Zweiphasenhypothese der Jurafaltung. *Eclogae Geol. Helv.*, 55: 1–22.

LAUBSCHER, H. P., 1965. Ein kinematisches Modell der Jurafaltung. *Eclogae Geol. Helv.*, 58: 231–318.

LIENHARDT, G., 1962. Géologie du bassin houiller Stéphanien du Jura et de ses morts-terrains. *Mém. Bur. Rech. Géol. Minières*, 9: 449 pp.

LUGEON, M., 1941. Une hypothèse sur l'origine du Jura. *Bull. Lab. Géol. Mineral. Géophys. Musée Géol. Univ. Lausanne*, 73: 14 pp.

PAVONI, N., 1961. Faltung durch Horizontalverschiebung. *Eclogae Geol. Helv.*, 54: 515–534.

RUBEY, W. W. and HUBBERT, M. K., 1959. Role of fluid pressure in mechanics of overthrust faulting. *Bull. Geol. Soc. Am.*, 70: 167–206.

SIGAL, S., 1962. Contribution à l'étude du faisceau occidental du Jura Bugeysan (Feuille de Belley au 50.000e). *Bull. Carte Géol. France*, 268: 77 pp.

UMBGROVE, J. H. F., 1950. *Symphony of the Earth.* Nijhoff, The Hague, 220 pp.

REFERENCES

WEGMANN, E., 1961. Anatomie comparée des hypothèses sur les plissements de couverture (le Jura plissé). *Bull. Geol. Inst. Uppsala*, 40: 196–182.

WEGMANN, E., 1963. Le Jura plissé dans la perspective des études sur le comportement des socles. In: M. DURAND DELGA (Editor), *Livre à la Mémoire du Professeur P. Fallot. II.* Soc. Géol. France, Paris, pp. 99–104.

WINNOCK, E., 1961. Résultats géologiques du forage du Risoux 1. *Bull. Ver. Schweiz. Petrol. Geol. Ingr.*, 28: 17–26.

CHAPTER 16

Alpine Europe: Betic Cordillera, Iberian Chains and Catalanides

INTRODUCTION

As stated on p.171 and p.172 the main Alpine elements of Spain, apart from the southern flanks of the Pyrenees, are the Iberian or Celtiberian[1] chains, the Catalanides and the Betic Cordillera. Of these the Iberian chains and the Catalanides, which border the Ebro Basin to the southwest and southeast (Fig.252), are minor chains, comparable in structure to the Jura Mountains. The Betic Cordillera, on the other hand, is a major element, and, at least in part, developed from a true geosynclinal basin. In view of this situation, we will begin our review with the Betic Cordillera.

THE BETIC CORDILLERA

The name Betic Cordillera embraces the vast mountainous area of southeastern Spain, some 500 km long by 100 km wide, bordering the Mediterranean. It stretches fom Cádiz and Gibraltar in the west to Carthagena and Alicante in the east.

Although forming an element of the Alpine chain, it is strikingly different from the Alps proper in several ways. One of these is its general topography. Topographically, this area is occupied, not by a system of elongated moutnain chains, but by a more or less irregular mosaic of individual sierras, high and steep, but rather short, separated by quite extensive plateaus, both high and low. Its major element, for instance, the Sierra Nevada, culminating in the Mulhacén of 3,481 m of altitude, is undoubtedly of alpine dimensions; its length, however, is but 75 km, and as such it forms less than one sixth of the Betic Cordillera.

This topography of an irregular network of smaller individual sierras alternating with plateaus is due to the development of younger, Neogene, intramontane basins. Instead of the larger external, more or less continuous, molasse basins of the Alps, the Betic Cordillera is characterized by these smaller, internal and discontinuous molasse basins.

The result is that on topographic maps one will only find the names of these separate sierras. Many of these form various subunits of the Betic Cordillera, and the latter name is essentially geological in nature. It is used to indicate the totality of the Alpine chains of southeastern Spain. As such, I prefer to use it in the singular.

[1] "Celtiberian" is the better name, because the Celtiberes lived southwest of the *Iberis*, the present river Ebro. For shortness the former name will be used.

Into this region two types of geologists have been working, e.g., alpine and non-alpine. The alpine geologists have tried to interpret the Betic Cordillera as similar to, or even forming a part of, the Alps. Their studies have been admirably summarized by FALLOT (1948), to which paper the reader is also referred for the bibliography of the older literature, upon which this summary is based. Apart from earlier investigations, most of the field work has been done by the Swiss geologist M. M. Blumenthal, by Fallot himself, and by the Dutch school of H. A. Brouwer. It has been synthesized into the classic picture of the Alpine chain by STAUB (1934).

An etymological peculiarity of the passage of the Dutch school is the retention of two Dutch terms, often in their German translation. These are, the "mengzone", or"Mischungszone", or in Spanish, the "llamada Mischungszone"; and the "konglomeratische mergel", or "Konglomeratischer Mergel". As we will see, both elements are found overlying the mica schists of the Sierra Nevada.

The non-alpine geologists, mostly represented by the German school, were more impressed by the individuality of the various sierras. Although conceding local overthrusting and nappe structures, they still think more in terms of autochthonous or par-autochthonous development.

Professor P. Fallot has always taken up an intermediate position. Although in favour of major nappe structures (cf. FALLOT, 1948), he was very much opposed to the detailed correlations with the Austrides and the Pennides, as proposed by R. Staub (FALLOT, 1954).

The latest developments in the Betic Cordillera are that teams of geologists from Spanish, French, Dutch and German universities are now engaged in detailed re-mapping and structural analysis. The results obtained so far indicate that the Betic Cordillera are extremely more complex than had originally been thought (SOCIÉTÉ GÉOLOGIQUE DE FRANCE, 1961; KONINKLIJK NEDERLANDS GEOLOGISCH MIJNBOUWKUNDIG GENOOTSCHAP, 1964).

THE MAIN DIVISIONS

Returning to the earlier, more schematized, pictures, the main divisions, as distinguished earlier, still persist. It is only in regard to their internal structure and their interpretation that newer results have been obtained.

These main divisions are indicated in Fig.239. South of the Hercynian uplands of the Spanish Meseta lies the Guadalquivir Basin. This shallows in easterly direction, and is replaced in that direction by an area of a thin epicontinental cover resting on Triassic, and tectonized in small, irregular, structures (BUSNARDO, 1961; GARCIA-RODRIGO, 1961). This zone, which presumably continues westwards underneath the Guadalquivir Basin, forms the Prebeticum.

A very first division covers a northern, external zone, the Subbetic, and a southern internal zone, the Betic. In the Subbetic the Triassic is developed in a Germanic facies, while the basement is nowhere exposed. In the Betic, on the other hand, the Triassic belongs to the Alpine facies, and the pre-Alpine basement has become incorporated wholesale into the Alpine structures.

The northernmost zone of the Betic Cordillera is formed by the Subbetic. This is often subdivided into a northern subzone, the Subbetic sensu stricto, and the more southerly Penibetic.

The main part of the Betic Cordillera is formed by its southern unit, the Betic itself. The Betic can, schematically, be subidvided into three main zones, i.e., the Sierra Nevada zone, the Málaga zone and the Alpujarride nappes. In its western part the divergent Rondaïdes must be mentioned, whereas still further west, in relation with an axial depression of the chain, the fiysch area of the Campo de Gibraltar is found.

Just as in the Alps, there is a striking difference between the external part of the Betic Cordillera, formed by the Subbetic, and the internal part, formed by the Betic. The difference is, if anything, even greater than in the Alps. The Subbetic consists of epidermis-type structures, sheared off over the evaporites of the Triassic in Germanic facies. The main nappe system of the Betic, on the other hand, the Alpujarrides, is characterized by geosynclinal Triassic in Alpine facies. Moreover, in both the Alpujarrides and in the Málaga Betic the pre-Alpine basement has become incorporated into the Alpine structures. In contrast to the two units mentioned, the Sierra Nevada Betic is in the main formed by a monotonous series of micaschists, several kilometers thick.

For STAUB (cf. 1934), the Alpujarrides, because they contain Alpine Triassic, must belong to the Austrides. And hence, the underlying micaschists of the Sierra Nevada must belong to the Pennide nappes, and form a large tectonic window, comparable to the Tauern. Although the Alpujarrides developed within the realm of the Alpine Triassic, there is no further evidence for this correlation. Nor is there any evidence for a correlation of the Nevada Betic with the Pennides, or even for the supposition that this forms a nappe and is not autochthonous (FALLOT, 1954).

The amount of schematization, incorporated in Fig.239, may be apparent from Fig. 240. In this figure the extent of the Neogene basins of subsidence within the central part

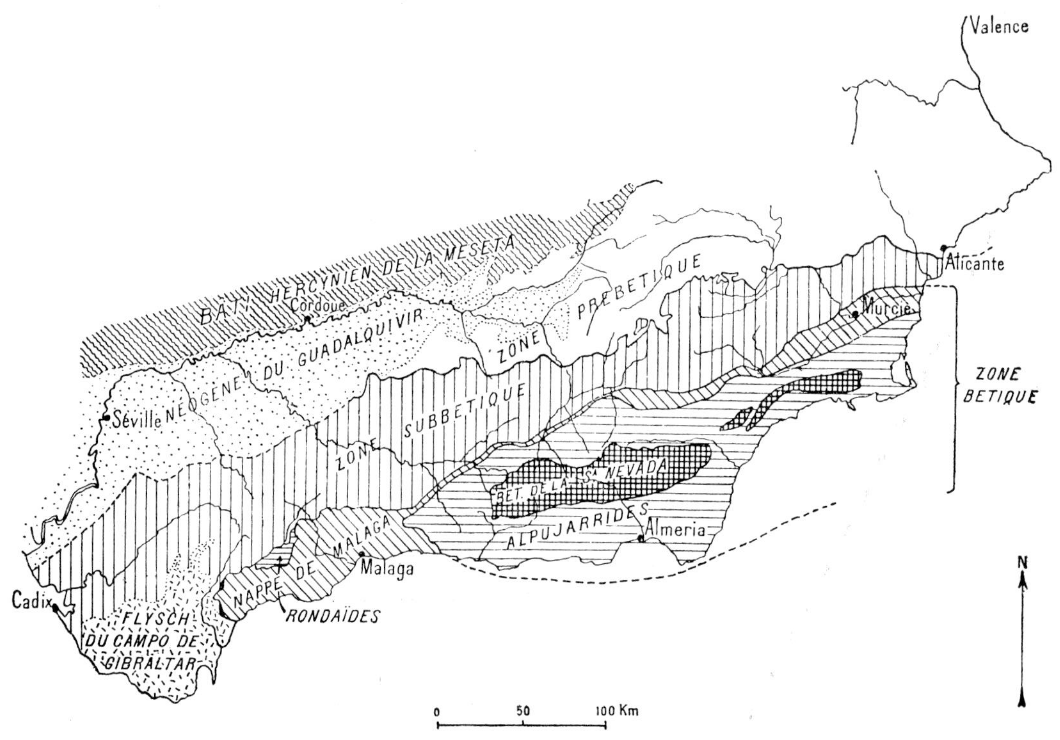

Fig.239. Generalized map of the Betic Cordillera, indicating the main divisions. (After FALLOT, 1948.)

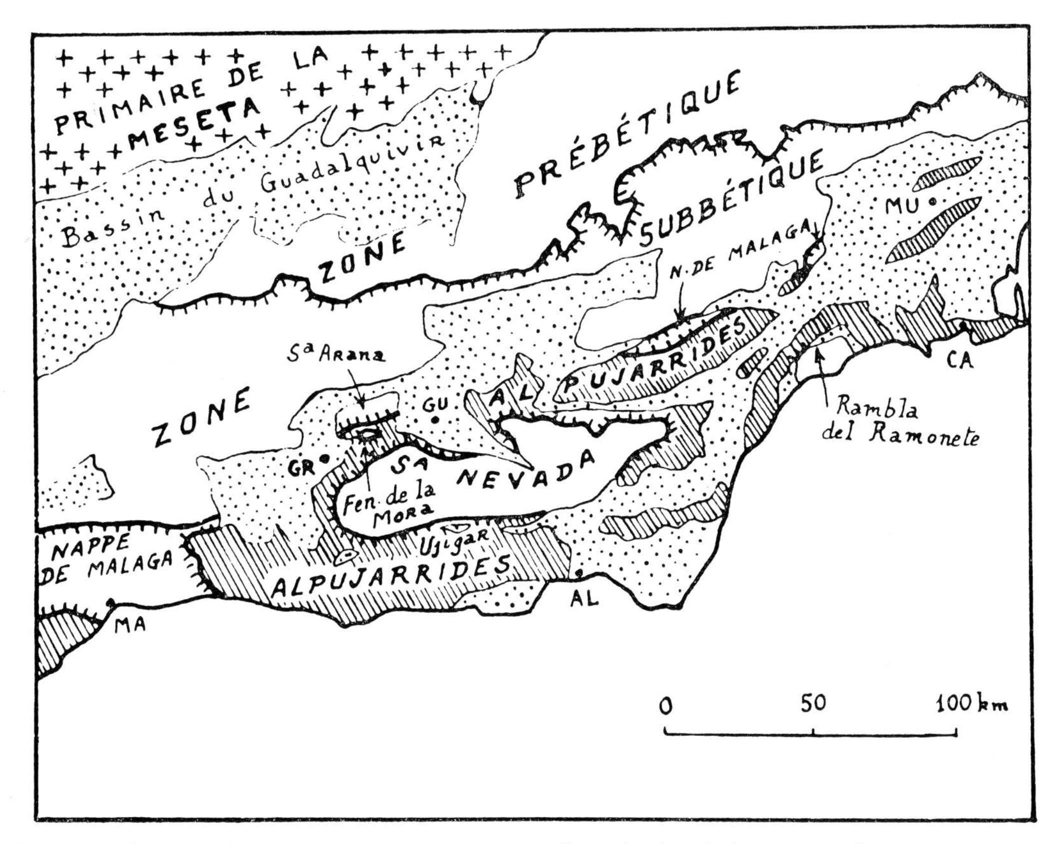

Fig.240. Outline map of the central area of the Betic Cordillera, showing the importance of post-orogenetic Neogene subsidence basins (stippled). Compare Fig.239.
GR = Grenada; *GU* = Guadix; *MA* = Málaga; *AL* = Almería; *CA* = Cartagena. (After DURAND DELGA and FONTBOTÉ, 1960.)

of the Betic Cordillera has been schematically indicated. There can be no doubt either that the present elevated position of the Sierra Nevada is due to young, post-orogenetic uplift, or that the Neogene basins are due to extensive and diversified graben movements of equally recent development. The vertical post-orogenetic movements, both positive and negative, have evidently not so much affected the chain as a whole, but have been of a more haphazard nature, producing this pattern of localized horsts and graben.

Before going into a more detailed account of the Subbetic and Betic, we must mention that FALLOT (1959) has once compared the basins of Po and Guadalquivir. As follows from the title of his paper, where he speaks of the deeper geology of the Po Basin, and the mystery of the Guadalquivir Basin, there is not much resemblance between the two, apart from the fact that both form Neogene basins of subsidence and sedimentation. Not only is the Guadalquivir Basin much the shallower of the two (PERCONIG, 1962), but, moreover, it is situated on the southern flank of the Hercynian uplands of the Meseta. In its position it is much more comparable to the molasse basin of northeastern Switzerland and southern Germany.

THE SUBBETIC

The epidermis type structures of the Subbetic, sheared off over the Triassic in Germanic

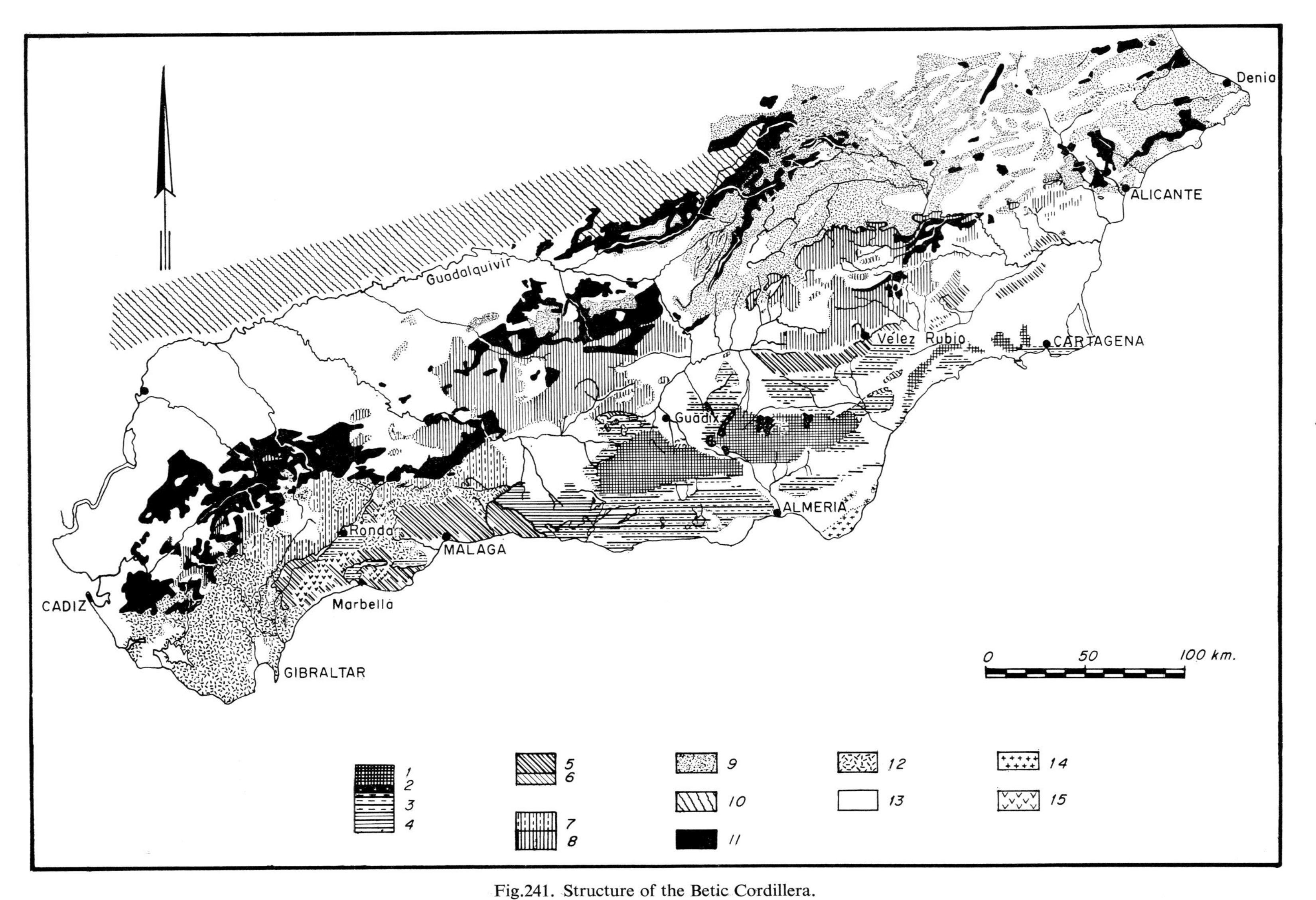

Fig.241. Structure of the Betic Cordillera.

1–10: Tectonic units. *1* = Sierra Nevada Betic; *2* = "Mischungszone"; *3, 4* = Alpujarrides (*3* = Lanjarón–Gador nappe; *4* = Lujar nappe); *5* = Málaga Betic; *6* = Rondaïdes; *7, 8* = Subbetic; *9* = Prebetic; *10* = Hercynian basement of the Meseta.

11–13: Stratigraphic units. *11* = Triassic in Germano-Andalusian facies; *12* = Senonian and Lower Tertiary flysch; *13* = Neogene and Quaternary basins; *14, 15* = volcanics. *14* = younger volcanics of the Cabo de Gata; *15* = Serranea de Ronda peridotites. (After FALLOT, 1948.)

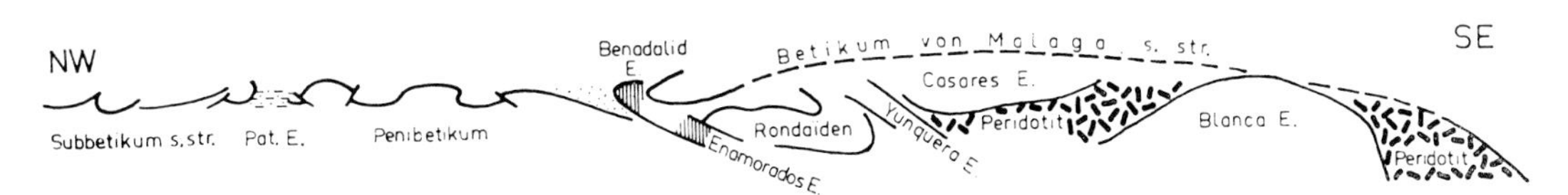

Fig.242. Schematic cross-section over the westerly part of the Betic Cordillera, showing the relation between the Subbetic and the Ronda and Málaga units of the Betic. (*E* = "Einheit" = unit; *Pat. E* = Paterna Einheit). (After HOEPPENER et al., 1964a, b.)

facies, are characterized by a relatively thin carbonate series of the higher Mesozoic, in an epicontinental development, reaching into the Lower Tertiary.

The Triassic, though in a general way developed in Germanic facies, is characterized by thick gypsum deposits in its lower part. These are followed by a "Muschelkalk", which may contain fossils of Anisian and Ladinian age, showing a certain transition towards the Alpine facies of the Triassic. The higher Triassic may be developed again in a continental facies, but on the other hand a discontinuous carbonate series ranging from the Muschelkalk into the lowermost Jurassic is also found.

This development of the Triassic, which is slightly at variance with the normal Germanic facies, has been indicated as the Andalusian facies. As follows from Fig.241, it is not restricted to the Subbetic, but occurs in the Prebetic as well, and even much further north, over most of Spain.

Tectonically, the important thing in this Germano-Andalusian facies of the Triassic is the development of thick gypsum deposits in the lower part of the series. These serve as "décollement" zone, and may well develop strong local halokinesis.

The reason why, incidentally, such large areas of Spain are so fertile, lies mainly in the gypsum in the Germano-Andalusian facies of the Triassic and in the Miocene. No matter how saline irrigation water may be, gypsum normally predominates over salt. This results in a nice, crumbly structure of the clays in the irrigated fields.

Within the Subbetic, the more northerly Subbetic sensu stricto shows a more marly development than the southern subunit, the Penibetic. This can be correlated with the influx of terrigenous material derived from the north from the Hercynian borderlands.

In the literature much stress has always been laid on the irregularity of the pattern, both of paleogeography and facies, and of the structure of the Subbetic. In this respect the southerly Penibetic, with fewer marls, and more, and thicker pelagic limestones of Jurassic and Cretaceous age, shows a more regular pattern both of facies changes and of structure (cf. BUSNARDO and DURAND DELGA, 1961).

A very schematic idea of the tectonic style of the Subbetic can be gleaned from the outline section of Fig.242. For more details, the reader is referred to the regional literature (cf. FALLOT, 1945; BUSNARDO, 1962, 1964; FOUCAULT, 1962, 1964; HOEPPENER et al., 1964a, b). There is, in my opinion, no evidence for nappes in the Subbetic. The structures presumably are autochtonous or par-autochthonous. The tectonic style resembles that of the Subalpine chains.

THE BETIC

Of old, the three main units of the Betic, each of which may consist of a number of individual nappes, are the Málaga Betic, the Alpujarrides and the Sierra Nevada Betic.

The Rondaïdes, found around Ronda in the western part of the chain, were formerly thought of as a local, aberrant element. More recent research, however, indicates the possibility that they are more important. They might also form a major element, and even represent the westerly replacement of the Alpujarrides.

The sequence, as cited above, indicates the most generally accepted tectonic position. That is, the highest element is formed by the Málaga Betic, the lowest by the Sierra Nevada Betic, whereas the Alpujarrides and Rondaïdes are thought to occupy an intermediate position.

The Málaga Betic

The Málaga Betic is characterized in its stratigraphy by the predominance of pre-Alpine series. These comprise both an older group of high-grade metamorphic rocks—gneisses, schists and "cipolins" — and a younger, hardly metamorphic series, containing clastic rocks in which fossils ranging from the Silurian to the Carboniferous have been found.

The post-Hercynian series start with the transgression of the Buntsandstein, and is further represented by individualized complexes of the higher Mesozoic. Generalized sedimentation sets in only during Late Cretaceous and Early Tertiary, with the formation of flysch series.

As follows from Fig.239 the geographical extent of the Málaga Betic is peculiar. Its main area is in the westerly part of the chain, around Málaga itself. But from there it can be followed in a thin zone towards the northeast, stretching all the way along the northern border of the Betic.

In its southwesterly area, the Málaga Betic is thought to dip into the Mediterranean, along the whole coast in the Marbella–Málaga region. From there it is thought to override northwards, not only all of the Betic, but in its frontal part even a substantial zone of the Subbetic. This interpretation was first proposed by M. Blumenthal, and has been followed by P. Fallot (Fig.243). It is also adhered to by R. Hoeppener (Fig.242).

A quite different interpretation has recently been revived for the thin zone trending in northeasterly direction along the northern front of the Betic by the Amsterdam school.

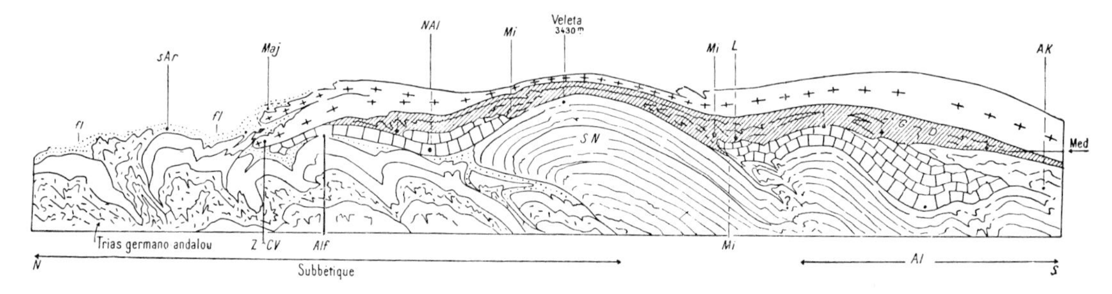

Fig.243. Section through the western part of the Betic Cordillera, in the interpretation of Blumenthal. Showing the Málaga Betic (elongated crosses), overriding all other tectonic units.

AK = Almunecar zone; *Al* = Alpujarride nappes; *Alf* = Alfacar window; *fl* = flysch; *GD* = Gádor nappe; *La* = Lanjarón; *Maj* = Majalijar digitations; *Med* = Mediterranean sea level; *Mi* = "Mischungszone"; *NAL* = northern Alpujarrides; *SAR* = Sierra Arana; *SN* = Sierra Nevada; *ZCV* = Cogollos Vega zone, the frontal element of the Málaga Betic. (After M. Blumenthal, in: FALLOT, 1948.)

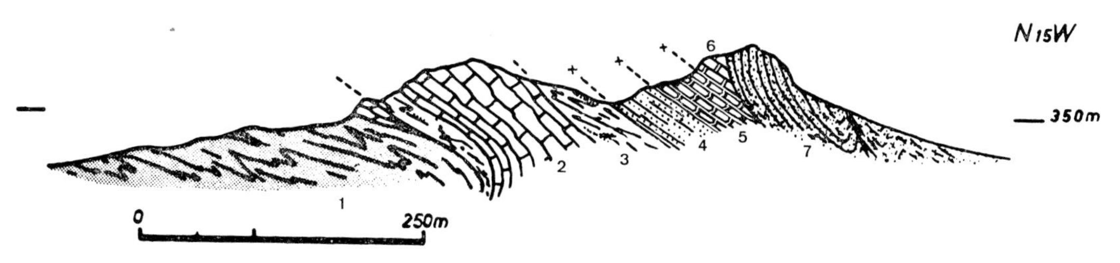

Fig.244. Section through the Cerro del Medio in the western Sierra Cabrera, situated in the southeasterly part of the Betic Cordillera. Showing small elements of the Málaga Betic (*4*, *6*) pinched in the Alpujarrides (*1–3*, 5). This forms an argument for a southerly thrust of the Málaga Betic.
1 = older micaschist; *2* = Triassic limestone and dolomite; *3* = Permo-Triassic or older phyllite and quartzite; *4* = Paleozoic sandstone; *5* = Triassic gypsum; *6* = Jurassic limestone and dolomite; *7* = Neogene. (After RONDEEL, 1965.)

They recently came up with a detailed analysis (MACGILLAVRY, 1964; MACGILLAVRY et al., 1964) of the area around Vélez Rubio. According to their view, this zone is not only subautochthonous, but represents a sort of root zone, a cicatrice, from which elements have been overthrust in southerly direction over the Alpujarrides. Such small remnants of Málaga Betic within the area of the Alpujarrides have previously been described by SIMON (1963) and also by RONDEEL (1965), see Fig.244. Similar views are held by DURAND DELGA (1963).

This interpretation results in the assumption of an earlier northward thrust during the Early Oligo-Miocene, when the Alpujarrides were formed. It was followed by a southward thrust of the Málaga Betic over the Alpujarrides during later Oligo-Miocene, and again by a northward thrust of the Subbetic during Early Mio-Pliocene. Such a sequence of orogenetic movements is, to my mind, difficult to visualize. Such interpretation may well be another result of a too strict an application of the facies concept of the classic picture of the Alps.

The Alpujarrides

This unit of the Betic derives its name from the Sierra de las Alpujarras, situated between the Sierra Nevada and the Mediterranean. In area, it is the most important part of the Betic, as may be seen from Fig.239.

The Alpujarrides are built up by two distinct rock series, an older and a younger. But not all tectonic structures contain both series, in which case one normally only finds exposed the younger series, which has been sheared off from the older.

The lower series consists mainly of medium grade schists with intercalated "cipolins". Their age has always been indicated as "Paleozoic" in earlier literature. However, upon the admittedly slight evidence of lithofacial resemblance of some intermediate graywackes found in the Alpujarrides to the Paleozoic rocks of the Málaga Betic, EGELER (1964) assumes a pre-Silurian age for the older series of the Alpujarrides.

The younger series consists of a massive, thick, predominantly dolomitic carbonate series. Fossils of Middle and Upper Triassic age have been found, and both lithology and fossils point to a close relationship with the Alpine Triassic. This series starts with a gypsiferous shale horizon of Werfenian age, over which the "décollement" of the higher series of the Alpujarrides normally takes place.

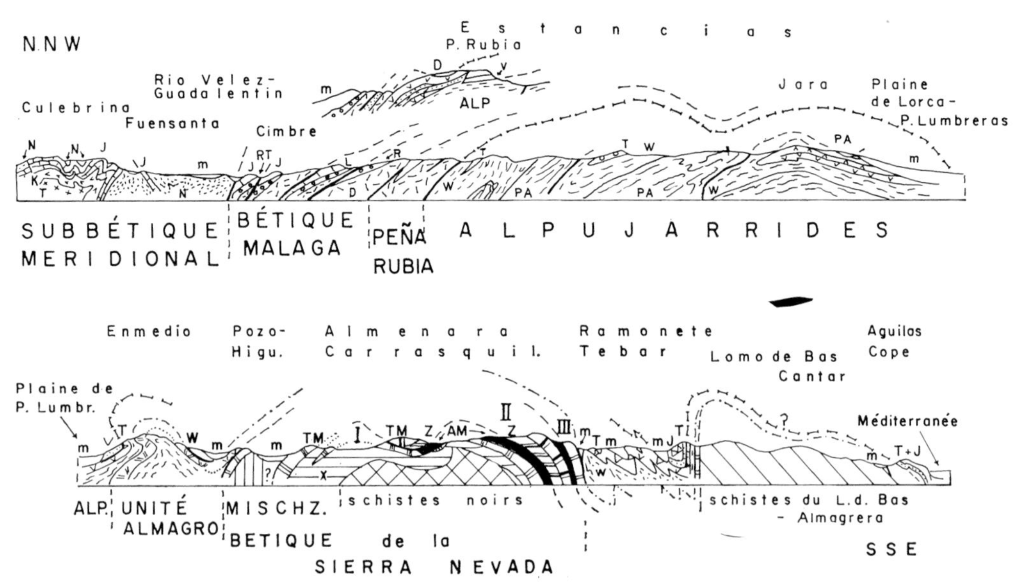

Fig.245. Section through the Betic Cordillera across the Sierra Nevada. The upper section forms the northern continuation of the lower section.
"Subbétique méridional" (= southern Subbetic): N = Upper Cretaceous–Lower Tertiary; J = Jurassic; T = Triassic.
Málaga Betic: J = Middle and Upper Jurassic and Lower Tertiary; L = Upper Triassic and Lias; RT = Permo-Triassic and Triassic; P = Permo-Triassic; D = Devonian–Carboniferous.
Complex unit of Pena Rubia: V = Permo-Triassic.
Alpujarrides: T = Middle and Upper Triassic; W = Permian and Werfenian; PA = older schists.
Sierra Nevada Betic: TM = Trias marble; AM = amphibolites and prasinites; X = micaschists; Z = garnet schists. (After FERNEX, 1964.)

No younger deposits, either of Upper Mesozoic, or of Lower Tertiary, are known from the Alpujarrides.

The Alpujarrides have been strongly tectonized, showing numerous major overthrusts. A first division into three nappes; from high to low the Guajar nappe, the Lanjarón–Gadór nappe and the Lujar nappe; has been introduced by H. A. Brouwer and his school (cf. SIMON, 1963). Not everybody has agreed, and moreover, recent work has proved the structure of the Alpujarrides to be much more complicated than had originally been thought. Nevertheless, these three early names still retain their value (cf. Fig.243), and are in a way comparable to the Roman numerals given by E. Argand to the Pennide nappes. A recent cross-section over the Sierra Nevada, with the Alpujarrides developed both to the north and to the south, is given in Fig.245.

The main folding period of the Alpujarrides, at first supposed to be Mesozoic by the school of H. A. Brouwer, then Eocene by M. Blumenthal and by P. Fallot, has since been rejuvenated further. It is now held to be either Late Oligocene or Early Miocene by DURAND DELGA and FONTBOTÉ (1960).

The Rondaïdes

Between Ronda and Málaga (Fig.246) an aberrant development of the Betic has long been known to exist. It resembles most closely the Alpujarrides, and has formerly been classed either with this unit, or with the Penibetic. Its structure is complex, and moreover

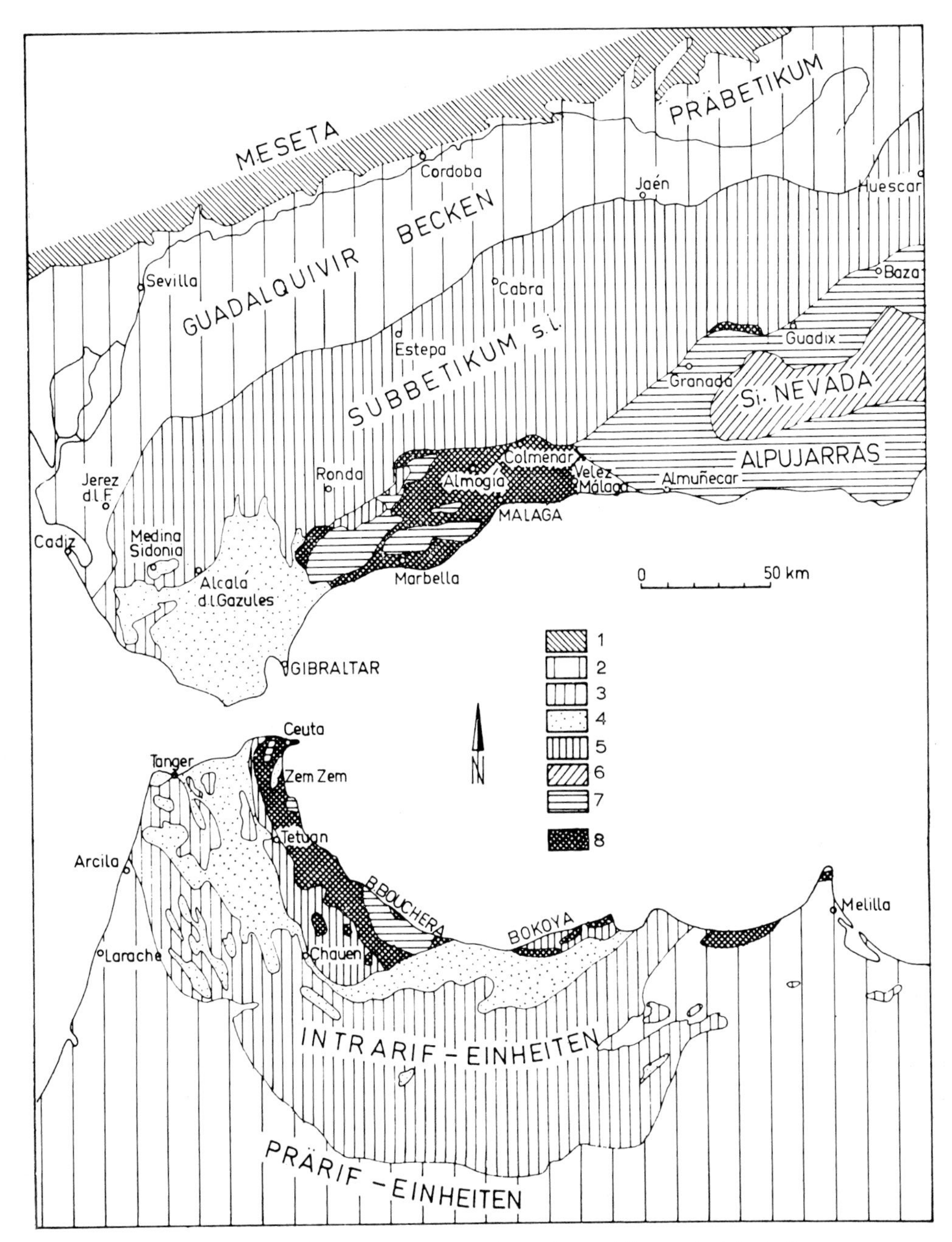

Fig.246. Map of the western part of the Betic Cordillera and of the Rif.
1 = Hercynian basement of the Meseta; *2* = Prebetic, Pre-Rif and post-orogenetic sediments; *3* = Subbetic and Intra-Rif; *4* = "Ultra-flysch", that is the flysch of the Campo de Gibraltar and of the Ultra-Rif; *5* = Rondaïdes; *6* = Sierra Nevada Betic; *7* = Alpujarrides; *8* = Málaga Betic. (After HOEPPENER et al., 1964a, b.)

difficult to untangle from elements of the overlying Málaga Betic. Perhaps because in later years a separate group, that guided by R. Hoeppener from Bonn, is working in this area, it has become customary nowadays to speak of the Rondaïdes as a separate unit of the Betic.

The stratigraphy of the Rondaïdes is very much restricted, and consists only of Upper Triassic and of Lower Jurassic. The Triassic is represented only by a massive series of

Norian dolomites, over 1 km thick. This represents the upper part of the Alpine Triassic and is, as such, comparable to the Alpujarride stratigraphy. Of the Jurassic only the lower part of the Lias is known, which is developed as several hundreds of meters of marls and limestones. Moreover, several intercalations of "cargneules" occur, the cellular dolomites related to evaporite series, which more normally occur in the Triassic (HOEPPENER et al., 1964a, fig.5).

The complicated structure of the Rondaïdes is schematically indicated in Fig.242.

Peridotites

Another extraneous element, occurring in much the same area, is formed by the extensive masses of peridotite. Formerly attributed to the Málaga Betic, they now seem to have been incorporated into the Alpujarrides by HOEPPENER et al. (1964a, b), Fig.246. Their enigmatic position follows from Fig.242.

The Betic "haut-fond", or Betic geanticline

As we saw, a characteristic of the major part of the Betic, that is of the Alpujarrides, the Rondaïdes and Sierra Nevada Betic, is the absence or near absence of higher Mesozoic, of Jurassic and Cretaceous. The Bonn school sees evidence in this for a first major orogenetic period during the Jurassic (cf. HOEPPENER et al., 1964a, b). MACGILLAVRY (1964) replies to this idea that if this had been so, it remains difficult to explain the absence of appropriate post-orogenetic detritus in the Málaga Betic.

The latter author even calls the absence of higher Mesozoic "one of the most vexing problems in the Betic area". It is, however, vexing only within the limits set by the classic model of Alpine geology. For in such a model a geosyncline has to subside more or less steadily through incipient tangential compression, ultimately leading to the orogenetic phase.

But as we saw during earlier discussions, orogenies do not ever conform to the classic picture. From the example of the Briançonnais zone of the French Alps, we know that "haut-fonds" of respectable duration and forming large areas of non-deposition may occur in geosynclines. I think we may better interpret the absence of higher Mesozoic over large parts of the Betic as due to a "haut-fond", instead of to an early major period of orogenesis.

The Sierra Nevada Betic

This unit of the Betic has, of course, been named after the Sierra Nevada. It is, however, not restricted to this mountain range, but can be followed much further east (Fig. 247). Its major element outside the Sierra Nevada is found in a part of the Sierra de los Filabres.

The Sierra Nevada Betic is extremely uniform. It consists of a lower, very thick, and extremely monotonous series of micaschists, covered by rocks of the "Mischungszone". This uniformity in the facies does not, however, apply to the structure of all of the Sierra Nevada Betic. In the Sierra Nevada itself it forms a single, elongated dome, and as such

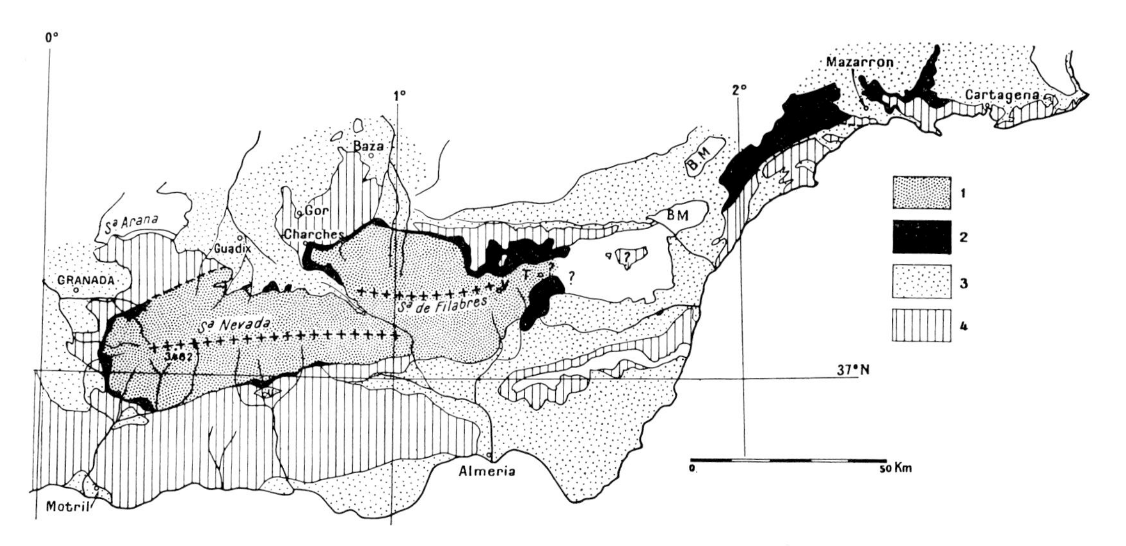

Fig.247. Schematic map of the Sierra Nevada Betic.
1 = Sierra Nevada micaschists; *2* = "Mischungszone"; *3* = Neogene and Quaternary; *4* = Alpujarrides. 0° longitude is as of Madrid. (After FALLOT et al., 1961.)

is the acme of tranquility. All more easterly elements of the Sierra Nevada Betic, from the Sierra de los Filabres on, on the other hand, are strongly tectonized.

The lower series of the Sierra Nevada Betic show (upper) mesozonal metamorphism. Apart from more sandy intercalations, nothing breaks the monotony of the 4 km of thickness of micaschists exposed in the cross-section through the Sierra Nevada (Fig.248, 249). Their age is unknown, beyond the fact that they must be pre-Alpine. EGELER (1964) has recently maintained that they must even be pre-Silurian, using an extension of the lithostratigraphic correlation between the Málaga Betic crystalline rocks and the Alpujarrides.

The younger schists, the "Mischungszone", form only a thin envelope around the core of the older schists in the Sierra Nevada itself. It thickens eastwards, reaching thicknesses of the order of 1 km. How much of this is due to tectonic repetition, which seems to play a part even in the Sierra Nevada envelope (J. Fontboté, in: FALLOT et al., 1961), has, however, not yet been elucidated.

Petrographically, the "Mischungszone" forms, of course, an extremely mixed zone. It contains many different elements, which at first sight seem more or less completely unrelated. It is this mixed character which is the main diagnostic of the "Mischungszone". First among its components are micaschists, which again are of mesozonal metamorphism, and show transitions towards albitic gneiss. The difference between the older and the

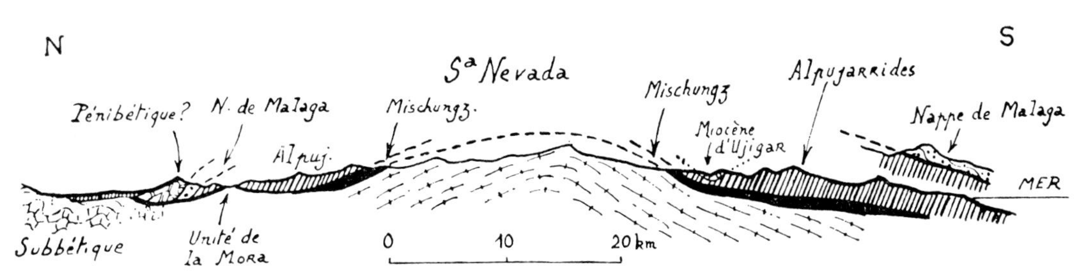

Fig.248. Schematic section through the Betic across the Sierra Nevada. (After DURAND DELGA and FONTBOTÉ, 1960.)

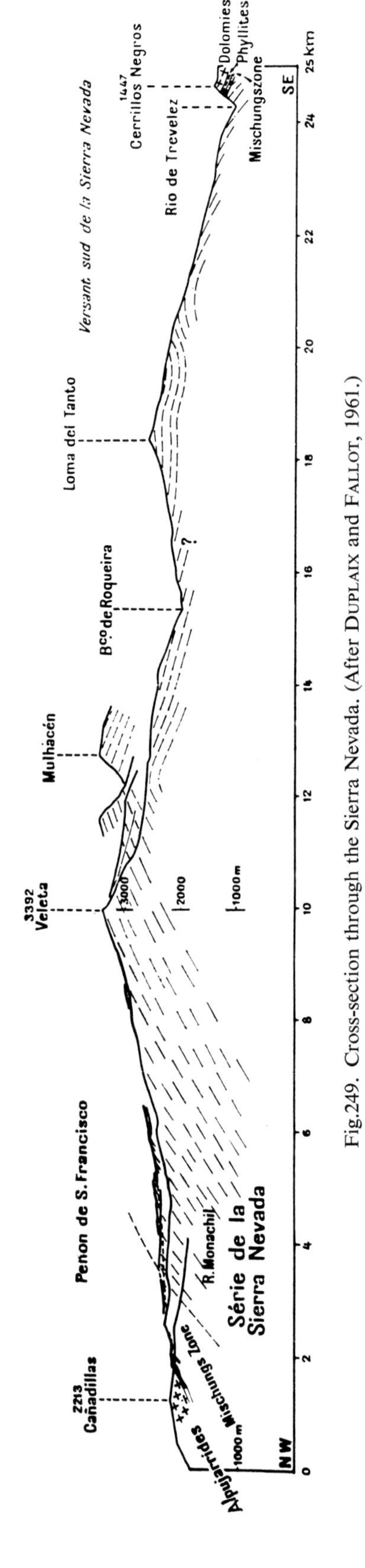

Fig.249. Cross-section through the Sierra Nevada. (After DUPLAIX and FALLOT, 1961.)

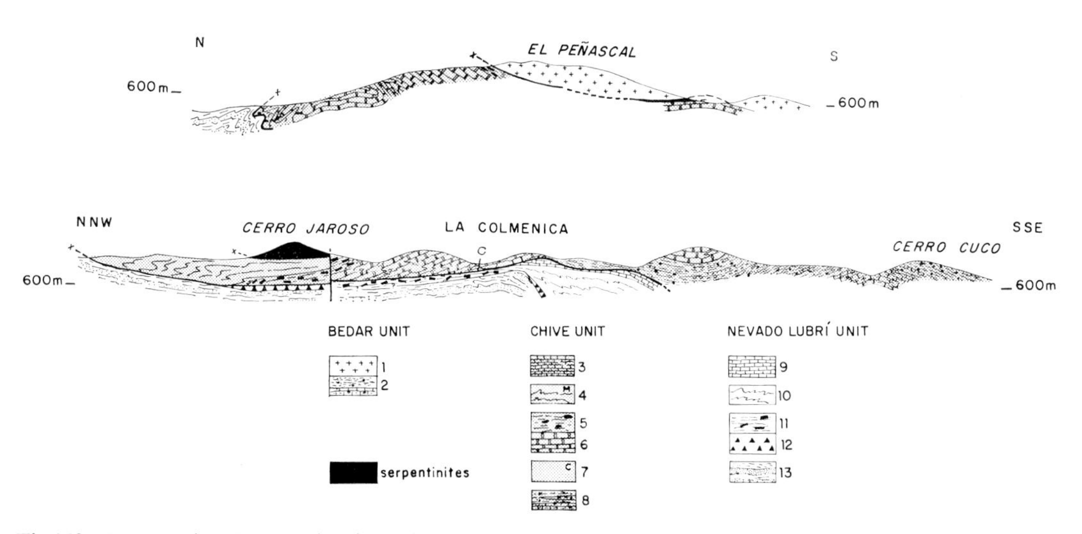

Fig.250. Cross-section through the Sierra de los Filabres. Showing the strong tectonization of the easterly elements of the Sierra Nevada Betic. (After NIJHUIS, 1964.)

1 = Tourmaline gneisses, metagranites; *2* = graphite schist-and-tourmaline gneisses complex; *3* = Las Casas marbles and schists; *4* = Muñoz amphibole-micaschists; *5* = predominantly albite–epidote amphibolites; *6* = Atalaya marbles and amphibolites, undifferentiated; *7* = Colmenica schists; *8* = graphite schist tourmaline gneiss complex; *9* = Las Casas marbles and schists; *10* = Muñoz amphibolite-micaschists; *11* = predominantly albite–epidote amphibolites; *12* = Huertecica brecciated marble zone, undifferentiated; *13* = Tahal schist complex.

younger schists is, that the latter are feldspathic. Much less important as to volume, but much more important for our understanding of the "Mischungszone", are intercalations of marble, of the "Konglomeratischer Mergel"[1] and even of gypsum.

There is general agreement now, I believe, that the "Mischungszone" represents metamorphosed Triassic in Germanic facies. In no series but the Germanic facies of the Triassic, is the gradient of the original facies, from fully marine, to saline, to evaporite, to continental, so great. As a consequence, it is felt that in no other facies could plurifacial metamorphism produce such a varied suite of rocks as is found in the "Mischungszone".

For one thing, the dating of the "Mischungszone" as Triassic, would prove mesozonal Alpine metamorphism. As we saw, in the other untis of the Betic, Alpine metamorphism has, in contrast, remained slight indeed.

For another, this correlation of the "Mischungszone" would mean that the Sierra Nevada Betic was in the realm of the Germanic facies during time of deposition. The major paleogeographic boundary of the Betic Cordillera during the Triassic consequently was situated between the Sierra Nevada Betic and the Alpujarrides. Hence, it does not coincide with the later major tectonic boundary, which lies between Subbetic and Betic.

The strong tectonization of the Sierra Nevada Betic in the Sierra de los Filabres and further east, tectonization of a style comparable to that of the Alpujarride nappes (Fig. 250), has recently induced EGELER (1964) and others to assume the existence of a lower nappe system in the Betic, named the "Nevado-Filabrides". They hereby extended earlier notions, such as of Brouwer and of Stille, that the Sierra Nevada is not autochthonous.

[1] The "Konglomeratischer Mergel" is neither a conglomerate, nor a marl. It is a residual rock, resembling "cargneules" (DUPLAIX and FALLOT, 1961). It is also found in the base of the Alpujarride nappe overriding the Sierra Nevada, which forms another argument to assign a Triassic age to the "Mischungszone".

With Fallot, who reviewed these earlier nappist interpretations (FALLOT, 1948, 1954), I think it safe to remember that no evidence is found for nappe structures in the Sierra Nevada. On the other hand, this mountain range is fairly reminiscent of the migmatite dome of Haller, and could well be autochthonous also. In that case the apparent narrow geographic relation, between the Sierra Nevada itself and the more easterly elements of the Sierra Nevada Betic, might indicate that the overthrusts within the latter area belong to a class of par-autochthonous movements.

Sierra Nevada morphology

The Betic Cordillera, as the most southerly European mountain chain with peaks of over 3,000 m in height, presents a quite striking, non-glaciated type of morphology, which has not developed in the other major chains of Alpine Europe. This morphology is, of course, best expressed in the uniform micaschist series of the Sierra Nevada dome. It is far less apparent in the higher sierras of the Alpujarrides, due to the influence on morphology exerted by lithological variations in these mountains.

Absence of glaciation in the Sierra Nevada is too strongly worded, because on the northern slopes of the higher peaks, such as the Veleta and the Mulhacén, small kar-type glaciers have indeed developed. These, however, only serve to underline the absence of glaciation elsewhere in the chain.

The prevalence of fluviatile, V-shaped valleys, and the absence of U-shaped valleys; the dendritic stream pattern developed all over the area, and the absence of hanging valleys, to quote but a few of the characteristic features, make the Sierra Nevada morphology unique in Europe. If one follows the road up to the Veleta, one of the major peaks of the Sierra Nevada, which with its 3,392 m altitude is the highest point which can be reached by car in Europe, one is continuously impressed with the decidedly tropical flavor of the morphology, up to about 3,000 m altitude. It is only above this level that the influence of small kar glaciers on the morphology becomes noticeable.

The Straits of Gibraltar

Alpine tectonicians have speculated widely on how to connect the Alpine elements of Spain, of the Mediterranean islands and of North Africa, with the Apennines and the Alps. Fig.251 gives an idea of the amount of theorizing that has gone on for 50 years.

In this text we are most interested in the westernmost connection, that across the Straits of Gibraltar. Is this a young feature, a graben cutting across the former continuation of the Betic Cordillera, curving southwards into Morocco? Or does the Betic Cordillera continue westwards, to be cut off by the Atlantic?

South of the Straits, in Morocco, a north–south trending fold belt is indeed found, the Rif (for a modern complete review article on the Rif, see DURAND DELGA et al., 1962). The Rif forms a separate element, and is not directly related to the more southerly, west–east trending chains of the Atlas Mountains sensu stricto. Moreover, it shows strong resemblance to part of the external elements of the Betic Cordillera, and, in part, to the Málaga Betic (Fig.246).

Notwithstanding these similarities, P. Fallot was of the opinion that the differences

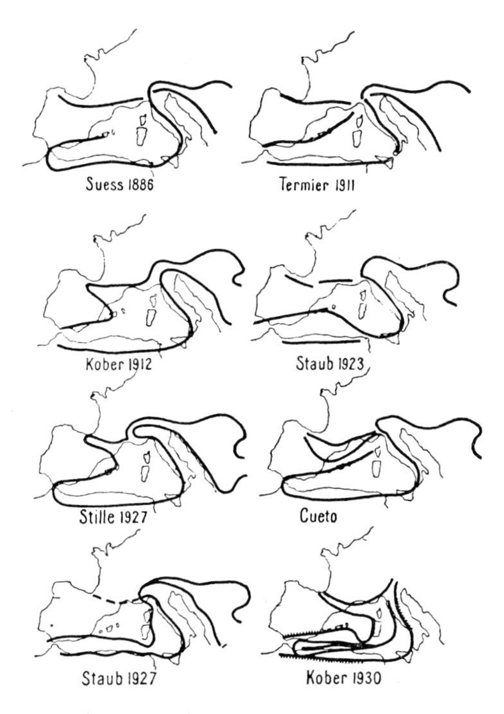

Fig.251. Various interpretations as to the connections between elements of the Alpine chain around the western Mediterranean, as proposed by major Alpine tectonicians. (After FALLOT, 1948.)

found between Rif and Betic Cordillera were so important, as to exclude the possibility that they formerly formed one single, curved fold belt.

In a way we meet here with the same mentality, found, for instance, in the Pyrenees. Variations across a gap, such as the axial zone of the Pyrenees, or the Straits of Gibraltar, are overrated, whereas variations that can be followed continuously, are underrated.

Recent research in the Campo de Gibraltar, in the Subbetic, and in the western part of the Rif has brought out more and more points of resemblance. Today all authors seem to consider the Rif to be the continuation of the Betic Cordillera (cf. HOEPPENER et al., 1964a, b; MATTAUER, 1964). This indicates a young, post-orogenetic age for the present Straits of Gibraltar. It must be kept in mind, however, that in the Rif only the "higher" elements of the Betic Cordillera, are exposed. Only the Subbetic and the highest element of the Betic, the Málaga Betic, are represented in the Rif. The axial dip, found in the Betic Cordillera westwards of the Sierra Nevada, evidently continues into the Rif.

IBERIAN CHAINS AND CATALANIDES

The area occupied by the Iberian chains and the Catalanides is shown in Fig.252, 253. They consist of an irregular network, in which the Hercynian basement outcrops, and is draped by an epicontinental series of Mesozoic, locally with some Eocene. These positive elements are separated by an equally irregular set of graben, in which quite varied, but always terrestrial, and normally strong, Neogene sedimentation has taken place. This was followed by a more generalised Plio-Pleistocene uplift and tilting.

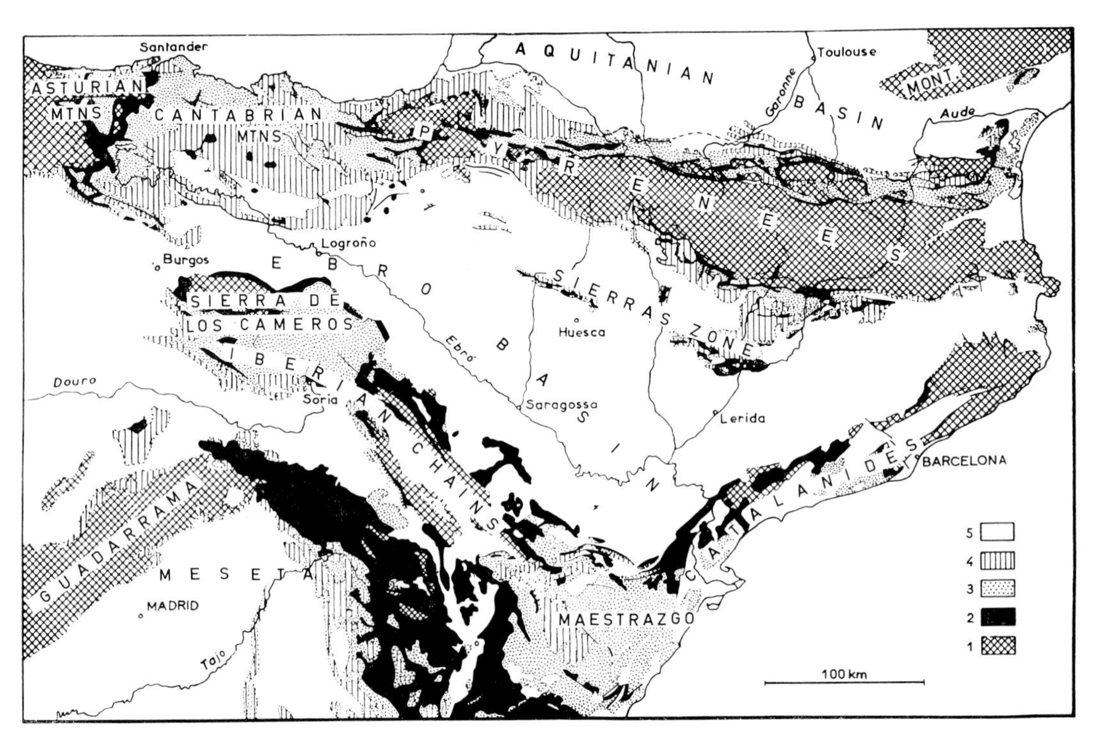

Fig.252. Map of northeastern Spain, with the Pyrenees, the Iberian chains and the Catalanides. *1* = Paleozoic; *2* = Triassic and Jurassic; *3* = Lower Cretaceous; *4* = Upper Cretaceous; *5* = Neozoic. Rat reported on a paleogeographic analysis of the Lower Cretaceous, and consequently grouped together Triassic and Jurassic, and Eocene with Neogene and Quaternary. Both Triassic and Eocene are indicated separately on Fig.253. (After RAT, 1964.)

Notwithstanding the epicontinental character of the Mesozoic, these chains have variously been interpreted as major orogenetic fold belts, comparable to, and of the same standing as, the Alps (Fig.251). Geologists of the Stille school have been very active in this respect. This reached a climax in the study of ASHAUER and TEICHMÜLLER (1935) on the Catalanides. In this paper, they were in the fortunate position of being able to place the majority of their "Doppelorogen" in the Mediterranean, between Cataluña and the island of Mallorca. The earlier German literature on these chains should consequently be used with caution.

The *Livre à la Mémoire du Professeur Fallot*, has, the same as for other Alpine chains, assembled the main modern studies on the area. These are predominantly directed towards much needed stratigraphic and paleogeographic analyses. For the Catalanides the best general account is still that of LLOPIS LLADO (1947).

STRATIGRAPHY

The Triassic is developed in Germanic facies all over the region. For the Catalanides we have the modern analysis by VIRGILI (1958). The Muschelkalk is well developed in Cataluña, and there are several gypsiferous horizons throughout the series (Fig.254). "Décollement" may take place over any or all of these.

The higher Mesozoic, the Jurassic and Cretaceous, and what is present of the Eocene, were predominantly developed in a carbonate facies. In the Lower Cretaceous, however,

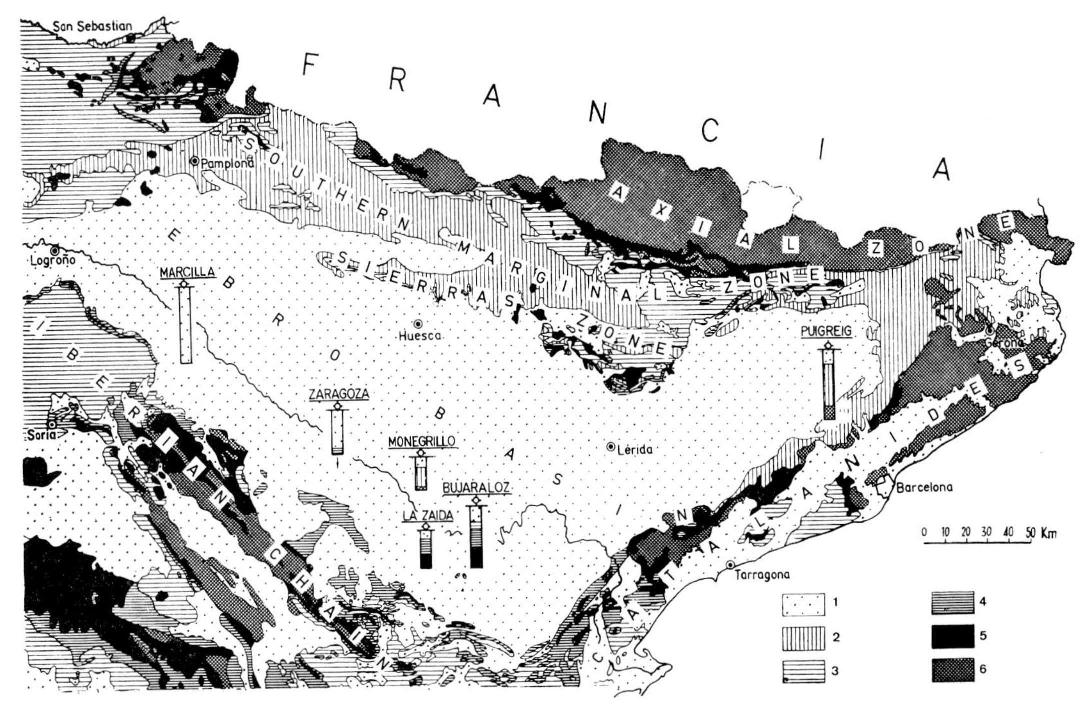

Fig.253. Map of the Ebro Basin and its surroundings. Compare Fig.252.
1 = Oligo-Miocene; *2* = Eocene; *3* = Cretaceous; *4* = Jurassic; *5* = Triassic; *6* = Paleozoic.
The Ebro Basin subsided most strongly in its northwestern part, with 3,415 m of Oligocene, without having reached the Cretaceous, in the Marcilla hole. The later uplift was also strongest in the northwest, resulting in the slight present tilting of the basin towards the southeast. (After ALMELA and RIOS, 1962.)

extensive areas were above sea level, leading to the deposition of rocks in Wealden facies. This followed a more generalized period of regression at the transition from Jurassic to Cretaceous (BRINKMANN, 1962; RAT, 1964).

Just as in the southern border of the Aquitanian Basin and in the northern marginal zone of the Pyrenees (Fig.227) both the rate of subsidence and the facies developed are strongly variable, both in place and in time. There is no apparent relation between the subsidence of a given area and its later sedimentary history. Nor is there a relation to the later block faulting movements, which led to the Alpine disturbances.

This is, for instance, quite apparent in the two basins of extra strong sedimentation at either end of the eastern Iberian chain, i.e., the Sierra de los Cameros and the Maestrazgo. Although these former basins, which by the way now stand out as mountaineous areas due to their thicker cover of Mesozoic, were, in a general way, subsiding strongly about the same time, they did so independently. There is no correlation between this extra subsidence in separate basins. A fact which, for instance, follows from the development of the relative sedimentary series. The Sierra de los Cameros has a continental Wealden facies, the Maestrazgo, on the other hand, a marine Urgo-Aptian.

STRUCTURE

The main character of the structure of the Iberian chains and of the Catalanides lies in its irregularity. There exist large numbers of short structures, which are more or less

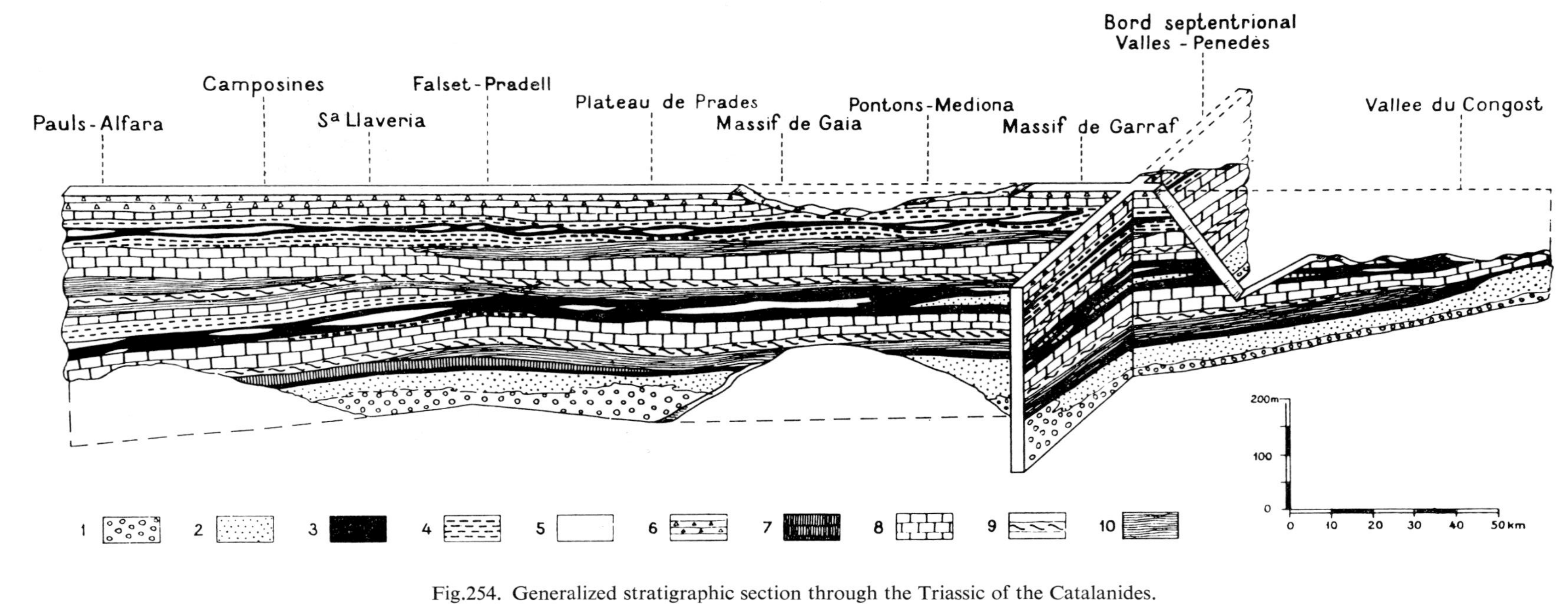

Fig.254. Generalized stratigraphic section through the Triassic of the Catalanides.
B = Buntsandstein; *M* = Muschelkalk; *K* = Keuper.
1 = conglomerate; *2* = red- and green sandstones, the "grès bigarrés" of the Buntsandstein; *3* = red iridescent clay and marl; *4* = grey and yellow marl; *5* = gypsum and anhydrite; *6* = "cargneule"; *7* = reddish dolomite; *8* = dolomite; *9* = *Fucoïdes* limestone; *10* = limestone. (After VIRGILI, 1962.)

aligned, but which nevertheless show an extremely irregular pattern. The "en échelon" arrangement often acquires a habit which is more reminiscent of the shape of a braided river, than that of a fold belt (Fig.255). More or less horizontal plateau areas, structurally situated either high or low, alternate with narrow, normally strongly asymmetric, folds, which show strong along-the-strike variation.

In the Catalanides the topographically higher of the horizontal or subhorizontal plateaus normally have the basement exposed. In those plateaus, the Buntsandstein invariably covers the Hercynian basement. Underneath the lowest gypsum horizon it has an extremely tranquil attitude, being either horizontal, or dipping only slightly. The folding of the higher elements of the series is entirely of the epidermis type. The basement and its thin cover of Buntsandstein shows only block faulting and tilting. Moreover, the movements individual blocks have effected are irregular and differ from block to block (RIBA and RÍOS, 1962).

This is, of course, quite normal for an epidermis type of fold belt. But in the Catalanides erosion has proceeded much further than in the epidermis type fold belts discussed so far. The average level of the erosion, measured against the upper surface of the basement, is perhaps about 1 km deeper in the Catalanides than in the Jura Mountains. The relationship between basement and epidermis consequently is far better visible in the latter chain than in the much advertized Jura Mountains.

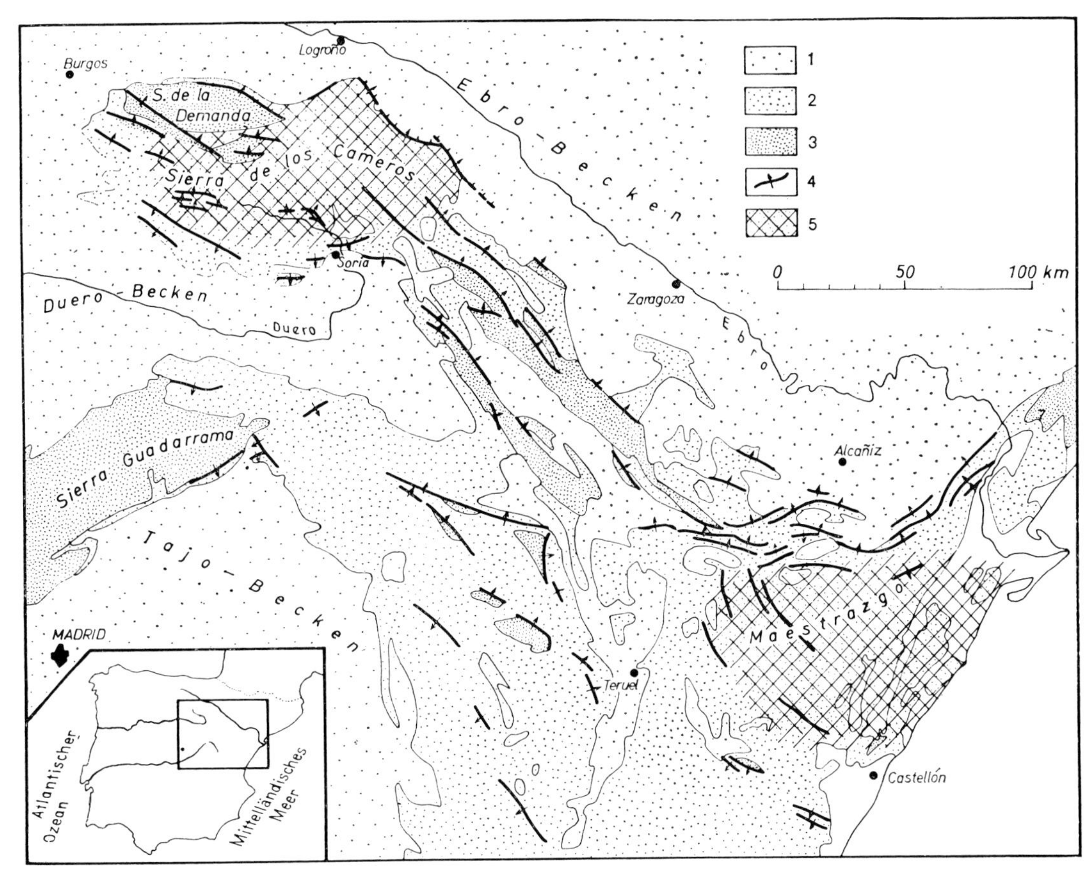

Fig.255. Tectonic map of the Iberian chains, showing the irregular distribution of fold axes. *1* = Tertiary; *2* = Mesozoic; *3* = Paleozoic; *4* = fold axes; *5* = zone of extra thick development of Wealden in the Sierra de los Cameros, and of Urgo-Aptian in the Maestrazgo. (After BRINKMANN, 1962.)

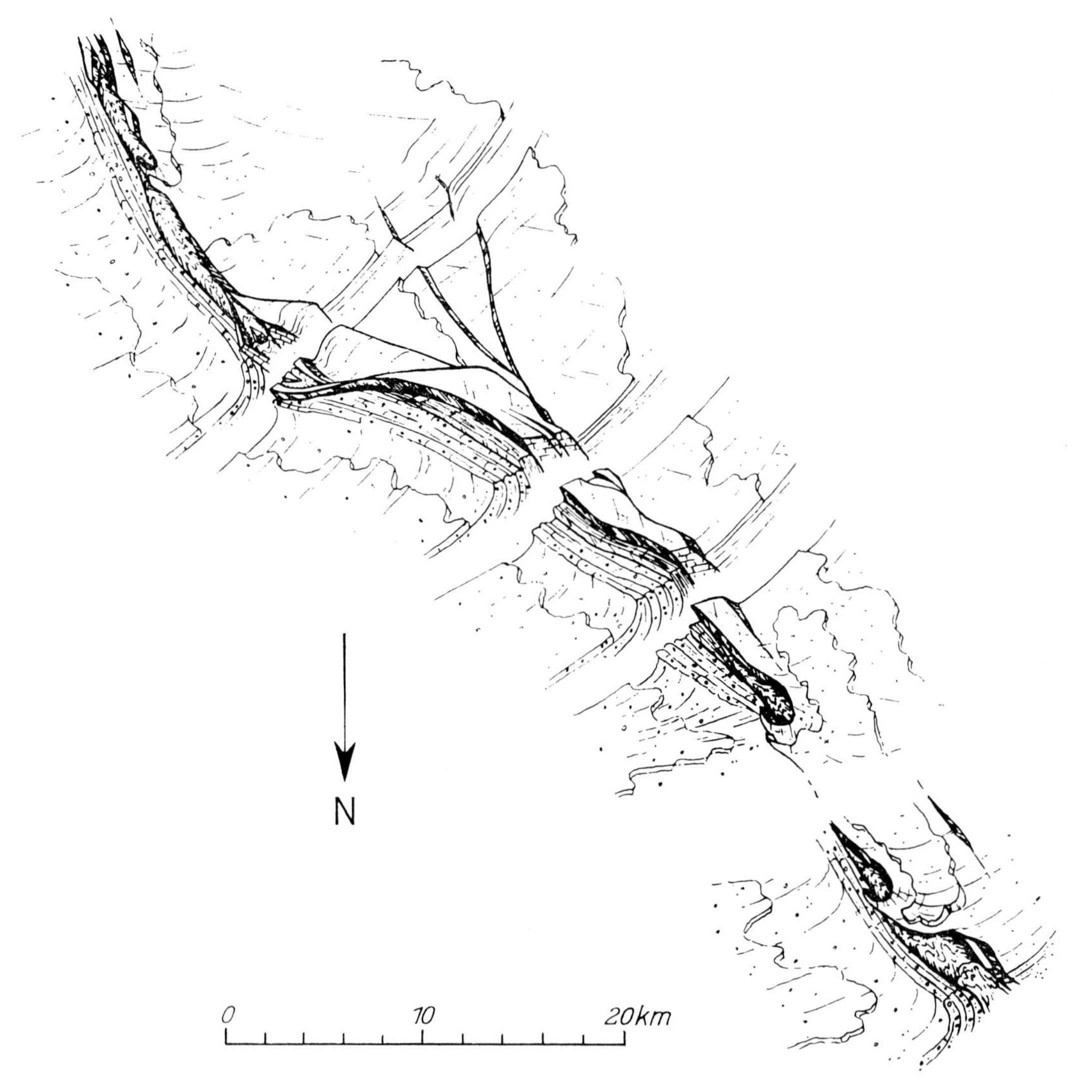

Fig.256. Block diagram of the northwestern border of the Sierra de los Cameros. Showing the draping of the epidermis over the border fault separating the block of the Sierra de los Cameros from the Ebro Basin. For location, compare Fig.255. (After G. Tischer, in: BRINKMANN, 1962.)

It seems to me that the folding of the epidermis is due either to gravitational sliding of epidermis blocks from rising areas, or to a draping over faults in the basement. A beautiful example of such draping is reproduced in Fig.256. The relation between tilting and faulting of the basement and folding of the epidermis follows from Fig.257. A miniature example of the "Branden einer Deckenstirn" is given in Fig.258.

GRAVITY TECTONICS

The three sections of Fig.258 are selected examples of structures favourable to an interpretation throughgravity tectonics. This is not found everywhere in the area under discussion. For one thing, the relation between basement and epidermis is hard to establish in areas with thicker epidermis sedimentation, in which the basement hardly outcrops, such as the

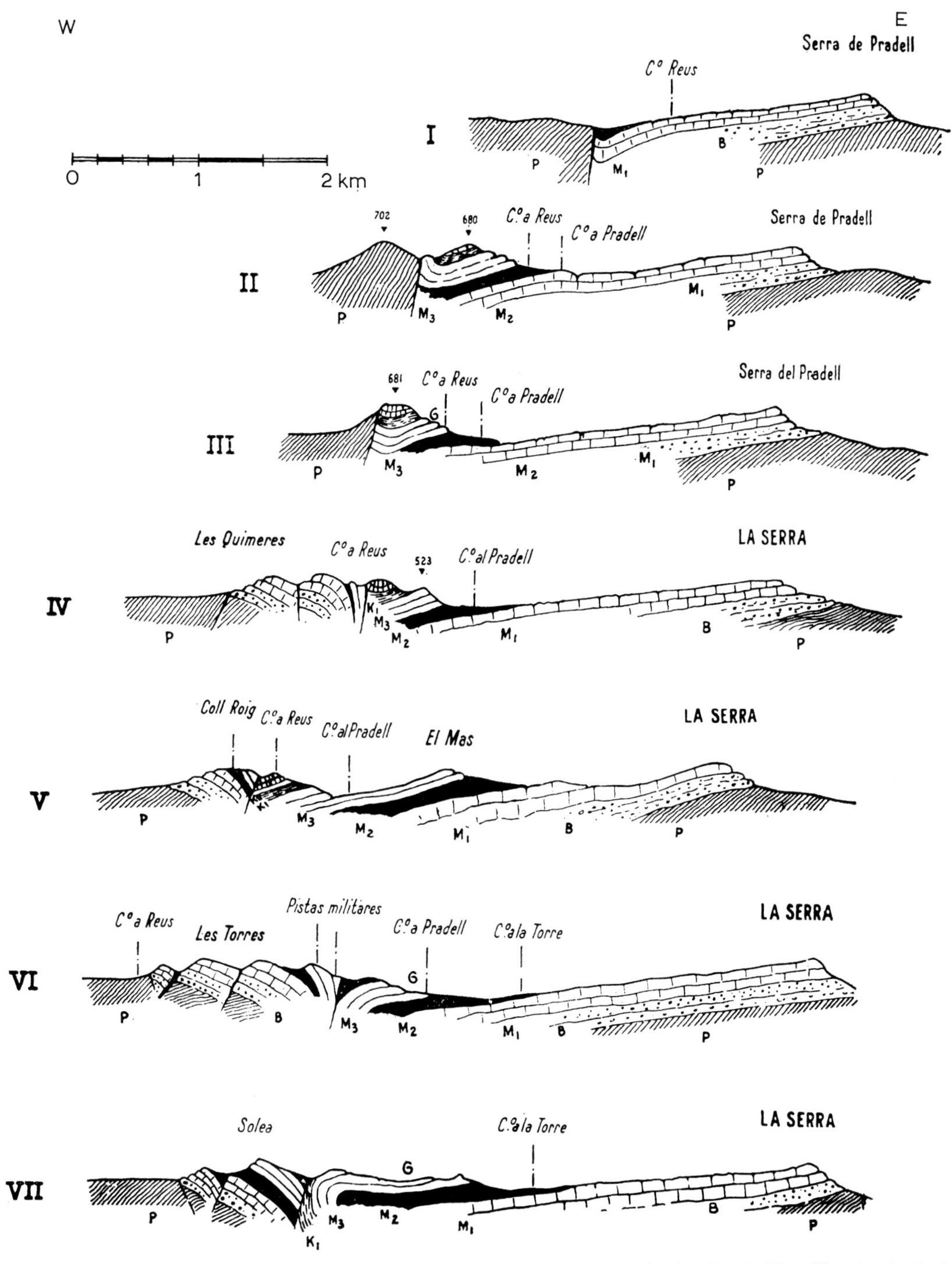

Fig.257. Serial sections through La Serra, west of Falset, near Tarragona in the Catalanides. Showing both the "décollement" in the higher Triassic, and the relation between basement faulting and tilting, and epidermis folding. P = Paleozoic; B = Buntsandstein; M_1, M_2, M_3 = Muschelkalk; K_1, K_2 = Keuper. (After VIRGILI, 1958.)

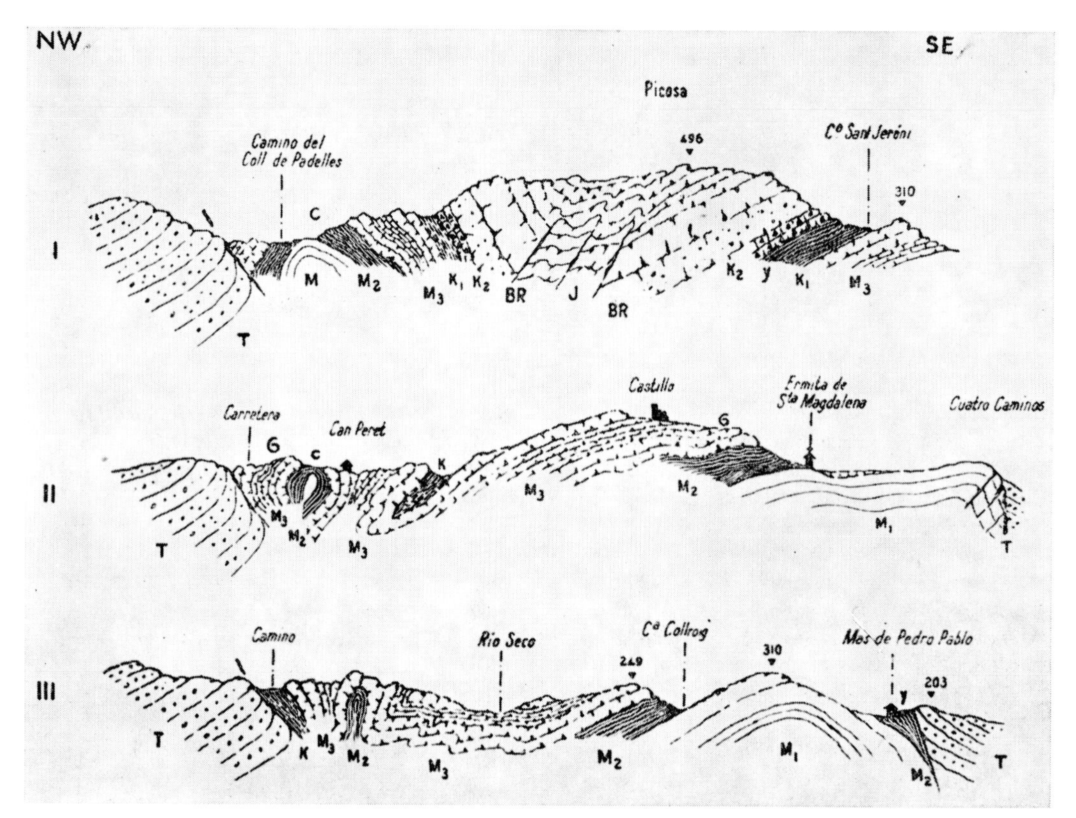

Fig.258. Three serial sections through the Camposines structure in the Catalanides. This is a miniature example of the "Branden einer Deckenstirn".
M_1, M_2, M_3 = Muschelkalk; K_1, K_2 = Keuper; *BR* = basal breccious limestone of the Jurassic; *J* = Jurassic; *T* = Tertiary. (After VIRGILI, 1958.)

Sierra de los Cameros and the Maestrazgo. Much more serious are, on the other hand, those structures in which the basement is exposed, but tilts the wrong way. That is, away from the "slided" blocks of epidermis, so that it seems as if gravity had pushed up those blocks from their original position. In these cases post-sliding movements have to be assumed, having tilted the basement away from the attitude it had when the "décollement" of the epidermis took place.

Such a hypothesis is quite acceptable, because we know of so many examples of basement blocks, in which vertical and tilting movements have shown one, or even more, reversals during their history. But it remains an ad hoc hypothesis for all those cases in which it is impossible to establish the exact history of the movements of the basement blocks. Worse still, the latter cases form the majority. And although it is quite normal in geology, to have more of that one knows nothing about, than of that one knows something of, this remains a serious diffculty.

I even believe that this necessity of an ad hoc assumption about a reversal of basement movements, every time we see a supposed "décollement" horizon dipping the wrong way, forms one of the major barriers against a more general acceptance of gravity as the motor in tectonics.

Also it seems as if this discussion on probabilities and improbabilities forms a good ending to the chapters on Alpine Europe.

REFERENCES

ALMELA, A. and RÍOS, J. M., 1962. Structure d'ensemble des Pyrénées aragonaises et découvertes récentes dans cette région. In: M. DURAND DELGA (Editor), *Livre à la Mémoire du Professeur Paul Fallot. I.* Soc. Géol. France, Paris, pp. 315–331.

ASHAUER, H. and TEICHMÜLLER, R., 1935. Die variscische und alpidische Gebirgsbildung Kataloniens. *Abhandl. Ges. Wiss. Göttingen, Math. Phys. Kl.*, 3 (16): 16–98.

BOULIN, J., 1963. Sur les Alpujarrides occidentales et leurs rapports avec la nappe de Málaga (Andalousie méridionale). *Bull. Soc. Géol. France*, 7 (4): 384–389.

BRINKMANN, R., 1962. Aperçu sur les chaînes ibériques du nord de l'Espagne. In: M. DURAND DELGA (Editor), *Livre à la Mémoire du Professeur Paul Fallot. I.* Soc. Géol. France, Paris, pp. 291–300.

BUSNARDO, R., 1961. Aperçu sur le Prébétique de la région de Jaén (Andalousie, Espagne). *Bull. Soc. Géol. France*, 7 (2): 324–328.

BUSNARDO, R., 1962. Regards sur la géologie de la région de Jaén. In: M. DURAND DELGA (Editor), *Livre à la Mémoire du Professeur Paul Fallot. I.* Soc. Géol. France, Paris, pp. 183–186.

BUSNARDO, R., 1964. Hypothèses concernant la position des unités structurales et paléogéographiques de la transversale Jaén–Grenade (Andalousie). *Geol. Mijnbouw*, 43: 264–267.

BUSNARDO, R. and DURAND DELGA, M., 1961. Données nouvelles sur le Jurassique et le Crétacé inférieur dans l'est des Cordillères bétiques. *Bull. Soc. Géol. France*, 7 (2): 278–287.

DE ROEVER, W. P. and NIJHUIS, H. J., 1964. Plurifacial Alpine metamorphism in the eastern Betic Cordilleras (SE Spain), with special reference to the genesis of glaucophane. *Geol. Rundschau*, 53: 324–336.

DUPLAIX, S. and FALLOT, P., 1961. Les "konglomeratische Mergel" des Cordillères bétiques. *Bull. Soc. Géol. France*, 7 (2): 308–317.

DURAND DELGA, M., 1963. Essai sur la structure des domaines émergés autour de la Méditerranée occidentale. *Geol. Rundschau*, 53: 534–535.

DURAND DELGA, M. and FONTBOTÉ, J. M., 1960. Le problème de l'âge des nappes alpujarrides d'Andalousie. *Rev. Géograph. Phys. Géol. Dyn.*, 3 (4): 181–269.

DURAND DELGA, M., HOTTINGER, L., MARÇAIS, J., MATTAUER, M., MILLARD, Y. and SUTER, G., 1962. Données actuelles sur la structure du Rif. In: M. DURAND DELGA (Editor), *Livre à la Mémoire du Professeur Paul Fallot. I.* Soc. Géol. France, Paris, pp.399–422.

EGELER, C. G., 1964. On the tectonics of the eastern Betic Cordilleras (SE Spain). *Geol. Rundschau*, 53: 260–269.

FALLOT, P., 1945. *Estudios Geológicos en la Zona Subbética entre Alicante y el Río Guadania Menor.* Consejo Super. Invest. Cient., Inst. Lucas Mallada, Madrid, 720 pp.

FALLOT, P., 1948. Les Cordillères bétiques. *Estud. Geol., Inst. Invest. Geol. "Lucas Mallada" (Madrid)*, 8: 83–172,

FALLOT, P., 1954. Comparaison entre Cordillères Bétiques et Alpes orientales. *Bol. Soc. Real. Españ. Hist. Nat., Tomo Extraordinario 80 Aniv. Prof. Ed. Hernández-Pacheco*, 1954: 259–279.

FALLOT, P., 1959. La géologie profonde du Bassin du Po et le mystère de celui du Guadalquivir. *Estud. Geol., Inst. Invest. Geol. "Lucas Mallada" (Madrid)*, 15: 155–162.

FALLOT, P., FAURE-MURET, A., FONTBOTÉ, J. M. and SOLÉ, L., 1961. Estudios sobre las series de Sierra Nevada y de la llamada Mischungszone. *Bol. Inst. Geol. Minero Españ.*, 71: 345–557.

FERNEX, F., 1964. Essai de corrélation des unités bétiques sur la transversale de Lorca–Aguilas. *Geol. Mijnbouw*, 43: 326–330.

FOUCAULT, A., 1962. Problèmes paléogéographiques et tectoniques dans le prébétique et le subbétique sur la transversale de la Sierra Sagra (Province de Grenade, Espagne). In: M. DURAND DELGA (Editor), *Livre à la Mémoire du Professeur Paul Fallot. I.* Soc. Géol. France, Paris, pp.175–182.

FOUCAULT, A., 1964. Sur les rapports entre les zones prébétiques et subbétiques entre Cazorla (prov. de Jaén) et Huescar (prov. de Grenade, Espagne). *Geol. Mijnbouw*, 43: 268–272.

GARCIA-RODRIGO, B., 1961. Sur la structure du Nord de la province d'Alicante (Espagne). *Bull. Soc. Géol. France*, 7 (2): 273–277.

HOEPPENER, R., HOPPE, P., DÜRR, S. and MOLLAT, H., 1964a. Ein Querschnitt durch die Betischen Kordilleren bei Ronda (SW Spanien). *Geol. Mijnbouw*, 43: 282–298.

HOEPPENER, R., HOPPE, P., MOLLAT, H., MUCHOW, S., DÜRR, S. and KOCKEL, F., 1964b. Über den westlichen Abschnitt der Betischen Kordillere und seine Beziehungen zum Gesamtorogen. *Geol. Rundschau*, 53: 269–296.

KONINKLIJK NEDERLANDS GEOLOGISCH MIJNBOUWKUNDIG GENOOTSCHAP (Editor), 1964. Symposium on the geology of the Betic Cordilleras. *Geol. Mijnbouw*, 43: 262–334.

LLOPIS LLADO, N., 1947. *Contribución al Conocimiento de la Morfoestructura de los Catalánides.* Consejo Super. Invest. Cient., Inst. Lucas Mallada, Madrid, 372 pp.

MACGILLAVRY, H. J., 1964. Speculations based upon a comparison of the stratigraphies of the different tectonic units between Vélez Rubio and Moratalla. *Geol. Mijnbouw*, 43: 299–309.

MacGillavry, H. J., Geel, T., Roep, T. B. and Soediono, H., 1964. Further notes on the geology of Málaga, the subbetic, and the zone between these two units, in the region of Vélez Rubio (southern Spain). *Geol. Rundschau*, 53: 233–259.

Mattauer, M., 1964. Le style tectonique des chaînes telliene et rifaine. *Geol. Rundschau*, 53: 296–313.

Nijhuis, H. J., 1964. *Plurifacial Alpine Metamorphism in the Southeastern Sierra de los Filabres, south of Lubrín, SE Spain.* Thesis, Univ. Amsterdam, Amsterdam, 151 pp.

Perconig, E., 1962. Sur la constitution géologique de l'Andalousie occidentale, en particulier du Bassin du Guadalquivir. In: M. Durand Delga (Editor), *Livre à la Mémoire du Professeur Paul Fallot. I.* Soc. Géol. France, Paris, pp.229–256.

Rat, P., 1964. Problèmes du Crétacé inférieur dans les Pyrénées et le nord de l'Espagne. *Geol. Rundschau*, 53: 205–218.

Riba, O. and Ríos, J. M., 1962. Observation sur la structure du secteur sud-ouest de la Chaîne ibérique (Espagne). In: M. Durand Delga (Editor), *Livre à la Mémoire du Professeur Paul Fallot. I.* Soc. Géol. France, Paris, pp.257–290.

Rondeel, H., 1965. *Geological Investigations in the Western Sierra Cabrera and Adjoining Areas, Southeastern Spain.* Thesis, Univ. Amsterdam, Amsterdam, 161 pp.

Simon, O. J., 1963. *Geological Investigations in the Sierra de Almagro, Southeastern Spain.* Thesis, Univ. Amsterdam, Amsterdam, 164 pp.

Simon, O. J., 1964. The Almagro unit: a new structural element in the Betic zone? *Geol. Mijnbouw*, 43: 331–334.

Société Géologique de France (Editor), 1961. Séance sur les Cordillères bétiques. *Bull. Soc. Géol. France*, 7 (2): 263–362.

Staub, R., 1934. Der Deckenbau Südspaniens in den Betischen Cordilleren. *Vierteljahresschr. Naturforsch. Ges. Zürich*, 79: 271–332.

Virgili, C., 1958. El Triássico de los Catalánides. *Bol. Inst. Geol. Minero Españ.*, 69: 856 pp.

Virgili, C., 1962. Le Trias du nord-est de l'Espagne. In: M. Durand Delga (Editor), *Livre à la Mémoire du Professeur Paul Fallot. I.* Soc. Géol. France, Paris, pp. 301–312.

CHAPTER 17

Lowlands and Low Plateaus

INTRODUCTION

The lowlands and low plateaus, which together form our third major subdivision of western Europe will only be treated very lightly. The reason is that these regions have received so much interest from stratigraphers, that there are now several general accounts, to which the reader may be referred. Principal amongst these are the stratigraphy textbooks by GIGNOUX (1950) and BRINKMANN (1960), and, of course, the various volumes of the *Lexique Stratigraphique International*, published by the Centre National de la Recherche Scientifique in Paris. Most regional publications, moreover, pay a lot of attention to the stratigraphy of the lowlands and low plateaus found in their area.

Lowlands and low plateaus form a mixed group, which, as we saw already in Chapter 1, comprises units of quite different structural histories. In regard to the lowlands, we may first distinguish those that are lowlands because, thought he crust is rising in their area, erosion is able to offset crustal rising, and maintain a peneplained surface of the earth. This type of lowlands is mainly found on the Fennoscandian Shield, and will not be considered further in this text. The opposite group of lowlands is formed by basins which are subsiding, but where subsidence is offset by sedimentation. Within this second group a further distinction may be made between those basins developing on the epicontinental part of Europe, and those developed post-orogenetically within the confines of Alpine Europe. The latter of the two subgroups is by far the more active, with sediment thicknesses of the order of 10 km, as against the order of 1 km for the first subgroup.

Most of the low plateaus are comparable to the basins developed on the epicontinental part of Europe, but subsidence and sedimentation have, in their case, been stopped some time ago, and after that some uplift has taken place.

Apart from the stratigraphy, there are, however, several other aspects of the lowlands and low plateaus, which merit attention. First of these is the individuality of separate basins, both in regard to the history of their subsidence and of their sedimentation. Second comes the compartmentality within individual basins. This feature weakens the image of regional uniformity which is the impression gained by a cursory examination of these basins.

This division of sedimentary basins is in many ways related to the rising of the crustal blocks which now form the Hercynian uplands. Several of the major faults have been active in both cases, but scissor like, with a hinge point separating the rising from the subsiding area. Moreover, elongated basins bordering the rising crustal blocks, the "Randtröge" have developed also. All of which shows that the stable part of continental Europe is not so stable after all.

And lastly, there is the influence of Permian and Triassic evaporites, which has led to the development of a pronounced form of salt tectonics called halokinesis.

VARIATIONS IN HISTORY

The variations in history of subsidence and sedimentation in separate basins are most readily seen, and consequently best known, in the low plateaus, where erosion has exposed the post-Hercynian series along the valley walls of the main rivers. Similar variations are, however, also found in the basins of the lowlands which are still subsiding. The latter have become better known in later years also, thanks to the increased drilling programs (see, for instance, for the lowlands of The Netherlands: HAANSTRA, 1963; KEIZER and LETSCH, 1963).

As examples, let us cite three basins, those of Swabia, Paris and Aquitaine. In the basin of the Swabian Jura in southern Germany (Fig.1), sedimentation stopped with the end of the Jurassic, as the name implies (GEYER and GWINNER, 1962). In the Paris and Aquitaine Basins, on the other hand, sedimentation continued up to the end of the Oligocene, but was of a quite different nature in the two basins. This is most apparent for the Lower Tertiary, which saw repeated transgressions and regressions in the Paris Basin, as against a single transgression in the Aquitaine Basin (cf. GIGNOUX, 1950, fig.110, 124). The Münster Basin in northwestern Germany, on the other hand, subsided strongly during a short period only, i.e., the Albian and the Late Cretaceous (ARNOLD, 1963; THIERMANN and ARNOLD, 1964).

So not only the onset and the end of subsidence and sedimentation vary in separate basins, but also their intervening history. I will not go into further detail here, but only stress the fact that most periods of transgression or regression occur at different times in each basin. Since the limits between stratigraphical zones were, from the beginning, placed at the obvious breaks in the succession formed either by a transgression or a regression, stage names from the onset were often limited to a single basin only. As, moreover, each nation has its own basin or basins, national sets of stage names have developed, which are difficult to correlate. Or, worse still, which do not indicate the same time period, when used in different basins.

At several points during the post-Hercynian history crustal movements of wider scope have, however, occurred, which can be traced over a number of separate basins. As an example, the regression at the end of the Jurassic may be cited, or the transgression of the "Cenomanian". But the former break in the succession does not "occur at the end of the Jurassic". Instead, reversely, "a regression found more or less at this point" has been used to define the boundary between the Jurassic and the Cretaceous. Modern, detailed paleontological studies trend to show again and again that such lithostratigraphic breaks due to regressions or transgressions are but the expression of events which occurred at various times in separate basins, or which even show a definite time lag within a single basin. Such a time lag has always been obvious for the transgression of the Late Cretaceous, which may range in time from the Cenomanian to the base of the Senonian.

Such variations found between separate basins, but also in different parts of a single basin, are, of course, common knowledge to the modern stratigrapher. In oil geology such variations are widely used in the search for oil traps. But in western Europe stratigraphy has a more venerable history. In its early days it was strictly limited to surface outcrops, whereas the difference between litho- and chronostratigraphy was hardly appreciated.

COMPARTMENTATION AND "RANDTRÖGE"

As a first example of the compartmentation found within the younger sedimentary

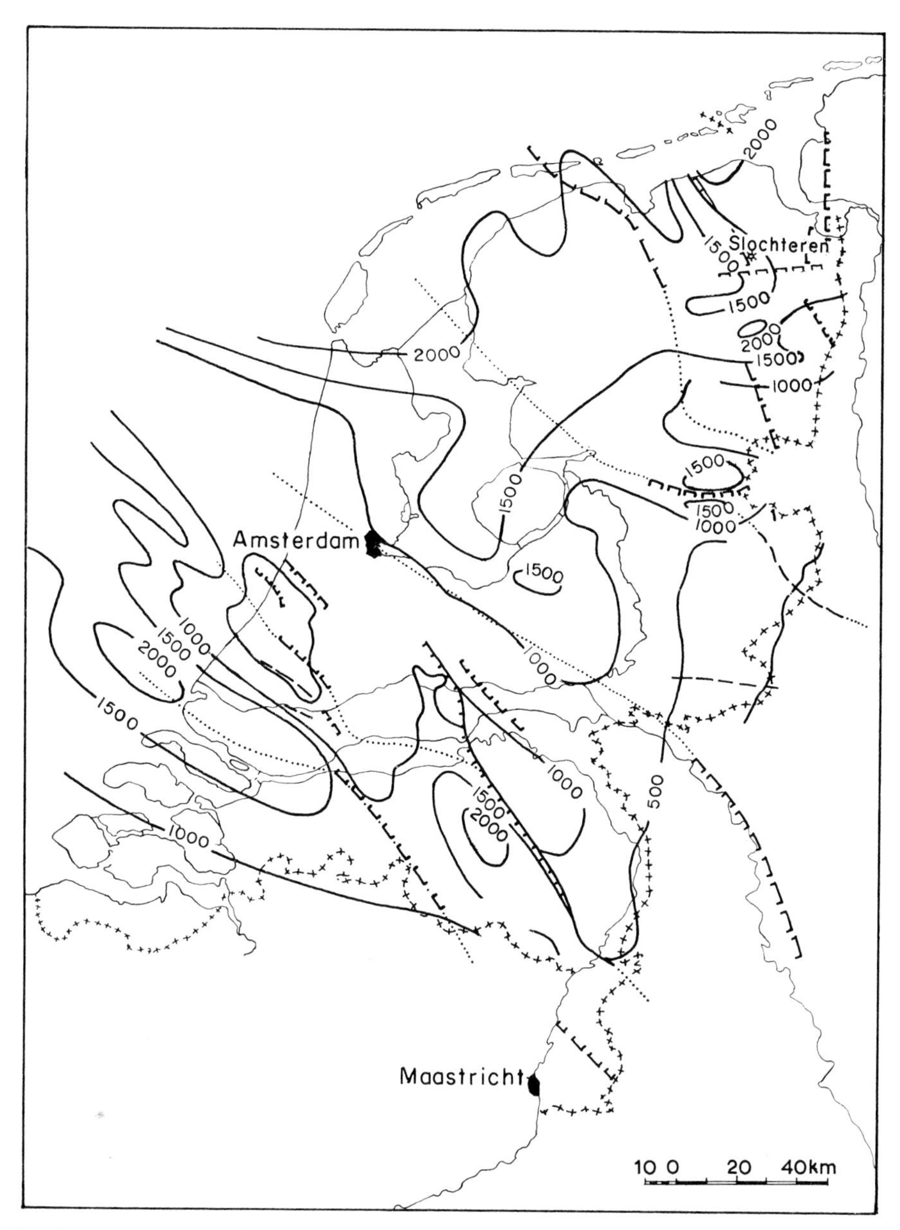

Fig.259. Depth-contour map of the base of the Tertiary in The Netherlands. Because of the low topographic elevation, this virtually is an isopach map for the Cenozoic, exept for the area near Maastricht. (After STHEEMAN, 1963).

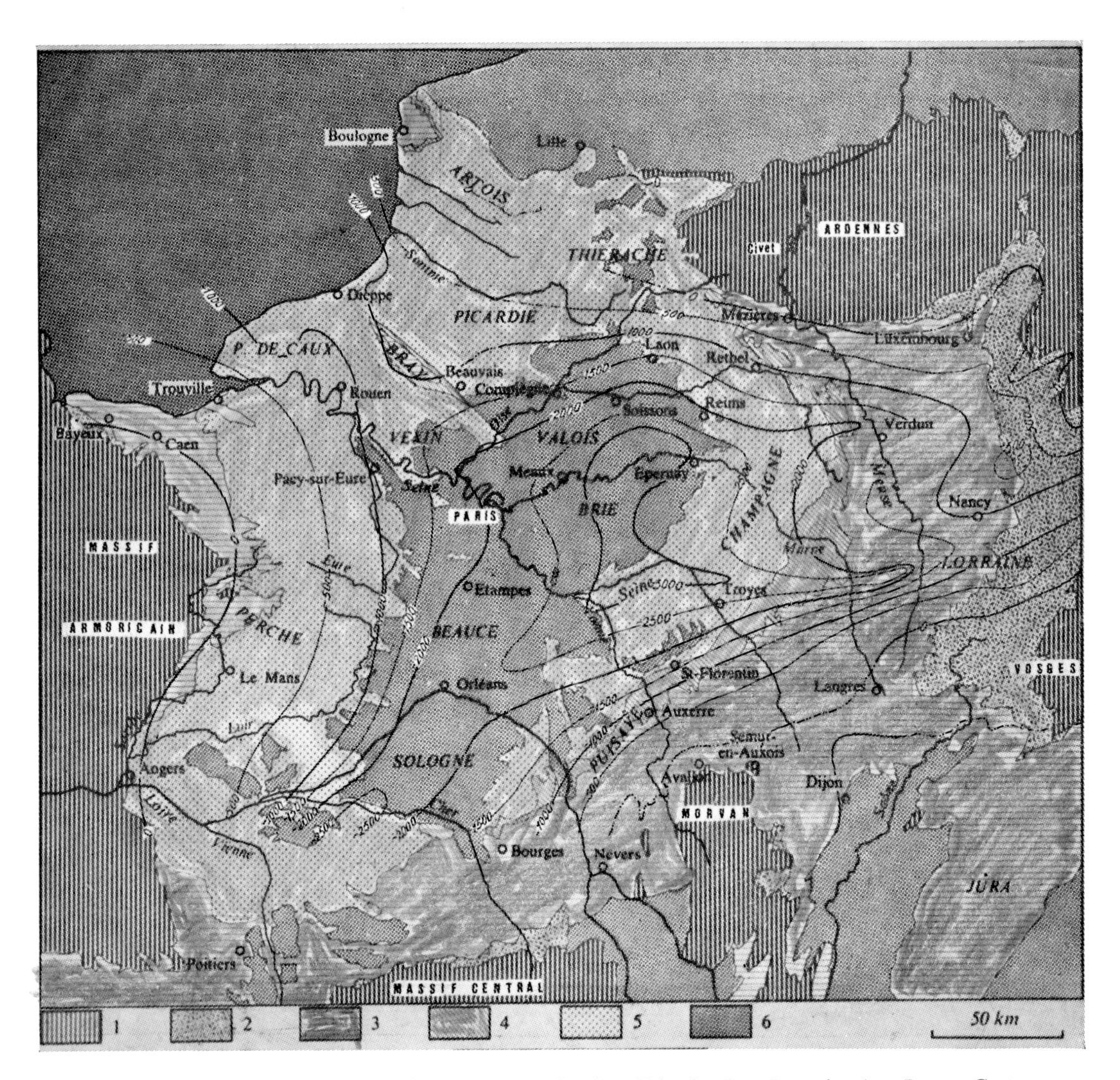

Fig.260. Geological map of the Paris Basin. *1* = Paleozoic; *2* = Triassic; *3* = Jurassic; *4* = Lower Cretaceous; *5* = Upper Cretaceous; *6* = Tertiary. (After DE LAPPARENT, 1964; depth-countour lines on top of pre- Permian basement after LIENHARDT, 1961.)

basins, a recent isopach map of the Cenozoic in The Netherlands is presented in Fig.259. The northwest–southeast trending horst and graben structures, which characterize the post-Hercynian blockfaulting tectonics in this part of western Europe, have evidently been active in the basin also, during the sedimentation of the Cenozoic.

Another example is supplied by the Paris Basin (compare, e.g., SOCIETÉ GÉOLOGIQUE DE FRANCE, 1966), which is for many the type epicontinental basin. Its beautiful saucer shape, evidenced by the concentric rings formed by successively younger formations from the Jurassic to the Oligocene, gives the impression of extreme regularity.

But this saucer-shaped form is not due to synsedimentary subsidence, but to the post-Oligocene uplift. That is, to a lesser uplift in the present centre of the basin, mirroring the stronger uplift of the Hercynian uplands along its margins; the Ardennes, the Vosges Mountains, the Massif Central with its promontory the Morvan, and the Armorican Massif (Fig.260).

Total post-Hercynian subsidence shows, however, a different orientation, as can be seen from the depth-contour lines on the top of the basement. It follows that the earlier subsidence was not conformable to the regular saucer-shaped pattern of the post-Oligocene uplift.

That this is not the case is well demonstrated by the well-known pictures of a paleogeographic sequence, i.e., that of the alternation of transgressions and regressions during the Early Tertiary in the Paris Basin. Subsidence during that time showed tilting towards the northwest, resulting in a number of transgressions always arriving from the northwest, and never reaching the southeastern part of the basin. Although everybody will be familiar with the relevant figures, I have reproduced them here (Fig.261, 262) to stress the difference between the movements of the Paris Basin during Early Tertiary subsidence and during later uplift.

Apart from these variations in subsidence and uplift, a good example of compartmentality is found in the northwest–southeast trending anticline of the Pays de Bray, northwest of Paris. PRUVOST had already in 1930 drawn attention to the irregularity of its history. The present anticline was in fact an area of strong subsidence and sedimentation, a partial trough, during the Jurassic. Crustal movements then reversed from the Cretaceous onwards

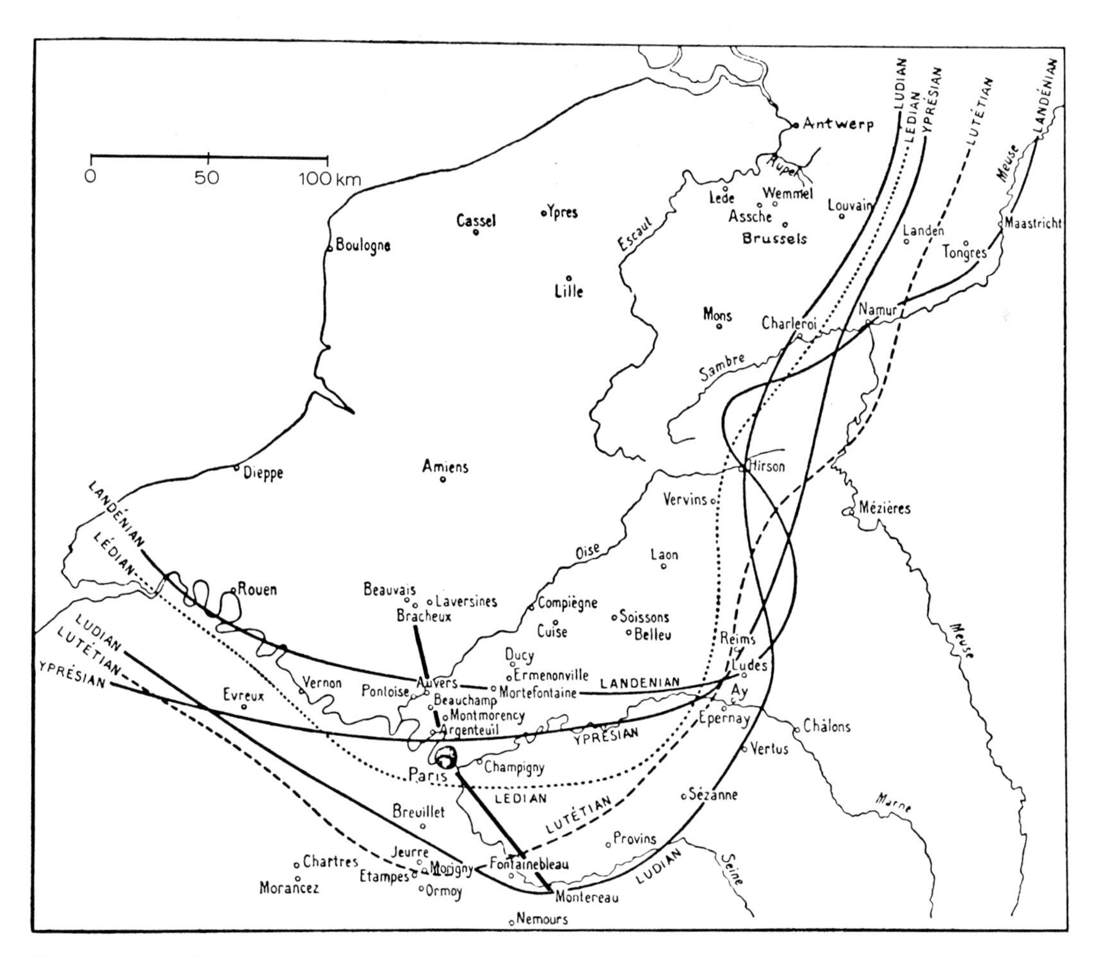

Fig.261. Limits of the marine transgressions in the Paris Basin during the Eocene. Thick line indicates location of section of Fig.262. (After GIGNOUX, 1950.)

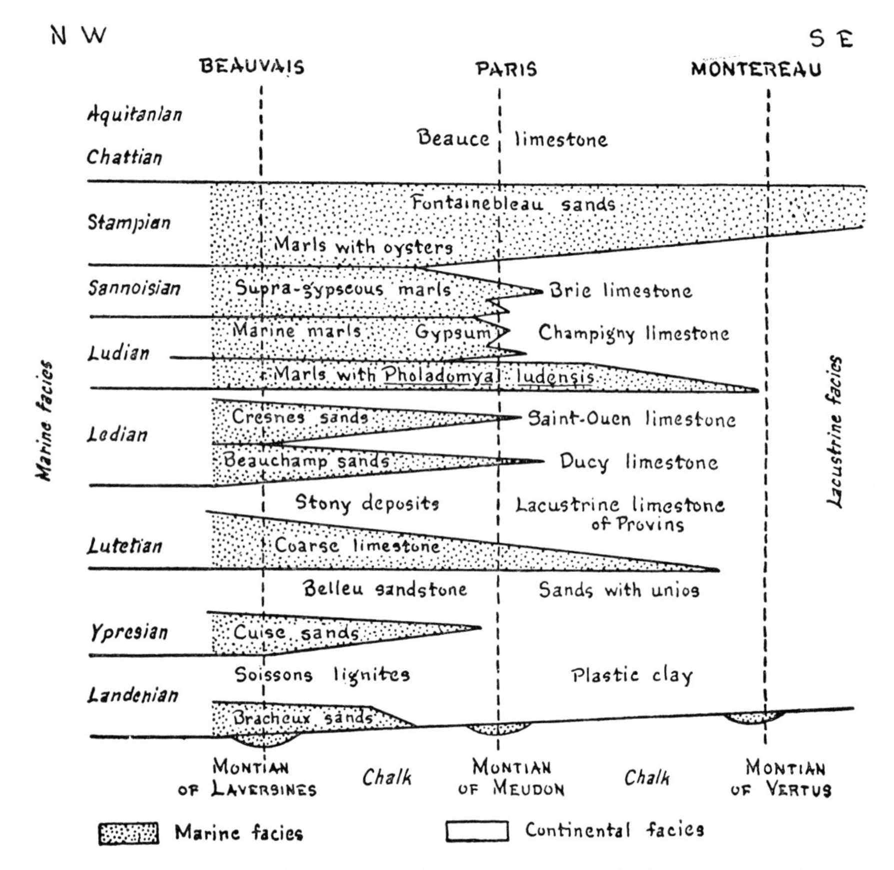

Fig.262. Stratigraphic section through the Lower Tertiary of the Paris Basin between Beauvais and Montereau. Showing the extent of the repeated marine invasions from the northwest. For location of section, compare Fig.261. (After GIGNOUX, 1950.)

(Fig.263). With VOIGT (1963), we may speak of an inverted trough. In fact, post-Hercynian total subsidence is hardly influenced by the Bray anticline, as can be seen from Fig.260.

Detailed studies of many younger sedimentary basins of northwestern Europe have led VOIGT (1963) to assume that such inversion of crustal movements of local, circumscribed troughs is a general rule in these basins.

One of the most widely cited of such more localized troughs within a larger sedimentary basin is the Niedersachsen Basin, north of the so-called Rheinische Masse. The latter term groups together the Rhienisches Schiefergebirge and its outliers covered by a thin veneer of younger sediments only (Fig.265). In contrast to the situation found in the Bray and Boulogne anticlines of Fig.263, the history of the Niedersachsen Basin is more complex. This results in a compartmentation that is even more pronounced.

During the Lias, subsidence and sedimentation were most active in a northwest–southeast trending zone situated about 100 km north of the Rheinische Masse (Fig.264 A). This zone forms the primary "Randtrog", comparable to the Bray anticline of the Paris Basin. During the Dogger more localized, and even stronger, sedimentation took place in still smaller troughs, whereas the remainder of the original trough already received much less sedimentation (Fig.264 B). Of these partial troughs the Gifhorn trough, well known for its oil production, is the most pronounced (for location see Fig.265). During the Malm, all of the primary "Randtrog" had already become inverted, and from that time onwards it forms the Pompekj sill or "Pompeckj'sche Schwelle", which during its further development only received limited sedimentation. Contrarily, the strongest sedimentation in the Niedersachsen Basin now took place in a new trough — a subsequent trough in Voigt's terminology — which developed between the primary "Randtrog" and the Rheinische Masse (Fig.264 C).

According to Voigt, such complex history of partial troughs seems to be the rule in the younger sedimentary basins of northwestern Europe. For some reason or other several of the boundaries between the rising and the subsiding crustal blocks are more active than others. It is along these active boundaries that the circumscribed troughs with stronger sedimentation, the primary "Randtröge", develop. These in time become inverted and develop into stable sills, which may develop even further into anticlinal structures. After the subsidence in the primary "Randtröge" ended, subsequent troughs may develop parallel to, and bordering upon, the primary troughs (Fig.265, 266).

After reviewing the history of these partial troughs in the various younger sedimentary basins of northwestern Europe, Professor Voigt stresses the extent of division, of mobility, of the crust in what is always called the stable part of Europe. He insists that these newer insights, based mainly on drilling programs executed during the latter decades, must necessarily lead to a re-evaluation of what in the German, Stille-type, nomenclature, is called

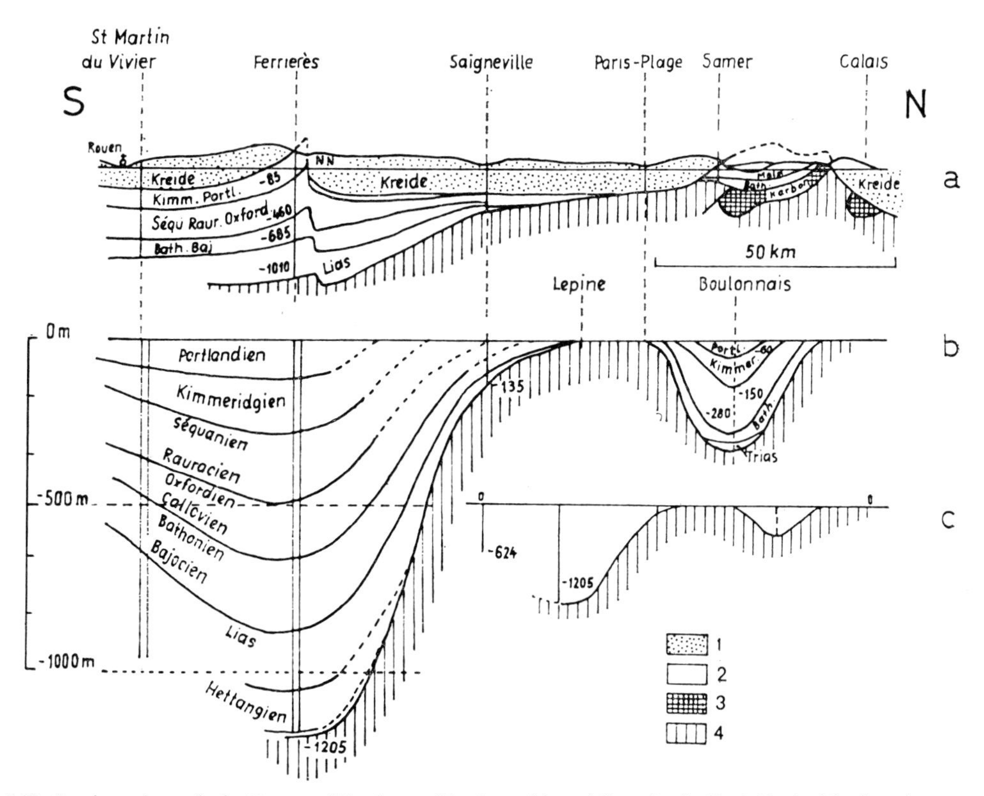

Fig.263. Sections through the Bray and Boulogne (Boulonnais) anticlines in the Paris Basin. For location, compare Fig.260. The Boulogne anticline forms the cross-channel continuation of the Wealden anticline in southern England. *a* = tectonic section; *b* = reconstruction of the Jurassic troughs, vertical scale strongly exaggerated; *c* = idem, but on a true scale.
1 = Cretaceous; *2* = Jurassic; *3* = Carboniferous; *4* = earlier Paleozoic. (After PRUVOST, 1930, simplified.)

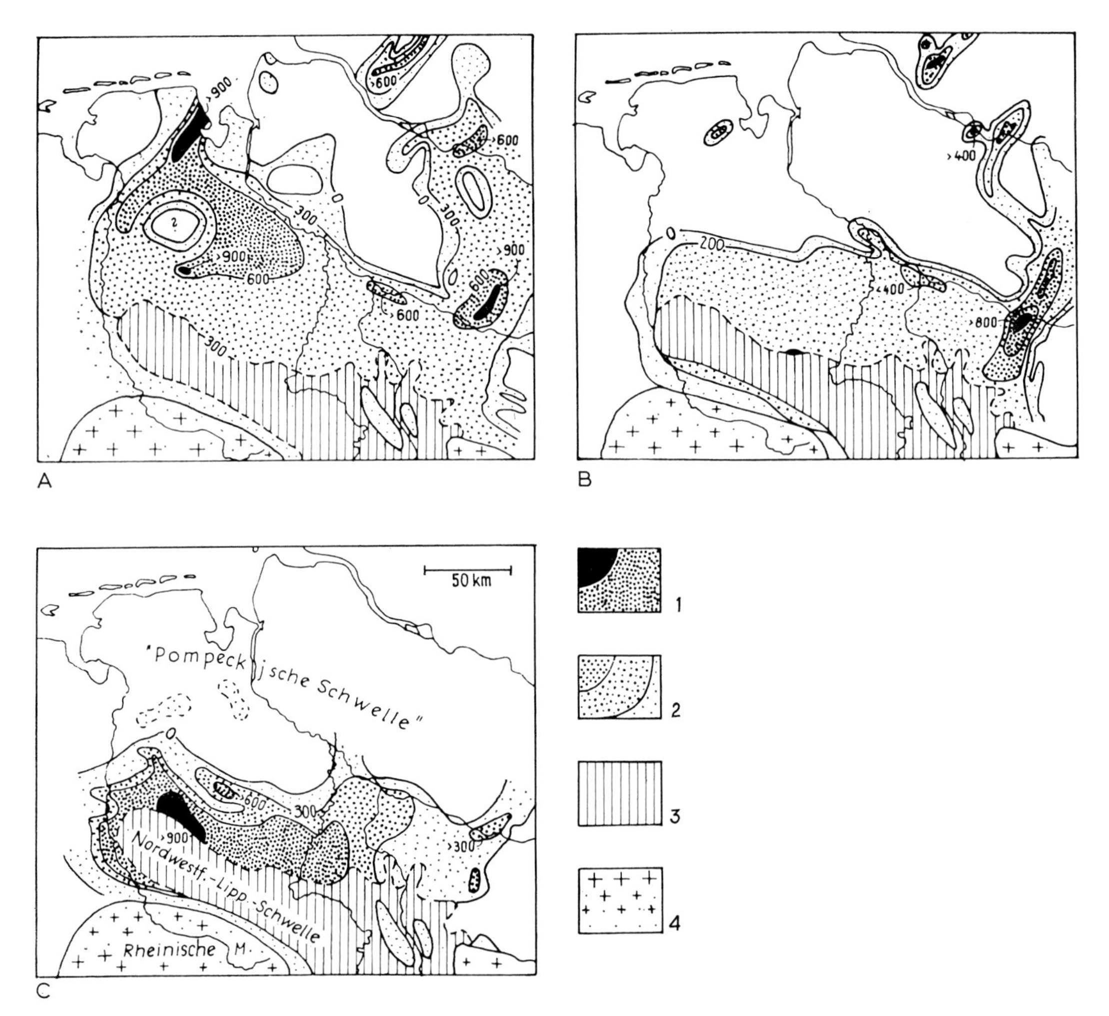

Fig.264. Sedimentation during the Jurassic in the Niedersachsen Basin of northern Germany. Showing the situation of the primary "Randtrog" of the Lias, in relation to the "Rheinische Masse" (A). Further, the development of more localized troughs during the Dogger (B). And lastly the inversion of the primary "Randtrog" during the Malm to form the Pompeckj sill (C), and the development of a subsequent trough between the primary trough and the "Rheinische Masse".

1 = thick sediment pile filling the troughs; *2* = thin cover of contemporaneous sediments; *3* = earlier post-Hercynian sediments, either at the surface or under a thin cover of Neozoic, uplifted through a reversal of basin subsidence; *4* = Paleozoic, partly under a thin sedimentary cover. (After VOIGT, 1963.)

"Germanotyp" and "Alpinotyp". That is of the difference between "epicontinental" and "geosynclinal", between "cratonic" and "orogenetic".

The answer to this question is that the difference still exists, but has become more qualified.

Sediment thicknesses of the order of 1 km, for shorter periods, are the rule, rather than the exception in the partial troughs within the sedimentary basins. For instance, 1 km for the Jurassic of the Pays de Bray, 1 km also for the Cenozoic of the northern Netherlands, and 2 km even for the Cretaceous in part of the Danish–Poland trough. But when subsidence stops, the trough becomes inverted, and total subsidence and sedimentation is not so much higher than in adjoining parts of the basin.

These then are evidently the limitations to this surprisingly strong subsidence in cratonic basins.

Although the rate of subsidence is similar to geosynclinal rates, the subsidence is strictly limited, both as to area and as to time. Holes may evidently fall into the otherwise stabilized shield of a continent, but the phenomenon never acquires the dimensions of a geosyncline. Nor is it proceeding for a time, sufficient to accumulate the material for a later fold belt. So, although the differences between craton and orogen still exist, there are convergent variations that approach each other to within an order of magnitude. Instead of the black and white antithesis, we now have a much more subtle distinction. Any geotectonic theory must at present take into account, why, nevertheless, as a whole the craton remained stable, and the orogen was mobilized.

The importance of these seesaw movements, which nowadays have been proved for sedimentary basins, even exceeds, in my view, the area of these younger sedimentary basins. In the preceding chapters we have repeatedly seen how certain structures of folding or overthrusting or nappe formation could be explained most elegantly by gravity tectonics, were it not for the fact that such structures had apparently "glided uphill".

To save gravity tectonics, ad hoc hypotheses about balancing movements of the basement had to be introduced. These would have been elevated to their present high position, but only after nappe formation, the regions which formerly were low and consequently could accumulate gravitative nappes.

But apart from the "rétrocharriage" in the internal zones of the French Alps, such seesaw movements of the basement were difficult to detect. The reason being that the folded and overthrust epidermis of younger sedimentary series effectively masks the basement in almost all fold belts.

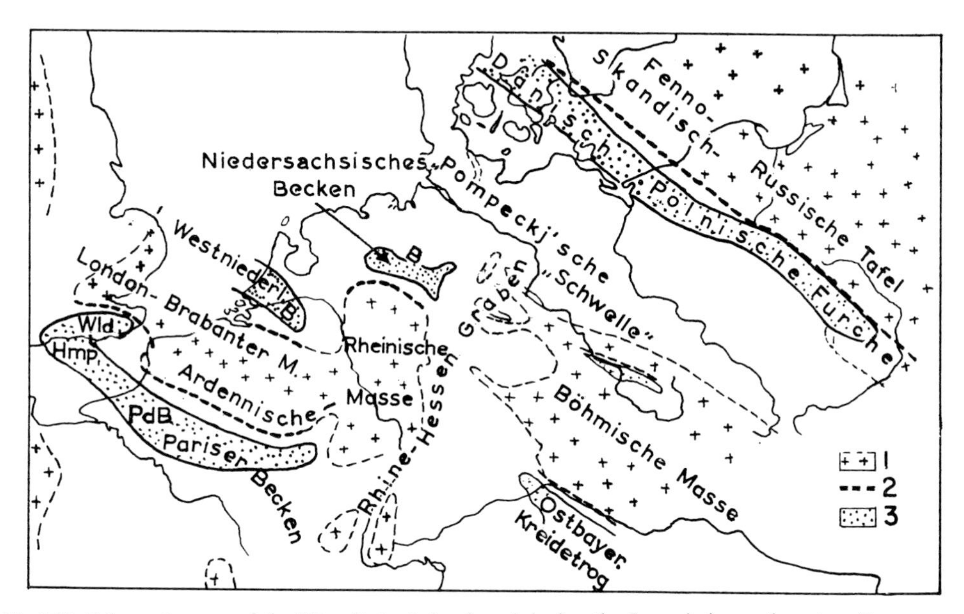

Fig.265. Schematic map of the "Randtröge" developed during the Jurassic in northwestern Europe. *1* = crustal blocks with rising tendencies, Hercynian uplands with their outliers; *2* = active boundaries in the younger sedimentary basins; *3* = "Randtröge".
Hmp = Hampshire; *Wld* = Wealden; *PdB* = Pays de Bray Basin; *GF* = Gifhorn Basin; *H* = Harz Mountains. (After VOIGT, 1963.)

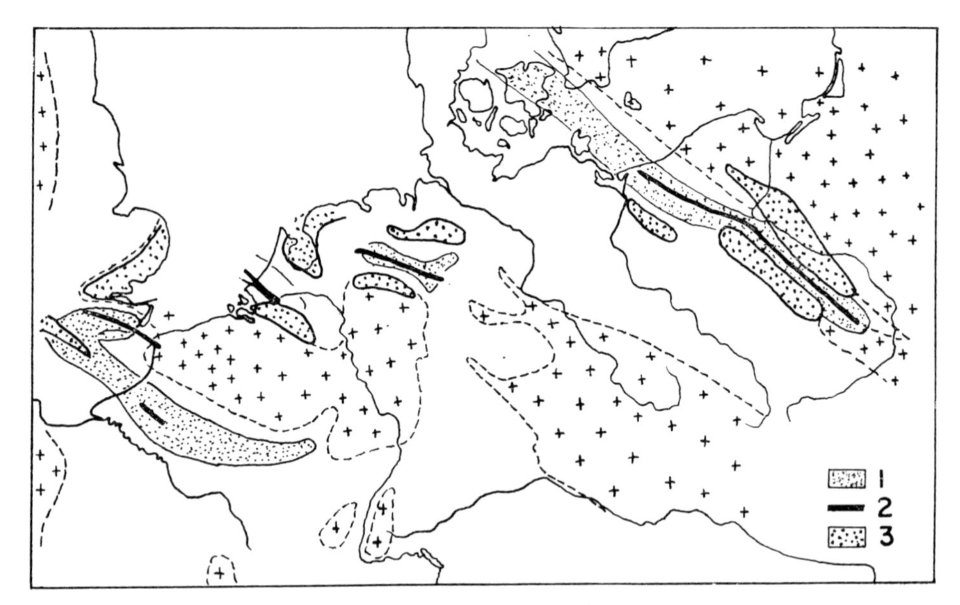

Fig.266. Development of subsequent troughs during Cretaceous and Tertiary times, parallel to some of the primary "Randtröge" of the Jurassic. Compare Fig.265.
1 = inverted "Randtröge" of the Jurassic; *2* = anticlinal structures; *3* = subsequent troughs. (After VOIGT, 1963.)

However, we now have proof that in the younger sedimentary basins similar seesaw movements effectively occurred. Movements, moreover, which appear to be of the order of magnitude required for the explanation by gravity tectonics of the genesis of major fold belts. That is, vertical movements of the order of one to several kilometers, extending over periods of the order of some 50 million years, before reversing in sign.

This knowledge, gained from the younger sedimentary basins, is, in itself, no proof that similar movements have occurred underneath the fold belts too. But such knowledge gained from areas which hitherto had been regarded as stable, will nevertheless be helpful in validating the ad hoc assumptions as to the existence of similar movements in the more mobile fold belts.

HALOKINESIS

Halokinesis, a term coined by TRUSHEIM (1957, 1960) means salt movement, and hence salt tectonics. The latter term does, however, cover a much wider field than halokinesis, and is, for instance, also applied to the "décollement" of an epidermis type of fold belt over an evaporite series. The term halokinesis was introduced to describe the salt tectonics of sedimentary basins which were not subjected to later orogenetic movements. In those basins, the movements due to salt tectonics are limited to the formation of salt domes, salt walls and comparable structures, to their surrounding basins, and even, in extreme cases, to pseudo-anticlines.

The main result of the studies of the related phenomena which have been brought under the term halokinesis is that the development of salt domes is a complex event. The rise of the salt plug is accompanied by the development of a marginal basin around the salt structure, due to the movement of salt in the underlying salt layer towards the plug. In these marginal basins sedimentary thicknesses are thicker than elsewhere in the basin, offsetting the thinning of the salt layer. These locally thicker sediments form as long as the movement of the salt continues.

When the salt structure develops from the initial pillow to a diapiric plug, its surrounding basin may at first become much more acute. But when, in an even later stage, its surrounding sediments become dragged upwards, a younger marginal basin may develop outwards (Fig.267).

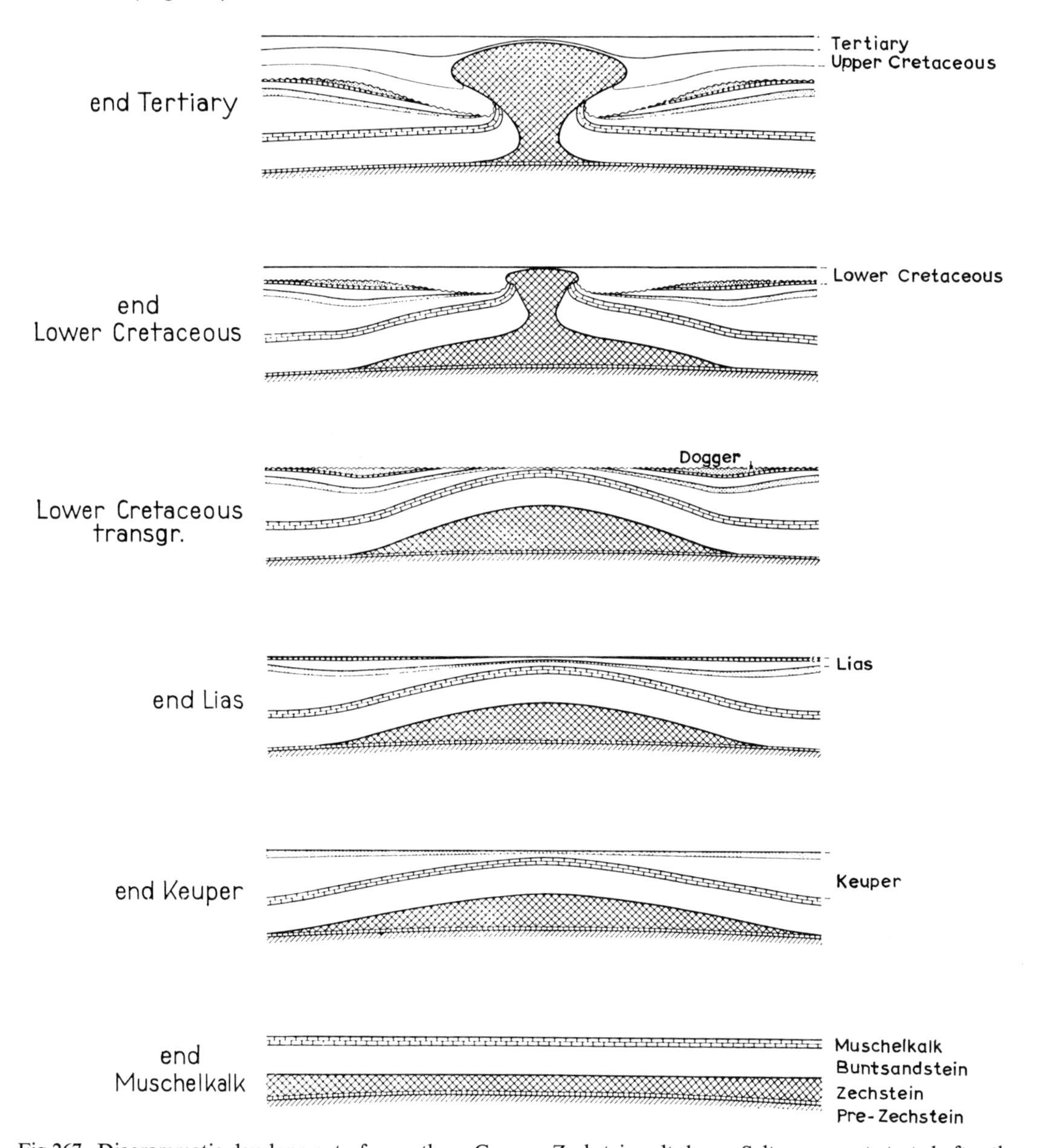

Fig.267. Diagrammatic development of a northern German Zechstein salt dome. Salt movement started after the end of the Muschelkalk and a marginal basin developed during the Keuper and all of the Jurassic. Further diapiric rise of the salt plug led to the well-known tectonization of the sediments surrounding the salt dome. Apart from the Albian transgression (fourth section from below), which is a general event in northwestern Europe, salt domes develop quite individually. According to Trusheim, they do not follow certain specific dates; there is no apparent control by orogenetic phases. (After TRUSHEIM, 1960.)

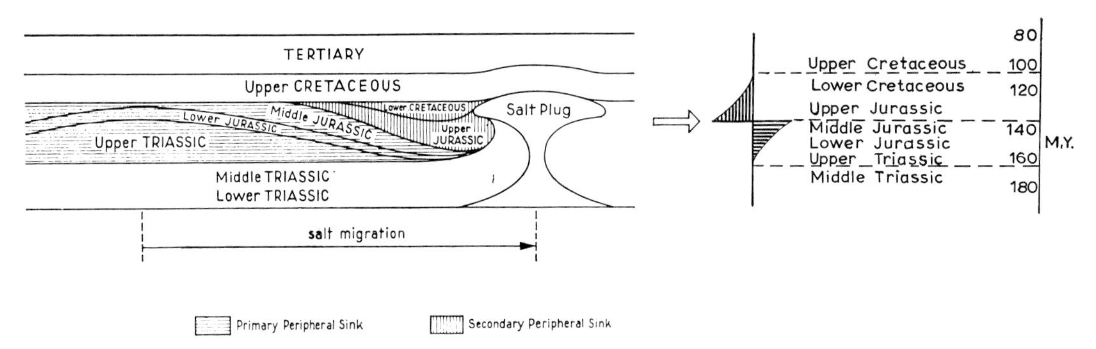

Fig.268. Diagram illustrating usage of the thicker sediments in a marginal syncline around a growing salt dome to calculate salt flow rate. The amount of extra sediment, as derived from the section, is integrated in the graph at right. (After SANNEMANN, 1963.)

The extra thickness of sediments of a certain age around a salt dome conversely dates a phase of movement of the salt. It can also be used to calculate the rate of salt flow, if one equalizes the amount of extra sediment with the amount of salt which must has flowed towards the salt dome during the same time (Fig.268). This rate of salt flow has been estimated at 0.3 mm/year (SANNEMANN, 1963).

A salt dome reaches maturity when all salt that can be mobilized in its surroundings has flowed into the salt structure. In this case second and third generation salt domes may develop further out. These can be distinguished from the original master salt dome because their surrounding basins only began to develop when the basin around the master salt dome had stopped its subsidence. The extra-thick sediments around a first generation salt dome consequently are older than those around a second generation salt dome, etc. (Fig.269).

In this way a history of the movements due to halokinesis could be established for most of the salt dome territory of northwestern Europe (Fig.270).

Trusheim stresses the fact that the growth of individual salt domes and the development of their surrounding basins varies from one structure to the next, or from salt dome family to salt dome family. Halokinesis is, so he says, not "phase-bound". By this is meant

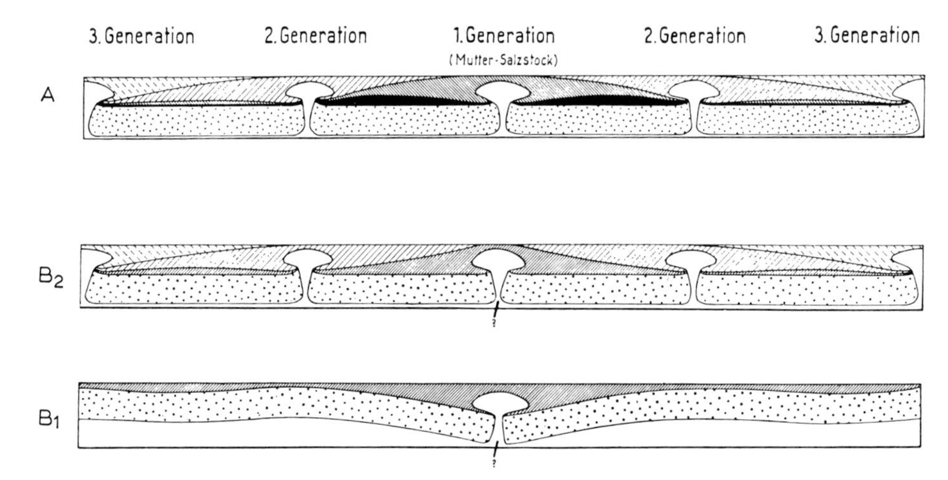

Fig.269. Diagram showing the development of salt dome families.
Cross hatching schematically indicates the thicker sediments of the marginal synclines. The second and third generation daughter salt domes begin to develop only after the growth of the first generation master salt dome has come to an end. (After SANNEMANN, 1963.)

that it is not bound to, not controlled by, distinct orogenetic phases. Or, because Trusheim is a German geologist, more precisely still, not controlled by the orogenetic phases of the Stille codex.

To a German geologist, bottle-fed on the ideas of this codex, this must have seemed an unprecedented departure from a well-established law. To those of us, who have seen so

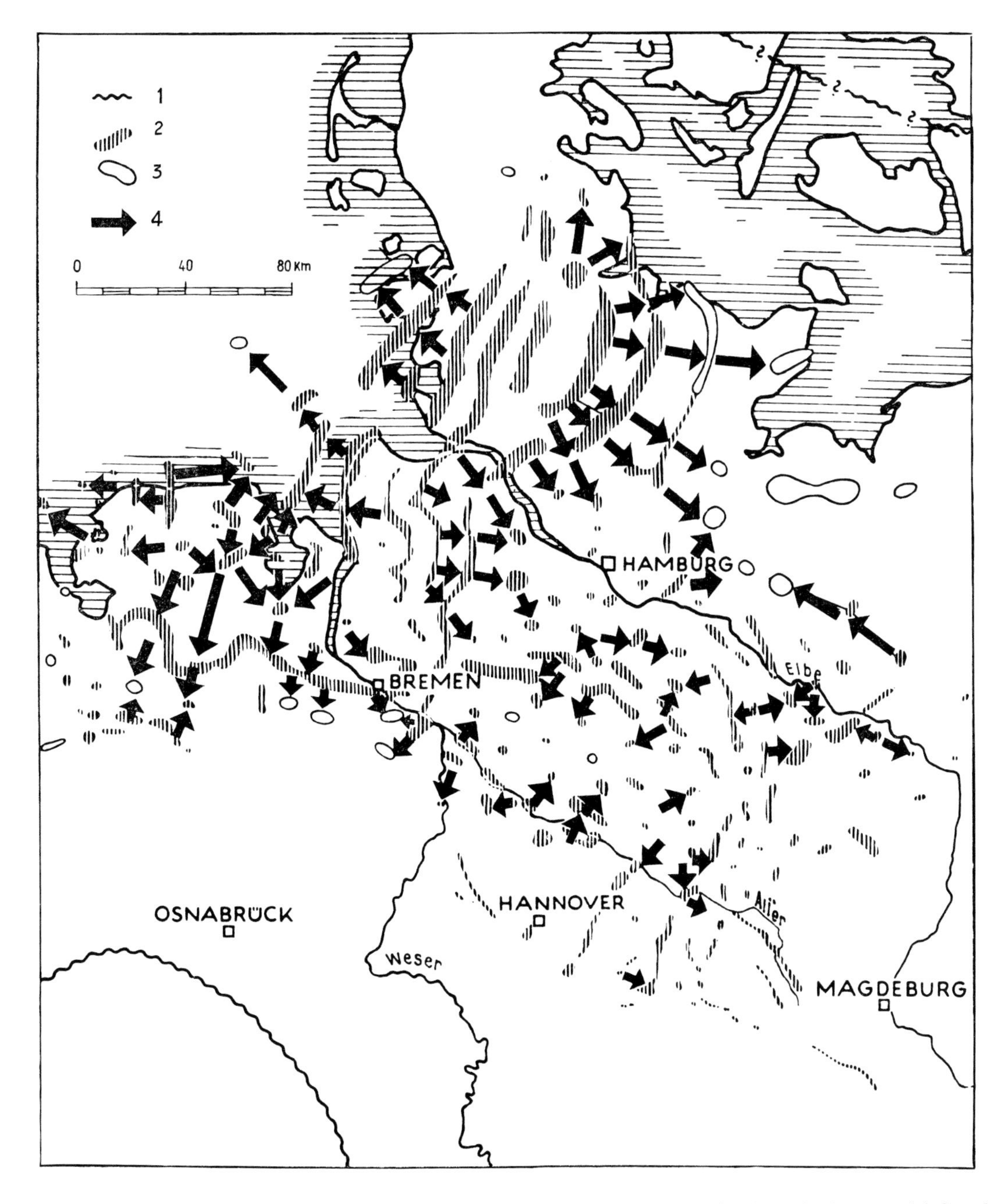

Fig.270. Schematic map of the Zechstein salt basin of northwestern Europe. Showing salt domes and inferred directions of salt movement.

1 = limits of the marine basin of the Zechstein; *2* = salt dome; *3* = salt pillow; *4* = direction of salt movement. (After SANNEMANN, 1963.)

many examples of the relativity of timing of orogenetic movements, it can hardly be surprising.

Swinging the pendulum a little too far, Trusheim denies any tectonic influence on halokinesis, which is thought to be an anorogenetic process (TRUSHEIM, 1957, p.119). Contrarily, I think we may safely assume that basement movements, either faulting, or tilting, or both, must have triggered halokinetic reactions. SANNEMANN (1963) also indicates a hypothetical fracture, at least below the master salt dome of a salt dome family (cf. Fig.269).

Such tectonic control triggering halokinetic movements must almost certainly have taken place in the case of the immense straight salt walls, found in Holstein and extending towards the Danish–German border (Fig.270).

In the same vein TRUSHEIM (1957) himself states that halokinesis on the Pompeckj sill and in the Niedersachsen Basin is markedly different. For him these differences are of a secondary nature. It seems more sensible, however, to attribute them to the primary differences in subsidence and sedimentation, as described in the preceding paragraph.

It is, however, beyond any question, that the effects of halokinesis exceed by far the primary movements that triggered them. It is also beyond doubt that the lucid presentation of all the various effects of salt dome formation and of horizontal salt flow, which together make up halokinesis, has greatly clarified our insight into the structure of the North German Basin.

Comparing for a moment the situation around the Canadian Shield with that found in Europe around the Fennoscandian Shield, one might ask the question: "why is there no stable interior around the Fennoscandian Shield?" The answer being: "because of halokinesis".

In the North German Basin halokinesis has intensified the normal draping phenomena of a post-Hercynian sedimentary cover over a block-faulted basement, to such an effect, that a special type of orogeny, the Saxonic (cf. LOTZE, 1948; METZ, 1957), is often distinguished. Niedersachsen is the type area for this supposedly special fold belt type. In the latest edition of the well-known *Abriss der Geologie* of Kayser and Brinkmann (BRINKMANN, 1961), for instance, it is called a "Bruchfaltengebirge", a faulting-and-folding belt, as distinct from both a "Blockgebirge", a blockfault belt, and a "Falten- und Deckengebirge", a folding-and-nappe belt. Such distinction seems meaningful only in an area so dominated by halokinesis as the North German Basin. It can hardly be extrapolated to other, non-saliferous areas, as a major group of tectonic movements.

REFERENCES

ARNOLD, H., 1963. Regionalgeologische Betrachtungen zum Kreideprofil der Bohrung Münsterland 1. *Fortschr. Geol. Rheinland Westfalen*, 11: 459–468.

BRINKMANN, R., 1960. *Geologic Evolution of Europe*. Enke, Stuttgart, 161 pp.

BRINKMANN, R., 1961. *Abriss der Geologie*. Enke, Stuttgart, 280 pp.

DE LAPPARENT, A. F., 1964. *Région de Paris*. Hermann, Paris, 195 pp.

GEYER, O. F. and GWINNER, M. F., 1962. Der Schwäbische Jura. *Sammlung Geol. Führer*, Borntraeger, Berlin, 40: 452 pp.

GIGNOUX, M., 1950. *Stratigraphic Geology*. Freeman, San Francisco, Calif., 682 pp.

HAANSTRA, U., 1963. A review of Mesozoic geological history in The Netherlands. *Verhandel. Koninkl. Ned. Geol. Mijnbouwk. Genoot., Geol. Ser.*, 21 (1): 35–56.

REFERENCES

KEIZER, J. and LETSCH, W. J., 1963. Geology of the Tertiary in The Netherlands. *Verhandel. Koninkl. Ned. Mijnbouwk. Genoot., Geol. Ser.*, 21 (2): 147–172.

LIENHARDT, M. J., 1961. Étude stratigraphique, pétrographique et structurale du socle anté-permien du bassin de Paris. *Ann. Soc. Géol. Nord*, 81: 223–241.

LOTZE, F., 1948. 100 Jahre Forschung in der saxonischen Tektonik. *Z. Deut. Geol. Ges.*, 100: 321–337.

METZ, K., 1957. *Lehrbuch der tektonischen Geologie*. Enke, Stuttgart, 294 pp.

PRUVOST, P., 1930. Sédimentation et subsidence. *Livre Jubilaire Soc. Géol. France*, 2: 545–564.

SANNEMANN, D., 1963. Über Salzstock-Familien in NW-Deutschland. *Erdoel Z.*, 11: 3–10.

SOCIÉTÉ GÉOLOGIQUE DE FRANCE, 1966. Séance sur la géologie du Bassin de Paris. *Bull. Soc. Géol. France*, 7 (7): 179–340.

STHEEMAN, H. A., 1963. Petroleum development in The Netherlands (with special reference to the origin, subsurface migration and geological history of the country's oil and gas resources). *Verhandel. Koninkl. Ned. Geol. Mijnbouwk. Genoot., Geol. Ser.*, 21 (1): 57–96.

THIERMANN, A. and ARNOLD, H., 1964. Die Kreide in Münsterland und Nordwestfalen. *Fortschr. Geol. Rheinland Westfalen*, 7: 691–724.

TRUSHEIM, F., 1957. Über Halokinese und ihre Bedeutung für die strukturelle Entwicklung Norddeutschlands. *Z. Deut. Geol. Ges.*, 109: 111–157.

TRUSHEIM, F., 1960. Mechanism of salt migration in northern Germany. *Bull. Am. Assoc. Petrol. Geologists*, 44: 1519–1540.

VOIGT, E., 1963. Über Randtröge von Schollenrändern und ihre Bedeutung im Gebiet der mitteldeutsche Senke und angrenzender Gebiete. *Z. Deut. Geol. Ges.*, 114: 378–418.

CHAPTER 18

Younger Volcanics

INTRODUCTION

The younger volcanics of western Europe all belong to the Cenozoic. Within this Period volcanic activity has been the most active during the Mio-Plio-Pleistocene Epochs. It has waned since to such an extent that the only active volcanoes of western Europe are now found in Italy, between Vesuvius and Etna. During the later half of the Pleistocene, and perhaps even in Early Holocene times, volcanic activity must still have been much more wide-spread than at present. Our ancestors must, for instance, have seen volcanic eruptions, not only in the Auvergne in central France, but also as far north as the Eifel and the Siebengebirge in Germany.

Parenthetically, it might be mentioned here that a theory has been proposed according to which the Upper Pleistocene Alleröd Horizon was caused by a volcanic eruption. This charcoal horizon, found all over the sedimentary basin of northwestern Europe has been dated again and again at $\pm$ 11,000 y B.P. It is assumed that a climatic variation during the Holocene had rendered most of the vegetation of the contemporary primeval forests rathei unstable. A volcanic eruption, probably in the Laacher See area in the Eifel, would then have ignited the forest, causing almost continent-wide forest fires. If this would be true, this would have been an after-effect surpassing all other known volcanic calamities. It would even offer an example of an event which could hardly be classed under actualistic philosophy. This deduction would, of course be the more interesting, in view of the very short time which has elapsed since the formation of the Alleröd Horizon. It would be quite unexpected, if it could be proved how, already so short a time ago, actualistic extrapolations broke down. Unfortunately, the volcanic eruption theory as to the genesis of the Alleröd has not been proved. It is just as well possible that, due to climatic change, sporadic forest fires developed during a certain period, whilst, completely unrelated to these fires, one or more volcanic eruptions took place in the Eifel.

Another example of an after-effect of the youthful volcanic activity lies in the success of heavy mineral analysis of Pleistocene sediments in the northwestern European Basin. Index minerals derived from the Eifel volcanism, mainly in the form of hornblende crystals of different hues, offer excellent time markers.

With the exception of northern Europe, it is therefore a fact that volcanism has played an important part during the younger, and even the youngest history of western Europe. Thus, it is appropriate to devote the final chapter to this subject.

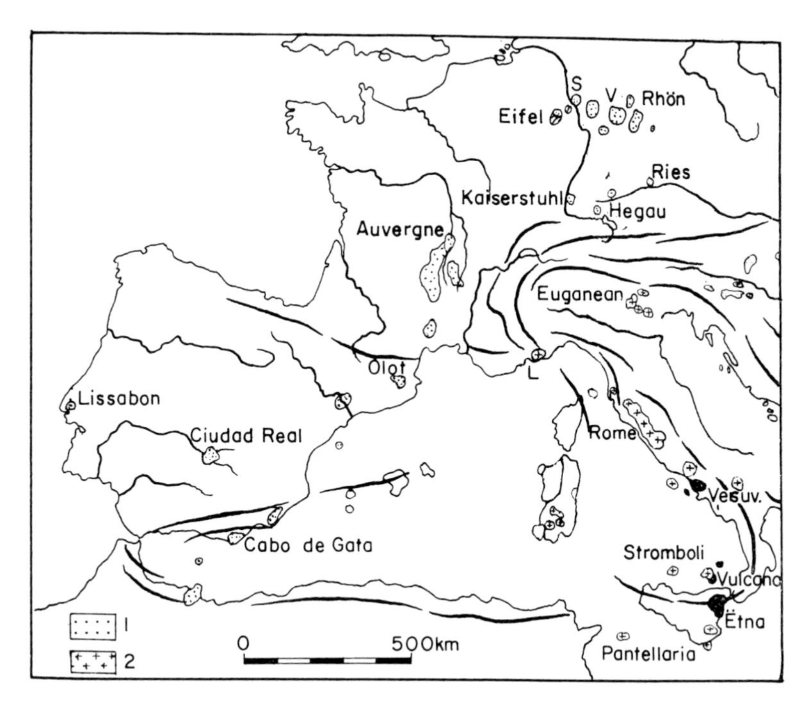

Fig.271. Generalized map of the younger volcanics of western Europe.
1 = outer-Alpine volcanics; *2* = inner-Alpine volcanics; *L* = Ligurian province, near Cannes; *S* = Siebengebirge; *V* = Vogelsberg. (After SCHWARZBACH, 1961b.)

A glance at the map of Fig.271 shows that the younger volcanics fall into two categories. The first is outer-Alpine, the second inner-Alpine. The first group is related to block-faulting movements. The second, though Alpine, is post-orogenetic to the folding, and consequently may well be related to faulting too. But in this case faulting may still be seen as an after effect of the Alpine orogeny, and we may assume that the relation of these volcanoes to the Alpine chain forms their primary control. In contrast, in the first group there is no apparent influence by the Alpine orogeny, and the correlation with block-faulting movements seems to be the only control.

ECONOMIC IMPORTANCE

The economic importance of the younger volcanics has been great, and still is, in the realm of building material. Lava has been extensively used as building stone, as many monuments still attest. The black basaltic lava of, for instance, the cathedrals of Clermont Ferrand and Agde and the Black Gate of Trier, has a charm of its own. In Italy considerable amounts of tuff stone, "tuffo lithico", which can be exploited by a hand saw, or at present with a saw armed by an auxiliary bicycle engine, have been used. In post-war times a large amount of the re-building of Germany was based on the "Trass" of the Laacher See area.

And even after the importance of lavas as building stone has waned, owing to excessive labour cost, several smaller scoriae volcanoes have by now almost dissappeared, because this material is well suited for road bed fills, not being subject to thaw collapse even after prolongued periods of freezing.

The exploitation of steam for energy is, on the other hand, often overrated. The best known field of so-called "volcanic" steam exploitation, Larderello in northern Italy, seems not so much bound to volcanic activity, as to a shallow igneous body, whose heat is effectively sealed off by the Tertiary cap rock of clayey shales, the "argilia scaglioso" of the Apennines (MARINELLI, 1963).

Returning to the importance of the younger volcanics as building material, one might even hold that the invention by the Romans of their special type of brick masonry rests on the fact that this race originally occupied the area around the Bay of Naples. Roman masonry, consisting of a double wall of thin brick, filled with concrete, might well have originated from the fact that they had at hand the natural cement pozzolan. This special type of tephra is named after the coastal town of Pozzuoli, situated in the Phlegrean Fields west of Naples. The Etruscans, which at that time occupied the area around Rome, built their mausoleums out of the tuff stone of the volcanoes north of Rome, and, not having at hand a natural cement, never practised the art of masonry.

OUTER-ALPINE YOUNGER VOLCANICS

The main provinces of the outer-Alpine younger volcanics are found in Germany and in France. From southern France they can be followed southwards, but in much smaller quantity, and very much interrupted, through eastern Spain (Olot, Cabo de Gata), and even to the northern coast of Africa, in the neighbourhood of Melilla (Fig.271).

In Germany the younger volcanics occur in a northern and a southern province. The northern group begins in the west with the Eifel volcanoes, and continues with the Laacher See and Neuwied Basin. On the other side of the Rhine follow the Siebengebirge, the Westerwald, the Vogelsberg with the related volcanism in and around the Hessen graben, the Rhön and some smaller volcanic points still further east (Fig.272).

In the western part of this province, from the Eifel up to and including the Siebengebirge, the composition of the volcanic products varies strongly, as a whole, and although numerous basalt volcanoes are present, trachytes and related acid rocks predominate as to volume. The easterly volcanism of this northern province is, on the other hand, predominantly basaltic.

The southern German province contains much smaller units than the northern. It begins in the west with the complex Kaiserstuhl, situated in the Upper Rhine graben, and largely covered by loess and vineyards. Then follow the Hegau volcanoes, mostly topped by medieval castles, and the scattered, small tuff dikes of the Schwäbische Alb. Further east lies the small Steinheim Basin, and the province ends with the large Ries Basin, which is now interpreted as an astrobleme.

In France a northeastern province is delineated, composed of small, scattered basalt volcanoes in the Vosges Mountains (JUNG and BROUSSE, 1962). Were it not for the French–German border, these would have been grouped with the southern German province (Fig.273).

The main area of younger volcanics in France lies in and around the eastern part of the Massif Central. Within this province we find the two large central volcanoes of the Cantal and the Mont Dore, the Auvergne and the Velay districts, and several other subsidiary districts (Fig.280).

From the southeastern edge of the Massif Central the volcanism continues straight south, through the Causses and into the Bas Languedoc, down to the Agde Volcano on the coast of the Mediterranean. South from Agde, it is possible, as we saw, to see a further prolongation of this line in the scattered volcanism of eastern Spain and as far as the southern coast of the Mediterranean, near Melilla.

"HEBUNG–SPALTUNG–VULKANISMUS"

"Hebung–Spaltung–Vulkanismus", rising–fissuring–volcanism, is the classic title of the paper by CLOOS (1939), in which the relation between tectonics and volcanism in outer-Alpine and related regions was set forth. Fig.274 gives the frontispiece of this publication. Hackneyed though this drawing may be by now, it still has to find its place in this text,

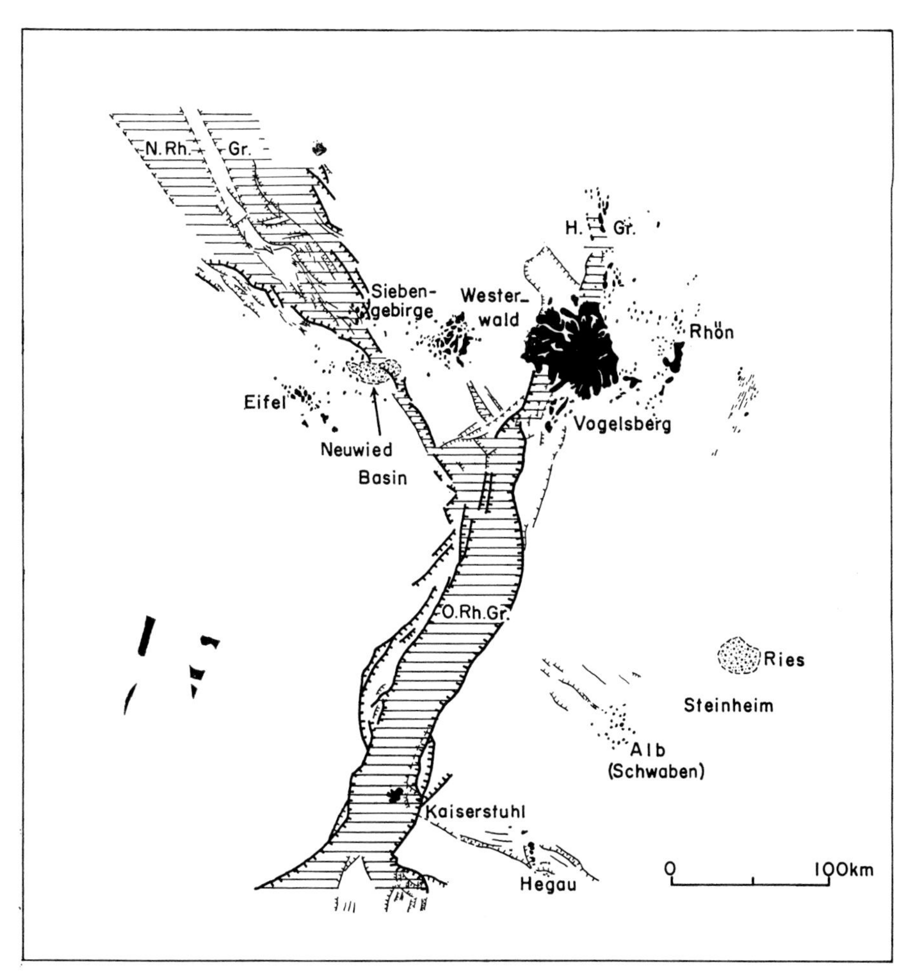

Fig.272. Generalized map of the younger volcanics of Germany. Showing the relationship between major faulting and volcanism. For the relation between volcanism and smaller faults in, say, the Eifel or the Hegau, consult the relevant maps in the original publication.
N.Rh.Gr. = "Nieder" or Lower Rhine graben; *O.Rh.Gr.* = "Ober" or Upper Rhine graben; *H.Gr.* = Hessen graben. (After CLOOS, 1939.)

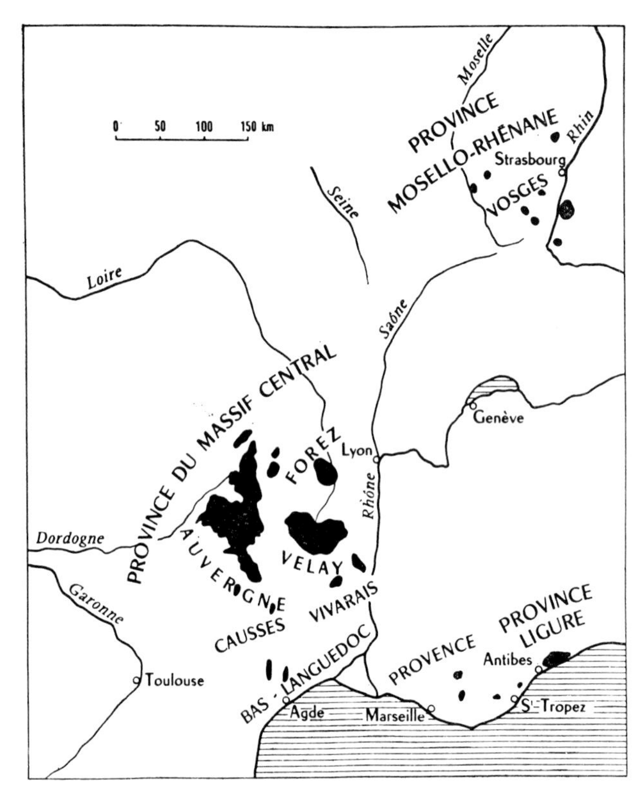

Fig.273. The younger volcanics of France. The Ligurian province is inner-Alpine, the others are outer-Alpine. (After JUNG and BROUSSE, 1962.)

because relations such as those thrashed out by Cloos form the control of the outer-Alpine volcanism of western Europe.

An important fact, often too little stressed, is that the major volcanism does normally not occur within, or along the borders of, the major graben. Apparently it is the fissures along which no larger movements took place that remained open and offered the better

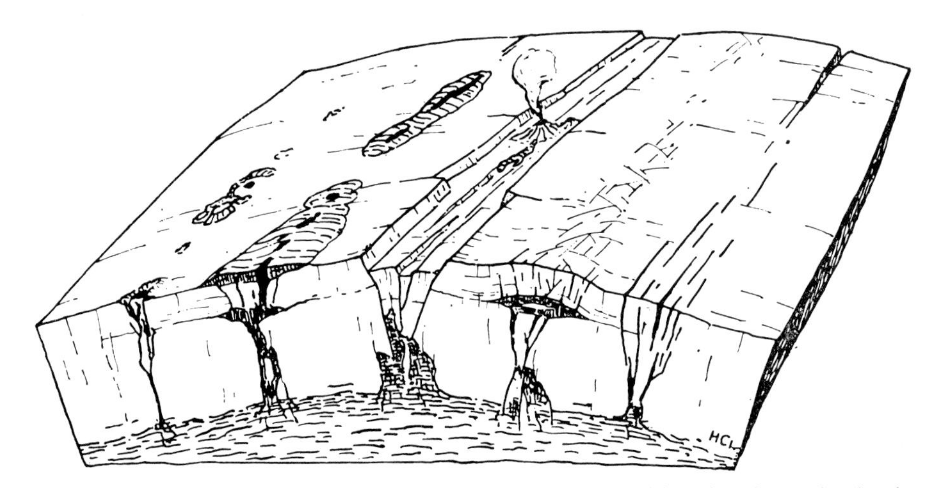

Fig.274. Schematic block diagram, showing the relation between rising, fissuring and volcanism. (After CLOOS, 1939.)

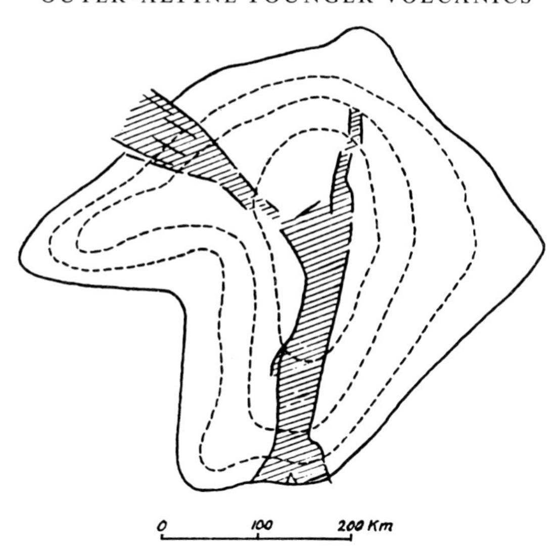

Fig.275. Schematized map of the updomed shield of the "Rheinische Masse" and of its major grabens. Full line: approximate position of the coast line at the beginning of the Upper Cretaceous. Dotted lines: schematized isohypses of the base of the (eroded) sedimentary cover, as constructed from the centripetal dip of the Mesozoic in the surrounding basins. (After CLOOS, 1939.)

communication with the underlying magma, whereas the repeated movements along the major graben faults tended, on the other hand, to seal them off.

The relation between rising and fissuring for the "Rheinische Masse" (p.422) follows from Fig.275, chosen amongst several figures — some of them much more detailed — supplied by CLOOS (1939). There is no doubt that this correlation exists. But Cloos' explanation of this correlation will, I think, satisfy most geologists no more at present.

Cloos imagined tangential forces which very gradually squeezed up these large dome-like structures over rather long periods. Through the updoming, lengthening occurred as a secondary feature, leading to the development of the centrally situated major graben structures by tension. With the newer insight as to the relatively minor importance of tangential compression for tectonics, and as to the much larger importance of vertical instability of the crust, we would now tend to consider the general rising of the dome and the formation of antithetic graben structures as due to differential vertical movements of the crust. This does, of course, not offer any explanation as to why these vertical crustal movements have occurred, nor why they have occurred in so regular a pattern.

COMPOSITION OF THE OUTER-ALPINE VOLCANICS

From the enumeration of the northern province of the volcanism of Germany, it follows already that the magmatic composition varies widely. The more acid magmas are thought to be anatexitic, whereas the basaltic magmas are often thought to be formed mainly from primary, subcrustal material. But even so, it appears as if most of the more basic magmas are sclerotic, and also contain re-melted crustal material in varying amounts.

The tendency is, of course, to alkaline, and varies from strongly to mildly atlantic. For the French volcanics a few more indications will be supplied (pp.444, 446, 447), taken from JUNG and BROUSSE (1962), but for more precise information the reader is referred to the regional literature.

YOUNGER VOLCANICS OF GERMANY

In view of the interrupted occurrence of the volcanics of Germany, in a northern and a southern province, and moreover in separate districts within these provinces, a short description of the various districts seems in order.

NORTHERN PROVINCE

Starting with the more important northern province, the westernmost district is that of the Eifel. We note that in this district the volcanism is added to the much earlier Hercynian history. It is developed as a great number of small units, which either stand out as small steep hills on top of the post-Hercynian peneplain or form smallish, deep holes, mostly occupied by lakes, in this same peneplain.

The hills are formed by basalt volcanoes, which require no further comment. The holes in the peneplain, on the other hand, are the famous Eifelmaare, which have given their name to a group of volcanic phenomena.

Maar in the local dialect means "lake" (Dutch: *meer*), so the word does no more than describe the small, roundish lakes so typical of a certain zone of the Eifel. These lakes are, however, without exception, due to volcanism. They must have resulted from an explosive activity by a magma that was so viscous that, apart from an encircling tuff wall which sometimes developed, no volcano was formed. A blast hole was all that remained.

Magmas of such high viscosity are only found amongst the very acid ones, as is well known from volcanology. Ironically enough, most of the Eifelmaare have also served as eruption centres for basaltic magma, a duality of a single volcanic structure, also well known from other volcanic districts. In the Eifel, however, the fact that what one sees at the border of the Maare is the basalt, whereas very little is found of the acid tuffs from the eruption which originally must have formed the holes, has led to quite an extensive discussion as to their origin.

The separation of basic and acid magmas is in part geographic too. The main branch of the basalt volcanoes trends northwest–southeast, that is radial to the dome of the Rheinische Masse. The main branch of the Maare, on the other hand, trends almost north–south (cf. Fig.276 and HOPMANN et al., 1959).

The next area, that of the Laacher See and the Neuwied Basin, is similar to the Eifel volcanism in the occurrence of acid and basic eruptions. But in contrast with the Eifel the production of acid volcanics has been much larger than that of basaltic rocks. The Laacher See, whose explosive activity persisted up to the Alleröd time, forms a beautiful caldera, and even the Neuwied Basin might well be, at least in part, of volcanotectonic origin (Fig.277, and FRECHEN, 1953).

The next district is that of the Siebengebirge (FRECHEN, 1962). The seven mountains from which this district takes its name are but the most prominent of the many basaltic volcanic structures. As in the Eifel, many of these stand on top of the post-Hercynian peneplain. There are, however, also a number of sills and necks, now uncovered by post-volcanic erosion, or by quarrying. One of the best known examples is the composite sill of the Grosser Weilberg (CLOOS, 1947). Apart from this, there has occurred acid volcanic activity too, resulting in widely spread trachytic tuffs. The most widely known elements of the Siebengebirge do, however, not belong to the basalt cones. They are two smaller structures, situated on either side of the Rhine, the Rodderberg and the Drachenfels (Fig.278).

The Rodderberg (RICHTER, 1942), which stands on a Pleistocene terrace, is so youthful that this scoria volcano has still a perfectly preserved morphology. A low ringwall entirely surrounds a large crater bottom.

The Drachenfels is formed by the remnants of a trachytic neck, which rises steeply from the eastern side of the Rhine channel. It is related to the trachytic tuffs mentioned above. Its large, tabloid sanidine crystals are so strictly orientated, that a rockwall will light up all at once, when the sun catches the surfaces of the sanidines, as they all reflect the light at the same angle. H. Cloos and E. Cloos (cf. CLOOS, 1936) used this orientation of

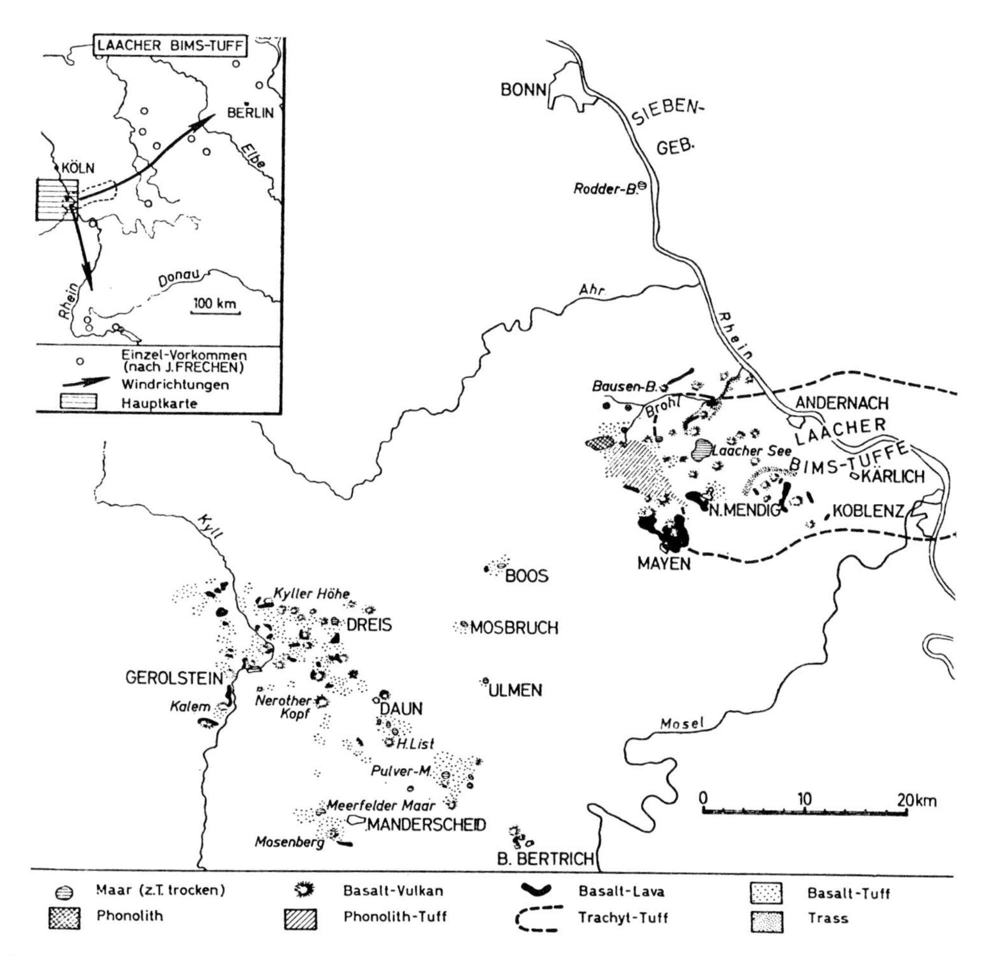

Fig.276. Schematic map of the younger volcanics of the Eifel and of the Laacher See and Neuwied Basin. In the Eifel the main branch of the basaltic volcanoes trends northwest–southeast, from Kyller Höhe to Bad Bertrich. That of the Maare, on the other hand, trends almost north–south, from Boos Maar to Meerfelder Maar. Several Maare do, however, accompany the main branch, such as the Daun and Dreis Maare. The Neuwied Basin, south of Andernach, in part possibly a volcanotectonic depression, is filled with tuffs of the Laacher See eruption. When windblown, they form layers of pumice, the *Bimstuffe*. The majority is, however, waterlaid or re-deposited by water; the so-called *Trass*. It is the latter, which, with varying additions of lime, is now synthesized to the almost universal building "brick" of western Germany. Windblown Laacher See ashes have been found over a much greater area, even beyond Berlin (see inset map). They have, however, no economic importance. (After SCHWARZBACH, 1961.)

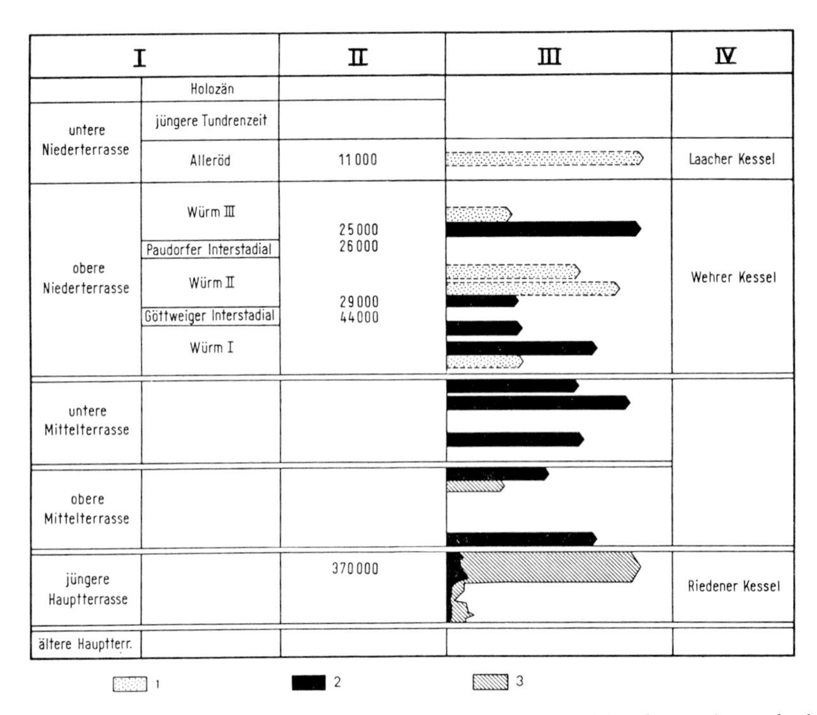

Fig.277. List of the major eruptions of the Laacher See Volcano, as dated by the K–Ar method. *I* = stages; *II* = K–Ar dates (B.P.); *III* = volcanic activity; *IV* = eruption centres; *1* = pumice ("Bimstuffe"); *2* = basaltic tuff and basalt; *3* = phonolitic tuffs and phonolites. (After FRECHEN, 1959.)

the phenocrysts in one of the earliest "microtectonic" studies, to map the flow lines of the viscous magma during the intrusion (Fig.279).

The more easterly districts of the northern province, the Westerwald, the Vogelsberg (SCHOTTLER, 1937), and the Röhn, are predominantly basaltic, and will not be further detailed here.

As a general remark, it may, however, be said that these complexes are built up from a great number of individual eruption centres, whereas individual lava flows have in general been of rather small dimensions. Moreover, necks, sills and in general "Subvulkane" (CLOOS, 1936), subvolcanoes, abound (see also SCHENK, 1964, who uses the new term "Subfusion" for an intrusion). It is not a plateau-basalt area, in which lavas make up almost the total volume of the volcanic products.

I believe this is the main reason, why, in Germany, the volcanic nature of basalts has for such a longtime been contested. Whereas Nicolas Desmarest in 1777 had already described the volcanic nature of basalts in the Auvergne, Abraham Gottlob Werner could, in Germany, until his death, forty years later, succesfully defend his theory on the aquatic origin of basalt. This also explains how it is possible that Leopold von Buch, after having visited the active volcanoes of Italy and the extinct volcanoes of the Auvergne, still could maintain a neptunist origin of the basalts of Germany, as had been proclaimed by his teacher Werner (cf. GEIKIE, 1960).

The volcanic nature of basalts has of course been generally accepted since. This is not the place to go into the battle between neptunists and plutonists. But, curiously enough, this same character of the volcanics of Germany, in which the relation between volcanoes and basalt is not very apparent, has, in later years, led KLÜPFEL (cf. 1953) to assume that

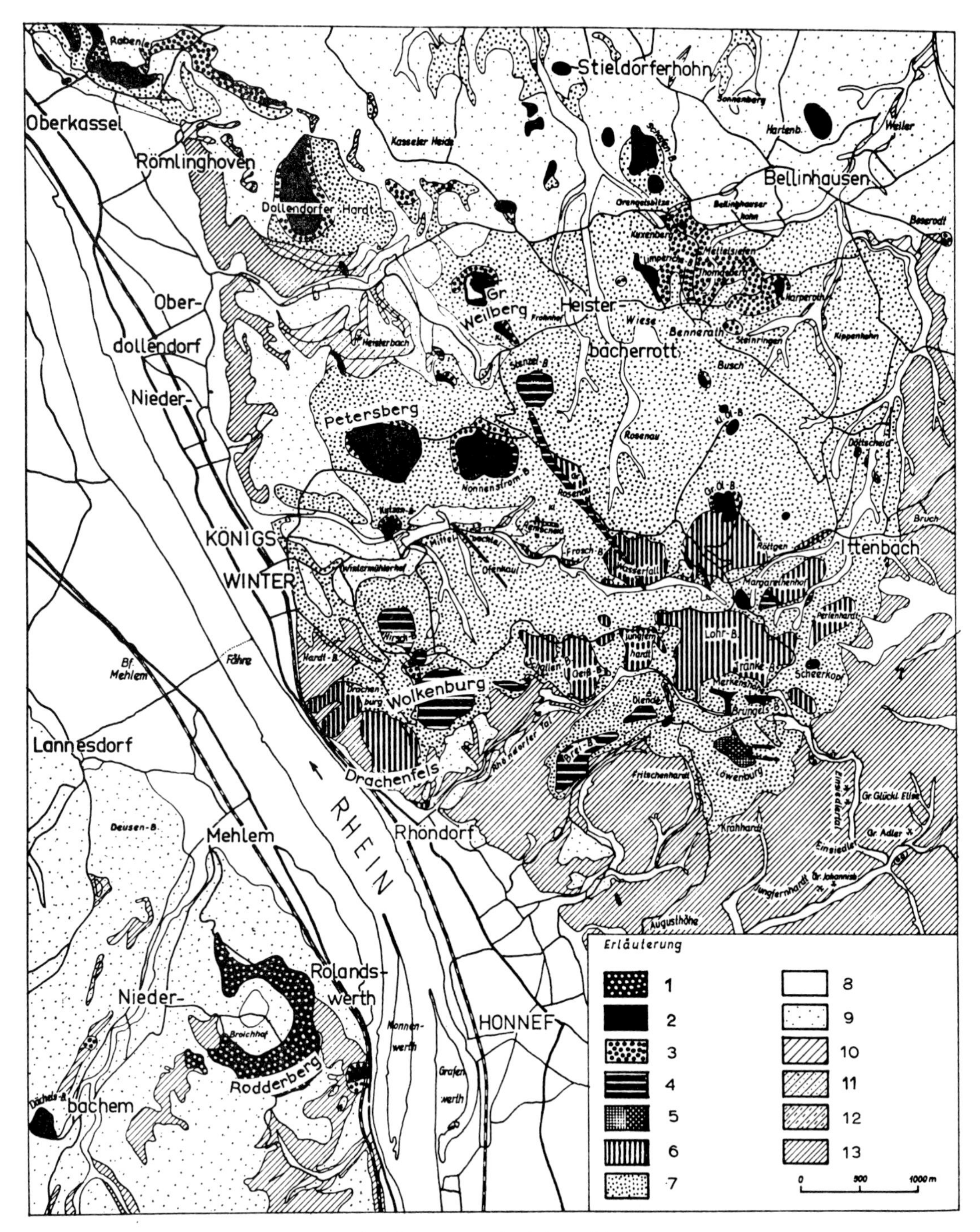

Fig.278. Map of the Siebengebirge.
1–7: volcanics. *1* = tephritic basalt tuff; *2* = alkalibasalt; *3* = basalt tuff; *4* = latite; *5* = nepheline–latite; *6* = trachyte; *7* = trachyte tuff.
8–13: sediments. *8* = Recent; *9* = Pleistocene Rhine terraces; *10* = clayey Tertiary deposits; *11* = sandy Tertiary deposits; *12* = post-volcanic Tertiary; *13* = Lower Devonian, basement of the Rheinisches Schiefergebirge. (After FRECHEN, 1962.)

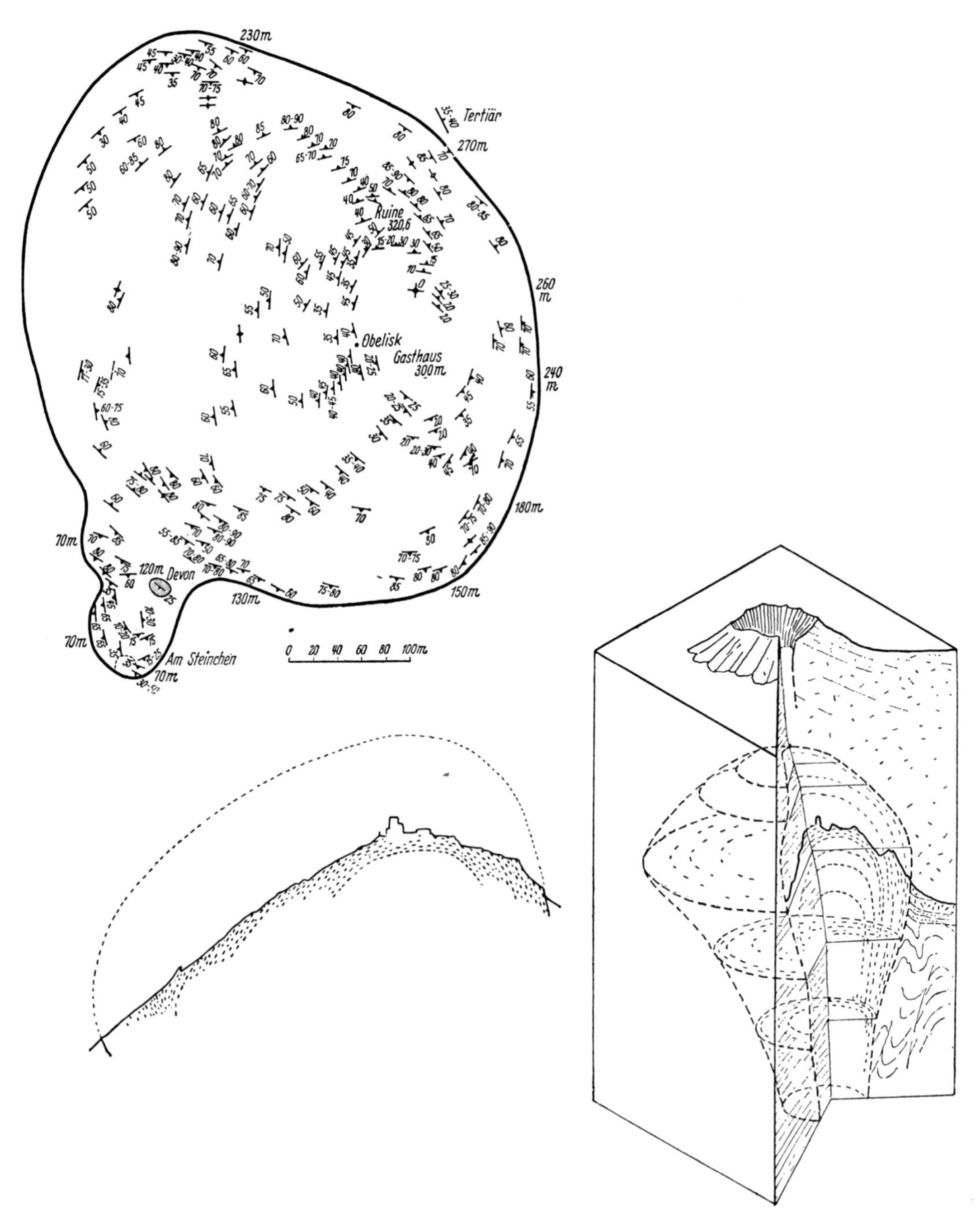

Fig.279. Map, section and block diagram of the trachytic intrusion of the Drachenfels. Showing the flow lines of the viscous magma during the intrusion, as derived from the regular orientation of the tabloid sanidine phenocrysts. The interpretation given in the block diagram of a former volcano situated very high above the present surface is doubtful. It is not easy to see how erosion should have been able to remove so much of the superstructure. An interpretation as a neck or dome seems more probable. (After CLOOS, 1936.)

there are no Tertiary basaltic lavas at all. A generalization as false as the earlier trend to interpret every subhorizontal basalt as a lava.

SOUTHERN PROVINCE

We can be shorter in our description of the southern province of the younger volcanics of Germany, because there is so much less of them.

This is already clear in its westernmost element, the Kaiserstuhl (GEOLOGISCHES LANDESAMT IN BADEN–WÜRTTEMBERG, 1957, 1959; WIMMENAUER, 1957, 1959, 1962a, b, c). The Kaiserstuhl occupies a singular position, because with the adjoining smaller Limburg volcano, it is the only one situated on the floor of the Upper Rhine graben. It is, moreover, quite complex and consists of quite peculiar products of magmatic differentiation and/or assimilation, which have won its fame. But, as stated already in the introductory remarks, it is almost totally covered by loess. If quarrying goes on at the present rate in the few exposures left, it will not be long before more of the Kaiserstuhl rocks are to be found in the petrological musea, than in the field.

The next district, that of the Hegau basaltic and phonolitic volcanoes, is quite commonplace. Although it must be conceded that the dark basalt hills, many of which are thought to be sills or necks originally, form a spectacular landscape, being offset by the white limestones of the Jurassic. This brings us to the tuff dikes of the Schwäbische Alb, southeast of Stuttgart. These tuff dikes, the "Tuffschlote" of CLOOS (1941), over 160 of them, occur in an oval area of about 30 km by 40 km. They contain tuffbreccia of very variable grain-size, from fine-grained to blocky, with individual blocks of the order of several tens of meters in diameter. The latter are always formed by the white Jurassic limestone.

With Cloos, we may interpret the genesis of these "Tuffschlote" as due to a single volcanic event, his "Schwäbische Vulkan". It is supposed that a very gaseous magma only just arrived at piercing the surface in a larger number of narrow pipes. Moreover, we may assume with COE (1965) that fluidization was an important factor in the process.

Skipping the Steinheim Basin, the origin of which is questionable, we then arrive at the large basin of the Nördlinger Ries, which was formerly always cited as the easternmost element of the southern province of younger volcanics. It has now been proved to be an astrobleme by the occurrence of coesite and stiskovite (SHOEMAKER and SHAO, 1961; PREUSS, 1964).

PREUSS (1964) not only gives a full description of the Ries, but also cites the volcanic and meteoritic theories on its formation. The objections to the new interpretation take a form of circular reasoning. Because the Ries lies — more or less — in line with the volcanic structures of southern Germany, a "law" has formerly been deduced on the occurrence of volcanic events. Because of this law it has now become impossible to see the Ries as anything other than a volcanic event. But, as I see it, the only thing one can do, is to reproach the meteorite to have been so impolite as to fall at a point where a volcano could just as well have been. And, moreover, at a time (the Late Miocene) when the volcanism of southern Germany was particularly active.

OUTER-ALPINE YOUNGER VOLCANICS OF FRANCE

Only the main province, that on and around the eastern part of the Massif Central, will be dealt with here. Its volcanology has recently been summarized by BROUSSE (1961) and by JUNG and BROUSSE (1962), who also give maps and bibliographies.

Its backbone is formed by the volcanic part of the Auvergne. From north to south this is formed, first by the range of small volcanoes forming the Chaîne des Puys, then by the central volcano of the Mont Dore, thereupon the basalt lava plateau of the Cézallier, and finally, biggest of all, the central volcano of the Cantal. The main source of information on the volcanic Auvergne is still to be found in JUNG (1946).

Apart from the Auvergne a number of smaller districts make up the Massif Central province. Largest of these is the Velay, whose volcanic products coincide with the borders of the former province of that name. The others are either formed by a smaller or larger number of separate volcanic structures, or by basalt lava plateaus (Fig.280).

COMPOSITION OF THE VOLCANICS

With the exception of the lava plateaus, the composition of the volcanics varies widely. However, within a certain district, the variations follow either only one, or two patterns (JUNG and BROUSSE, 1962). Their chemical variation has been expressed in Jung's alkalinity parameter $\frac{K + Na}{K + Na + Ca} \cdot 100$, against Si.

In his group *I*, of which the Cascade Range has been taken as the type area, basalts are found to range to rhyolites via andesites and dacites. In group *II*, with Ascension Island as the type area, basalts range to rhyolites by way of trachyandesites and trachytes. In the other two groups basalts and basanites range via tephrites to phonolites (Fig.281).

It is clear that alkalinity increases from group *I* to group *IV*. But I think it a pity that every petrographer of note has to invent his own parameters. This leaves us, geologists, with the task of calculating, say, Niggli values, to compare with Jung values, or vice versa.

However, the result seems important, because it now transpires that the various districts differ in their differentiation. Only the types *II–IV* occur in the Auvergne. Type *II* occurs singly in the Chaîne des Puys, whereas type *III* occurs singly in a number of districts. Type *IV* only occurs together with type *III*, and then only in a small part of the Limagne. Types *II* and *III*, at last, occur together in the central volcanoes of the Cantal and the Mont Dore. These consist both of an older trachytic and a younger basaltic part (Fig.282).

Fig.280. Map of the twelve districts of the Massif Central province of the younger volcanics of France. (After BROUSSE, 1961.)
The districts are the following:

(*1*) volcanism along the "Sillon houiller"
(*2*) Sioule chain (northwest of Chaîne des Puys)
(*3*) Chaîne des Puys
(*4*) Mont Dore
(*5*) Cézallier and Margéride
(*6*) Cantal
(*7*) Aubrac
(*8*) Causses and Bas Languedoc
(*9*) Limagne
(*10*) Forez
(*11*) Velay
(*12*) Coirons

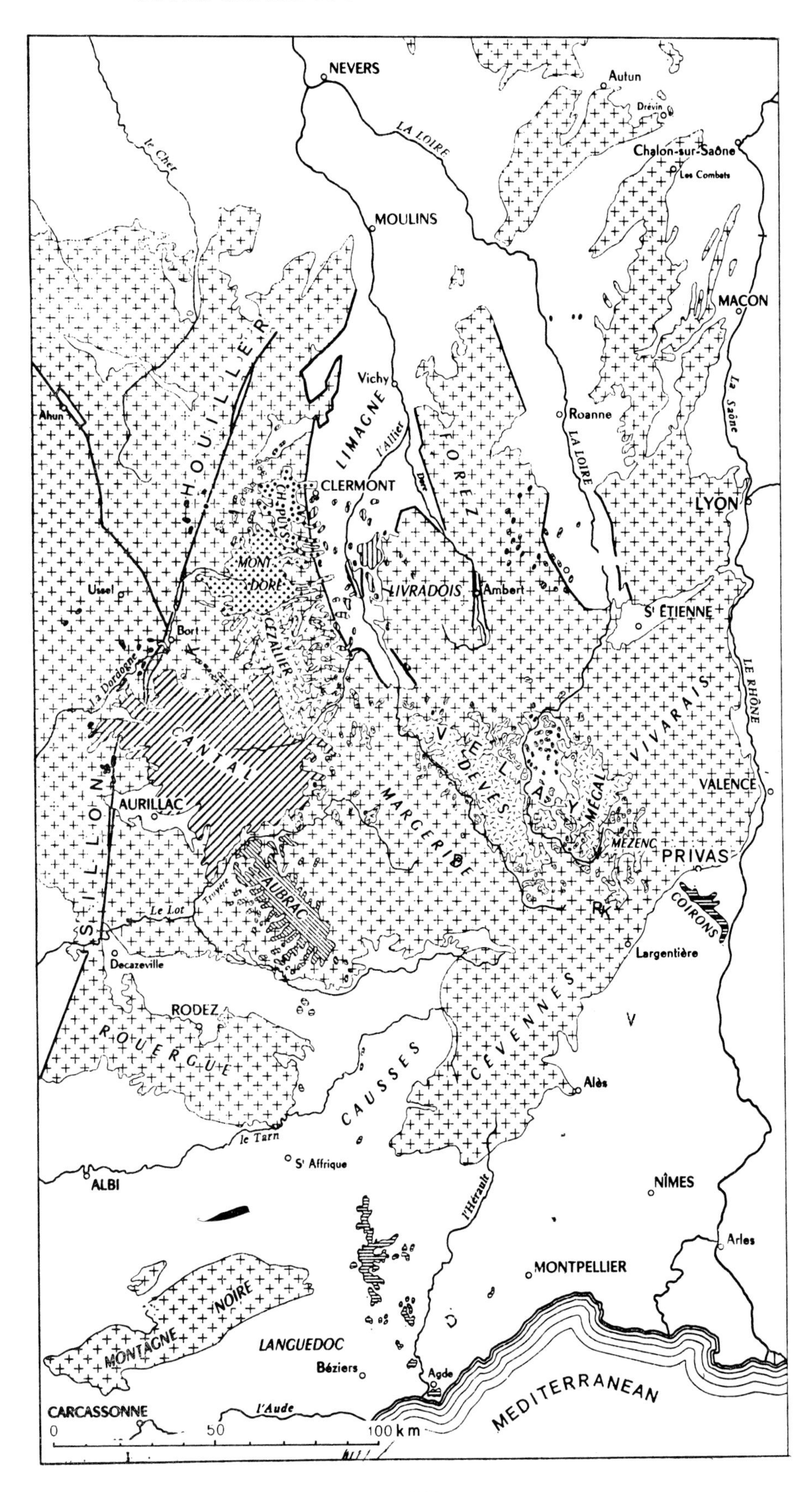
NEVERS
Autun
Drévin
LA LOIRE
le Cher
Chalon-sur-Saône
Les Combets
MOULINS
MACON
HOUILLER
Vichy
LIMAGNE
l'Allier
Ahun
Roanne
FOREZ
LA LOIRE
La Saône
CLERMONT
CH. PUYS
LYON
MONT DORE
Ussel
LIVRADOIS
Ambert
St ÉTIENNE
Bort
CEZALLIER
La Dordogne
LE RHÔNE
CANTAL
VELAY
DEVES
VIVARAIS
MÉGAL
SILLON
AURILLAC
MARGERIDE
VALENCE
MEZENC
PRIVAS
AUBRAC
Le Lot
Truyère
COIRONS
Decazeville
Largentière
CÉVENNES
RODEZ
ROUERGUE
CAUSSES
Alès
le Tarn
St Affrique
ALBI
l'Hérault
NÎMES
Arles
MONTPELLIER
MONTAGNE NOIRE
LANGUEDOC
Béziers
Agde
MEDITERRANEAN
CARCASSONNE
l'Aude
0
50
100 km

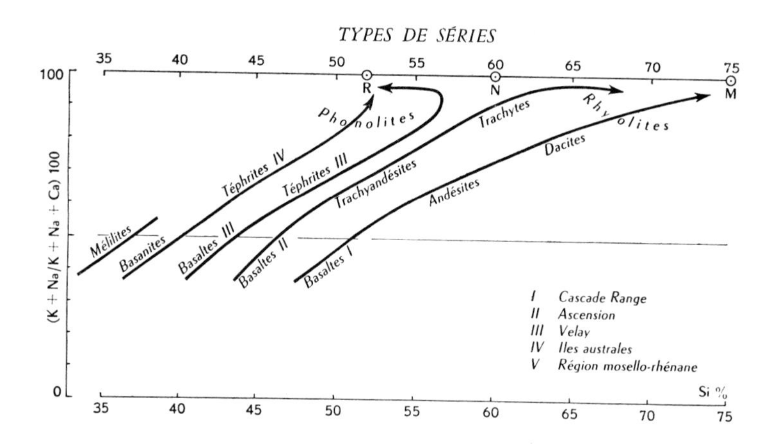

Fig.281. Differentiation of various volcanic areas, expressed in Jung's alkalinity diagram (After JUNG and BROUSSE, 1962.)

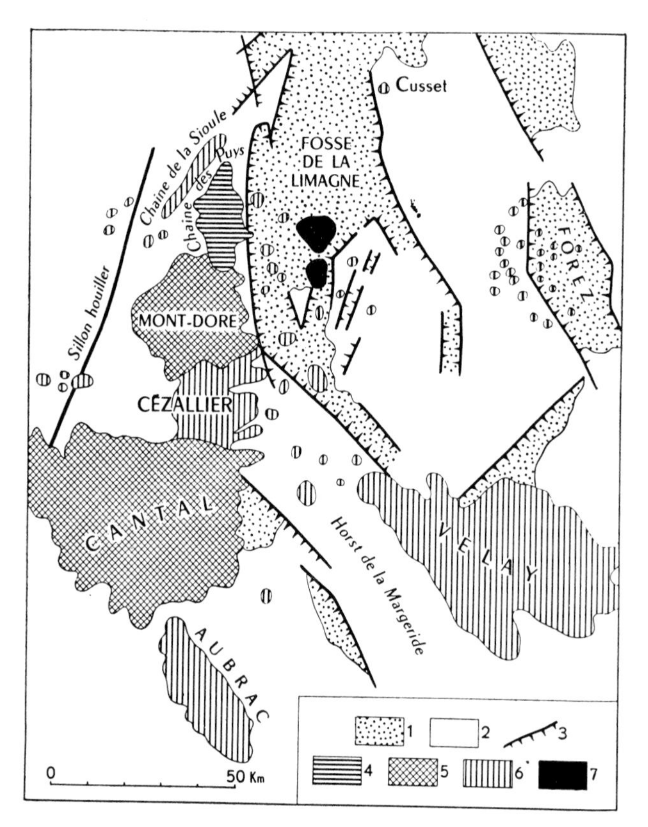

Fig.282. Schematic map of the younger volcanics of the Auvergne. Showing the types of differentiation of individual districts.

1 = Tertiary graben; *2* = Hercynian basement; *3* = faults; *4* = volcanics following differentiation type *II* (cf. Fig.281 only; *5* = same for types *II* + *III*; *6* = same for type *III* only; *7* = same for types *III* + *IV*. (After JUNG and BROUSSE, 1962.)

The macrovolcanological characteristics, which have been mainly used to delimit the various districts of this volcanic province, are consequently reflected by the chemical differentiation of the magma. Of course, the variations and similarities cited are purely descriptive. The lines of Fig.281 do not indicate that those acid rocks at one end of the line arose from the basaltic rocks at the other through crystal differentiation.

The large volume of acid rocks in several of the districts of the Auvergne even seems to indicate the probability of the existence of at least two magmas, one acid and anatexitic, the other basic and juvenile. The differentiation of the actual rocks within a given district would in that case be mainly determined by the amount of mixing. Whereas the differences found between the districts might well result from primary differences in the composition of the crust at or around that place, which would be reflected in the composition of the anatexitic magma.

THE AUVERGNE

When thinking of the Auvergne volcanism, one somehow always first gets in mind the Chaîne des Puys near Clermont-Ferrand. This may be legitimate for several reasons. For one thing, because this chain of small volcanoes dates only from the Quaternary, the morphology of its cones and craters, and the surface of its lavas, is still so well preserved that it seems as if volcanic activity had stopped only yesterday. And did not Blaise Pascal in 1646 have the experiments with a mercury barometer, first carried out on a Paris church spire, repeated on the Puy de Dôme proving that air was thinner at higher altitudes? And did not Alexander Von Humboldt convince himself only from the example of the Chaîne des Puys much more than from the active volcanoes of Italy, that volcanism clearly had occurred in earlier geologic periods? And that consequently the neptunist views of his teacher Abraham Gottlob Werner, if only in France, had to be wrong?

Nevertheless, it is good to remember that the Auvergne contains older and larger volcanic structures. The most important being the central volcanoes of the Cantal and the Mont Dore. It is therefore with these volcanoes that we will start our description of the Auvergne volcanics.

CANTAL AND MONT DORE

From afar both the Cantal and the Mont Dore (JUNG, 1946; ROQUES, 1958) give the impression of a shield volcano. Large, symmetric, round in form, with a low angle dip of the flanks of less than 5°, they have the same form as the bigger of the Icelandic shield volcanoes. But neither of them is still active, and on closer inspection one realizes that they represent no more than the ruins of former volcanoes. Nothing is left of the former craters, whereas the highest peaks of the present are chance remnants, left by the effects of fluviatile and glacial erosion. None of the present highest points is related to former eruption centres.

The outward resemblance with a shield volcano resides in the fact that the activity of both Cantal and Mont Dore, with the exception of smaller posthumous eruptions, terminated with the production of a large volume of basalt lavas during the Late Pliocene and Early Quaternary (Villafranchian). These lavas spread all over, and beyond the earlier volcanic structures. Upon cooling they assumed the typical low angle surface of basalt lavas. And so these volcanoes do really have an outer shell, which resembles a shield volcano. The basalt lavas now form the high altitude fertile plains, covered by lush meadows, from which originates the cantal cheese.

But although both Cantal and Mont Dore possess an outer shell similar to a shield volcano, they are quite different, because they form complex volcanoes. Their infrastructure is formed by acid tuffs and tuff breccias, the "cinérites" and the "cinérites à bloc" respectively, and by acid intrusions. This earlier acid volcanism started in the Middle Miocene, and continued all through the Early Pliocene.

The complexity of these two biggest elements of the Auvergne volcanism can be seen from Fig.283, 284 taken from the Cantal. The Mont Dore is constructed quite similarly, but is more complicated through the interference of the younger volcanism of the Chaîne des Puys and the Limagne districts and by the occurrence of a number of younger Maare which have punched the basalt plateaus (BROUSSE, 1961, fig.4).

A word must be added in regard to the cinerites. "Cinérites" and "cinérites à bloc" were formerly thought to be of different age, the finer cinerites being dated as the lower group. For the Mont Dore at least, this is not true, and fine-grained cinerites occur at all

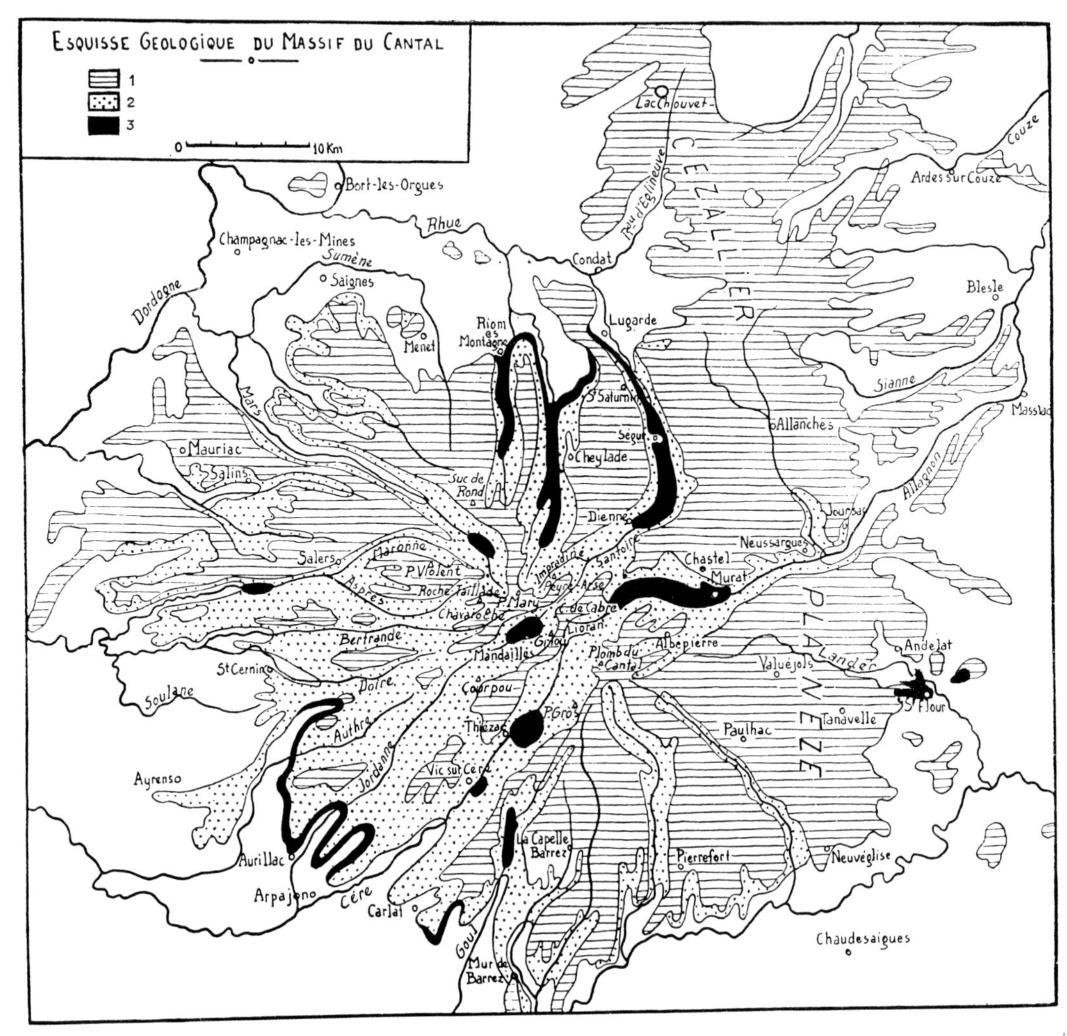

Fig.283. Schematic map of the Cantal.
1 = plateau basalts (Upper Pliocene–Villafranchian); *2* = blocky cinerites (Lower Pliocene); *3* = fine-grained cinerites (Miocene).
The outer form, so similar to a shield volcano, is due to the upper basalts. The radial stippled zones, so similar to the youngest lava flows on top of a shield volcano, are instead windows of the underlying cinerites, laid bare by the erosion of rivers and glaciers. (After JUNG, 1946.)

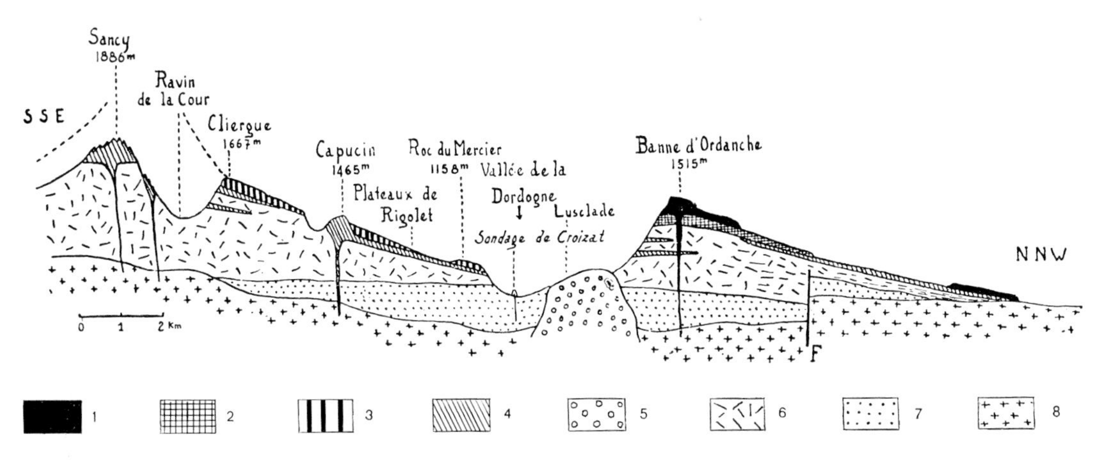

Fig.284. Strongly schematized section through the northwestern quadrant of the Cantal. *1* = basalt; *2* = ordanchite; *3* = andesite; *4* = trachyte; *5* = rhyolite; *6* = upper cinerite (= *2* of Fig.283); *7* = lower cinerite (= *3* of Fig.283); *8* = basement. (After JUNG, 1946.)

ages. The fine-grained cinerites are formed by water-laid, or re-deposited tephra, whereas the blocky cinerites are laharic breccias. It stands to reason that more of the latter developed during younger times, because the steep cones of the contemporaneous acid volcanoes provided the necessary potential energy for lahars to develop. Volcanism started in these regions on top of a well-developed peneplain, upon which no lahars could form (RUTTEN, 1962).

THE CHAÎNE DES PUYS

The Chaîne des Puys consists of a large number, almost a hundred, of small and smaller individual volcanoes, the "puys" or cones (Fig.285). They are normally formed by scoria, whereas the lavas, if produced, have breached a part of the cinder wall. The "puy" then has become "égueulé", or open mouthed (Fig.286). The larger of the scoriaceous puys stand between 100 m and 300 m above the surrounding plain, but many of them are much smaller. The composition of the volcanic products ranges all the way from basalts via trachyandesites to trachytes. They belong to Jung's differentiation type *II*, and consequently form the least alkaline district of the Auvergne volcanism.

One "puy" is aberrant, the Puy de Dôme (Fig.287). Standing more than 500 m above the plateau, reaching an altitude of almost 1,500 m, it projects several hundreds of meters above the normal skyline of the Chaîne des Puys. The reason is simple; the Puy de Dôme is formed by solid rock, instead of by scoria. It is a plug of trachyte of a special composition called domite.

An important aspect of the Chaîne des Puys is that its north–south alignment is clearly parallel to the western border fault of the Limagne graben, but is nevertheless situated about 5 km more westerly. In fact, only one volcano, the Puy de Gravenoire south of Royat, sits exactly on the border fault.

The Quaternary volcanic activity of the Chaîne des Puys continued until so recently that many of its lavas flowed from the plateau, upon which the chain stands, into adjoining basins through existing river valleys. There the lava flows can be dated according to which

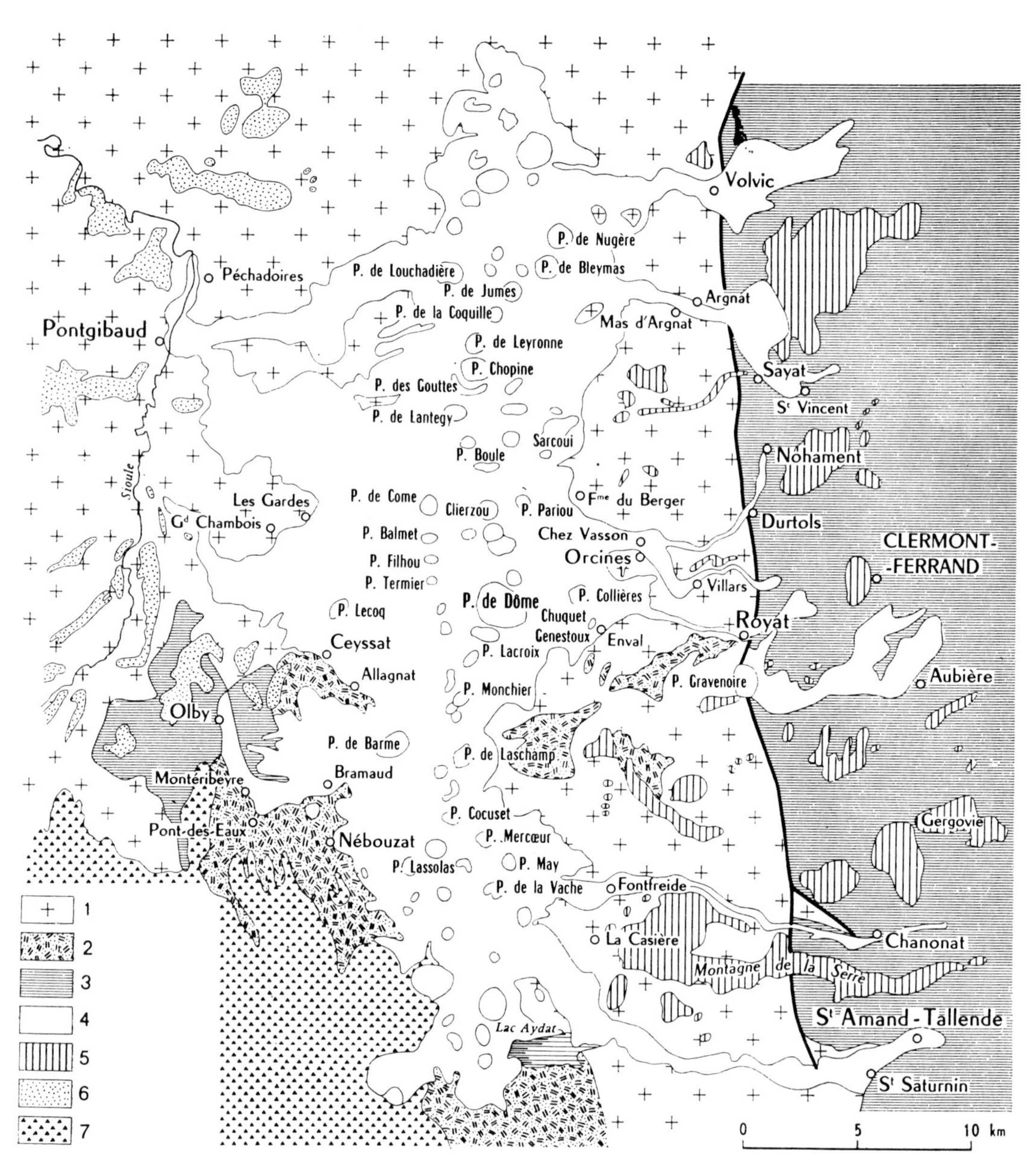

Fig.285. Map of the Chaîne des Puys.
1 = crystalline of the Hercynian basement; *2* = Paleozoic of the Hercynian basement; *3* = Tertiary of the younger graben; *4* = Chaîne des Puys; *5* = Limagne volcanism; *6* = Chaîne de la Sioule, or Little Chaîne des Puys; *7* = Mont Dore. (After BROUSSE, 1961.)

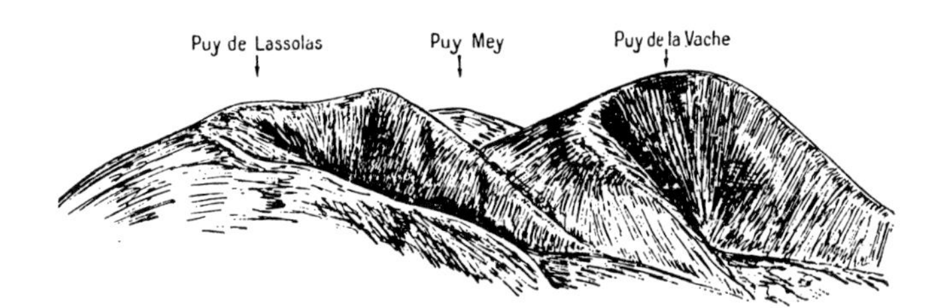

Fig.286. The twin scoriae volcanoes of the Puy de Lassolas and the Puy de la Vache in the southern part of the Chaîne des Puys. The lava produced by these "égueulé" craters flowed for over 20 km along a river valley into the Limagne Basin (cf. Fig.288). (After BENTOR, 1955.)

Fig.287. The Puy de Dôme, seen from the south. (After BENTOR, 1955.)

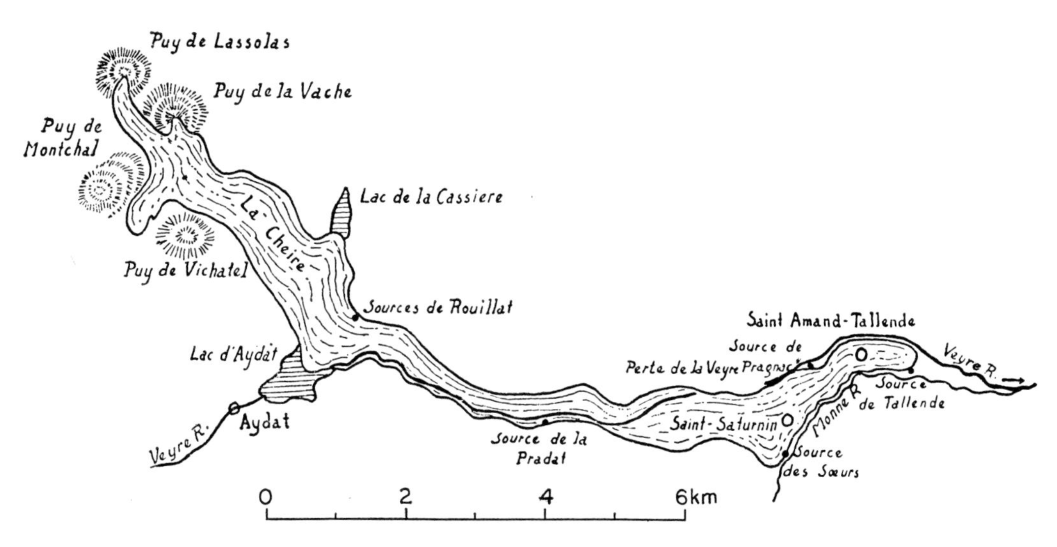

Fig.288. The "cheire" or Quaternary lava produced by the Puy de Lassolas and the Puy de la Vache, as it has flowed through the valley of the river Veyre into the Limagne Basin. Along its course it has dammed tributary valleys, to form the lakes of Aydat and Cassière. (After JUNG, 1946.)

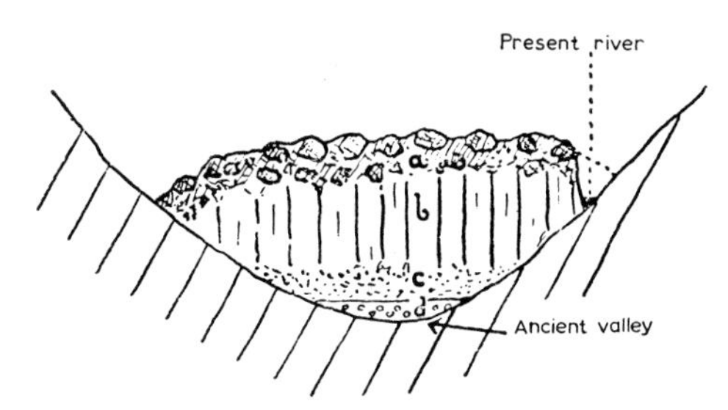

Fig.289. Schematic cross-section through a lava flow descending a valley from the plateau area of the Chaîne des Puys into an adjoining basin.
a = scoriaceous block lava; b = massif or columnar basalt; c = tephra; d = alluvial deposits. (After JUNG, 1946.)

of the fluvial terraces they have covered. Radiocarbon measurements on wood burnt by the lava of the Puy de la Vache and the Puy de Lassolas (Fig.288), gave an age of 7,650 $\pm$ 350 years.

In the deeply incised river valleys the lavas form a convex surface upon cooling, so that the river which formerly meandered over the valley floor is now pushed to one side, and runs close to the valley wall (Fig.289).

It follows that the Chaîne des Puys is an excellent area for the study of volcanic phenomena. But for further details we have to refer the reader to the literature cited.

LIMAGNE DISTRICT

The graben of the Limagne contains innumerable smaller eruption centres, dikes, sills and lava flows. Moreover extensive tuff layers are found, the peperites (MICHEL, 1953). Moreover, several lava flows of larger extension exist. Most important for the present day morphology is that the Limagne volcanism is rather old. It started during the Late Oligocene, and showed intermittent activity up till the Quaternary.

The intermittent volcanic activity in the sedimentary basin of the Limagne makes it possible to date some of the volcanics rather precisely. This has been the base for the dating of the reversals of the magnetic field of the earth by ROCHE (1953, 1963). A controversial point remains in the fact that the limnic deposits of the Limagne and other graben of the Massif Central are difficult to correlate with the normal stratigraphy of the Cenozoic.

Although of small size, and although very few volcanic structures have been preserved, the Limagne volcanics nevertheless are an important feature in the development of the morphology of the basin. Since the Miocene that graben has no longer subsided, and has even risen for several hunderds of meters. Which is but another example of the reversal of the vertical crustal movements in basin areas, cited in the preceding chapter.

The adjoining horsts, belonging to the Massif Central, rose much higher, so the graben still is a basin. But, nevertheless, the original basin floor has been attacked by erosion. It is only preserved where lava flows defended the underlying soft limnic sediments of the graben filling. This forms the origin of the many flat-topped hills which dot the Limagne Basin.

The largest of these flat-topped hills is the Plateau of Gergovia (*Gergovie* on Fig.280), south of Clermont-Ferrand. Its structure is seen in Fig.290.

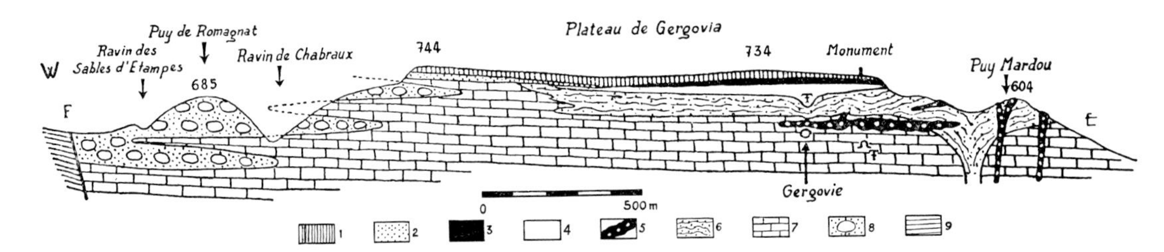

Fig.290. Section along the southern wall of the Gergovia Plateau south of Clermont-Ferrand. *1*, *2* = Upper Burdigalian (*1* = upper analcime basalt; *2* = felspar sand); *3*, *4* = Lower Burdigalian (*3* = lower analcime basalt; *4* = marls with *Melania escheri*); *5* = Aquitanian intrusions of olivine basalt; *6–8* = Upper Stampian (*6* = peperites; *7* = marly limestone with *Helix raymondi*; *8* = sandy marl with *Phryganes*); *9* = Lower Stampian, marls with *Cypris* and *Potamides lamarcki*. (After MICHEL, 1953.)

Apart for its complicated geology, the Gergovia Plateau is much more famed in history. It was on its flanks that Julius Caesar was defeated in 52 B.C. by the Gaulic chief Vercingetorix of the *Arverni*, the race after which the present Auvergne has been named.

VELAY DISTRICT

We will not touch upon the smaller districts, neither on those with many scattered, small single volcanoes, the Sioule and the Forez, nor on the basalt plateaus of the Cézallier and the Aubrac. Only a few words can be added on the Velay and the Causses districts.

The Velay is mainly known for the contrast between its younger basalt plateaus and its older phonolitic volcanism. The latter consists of remnants of necks, dikes and short, thick lava flows, easy to recognize by the attitude of the fluidal structure, well developed in these acid rocks. All three groups are, however, nowadays eroded to more or less rounded hills, projecting above their surroundings. These are the "sucs" or woman's breasts of the local dialect. The highest, and one of the largest, of these is the Mézenc, in the eastern part of the district, which forms one of the highest points of the Massif Central.

Apart from the volcanics mentioned, acid tuffs occur too. As a result of the high altitude of the Velay Plateau, periglacial climate reigned in this area, at least during the two last glaciations. Solifluction, consequently has been a widespread phenomenon. It is often difficult to distinguish primary volcanic deposits, basaltic scoria, cinerites or peperites, from layers re-deposited through periglacial solifluction (BOUT, 1960; RUTTEN, 1964).

CAUSSES AND BAS LANGUEDOC DISTRICT

This district is formed by one single fissure, stretching in a N–S direction over about 150 km (GÈZE, 1955). Its morphological development is, however, quite variable. It ranges from little points of basaltic eruption on top of the Mesozoic limestones of the Causses, at about 1 km altitude, to the much bigger Agde Volcano on the coast of the Mediterranean.

The volcanics produced are mainly basalt and related rocks, but tuff series occur too. Near the Mediterranean these may be developed as water laid, marine or limnic, tuffites (KLOOSTERMAN, 1960).

INNER-ALPINE YOUNGER VOLCANICS

Apart from several smaller volcanic centres further north, all inner-Alpine younger volcanics of western Europe are found on the Italian Peninsula and on islands to the west of this peninsula. The best known of these northerly inner-Alpine districts are the Ligurian province in southeastern France (Fig.273), and that northwest of Venice (PICCOLI, 1963, 1965).

Starting in the northern part of the Italian Peninsula, we find a row of major, extinct volcanoes, ranging from Tuscany to south of Rome. These will be combined to the Rome district in this text. Beyond this district follows the Naples district, whose main elements are the active Vesuvius and the all but extinct Phlegrean fields and the island of Ischia. South of the Naples district volcanic activity is interrupted. It can be picked up again in the Eolian Islands, from where it can be followed to the Etna and the Iblean region in eastern Sicily. After another interruption, we may then connect it with the small island of Pantellaria, half way between Sicily and Africa (Fig.271, 291).

The literature of the volcanics of Italy is vast and scattered, and, moreover, mostly in Italian. It contains, as is well known, the oldest volcanological paper, i.e., the description by the younger Pliny of the 79 eruption of Vesuvius.

History and volcanism have since remained interwoven in Italy. For instance, one of the Eolian Islands, Volcano, gave its name to the science, whereas the Monte Somma around Vesuvius is often cited as the type specimen of caldera walls.

On the occasion of the meeting of the Volcanological Association in Catania in 1961 a number of guide books were produced, which give a very good general view of the volcanics of Italy. I will draw heavily on these in my text, but I doubt if they have had as wide a circulation as they deserve.

The main aspect for many petrographers resides in the alkaline, mediterranean aspect of most Italian volcanics. In many petrographic studies the divisions have been drawn so fine that there are numbers of lava flows which are classed as different rock types in parts near to, or far away from the crater. As in every kind of systematics, there are lumpers and splitters.

If one is, however, more interested in the genetic problems of volcanics and volcanism, not in mere descriptions, one finds that the literature on the volcanics of Italy is dominated by one single name, that of A. Rittmann. His effect on Italian volcanology, and through this on general volcanology, has been of prime importance. First studying Vesuvius near Naples, and later adding to this his studies on Etna and other volcanic areas, he is, with his school, mainly responsible for all modern volcanological work in Italy. Rittmann's general ideas are set foreward in his *Vulkane und ihre Tätigkeit* (RITTMANN, 1960), so I can mainly confine myself to adding a few descriptive remarks in this text.

Before doing this, I want to interpolate, however, a few words on generalities. These are first about the main influence of Rittmann's ideas on volcanology. And second about some controversial issues on the mode of formation of certain volcanics, notably on ignimbrites and hyaloclastites.

The main influence of Rittmann has been to show, from the example of Monte Somma and Vesuvius, how the composition of volcanic products of a single volcanic centre could change gradually with time. And how, in the example chosen, the variation could be explained by the assimilation, by the magma in course of evolution, of material from the thick overlying carbonate series of the Alpine Triassic (RITTMANN, 1930). We now understand how a smaller volcano can be stronger mediterranean in its chemistry than a big one, such as Etna. The smaller magma volume assimilated relatively more material from the enveloping crust, than the bigger ones. So, normally the smaller magma had the bigger chance to become the more contaminated, the more sclerotic.

In this context the emphasis must be noted which Rittmann always laid on the fact that most volcanics, from their phenocrysts give a quite distorted picture of their chemical composition, because lighter elements tend to concentrate in the glass basis. For all rocks which are only studied petrographically, Rittmann insists on the prefix "pheno", such as pheno-basalt, or pheno-andesite.

The ideas on evolution of magmas through assimilation, as set forward by Rittmann

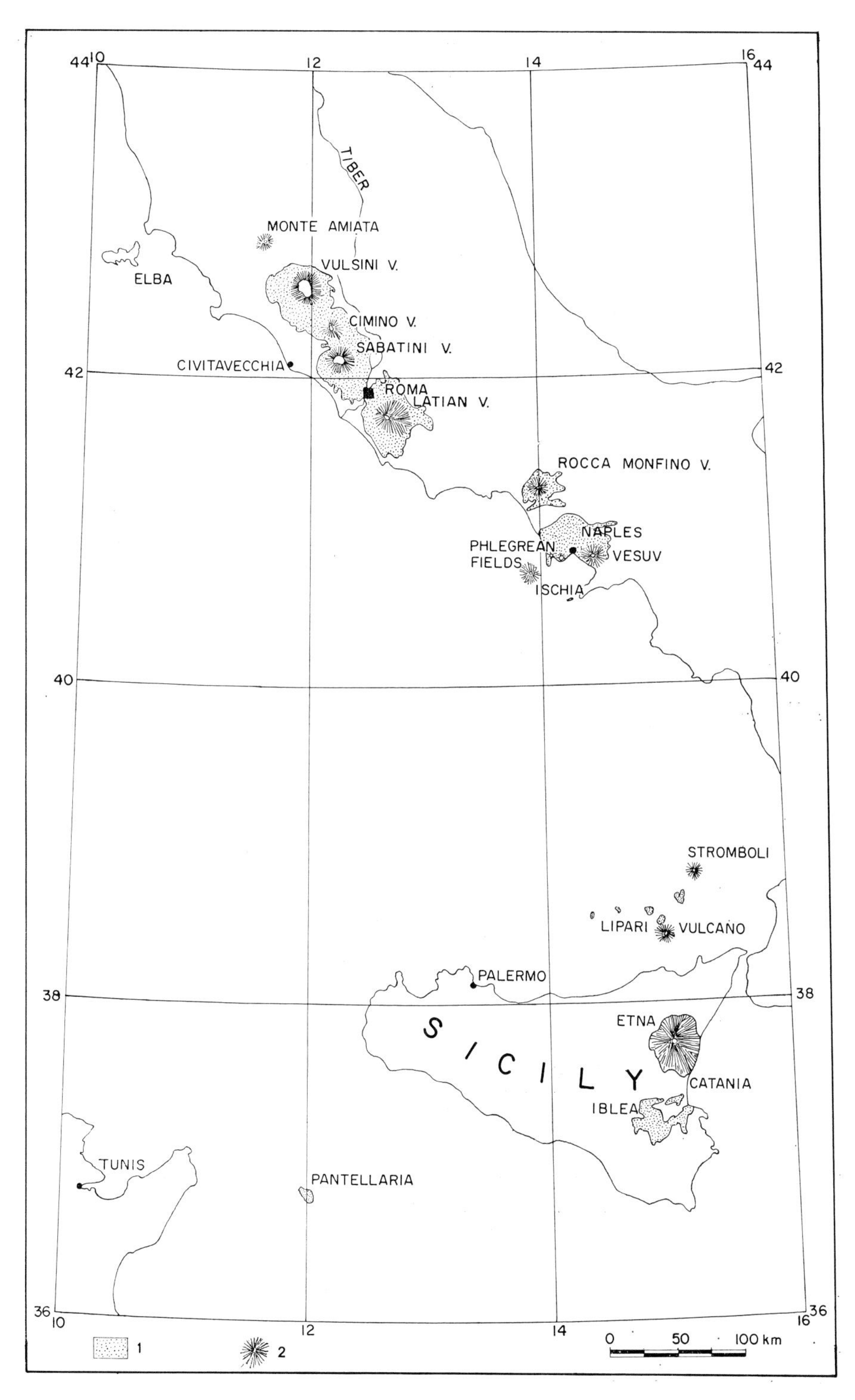

Fig.291. Schematic map of the younger volcanics of middle and southern Italy. *1* = volcanics, general; *2* = volcanoes, extinct and active.

have, by now, become generally accepted. This is not the case for the two problems of younger date, that of the genesis of certain tuffs called ignimbrites and certain breccias called hyaloclastites. The 1961 Catania symposium was even entirely centered upon these two problems (see *Bulletin Volcanologique*, 25 of 1963).

We have already touched upon the problems of ignimbrites in Chapter 3, in our discussion of the Permian volcanics of the Oslo graben. It now crops up again in relation to the younger volcanics of Italy. Though a certain repetition is unavoidable, it consequently merits discussion here. Ash flows or ignimbrites have to be welded to be accepted as such by a petrographer. Whereas for a field geologist its identifying characters lie in its massive, homogeneous structure, without the windsorted layers and the volcanic bombs that characterize air-fall tuffs, and equally without the strongly developed, contorted lamination due to fluidal texture of a viscous lava (VAN BEMMELEN, 1961).

Thus, to cite an example, from Monte Amiata RITTMANN (1958) described ignimbrites and rheo-ignimbrites which, for many field geologists were normal rhyolitic lavas. Whereas from the other volcanoes north of Rome, massive and extensive ash flows have come down pre-existing valleys, which consequently are ignimbrites to a field geologist (RUTTEN, 1959). But, because they are not welded, they are not accepted as such by many a petrographer.

There are to be found in Italy almost all transitions between lava flow and ash flow and air-fall tuff. So we may hope that eventually it may become possible to identify these groups on a sounder basis. Careful studies in which both the result of field mapping and of petrographic studies are incorporated, such as recently carried out by LOCARDI and MITTEMPERCHER (1965) and LOCARDI (1965) will eventually supply the answer. But the problem is at present still highly controversial.

Hyaloclastites, in the 1961 Catania Symposium definition, are submarine, shallow water deposits. If typical, such as in the Iblean region southwest of Catania in Sicily (CAMPIONE, 1961) they are coarsely clastic. They contain big, roundish to ellipsoidal inclusions, which one would tend to call volcanic bombs, if one forgets that they never have been flying through the air, as a bomb should have done.

These roundish inclusions, of the order of one or several decimeters in diameter are characterized by thick, glassy selvages. These are similar to, although not quite as thick, as those found in the truly subglacial Palagonite Breccia of Iceland. Such thick selvages indicate quick cooling, commensurate with submarine eruption. There are, however, many transitions. As always, it is in these border line cases that the problems of classification and of semantics arise (CUCUZZA SILVESTRI, 1963).

THE ROME DISTRICT

The Rome district consists of four major volcanoes north of Rome, of one major volcano south of Rome, and of several smaller volcanic areas accompanying the main chain to the west. The most important of the latter are those of San Vincenzo (BORSI, 1961) to the west of the Monte Amiata, and of Tolfa, northwest of the Sabatini Volcano. Although of importance, because they are older than the main chain, and moreover of different composition, they can not be accounted for here.

The northernmost volcano of the Rome district, the Monte Amiata, still forms a

TABLE XX

NAMES OF EXTINCT VOLCANOES AND CORRESPONDING CRATER LAKES IN THE ROME VOLCANIC DISTRICT

Volcanoes	*Crater lakes*
Monte Amiata	
Vulsini Volcano	Bolsena Lake
Cimino Volcano	Vico Lake
Sabatini Volcano	Bracciano Lake
Latian Volcano	Albano and Nemi Lakes

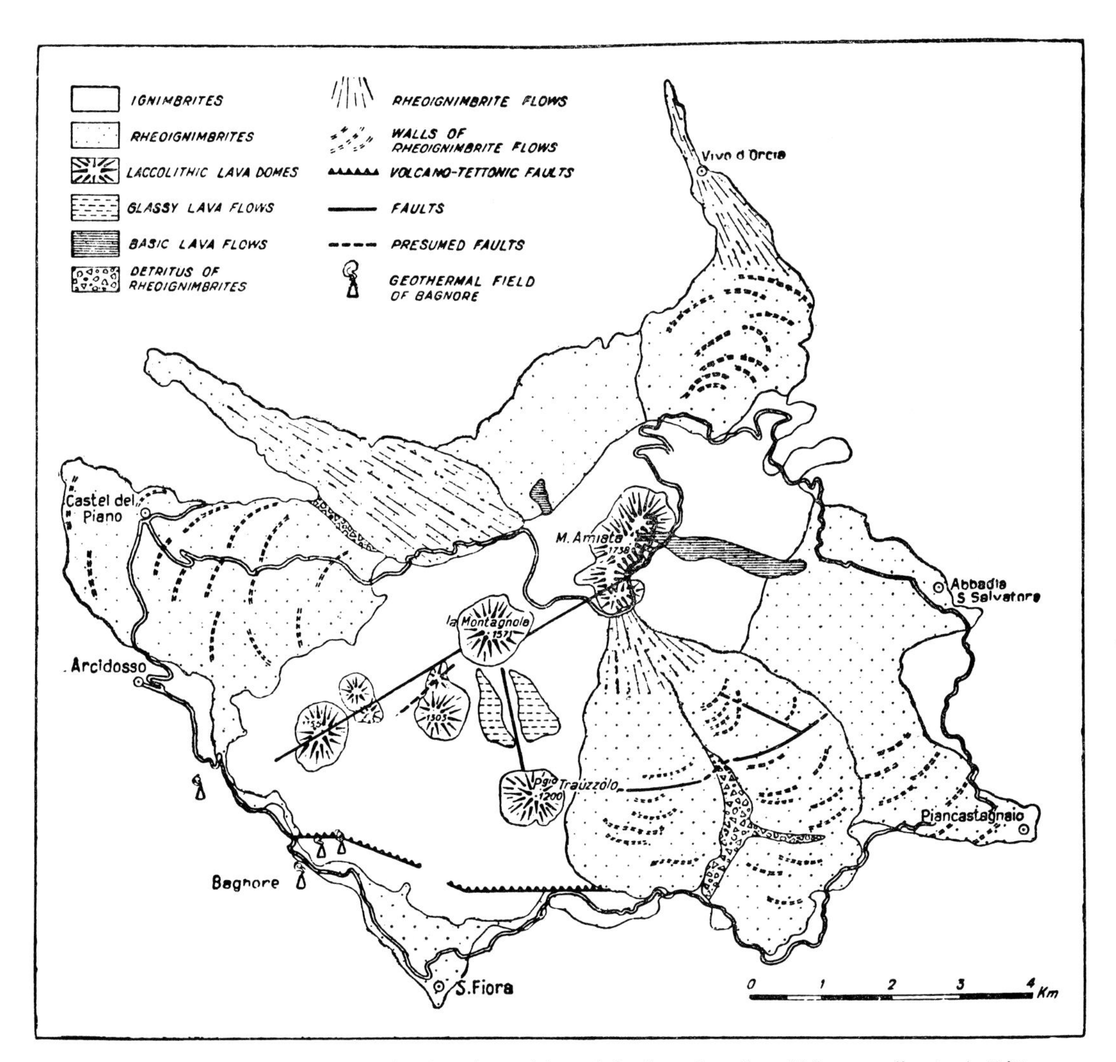

Fig.292. Map of the Monte Amiata. Showing the position of the "ruga" walls, which, according to A. Rittmann, are of rheo-ignimbritic origin. (After PRATESI and MAZZUOLI, 1961.)

beautiful cone. The others are all in a more or less pronounced caldera phase. They contain lakes, which are much better known than the volcanoes themselves. Unfortunately the names of the volcanoes and their crater lakes are not the same, so a synopsis has been given in Table XX.

The Monte Amiata (PRATESI and MAZZUOLI, 1961; MARINELLI, 1961), though forming a single cone, topographically, is still composed of a number of separated eruption centres (Fig.292). Because of the dogma stating that ignimbrites arise from fissure eruptions, Rittmann, who thinks that a part of the Monte Amiata is covered by ignimbrites, maintains that it is formed from a fissure eruption. A glance at the map of Fig. 292 will show that this can, in all objectivity, hardly be maintained.

More important from a general point of view are roughly tangentional steep walls of "lava" showing subvertical or strongly contorted fluidal texture. These "ruga", or wrinkles, several of them over 10 km long (Fig.292), were called rheo-ignimbritic walls by Rittmann. They are interpreted as formed by ash flow material which, after the eruption stopped, melted from its own heat at the base of the ash flow. Owing to the steep slopes on which these supposed ash flows came to rest, the melted base was mobilized by gravity and broke through the upper part of its own ash flow. The upper part did not melt and retained its tuffbreccia character. The pseudo-lava then cooled in the "ruga" walls. Other geologists, however, interpret the "rugas" as the front of normal viscous primary lava flows.

The three other volcanoes north of Rome (Fig.293) are quite similar in that they all have a crater lake, which may indicate caldera subsidence. No modern description of these volcanoes exists.

All three volcanoes are surrounded by extensive tuff flows. These can be sawed for building stone, and also, in weathering, produce very fertile soil. The discussion about the genesis of these tuffs, whether ignimbritic or not, has already been mentioned in the introduction of this chapter.

The volcano southeast of Rome, the Latium Volcano, which is also called the Colli Albani, is a much bigger structure than the volcanoes north of Rome. Its outer edge is formed by the horseshoe remnant of an old crater rim, about 10 km in diameter. This convex wall is called Tuscolo in the north and Artemisio in the south (Fig.294).

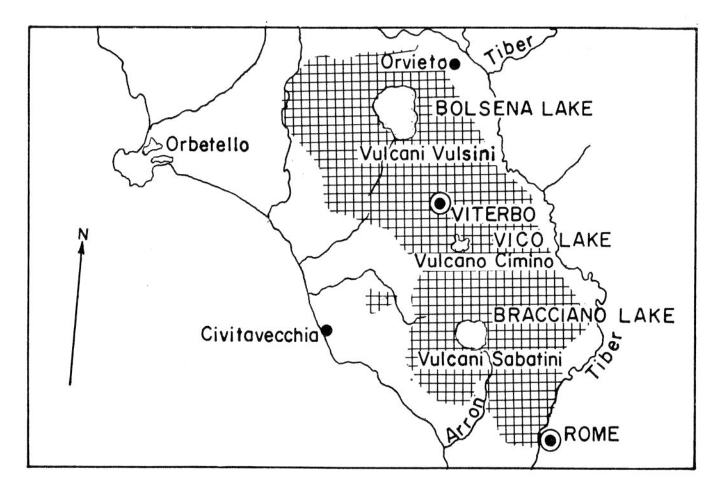

Fig.293. Map of the three major volcanoes north of Rome, with the surrounding volcanics and the crater lakes. (After MITTEMPERCHER, 1961.)

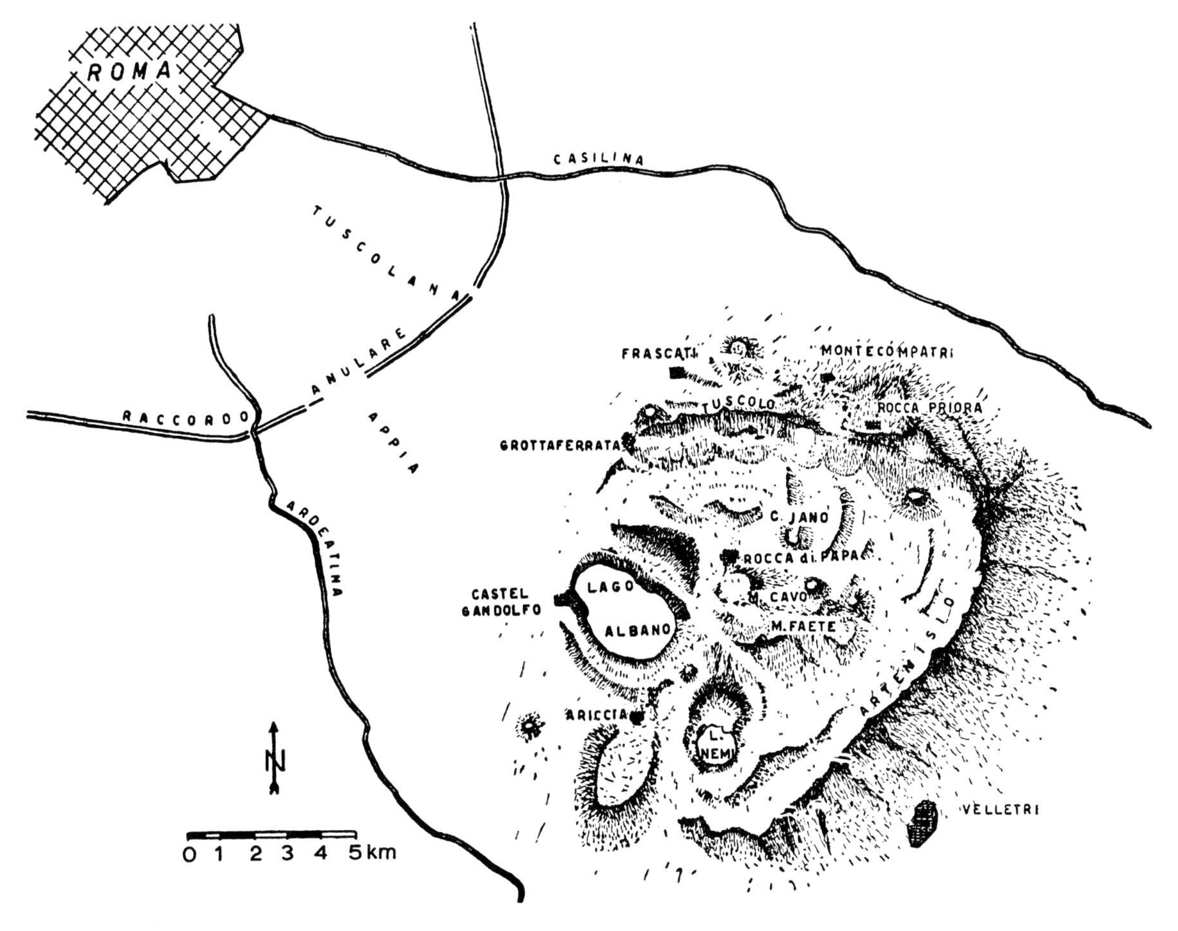

Fig.294. Map of the Latian Volcano (Colli Albani), southeast of Rome. (After MITTENPERCHER, 1961.)

Within the caldera three new volcanoes have been formed, the Nemi, Ariccia and Albano Volcanoes. Both the Nemi and the Albano Volcanoes have beautiful crater lakes. The first has become better known from Nero's pleasure boats, salvaged before, but destroyed during the last war. The second is known for the papal summer castle, which stands at its rim in Castel Gandolfo.

The eruptive history of the Latian Volcano belongs to the Pleistocene, to the Riss and Würm periods, according to VENTRIGLIA (1959).

THE NAPLES DISTRICT

The Naples district comprises three main units, the Vesuvius, the Phlegrean Fields and the island of Ischia. Moreover, there is the more or less isolated Roccamonfino Volcano, near Gaeta, half way between Rome and Naples.

The Vesuvius has been described so often, that I think I may refer the reader to RITTMANN's (1933, 1950, 1960) descriptions and to the map of Fig.295.

The Phlegrean fields (Fig.296) west of Naples form the other main volcanic unit on the mainland in the Naples district (GOTTINI, 1961a; RITTMANN et al., 1951). It is assumed that its volcanic history started with the formation of a big strato volcano, which blew up to form a caldera, 14 km in diameter. This produced the trachytic grey tuffs of the "Campano napolitano". It was followed by more eruptions, producing the yellow Napolitanean Tuff, and by more sinking and caldera formation. The present volcanic cones are the youngest in the district, and the Monte Nuova, a forerunner of Paricutín, was even formed in only two days during the historical eruption of 1538.

Several islands in the Tyrrhenian Sea further make up the Naples district. Monte di Procida and Ischia are the largest of this group.

Ischia (RITTMANN, 1930; GOTTINI, 1961b) consists of a basis formed by an older

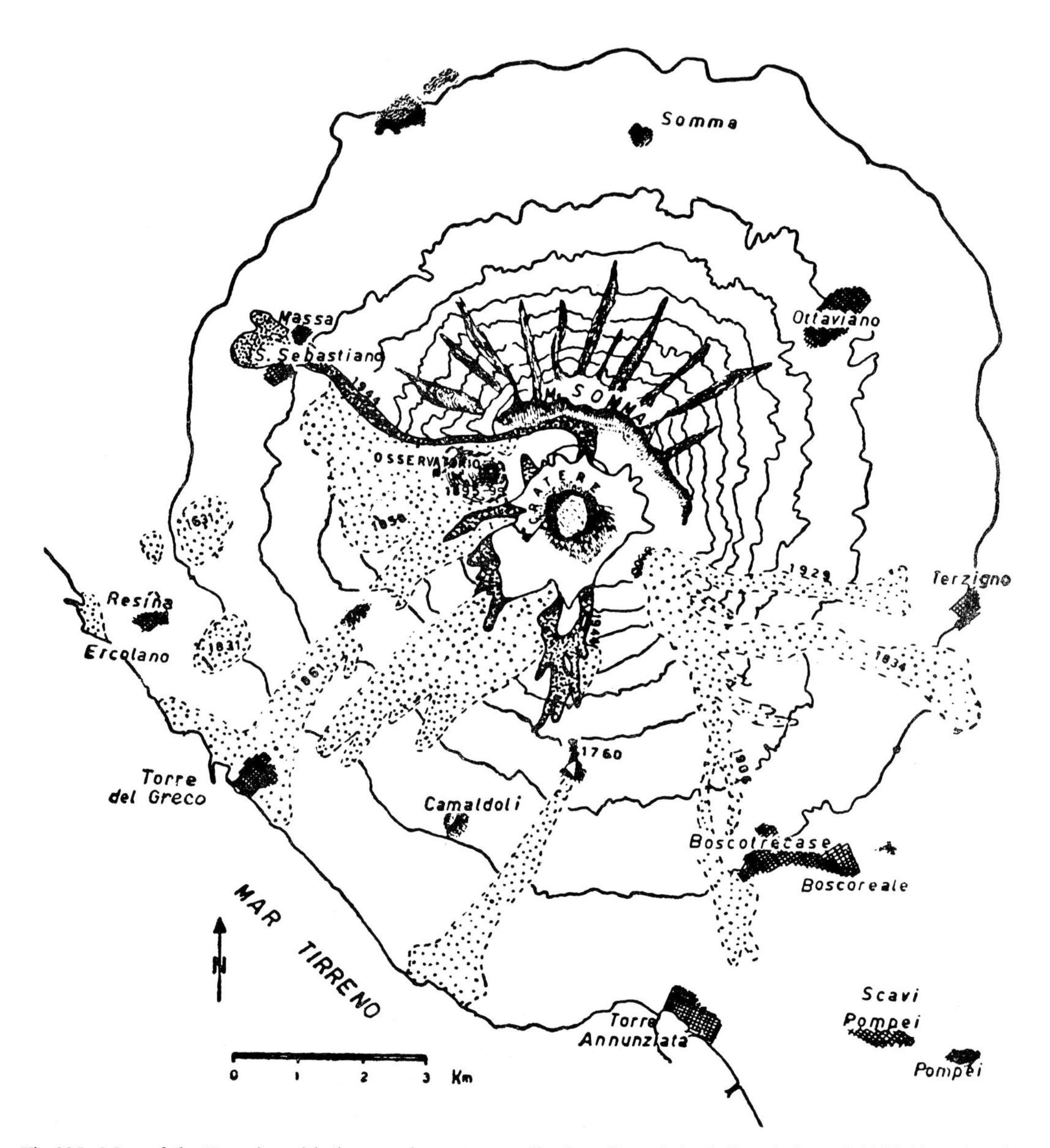

Fig.295. Map of the Vesuvius with the most important earlier lava flows (stippled) and that of 1944 (densely stippled).

Resina is the modern town built upon the ancient site of Herculaneum. Scavi Pompei is the old Pompei. Herculaneum was covered by a lahar, Pompei by ash and pumice during the 79 A.D. eruption. The Monte Somma north and east of the present crater is the only part left of the old caldera wall. (After CUCUZZA SILVESTRI, 1961a.)

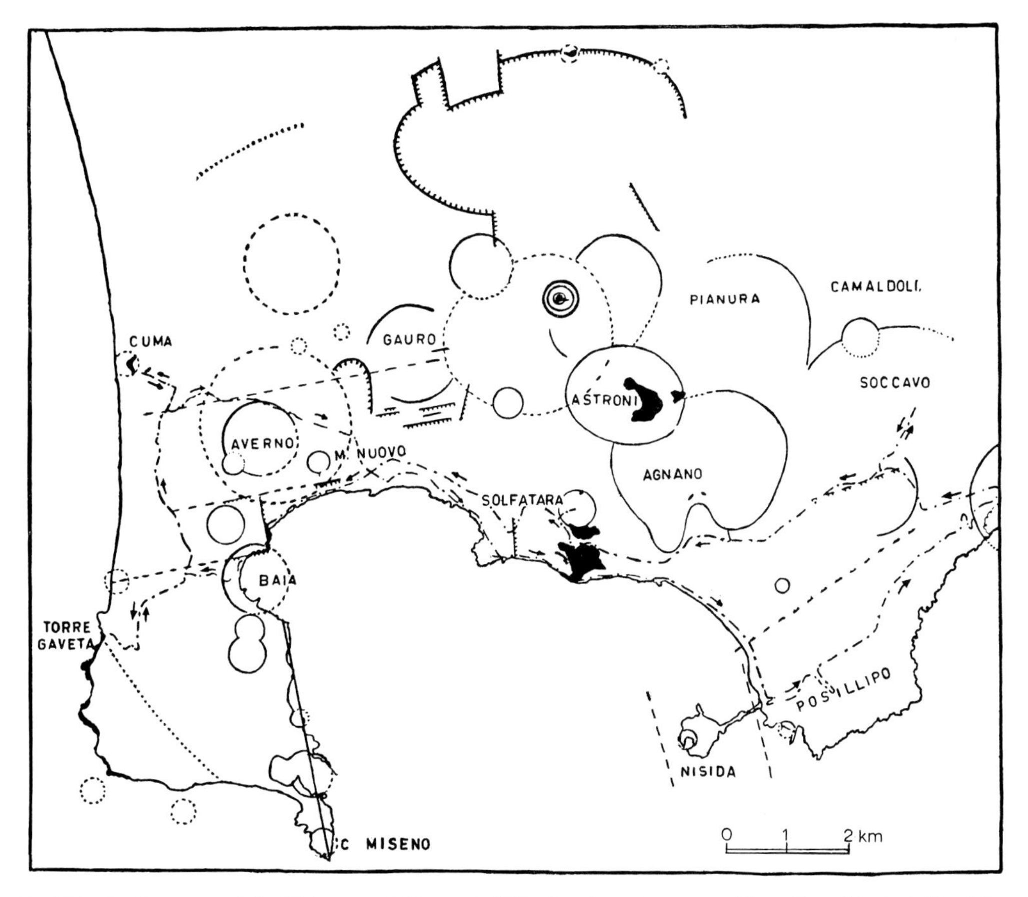

Fig.296. Outline map of the Phlegrean fields west of Naples. Lava domes and lava flows black, tuffs white.
The ringwall of the first caldera, the Archiphlegreus, 14 km in diameter, has been destroyed by later eruptions and by the formation of later calderas, except near Camaldoli. The many younger cones, which now dot the old caldera floor, are shown in outline. The Bay of Bahia is formed by further subsidence, at least in part of volcano-tectonic origin.
The town of Pozzuoli, from which the natural cement pozzolan draws its name, and where, from the marks of stone boring mussels on the pillars of its market place, Charles Lyell drew his conclusions on sea level oscillations, is situated on the lava promontory south of Solfatara Volcano. (After GOTTINI, 1961a.)

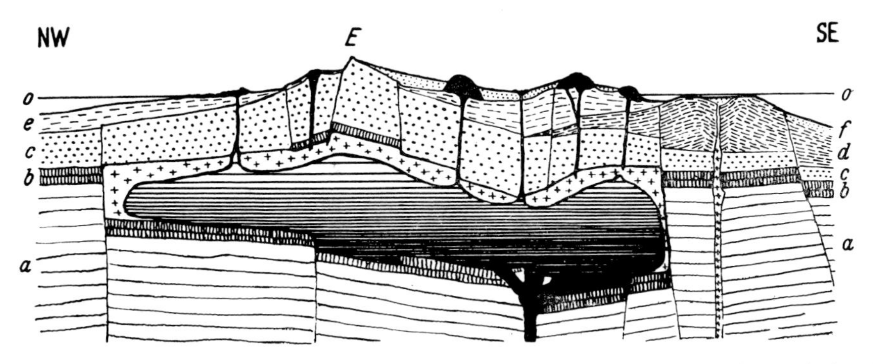

Fig.297. Cross-section through the island of Ischia (E = Monte Epomeo, the highest point, and also the highest tectonic block of the Ischia horst).
a = sedimentary basis; b = basalt and trachybasalt; c = green Epomeo trachyte tuff; d = Secca d'Ischia trachyte volcano; e = younger trachyte tuffs. (After RITTMANN, 1960.)

volcanic series, which is broken up by block faulting. Younger and smaller volcanic eruptions then took place along the faults.

Apart from some older basalts, most of the older volcanic basement is formed by the Epomeo Tuff, a greenish, massive, ignimbritic deposit. Its deposition is probably related to a violent explosion and caldera formation, such as of the Archiphlegreus. But neither the early stratovolcano, nor the caldera are known from Ischia. The Monte Epomeo is no more than the highest block of the Ischia Horst, where consequently the older Epomeo Tuffs crop out over large areas (Fig.297).

The later movements, which reversed the sign of the supposed caldera subsidence, and produced the Ischia Horst, are attributed by Rittmann to a hypothetical light trachytic magma chamber (Fig.297), which gave the crust a certain buoyancy. Ischia is called a "volcanotectonic horst" for this reason, but the relation between volcanic events and tectonic movements is much less obvious, than in the formation of a caldera after a major eruption. The formation of the Ischia horst could well be another example of "Hebung–Spaltung–Vulkanismus", in which the volcanic activity followed upon the tectonic block-faulting movements, and not the tectonics upon the volcanic event, such as the emptying of a magma chamber. It must be noted that the volume of the post-faulting volcanism is slight indeed, and does not warrant a magma chamber of the size as indicated by Rittmann.

THE EOLIAN ISLANDS

The Eolian Islands, north of Sicily, are all volcanic (Fig.298). The best known are those of Stromboli, Lipari and Volcano (Jakob, 1958).

Stromboli is, of course, well known from general volcanology, being the type volcano of persistent eruption. It forms a beautiful cone, over 900 m in altitude and not more than 5 km in diameter at its base. The sea around Stromboli is over 1 km deep, so the volcano actually forms a mountain over 2 km high. Its products are basic, without any acid members, an exception in this district. They range from basalt via trachybasalt to trachyandesite. The persistent eruptive activity is related to the basic composition of its magma.

The most conspicuous feature of Stromboli is the "Sciara del Fuoco", which, in Sicilian dialect, means "Fiery basalt scoriae", at its northwestern flank (Fig.299). This results from the fact that the present crater wall has become breached towards the sea, so that at every eruption an avalanche of red hot bombs and blebs of lava roll down the slope, sometimes even reaching the sea. At night it is always a grand sight, and because Stromboli lies on the route from Genoa to the Straits of Messina, passenger vessels often sail quite close to the Sciara del Fuoco.

Lipari (Fig.300), the largest of the Eolian Islands, has also a much more complex history than Stromboli. It is, on the other hand, also entirely formed by volcanic action during the Quaternary. Its formation started with the eruption of tuffs and lavas of basaltic to andesitic composition, comparable to those of present day Stromboli. It was only after this initial basic volcanism that the acid rocks, to which the island has given its name, began to be produced. They are formed both by rhyolitic lava flows and by massive deposits of pumice tuffs.

The rhyolitic lava flows are small, short and thick, and very steep fronted. They contain varying amounts of obsidian, which often becomes the predominant constituent of these lava flows. With their extremely strong fluidal texture, which follows the contortions of the convective flow of the viscous block lava, they may well stand as the type example of a viscous lava flow. As such they contrast with the ash flows of similar composition, not only because of the large volumes but also by the homogeneous structure of the latter, with only the faintest of subparallel flow lines, indicative of extreme fluidity of the erupting ash flow.

The pumice, which mainly covers the northeastern part of the island, is extensively quarried, and loaded in bulk in freighters off Puerto di Sparanello. Apart from its use in abrasives, it is now widely employed, very finely ground, as a base for beauty creams and other pharmaceutical products.

Volcano Island, south of Lipari, forms a composite volcano too, showing a similar

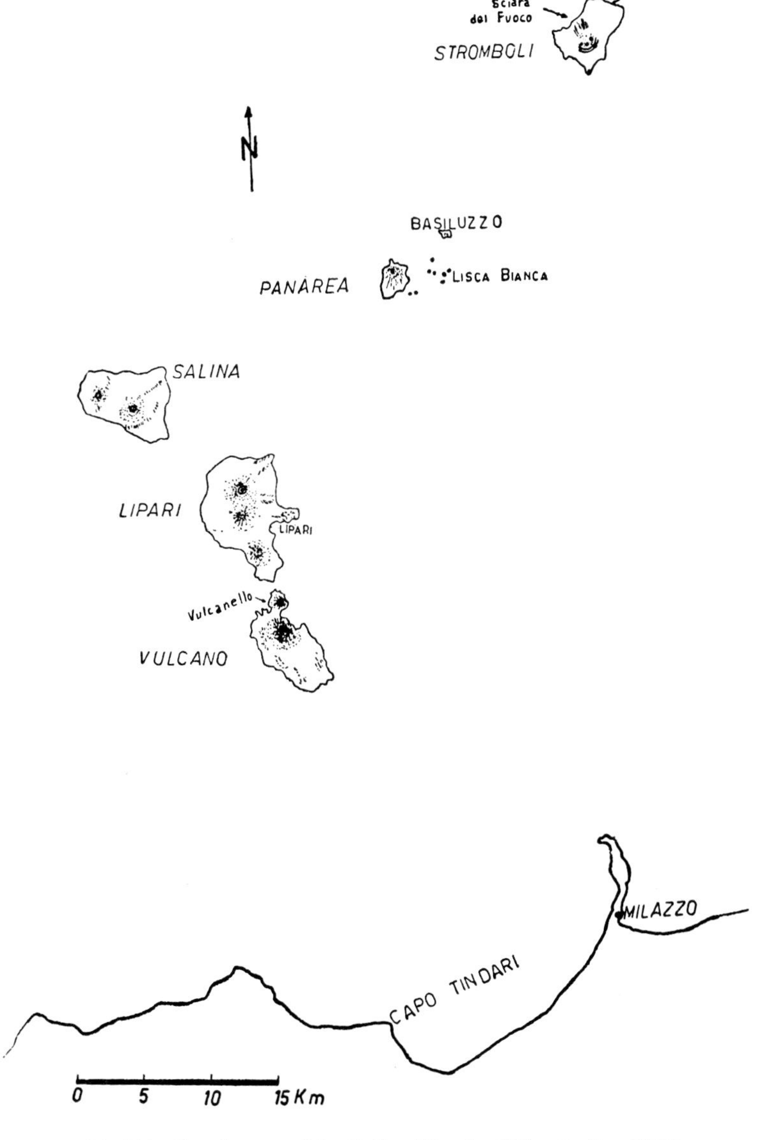

Fig.298. Sketch map of the Eolian Islands. (After DI RE, 1961.)

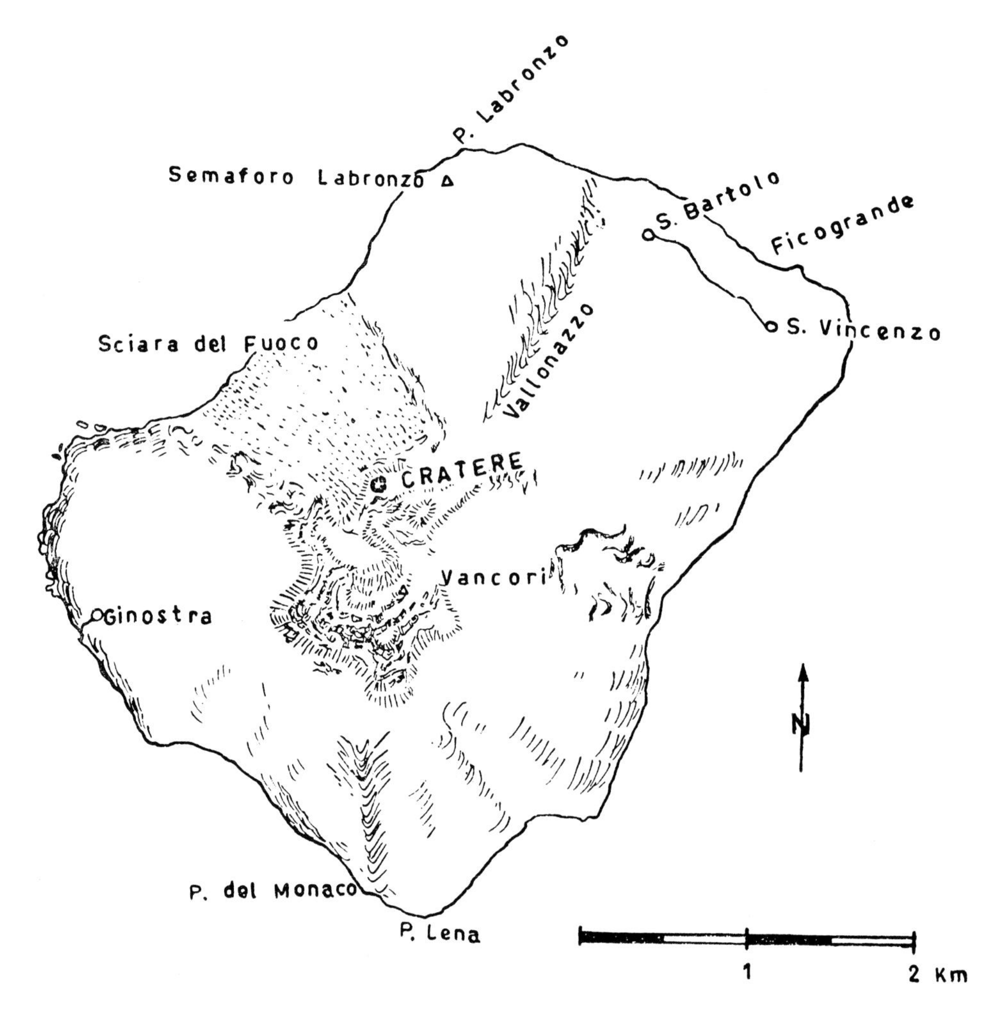

Fig.299. Sketch map of Stromboli, showing the position of the present crater and the *Sciara del Fuoco*. (After Di Re, 1961.)

evolution from a basaltic to a rhyolitic magma. The main volcanoes at present are Volcano I, Volcano II and Volcanello (Fig.301). Volcano II, although of small size, has had a dominantly explosive activity. Its eruptions are known from the 5th century B.C. until 1892. It is now in a dormant solfatara stage.

The relation between early history and volcanics is particularly strong in the Eolian Islands. As early as the 8th century B.C. Homer refers to these islands as the abode of Aeolus, the god of the winds. These he kept confined in a big cave on Strongyle, the present Stromboli. In the 4th century B.C. Aristotle, in his *Meteorics* (II, 8), describes volcanic eruptions on the island of Hiera, the present Volcano. Curiously enough, this is still thought to be the work of Aeolus. Because, in Aristotle's conception, volcanic eruptions are the result of air confined internally in the earth, breaking forth. The heat which forms part of volcanic phenomena is thought to be a secondary effect by Aristotle. The air became inflamed by the shock when it violently separated into the minutest fragments (Geikie, 1962).

More modern views, in which the fire was thought to be the primary correlation with volcanism, the strong winds a secondary effect, prevailed already in the 3rd century B.C. These opinions are found in Callimachus' *Hymn to Artemis* (46), and in the Sicilian writer

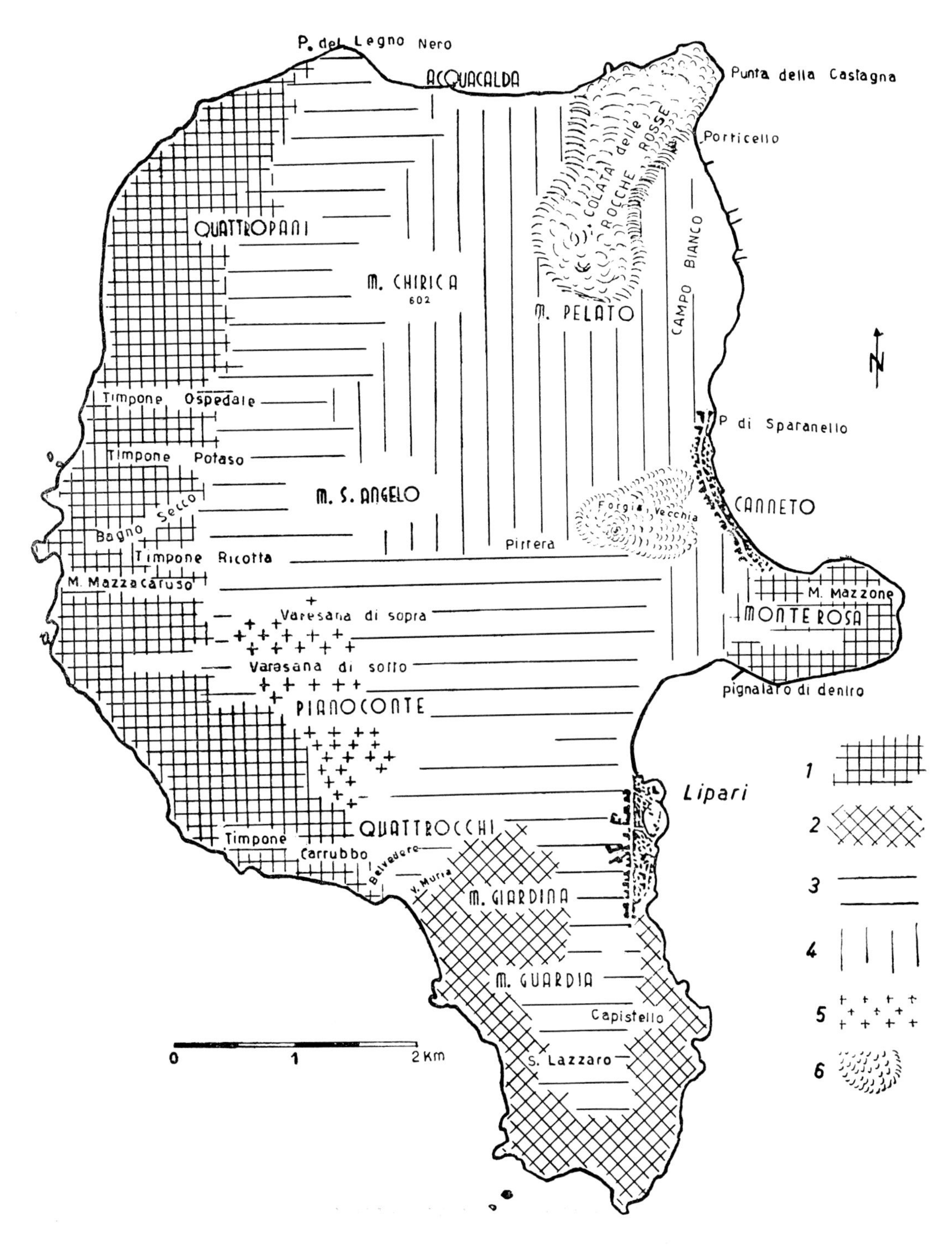

Fig.300. Sketch map of Lipari Island.
1 = basaltic and andesitic lavas and tuffs; *2* = older rhyolites; *3* = acid tuffs; *4* = rhyolitic pumice; *5* = quarz latite with cordierite; *6* = younger rhyolite with much obsidian. (After STURIALE, 1961.)

Theocritus (II: 133). From then onward the Eolian Islands not only are dedicated to Aeolus but are also believed to be the locus of the smithies of the Cyclopes and of Hephaestes. Or, in Roman times, of Volcan.

THE ETNA

Sicily has two quite different volcanic districts. One is the active Etna Volcano in its northeastern edge. The other are the Tertiary submarine volcanics of the Iblean region further south (CAMPIONE, 1961). Only the Etna will be described here.

The Etna forms a worthy subject indeed for the last paragraph of this book. Its ma-

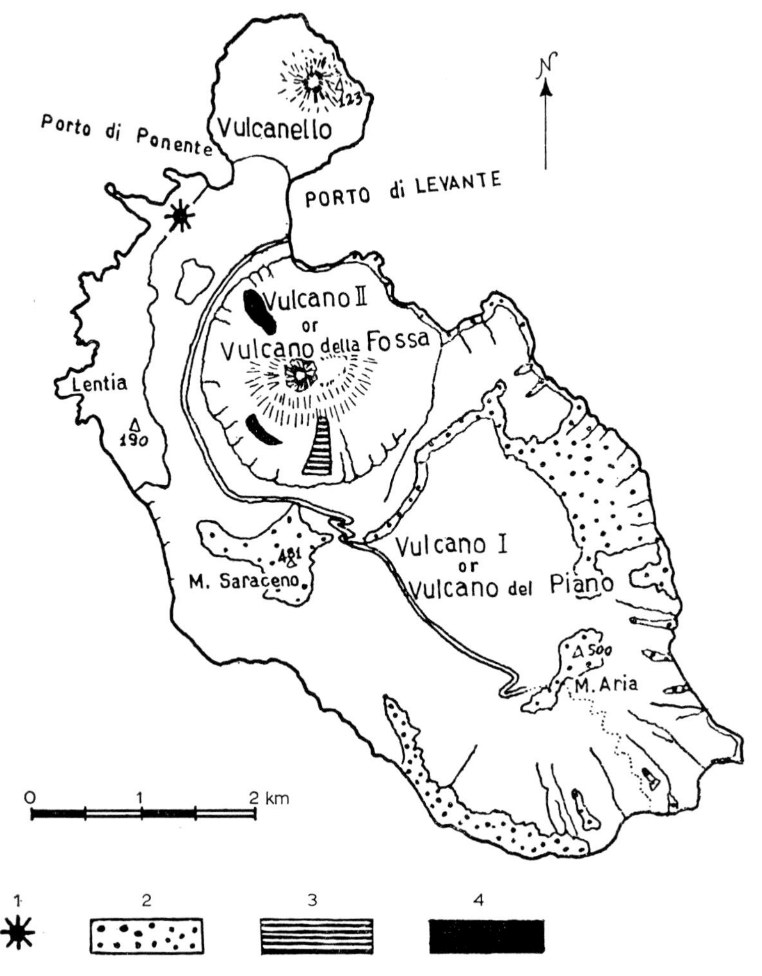

Fig.301. Sketch map of Volcano Island.

1 = quarry in obsidian, liparite, tuffs and breccias of the Mount Lentia series, forming the oldest part of the island; *2* = olivine trachybasalts and trachyandesites of Vulcano I or Vulcano del Piano (this oldest volcano is elsewhere strongly altered by fumarolic activity); *3* = trachytes of Vulcano II or Vulcano della Fossa; *4* = trachytic obsidian of Vulcano II or Vulcano della Fossa.

It is the explosive activity of this volcano that has attracted the attention of so many writers, since its earliest known description by Tucydides in the 5th century B.C.

Volcanello was born from the sea in the 2nd century (183?) B.C. Its products vary from tephritic trachyandesites to leucite basanites. It is now partly eroded by the sea and offers beautiful sections through a volcano. (After BALDANZA, 1961.)

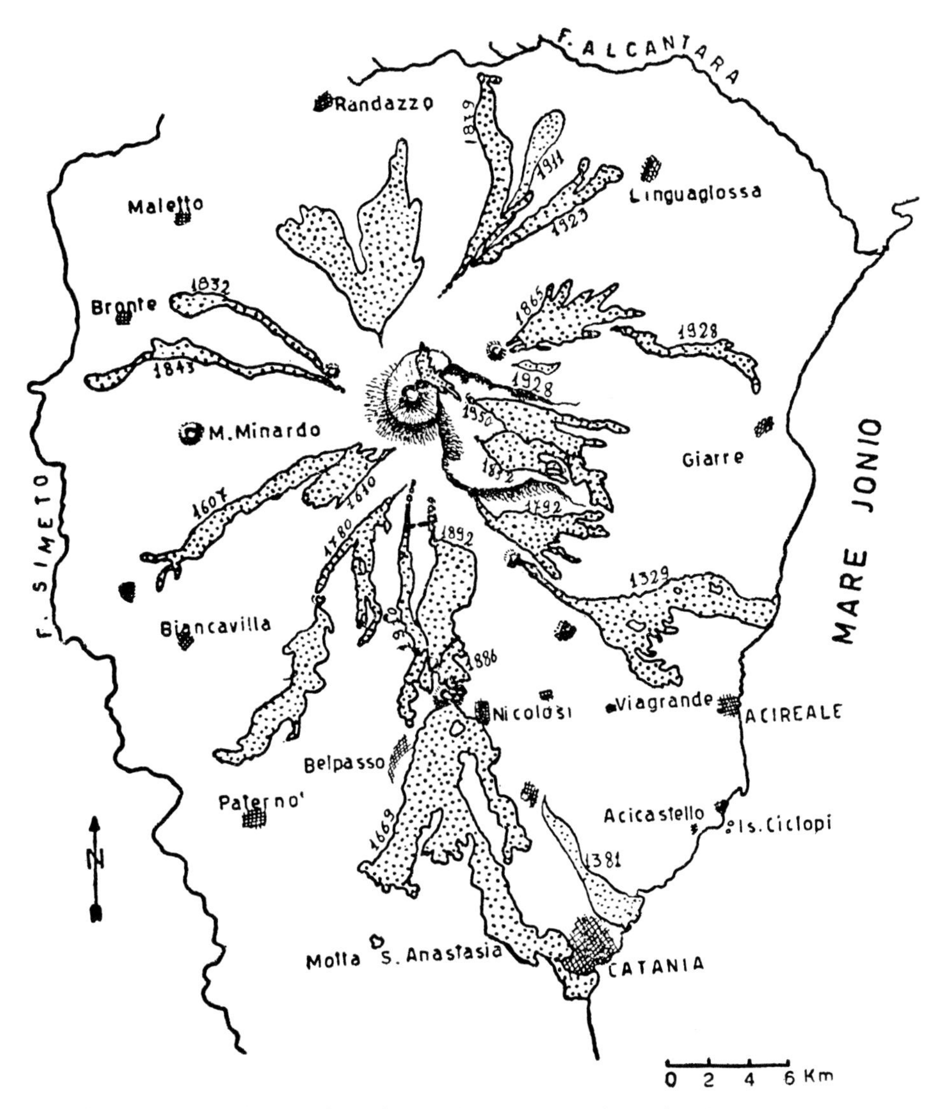

FIG.302. Sketch map of Etna, showing the main historic lava flows. The Valle del Bove, a collapsed former crater area, from which erupted the 1892 and 1950 flows, is found east of the present crater. The 1910 lava flow of Fig. 303 is situated due south of the present crater. (After CASTIGLIONE, 1961.)

jestic cone, the altitude of which changes with every major eruption, varying around 3,300 m, is one of the most beautiful mountains of western Europe.

The composition of the Etna products is mainly basaltic, but varies from 46% to 58% SiO_2 (CASTIGLIONE, 1961). Many rocks fall very near to the boundaries between basalts, andesine basalts, trachyandesites and nepheline tephrites. This mildly alkaline character may well indicate that the Etna magma is predominantly a primary magma, being less contaminated by crustal material than most of the other younger volcanics of Italy.

The building of Etna started at the beginning of the Quaternary with submarine basalt eruptions, which later continued above sea level. These pre-Etnean rocks are, for instance, found along the base of Etna on the sea coast north of Catania. Best known are the Cyclopean Islands, a set of dikes eroded by the sea, and supposed to be the legendary rocks the blinded Cyclops threw after departing Odysseus (*Is. Ciclopi* on Fig.302).

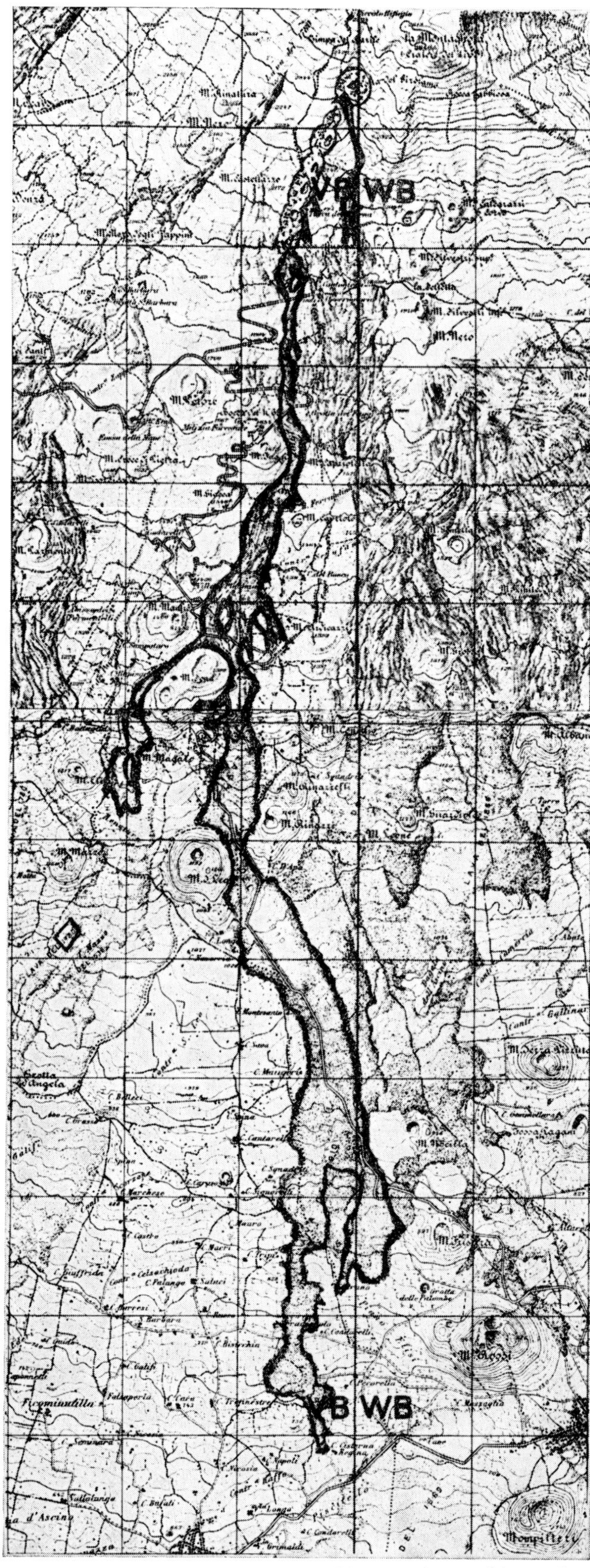

Fig.303. The 1910 lava flow of Etna, outlined on a topographic map of part of its southern flank. Each square is 1 km². The eruption took place from a radial row of adventitious cones in the two uppermost squares.

The present volcano was built up afterwards by a succession of numerous eruptions, many of them being historical (Fig.302). The present crater lies west from the former crater area, which collapsed to form the horse shoe shaped basin of the Valle del Bove.

A special characteristic of Etna, which can be correlated with the high altitude of its central crater, lies in the fact that eruptions often took place on its flanks. These lateral eruptions normally are not produced from a single eruption centre, but from a radial row of adventitious cones (see Fig.303 for the 1910, and Di Re, 1961a, for the 1780 eruptions). It is these radial rows of adventitious cones that largely determine the characteristic landscape of the flanks of Etna.

REFERENCES

Baldanza, B., 1961. *Guide to the Excursion to Vulcano.* Inst. Vulcanol., Catania, 13 pp.

Bentor, K., 1955. La Chaîne des Puys. *Bull. Serv. Carte Géol. France*, 252: 436 pp.

Borsi, S., 1961. *Guide for the Excursion to San Vincenzo (Tuscany).* Inst. Vulcanol., Catania, 9 pp.

Bout, P., 1960. *Le Villafranchien du Velay et du Bassin Hydrographique Moyen et Supérieur de l'Allier.* Thesis, Jeanne d'Arc, Le Puy, 344 pp.

Bout, P., 1966. Histoire géologique et morphogénétique du système Velay SE–Boutières–Coiron. *Rev. Géograph. Phys. Géol. Dyn., Sér. 2*, 8: 225–251.

Brousse, R., 1961. Recueil des analyses chimiques des roches volcaniques tertiaires et quaternaires de la France. *Bull. Serv. Carte Géol. France*, 263: 140 pp.

Campione, A., 1961. *Guide for the Iblean Region.* Inst. Vulcanol., Catania, 13 pp.

Castiglione, M., 1961. *Guide to the Etna Excursion.* Inst. Vulcanol., Catania, 12 pp.

Cloos, H., 1936. *Einführung in die Geologie.* Borntraeger, Berlin, 503 pp.

Cloos, H., 1939. Hebung–Spaltung–Vulkanismus. *Geol. Rundschau*, 30: 401–519; 637–640.

Cloos, H., 1941. Bau und Tätigkeit von Tuffschloten. *Geol. Rundschau*, 32: 708–800.

Cloos, H., 1947. Der Basaltstock des Weilberges im Siebengebirge. *Geol. Rundschau*, 35: 33–35.

Cloos, H. and Cloos, E., 1927. Die Quellkuppe des Drachenfels am Rhein. Ihre Tektonik und Bildungsweise. *Z. Vulkanol.*, 11: 33–40.

Coe, K., 1966. Intrusive tuffs of west Cork, Ireland. *Quart. J. Geol. Soc. London*, 122: 1–28.

Colin, F., 1963. Le volcanisme du sud de l'Aubrac. *Congr. Soc. Savantes, 88e, Clermont-Ferrand, Compt. Rend.*, 2: 115–123.

Cucuzza Silvestri, S., 1961a. *Guide for the Excursion to Vesuvius.* Inst. Vulcanol., Catania, 22 pp.

Cucuzza Silvestri, S., 1961b. *Guide for the Excursion to the Latian Volcano.* Inst. Vulcanol., Catania, 18 pp.

Cucuzza Silvestri, S., 1963. Proposal for a genetic classification of hyaloclastites. *Bull. Volcanol.*, 25: 315–321.

Di Re, M., 1961a. L'eruzione dell'Etna del 1780 ed i suo prodotti. *Boll. Accad. Gioenia Sci. Nat. Catania*, 4 (6): 283–302.

Di Re, M., 1961b. *Guide for the Excursion to Stromboli.* Inst. Vulcanol., Catania, 10 pp.

Frechen, J., 1953. *Der Rheinische Bimsstein.* Fischer, Wittlich, 96 pp.

Frechen, J., 1959. Die Tuffe des Laacher Vulkangebietes als quartärgeologische Leitgesteine und Zeitmarken. *Fortschr. Geol. Rheinland Westfalen*, 4: 363–370.

Frechen, J., 1962. *Führer zu vulkanologisch-petrographischen Excursionen im Siebengebirge am Rhein, Laacher Vulkangebiet und Maargebiet der Westeifel.* Schweizerbart, Stuttgart, 151 pp.

Geikie, A., 1962. *Founders of Geology.* Reprint of 2nd ed. (1905). Dover, New York, N.Y., 486 pp.

Geologisches Landesamt in Baden–Württemberg, 1957. *Geologische Exkursionskarte des Kaiserstuhls, 1:25,000.* Geol. Landesamt Baden–Württemberg, Freiburg i. Brsg.

Geologisches Landesamt in Baden–Württemberg, 1959. *Erläuterungen zur geologischen Exkursionskarte des Kaiserstuhls, 1:25,000.* Geol. Landesamt Baden–Württemberg, Freiburg i. Brsg., 139 pp.

Gèze, B., 1955. Le volcanisme des Causses et du Bas-Languedoc (France). *Bull. Volcanol.*, 17: 73–89.

Gottini, V., 1961a. *Guide to the Excursion to the Phlegrean Fields.* Inst. Vulcanol., Catania, 15 pp.

Gottini, V., 1961b. *Guide for the Excursion to Ischia.* Inst. Vulcanol., Catania, 19 pp.

Hopman, M., Frechen, J. and Knetsch, G., 1959. *Die vulkanische Eifel*, 2nd ed. Fischer, Bonn, 143 pp.

Jakob, R., 1958. *Zur Petrographie von Vulcano, Vulcanello und Stromboli.* Vulkaninst. Friedländer, Zürich, 117 pp.

JUNG, J., 1946. Géologie de l'Auvergne et de ses confins bourbonnais et limousins. *Mém. Carte Géol. France*, 1946: 372 pp.

JUNG, J. and BROUSSE, R., 1962. Les provinces volcaniques néogènes et quaternaires de la France. *Bull. Serv. Carte Géol. France*, 267: 61 pp.

KLOOSTERMAN, J. B., 1960. Le volcanisme de la région d'Agde. *Geol. Ultraiectina*, 6: 79 pp.

KLÜPFEL, W., 1953. Basaltgeologie. *Z. Deut. Geol. Ges.*, 104: 326–353.

LOCARDI, E., 1965. Tipi di ignimbrite di magmi mediterranei. Le ignimbriti del vulcano di Vico. *Atti Soc. Tosc. Sci. Nat.*, A,72: 55–173.

LOCARDI, E. and MITTEMPERGHER, M., 1965. Study of an uncommon lava sheet in the Bolsena district (central Italy). *Bull. Volcanol.*, 28: 1–6.

MARINELLI, G., 1961. Les anomalies thermiques et les champs géothermiques dans le cadre des intrusions récentes en Toscane. *U.N. Conf. New Sources Energy, Rome, 1961, Proc.*, 2: 288–291.

MARINELLI, G., 1961. Genesi e classificazione delle vulcaniti recenti Toscane. *Atti Soc. Tosc. Sci. Nat.*, A,68: 74–116.

MARINELLI, G., 1963. L'Energie géothermique en Toscane. *Ann. Soc. Géol. Belg., Bull.*, 85: 417–438.

MICHEL, R., 1953. Contribution à l'étude pétrographique des pépérites et du volcanisme tertiaire de la Grande Limagne. *Publ. Fac. Sci. Univ. Clermont*, 1: 140 pp.

MITTEMPERGHER, M., 1961. *Guide for the Excursion from Arcidosso to Rome*. Inst. Vulcanol., Catania, 7 pp.

PICCOLI, G., 1963. Cenni di geologia euganea. In: *Colli Euganea, Guide Alpin.* Turistica, Padova, pp. 1–16.

PICCOLI, G., 1965. Rapporto tra gli allineamenti dei centri vulcanici paleogenici e le structture tettoniche nei Lessini. *Boll. Soc. Geol. Ital.*, 84: 1–19.

PRATESI, M. and MAZZUOLI, R., 1961. *Guide for the Excursion to Mt. Amiata (Tuscany)*. Inst. Vulcanol., Catania, 10 pp.

PREUSS, E., 1964. Das Ries und die Meteoriten-Theorie. *Fortschr. Mineral.*, 41: 271–312.

RICHTER, M., 1942. Geologie des Rodderberges südlich von Bonn. *Decheniana*, 101: 1–24.

RITTMANN, A., 1930. Geologie der Insel Ischia. *Z. Vulkanol.*, 6: 265 pp.

RITTMANN, A., 1933. Die geologisch bedingte Evolution und Differentiation des Somma-Vesuvmagmas. *Z. Vulkanol.*, 15: 94 pp.

RITTMANN, A., 1951. Sintesi geologico dei Campo Flegrei. *Boll. Soc. Geol. Ital.*, 69: 117–362.

RITTMANN, A., 1958. Sul meccanismo dell'attività vulcanica persistente. *Boll. Accad. Gioenia Sci. Nat., Catania*, 4 (4): 352–360.

RITTMANN, A., 1958. Cenni sulle colate di ignimbriti. *Bol. Accad. Gioenia Sci. Nat. Catania*, 4 (4): 524–533.

RITTMANN, A., 1960. *Vulkane und ihre Tätigkeit*, 2nd ed. Enke, Stuttgart, 336 pp. (English translation: Wiley, New York, N.Y., 1962.)

ROCHE, A., 1953. *Étude sur l'Aimantation des Roches Volcaniques Tertiaires et Quaternaires d'Auvergne et du Velay.* Thesis. Univ. Paris, manuscript.

ROCHE, A., 1963. Elaboration d'une stratigraphie paléomagnétique des formations volcaniques. *Bull. Soc. Géol. France*, 7 (5): 182–187.

ROQUES, M., 1958. Itinéraires géologiques en Auvergne. In: *Le Touriste en Auvergne*. Bussac, Clermont-Ferrand, 26: 1–42.

RUDEL, A., 1963. *Les Volcans d'Auvergne*. Éd. Volcans, Clermont-Ferrand, 160 pp.

RUTTEN, M. G., 1959. Ignimbrites or fluidised tuff flows on some mid-Italian volcanoes. *Geol. Mijnbouw*, 21: 396–399.

RUTTEN, M. G., 1962. Cinérites of the Mont Dore, central France. *Geol. Mijnbouw*, 41: 351–355.

RUTTEN, M. G., 1964. Three examples of periglacial solifluction in the southwestern Plateau Central France. *Geol. Mijnbouw*, 43: 1–9.

SCHENK, E., 1964. Die geologischen Erscheinungen der Subfusion des Basaltes. *Abhandl. Hess. Landesamtes Bodenforsch.*, 46: 31 pp.

SCHOTTLER, W., 1937. Der Vogelsberg. *Notizbl. Hess. Geol. Landesamtes Darmstadt*, 5 (18): 3–86.

SCHWARZBACH, M., 1961a. Eine Karte des quartären und tertiären Vulkanismus in Europa. *Eiszeitalter Gegenwart*, 12: 5–8.

SCHWARZBACH, M., 1961b. Der junge Vulkanismus in den Rheinlanden. In: *Köln und die Rheinlande*. Steiner, Wiesbaden, pp. 231–235.

SHOEMAKER, E. M. and CHAO, E. C. T., 1961. New evidence for the impact origin of the Ries Basin, Bavaria, Germany. *J. Geophys. Res.*, 66: 3371–3378.

STURIALE, C., 1961. *Guide for the Excursion to Lipari*. Inst. Vulcanol., Catania, 12 pp.

VAN BEMMELEN, R. W., 1961. Volcanology and geology of ignimbrites in Indonesia, North Italy and the U.S.A. *Geol. Mijnbouw*, 40: 399–411.

VUITTENEZ, H., 1964. *Géologie Cantalienne*. Imp. Moderne, Aurillac, 195 pp.

WIMMENAUER, W., 1957. Beiträge zur Petrographie des Kaiserstuhls. *Neues Jahrb. Mineral., Abhandl.*, 91: 131–150.

REFERENCES

WIMMENAUER, W., 1959. Beiträge zur Petrographie des Kaiserstuhls. I, Schluss, II und III. *Neues Jahrb. Mineral., Abhandl.*, 93: 133–173.

WIMMENAUER, W., 1962a. Beiträge zur Petrographie des Kaiserstuhls. IV, V. *Neues Jahrb. Mineral., Abhandl.*, 98: 367–415.

WIMMENAUER, W., 1962b. Beiträge zur Petrographie des Kaiserstuhls. VI, VII. *Neues Jahrb. Mineral., Abhandl.*, 99: 231–276.

WIMMENAUER, W., 1962c. Zur Petrogenese der Eruptivgesteine und Karbonatite des Kaiserstuhls. *Neues Jahrb. Mineral. Abhandl.*, 1962: 1–11.

Addendum

Owing to several circumstances much time has elapsed since the manuscript for this book was closed. Several important newer publications have been added as footnotes. There is one group of papers wich might be cited here, i.e., those presented to the XXIIIth International Geological Congress in Prague in 1968. On the basis of the abstract volume, published in May 1968, the following papers will be of importance:

Chapter 2

LOBACH-ZHUCHENKO, S. B., KRATTS, K. O., GERLING, E. K. and GOROKHOV, I. M., The Precambrian chronology of the Baltic Shield. Section 4.

SEMENENKO, N. P., Geochronology of the Precambrian of the East European Platform and its adjoining areas. Section 6.

Chapter 4

BOIGK, H., CLOSS, H., DOEBL, F., HAHN, A. and PLAUMANN, S., Geologisch–geophysikalisches Profil durch den Oberrheingraben. Section 1.

Chapter 5

CHARLIER, R. H. and LECOMPTE, M., Geographical distribution of the Devonian in Belgium with emphasis on the Frasnian and the Famennian. Section 9.

Chapter 9

PEIVE, A. V., The main tectonic features of the Alpine belt of Europe. Section 3.

Chapter 10

STEIN, A., HINZ, K. and PLAUMANN, S., Die Krustenstruktur der Poebene und des südlichen Alpenrands nach seismischer Tiefensondierung und Auswertung von Schweremessungen. Section 1.

Chapter 11

ANDRUSOV, D. and SCHEIBNER, E., Classification of "Klippes" and "Klippen". Section 3.

MILNES, A. G., Strain analysis of the basement nappes in the Simplon region, northern Italy. Section 3.

ROST, F., Über die Fazies-Einstufung orogenotyper Peridotite. Section 1.

Chapter 13

OXBURGH, E. R., Recent Rb/Sr and K/Ar age determinations and the tectonic history of the Central Eastern Alps. Section 3.

Chapter 14

PAQUET, J., Les différentes phases orogéniques des Cordillères bétiques dans l'ouest de la Province de Murcie, Espagne. Section 3.

Sources for Geological Information in West European Countries

AUSTRIA

Geological survey: Geologische Bundesanstalt, Wien.

General map: P. Beck-Mannagatta, 1964: *Geologische Übersichtskarte der Republik Österreich mit tektonischer Gliederung, 1:1.000.000.* Geologische Bundesanstalt, Wien.

General description: P. Beck-Mannagatta, in press: *Erläuterungen zur geologischen und zur Lagerstätten-Karte, 1:1.000.000, von Österreich* (with translations in English and French). Verhandlungen der Geologischen Bundesanstalt, Wien.

F. X. Schaffer (Editor), 1951: *Geologie von Österreich*, 2nd ed. Deuticke, Wien, 810 pp.

Society: Geologische Gesellschaft, Wien.

Journals: *Mitteilungen der Abteilung für Bergbau, Geologie und Paleontologie des Landesmuseums "Joanneum"*, Graz.

Mitteilungen der Geologischen Gesellschaft, Wien.

Mitteilungen der Gesellschaft der Geologie- und Bergbaustudenten in Wien.

Tschermaks Mineralogische und Petrographische Mitteilungen, Wien.

BELGIUM

Geological survey: Service Géologique de Belgique, 13 Rue Jenner, Parc Léopold, Brussels 4.

General map: (*See* General descriptions.)

General descriptions: P. Fourmarier, 1934: *Vue d'Ensemble sur la Géologie de la Belgique.* Annales de la Société Géologique de Belgique, Mémoires, 4: 200 pp.

P. Fourmarier (Editor), 1954: *Prodrome d'une Description géologique de la Belgique* (with geological map, 1:500,000). Société Géologique de Belgique, 826 pp.

E. Mailleux, 1933: *Terrains, Roches et Fossiles de la Belgique*, 2nd ed. Musée Royal de l'Histoire Natural Belge, 217 pp.

Societies: Société Belge de Géologie, de Paléontologie et d'Hydrologie, Bruxelles.

Société Géologique de Belgique, Liège.

Journals: *Annales de la Société géologique de Belgique, Bulletin.*

Annales de la Société géologique de Belgique, Mémoires.

Bulletin de la Société belge de Géologie, de Paléontologie et d'Hydrologie, Bruxelles.

Mémoires de la Société belge de Géologie, de Paléontologie et d'Hydrologie, Bruxelles.

Mémoires de l'Institute géologique de l'Université de Louvain, Louvain.

Publication de l'Association pour l'Etude de la Paléontologie et de la Stratigraphie Houillère, Bruxelles.

Publication du Centre National de Géologie Houillère, Bruxelles.

Publication du Centre National de Vulcanologie, Bruxelles.

DENMARK

Geological survey: Danmarks Geologiske Undersøgelse, Raadhusvej 36, Charlottenlund.

Grønlands Geologiske Undersøgelse, Oster Voldgade 7, Copenhagen K.

General map: (*See* General description.)

General description: V. Nordman and M. Nordman (Editors), 1928: Summary of the Geology of Denmark (with a small map). *Danmarks Geologiske Undersøgelse*, 5 (4): 1–219.

Society: Dansk Geologisk Forening, Oster Voldgade 5–7, Copenhagen K.

FINLAND

Geological survey: Geologinen Tutkimuslaitos, Otaniemi.
General map: *see* General description.
General description: A. Simonen, 1960: Pre-Quaternary Rocks in Finland (with map 1:2,000,000). *Bulletin de la Commission Géologique de Finlande*, 191: 1–49.
Society: Suomen Geologinen Seura, Otaniemi.
Journal: *Bulletin de la Commission Géologique de Finlande*, Helsinki.

FRANCE

Geological surveys: Bureau de Recherches Géologiques et Minieres (B.R.G.M.), Paris (15e).
Service de la Carte Géologique d'Alsace et de Lorraine, Strasbourg.
Service de la Carte Géologique de France, 62 Boulevard St. Michel, Paris (6e)
General map: *Carte géologique de la France, 1:1.000.000*. Service de la Carte Géologique de France.
General descriptions: R. Abrard, 1940: *Géologie de la France*. Payot, Paris, 607 pp.
L. Bertrand, 1946: *Histoire géologique du Sol français*. Flammarion, Paris, Part I, 356 pp.; Part II, 369 pp.
Societies: Société Géologique de France, Paris (6e).
Société Géologique du Nord, Lille.
Journals: *Annales de la Faculté des Sciences de l'Université de Clermont; Géologie et Mineralogie.*
Annales de la Société géologique du Nord, Lille.
Annales de Paléontologie, Paris.
Annales Scientifiques de l'Université de Besancon.
Bulletin de la Société Géologique de France, Paris.
Mémoires de la Société Géologique de France, Paris.
Sciences de la Terre, Nancy.
Travaux du Laboratoire de Géologie de la Faculté des Sciences de l'Université d'Aix, Marseille.
Travaux du Laboratoire de Géologie de la Faculté des Sciences de l'Université de Grenoble, Grenoble.

ITALY

Geological survey: Servizio Geologico d'Italia, Rome.
General map: *Carta Geologica d'Italia, 1:1.000.000*. 2nd ed., 1961, 2 sheets. Servizio Geologico, Rome.
Societies: Società Geologica Italiana, Rome.
Società Palaeontologica Italiana, Rome.
Journals: *Atti dell' Istituto di Geologia della Università di Genova.*
Atti dell' Istituto Geologico della Università di Pavia.
Bullettino della Società Geologica Italiana.
Geologica Romana, Rome.
Memoria dell' Istituto Geologico dell' Università di Padova.
Rivista Italiana di Palaeontologia e Stratigrafia, Milano.

NORWAY

Geological surveys: Geofysisk Avdeling, Trondheim.
Norges Geologiske Undersøkelse (N.G.U.), Josefinesgatan 34, Oslo NV.
General map: O. Holtedahl and J. Dons, 1960: *Geologiske Kart over Norge, 1:1.000.000*. 2nd ed. Norges Geologiske Undersøkelse.
General description: O. Holtedahl (Editor), 1960: *Geology of Norway*. Norges Geologiske Undersøkelse, 208: 540 pp.
Society: Norsk Geologisk Forening, Sarsgatan 1, Oslo.
Journal: *Norsk Geologisk Tidsskrift*, Bergen.

PORTUGAL

Geological survey: Servicos Geológicos de Portugal, Lisbon.
General map: (Included in the map of Spain.)

Society: Sociedade Geológica de Portugal, Lisbon.

Journals: *Boletim da Sociedade Geológica de Portugal*, Lisbon.
Boletim do Museu e Laboratorio Mineralógico e Geológico da Universidade de Lisboa, Lisbon.
Estudos de Geologia e Paleontologia, Lisbon.

SPAIN

Geological surveys: Instituto de Investigaciones Geologicas "Lucas Mallada", Madrid.
Instituto Geologico y Minero de España, Madrid.

General map: *Mapa Geologico de Espana, 1:1.000.000.* 2nd ed., 1936, 2 sheets. Instituto Geologico y Minero de España, Madrid.

General descriptions: M. San Miquel de la Camara, 1946: Historia de la formacion del suelo español. *Boletin de la Universidad de Granada*, 81: 14 pp.
W. Von Seidlitz, 1946: *La Construcion Geologica de España y de la Region del Mediterraneo Occidental.*

Journals: Estudios Geologicos, Madrid.
Monografias Geologicas, Oviedo.

SWEDEN

Geological survey: Sveriges Geologiska Undersökning (S.G.U.), Frescati, Stockholm (50).

General maps: G. Lundqvist, 1958: *Karta över Sveriges Jordarter (Quaternary deposition), 1:1.000.000, Ba 17.* 3 sheets. Sveriges Geologiska Undersökning, Stockholm.
N. H. Magnusson, 1958: *Karta över Sveriges Berggrund (pre-Quaternary), 1:1.000.000, Ba 16.* 3 sheets. Sveriges Geologiska Undersökning, Stockholm.

General descriptions: G. Lundqvist, 1959: *Description to accompany the Map of the Quaternary Deposits of Sweden, Ba 17.* Sveriges Geologiska Undersökning, Stockholm.
N. H. Magnusson (Editor), 1960: *Description to accompany the Map of the pre-Quaternary Rocks in Sweden, Ba 16.* Sveriges Geologiska Undersökning, Stockholm.

Society: Geologiska Föreningen i Stockholm, Stockholm 50.

Journals: *Arkiv för Mineralogi och Geologi*, Stockholm.
Bulletin of the Geological Institution of Uppsala.
Geologiska Föreningens i Stockholm Förhandlingar, Stockholm.
Stockholm Contributions in Geology.

SWITZERLAND

Geological survey: Geologische Kommission der Schweizerische Naturforschende Gesellschaft, Basel Bernoullianum.

General map: *Geologische Generalkarte der Schweiz, 1:200.000.* 8 sheets. Kimmerley und Frey, Bern.

General descriptions: H. Badoux, in press: *Géologie abrégée de la Suisse.*
E. Gagnebin and P. Christ, 1934: *Geologischer Führer der Schweiz.* Wepf, Basel, 14 vol.
A. Heim, 1919–1922: *Geologie der Schweiz.* Tauchnitz, Leipzig,

Society: Schweizerische Geologische Gesellschaft, Luzern.

Journals: *Bulletin des Laboratoires de Géologie, Mineralogie, Géophysique et du Musée géologique de l'Université de Lausanne*, Lausanne.
Eclogae geologicae Helvetiae, Lausanne.
Schweizerische Mineralogische und Petrographische Mitteilungen, Zürich.

THE NETHERLANDS

Geological survey: Geologische Dienst, Spaarne 17, Haarlem.

General map: *Kleine Geologische Overzichtskaart van Nederland, 1:600.000.* Geologische Stichting, Afd. Geologische Kaarten, 1947.

General description: A. J. Pannekoek (Editor), 1956: *Geologic History of The Netherlands.* Staatsdrukkerij, Den Haag, 154 pp.

Society: Koninklijk Nederlands Geologisch Mijnbouwkundig Genootschap, The Hague.

Journals:

Geologica Ultraiectina, Geologisch Instituut, Utrecht.
Geologie en Mijnbouw, Den Haag.
Leidse Geologische Mededelingen, Geologisch Instituut, Leiden.
Verhandelingen van het Koninklijk Nederlands Geologisch Mijnbouwkundig Genootschap, Mouton, Den Haag.

WEST GERMANY

Geological survey:

Bayerisches Geologisches Landesamt, München.
Bundesanstalt für Bodenforschung, Wiesenstrasse 1, Hannover.
Geologisches Landesamt in Baden-Wurttemberg, Freiburg in Breisgau.
Geologisches Landesamt für Nordrhein-Westfalen, Krefeld.
Geologisches Landesamt, Rheinland-Pfalz, Mainz.
Geologisches Landesamt des Saarlandes, Triererstrasse 1, Enshein über Saarbrücken.
Hessisches Landesamt für Bodenforschung, Wiesbaden.
Niedersachsischer Landesamt für Bodenforschung, Hannover.

General maps:

No general map 1:1,000,000.
Geologische Karte von Bayern, 1:500.000. Bayerisches Geologisches Landesamt, München, 1954.
Geologische Karte von Nordwestdeutschland, 1:300.000. Amt für Bodenforschung, Hannover, 1951, 4 sheets.
Geologische Überschichtskarte von Hessen, 1:1.000.000. 1958; and
Geologische Übersichtskarte von Hessen, 1:300.000. 1960, Hessisches Landesamt für Bodenforschung, Wiesbaden.
Geologische Übersichtskarte von Nordrhein-Westfalen, 1:500.000, 2nd ed. August Bagel, Düsseldorf, 1956.
(*See also* General descriptions.)

General descriptions:

P. Dorn, 1960: *Geologie von Mitteleuropa*, 2nd ed. Schweizerbart, Stuttgart, 488 pp., 1 map.
G. Kuetsch, 1963: *Geologie von Deutschland und einigen Randgebieten.* Enke, Stuttgart, 386 pp., 1 map.

Societies:

Deutsche Geologische Gesellschaft, Hannover.
Geologische Rundschau, Bonn.
Geologische Vereinigung, Bonn.

Journals:

Abhandlungen zur Geotektonik, Berlin.
Arbeiten aus dem Geologisch-Palaeontologischen Institut der Technischen Hochschule, Stuttgart, Neue Folge.
Erlanger Geologische Abhandlungen, Erlangen.
Fortschritte der Geologie und Palaeontologie, Berlin.
Fortschritte der Mineralogie, Stuttgart.
Geologie (Berlin).
Geologie, Freberg.
Geologische Mitteilungen, Aachen.
Geologische Rundschau, Bonn.
Geotektonische Forschungen, Stuttgart.
Hallesches Jahrbuch für Mitteldeutsche Erdgeschichte, Halle.
Heidelberger Beiträge zur Mineralogie und Petrographie, Berlin.
Mitteilungen aus dem Geologischem Statsinstitut in Hamburg.
Neues Jahrbuch für Mineralogie, Geologie und Palaeontologie, Stuttgart.
Palaeontographica, Stuttgart.
Palaeontologische Abhandlungen, Berlin.
Palaeontologische Zeitschrift, Stuttgart.
Zeitschrift der Deutschen Geologischen Gesellschaft, Berlin.
Zeitschrift für Angewandte Geologie, Berlin.
Zeitschrift für das Gesamtgebiet der Geologie und Mineralogie sowie der Angewandten Geophysik.

References Index[*]

[*] The names in capitals and small capitals refer to authors mentioned in the text (page numbers not in italics) and references lists (page numbers in italics); the names in capitals and lower case refer to persons mentioned in text only.

*Subject Index**

* Page numbers printed in italics refer to matter included in illustrations rather than in the text.